Springer Proceedings in Earth and Environmental Sciences

The series Springer Proceedings in Earth and Environmental Sciences publishes proceedings from scholarly meetings and workshops on all topics related to Environmental and Earth Sciences and related sciences. This series constitutes a comprehensive up-to-date source of reference on a field or subfield of relevance in Earth and Environmental Sciences. In addition to an overall evaluation of the interest, scientific quality, and timeliness of each proposal at the hands of the publisher, individual contributions are all refereed to the high quality standards of leading journals in the field. Thus, this series provides the research community with well-edited, authoritative reports on developments in the most exciting areas of environmental sciences, earth sciences and related fields.

Olga V. Frank-Kamenetskaya · Dmitry Yu. Vlasov ·
Elena G. Panova · Tatiana V. Alekseeva
Editors

Biogenic—Abiogenic Interactions in Natural and Anthropogenic Systems 2022

Springer

Editors
Olga V. Frank-Kamenetskaya
Department of Crystallography
Saint Petersburg State University
Saint Petersburg, Russia

Dmitry Yu. Vlasov
Department of Botany
Saint Petersburg State University
Saint Petersburg, Russia

Elena G. Panova
Department of Geochemistry
Saint Petersburg State University
Saint Petersburg, Russia

Tatiana V. Alekseeva
Institute of Physical Chemical
and Biological Problems of Soil Science
RAS
Pushchino, Russia

ISSN 2524-342X ISSN 2524-3438 (electronic)
Springer Proceedings in Earth and Environmental Sciences
ISBN 978-3-031-40469-6 ISBN 978-3-031-40470-2 (eBook)
https://doi.org/10.1007/978-3-031-40470-2

This Springer imprint is published by the registered company Springer Nature Switzerland AG
The registered company address is: Gewerbestrasse 11, 6330 Cham, Switzerland

Paper in this product is recyclable.

Preface

The book represents a collection of papers presented at VII International Symposium *Biogenic-Abiogenic Interactions in Natural and Anthropogenic Systems* which was held on September 26–29, 2022 in Saint Petersburg, Russia. The previous symposiums on this topic took place in St. Petersburg in 2002, 2004, 2007, 2011, 2014, and 2018. Organizers of the symposium were Saint Petersburg State University and Saint Petersburg Society of Naturalists.

Saint Petersburg State University, which was founded by Peter the Great in 1724, is the oldest University in Russia. The University is not only the largest educational center of Russia, but it also is a great scientific center.

The Saint Petersburg Society of Naturalists—one of the oldest natural science societies of Russia, being founded in 1868 by the Emperor Alexander II. From its foundation through the present day, the society has been strongly associated with Saint Petersburg State University. The first president of the society, Karl Kessler, was the rector of Saint Petersburg State University. Presidents of the Saint Petersburg Society of Naturalists have included famous scientists such as Prof. Beketov, Academician Vernadsky, Academician Ukhtomsky, and others.

The priorities of the society currently include complex interdisciplinary study in the different fields of earth science and life science, support of scientific communication among scientists, and support for young researchers of nature.

According to the results of the previous symposiums, monographs were published by Springer International Publishing in 2016 and 2020. The present book continues to discuss wide range of issues connecting with biogenic and abiogenic interactions in lithosphere, biosphere, and technosphere.

The book contains seven parts:

1. Biomineralization in living organisms and nature-like materials
2. Biomineralization in geosystems
3. Geochemistry of biogenic–abiogenic systems
4. Biomineral interactions in soil
5. Interactions of microorganisms with natural and artificial materials
6. Medical geology
7. Philosophical aspects.

The part *Biomineralization in Living Organisms and Nature-Like Materials* start with two reviews, one on to mechanisms, risks, and management of ectopic mineralization following liver transplantation (Radhakrishnan et al.) and the other on to crosslinking mechanisms in the designing of scaffolds for biomedical applications (Martin et al.). In work (Pikhur et al.) the results of the original experimental studies of odontomas (benign tumors of the maxillofacial region) of contemporary humans and animals are regarded. The work of Bespalov et al. is devoted to theoretical modeling of the structure and IR spectra of calcium and magnesium glutamate. In a series of works the synthesis conditions, stability, properties, and mechanisms of formation of biological apatites and apatite-organic composites are considered: hydroxyapatites synthesized from solutions modeling human joint synovia fluid (Gerk and Golovanova), Co-bearing hydroxyapatite (Korneev et al.), hydroxyapatite-chitosan biopolymer composite materials (Golovanova). The possibilities of application of the obtained results are discussed.

The part *Biomineralization in Geosystems* focuses on the study of natural objects formed and/or transformed with the participation of living organisms and the environment. Here, first of all, it should be noted the results of a study of chronic maxillary sinusitis in ancient populations (Zubova et al.). Magnetic properties and composition of inclusions in foraminifera shells at the Mid-Atlantic Ridge (Sergienko et al.), calcite microspherulites reflecting the relationship between abiotic biological processes (Antoshkina), the contribution of microorganisms to formation of framboidal pyrites in various geological settings (Leonova et al.) and the role of biogenic processes in formation of Vendian Oncolites (Litvinova, Kolosov) are also discussed.

Geochemical aspects of biogenic–abiogenic interactions have been considered in the part *Geochemistry of Biogenic–Abiogenic Systems*. The results of the study of the carbon isotopic composition of the Pai-Khoi amber-like fossil resin (NW Asia) (Zhuravlev and Astakhova) and bioavailable Sr-isotope ratio in the Caspian catchment basin (Kuznetsov and Gavrilova) are regarded. Some papers provide data on the accumulations of chemical elements, as well as the seasonal transformation of organic molecules in plants (Kosheleva et al.; Shtangeeva et al.; Chelibanov et al.). Chemical and biological weathering of black shales (Panova et al.) as well as effect of bryophyte vegetation on the chemical composition of sandstone (Smirnova et al.) are discussed.

The part *Biomineral Interactions in Soil* cover the wide range of classical and actual problems of soil science. In a comparison with previously published books of this series (2016 and 2020) the topics of this part have more applied character being related to the influence of climate change and human impact on biomineral interactions in soil. In work (Aparin et al.) the microbiome analysis was performed in different aspects. It is concluded that climate change globally leads to the spread of invasive earthworm species in both agrocenoses and natural ecosystems. Golovanova et al. experimentally showed that earthworm invasions decrease the amount of humus in tested meadow-chernozem soils and thus lead to soil degradation and soil fertility loss. The technology of soil geotechnical properties (filtration coefficient, compaction depth) improvement based on microbially induced calcite precipitation approach

(MICP) is tested in a paper of Golovkina et al. The authors propose the protocol for soil stabilization by calcium carbonate biomineralization. The effect of wildfires on soil properties (Podzol in Russian North-West, Novgorod region): dynamics of biogenic elements (P, N, K), vegetation cover and soil density are described in a paper of E. Yu. Chebykina et al. The impact of urbanization on soil properties—pH, organic carbon, $CaCO_3$, mobile phosphorus compounds, heavy metals (Pb, Cu, Zn), soil texture is reported by Bakhmatova et al. The data are based on the study of a series of urban soils from the central part of Saint Petersburg. Was the pedogenesis of the geological past of our planet specific taking in mind the changes which the Earth biosphere (vegetation cover, atmosphere, etc.) underwent? This problem is discussed in the paper of Alekseeva et al. based on the interdisciplinary study of fossil soils of Devonian age. Rozanov et al. studied the clay mineralogy as a marker of volcanic biogeosystem evolution in Laetoli, Tanzania. The authors were able to show that following the deposition and consolidation of the tuffs this biogeosystem experienced a complex evolution which led to the formation of the present-day soil cover.

In part *Interactions of Microorganisms with Natural and Artificial Materials* the different aspects of microbial effect on natural and artificial substrates are discussed. Special attention is paid to the problem of biological damage of cultural heritage objects in various ecological conditions (Lobzova et al.; Vlasov et al.). In these works the role of microorganisms in the transformation of different bedrocks is shown. The influence of biofilms on the surface properties of carbonates in speleothems in karst caves (Sofinskaya et al.) and the role of microorganisms at various stages of stromatolite formation (Litvinova) are discussed. Also of interest are studies on the adaptation of microorganisms to extreme conditions, such as ionizing radiation (Kirtsideli et al).

Different aspects of geological factors effect on human health are discussed in the part *Medical Geology*. The state and prospects of medical geology in Russia and in the world are considered (Volfson et al). The influence of the environment on the elemental blood composition in the human body in different regions is estimated (Drogobuzhskaya et al.; Semenova). Works on the composition of drinking water, as well as monitoring and modeling the impact of water composition on human health are relevant (Mazukhina et al.).

In part *Philosophical Aspects* the fundamental philosophical problem of the ratio of simple and complex in evolution is discussed (Sumina and Sumin). Comparative analysis of approaches to the species concept in biology and mineralogy is considered from interdisciplinary positions (Krivovichev and Borovichev).

The present book devoted to processes and phenomena on the boundary between biogenic and abiogenic nature should be of interest to a wide range of readers.

Saint Petersburg, Russia — Olga V. Frank-Kamenetskaya
Saint Petersburg, Russia — Dmitry Yu. Vlasov
Saint Petersburg, Russia — Elena G. Panova
Pushchino, Russia — Tatiana V. Alekseeva

Contents

Biomineralization in Living Organisms and Nature-Like Materials

Ectopic Mineralization Following Liver Transplantation—Mechanisms, Risks, and Management: A Review

Subathra Radhakrishnan, Catherine Ann Martin, Geethanjali Dhayanithy, Koustav Jana, Dinesh Jothimani, Alina R. Izatulina, Narayana Subbaraya Kalkura, and Mohamed Rela

Abstract A progressive calcification/mineralization that occurs ectopically (other than teeth, cartilage, and bone) following orthotopic liver transplantation is an undervalued condition but can turn out to be a fatal one, especially if it is a metastatic pulmonary calcification. Recent advancements in the epidemiology of the disease revealed new targets for inhibition and consequently recovery of the transplanted patients from massive calcification. Additionally, it is more conceivable that developments in surgical techniques and peri-transplant procedures might considerably reduce the occurrence of this disease condition. This review on ectopic mineralization following liver transplantation describes the pathomechanisms underlying the disease, the effects of ectopic calcification, and treatments available to overcome the ailing. Moreover, we address the importance of these complications (pulmonary restriction) and herein some aspects of precautionary aspects (monitoring plasma ionized levels of calcium and citrate metabolism) to prevent the disease.

Keywords Ectopic mineralization · Liver transplantation · Inorganic pyrophosphate · Calciphylaxis

---

S. Radhakrishnan · C. A. Martin
National Foundation for Liver Research, CLC Works Road, Chromepet, Chennai 600044, India

G. Dhayanithy · N. S. Kalkura
Crystal Growth Centre, Anna University, Guindy, Chennai 600025, India

K. Jana · D. Jothimani · M. Rela (✉)
Dr. Rela Institute and Medical Centre, Chromepet, Chennai 600044, India
e-mail: mohamed.rela@relainstitute.com

A. R. Izatulina
Institute of Earths Sciences, St. Petersburg State University, St. Petersburg, Russia

O. V. Frank-Kamenetskaya et al. (eds.), *Biogenic—Abiogenic Interactions in Natural and Anthropogenic Systems 2022*, Springer Proceedings in Earth and Environmental Sciences, https://doi.org/10.1007/978-3-031-40470-2_1

1 Introduction

The world's first successful liver transplantation (LT), a highly complicated surgical procedure was performed by Thomas Starzl in the year 1967 at Pittsburgh, followed by Sir Roy Yorke Calne in the year 1968 at Cambridge, both surgeons pioneered the path to enlightenment (Fricker 2017) in the field of liver transplantation. Now, LT is considered as the best therapeutic option for patients with end-stage liver disease of varied etiology, selected cases of hepatocellular carcinoma, confined metastatic tumor to the liver, and acute liver failure. Usually, LT could be carried out in two ways (i) deceased donor liver transplant (DDLT) (ii) living donor liver transplant (LDLT). The new donor liver/graft can be placed either heterotopically, where the recipient's liver is kept in situ and the graft is placed in a different site, or orthotopically, where the native liver is completely replaced with graft at the same site. Due to the high morbidity and mortality associated with heterotopic LT, orthotopic liver transplantation (OLT) remains the standard of care (de Rave et al. 2005). Even though the current LT survival statistics revealed that the 5-year survival rate is more than 75%, the complications of the procedure such as bleeding, infection, graft rejection, and bile duct complications are perilous. Nevertheless, over the past two decades, there was a massive improvement in the results of LT. Yet another major challenge faced by hepatologists is the metabolic complications after OLT which also affect the overall survival rate (Moreno and Berenguer 2006).

The liver is a metabolic organ that interacts with almost all the organ systems in the human body, and it is evident that patients who underwent liver transplants with graft cope with enormous physiological changes. The changes that affect the graft in OLT patients occur in two phases namely, the early postoperative phase and the late postoperative phase. During the early postoperative phase, the major challenges faced by the grafts are sepsis, hypotension, ischemia, hemodynamic instability, haemorrhage, dysregulated coagulation, and drug-related toxicity (Gray et al. 1986; Moreno and Berenguer 2006). Besides, the liver is a vascularized organ and since transplantation involves two genetically distinct individuals it highly necessitates the use of immunosuppression drugs. In the late postoperative phase, graft rejection, infection, new onset of diabetes, dyslipidemia, kidney failure, arterial hypertension, osteoporosis and side effects from the immunosuppressive drugs are the major complications (Jiménez-Pérez et al. 2016; Moreno and Berenguer 2006). The potential complications of LT have been listed in the Table 1. Overall, the clinical outcome of LT greatly depends on the preoperative patient's clinical condition, quality of the donor organ, intraoperative events, early identification of aforementioned factors in two phases, and the management of developing clinical events (e.g., neurological/cardiovascular).

Ectopic mineralization (EM) is a pathological state where the deposition of calcium phosphates in abnormal locations other than bone, cartilage, and teeth (Li et al. 2013). It is an incorrect course of the normal physiological process which endows the unique mechanical property of the teeth and bone. EM has been linked to several disorders such as aging, cancer, and autoimmune diseases, and has been

Table 1 Potential complications in liver transplant surgery; Liver transplant LT

LT short-term post-operative phase	LT Long-term post-operative phase
Haemorrhage	Chronic rejection
Biliary complications e.g., stricture, leak	Infection
Vascular complications e.g., thrombosis, stenosis	Recurrence of cancer
Sepsis	Diffuse abnormal parenchyma (hepatitis, cholangitis
Coagulation disorders	Ectopic mineralization/calcification
Hemodynamic instability	Side effects from immunosuppressant
Thrombocytopenia	Diabetes, obesity, and dyslipidemia
Renal dysfunction	Arterial hypertension
Acute rejection	Renal failure
Hypovolemia	Neurological complications

correlated with morbidity and mortality. Previous clinical reports demonstrated that numerous OLT patients acquiring EM, especially, pseudoxanthoma elasticum and calcinosis cutis (Gray et al. 1986; Neau-Cransac et al. 2005; Bercovitch et al. 2011; Munoz et al. 1988; Larralde et al. 2003; Eshani et al. 2006; Lisbon et al. 1993). Geographic graft calcification after LT has also been reported in two independent studies (Jeng et al. 2017; Tzimas et al. 2004). Based on their etiology, ectopic calcification can be classified as dystrophic and metastatic calcification. Calcification is said to be dystrophic if the serum calcium and phosphorous levels are normal. Calcification occurrence along with the abnormal level of calcium and phosphorous resulted in metastatic calcification. Being benign and asymptomatic, EM can be resolved with close follow-up and without treatment. But metastatic pulmonary and corneal calcification needs early diagnosis and appropriate treatment (Winter et al. 1995). The severity of the EM is highly variable. Irrespectively, ectopic mineralization/calcification is considered clinically significant when occurs in the Lungs, Cornea, Soft tissue, Kidneys, and intestine. The pathomechanism of ectopic mineralization is still not clearly understood. EM, a multifactorial induced disease depends on numerous factors which involve genetics (e.g. ABCC6, NT5E) molecular signal transduction pathways (e.g. BMP2 pro-osteogenic), systemic mineral imbalance (e.g. serum calcium), and inflammatory pathway (e.g. NF-KB).

Here, in this chapter, we review ectopic mineralization/ calcification that occurs in post-liver transplant patients. We have perused the related prospective and retrospective studies; and envisaged to understand and discuss the potential pathological mechanism and the therapeutic interventions related to this disorder. Furthermore, we have also highlighted the recent advancements in this field.

2 Organ Transplantation and Ectopic Calcification

Organ transplantation involves surgical removal/resecting of an organ from one person (donor)/location of the same body (self) to another person (recipient)/ other location of the body (self). The transplanted organ replaces the damaged or non-functional organ/irreversibly damaged organ. The liver, kidney, heart, lungs, pancreas, and intestines are the organs that are most often transplanted. Even though the skin flaps were used to replace organs early before the birth of Christ (BC), the first successful skin transplant took place in the sixteenth century by Italian surgeon Gaspare Tagliacozzi (Tagliacozzi 1597). He was the first one to describe graft-based immunological reactions which paved way for successful organ transplantation. Even today, his pioneering technique, vascular anastomosis remains a vital turn-point in xenotransplant research and surgery (Rodger et al. 2022). The very first attempt at organ transplantation dates to 1906 when Mathieu Jaboulay tried to transplant a pig kidney for one and a goat kidney for another in a human being (Baker et al. 2013). The first human-to-human kidney transplant was carried out in 1933, by a Soviet surgeon Yu Yu Voronoy. Solid organ transplantation is widely practiced now and is one of the best treatment options for chronically ill patients (Bezinover et al. 2019). Around 1,29,681 organ transplants took place worldwide in 1920. Recently, a genetically modified pig heart has been transplanted into a 57-year-old man at the University of Maryland Medical Centre (UMMC), Baltimore, Maryland is now considered a benchmark in the field of xenotransplantation (Reardon 2022).

Since the inception of organ transplantation (i.e., the mid-1950s), it has been considered as the precious gift of life to end-stage organ failure patients. The science of organ transplantation faces several challenges such as the availability of donors, and postoperative complications such as graft rejection and immunosuppression (Fox et al. 1974). However, present-day modifications of surgical technique and innovative peri-transplant strategies increased the volume of donor liver which was once considered marginal livers (Reddy et al. 2013). The graft rejection is also well managed with a multidisciplinary approach (e.g., active T-Cell monitoring, administration of immunosuppression) by clinicians and surgeons. Overall, the clinical outcome of solid organ transplantation is analysed by four factors namely, patient survival rate, reduction in the risks associated with co-morbidities, improvement in the patient's lifestyle, and management with minimal medications (Grinyó et al. 2013).

Post-transplantation abnormalities such as abnormal mineral metabolism are not uncommon. Elevation of phosphorous and calcium levels is found in serum during early post-transplantation. This abnormality in mineral levels could also be attributed to hyperparathyroidism medication and baseline cardiovascular diseases (Egbuna et al. 2007). On the contrary, fluctuation in mineral levels also could lead to cardiovascular disorders.

Liver transplantation incurs a certain degree of ischemic reperfusion-induced injury to the implanted graft. Several morphological and cellular events have been identified post-liver transplantation (Kalantari et al. 2007). A certain degree of cellular apoptosis or necrosis has been identified because of I/R-induced ischemia.

Moreover, several clinical reports have also suggested the development of vacuoles and calcification of the liver following extensive ischemia (Tzimas 2004). Hypocalcemia remains an important metabolic disorder demonstrated in OLT patients. This occurs because of calcium binding to the citrate content of the plasma during operative procedures. To nullify the calcium depletion external calcium supplies are administered. Though this external calcium supply prevents secondary hyperparathyroidism; it leads to calcium deposition in soft tissues. This kind of calcification may lead to respiratory malfunction and bone disorders (Munoz et al. 1988). Moreover, rapid pulmonary failure may also occur due to such mineral deposition in the tissues post-liver transplantation. Though there are several underlying reasons for such mineralization, the alkaline environment created by assisted ventilation and decreased blood flow may also contribute to mineralization (Winter et al. 1995).

One of the major causes of fatality in renal transplant recipients is coronary artery disease. Also, facts suggest that there is a positive correlation between coronary artery disease and calcification and further cardiovascular morbidity. Coronary artery calcification is reported in approximately two-thirds of renal transplant recipients. (Oschatz et al. 2006). Calcification of renal allografts has been described in patients with renal grafts. Calcification occurs within the first week after transplantation. Also, there are instances of increased calcification in patients with combined pancreatic and renal transplantation. These instances of renal transplantation occur in the tubular lamina and less frequently at the medullar tubulointerstitium (Gwinner et al. 2005).

Pulmonary calcification is also observed in post-cardiac transplantation patients. In this case, calcifications are observed in alveolar septa, blood vessels, and bronchial areas. Though Metastatic calcification is also common in other regions like the stomach and kidney, the lungs remain as the most common site. Corneal calcification is also found during Amniotic membrane transplantation. A crucial risk factor associated with such corneal calcifications is the usage of phosphate-containing eye drops (AlNuaimi et al. 2020).

Ectopic calcification results in the formation of hydroxyapatite in many non-skeletal tissues. Metastatic and dystrophic calcification are two major types of calcifications. Metastatic calcification occurs generally due to elevated levels of calcium and phosphate whereas dystrophic calcification occurs due to the deposition of hydroxyapatite in the organic matrix of ECM proteins (Kalantari et al. 2007). This kind of calcification in post-liver transplantation has been noted to occur in several internal organs and sometimes in the skin (Lateo et al. 2005a). Though the pathogenesis of such calcification is uncertain, it is evident that intravenous calcium required to correct hypocalcemia may play an important role. This kind of calcification does not occur only in post-transplantation patients but also in patients with renal disorders.

3 Ectopic Mineralization After Orthotopic Liver Transplantation (OLT)

OLT is a complex surgery with distinct stages viz., removal of unhealthy liver from the recipient, implantation of the healthy graft (from the deceased or live donor) revascularization (reperfusion stage), and bile duct reconstruction. (Makowka et al. 1988; Gray et al. 1986). The anhepatic phase begins from the resection of the diseased native liver from the recipient and ends when the implanted graft is reperfused. During this phase, the hemorrhage, hemodynamic instability, coagulopathy, hypocalcemia, and hyperfibrinolysis necessitate blood and blood products transfusion, calcium infusion, magnesium supplementation, and antifibrinolytics administration to prevent coagulopathy and cardiac depression (Fabbroni et al. 2006; Gray et al. 1986; Makwana et al. 2010). These transfusions increase the citrate load on the new graft liver. Nevertheless, during the anhepatic phase citrate metabolism is unlikely to occur. Previous studies substantiate that there was a 40% drop in the calcium content and an utmost rise in the citrate levels during the anhepatic phase (Gray et al. 1986; Merritt 2000).

Altogether, ectopic mineralization is an intricate process of progressive deposition of calcium and phosphates in the extracellular matrices of various organ systems, arterial blood vessels, soft tissues, and elastic fibres. In post-liver transplantation patients, the first case of ectopic mineralization was observed in the lungs (Lateo et al. 2004). Other parts that may have EM include the liver graft itself, colon, vascular walls, kidneys, adrenal glands, and gastric mucosa. Interestingly, Deguchi et al. (2016) reported systemic disseminated metastatic tissue calcification, including massive myocardial calcification after orthotopic liver transplantation in a 20-year-old man. Pulmonary calcification following OLT has been equally reported in adults and children. These congestions may be detrimental to the patient. It is a progressive disorder that could be radiologically diagnosed. Dystrophic and metastatic calcification occurs in damaged and normal lungs, respectively (Winter et al. 1995; Libson et al. 1993). Rather, calcifications in the liver graft itself have been scarcely reported. The graft calcification might be caused because of liver ischemia/reperfusion (I/R) induced cellular events and in the case of living-related donor liver transplantation, the graft is also affected rarely because of the hepatic vein torsion. (Jeng et al. 2017; Tzimas et al. 2004). Furthermore, the graft microcalcification resulting in vacuole formation and cellular necrosis initiated by the consequence of two pathological mechanisms viz., I/R generated osteoblast-like cell lineage derived from activated myofibroblasts and intracellular accumulation of hydroxyapatite deposits in hepatocytes. Progressive graft calcification may cause liver dysfunction (Kalantari et al. 2007).

There are few case reports on evidence of calcification of the skin in post-liver transplant recipients. The calcium salt deposition especially in the skin and the subcutaneous tissue is known as calcinosis cutis. After transplantation, the reported cases develop CC within 10–12 days and it can be dystrophic, metastatic, or idiopathic CC (Jucgla et al. 1995). In dystrophic calcification calcium deposits in injured skin tissue with normal sera levels of calcium and phosphorus. Conversely, in metastatic calcinosis, due to solubility products of calcium and phosphorous, there is an abnormal level of calcium and phosphorus may result in chronic kidney injury (Lateo et al. 2005b; Jiménez-Gallo et al. 2015). Pseudoxanthoma elasticum (PXE), another genetic metabolic disease has been often reported along with the minimal presence of anti-mineralization factors in circulation. PXE causes elastic tissue in the body to get mineralized. Besides, calcium deposition occurs in the eyes, cardiovascular system, and gastrointestinal tract (Bercovitch et al. 2011). Small yellow papules are seen at the back and sides of the neck, which may be the first clinical symptom. Calcification in the retina and the arterial walls may result in blindness and peripheral heart disease, respectively. Dystrophic calcification in the elastic fibres of the skin is also seen (Germain 2017). Another disorder presented after OLT is generalized arterial calcification of infancy (GACI), a very rare genetic disorder that forms abnormal clumps of calcium in the blood vessels. The most common features are calcification of the arteries and ectopic mineralization of extravascular tissues. This condition can be fatal if not treated within 6 months (Kawai et al. 2022). Arterial Calcifications due to Deficiency in CD73 (ACDC) is a genetic condition where calcium forms deposits in the arteries as well as in the lower extremities of the body. Unlike GACI, this is mostly seen in adult patients following liver transplantion. Rarely calcification can be found in the joints of the fingers, wrists, ankles, and feet (Gutierrez et al. 2016).

Taken together, the ectopic mineralization following OLT needs specific attention in the early postoperative phase as well as during follow-up. The cases of EM after liver transplantation reported have been tabulated in Table.2.

4 Pathological Mechanisms of Ectopic Mineralization After Liver Transplantation

The liver is the largest internal organ composed of 80% of hepatocytes (parenchymal cells) and 20% non-parenchymal cells. It is the largest solid organ that accomplishes numerous biosynthetic and metabolic processes especially involved in detoxification, regulation of mineralization, balancing the chemical levels in the circulation, and bile secretion (Bercovitch et al. 2011). During the anhepatic phase and early post-liver transplantation, the capacity of these processes significantly declined. It is not surprising that the perturbations cause abnormalities in coagulation, metabolism, and homeostasis (Gray et al. 1986).

The pathogenesis of ectopic mineralization post-liver transplantation is multifactorial. Haemorrhage during liver transplantation also requires large volumes of blood replacement or the use of automatic blood transfusion technology. The anticoagulant citrate is the most content, approximately 3 g per unit of red blood cells (RBC) (Li et al. 2015). The citrate metabolism which is a highly oxygen-consuming process physiologically occurs in huge mitochondria-bearing organs especially, the liver. A healthy liver can metabolize the 3 g of citrate in 5 min which is greatly reduced during the anhepatic phase and enhance citrate elevation. The accumulation of citrate chelates the calcium ions and decreased the ionized calcium level in circulation

Table 2 Ectopic mineralization/ Calcification (Calcium deposition) Cases reported after orthotopic liver Metastatic calcification occurs generally due to elevated levels of calcium and phosphate whereas dystrophic calcification occurs due to the deposition of hydroxyapatite (like bone mineralization) in the organic matrix of ECM proteins

S.no.	Calcium deposition/ accumulation in ectopic mineralization	Organ	Type	Physiological impact	References
1	Calcium deposition	Soft tissues	Dystrophic and metastatic soft tissue calcification	Renal failure, restricted pulmonary function, osteopenia	Munoz et al. (1988)
2	Calcium deposits	Dermal collagen bundles in the reticular dermis	Dystrophic tissue calcification	Plaque formation	Ehsani et al. (2006)
3	Congestion necrosis with geographic graft calcification	Transplanted graft	Not mentioned about serum calcium and phosphorus	Parenchymal hepatocytes damage: Patient died because of infection	Jeng et al. (2017)
4	Calcinosis cutis	White papules on an erythematous base in linear and rosette configurations that developed in the abdominal and lumbar areas	Dystrophic tissue calcification	Severe metabolic problems and later improved	Larralde et al. (2003)
5	Calcinosis cutis	Arm become inflamed and indurated; Central necrosis has been noted; dermal calcification	Dystrophic tissue calcification	Plaque formation with hyperpigmentation resolved after 4 months with scarring	Lateo et al. (2005a)
6	Calcium deposits	Pulmonary parenchyma	Dystrophic/ metastatic	Pleural effusion on chest radiograph; subcapsular liver calcification; pulmonary consolidation	Lisbon et al. (1993)

(continued)

Table 2 (continued)

S.no.	Calcium deposition/ accumulation in ectopic mineralization	Organ	Type	Physiological impact	References
7	Calcium and other minerals; acquired Pseudoxantomona elasticum	Elastic fibres all around the body	Dystrophic/ metastatic	Skin; peripapillary angioid streaks; Cutaneous PXE	Bercovitch et al. (2011)
8	Calcification	Pulmonary Calcification	Dystrophic/ metastatic	Progressive pulmonary calcification and restricted pulmonary function; the patient passed away	Isaccs et al. (2021)
9	Calcification	Pulmonary calcification	Dystrophic/ metastatic	Tachypnoea persisted; restricted pulmonary function	Winter et al. (1995)
10	Calcinosis cutis	Skin, elastic fibres	Dystrophic/ metastatic	Severe chronic liver graft dysfunction, and hypercalcemia/ hyperphosphatemia, renal Failure	Neau-Cransac et al. (2005)
11	Calcification	Transplanted graft	Dystrophic/ metastatic	Primary graft dysfunction: both patients died secondary to the sequelae of the graft dysfunction	Tzimas et al. (2004)
12	Calcification	Abdominal aortic calcification (AAC)	Not mentioned serum calcium and phosphorus	Diabetes and associate cardiovascular disease, affecting overall survival	Imaoka et al. (2019)
13	Calcium	Non- uremic Calciphylaxis	Not mentioned serum calcium and phosphorus	Subcutaneous nodules with ulceration and eschar formation of overlying skin involving the medial side of the thighs bilateral	Prabhakar et al. (2018)

(continued)

Table 2 (continued)

S.no.	Calcium deposition/ accumulation in ectopic mineralization	Organ	Type	Physiological impact	References
14	Calcium	Non- uremic Calciphylaxis; calcification in the media of small-sized and medium-sized arterial walls with endothelial proliferation and luminal obliteration	Not mentioned serum calcium and phosphorus	Painful ulcerative rash on her thighs and right buttock; fibrin deposition and fat necrosis	Frunza-Stefan et al. (2018)
15	Calcium	Calciphylaxis; in OLT patients with acute kidney failure	Metastatic (hypercalcemia)	Calcium deposits, and thrombotic vasculopathy; painful, non-ulcerating, erythematous plaques over her shins and thighs	Nseir et al. (2021)
16	Calcium	Metastatic Pulmonary Calcification	Metastatic calcification	Bilateral anterior ground glass opacities; Chest pain; myxoid interstitial fibrosis and calcification	Charokopos et al. (2021)
17	Calcium	Calciphylaxis; in small blood vessels; calcific uremic arteriolopathy	Not mentioned about serum calcium and phosphorus	Subcutaneous calcification and fibrosis, and small-medium size blood vessel calcification and thrombosis; 1 patient died	Bohorquez et al. (2015)

which affects cardio muscular contractility. In addition, the acid-base imbalance in the plasma also causes metabolic alkalosis. The pH of the tissues, certain drugs (e.g., Pitressin tannate), and the alkalotic environment generated by assisted ventilation promote the precipitation of calcium salts. Aforesaid, the chelation of calcium necessitates large volumes of calcium substitution, which causes cutaneous calcifications and ectopic mineralization when exceeds the metabolic need (Li et al. 2015; Gray et al. 1986; Jucgla et al. 2006; winter et al. 1995). Such impaired citrate metabolism that can lead to citrate toxicity could also be seen in acute liver failure patients and acute-on-chronic liver disease patients substantiating the correlation between liver dysfunction and low ionized calcium levels in the plasma. Fetuin-A (α2-Heremans–Schmid glycoprotein), a 64 K Da glycoprotein, is predominantly secreted by the liver, involved in calcified matrix metabolism, and prevents pathological calcification in patients (Hendig et al. 2006). Nonetheless, the direct involvement of ectopic mineralization in post-liver transplant patients has not been reported, previous clinical study has demonstrated a deficiency of ABCC6 (multidrug resistance-associated protein 6) alterations leads to declined Feutin A in the blood which in turn causes mineralization in elastic fibres i.e., PXE (Herrmann et al. 2012).

Inorganic pyrophosphate (PPi) is a potent mineralization inhibitor in circulation. The major source of PPi is the liver. Functionally, PPi inhibits the precipitation of calcium and phosphorous (inorganic) which leads to the formation of hydroxyapatite crystals (Jansen et al. 2014; Shimada et al. 2021). The four predominant genes viz., ABCC6, ectonucleotide pyrophosphatase/phosphodiesterase 1 (ENPP1), ecto-5-prime-nucleotidase (NT5E) and Alkaline Phosphatase, Biomineralization Associated (ALPL) associated with the liver are involved in the regulation of purinergic metabolism, intricately regulates the production of systemic PPi. Inside the liver sinusoid, the ABCC6 belongs to the family of unidirectional ATP-driven transmembrane transporters that release ATP which is converted by ENPP1 to AMP and PPi in the liver vasculature. Mutations within these ABCC6 and ENPPI genes have been reported to cause PXE and artery calcifications, respectively. (Vanakker et al. 2013; Jansen et al. 2014; Ma et al. 2018). In the periphery, the AMP is further broken into adenosine and Pi by functional NT5E, where adenosine negatively regulates ALPL/TNAP, which successively cleaves PPi to Pi. Non-functional ecto-5-prime-nucleotidase not surprisingly, could not inhibit TNAP, leading to arterial calcification and Calcification of joints and arteries (CALJA) (Quaglino et al. 2020; Shimada et al. 2021). Furthermore, during the anhepatic phase, due to retarded function of ABCC6, the central gene in purinergic metabolism, there is an upregulation in the pro-osteogenic pathways such as WNT, TGFβ, and BMP2 that eventually leads to ectopic mineralization (De Vilder et al. 2015). Altogether, the key role of ABCC6 and circulating PPi in ectopic mineralization in post-liver transplantation is well demonstrated in humans. In the liver, additional metabolic activity is involved in ectopic mineralization inhibition through carboxylation of vitamin K-dependent coagulation factors and matrix GLA proteins (MGP). The gamma-glutamyl carboxylase (GCCX), a multipass transmembrane protein present in the endoplasmic reticulum post-translationally carboxylates clotting factors and MGP and activates them. The MGP potentially inhibits BMP2 signalling involved in ectopic mineralization. Impairment in GCCX function directs pathological mineralization through downstream genes of BMP2 signalling (De Vilder et al. 2015; Li et al. 2012).

Besides, studies have reported that liver transplantation patients who had earlier kidney failure/chronic kidney disease associated with increased parathyroid hormone levels developed calcifications. In such conditions, the production of vitamin D is retarded which leads to hypocalcaemia and secondary hyperparathyroidism. Clinical studies revealed that hyperparathyroidism could cause ectopic mineralization

(Larralde et al. 2003; Cunningham et al. 2011). Pulmonary calcification mostly occurs in OLT patients with chronic renal failure and hyperparathyroidism. Besides, post-operatively, these patients exhibited high serum levels of calcium and phosphorus and received citrate-containing frozen blood plasma and platelets intraoperatively. In a retrospective review, metastatic calcification has been reported. The autopsy study was carried out with 25 patients who underwent OLT. 84% of the patients exhibited pulmonary calcification identified with microscopic foci and hypercalcemia (Winter et al. 1995).

The I/R -induced parenchymal cells' apoptosis and necrosis result in primary graft dysfunction because of calcification. A very recent study demonstrated the molecular events that followed cell death. The extracellular DNA released during cell death has the capacity to chelate the cations in the protein moieties present in the extracellular matrix. The agglomeration endows protection to the extracellular DNA from being degraded. Such DNA agglomerates can cause pathological calcification (Tzimas et al. 2004; Shen et al. 2022). Taken together, the molecular and sub molecular level clinical reports regarding pathomechanisms of ectopic mineralization/calcification are intriguing multi-cascades, intricately regulated by the liver. A pictorial representation of pathomechanisms involved in ectopic mineralization is shown in Fig. 1.

5 The Therapeutic Modalities for Ectopic Mineralization

The management of the anhepatic phase and intensive care following liver transplantation, especially the immunological, microbial, molecular, and biochemical parameters is crucial. Numerous studies reported that the post-liver transplant patient acquiring EM, especially, PXE/CC (Gray et al. 1986; Neau-Cransac et al. 2005; Bercovitch et al. 2011; Munoz et al. 1988; Larralde et al. 2003). The treatment followed for EM in post-transplant patients also highlights the therapeutic modes for PXE and CC. Complications associated with ectopic mineralization in post-transplant patients form a small part of this huge treatment regime which sometimes needs to be carefully monitored. Some patients with EM remain asymptomatic and recovered well with close follow-up and no specific treatment. Besides, treatment becomes mandatory when metastatic and persistent pulmonary calcification is seen after OLT (Winter et al. 1995).

In the cases where treatment is needed, currently, there are six types of treatment modalities available (Fig. 2). They are (i) hypocalcaemia and hyperkalaemia targeted (ii) ABCC6 Pathway targeted (iii) apoptosis or necrosis targeted (iv) molecular-based targets (v) BMP2 signalling targeted pathway and (iv) extracellular DNA targeted. The targeted therapies are either given independently or in combination.

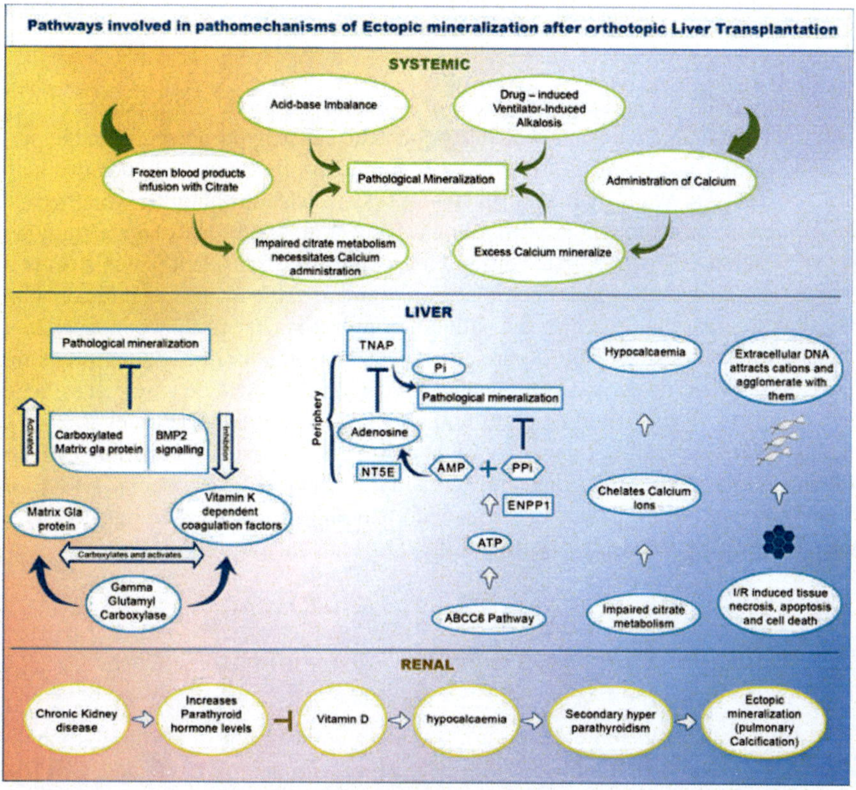

Fig. 1 Path mechanisms of EM after OLT. Major pathways involved pertaining to the Liver, Kidney and circulation are exhibited. EM: Ectopic mineralization; OLT: Orthotopic liver Transplantation

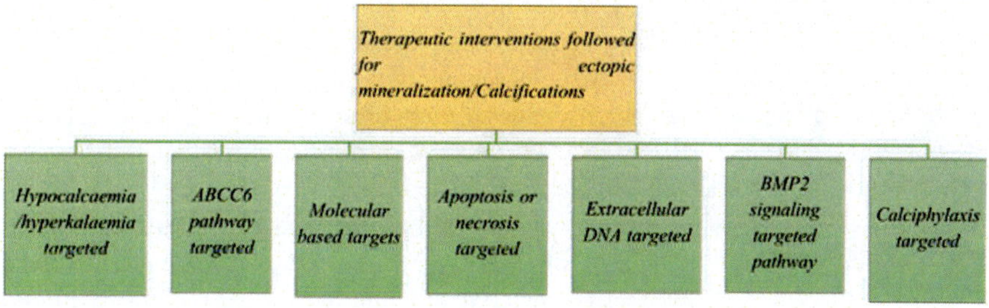

Fig. 2 The therapeutic intervention followed for ectopic mineralization/Calcification followed by the liver transplant

5.1 Systemically Targeted Therapeutic Modalities

Substantial amounts and rapid infusion of stored blood or plasma products during liver transplantation, due to its high citrate content, cause hypocalcaemia and hyperkalaemia, a hemodynamic instability. Consequently, the created hypotension in patients will be reversed by infusing a 10 mM bolus of calcium chloride which raised the plasma ionized calcium concentrations (Gray et al. 1986). However, a study by Jawan et al. (2003) with 39 paediatric living-donor liver transplant patients (in two groups) revealed that in spite of complete replacement of blood product infusion with 5% albumin it couldn't recover the patients from hypocalcaemia. Two independent studies have demonstrated the importance of systematic monitoring of serum calcium levels intra- operatively and consistent rectification along with accurate anesthetic management will prevent mineral instability (Rando et al. 2014; Jawan et al. 2003). Such regulation avoids the necessity of calcium infusion thereby impeding ectopic mineralization. If the excess calcium infusion resulted in EM, especially PXE, it could be treated using EDTA-encapsulated albumin nanoparticles conjugated with anti-elastin antibodies which facilitates the release of calcium-affected sites.

5.2 Molecular Pathways Targeted Therapeutic Modalities.

Liver cells are considered the predominant production site of inorganic pyrophosphate (PPi), a potent inhibitor of mineralization through ABCC6 and ENPP1-dependent pathways. Any disruption in PPi production might cause EM. During the anhepatic and early post-transplant phases, the production of PPi is diminished. Bisphosphonates, a chemically stable derivative of PPi, were originally used for the treatment of bone diseases. Oral formulations of bisphosphonates in post-liver transplants have shown to reduce fracture incidence and bone loss (Ho et al. 2021). The bisphosphonates consist of two classes, namely amino bisphosphonates, and non-amino bisphosphonates which differ in their efficacy and action mechanism (Furman 2007). Li et al. (2015) in their extensive study demonstrated that both classes of bisphosphonates could prevent the EM caused because of ABCC6 mutation in PXE and GACI. In addition, a previous study determined that the administration of biphosphonates in the treatment of EM not only reduces the size of calcifications but also exhibited anti-inflammatory roles (Neau-Cransac et al. 2005). Furthermore, a randomized clinical trial of treating PXE with 74 patients significantly reduced arterial calcification when compared to a placebo (Kranenburg et al. 2018). Administration of bisphosphonates might be a promising therapy to prevent EM. A previous PXE case report substantiates that oral administration of disodium pyrophosphate prevents re-occlusion after surgery (Väärämäki et al. 2019; Dedinszki et al. 2017). An evaluation of the efficacy of oral administration of PPi salts is carried out in a phase II-controlled trial against a placebo in PXE patients, against retarded ABCC6 pathway (RCT: NCT04868578:ongoing). Other supplementary treatments that could be considered for the prevention/inhibition of EM are dietary magnesium, Phytic acid, Sodium thiosulphate, Phosphate binder, Feutin A, and vitamin K (Shimada et al. 2021). By competing with calcium, the supplemented dietary magnesium form precipitates with inorganic phosphates which are readily soluble and resulted in less hydroxyapatite structure. Hence, Dietary magnesium significantly decreased ectopic mineralization (Luo et al. 2020).

5.3 *Apoptotic, Necrosis and Cell Death Targeted Modalities*

Developing effective therapies for systematic manifestations of EM remains to be a challenge since certain therapeutic agents such as inositol hexaphosphate only can prevent newly formed mineralization/calcification but not effective in reversing the existing pathological effects. In a double-blind, placebo-controlled phase 2b trial, the intravenous administration of myoinositol hexaphosphate reduced the progression of cardiovascular calcification, but further studies in a randomized manner are needed to prove the reversal of detrimental cardiovascular events (Raagi et al. 2018). The EM occurring in post-transplant patients might be the outcome of apoptotic cell death which displays phosphatidyl serine in the ECM and causes mineralization by binding to calcium or by intersecting the apoptotic pathway of vascular smooth muscle cells which elevates reactive oxygen species. Apoptotic-mediated crystal deposition occurs predominantly in cartilage and blood vessels (Boraldi et al. 2021; De Vilder et al. 2015). Exploring the relationships between apoptosis and EM and targeting the pathway is still in its nascent stage. Unlike other therapeutic approaches targeting apoptosis or necrotic pathway in treating EM will reduce negative interference (naturally occurring mineralization process).

Molecularly targeted drugs are very precise in their mode of action. ENPP1 (Ectonucleotide Pyrophosphatase/Phosphodiesterase 1) replacement therapy has been achieved by the administration of recombinant ENPPI protein which breaks the ATP into AMP and PPi. In animal studies, the therapy reduces the mortality rate by reducing vascular calcification. Systemic Administration of Recombinant human ENPP1-Fc protein in an animal model proved that it is effective against calcification and arterial stenoses (Nitschke et al. 2018; Luo et al. 2020). TNAP inhibitors are used to treat ectopic calcification in ABCC6 mutant mice. A recent study on SBI-425, an aryl sulfonamide, selective inhibitor of TNAP used in treated disease model mice. It is considered as a potential therapeutic target for treating PXE. The oxidative stress and ROS cause soft tissue mineralization through the activation of the NF-KB pathway and upregulation of the pro-osteogenic BMP2 pathway. Interleukin-37 and SMAD 6 act as an inhibitor of BMP2, suppresses the level of alkaline phosphatase, and play a protective role in calcific aortic valve disease (Song et al. 2022). As mentioned before, the extracellular DNA released in the ECM because of cell death and apoptosis is the major cause of collagen intrafibrillar mineralization, a pathogenetic aspect of EM. It was hypothesized that the use of the enzyme DNase at the intended tissue site will neutralize the extracellular DNA and reduce the EM (Shen et al. 2022).

Unlike the administration of calcium chloride, biphosphonates, pyrophosphates and other supplementary methods all other treatment modules are in their infancy. Some of the modules like aryl sulfonamides are validated in the animal model, yet to undergo clinical trials in all phases. The preclinical study design with disease model animals (e.g., MGP deficient mice) is very crucial in justifying a therapeutic agent/molecule. Future work needs to be focused on the development of effective therapies for EM in vulnerable post-transplant patients.

6 Concluding Remarks: Recent Advancements

Ectopic calcification although described in the nineteenth century is now gaining recognition due to the expanding knowledge of the mechanisms of ectopic calcification. Recent studies have shown significant progress in the understanding of molecular mechanisms, inhibitors, proteins, and cellular processes that are causes or consequences of ectopic calcification. Intriguingly, a report by Ilona et al. (2021) has linked ectopic calcification in aortic valves to the telomere length. However, this novel finding is yet to define if this is a cause or consequence. Similarly, another study has shown that extracellular DNA is associated with ectopic mineralization and the reaction can take place as early as 24 h (Shen et al. 2022). Although calcium and phosphate ions are known to play key roles in early arterial calcification, the role of thermodynamics in calcification, a new and exciting concept was put forward by Millán et al. (2021). They determine three thermodynamic parameters to be attributed to ectopic calcification, which are (i) the Ca^{2+} and PO_4^{3-} functions that result in enhanced precipitation (ii) focally defective balance of the biomolecules present as promoters and inhibitors, and (iii) the irreversible hydroxyapatite crystallization that continues despite the normal levels of Ca^{2+} and PO_4^{3-}. In parallel, medial vascular calcification is caused due to the increased levels of calcium deposition in the arteries with emerging chronic kidney disease and diabetes. Another interesting report by Phadwal et al. (2021) describes the involvement of mitochondrial dysfunction in ectopic mineralization. This study highlights the effect of mitochondrial dysfunction due to increased ROS production and mineral dysregulation. Even though these are advanced findings the question of whether it is because of the cause or consequence of ectopic mineralization remains elusive. Similarly, another recent study by Laurain et al. (2022) revealed that liver cirrhosis is associated with arterial calcification. It is reported that the low levels of pyrophosphate production in the liver cause this ectopic calcification. This study was carried out in patients before and after liver transplantation and surprisingly three months post-transplant, the levels of hepatic pyrophosphate are normalized, and the condition is resolved. The very recent recognition that inflammation drives various channels of ectopic mineralization has opened new avenues to enable the development of novel strategies to treat this debilitating condition (Song et al. 2022). Excitingly, Peeters et al. (2020) have introduced sex difference, a new concept to deal with in ectopic calcification. Herein, they have reported biomarkers that are associated with inflammation and aortic valve calcification to be expressed in males, while in females' makers involved in fibrosis are expressed more. This could lead to the development of sex-dependent precision treatment strategies.

Despite the extensive studies and emerging evidence on the mechanisms of action and the pathomechanisms of ectopic calcification, there remain two critical challenges. First, the need for biomarkers to detect early calcification, and then, effective precision treatment strategies that could reverse the onset of calcification and treat chronic conditions in individuals.

References

AlNuaimi D, ElKaissi M, Wahla A, Shafiq I, Abdulghaffar S (2020) Metastatic pulmonary calcifications postcardiac transplant in a 15-year-old patient: A case report and review of literature. Radiol Case Rep. Aug 20;15(10):1978–1982. https://doi.org/10.1016/j.radcr.2020.07.061. PMID: 32874395; PMCID: PMC7452014.

Barker CF, Markmann JF (2013) Historical overview of transplantation. Cold Spring Harb Perspect Med. 3(4):a014977. https://doi.org/10.1101/cshperspect.a014977. PMID: 23545575; PMCID: PMC3684003.

Bercovitch L, Martin L, Chassaing N, Hefferon TW, Bessis D, Vanakker O, Terry SF (2011) Acquired pseudoxanthoma elasticum presenting after liver transplantation. J. Am. Acad. Dermatol. 64(5):873–878. https://doi.org/10.1016/j.jaad.2010.03.030.

Bezinover D, Saner F (2019) Organ transplantation in the modern era. BMC Anesthesiol. 19(1):32. https://doi.org/10.1186/s12871-019-0704-z. PMID: 30832567; PMCID: PMC6399965.

Bohorquez HE, Chamorro N, Garces J, Cohen AJ, Reichman TW, Davis NK, Vincent B, Bruce DA, Carmody IC, Moiz A, Staffeld C, Loss GE (2015) Calciphylaxis in simultaneous liver-kidney transplantation. Am J Transplant. 15(4):1105-9. https://doi.org/10.1111/ajt.13082. Epub 2015 Feb 25. PMID: 25716643.

Boraldi F, Lofaro FD, Quaglino D (2021) Apoptosis in the Extraosseous Calcification Process. Cells, 10(1):131. https://doi.org/10.3390/cells10010131.

Charokopos A, Dua S, Beasley MB, Chung MS, Bernheim A, Padilla M, Mathur A (2021) Radiographically Progressive Metastatic Pulmonary Calcification in a Three-Time Liver Transplant Recipient Requiring Intraoperative Resuscitation. Am J Respir Crit Care Med. 203(4):502-503. https://doi.org/10.1164/rccm.202001-0027IM. PMID: 33091307.

Cunningham J, Locatelli F, Rodriguez M (2011) Secondary hyperparathyroidism: pathogenesis, disease progression, and therapeutic options. Clin J Am Soc Nephrol. 6(4):913-21. https://doi.org/10.2215/CJN.06040710. Epub 2011 Mar 31. PMID: 21454719.

De Vilder EY, Vanakker OM (2015) From variome to phenome: Pathogenesis, diagnosis and management of ectopic mineralization disorders. World J Clin Cases. 3(7):556-74. https://doi.org/10.12998/wjcc.v3.i7.556. PMID: 26244149; PMCID: PMC4517332.

Dedinszki D, Szeri F, Kozák E, Pomozi V, Tőkési N, Mezei TR, Merczel K, Letavernier E, Tang E, Le Saux O, Arányi T, van de Wetering K, Váradi A (2017) Oral administration of pyrophosphate inhibits connective tissue calcification. EMBO Mol Med. 9(11):1463-1470. https://doi.org/10.15252/emmm.201707532. PMID: 28701330; PMCID: PMC5666306.

Deguchi K, Ueno T, Matsuura R, Yamanaka H, Nara K, Uehara S, Tazuke Y, Bessho K, Okuyama H (2016) Disseminated Metastatic Tissue Calcification After Orthotopic Liver Transplantation: A Case Report. Transplant Proc. 48(1):251-4. https://doi.org/10.1016/j.transproceed.2015.12.033. PMID: 26915877.

Egbuna OI, Taylor JG, Bushinsky DA, Zand MS (2007) Elevated calcium phosphate product after renal transplantation is a risk factor for graft failure. Clin Transplant. 21(4):558-66. https://doi.org/10.1111/j.1399-0012.2007.00690.x. PMID: 17645720.

Ehsani AH, Abedini R, Ghiasi M, Hoseini MS (2006) Calcinosis cutis complicating liver transplantation. Dermatol Online J. 12(7):23. PMID: 17459309.

Fabbroni D, Bellamy M (2006) Anaesthesia for hepatic transplantation, Continuing Education in Anaesthesia Critical Care & Pain, Volume 6, Issue 5, October 2006, Pages 171–175, https://doi.org/10.1093/bjaceaccp/mkl040.

Fricker J (2017) Thomas Starzl. BMJ, j1806. https://doi.org/10.1136/bmj.j1806

Frunza-Stefan S, Poola-Kella S, Silver K (2018) Non-uraemic calciphylaxis (NUC) postliver transplantation. BMJ Case Reports. bcr-2018–226537. https://doi.org/10.1136/bcr-2018-226537. PMID: 30361453; PMCID: PMC6203001.

Furman BL (2007) Bisphosphonates. xPharm: The Comprehensive Pharmacology Reference, 1–3. https://doi.org/10.1016/b978-008055232-3.61038-4.

Germain DP (2017) Pseudoxanthoma elasticum. Orphanet J Rare Dis. 12(1):85. https://doi.org/10.1186/s13023-017-0639-8. PMID: 28486967; PMCID: PMC5424392.

Gray TA, Buckley BM, Sealey MM, Smith SCH, Tomlin P, Mcmaster P (1986) Plasma Ionized Calcium Monitoring During Liver Transplantation1. Transplantation 41(3):335–339. https://doi.org/10.1097/00007890-198603000-00010.

Grinyó JM (2013) Why is organ transplantation clinically important? Cold Spring Harb Perspect Med. 3(6):a014985. https://doi.org/10.1101/cshperspect.a014985. PMID: 23732857; PMCID: PMC3662355.

Gutierrez LB, Link T, Chaganti K, Motamedi D (2016) Arterial calcification due to CD73 deficiency (ACDC): imaging manifestations of ectopic mineralization. Skeletal Radiol. 45(11):1583-7. https://doi.org/10.1007/s00256-016-2465-9. Epub 2016 Aug 26. PMID: 27562569.

Gwinner W, Suppa S, Mengel M, Hoy L, Kreipe HH, Haller H, Schwarz A (2005) Early calcification of renal allografts detected by protocol biopsies: causes and clinical implications. Am J Transplant. 5(8):1934-41. https://doi.org/10.1111/j.1600-6143.2005.00938.x. PMID: 15996242.

Hendig D, Schulz V, Arndt M, Szliska C, Kleesiek K, Götting C (2006) Role of Serum Fetuin-A, a Major Inhibitor of Systemic Calcification, in Pseudoxanthoma Elasticum, Clinical Chemistry, 52(2):227–234, https://doi.org/10.1373/clinchem.2005.059253.

Herrmann M, Kinkeldey A, Jahnen-Dechent W (2012) Fetuin-A function in systemic mineral metabolism. Trends Cardiovasc Med. 22(8):197-201. https://doi.org/10.1016/j.tcm.2012.07.020. Epub 2012 Aug 16. PMID: 22902178.

Ho OTW, Ng WCA, Ow ZGW, Ho YJ, Lim WH, Yong JN, Wang RS, Wong KL, Ng CH, Muthiah MD, Teo CM (2021) Bisphosphonate therapy after liver transplant improves bone mineral density and reduces fracture rates: an updated systematic review and meta-analysis. Transpl Int. 34(8):1386-1396. https://doi.org/10.1111/tri.13887. Epub 2021 Jul 1. PMID: 33884669.

Ilona S, Athanase B, Carlos L, Anders FC, Magnus B, Simon T (2021) Telomere Length in Valve Tissue Is Shorter in Individuals with Aortic Stenosis and in Calcified Valve Areas, Front. Cell Dev. Biol. vol.9, DOI https://doi.org/10.3389/fcell.2021.618335.

Imaoka Y, Ohira M, Nakano R, Shimizu S, Kuroda S, Tahara H, Ide K, Kobayashi T, Ohdan H (2019) Impact of Abdominal Aortic Calcification Among Liver Transplantation Recipients. Liver Transpl. 25(1):79-87. https://doi.org/10.1002/lt.25311. PMID: 30021054.

Isaacs T, Henderson C, Cole MC, Das S (2021) TP070 Diffuse lung disease and rare disease/ Thematic Poster Session Rapid Pulmonary Calcification in a Liver Transplant Patient

Jansen RS, Duijst S, Mahakena S, Sommer D, Szeri F, Váradi A, Plomp A, Bergen AA, Oude Elferink RP, Borst P, van de Wetering K (2014) ABCC6-mediated ATP secretion by the liver is the main source of the mineralization inhibitor inorganic pyrophosphate in the systemic circulation-brief report. Arterioscler Thromb Vasc Biol. 34(9):1985–9. https://doi.org/10.1161/ATVBAHA.114.304017. Epub 2014 Jun 26. PMID: 24969777; PMCID: PMC6743317

Jawan B, de Villa V, Luk HN, Chen YS, Chiang YC, Wang CC, Wang SH, Cheng YF, Huang TL, Eng HL, Liu PP, Chen CL (2003) Ionized calcium changes during living-donor liver transplantation in patients with and without administration of blood-bank products. Transpl Int. 16(7):510-4. https://doi.org/10.1007/s00147-003-0578-2. Epub 2003 Apr 10. PMID: 12687324.

Jeng KS, Huang CC, Lin CK, Lin CC, Chen KH (2017) Graft calcification caused by a torsion of the hepatic vein after a living-related donor liver transplantation. Ann Hepatol. 16(1):164-168. https://doi.org/10.5604/16652681.1226954. PMID: 28051807.

Jiménez-Gallo D, Ossorio-García L, Linares-Barrios M (2015) Calcinosis Cutis and Calciphylaxis. Actas Dermo-Sifiliográficas (English Edition), 106(10):785–794. https://doi.org/10.1016/j.adengl.2015.10.018.

Jiménez-Pérez M, González-Grande R, Omonte Guzmán E, Amo Trillo V, Rodrigo López JM (2016) Metabolic complications in liver transplant recipients. World J Gastroenterol. 22(28):6416-23. https://doi.org/10.3748/wjg.v22.i28.6416. PMID: 27605877; PMCID: PMC4968123

Jucglà A, Sais G, Curco N, Marcoval J, Moreno A, Peyri J (2006) Calcinosis Cutis Following Liver Transplantation: A Complication Of Intravenous Calcium Administration. Br J Dermatol 132(2):275–278. https://doi.org/10.1111/J.1365-2133.1995.Tb05026.X.

Kalantari F, Miao D, Emadali A, Tzimas GN, Goltzman D, Vali H, Chevet E, Auguste P (2007) Cellular and molecular mechanisms of abnormal calcification following ischemia-reperfusion injury in human liver transplantation. Mod Pathol. 20(3):357-66. https://doi.org/10.1038/modpathol.3800747. PMID: 17334330.

Kawai K, Sato Y, Kawakami R, Sakamoto A, Cornelissen A, Mori M, Ghosh S, Kutys R, Virmani R, Finn AV (2022) Generalized Arterial Calcification of Infancy (GACI): Optimizing Care with a Multidisciplinary Approach. J Multidiscip Healthc. 15:1261-1276. https://doi.org/10.2147/JMDH.S251861. PMID: 35677616; PMCID: PMC9167688.

Kranenburg G, de Jong PA, Bartstra JW, Lagerweij SJ, Lam MG, Ossewaarde-van Norel J, Risseeuw S, van Leeuwen R, Imhof SM, Verhaar HJ, de Vries JJ, Slart RHJA, Luurtsema G, den Harder AM, Visseren FLJ, Mali WP, Spiering W (2018) Etidronate for Prevention of Ectopic Mineralization in Patients With Pseudoxanthoma Elasticum. J Am Coll Cardiol. 71(10):1117-1126. https://doi.org/10.1016/j.jacc.2017.12.062. PMID: 29519353.

Larralde M, Giachetti A, Kowalczuk A, D'Agostino D, Galimberti R (2003) Calcinosis cutis following liver transplantation in a pediatric patient. Pediatr Dermatol. 20(3):225-8. https://doi.org/10.1046/j.1525-1470.2003.20308.x. PMID: 12787271.

Lateo S, Charlton F, Hudson M, Lawrence CM (2005a) Cutaneous calcification following liver transplantation. Clin Exp Dermatol. 30(5):484-6. https://doi.org/10.1111/j.1365-2230.2005.01793.x. PMID: 16045672.

Lateo S, Charlton F, Hudson M, Lawrence CM (2005b) Cutaneous calcification following liver transplantation. Clin. Exp. Dermatol. 30(5):484-486.

Laurain A, Rubera I, Razzouk-Cadet M, Bonnafous S, Albuquerque M, Paradis V, Patouraux S, Duranton C, Lesaux O, Lefthériotis G, Tran A, Anty R, Gual P, Iannelli A, Favre G (2022) Arterial Calcifications in Patients ith Liver Cirrhosis Are Linked to Hepatic Deficiency of Pyrophosphate Production Restored by Liver Transplantation. Biomedicines. 10(7):1496. https://doi.org/10.3390/biomedicines10071496. PMID: 35884801; PMCID: PMC9312703.

Li Q, Uitto J (2013) Mineralization/anti-mineralization networks in the skin and vascular connective tissues. Am J Pathol. 183(1):10–8. https://doi.org/10.1016/j.ajpath.2013.03.002. Epub 2013 May 8. PMID: 23665350; PMCID: PMC3702739.

Li K, Xu Y (2015) Citrate metabolism in blood transfusions and its relationship due to metabolic alkalosis and respiratory acidosis. Int J Clin Exp Med. 8(4):6578–84. PMID: 26131288; PMCID: PMC4483798.

Li Q, Uitto J (2012) Heritable ectopic mineralization disorders: the paradigm of pseudoxanthoma elasticum. J Invest Dermatol. 132(E1):E15-9. https://doi.org/10.1038/skinbio.2012.5. PMID: 23154625; PMCID: PMC4006979.

Libson E, Wechsler RJ, Steiner RM (1993) Pulmonary calcinosis following orthotopic liver transplantation. J Thorac Imaging. 8(4):305-8. https://doi.org/10.1097/00005382-199323000-00007. PMID: 8246329.

Luo H, Li Q, Cao Y, Uitto J (2020) Therapeutics Development for Pseudoxanthoma Elasticum and Related Ectopic Mineralization Disorders: Update 2020. J Clin Med. 10(1):114. https://doi.org/10.3390/jcm10010114. PMID: 33396306; PMCID: PMC7795895.

Ma H, Wang P, Jin D, Jia T, Mao H, Zhang J, Zhao S (2018) The hepatic ectonucleotide pyrophosphatase/phosphodiesterase 1 gene mRNA abundance is reduced by insulin and induced by dexamethasone. Braz J Med Biol Res. 51(4):e6980. https://doi.org/10.1590/1414-431X20176980. PMID: 29513794; PMCID: PMC5856437.

Makowka L, Stieber AC, Sher L, Kahn D, Mieles L, Bowman J, Marsh JW, Starzl TE (1988) Surgical technique of orthotopic liver transplantation. Gastroenterol Clin North Am. 17(1):33–51. PMID: 3292430; PMCID: PMC3228369.

Makwana J, Paranjape S, Goswami J (2010) Antifibrinolytics in liver surgery. Indian J Anaesth. 54(6):489-95. https://doi.org/10.4103/0019-5049.72636. PMID: 21224964; PMCID: PMC3016567.

Merritt WT (2000) Metabolism and liver transplantation: review of perioperative issues. Liver Transpl. 6(4 Suppl 1):S76-84. https://doi.org/10.1002/lt.500060515. PMID: 10915196.

Millán Á, Lanzer P, Sorribas V (2021) The Thermodynamics of Medial Vascular Calcification. Front Cell Dev Biol. 9:633465. https://doi.org/10.3389/fcell.2021.633465. PMID: 33937234; PMCID: PMC8080379.

Moreno R, Berenguer M (2006) Post-liver transplantation medical complications. Ann. Hepatol. 5(2): 77–85. https://doi.org/10.1016/s1665-2681(19)32022-8.

Munoz SJ, Nagelberg SB, Green PJ, Angstadt JD, Yang SL, Jarrell BE, Maddrey WC (1988) Ectopic soft tissue calcium deposition following liver transplantation. Hepatology. 8(3):476-83. https://doi.org/10.1002/hep.1840080307. PMID: 3286457.

Neau-Cransac M, Lepreux S, Labreze C, Quinart A, Revel P, Saric J, Bioulac-Sage P (2005) Uncommon Localization of Calcinosis Cutis after Liver Transplantation. Dermatology 211(4):379–381. https://doi.org/10.1159/000088516.

Nitschke Y, Yan Y, Buers I Kintziger K, Ask,ew K Rutsch F (2018) ENPP1-Fc prevents neointima formation in generalized arterial calcification of infancy through the generation of AMP. Exp Mol Med 50:1–12 https://doi.org/https://doi.org/10.1038/s12276-018-0163-5.

Nseir V, Bradauskaite G, Pedroza M, Minimo C, Zaki R, Chewaproug D (2021) A Rare Case of Calciphylaxis in an Orthotopic Liver Transplant Recipient with Acute Kidney Injury. Exp Clin Transplant. 19(4):382-385. https://doi.org/10.6002/ect.2017.0123. Epub 2018 Apr 9. PMID: 29633925.

Oschatz E, Benesch T, Kodras K, Hoffmann U, Haas M (2006) Changes of coronary calcification after kidney transplantation. Am J Kidney Dis. 48(2):307-13. https://doi.org/10.1053/j.ajkd.2006.04.066. PMID: 16860198.

Peeters FECM, Dudink EAMP, Weijs B, Fabritz L, Chua W, Kietselaer BLJH, Wildberger JE, Meex SJR, Kirchhof P, Crijns HJGM, Schurgers LJ (2020) Biomarkers Associated With Aortic Valve Calcification: Should We Focus on Sex Specific Processes? Front Cell Dev Biol. 8:604. https://doi.org/10.3389/fcell.2020.00604. PMID: 32754594; PMCID: PMC7366171.

Phadwal K, Vrahnas C, Ganley IG, MacRae VE (2021) Mitochondrial Dysfunction: Cause or Consequence of Vascular Calcification? Front Cell Dev Biol. 16(9):611922. https://doi.org/10.3389/fcell.2021.611922. PMID: 33816463; PMCID: PMC8010668

Prabhakar S, Tuffaha AM (2018) Non-Uremic Calciphylaxis Post Liver Transplantation: A Case Report and Literature Review of an Unusual Presentation of a Rare Disease. Am. J. Med. Case Rep. 19:118-122. https://doi.org/10.12659/ajcr.906981. PMID: 29386500; PMCID: PMC5804298.

Quaglino D, Boraldi F, Lofaro FD (2020) The biology of vascular calcification. Int Rev Cell Mol Biol. 354:261-353. https://doi.org/10.1016/bs.ircmb.2020.02.007. Epub 2020 Apr 6. PMID: 32475476.

Rando K, Vázquez M, Cerviño G, Zunini G (2014) Hypocalcaemia, hyperkalaemia and massive haemorrhage in liver transplantation. Rev. Col. Anest. 42(3):214–219. https://doi.org/10.1016/j.rcae.2014.04.003.

De Rave S, Hansen BE, Groenland TH, Kazemier G, de Man RA, Metselaar HJ, Terpstra OT, Tilanus HW, Ijzermans JH, Schalm SW (2005) Heterotopic vs. orthotopic liver transplantation for chronic liver disease: a case-control comparison of short-term and long-term outcomes. Liver Transpl. 11(4):396–401. https://doi.org/10.1002/lt.20376. PMID: 15776459.

Reardon S (2022) First pig-to-human heart transplant: what can scientists learn? Nature. 601(7893):305-306. https://doi.org/10.1038/d41586-022-00111-9. PMID: 35031782.

Reddy MS, Varghese J, Venkataraman J, Rela M (2013) Matching donor to recipient in liver transplantation: Relevance in clinical practice. World J Hepatol. 5(11):603-11. https://doi.org/10.4254/wjh.v5.i11.603. PMID: 24303088; PMCID: PMC3847943.

Rodger D, Hurst DJ (2022) Mathieu Jaboulay's (1860-1913) contribution to xenotransplantation. Xenotransplantation. 29(5):e12765. https://doi.org/10.1111/xen.12765. Epub 2022 Jun 13. PMID: 35695309.

Shen MJ, Jiao K, Wang CY, Ehrlich H, Wan MC, Hao DX, Li J, Wan QQ, Tonggu L, Yan JF, Wang KY, Ma YX, Chen JH, Tay FR, Niu LN (2022) Extracellular DNA: A Missing Link in the Pathogenesis of Ectopic Mineralization. Adv Sci (Weinh). 9(5):e2103693. https://doi.org/10.1002/advs.202103693. Epub 2021 Dec 23. Erratum in: Adv Sci (Weinh). 2022 Apr;9(11):e2201368. PMID: 34939364; PMCID: PMC8844461.

Shimada BK, Pomozi V, Zoll J, Kuo S, Martin L, Le Saux O (2021) ABCC6, Pyrophosphate and Ectopic Calcification: Therapeutic Solutions. Int J Mol Sci. 22(9):4555. https://doi.org/10.3390/ijms22094555. PMID: 33925341; PMCID: PMC8123679.

Song JH, Liu MY, Ma YX, Wan QQ, Li J, Diao XO, Niu LN (2022) Inflammation-associated ectopic mineralization, Fundamental Research, ISSN 2667-3258. https://doi.org/10.1016/j.fmre.2022.04.020

Tagliacozzi G (1597) Decurtorum cirugia per insitionum. Bindonum, Venice, Italy.

Tzimas GN, Afshar M, Emadali A, Chevet E, Vali H, Metrakos PP (2004) Correlation of cell necrosis and tissue calcification with ischemia/reperfusion injury after liver transplantation. Transplant Proc. 36(6):1766-8. https://doi.org/10.1016/j.transproceed.2004.06.013. PMID: 15350473.

Väärämäki S, Pelttari S, Uusitalo H, Tökési N, Váradi A, Nevalainen PI (2019). Pyrophosphate Treatment in Pseudoxanthoma Elasticum (PXE)-Preventing Re-Occlusion After Surgery for Critical Limb Ischaemia. Surgical Case Reports, 2(4):1-3.

Vanakker OM, Hosen MJ, Paepe AD (2013) The ABCC6 transporter: what lessons can be learnt from other ATP-binding cassette transporters? Front Genet. 4:203. https://doi.org/10.3389/fgene.2013.00203. PMID: 24137173; PMCID: PMC3797522.

Winter EM, Pollard AJ, Chapman S, Kelly D, Spencer D (1995) Pulmonary calcification after liver transplantation in children. Brit. J. Radiol., 68(812):923-925.

Crosslinking Mechanisms in the Designing of Scaffolds for Biomedical Applications: A Review

Catherine Ann Martin, Subathra Radhakrishnan, Josette Caroline Martin, Nivethaa EAK, Shanthini GM, Geethanjali Dhayanithy, Olga V. Frank-Kamenetskaya, Narayana Subbaraya Kalkura, and Mohamed Rela

Abstract Biomaterials are designed to have direct interaction with the living system and are aiming to reinstate the paradigmatic tissue grafting techniques by augmenting natural functions. Metals, ceramics, and polymers are often used in combination to develop smart and innovative biomaterials. However, their low mechanical stiffness and degradation are of significant concern. In this regard, crosslinkers are being used to overcome these limitations. Although crosslinkers increase mechanical stiffness, they also result in some adverse reactions and cause toxicity. A variety of crosslinkers and methods have been used in the last two decades to synthesize scaffolds for tissue engineering. The classical crosslinker, glutaraldehyde, which can exist at least in 13 forms, has been used in numerous scaffolds, despite its subtle toxicity. Researchers and scientists have tried to replace discordant chemicals with green chemicals, plant-derived chemicals, enzymatic, non-enzymatic chemicals, and physical methods. The classification of crosslinkers based on their applications has been discussed. This chapter summarizes the types of crosslinkers, and the mechanism of crosslinking with a comparison of their advantages, efficiency, efficacy, and limitations. The

C. A. Martin · S. Radhakrishnan · M. Rela (✉)
National Foundation for Liver Research, CLC Works Road, Chromepet, Chennai 600044, India
e-mail: mohamed.rela@relainstitute.com

J. C. Martin
Sri Venkateshwaraa Medical College Hospital and Research Centre, Pondicherry 605102, India

N. EAK
B.S. Abdur Rehman Crescent Institute of Science and Technology, Chennai 600048, India

S. GM
Bharath Institute of Higher Education and Research, Chennai 600126, India

G. Dhayanithy · N. S. Kalkura
Crystal Growth Centre, Anna University, Guindy, Chennai 600025, India

O. V. Frank-Kamenetskaya
Institute of Earths Sciences, St. Petersburg State University, St. Petersburg, Russia

O. V. Frank-Kamenetskaya et al. (eds.), *Biogenic—Abiogenic Interactions in Natural and Anthropogenic Systems 2022*, Springer Proceedings in Earth and Environmental Sciences, https://doi.org/10.1007/978-3-031-40470-2_2

pursuit of a perfect biomaterial that flawlessly matches the surrounding microenvironment is a never-ending challenge within biomedical research. This chapter, however, covers the recent advances in crosslinking methodologies, providing scope for further innovations in the field of tissue engineering.

Keywords Tissue engineering · Scaffolds · Crosslinking mechanisms

1 Introduction

Extensive research in biomaterials and tissue regeneration has been carried out in the past two decades, which has revolutionized the field of tissue engineering. Polymer-based scaffolds or hydrogels are requisites for tissue engineering and regenerative medicine. These polymeric materials support the regeneration/repair of damaged tissues or organs by mimicking the natural ECM and delivery of bioactive molecules such as drugs, growth factors, proteins, peptides, etc. In addition, they also provide hydration to the cells (Aguero et al. 2021; Guo et al. 2019). The designing of these scaffolds involves various strategies and numerous criteria that need to be satisfied related to the structure and function of the target site. The four main criteria for a successful scaffold include (i) biocompatibility: the most vital feature which involves cell adhesion, proliferation, and migration. The very first step towards regeneration. (ii) biodegradability: this functions in coherence with regeneration. An ideal scaffold degrades as tissue growth occurs, allowing the rejuvenation of the diseased tissue. (iii) mechanical properties: the mechanical stiffness of the scaffolds should be equivalent to that of the injured site. In some cases, like bone replacement materials mechanical stability also refers to surgical handling when implanted in the diseased site. (iv) scaffold architecture: last but not the least, an interconnected porous structure that plays an important role in cell attachment and exchange of nutrients and gases. Recent advances highlight that even micropores influence cell adhesion (Fergal et al. 2011; Pina et al. 2019). Polymeric materials can be synthesized in various forms like injectable hydrogels, thermo-responsive hydrogels, electrospun mats, 3D porous scaffolds, nanoparticles/microspheres, core-shell structures, etc. The production of the scaffolds in accordance with the application depends on the fabrication technique and effective crosslinking mechanisms. Thus, it is clearly evident that crosslinking plays a vital role in the synthesis of scaffolds for various biomedical applications.

Scaffolds for biomedical applications are usually synthesized using a combination of natural and synthetic polymers. Nevertheless, in aqueous environments, they lack the mechanical strength which is an essential prerequisite for medical applications. Generally, a crosslinker is used to enhance the mechanical property of the polymers used. Hence, Crosslinking is a fundamental process of tissue engineering which initiates physical, chemical, or enzymatic bonds between polymer chains to create a 3D matrix with enhanced degradability, biological, mechanical, and thermal stability (Jiang et al. 2022; Oryan et al. 2018). The crosslinkers contain instantly reactive chemical groups which link themselves to the functional groups present in the protein/

solid surfaces/drugs/other molecules of the scaffold thereby predominantly varying the stiffness and stability. In addition, the crosslinking agent largely determines the pore size and texture of the material synthesized.

Crosslinkers are classified into five types based on their functions as illustrated in Fig. 1. Zero-length crosslinkers are the small group of molecules that aids in the coupling of two polymers/molecules by creating a covalent bond, without insertion of any additional spacer molecules e.g., Carbodiimide, Woodward's reagent K, N-N-carbonyl diimidazole (Nadeau 2005; Hermanson 2013). As the name suggests homobifunctional crosslinkers have two identical spacer arms. They are symmetrical in nature with carbon chain flanked by identical reactive ends e.g., homobifunctional NHS esters, homobifunctional sulfhydryl reactive crosslinkers, and homobifunctional photoreactive crosslinkers. Contrastingly, heterobifunctional crosslinkers have two entirely different spacers arms that functionally aim to couple two different polymers with two distinct bond types e.g., amine-reactive and sulfhydryl reactive crosslinkers, amine-reactive and photoreactive crosslinkers, arginine reactive and photoreactive crosslinkers. Trifunctional crosslinkers represent a very small and complex group of crosslinkers that possesses three entirely varied reactive groups that enable them to bioconjugate firmly e.g., 4-Azido-2-Nitrophenylbiocytin-4-Nitrophenyl Ester, sulfosuccinimidyl-2-[6-(biotinamido)-2-(p-azidobenzamido) hexanoamido]ethyl-1,3′-dithiopropionate (Sulfo-SBED). Finally, dendrimers and dendrons, are massive artificial macromolecules characterized by a very dense structure with huge numbers of varied functional groups. Dendrimers are used as novel crosslinkers e.g., poly(amidoamine) type (called PAMAM dendrimer) (Abbasi et al. 2014; Duan et al. 2005; Hermanson 2013). A pictorial representation is shown in Fig. 2.

Choosing an appropriate cross-linking agent enables the desirable polymerization with suitable physical properties. The biomaterials synthesized for medical applications for e.g., bone tissue engineering must possess shapes that fit anatomically in the host. The amalgamation of the properties of polymers/materials and crosslinkers used to synthesize a 3D matrix play a vital role in the development of desirable scaffolds for clinical applications. Some polymers such as thermoplastic elastomers need physical cross-linking. Unlike chemical, physical crosslinking is weak, mostly formed because of hydrogen bonds/electrostatic forces/ionic bonds. The strength and

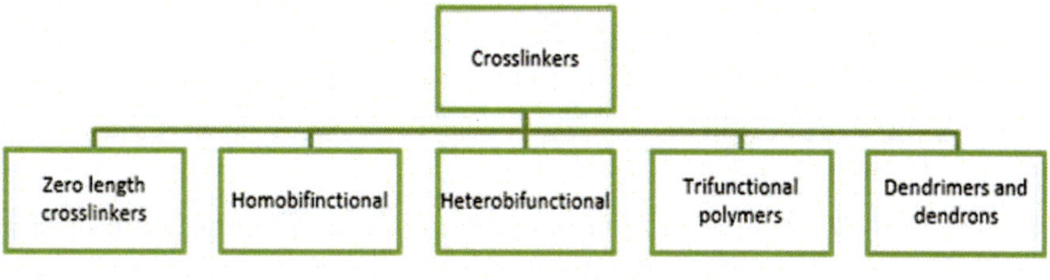

Fig. 1 Classification of crosslinkers. Adapted from Hermanson (2013)

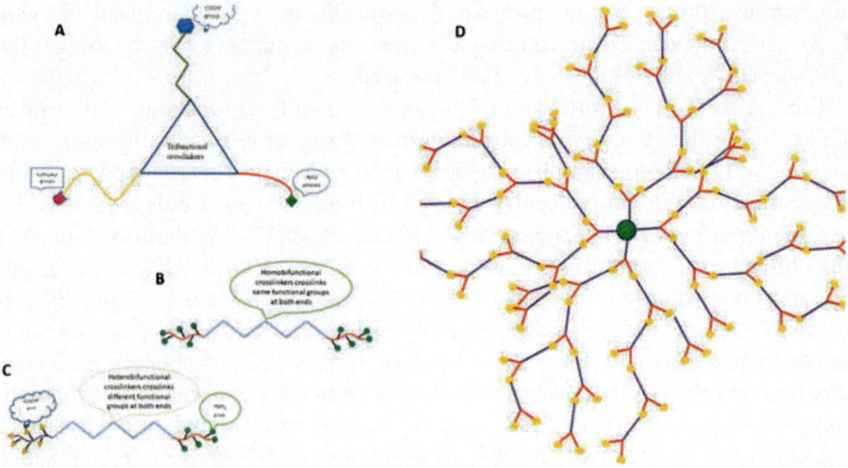

Fig. 2 Different types of Crosslinkers. Trifunctional Crosslinker (**a**). Homobifunctional crosslinkers (**b**). Heterobifunctional crosslinkers (**c**). Dendrimers (**d**)

durability are comparatively lesser than chemical cross-linking. Besides, intramolecular crosslinking is also used to stabilize the structure of polymers which can be either reversible or irreversible crosslinking. Reversible crosslinking includes dynamic bonds which could easily be disrupted and reorganized. This behaviour helps in improving the stability of the polymer and to modify the 3D structure according to the applications and environmental conditions (Chen et al. 2020a, b). This feature finds application in designing and fabricating stimuli-responsive biomaterials. While, in the case of irreversible crosslinking, the crosslinks will be stable in adverse conditions of pressure, stress, and similar environmental conditions. The irreversible crosslinking can be attained by click chemistry, metathesis, photodimerization, and coupling reactions.

Recently, the bioactive hydrogels are mildly crosslinked with biocompatible enzymatic crosslinking (e.g., Sortase A). The major advantage of enzymatic crosslinking is it can be controlled by physiological conditions such as PH and temperature (Heck et al. 2013). Herein, this chapter deals with the crosslinking mechanisms, applications, and recent advances.

2 Physical Crosslinking

As stated earlier, physical crosslinking is less strong compared to chemical crosslinking. But, it is one of the favourite choice of crosslinking, as it does not elicit any biological toxicity.

2.1 *Dehydrothermal (DHT) Method*

DHT is a physical type of crosslinking employed to improve the structural properties of polymers like collagen and to lower the degree of swelling (Chen et al. 2020a, b). This process is carried out at a temperature above 100 °C, under vacuum that results in dehydration to form an intermolecular bond between the functional groups like carboxyl and amine or hydroxyl groups. The optimized temperature range is reported to be between 130 °C to 140 °C for collagenous materials (Salvatore et al. 2021). This method could facilitate both crosslinking and sterilization concurrently, based on the exposure time and temperature which in turn reduces the immune response and increases cell activity. Chen et al. reported that DHT treatment of collagen films for a period of 1 week at a temperature range of 85–145 °C resulted in an increase in the crosslinking density with temperature and time until 3 days, after which the films started to denature resulting in decreased mechanical properties of crosslinked collagen films (Chen et al. 2020a, b). Higher temperature exposure improves the crosslinking density of collagen–GAG scaffolds, however, no considerable change in crosslinking is observed with respect to time. One of the major disadvantages of this technique is that in the higher temperature range and longer duration of exposure, collagen loses its structure due to denaturation (Haugh et al. 2009). The quantity of free residues in collagen structure has reduced considerably with dehydrothermal treated collagen scaffolds as these free acidic and basic residues are involved in crosslinking of the collagen scaffold by a condensation reaction (Silver et al. 1979). This method is highly favourable in improving the porosity, mechanical property, and aqueous stability of collagen-based biomaterials (Zhang et al. 2018). The major advantage of this process is that they do not involve any harsh chemicals that could be toxic to cells, which makes DHT a much-preferred technique.

2.2 *Irradiation Application*

Radiation crosslinking is done using beta and gamma radiation. Here the polymers are subjected to a predetermined dosage of radiation, which is absorbed by the material, causing the breakage of chemical bonds and the creation of free radicals which in turn facilitate the formation of new chemical bonds. Other forms of radiation like UV irradiation and photo crosslinking using visible light have also been reported for collagen-based scaffolds and chitosan nanoparticles respectively (Burgstaller et al. 2020).

2.3 Electron Beam Application

As the electron beam is passed through the accelerator, factors such as dosage, dose rate, voltage, and current control the crosslinking density. With the modification in these parameters, the ratio of free radical formation varies which further enhances crosslinking of the polymer chains by forming additional bonds between them. The major advantage of this technique is that it preserves the structural morphology of the exposed/treated material. But, in the case of higher dosage of electron beams, the polymer chains break down and result in weight loss of the material (Stelescu et al. 2018).

2.4 Plasma Method

In this method, a single-step plasma process is used to initiate the crosslinking of polymers that are water soluble in the liquid phase. The plasma treatment leads to the formation of radicals that recombine producing crosslinks. As plasma cannot penetrate through the material this is considered as a surface modification technique and has been employed for crosslinking polymers like gelatin (Liguori et al. 2016) and starch (Taslikh et al. 2022).

2.5 UV Light Application

UV irradiation-based physical crosslinking technique utilizes a highly energetic UV light that is passed on to the scaffolds to initiate photocrosslinking. In this technique, inter and intramolecular photodimerization aids in controlling the crosslinking density. The crosslinking density enhances with an increase in UV exposure until saturation of crosslinking density above which the UV light will not cause any further crosslinking. Researchers also experienced lesser water stability of biomaterials crosslinked with a longer duration of time, while materials exposed for a time interval of 5–10 min are stable for an extended period of time. The problem with utilizing UV irradiation for crosslinking is that it cannot reach the complete depth of the samples, resulting in uneven crosslinking (or) just crosslinking the surface of the material. Also, the UV irradiation-based crosslinking technique necessitates the presence of photoinitiators for the initiation of cross-linking. Generally, acrylates are used as initiators of photocrosslinking. Aryl azides and diazirines are also used as initiators. But post crosslinking, excess unreacted acryl residues may be toxic to cells or may elicit an immune response and in some instances might even cause genetic mutation. In order to overcome these drawbacks non-acrylate-based UV crosslinking can be carried out by using riboflavin i.e., vitamin B2. Riboflavin is a natural source of

UV curing material that can efficiently mediate and crosslink biomaterials by undergoing a radical polymerization reaction (Choi and Cha 2019). Combining riboflavin and gelatin with UV initiation, results in the formation of hydrogels that possess good rheological and mechanical properties. The crosslinked gelatin scaffold has been reported to have enhanced thermal properties with approximately 16 °C rise in transition temperature (Galdoporpora et al. 2019). A recent study using zein-poly ethylene oxide composite was crosslinked using UV irradiation. The improvement in tensile strength and Young's modulus was observed post-crosslinking which is due to the conformational changes of helical structure to beta sheets as the effect of UV treatment (Surendranath et al. 2022). In addition, collagen-based biomaterials are crosslinked with EDC by modifying the amino acid side chains of collagen which is responsible for integrin-assisted cell attachment. UV-assisted crosslinking helps in modifying the aromatic amino acids of collagen. The combined effect of EDC and UV on crosslinking exercises its effect on integrin binding which in turn exerts highly responsive cellular activity (Bax et al. 2019).

3 Chemical Crosslinking

This is the most commonly used crosslinking methods wherein crosslinking occurs due to the formation of a covalent chemical bond between the polymer chains. These crosslinks form by chemical reactions that can be initiated through heat, pressure, change in pH, or irradiation. Chemical crosslinkers are commonly used for improving the mechanical and degradation properties of polymeric scaffolds. The molecular structure of the various crosslinkers is shown in Fig. 3 and the reactive groups are listed in Table 1.

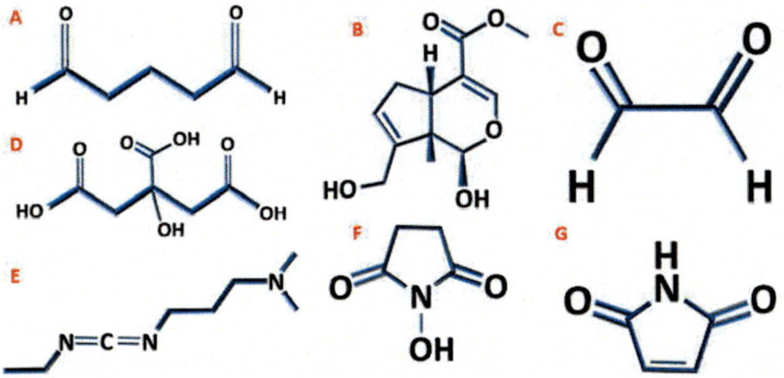

Fig. 3 Molecular structure of different crosslinkers. Glutaraldehyde (**a**). Structure of Genipin (**b**). Structure of Glyoxal (**c**). Structure of citric acid (**d**). Structure of EDC (**e**). Structure of NHS (**f**). Structure of Maleimide

Table 1 The reactive functional group of crosslinkers

S. no.	Chemical	Reactive group
1	Carbodiimide (e.g., EDC)	Carboxyl-to-amine reactive groups
2	Maleimide Haloacetyl (Bromo- or Iodo-) Pyridyldisulfide Thiosulfonate Vinylsulfone	Sulfhydryl-reactive groups
3	NHS ester Imidoester Pentafluorophenyl ester Hydroxymethyl phosphine	Amine-reactive groups
4	Diazirine Aryl azide	Photoreactive groups
5	Hydrazide Alkoxyamine	Aldehyde-reactive groups
6	Isocyanate	Hydroxyl (nonaqueous)-reactive groups

3.1 Small Molecule Application

3.1.1 Glutaraldehyde (GTA)Use

It is one of the most commonly used crosslinkers that chemically crosslink biopolymeric tissue scaffolds, composites, and hydrogels. It significantly enhances the durability and improves the mechanical properties of biomaterials. Intra and intermolecular bonds are formed through Schiff's base reaction by interacting with the amine and hydroxyl groups of polymers and proteins. Glutaraldehyde is widely used for chitosan-based scaffolds as it enhances biostability. In addition, the usage of glutaraldehyde as a crosslinker has impacted the morphology of the scaffold. One of the major disadvantages is due to the aldehyde group that has been reported to be toxic to the cells and causes inflammation in the body, limiting its use in various applications as a crosslinking agent (Barbosa et al., 2014; Hoffman et al. 2009; Wu et al., 2007; Ma et al. 2014; Liu et al. 2022a, b). In spite of the various limitations, GTA is still used as an effective crosslinker.

3.1.2 Carbodiimide-Containing Agent Use

These consist of a C=N=C bond. EDC 1-Ethyl-3-(3-dimethyl aminopropyl)-carbodiimide is a water-soluble carbodiimide which is a zero-length crosslinker that causes direct crosslinking of carboxylates (–COOH) to primary amines ($-NH_2$) without becoming part of the final crosslink between the target molecules. It can also react with hydroxyl and sulfhydryl functional groups. Even though the greatest crosslinking activity of EDC is obtained at a pH of 4.5 (acidic conditions) and in a buffer that is free of extraneous carboxyl and amine groups (e.g. 4-morpholinoethanesulfonic acid), this reaction can also happen at neutral pH (pH < 7.2) in a solution like PBS (Phosphate buffer saline). The residues of this method are all water soluble thus making it possible to remove these residues by merely washing the scaffold with distilled water. One notable property of carbodiimide cross-linked scaffolds is that it exhibits good cell adhesion potential (Cornwell et al. 2007). EDC-crosslinked materials exhibit poor biomechanical properties and more rapid biodegradation profiles when compared to GTA-crosslinked scaffolds (Ma et al. 2014; Yang et al. 2018).

3.1.3 Epoxy Compounds Use

These are three-membered rings with two carbon atoms and an oxygen atom. These compounds are capable of reacting with amine, hydroxyl, and carboxyl groups which renders them useful in preparing scaffolds for bone tissue engineering. It is also reported that epoxy cross-linked materials have increased elasticity, but exhibit cytotoxicity of the same order as that of glutaraldehyde which limits their application. A study by Poursamar et al. shows that an enhanced water-absorbing scaffold with uniform microstructure is obtained when epoxy crosslinking is used. In addition, a bi-functional epoxy crosslinker was employed to synthesize collagen-based scaffolds which demonstrated higher thermal and biocompatibility to human corneal and neural progenitor cells when compared to EDC-NHS crosslinking (Koh et al. 2013). Similarly, epoxied chitosan chloride combined with porcine acellular dermal matrix displayed higher mechanical and thermal stability which impacted cell adhesion and proliferation (Zheng et al. 2021).

3.1.4 Genipin Use

It is a safe, natural crosslinking agent extracted from the fruit of Gardenia Jasminoides Ellis. It possesses various active groups such as hydroxyl and carboxyl groups. Apart from this, genipin has anti-inflammatory, neuroprotective, neurogenic, and antidepressant effects which give this compound therapeutic potential for diseases of the central nervous system. It has been used for crosslinking polymers like gelatin, collagen, and chitosan (Liu et al. 2022a, b; Výborný et al. 2019). A study involving chitosan incorporated with hydroxyapatite nanocrystals and L-arginine crosslinked with genipin showed an interconnected porous network for osteoinduction (Zafeiris et al. 2021). A recent study by Cassimjee et al. (2022) has reported a scaffold synthesized using gelatine, and chitosan in addition to polysaccharides and crosslinked with genipin displayed remarkable properties, proving to be suitable for neural regeneration as well as drug delivery vehicles.

3.1.5 Citric Acid Use

It is a naturally occurring organic acid that is usually extracted from citrus fruits via fermentation. This non-toxic crosslinker is believed to form cyclic anhydrides which esterify the available hydroxyl groups forming polymer crosslinks (Mali et al. 2018). Although crosslinking of biomaterials with citric acid helps in increasing the hemocompatibility, balancing the hydrophilicity of the polymer network, and enhancing the availability of binding sites, the potency is reduced when compared to aldehydes (Salihu et al. 2021; Xu et al. 2015). Interestingly, it has been reported that residual citric acid may have a plasticizing effect on scaffolds (Sharmin et al. 2022). Similarly, methylcellulose crosslinked with citric acid formed stable thermoresponsive hydrogels that supported cell sheets, which dissociated by just altering the temperature. This novel hydrogel could replace the use of plastic substrates for tissue culture (Bonetti et al. 2021). Various reports have suggested the use of citric acid with polymers like cellulose, chitosan, and starch to achieve effectively crosslinked materials.

3.1.6 Glyoxal Use

It is an organic compound, used as a crosslinker for various polymers like collagen, chitosan, and cellulose. Glyoxal reacts with the hydroxyl and amino groups to form polymer chains (Yang et al. 2005). In an innovative study, chitosan/gelatin hydrogels crosslinked with glyoxal were encapsulated with human platelet lysate where the release stimulated the growth and migration of HUVECs. It exhibits lower cytotoxicity when compared to glutaraldehyde in spite of it being an aldehyde. An interesting study utilized decellularized articular cartilage ECM crosslinked with glyoxal which resulted in the formation of a porous, highly elastic matrix with osteointegration properties. In addition, this study also elucidated the compatibility of the scaffold with human macrophages without an immune response (Browe et al. 2019).

Apart from these, crosslinkers can also be classified based on the functional groups to with which they are reactive as amine-reactive crosslinkers, sulfhydryl-reactive crosslinker groups, carbonyl reactive crosslinkers, and photoreactive crosslinkers.

Among the amine-reactive crosslinkers, N-Hydroxysuccinimide esters and imidoesters are the most commonly studied.

3.2 NHS Esters Application

These are reactive groups formed by the EDC-activation of carboxylate molecules. EDC activates carboxyl groups and forms an amine-reactive O-acylisourea intermediate which spontaneously reacts with primary amines to form an amide bond and an isourea by-product. NHS ester-activated crosslinkers react with primary amines at a pH 7.2–8.5 (alkaline) to yield stable amide bonds. Although *N*-hydroxysuccinimide is released during the reaction, it can be easily removed by dialysis or desalting. NHS-ester crosslinking reactions are usually performed in phosphate buffer at pH 7–9.0 for about 30 min to 4 h at room temperature or 4 °C. A recent study with EDC/NHS cross-linked silk fibroin scaffold showed enhanced mechanical, thermal and biocompatibility with olfactory ensheathing cells that makes it suitable as a neural repair material (Li et al. 2023). Similarly, collagen, chitosan, and silk fibroin were combined with EDC/NHS to form porous scaffolds that had improved physio-chemical properties which also displayed cytocompatibility with osteoblast cells leading to the development of a bone repair material (Grabska-Zielińska et al. 2021).

Imidoesters have been known to crosslink agents as early as 1977, when bifunctional imido esters were used to crosslink proteins (Schramm et al. 1977). Imidoester crosslinkers react with primary amines to form amidine bonds. The imidoester reactions are carried out at pH of ~10 using buffers like borate buffer in an amine free environment to ascertain specificity for primary amines as the crosslink has a positive charge at physiological pH due to the protonation of the amidine bond. The amidine bonds formed are reversible at high pH, causing a reduction in usage of imido esters in most applications. In spite of their limitations, dimethylsuberimidate (DS) has been used to crosslink decellularized human umbilical veins in comparison with EDC and glutaraldehyde (GTA). It was found that DS crosslinked scaffolds showed better physical and biomechanical properties which could be used as a substitute for EDC and GTA (Narayani et al. 2017).

3.3 Sulfhydryl-Reactive Groups Application (Case of Maleimides)

Maleimide-activated crosslinkers react specifically with sulfhydryl groups (–SH) to form stable thioether linkages at pH ranging in between 6.5 and 7.5. Homobifunctional maleimide crosslinkers convert disulfide bridges in protein structures to be to permanent, irreducible linkages between cysteines. Apart from this, maleimide is used in combination with NHS-ester in the form of heterobifunctional crosslinkers to bring about a controlled, two-step conjugation of purified peptides and proteins (Partis et al. 1983). Maleimide crosslinked PEG hydrogels were synthesized with MMP degradable peptides with increased cytocompatibility when tested with mesenchymal stem cells (Yu et al. 2016). Another recent study showed that maleimide-modified hyaluronic acid, gelatin, and PEG hydrogel had enhanced protein release, and excellent biocompatibility and could be a potential candidate for regenerative medicine and drug delivery (Yoo et al. 2021a, b).

Haloacetyls These crosslinkers mostly contain an iodoacetyl or a bromoacetyl group. Haloacetyls react with sulfhydryl groups to form stable thioether linkages at pH 7.2–9. Iodoacetyl reactions are generally performed in dark conditions, to limit free iodine generation.

Pyridyl disulfides react with sulfhydryl groups to form disulfide bonds over a broad pH range. Conjugates prepared using these crosslinkers can be cleaved easily by using disulfide reducing agents, such as dithiothreitol (DTT).

3.4 *Carbonyl-Reactive Groups Applications*

Hydrazides Mild oxidation of some sugar glycols using sodium meta-perodate produces carbonyls (ald oxidation o ehydes and ketones) in glycoproteins and other polysaccharide-containing. Hydrazide-activated crosslinkers then conjugate with these carbonyls, leading to the formation of hydrazone bonds at pH 5–7.

Alkoxyamine compounds also conjugate to carbonyls (aldehydes and ketones) in much the same manner as hydrazides.

4 Enzyme Crosslinking

Enzyme based crosslinkers have been of great interest to scientists worldwide as it mimics the natural crosslinking mechanism of the human body. Transglutaminases are enzymes produced by the cells during wound healing. These enzymes initiate the formation of a blood clot, an insoluble protein network, during traumatic injury by crosslinking fibrin (Franz et al. 2007). Transglutaminase is a highly conserved protein from prokaryotes to vertebrates. Besides blood clotting, transglutaminase also plays a role in other biological functions such as epidermal keratinization and regulation of erythrocyte membranes (Nonaka et al. 1989). Factor VIII is the most frequently used transglutaminase which is used to crosslink acyl and amine groups found on lysine and glutamine. Consequently, Ehrbar et al. (2007) has crosslinked PEG synthetic polymers that contain lysine and glutamine with factor VIII forming a hydrogel that can degrade during regeneration. The non-mammalian transglutaminase, microbial transglutaminase (mTGase) was first isolated from the culture medium of Streptoverticillium, a variant of Streptoverticillium mobaraense (sp. S-8112) (Halloran et al. 2006). mTGase is functionally similar to human transglutaminase and crosslinks protein by forming an acyl transfer between γ-carboxamide group of the glutamine and the ε-amino groups of lysine or any primary amino groups (Motoki et al. 1998). The major advantages of mTGase is the Ca^{2+} independence, smaller size, higher reaction rate and broader specificity, which makes it better for tissue engineering and drug delivery applications (Collighan et al. 2002). Broderick et al. (2005) showed that crosslinking gelatin based materials with mTGase has improved their mechanical stability. They have also proved that 3-D hydroxyapatite/collagen composites have increased mechanical stability along with high thermal stability. Biomaterials crosslinked with transglutaminase have exquisite properties that support enhanced adhesion, proliferation and differentiation of cells (Ciardelli et al. 2010). Cui et al. (2017) in his study showed that casein fibres exhibited better tensile strength and had slow release of the drug when crosslinked with transglutaminase. Hyaluronic acid nanogels loaded with doxorubicin were crosslinked with transglutaminase, that has been tested both *in vitro* and *in vivo* demonstrating increased antitumor activity (Yang et al. 2015). Another study by Yin et al. (2012) with casein hydrogels demonstrated controlled release of docetaxel. Altogether, advanced materials both for tissue engineering and drug delivery crosslinked with transglutaminases have shown promising results.

Other enzyme based crosslinkers like horseradish peroxidase (HRP) and hydrogen peroxidase are also being used for tissue engineering and drug delivery applications. HRP which is a heme containing enzyme is found in the root of horseradish (Armoracia rusticana) crosslinks polymer phenol compounds in the presence of H_2O_2 (Bae et al. 2015). Extensive studies have been carried out in the mechanism of HRP—catalysed crosslinking. Recent studies have shown that HRP-catalysed in situ-forming hydrogels have superior characteristics where their gelation time, rate of degradation and tensile properties can be altered according to the concentration of HRP and H_2O_2 (Lee et al. 2015).

The most significant advantages of using enzyme crosslinking is that it occurs under mild physiologic conditions without any harsh temperature change, toxic chemicals or any detrimental radiation and can be carried out even for opaque materials.

The various crosslinkers with their applications are listed in Table 2.

5 Conclusion: Recent Advancements

Recent developments in chemistry, biology and engineering have led to the evolution of hybrid biomaterials like complex hydrogels, micro/nanogels incorporating newer crosslinking mechanisms for tissue engineering and drug delivery applications with better outcomes are gaining popularity in contrast to the contemporary methods of physical, chemical and enzymatic or biological crosslinking that have been previously discussed. The hybrid or complex hydrogels/nanogels are synthesized using synthetic and biopolymers with nanoparticles or microstructures and bioactive substances like peptides or proteins (Mao-Hua et al. 2021). Photocrosslinking is one such method used to crosslink polymers forming a biodegradable network (Matsuda et al. 2002). Oligomers can be photocrosslinked if they possess the following photo polymerizable groups. Oligomers with end groups like (i) cinnamate-, coumarin- or thymine (ii) inter- and intramolecular crosslinking with phenyl azide-, dithiocarbamate- and benzophenone end groups and (iii) styryl-, fumarate- or (meth)acrylate end-groups can be used (Mizutani et al. 2002; Matsuda et al. 2000; Grijpma et al. 2005). Star shaped Poly (D, L-lactide) (PDLLA) was synthesized without reactive diluents using stereolithography, a type of photopolymerization

Table 2 Crosslinkers with their application

Crosslinking agent	Polymers used	Application	Method of fabrication	References
Citric acid	PVA Citric acid modified Cotton cellulose, gelatin Collagen, cellulose, gelatin, chitosan, PVA, starch Collagen, chondroitin sulfate, PVA, Hyaluronic acid Carboxymethyl cellulose Alginate/pectin and cellulose nanofibrils Polyvinyl alcohol Bacterial cellulose sponges crosslinked with combination of citric acid and glucose	Tissue engineering, medical applications Bone tissue engineering Bone tissue engineering, cancer therapy, wound dressings Adhesion and proliferation of HEK cells Bone tissue engineering Development of stem cells Vascular tissue engineering Non inflammatory response in macrophages	Electrospinning, microfibrous fabrication, freeze drying, hydrogel, porous sponge formation	Nataraj et al. (2020) Singh et al. (2020) Salihu et al. (2021) Delgado-Rangel et al. (2019) Priya et al. (2021) Lungu et al. (2021) Velutheril et al. (2019) Frone et al. (2020)
Maleimide	Thiol modified/bismaleimide Thiolated hyaluronic acid/chitosan/cyclodextrin/PEG/dextran Maleimide functionalized Gelatin Hyaluronan/maleimide Furyl-modified hydroxypropyl chitin/maleimide terminated PEG Maleimide-modified hyaluronic acid/gelatin/bifunctional thiolated PEG	Biomedical applications Diagnostics, matrix for protein and drug delivery, tissue engineering Artificial stem cell niches, tunable biophysical properties, intrinsic cell interaction motifs Injectable cartilage repair filler 3D cell culture and tissue repair Well characterized ECM based platform: regenerative medicine tissue engineering	Hydrogels are formed by Michael addition reaction Crosslinking using sulfhydryl groups Orthogonal addition of bioactive crosslinks 1st degree crosslinking with thiol click chemistry and 2nd degree crosslinking with thiol oxidation reaction Integration of diels adler click reaction Hydrogel formation	Guaresti et al. (2019) Summonte et al. (2021) Gilchrist et al. (2021) Yao et al. (2020) Bi et al. (2019) Yoo et al. (2021a, b)

(continued)

Table 2 (continued)

Crosslinking agent	Polymers used	Application	Method of fabrication	References
Transglutaminase	Gelatin Gelatin/mTGase Alginate dialdehyde/gelatin Sodium caseinate/starch/tricalcium phosphate Gelatin/alginate Marine collagen	Tunable bioactivity of collagen constructs Bone regeneration Cartilage, bone, blood vessel engineering Bone tissue engineering Bone tissue engineering Skin/bone/cartilage reconstruction	Hydrogel 3D IPN scaffolds Ionic calcium and mTGase crosslinked scaffolds mTGase crosslinked flexible sponges Enzyme based in-situ externally crosslinked hybrid hydrogels TGase/genipin hybrid crosslinking	Nair et al. (2020) Echave et al. (2019) Distler et al. (2020) Sengor (2022) Sood et al. (2022) Liu et al. (2022a, b)
Dehydrothermal Treatment	Collagen/hyaluronic acid Nanofibrillated cellulose/carboxymethyl cellulose Epigallocatechin gallate-modified gelatin Ammonia treated Collagen Decellularised meniscus ECM	Tissue engineering 3D bioprinting, bone regeneration Osteo regenerative therapy Tissue engineering Meniscus regeneration	Physical crosslinked scafflold Self-standing dual porous scaffold Vacuum heating and DHT DHT combined with GTA crosslinked scaffolds Porous scaffolds	Bavaresco et al. (2020) Mohan et al. (2020) Honda et al. (2018) Chen et al. (2021) Ding et al. (2022)

(continued)

Table 2 (continued)

Crosslinking agent	Polymers used	Application	Method of fabrication	References
Genipin	Chitosan/hyaluronic acid/ gelatin Fucoidan adsorbed nanohydroxyapatite/ hydroxypropyl chitosan Gelatin Chitosan/polyethylene oxide Collagen/chitosan/ gelatin Nanographene oxide/adipose stem cell derived ECM Human umbilical cord ECM Decellularized nerve ECM/ chitosan	Drug delivery vehicles, neural tissue engineering constructs Bone tissue engineering Tissue engineering Articular cartilage regeneration Bone, cartilage, tendon regeneration Skin tissue engineering Neural tissue regeneration Nerve repair	IPN scaffolds Hybrid nanocomposite scaffold 3D scaffold Electrospun mats 3D scaffolds Sponge scaffold Genipin/EDC crosslinked scaffold Porous moldable scaffold	Cassimjee et al. (2022) Lu et al. (2019) Amirul et al. (2019) Ching et al. (2021) Wang et al. (2020) Vyborny et al. (2019) Zhang et al. (2021)
Plasma	Gelatin/polyvinylidene fluoride/polyaniline/graphene nanoparticles	Neural tissue engineering	Cold atmospheric plasma treated electrically conductive scaffold	Sahrayi et al. (2022)
Glutaraldehyde	Chitosan/calcium phosphate granules Procyanidin Decellularized ECM from rat kidney Bombyx mori silk fibroin/ gelatin nanofibres Collagen/silk	Osteoblastic response/ regeneration Soft tissue engineering Renal tissue engineering Tissue engineering Tissue engineering	Freeze drying 3D porous scaffolds Decellularized matrix Electrospun mats Electrospun mats	Pinto et al. (2020) Yang et al. (2017) Yu et al. (2017) Mohammadzaehmoghadam et al. (2019) Zhu et al. (2017)

(continued)

Table 2 (continued)

Crosslinking agent	Polymers used	Application	Method of fabrication	References
EDC/NHS/ Carbodiimide	Gelatin/carboxymethyl cellulose	Wound healing	Freeze dried scaffold	Chaijit et al. (2020)
	Recombinant human collagen polypeptide/chitosan	Soft tissue engineering	Freeze dried scaffold	Yang et al. (2021)
	Collagen	Corneal tissue engineering	EDC/GTA dual crosslinked hydrogel	Islam et al. (2021)
	Gum Arabic/gelatin	Cartilage/bone tissue engineering	Freeze dried matrix	Edwin et al. (2019)
	Collagen	Controlled cell adhesion	UV/EDC dual crosslinked films	Bax et al. (2019)
Glyoxal	Solublised ECM	Orthopaedic tissue engineering	Glyoxal/DHT dual treated porous scaffold	Browe et al. (2019)
	Silk fibroin/collagen/chitosan	Bone tissue regeneration	Lyophilized scaffolds	Grabska-Zielinska et al. (2020)
	Collagen/chitosan/ Nanohydroxyapaptite	Bone regeneration	3D porous scaffold	Karakecili et al. (2022)
	Chitosan/keratin	Soft tissue engineering	3D porous scaffold	Adil et al. (2022)
	Chitosan/melatonin cultured collagen	Skin tissue engineering	3Dbiomaterials	Kaczmarek-szczepanska et al. (2022)
	Gelatin/hydroxyapatite	Periodontal tissue engineering	Electrospun mats	Dieterle et al. (2022)
	Cissus quandrangularis extract/ chitosan/ collagen	Bone regeneration	Porous scaffolds	Nair et al. (2021)
	Collagen	Suture healing	Microfibres	Dasgupta et al. (2021)
	Chitosan/PVA nanofibres/ halloysite nanotubes	Skin tissue regeneration	Electrospun/hydrogel	Koosha et al. (2019)

was first reported by Melchels et al. (2009) for tissue engineering. Gelatin and hyaluronic acid, natural polymers extensively used for biomedical applications are also photocrosslinked to improve the mechanical properties by modifying lysine unit or glutamic acid/aspartate unit on gelatin. Gelatin methacryloyl (GelMA), a crosslinkable form of gelatin has found widespread applications due to the excellent biocompatibility (Yue et al. 2015; Piao et al. 2021). GelMA was first synthesized by Van den Bulcke et al. (2000) and has been optimized by various researchers worldwide for the past two decades. This form overcomes the existing limitations of the classic gelatin hydrogel in terms of solubility at room temperature and cytotoxicity as a consequence of glutaraldehyde crosslinking. GelMAcould be synthesized in different forms like 3D scaffolds, injectable hydrogels, electrospun fibres or 3D printed materials for various applications including bone, liver, skin, tendon, cartilage tissue engineering and as a wound dressing material (Eke et al. 2017; Levato et al. 2017). In addition to photocrosslinked GelMA, dynamic hydrazone-crosslinked hyaluronic acid (HA-HYD) has been combined to form a double network hydrogel which has been synthesized using two crosslinking mechanisms. This has been used as a bioink for 3D printing scaffolds exhibiting better mechanical strength thereby supporting the proliferation and differentiation of bone marrow mesenchymal stem cells (Wang et al. 2022). Some other polymers that can be photocrosslinked include poly(e-caprolactone) (PCL), poly(trimethylene carbonate) (PTMC) and poly(ethylene carbonate) (PEC) (Helminen et al. 2003; Matsuda et al. 2004; Cornacchione et al. 2012). Photocrosslinked biomaterials are an interesting group of materials to scientists as they are easy to synthesize, can entrap a vast range of polymers and bioactive substances and the ease of temporal and spatial control in polymerization, allowing the formation of complex structures. Another, fascinating strategy of crosslinking is click chemistry that is defined as rapid extremely selective reactions that are high yielding in connecting molecular components. This was first described by Kolb et al. (2001). Copper catalyzed azidealkyne cycloaddition (CuAAC) was the first to be termed as click chemistry. Later, Michael additions, photo-initiated thiol-ene reactions, Staudinger ligation, and strain promoted azide-alkyne cycloaddition (SPAAC) satisfied the concept of click chemistry (Hoyle et al. 2010; Sletten et al. 2009). Click chemistry has found numerous applications in tissue engineering for the synthesis of scaffolds. Poly(ε-caprolactone) was synthesized as an injectable, hyperbranched hydrogel using SPAAC reaction. Besides being used as bone defect repair material, this can also be used as a stiff matrix for tissue engineering (Liu et al. 2016). Similarly, hydroxyethyl cellulose modified with citric acid, incorporated with azide and alkyne moieties was synthesized using the bioorthogonal click chemistry mechanism, which was further lyophilized to obtain a porous interconnected scaffold. This scaffold had better stability and mechanical properties which favours chondrogenic cell adhesion and proliferation making it a suitable material for cartilage tissue engineering (Nouri-Felekori et al. 2021). Another latest study by Sousa et al. (2022) showed that chondroitin sulfate (CS) and polyethylene glycol (PEG) could be crosslinked with PEG-Norbornene (A-PEG-N) to generate a catalyst free click chemistry based self standing hydrogel. This hydrogel exhibited increased blood perfusion, enhanced the number of blood vessels, had fibrous matrix orietnatation and normal collagen deposition when tested in vivo. Thus, this hydrogel with promising features can be used as a substitute for skin regeneration. Normally, for organic reactions, solvents or catalysts that are toxic with heat and pressure are used for crosslinking. However, crosslinking in the absence of toxic catalysts has been made possible by copper-free click chemistry. The hydrogels and nanoparticles synthesized based on this method is used for drug delivery, particles for cell tracking (cellular imaging), bone tissue engineering and cardiac tissue engineering applications (Yoon et al. 2022). Recent reports show that sequential patterning and depatterning of biologically relevant cues have been achieved through click chemistry for human mesenchymal stem cells that leads to cell delivery applications (DeForest et al. 2009). Click chemistry has transformed the field of tissue engineering over the past decade due to the convenient, versatile and bio-orthogonal nature in generating complex scaffolds for regenerative medicine. In parallel, hyaluronic acid has been till date crosslinked using traditional methods, but due to their limitations, researchers have attempted to synthesize hyaluronic acid formulations under mild physiological conditions that have found applications in tissue engineering (Pérez et al. 2021). Another rarely used crosslinker, tannic acid is now gaining importance due to its variable properties. Tannic acid is a plant based crosslinker called hydrolysable plant tannin that crosslinks through hydrogen bonding mechanisms. Tannic acid is reported to have anti tumor activity that makes it an interesting crosslinker as it can perform dual roles of crosslinking and drug delivery. TA was used to crosslink collagen scaffolds as an anticancer agent, where it limited the proliferation of MCF-7 breast cancer cells with slight toxicity to mesenchymal stem cells (Cass et al. 2012). Tannic acid has been used through the years to crosslink numerous polymers like hydroxyapatite, collagen, gelatin, alginate, silk fibroin, chitosan, poly vinyl alcohol (PVA), poly ethylene glycol (PEG) and graphene for bone and tissue engineering, wound dressings and in nanoparticles for drug delivery. The field of tannic acid based crosslinking materials is still expanding and has scope for further research which interests scientists and biomedical researchers (Chen et al. 2022). The requisites for various tissue engineering applications are elabotared in Table 3. The quest for an ideal biomaterial perfectly matching the microenvironment of the surrounding tissues and cells is an endless challenge within biomedical research, in addition to integrating this with a facile and sustainable technology for its preparation.

Table 3 Requisites for successful 3D implants

Types of tissue engineering Requisites for scaffolds	Neural tissue engineering	Liver tissue engineering	Bone tissue engineering	Corneal tissue engineering	Skin tissue engineering	Vascular tissue engineering	Cardiac tissue engineering
Transparent	No	No	No	Yes	No	No	No
Suturable	Yes	Yes	Yes	Yes	No	Yes	No
Biodegradable	Yes	Yes	Yes	Yes	Yes	Yes	Yes
Nutrient/waste exchange	Yes	Yes	Yes	Yes	Yes	Yes	Yes
Mechanical stiffness	Moderate/less	Yes	Yes	No	No	No	No
Fibrous	No	Yes	No	No	No	No	No
Elasticity	Yes	Yes	No	No	Yes	No	Yes
Dome shaped	No	No	No	Yes	No	No	No
Rod/fibril shaped	Yes	No	Yes	No	No	Yes	No
Conductive materials	Yes	No	No	No	No	No	Yes

References

Aadil KR, Gupta P (2022) Fabrication and Characterization of Hair Keratin–Chitosan-Based Porous Scaffolds for Biomedical Application. InRe-Use and Recycling of Materials 2022 Sep 1 (pp. 313–325). River Publishers.

Abbasi E, Aval SF, Akbarzadeh A, Milani M, Nasrabadi HT, Joo SW, Hanifehpour Y, Nejati-Koshki K, Pashaei-Asl R (2014) Dendrimers: synthesis, applications, and properties. Nanoscale Res Lett. 9(1):247. https://doi.org/10.1186/1556-276X-9-247. PMID: 24994950; PMCID: PMC4074873.

Aguero L, Alpdagtas S, Ilhan E, Zaldivar-Silva D, Gunduz O (2021) Functional role of crosslinking in alginate scaffold for drug delivery and tissue engineering: A review. European Polymer Journal. 160:110807, ISSN 0014-3057, https://doi.org/10.1016/j.eurpolymj.2021.110807.

Bae JW, Choi JH, Lee Y, Park KD (2015) Horseradish peroxidase-catalysed in situ-forming hydrogels for tissue-engineering applications. J Tissue Eng Regen Med. 9(11):1225–32. https://doi.org/10.1002/term.1917. Epub 2014 Jun 11. PMID: 24916126.

Barbosa O, Ortiz C, Berenguer-Murcia Á, Torres R, Rodrigues RC, Fernandez-Lafuente R (2014) Glutaraldehyde in bio-catalysts design: a useful crosslinker and a versatile tool in enzyme immobilization. RSC Adv. 4(4):1583–1600. https://doi.org/10.1039/c3ra45991h

Bavaresco B, Comín R, Salvatierra NA, Cid MP (2020) Three-dimensional printing of collagen and hyaluronic acid scaffolds with dehydrothermal treatment crosslinking. Compos. Commun. 19:1–5.

Bax DV, Davidenko N, Hamaia SW, Farndale RW, Best SM, Cameron RE (2019) Impact of UV-and carbodiimide-based crosslinking on the integrin-binding properties of collagen-based materials. Acta Biomater. 100:280–91.

Bi B, Ma M, Lv S, Zhuo R, Jiang X (2019) In-situ forming thermosensitive hydroxypropyl chitin-based hydrogel crosslinked by Diels-Alder reaction for three dimensional cell culture. Carbohydr. Polym. 212:368–77.

Broderick EP, O'Halloran DM, Rochev YA, Griffin M, Collighan RJ, Pandit AS (2005) Enzymatic stabilization of gelatin-based scaffolds. J Biomed Mater Res B Appl Biomater 72B (1):37–42.

Browe DC, Mahon OR, Díaz-Payno PJ, Cassidy N, Dudurych I, Dunne A, Buckley CT, Kelly DJ (2019) Glyoxal cross-linking of solubilized extracellular matrix to produce highly porous, elastic, and chondro-permissive scaffolds for orthopedic tissue engineering. J Biomed Mater Res A. 107(10):2222–34.

Burgstaller C, Hoftberger T, Wagner BG, Stadlbauer W (2020) Effects of radiation type and dose on the properties of selected polymers. Polym Eng Sci. 61(1): 39–54.

Cass CA, Burg KJ (2012) Tannic acid cross-linked collagen scaffolds and their anti-cancer potential in a tissue engineered breast implant. J Biomater Sci Polym Ed. 23(1–4):281–98. https://doi.org/10.1163/092050610X550331. Epub 2011 Jan 18. PMID: 21244722.

Cassimjee H, Kumar P, Ubanako P, Choonara YE (2022) Genipin-Crosslinked, Proteosaccharide Scaffolds for Potential Neural Tissue Engineering Applications. Pharmaceutics. 14(2):441.

Chaijit S, Wiwatwongwana F (2020) Strengthening Scaffold by Using Carbodiimide Crosslinking on Gelatin/carboxymethylcellulose from Waste Product. Chem. Eng. Trans. 80:331–6.

Chen J, Garcia ES, Zimmerman SC (2020a) Intramolecularly cross-linked polymers: From Structure to Function with Applications as Artificial Antibodies and Artificial Enzymes. Acc. Chem. Res. 53(6):1244–1256.

Chen X, Zhou L, Xu H, Yamamoto M, Shinoda M, Kishimoto M, Tanaka T, Yamane H (2020b) Effect of the Application of a Dehydrothermal Treatment on the Structure and the Mechanical Properties of Collagen Film. Materials 13.

Chen X, Meng J, Xu H, Shinoda M, Kishimoto M, Sakurai S, Yamane H (2021) Fabrication and properties of electrospun collagen tubular scaffold crosslinked by physical and chemical treatments. Polymers. 13(5):755.

Chen C, Yang H, Yang X, Ma Q (2022) Tannic acid: a crosslinker leading to versatile functional polymeric networks: a review. RSC Adv. 12:7689–7711

Ching KY, Andriotis O, Sengers B, Stolz M (2021) Genipin crosslinked chitosan/PEO nanofibrous scaffolds exhibiting an improved microenvironment for the regeneration of articular cartilage. J. Biomater. Appl. 36(3):503–16.

Choi G, Cha HJ (2019) Recent advances in the development of nature-derived photocrosslinkable biomaterials for 3D printing in tissue engineering. Biomater Res. 23:18 https://doi.org/10.1186/s40824-019-0168-8

Ciardelli G, Gentile P, Chiono V, Mattioli-Belmonte M, Vozzi G, Barbani N, Giusti P (2010) Enzymatically crosslinked porous composite matrices for bone tissue regeneration. J Biomed Mater Res A 92A(1):137–151.

Collighan R, Cortez J, Griffin M (2002) The biotechnological applications of transglutaminases. Minerva Biotechnol 14:143–148.

Cornacchione LA, Qi B, Bianco J, et al. Photo-cross-linked poly (ethylene carbonate) elastomers: synthesis, in vivo degradation, and determination of in vivo degradation mechanism. Biomacromolecules. 2012;13:3099–3107.

Cornwell KG, Lei P, Andreadis ST, Pins GD (2007) Crosslinking of discrete self-assembled collagen threads: Effects on mechanical strength and cell-matrix interactions. J Biomed Mater Res A. 80(2):362–71. https://doi.org/10.1002/jbm.a.30893. PMID: 17001644.

Cui L, Reddy N, Xu H, Fan X, Yang Y (2017) Enzyme-modified casein fibers and their potential application in drug delivery. Fibers Polym 18(5):900–906

Dasgupta A, Sori N, Petrova S, Maghdouri-White Y, Thayer N, Kemper N, Polk S, Leathers D, Coughenour K, Dascoli J, Palikonda R (2021) Comprehensive collagen crosslinking comparison of microfluidic wet-extruded microfibers for bioactive surgical suture development. Acta Biomater. 128:186–200.

DeForest CA, Polizzotti BD, Anseth KS (2009) Sequential click reactions for synthesizing and patterning three-dimensional cell microenvironments. Nature Mater. 8(8):659–664. https://doi.org/10.1038/nmat2473

Delgado-Rangel LH, Hernández-Vargas J, Becerra-González M, Martínez-Torres A, Prokhorov E, González Campos JB (2019) Development of collagen/poly (vinyl alcohol)/chondroitin sulfate and collagen/poly (vinyl alcohol)/HA electrospun scaffolds for tissue engineering. Fibers Polym. 20(12):2470–84.

Dieterle MP, Steinberg T, Tomakidi P, Nohava J, Vach K, Schulz SD, Hellwig E, Proksch S (2022) Novel In Situ-Cross-Linked Electrospun Gelatin/Hydroxyapatite Nonwoven Scaffolds Prove Suitable for Periodontal Tissue Engineering. Pharmaceutics 14(6):1286.

Ding Y, Zhang W, Sun B, Mo X, Wu J (2022) Cyclic freeze–thaw grinding to decellularize meniscus for fabricating porous, elastic scaffolds. J Biomed Mater Res A. 110(11):1824–39.

Distler T, McDonald K, Heid S, Karakaya E, Detsch R, Boccaccini AR (2020) Ionically and enzymatically dual cross-linked oxidized alginate gelatin hydrogels with tunable stiffness and degradation behavior for tissue engineering. ACS Biomater. Sci. Eng. 6(7):3899–914.

Duan X, Sheardown H (2005) Crosslinking of collagen with dendrimers. J Biomed Mater Res A. 75(3):510–8. https://doi.org/10.1002/jbm.a.30475. PMID: 16108047.

Edwin BT, Dhanya H, Nair PD, Kassem M (2019) In Vitro and In Vivo Evaluation of 1-(3 Dimethylaminopropyl)-3-Ethyl Carbodiimide (EDC) Cross-Linked Gum Arabic–Gelatin Composite as an Ideal Porous Scaffold for Tissue Engineering. InBiomaterials in Orthopaedics and Bone Regeneration 131–145 Springer, Singapore.

Ehrbar M, Rizzi SC, Schoenmakers RG, Miguel BS, Hubbell JA, Weber FE, Lutolf MP (2007) Biomolecular hydrogels formed and degraded via site-specific enzymatic reactions. Biomacromolecules 8(10):3000–7. https://doi.org/10.1021/bm070228f. Epub 2007 Sep 21. PMID: 17883273.

Eke G, Mangir N, Hasirci N, MacNeil S, Hasirci V (2017) Development of a UV crosslinked biodegradable hydrogel containing adipose derived stem cells to promote vascularization for skin wounds and tissue engineering. Biomaterials 129:188–198. https://doi.org/10.1016/j.biomaterials.2017.03.021 https://doi.org/10.1016/j.biomaterials.2017.03.021

Frone AN, Panaitescu DM, Nicolae CA, Gabor AR, Trusca R, Casarica A, Stanescu PO, Baciu DD, Salageanu A (2020) Bacterial cellulose sponges obtained with green cross-linkers for tissue engineering. Mater. Sci. Eng. C. 110:110740.

Galdopórpora JM, Perez CJ, Tuttolomondo MV, Desimone MF (2019) Riboflavin-UVA gelatin crosslinking: Design of a biocompatible and thermo-responsive biomaterial with enhanced mechanical properties for tissue engineering. VBRI Press; Advanced Materials Letters. 10 (5): 324–328.

Gilchrist AE, Serrano JF, Ngo MT, Hrnjak Z, Kim S, Harley BA (2021) Encapsulation of murine hematopoietic stem and progenitor cells in a thiol-crosslinked maleimide-functionalized gelatin hydrogel. Acta Biomater. 131:138–48.

Grabska-Zielińska S, Sionkowska A, Coelho CC, Monteiro FJ (2020) Silk fibroin/collagen/chitosan scaffolds cross-linked by a glyoxal solution as biomaterials toward bone tissue regeneration. Mater. 13(15):3433.

Grijpma DW, Hou Q, Feijen J (2005) Preparation of biodegradable networks by photo-crosslinking lactide, epsilon-caprolactone and trimethylene carbonate-based oligomers functionalized with fumaric acid monoethyl ester. Biomaterials 26:2795–2802

Guaresti O, Basasoro S, González K, Eceiza A, Gabilondo N (2019) In situ cross–linked chitosan hydrogels via Michael addition reaction based on water–soluble thiol–maleimide precursors. Eur. Polym. J. 119:376–84.

Guo JL, Kim YS, Xie VY, Smith BT, Watson E, Lam J, Pearce HA, Engel PS, Mikos AG (2019) Modular, tissue-specific, and biodegradable hydrogel cross-linkers for tissue engineering. Sci Adv. 5(6):eaaw7396. https://doi.org/10.1126/sciadv.aaw7396. PMID: 31183408; PMCID: PMC6551165.

Halloran DM, Collighan RJ, Griffin M, Pandit AS (2006) Characterization of a microbial transglutaminase cross-linked type II collagen scaffold. Tissue Eng 12 (6):1467–1474.

Haugh MG, Jaasma MJ, O'Brien FJ (2009). The effect of dehydrothermal treatment on the mechanical and structural properties of collagen-GAG scaffolds. J Biomed Mater Res A. 89(2):363–9. https://doi.org/10.1002/jbm.a.31955. PMID: 18431763.

Heck T, Faccio G, Richter M, Thöny-Meyer L (2013) Enzyme-catalyzed protein crosslinking. Appl Microbiol Biotechnol. 97(2):461–75. https://doi.org/10.1007/s00253-012-4569-z. Epub 2012 Nov 25. PMID: 23179622; PMCID: PMC3546294.

Helminen AO, Korhonen H, Seppala JV (2003) Crosslinked poly(ester anhydride)s based on poly(epsilon-caprolactone) and polylactide oligomers. J Polym Sci A Polym Chem. 41:3788–3797.

Hermanson GT (2013) Chapter 1 - Introduction to Bioconjugation, in: G.T. Hermanson (Ed.) Bioconjugate Techniques (Third Edition), Academic Press, Boston 1–125.

Hoffmann B, Seitz D, Mencke A, Kokott A, Ziegler G (2009) Glutaraldehyde and oxidised dextran as crosslinker reagents for chitosan-based scaffolds for cartilage tissue engineering. J Mater Sci Mater Med. 20(7):1495–503. https://doi.org/10.1007/s10856-009-3707-3. Epub 2009 Mar 4. PMID: 19259790.

Honda Y, Takeda Y, Li P, Huang A, Sasayama S, Hara E, Uemura N, Ueda M, Hashimoto M, Arita K, Matsumoto N (2018) Epigallocatechin gallate-modified gelatin sponges treated by vacuum heating as a novel scaffold for bone tissue engineering. Molecules. 23(4):876.

Hoyle CE, Bowman CN (2010) Thiol-ene click chemistry. Angew Chem Int Ed 49(9):1540–73. https://doi.org/10.1002/anie.200903924. PMID: 20166107.

Islam MM, AbuSamra DB, Chivu A, Argüeso P, Dohlman CH, Patra HK, Chodosh J, González-Andrades M (2021) Optimization of collagen chemical crosslinking to restore biocompatibility of tissue-engineered scaffolds. Pharmaceutics 13(6):832.

Jiang YH, Lou YY, Li TH, Liu BZ, Chen K, Zhang D, Li T (2022) Cross-linking methods of type I collagen-based scaffolds for cartilage tissue engineering. Am J Transl Res. 14(2):1146–1159. PMID: 35273719; PMCID: PMC8902548.

Kaczmarek-Szczepańska B, Pin JM, Zasada L, Sonne MM, Reiter RJ, Slominski AT, Steinbrink K, Kleszczyński K (2022) Assessment of Melatonin-Cultured Collagen/Chitosan Scaffolds Cross-Linked by a Glyoxal Solution as Biomaterials for Wound Healing. Antioxidants 11(3):570.

Karakeçili A, Korpayev S, Orhan K (2022) Optimizing Chitosan/Collagen Type I/ Nanohydroxyapatite Cross-linked Porous Scaffolds for Bone Tissue Engineering. Appl. Biochem. Biotechnol. 11:1–7.

Koh LB, Islam MM, Mitra D, Noel CW, Merrett K, Odorcic S, Fagerholm P, Jackson WB, Liedberg B, Phopase J, Griffith M (2013) Epoxy cross-linked collagen and collagen-laminin Peptide hydrogels as corneal substitutes. J Funct Biomater. 4(3):162–77. https://doi.org/10.3390/jfb4030162. PMID: 24956085; PMCID: PMC4030907.

Kolb HC, Finn MG, Sharpless KB (2001) Click Chemistry: Diverse Chemical Function from a Few Good Reactions. Angew Chem Int Ed Engl. 40(11):2004–2021. https://doi.org/10.1002/1521-3773(20010601)40:11<2004::aid-anie2004>3.3.co;2-x. PMID: 11433435

Koosha M, Raoufi M, Moravvej H (2019) One-pot reactive electrospinning of chitosan/PVA hydrogel nanofibers reinforced by halloysite nanotubes with enhanced fibroblast cell attachment for skin tissue regeneration. Colloids Surf. B. 179:270–9.

Lee F, Bae KH, Kurisawa M (2015) Injectable hydrogel systems crosslinked by horseradish peroxidase. Biomed Mater. 11(1):014101. https://doi.org/10.1088/1748-6041/11/1/014101. PMID: 26694014

Levato R, Webb WR, Otto IA, Mensinga A, Zhang Y, van Rijen M, van Weeren R, Khan IM, Malda J (2017) The bio in the ink: cartilage regeneration with bioprintable hydrogels and articular cartilage-derived progenitor cells. Acta Biomater. 61:41–53. https://doi.org/10.1016/j.actbio.2017.08.005. Epub 2017 Aug 4. PMID: 28782725; PMCID: PMC7116023.

Li S, Wu P, Ji Z, Zhang Y, Zhang P, He Y, Shen Y (2023) *In vitro* biocompatibility study of EDC/NHS cross-linked silk fibroin scaffold with olfactory ensheathing cells. Journal of Biomaterials Science, Polymer Edition. 34(4):482–496, https://doi.org/10.1080/09205063.2022.2135076

Liguori A, Bigi A, Colombo V, Focarete ML, Gherardi M, Gualandi C, Oleari MC, Panzavolta S (2016) Atmospheric Pressure Non-Equilibrium Plasma as a Green Tool to Crosslink Gelatin Nanofibers. Sci. Rep. 6:38542.

Liu S, Lau CS, Liang K, Wen F, Teoh SH (2022a) Marine collagen scaffolds in tissue engineering. Curr. Opin. Biotechnol.74:92–103.

Liu S, Zhang H, Ahlfeld T, Kilian D, Liu Y, Gelinsky M, Hu Q (2022b) Evaluation of different crosslinking methods in altering the properties of extrusion-printed chitosan-based multi-material hydrogel composites. Bio-Des. Manuf.

Liu X, Miller AL, Fundora KA, Yaszemski MJ, Lu L (2016) Poly(ε-caprolactone) Dendrimer Cross-Linked via Metal-Free Click Chemistry: Injectable Hydrophobic Platform for Tissue Engineering. ACS Macro Letters, 5(11):1261–1265. https://doi.org/10.1021/acsmacrolett.6b0073

Lu HT, Lu TW, Chen CH, Mi FL (2019) Development of genipin-crosslinked and fucoidan-adsorbed nano-hydroxyapatite/hydroxypropyl chitosan composite scaffolds for bone tissue engineering. Int. J. Biol. Macromol. 128:973–84.

Lungu A, Cernencu AI, Dinescu S, Balahura R, Mereuta P, Costache M, Syverud K, Stancu IC, Iovu H (2021) Nanocellulose-enriched hydrocolloid-based hydrogels designed using a Ca2+ free strategy based on citric acid. Mater. Des. 197:109200.

Ma B, Wang X, Wu C, Chang J (2014) Crosslinking strategies for preparation of extracellular matrix-derived cardiovascular scaffolds. Regen. Biomater. 1:81–89.

Mali KK, Dhawale SC, Dias RJ, Dhane NS, Ghorpade VS (2018) Citric Acid Crosslinked Carboxymethyl Cellulose-based Composite Hydrogel Films for Drug Delivery. Indian J Pharm Sci. 80(4):657–667

Mao-Hua C, Xiao-Yi C, Luo-Qin F, Wen-Lin D, Xue Y, Xiao-Zhou M, Pei-Yang H (2021) Design and Development of Hybrid Hydrogels for Biomedical Applications: Recent Trends in Anticancer Drug Delivery and Tissue Engineering. Front. bioeng. biotechnol. 9. DOI https://doi.org/10.3389/fbioe.2021.630943.

Matsuda T, Magoshi T (2002) Preparation of vinylated polysaccharides and photofabrication of tubular scaffolds as potential use in tissue engineering. Biomacromolecules 3: 942–950

Matsuda T, Mizutani M, Arnold SC (2000) Molecular design of photocurable liquid biodegradable copolymers. 1. Synthesis and photocuring characteristics. Macromolecules 33:795–800.

Matsuda T, Kwon IK, Kidoaki S (2004) Photocurable biodegradable liquid copolymers: Synthesis of acrylate-end-capped trimethylene carbonate-based prepolymers, photocuring, and hydrolysis. Biomacromolecules 5:295–305.

Melchels FPW, Feijen J, Grijpma DW (2009) A poly(d,l-lactide) resin for the preparation of tissue engineering scaffolds by stereolithography. Biomaterials 30(23-24):3801–3809. https://doi.org/10.1016/j.biomaterials.2009.03.055 https://doi.org/10.1016/j.biomaterials.2009.03.055

Mizutani M, Matsuda T (2002) Liquid photocurable biodegradable copolymers: in vivo degradation of photocured poly(epsilon-caprolactone-co-trimethylene carbonate). J Biomed Mater Res. 61:53–60.

Mohammadzadehmoghadam S, Dong Y (2019) Fabrication and characterization of electrospun silk fibroin/gelatin scaffolds crosslinked with glutaraldehyde vapor. Front. Mater. Sci. 6:91.

Mohan T, Dobaj Štiglic A, Beaumont M, Konnerth J, Gürer F, Makuc D, Maver U, Gradišnik L, Plavec J, Kargl R, Stana Kleinschek K (2020) Generic method for designing self-standing and dual porous 3D bioscaffolds from cellulosic nanomaterials for tissue engineering applications. ACS Appl. Bio Mater. 3(2):1197–209.

M. Motoki, K. Seguro, Transglutaminase and its use for food processing. Trends Food Sci. Technol. 9 (5) (1998) 204–210.

Muhammad Mior Amirul A, Mohd Heikal MY, MH Busra F (2019) Genipin-crosslinked gelatin scaffold in tissue engineering: a systematic review. Medicine & Health 14(2).

Nadeau OW (2005) Protein Interaction Analysis: Chemical Cross-Linking. In: Encyclopedic Reference of Genomics and Proteomics in Molecular Medicine. Springer, Berlin, Heidelberg. https://doi.org/10.1007/3-540-29623-9_3540.

Nair M, Johal RK, Hamaia SW, Best SM, Cameron RE (2020) Tunable bioactivity and mechanics of collagen-based tissue engineering constructs: A comparison of EDC-NHS, genipin and TG2 crosslinkers. Biomaterials 254:120109.

Nair PR, Sreeja S, Sailaja GS (2021) Early biomineralizing chitosan–collagen hybrid scaffold with Cissus quadrangularis extract for regenerative bone tissue engineering. New J. Chem. 45(42):19733–45.

Narayani R, Rajaram R (2017) Decellularised and Chemically Crosslinked Human Umbilical Vein As a Small Caliber Conduit: Scaffold for Vascular Tissue Engineering. Trends Biomater Artif Organs 31:90–96.

Nataraj D, Reddy R, Reddy N (2020) Crosslinking electrospun poly (vinyl) alcohol fibers with citric acid to impart aqueous stability for medical applications. Eur. Polym. J. 124:109484.

Nonaka M, Tanaka H, Okiyama A, Motoki M, Ando H, Umeda K, Matsuura A (1989) Polymerization of several proteins by Ca2+-independent transglutaminase derived from microorganisms. Agric Biol Chem 53(10):2619–2623.

Nouri-Felekori M, Nezafati N, Moraveji M, Hesaraki S, Ramezani T (2021) Bioorthogonal hydroxyethyl cellulose-based scaffold crosslinked via click chemistry for cartilage tissue engineering applications. Int J Biol Macromol. 183:2030–2043. https://doi.org/10.1016/j.ijbiomac.2021.06.005. Epub 2021 Jun 4. PMID: 34097959.

Oryan A, Kamali A, Moshiri A, Baharvand H, Daemi H (2018) Chemical crosslinking of biopolymeric scaffolds: Current knowledge and future directions of crosslinked engineered bone scaffolds. Int. J. Biol. Macromol. 107 (2018) 678–688.

Partis MD, Griffiths DG, Roberts GC, Beechey RB (1983) Cross-linking of protein by ω-maleimido alkanoylN-hydroxysuccinimido esters. J. Protein Chem. 2:263–277.

Pérez LA, Hernández R, Alonso JM, Pérez-González R, Sáez-Martínez V (2021) Hyaluronic Acid Hydrogels Crosslinked in Physiological Conditions: Synthesis and Biomedical Applications. Biomedicines 9(9):1113. https://doi.org/10.3390/biomedicines9091113. PMID: 34572298; PMCID: PMC8466770.

Piao Y, You H, Xu T, Bei HP, Piwko IZ, Kwan YY, Zhao X (2021) Biomedical applications of gelatin methacryloyl hydrogels. Engineered Regeneration, 2, 47–56. https://doi.org/10.1016/j.engreg.2021.03.002 https://doi.org/10.1016/j.engreg.2021.03.002

Pina S, Ribeiro VP, Marques CF, Maia FR, Silva TH, Reis RL, Oliveira JM (2019) Scaffolding Strategies for Tissue Engineering and Regenerative Medicine Applications. Materials (Basel). 12(11):1824. https://doi.org/10.3390/ma12111824. PMID: 31195642; PMCID: PMC6600968.

Pinto RV, Gomes PS, Fernandes MH, Costa ME, Almeida MM (2020) Glutaraldehyde-crosslinking chitosan scaffolds reinforced with calcium phosphate spray-dried granules for bone tissue applications. Mater. Sci. Eng. C. 109:110557.

Priya G, Madhan B, Narendrakumar U, Suresh Kumar RV, Manjubala I (2021) In vitro and in vivo evaluation of carboxymethyl cellulose scaffolds for bone tissue engineering applications. ACS Omega 6(2):1246–53.

Sahrayi H, Hosseini E, Ramazani Saadatabadi A, Atyabi SM, Bakhshandeh H, Mohamadali M, Aidun A, Farasati Far B (2022) Cold atmospheric plasma modification and electrical conductivity induction in gelatin/polyvinylidene fluoride nanofibers for neural tissue engineering. Artif. Organs.

Salihu R, Abd Razak SI, Ahmad Zawawi N, Rafiq Abdul Kadir M, Izzah Ismail N, Jusoh N, Riduan Mohamad M, Hasraf Mat Nayan N (2021) Citric acid: A green cross-linker of biomaterials for biomedical applications. Eur. Polym. J. 146:110271.

Salvatore L, Calò E, Bonfrate V, Pedone D, Gallo N, Natali ML, Sannino A, Madaghiele M (2021) Exploring the effects of the crosslink density on the physicochemical properties of collagen-based scaffolds. Polym. Test. 93:106966.

Schramm HJ, Dülffer T (1977) Synthesis and application of cleavable and hydrophilic crosslinking reagents. Adv Exp Med Biol. 86A:197–206. https://doi.org/10.1007/978-1-4684-3282-4_11. PMID: 335837.

Sengor M (2022) Transglutaminase crosslinked sodium caseinate/starch/tri calcium phosphate based flexible sponge grafts. Mater. Lett. 326:132943.

Sharmin N, Rosnes JT, Prabhu L, Böcker U, Sivertsvik M (2022) Effect of Citric Acid Cross Linking on the Mechanical, Rheological and Barrier Properties of Chitosan. Molecules. 27(16):5118. https://doi.org/10.3390/molecules27165118. PMID: 36014369; PMCID: PMC9415850.

Silver FH, Yannas IV, Salzman EW (1979) In vitro blood compatibility of glycosaminoglycan-precipitated collagens. J Biomed Mater Res. 13(5):701–16. https://doi.org/10.1002/jbm.820130504. PMID: 479217.

Singh S, Dutt D, Mishra NC (2020) Cotton pulp for bone tissue engineering. J. Biomater. Sci. Polym. Ed. 31(16):2094–113.

Sletten EM, Bertozzi CR (2009) Bioorthogonal chemistry: fishing for selectivity in a sea of functionality. Angew Chem Int Ed Engl. 48(38):6974–98. https://doi.org/10.1002/anie.200900942. PMID: 19714693; PMCID: PMC2864149.

Sood A, Ji SM, Kumar A, Han SS (2022) Enzyme-Triggered Crosslinked Hybrid Hydrogels for Bone Tissue Engineering. Materials 15(18):6383.

Sousa GF, Afewerki S, Dittz D, Santos FEP, Gontijo DO, Scalzo SRA, Santos ALC, Guimaraes LC, Pereira EM, Barcelos LS, Do Monte SJH, Guimaraes PPG, Marciano FR, Lobo AO (2022) Catalyst-Free Click Chemistry for Engineering Chondroitin Sulfate-Multiarmed PEG Hydrogels for Skin Tissue Engineering. J Funct Biomater. 13(2):45. https://doi.org/10.3390/jfb13020045. PMID: 35466227; PMCID: PMC9036249.

Stelescu MD, Airinei A, Manaila E, Craciun G, Fifere N, Varganici C, Pamfil D, Doroftei F (2018) Effects of electron beam irradiation on the mechanical, thermal and surface properties of some EPDM/Butyl Rubber Composites. Polymer. 10(11): 1206.

Summonte S, Racaniello GF, Lopedota A, Denora N, Bernkop-Schnürch A (2021) Thiolated polymeric hydrogels for biomedical application: Cross-linking mechanisms. J Control Release. 330:470–82.

Surendranath M, Rajalekshmi R, Ramesan RM, Nair P, Parameswaran R (2022) UV-Crosslinked Electrospun Zein/PEO Fibroporous Membranes for Wound Dressing. ACS Appl. Bio Mater. 5 (4): 1538–1551

Taslikh M, Abbasi H, Mortazavian AM, Ghasemi JB, Naeimabadi A, Nayebzadeh K (2022) Effect of Cold Plasma Treatment, Cross-Linking, and Dual Modification on Corn Starch, Starch - Stärke, 74:2200008.

Van Den Bulcke AI, Bogdanov B, De Rooze N, Schacht EH, Cornelissen M, Berghmans H (2000) Structural and rheological properties of methacrylamide modified gelatin hydrogels. Biomacromolecules 1(1):31-8. https://doi.org/10.1021/bm990017d. PMID: 11709840.

Velutheril Thomas L, Nair PD (2019) An electrospun citric acid modified polyvinyl alcohol scaffold for vascular tissue engineering. Journal J. Bioact. Compat. Polym. 34(3):263–79.

Výborný K, Vallová J, Kočí Z, Kekulová K, Jiráková K, Jendelová P, Hodan J, Kubinová Š (2019) Genipin and EDC crosslinking of extracellular matrix hydrogel derived from human umbilical cord for neural tissue repair. Sci. Rep. 9(1):1–5.

Wang Z, Liu H, Luo W, Cai T, Li Z, Liu Y, Gao W, Wan Q, Wang X, Wang J, Wang Y (2020) Regeneration of skeletal system with genipin crosslinked biomaterials. J. Tissue Eng. 11:2041731420974861.

Wang Y, Chen Y, Zheng J, Liu L, Zhang Q (2022) Three-Dimensional Printing Self-Healing Dynamic/Photocrosslinking Gelatin-Hyaluronic Acid Double-Network Hydrogel for Tissue Engineering. ACS Omega 7(14):12076–12088 Publication Date: March 29, 2022, https://doi.org/10.1021/acsomega.2c00335.

Wu X, Black L, Santacana-Laffitte G, Patrick CW Jr. (2007) Preparation and assessment of glutaraldehyde-crosslinked collagen-chitosan hydrogels for adipose tissue engineering. J Biomed Mater Res A. 81(1):59–65. https://doi.org/10.1002/jbm.a.31003. PMID: 17109417.

Yang C, Wang X, Yao X, Zhang Y, Wu W, Jiang X (2015) Hyaluronic acid nanogels with enzyme-sensitive cross-linking group for drug delivery. J Control Release 205:206–217

Yang Y, Ritchie AC, Everitt NM (2017) Comparison of glutaraldehyde and procyanidin cross-linked scaffolds for soft tissue engineering. Mater. Sci. Eng. C. 80:263–73.

Yang G, Xiao Z, Long H, Ma K, Zhang J, Ren X, Zhang J (2018) Assessment of the characteristics and biocompatibility of gelatin sponge scaffolds prepared by various crosslinking methods. Sci Rep. 8:1616.

Yang Y, Ritchie AC, Everitt NM (2021) Using type III recombinant human collagen to construct a series of highly porous scaffolds for tissue regeneration. Colloids Surf. B. 208:112139

Yang Q, Dou F, Liang B, Shen Q (2005) Studies of cross-linking reaction on chitosan fiber with glyoxal. Carbohydrate Polymers. 59 (2): 205–210, ISSN 0144-8617, https://doi.org/10.1016/j.carbpol.2004.09.013.

Yao Y, Wang P, Li X, Xu Y, Lu G, Jiang Q, Sun Y, Fan Y, Zhang X (2020) A di-self-crosslinking hyaluronan-based hydrogel combined with type I collagen to construct a biomimetic injectable cartilage-filling scaffold. Acta Biomater. 111:197–207.

Yin W, Su R, Qi W, He Z (2012) A casein-polysaccharide hybrid hydrogel cross-linked by transglutaminase for drug delivery. J Mater Sci 47(4):2045–2055.

Yoo KM, Murphy SV, Skardal A (2021a) A Rapid Crosslinkable Maleimide-Modified Hyaluronic Acid and Gelatin Hydrogel Delivery System for Regenerative Applications. Gels 7(1):13. https://doi.org/10.3390/gels7010013. PMID: 33535669; PMCID: PMC7931058

Yoo KM, Murphy SV, Skardal A (2021b) A Rapid Crosslinkable Maleimide-Modified Hyaluronic Acid and Gelatin Hydrogel Delivery System for Regenerative Applications. Gels. 7(1):13. https://doi.org/10.3390/gels7010013. PMID: 33535669; PMCID: PMC7931058.

Yoon HY, Lee D, Lim DK, Koo H, Kim K (2022) Copper-Free Click Chemistry: Applications in Drug Delivery, Cell Tracking, and Tissue Engineering. Adv Mater. 2022 34(10):e2107192. https://doi.org/10.1002/adma.202107192. Epub 2022 Jan 28. PMID: 34752658.

Yu J, Chen F, Wang X, Dong N, Lu C, Yang G, Chen Z (2016) Synthesis and characterization of MMP degradable and maleimide cross-linked PEG hydrogels for tissue engineering scaffolds. Polym. Degrad. Stab. 133:312–320. https://doi.org/10.1016/j.polymdegradstab.2016.09.008

Yu Y, Liu L, Zhang J, Wei Z, Mei J (2017) Glutaraldehyde Cross-linking Modification of Decellularized Rat Kidney Scaffolds. In Decellularized Scaffolds and Organogenesis 111–119. Humana Press, New York, NY.

Yue K, Trujillo-de Santiago G, Alvarez MM, Tamayol A, Annabi N, Khademhosseini A (2015) Synthesis, properties, and biomedical applications of gelatin methacryloyl (GelMA) hydrogels. Biomaterials 73:254–271.

Zafeiris K, Brasinika D, Karatza A, Koumoulos E, Karoussis IK, Kyriakidou K, Charitidis CA, (2021) Additive manufacturing of hydroxyapatite–chitosan–genipin composite scaffolds for bone tissue engineering applications. Materials Science and Engineering: C.119:111639, ISSN 0928-4931, https://doi.org/10.1016/j.msec.2020.111639.

Zhang F, Zhang N, Xu Q, Zhang L, Zhang C, Liu H, Yu Z, Zhou S, Feng G, Huang F (2021) Decellularized nerve extracellular matrix/chitosan crosslinked by genipin to prepare a moldable nerve repair material. Cell Tissue Bank. 22(3):419–30.

Zhang Z, Ma Z, Zhang Y, Chen F, Zhou Y, An Q (2018) Dehydrothermally crosslinked collagen/hydroxyapatite composite for enhanced in vivo bone repair. Colloids and Surfaces B: Biointerfaces. 163:394–401, ISSN 0927-7765, https://doi.org/10.1016/j.colsurfb.2018.01.011.

Zheng X, Chen Y, Dan N, Dan W, Li Z (2021) Highly stable collagen scaffolds crosslinked with an epoxidized natural polysaccharide for wound healing. International Journal of Biological Macromolecules. 182: 1994–2002, ISSN 0141-8130, https://doi.org/10.1016/j.ijbiomac.2021.05.189.

Zhu B, Li W, Chi N, Lewis RV, Osamor J, Wang R (2017) Optimization of glutaraldehyde vapor treatment for electrospun collagen/silk tissue engineering scaffolds. ACS Omega. 2(6):2439–50.

Odontomas of Contemporary Humans and Animals: The Morphology and Composition

Oksana L. Pikhur, Yulia V. Plotkina, Alexander M. Kulkov, Denis S. Tishkov, and Alexander L. Gromov

Abstract Odontoma is one of the most common benign tumors of the maxillofacial region of contemporary humans and animals. The etiology and pathogenesis of odontoma have not been studied well enough. The morphology of human and animal odontomas (dogs and hedgehogs) was studied using optical microscopy, scanning electron microscopy and X-ray microtomography. It was found that the morphological structure of the odontomas is characterized by heterogeneity. The methods of morphological investigation we used allowed to differential diagnosis of odontoma and to clarify its classification type, as well as to study in detail the unique morphological features of each sample. The chemical composition of the studied hard tissues of the odontomas in most cases corresponds to the composition of hard tissues of the teeth of contemporary humans and animals.

Keywords Odontoma · Benign odontogenic tumor · Odontogenic neoplasm · Odontoid · Hard tooth tissue · Enamel · Dentin · X-ray computed microtomography

1 Introduction

An odontoma is a benign odontogenic tumor that consists of elements of dental tissues and develops in the jaw bones. The frequency of occurrence of odontogenic tumors ranges from 0.8 to 3.7% of all tumors of the maxillofacial region. Odontoma accounts for 22.0–34.6% of all odontogenic tumors and shares the first place in

O. L. Pikhur (✉) · D. S. Tishkov · A. L. Gromov
Kursk State Medical University, Kursk, Russia
e-mail: oruhk@mail.ru

Y. V. Plotkina
Institute of Precambrian Geology and Geochronology RAS, St. Petersburg, Russia

A. M. Kulkov
Research Center for X-Ray Diffraction Studies, St. Petersburg State University, St. Petersburg, Russia

O. V. Frank-Kamenetskaya et al. (eds.), *Biogenic—Abiogenic Interactions in Natural and Anthropogenic Systems 2022*, Springer Proceedings in Earth and Environmental Sciences, https://doi.org/10.1007/978-3-031-40470-2_3

frequency of occurrence with ameloblastoma, so this problem can be called quite relevant in dentistry (Hidalgo-Sanchez et al. 2008).

Odontoma is a formation that is the result of an anomaly in the development of dental tissues, the tumor is organ-specific and develops only in the jaw bones (Strukov and Serov 2015). It is mainly localised in the area of large molars, according to the position of the wisdom tooth, and is located either in the area of the alveolar process, or centrally—inside the body of the jaw (Akerzoul et al. 2017). Odontoma often affects the bone in the area of premolars and molars, both on the maxilla and mandible. Moreover, an odontoma may be localised on the maxilla in the area of the anterior group of teeth (De Vasconcellos Machado et al. 2015). It is possible, both unilateral (more often) and bilateral damage to the jaws by a tumor (Sun and Sun 2016). Odontoma, as a rule, has an irregular knobby shape and varies in size from a few millimeters to several centimeters. The symptoms of the disease depend on the odontoma's location, its size and the severity of inflammatory phenomena in the surrounding tissues (Vengal et al. 2007; Gupta et al. 2014). Most often, the disease proceeds slowly, gradually, asymptomatically or oligosymptomatically (Afanasyev 2019).

Currently, the causes of odontoma are not fully understood. There are several main factors contributing to the appearance and development of a tumor: injuries, bruises, cracks of the jaw bone; chronic infections of the jaw, oral cavity and nasopharynx (for example, periodontitis, periodontal disease, tonsillitis, osteomyelitis of the jaw bone); genetic predisposition to such neoplasms (Willis 2000; Sviridov et al. 2019).

Odontomas are found in people of different ages, but more often in childhood (Friedrich et al. 2010; Altay et al. 2016) and young age (up to 20 years) with equal probability in persons of both sexes, however, it can be observed in adulthood, and sometimes in old age. The growth's maximum of the tumor is observed either at the age of 6–11 years, which is associated with the period of eruption of permanent teeth (Snetkov et al. 2017), or at a later age—wth the eruption of the third molars. Its growth slows down or stops with the end of the formation of the dental rudiment, therefore, in adults, the tumor is more often detected accidentally during X-ray examination or with the development of the inflammatory process in the tissues of the neoplasm (Kaneko et al. 1998).

In literature, there are data on the detection of odontomas related to animals (Savina 2007). A number of publications describe clinical cases of odontoma detection related to animals, more often to dogs (Boyd 2002; Felizzola et al. 2003; Hoyer et al. 2016; Huang et al. 2019), cats and horses (Knowles et al. 2010; Andrews et al. 2014). Etiopathogenesis, clinical manifestations and signs of classification of human and mammalian odontomas are similar, as well as diagnosis and treatment (Starchenkov 2013; Klima and Goldstein 2007; De Oliveira et al. 2001, Walker et al. 2009). However, the morphology of odontomas found related to animals has not been studied enough.

The present investigation is a continuation of the early studies of dog's odontoma using a number of methods for studying morphology and chemical composition (Plotkina et al. 2018; Pikhur et al. 2020). The collection of samples of this study includes odontomas of different classification types belonging to various biological

Table 1 Samples of odontomas for studying

No.	Identity	Gender	Age	Localization in the maxillofacial region
1	Human	Male	7 years old	Central upper left incisor
2	Human	Female	20 years old	Third lower left molar
3	Dog	Male	7 months old	First lower left and right molars
4	Hedgehog	Male	5 years old	First upper left molar

species (contemporary humans and animals) and to different age groups. A detailed study of these odontomas will contribute to the elucidation of the etiology and pathogenesis, as well as the establishment of methods for the prevention and treatment of this pathology.

2 Materials and Methods

2.1 Samples

Samples for the study (Table 1) were obtained by removing, for medical reasons, odontomas of people from different age groups and some animals (dog and hedgehog). An odontoma removed from a 7-year-old boy patient in the area of the central upper left incisor and an odontoma removed from a 20-year-old woman patient in the area of the third lower molar on the left were examined. These odontomas were found in patients during X-ray dental examination. Odontomas were removed for medical reasons from young (7 month) male dog of the Black Russian Terrier during a period of mixed occlusion. The odontoma of a 5-year-old hedgehog (Erinaceus Europaeus) located on the left side of the upper jaw and removed in the veterinary clinic of the Zoo for medical reasons (inflammatory process) was studied.

2.2 Methods

Morphology of odontoma's samples was studied by optical microscopy, scanning electron microscopy (SEM) and X-ray microtomography (μCT). The chemical composition of the odontomas was studied by Energy-Dispersive Spectroscopy (EDS).

Optical Microscopy

The external appearance of the odontomas was studied by stereomicroscope «Olympus SZ61» (Japan).

Scanning Electron Microscopy

The surface and split morphology of the odontomas was studied by scanning electron microscopy TESCAN VEGA 3 (Czech Republic). Scanning parameters: accelerating voltage—10 kV, intensity—19, working distance—6.5 mm.

X-ray Computed Microtomography

To study the morphology of pathological formations in human and animal bodies, we used the method of X-ray computed microtomography. The method is a combination of tomographic algorithms and X-ray microscopy (Sasov and Van Dyck, 1998; De Clerck et al. 2003), which makes it possible to obtain visual information about the microstructure of hard tissues without their destruction based on an assessment of physical density. As a result of studies by this method, 3D models reflecting the internal structure of the studied samples were obtained. The undoubted advantage of this method is the absence of special preparation of samples, as well as their complete preservation during the study.

The odontoma's samples were scanned by Skyscan 1172 (Belgium). In all cases used conditions are source voltage—100 kV, source current—100 μA, Al filter 0.5 mm, pixel size—4.69 μm, frame averaging—3, rotation step—0.4°.

Energy-Dispersive Spectroscopy

The main element content in the odontomas was determined using a Hitachi S-3400N scanning electron microscope with Oxford Instruments X-Max20 EDS. The spectra were processed using the AzTec Energy software package (ver. 2.2) using the TrueQ method. Shooting parameters: accelerating voltage—20 kV, probe current—1.7 nA, working distance—10 mm, time of spectrum accumulation at a point (in point mode)—30 s. The quantitative calculation of the spectra was carried out using standard samples of natural and synthetic compounds.

3 Results

The study of macro- and micromorphology of samples allows to establish an accurate diagnosis of "odontoma" and determine the classification type of odontoma. As a result of the study, it was revealed that all samples are odontomas of various classification types (Table 2).

Table 2 Classification of odontomas samples

No.	Identity	Classification type of odontoma
1	Human (M)	Solid simple odontoma
2	Human (F)	Solid complex composite odontoma
3	Dog	Solid complex composite odontoma
4	Hedgehog	Solid complex composite odontoma

Basing on the morphological structure, soft and hard (solid) odontomas are distinguished. The soft odontoma (odontoma molle) is considered as the initial stage of the development of a hard odontoma. There are three forms of solid odontoma: simple (complete and incomplete); complex (composite and mixed); cystic. The solid odontoma (odontoma compactum) consists of petrified highly differentiated dental structures. The composition of this tumor may include different parts of the tooth: enamel, dentin, cement, periodontal, pulp. These tissues are calcified and may be at different stages of development.

It is known that there are simple and complex odontomas (Fletcher 2013, El-Naggar et al. 2017). Complex, in turn, are divided into composite and mixed. A simple odontoma (odontoma simplex) is an odontoma formed by a small number of randomly mixed hard tissues sufficient to form only one tooth. Macromorphologically, such a lesion looks like an encapsulated nodule with an underdeveloped or deformed tooth. A simple odontoma is associated with a malformation of one dental germ. A complex odontoma (odontoma complexu) is an odontoma formed by a significant amount of hard tissue sufficient to build several teeth. A complex odontoma is associated with a violation of the development of several rudiments of teeth, therefore it contains hard tissues at different stages of development (Barros et al. 2010). Composite odontoma (odontoma compositum) is a complex odontoma of several correctly formed but deformed teeth (tooth-like elements or odontoids) with pulp centrally located in them (Kalra et al. 2018). Morphologically, it is an encapsulated node of several odontoids, easily separated or tightly soldered together by connective tissue.

Various morphological species of odontomas of contemporary humans and animals determine the classification type of odontoma. The odontomas we studied, according to the classification, turned out to be solid odontomas. At the same time, we have identified a solid simple odontoma and solid complex odontomas. All the solid complex odontomas that we studied were composite.

Morphology and Chemical Composition of Human Solid Simple Odontoma

The solid simple odontoma removed from a 7-year-old patient in the area of the central upper left incisor has a size of 0.6 × 0.9 cm and is formed by dentin covered with an uneven layer of enamel, sometimes almost completely disappearing (Fig. 1). There is a cone-shaped cavity inside the odontoma.

SEM investigation of human solid simple odontoma showed that the enamel surface is heterogeneous. It is represented by areas with smooth areas similar to the enamel surface of intact human teeth, and areas with a broken structure (Fig. 2e, f). Extensive surface defects include areas of enamel with immature mineralization, which are, firstly, clearly defined regular layers (growth or dissolution) of a wave-like shape (Fig. 2a); secondly, areas with a clear mesh texture (Fig. 2b, c). The mesh texture was formed as a result of the dissolution and thinning of the enamel up to the exposure of enamel prisms (Fig. 2a). Significant areas with open enamel prisms are observed on the surface of the odontoma (Fig. 2d, f).

3D model obtained as a result of a microtomographic study demonstrates a conical cavity formed by enamel and dentin. The thickness of the enamel varies from 0.1

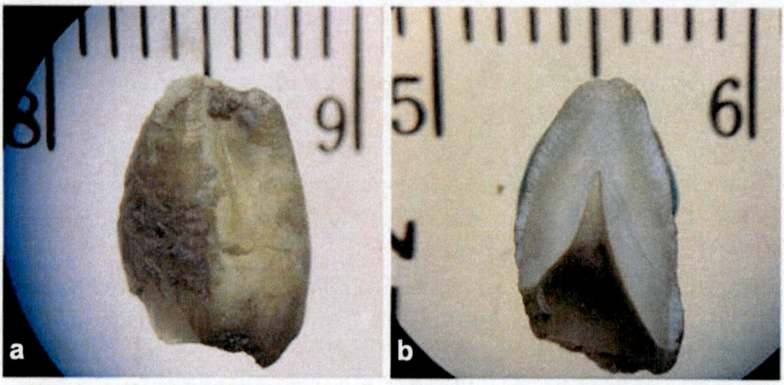

Fig. 1 Optical microscopy images of human solid simple odontoma: **a** the exterior; **b** longitudinal cleavage

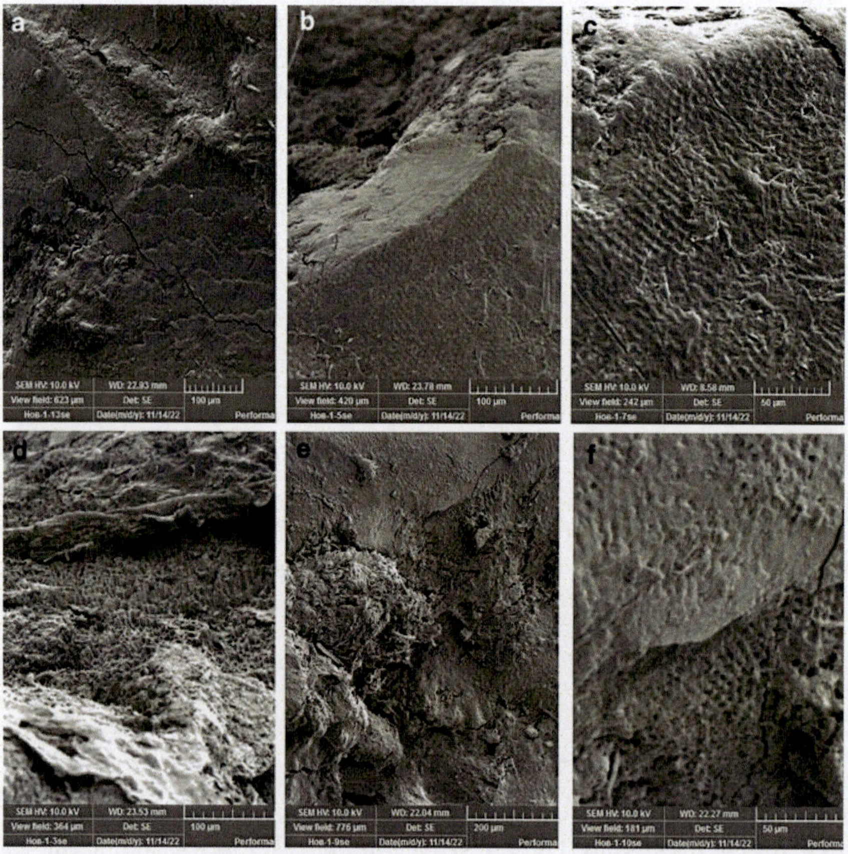

Fig. 2 SEM images of human solid simple odontoma: **a** wavelike layers on the enamel surface; **b**, **c** mesh texture of enamel; **d**, **f** exposed enamel prisms; **e** defective enamel surface

to 0.9 mm, dentin—from 0.1 to 21.0 mm. The enamel surface is uneven, with many defects in the form of scratches, depressions and dents (Fig. 3).

According to microprobe analysis, the Ca/P coefficient ($K_{Ca/P}$) calculated for enamel is 1.66–1.68. In the area of the enamel-dentine border, $K_{Ca/P} = 1.68$. In dentine, $K_{Ca/P} = 1.44$. The chemical composition of enamel includes Na (0.46–0.66 wt %), Mg (0.20–0.61 wt %), Cl (0.28–0.35 wt %); dentine—Na (0.29 wt %), Mg (0.82 wt %).

Morphology and Chemical Composition of Human Solid Complex Composite Odontoma

Optical investigation of the tumor (Fig. 4), removed by surgical intervention in a 20-year-old women patient in the area of the third lower molar on the left, revealed that it is a solid complex composite odontoma measuring 0.5 × 1.0 cm.

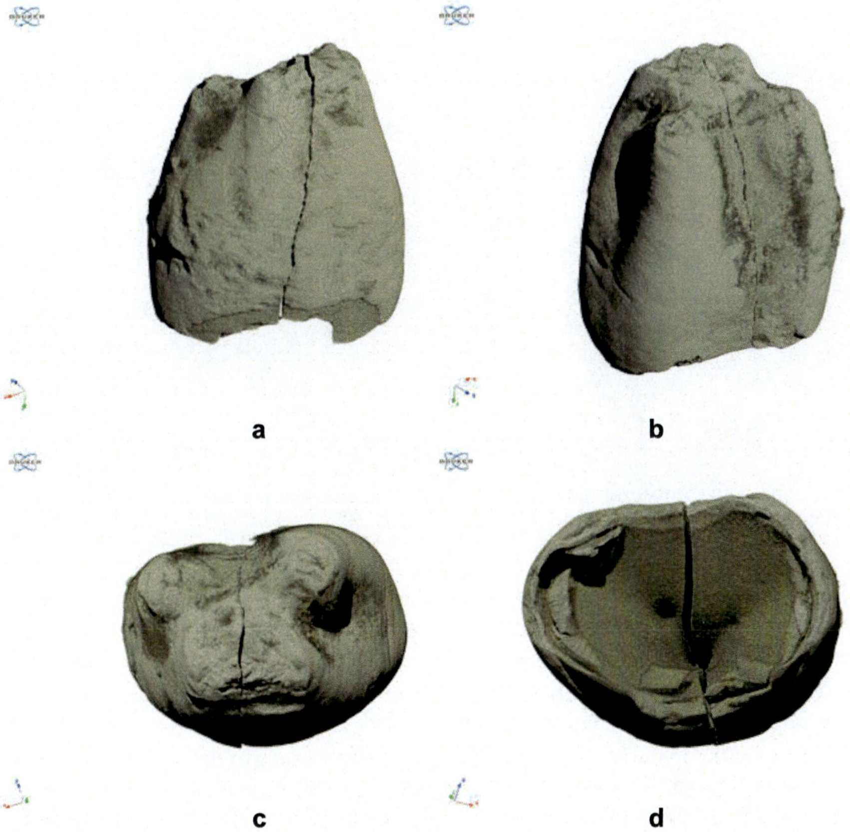

Fig. 3 Micro CT images of human solid simple odontoma: **a**, **b** exterior on the right and left; **c**, **d** exterior from above and below; **e**, **f** 3D model of the internal structure

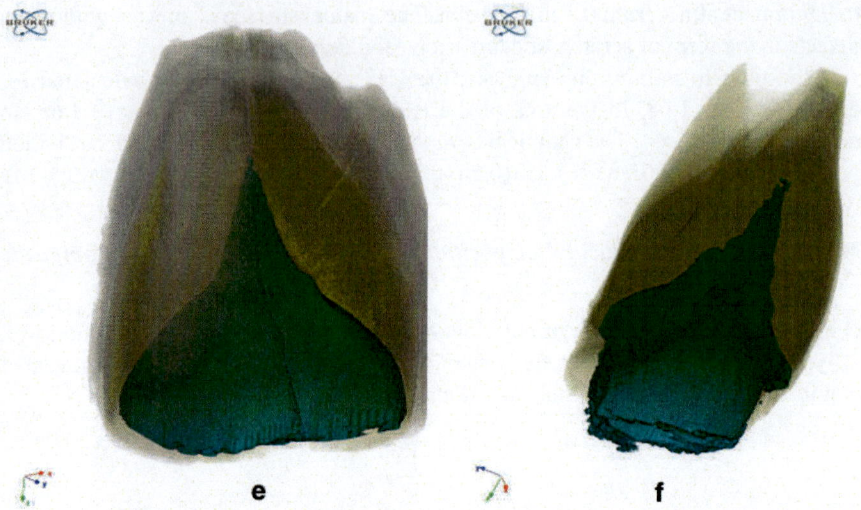

Fig. 3 (continued)

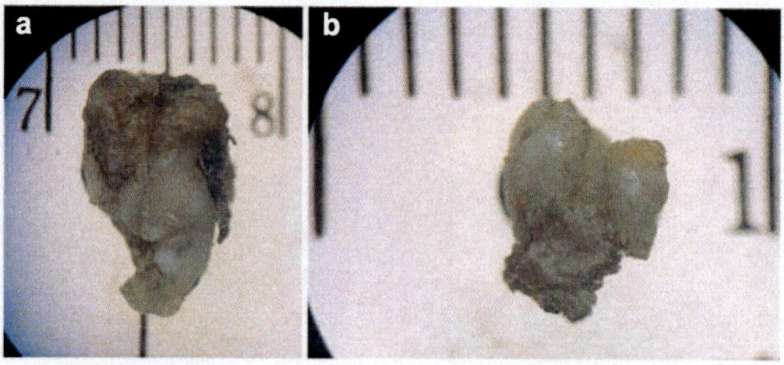

Fig. 4 Optical microscopy images of human solid complex composite odontoma: **a**, **b** the exterior of the tumor from different sides

SEM study of a large odontoid in the human solid complex composite odontoma showed that the enamel surface is uneven—coarse and fine-grained areas (Fig. 5a–c) are interspersed with smooth ones. Large cavities of various volumes are present in the enamel (Fig. 5a, b). Enamel defects on the surface are represented by well-mineralized irregularly shaped tubercles (Fig. 5c) with a size of 0.1–0.5 mm. The second odontoid (smaller) has a clear morphological zoning of the enamel. One third of the odontoid on the side of the enamel chewing surface is completely formed by separate irregularly shaped tubercles of different sizes (Fig. 5d–f). Two thirds of the odontoid, located below, have a relatively smooth surface (Fig. 5d).

This solid complex composite odontoma was studied by X-ray computed microtomography. The odontoma consists of enamel, dentin and cement arranged in order relative to each other like intact teeth. The morphology of the odontoma is determined

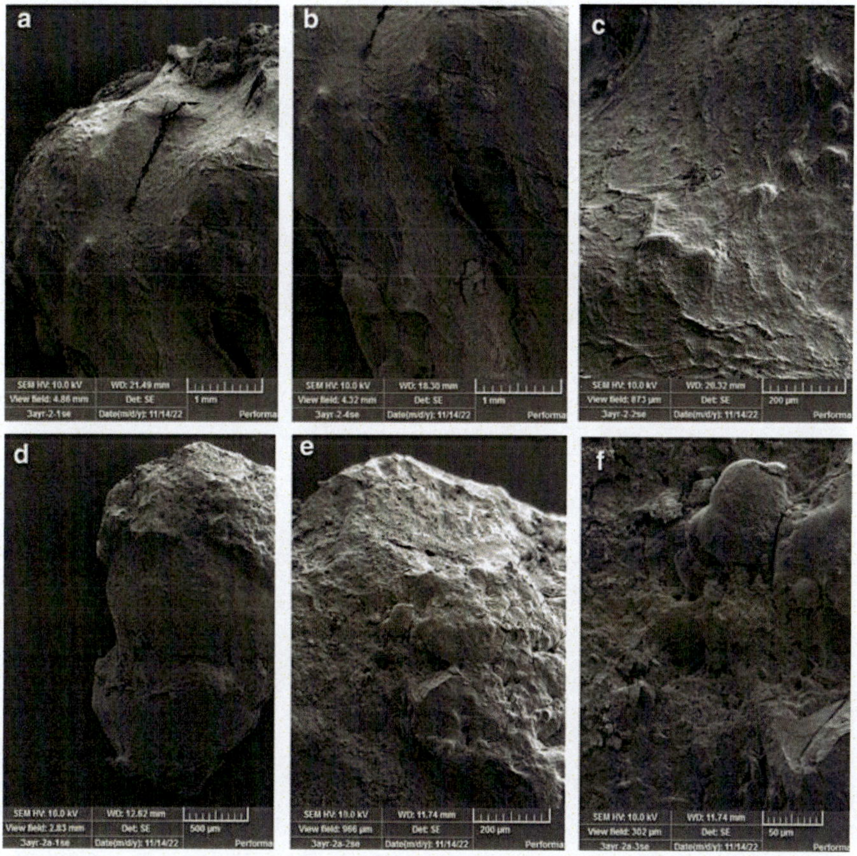

Fig. 5 SEM images of human solid complex composite odontoma: **a** a large odontoid on the chewing surface; **b**, **c** tubercles and cavities of the enamel surface; **d** a smaller odontoid; **e**, **f** a bumpy enamel surface

by the number and location of the odontoids that make up it. This odontoma is represented by 2 great odontoids measuring 4–6 × 5–9 mm (Fig. 6a) and 6 small odontoids measuring 1–2 × 2–3 mm (Fig. 6d–f). Great odontoids are spatially deployed 180° relative to each other. Small odontoids are located chaotically between them and are held by connective tissue. There are numerous cavities and root channels inside the formation (Fig. 6b, c). The volume of root canals varies in great odontoids from 2.00 to 2.58 mm^3, and in small odontoids—from 0.008 to 0.297 mm^3. Anastomoses are present between the root canals of large volume (Fig. 6g, h).

According to the microprobe analysis of the Ca/P coefficient in different areas of enamel varies within 1.37–1.81. At the same time, the maximum values of the coefficient are observed in areas with a bumpy shape, and the minimum values are in areas with a mesh texture. Areas of enamel with a smooth homogeneous surface

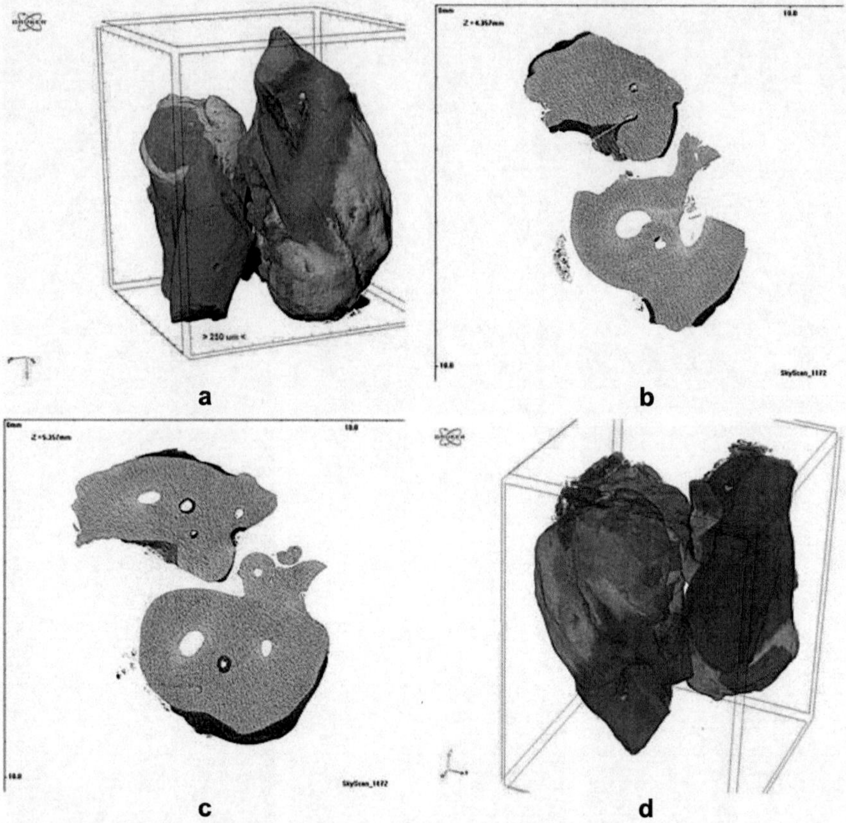

Fig. 6 Micro CT images of human solid complex composite odontoma: **a**, **d** the exterior; **b**, **c** cross-sections of sample; **e** longitudinal section; **f** cross section; **g**, **h** 3D model of internal channels and cavities

have the Ca/P coefficient of 1.57–1.65, which is characteristic of the enamel of intact teeth. The enamel contains Mg (0.92–1.11 wt %) and Na (0.42–0.88 wt %).

Morphology and Chemical Composition of Animals' Solid Complex Composite Odontoma

Odontomas extracted from dogs and hedgehogs were classified as solid complex composite odontomas.

Dog's Odontoma

The dog's odontoma has a size of 10–12 mm and is formed by randomly arranged 10 odontoids ranging in size from 2–3 to 7–8 mm and interconnected by connective tissue (Figs. 7 and 8). The odontoma consists of different constituent parts: enamel, dentin, cement, pulp of tooth and fibrous tissue. All the odontoids that make up the odontoma are oriented according to the direction of the intact tooth—from the root

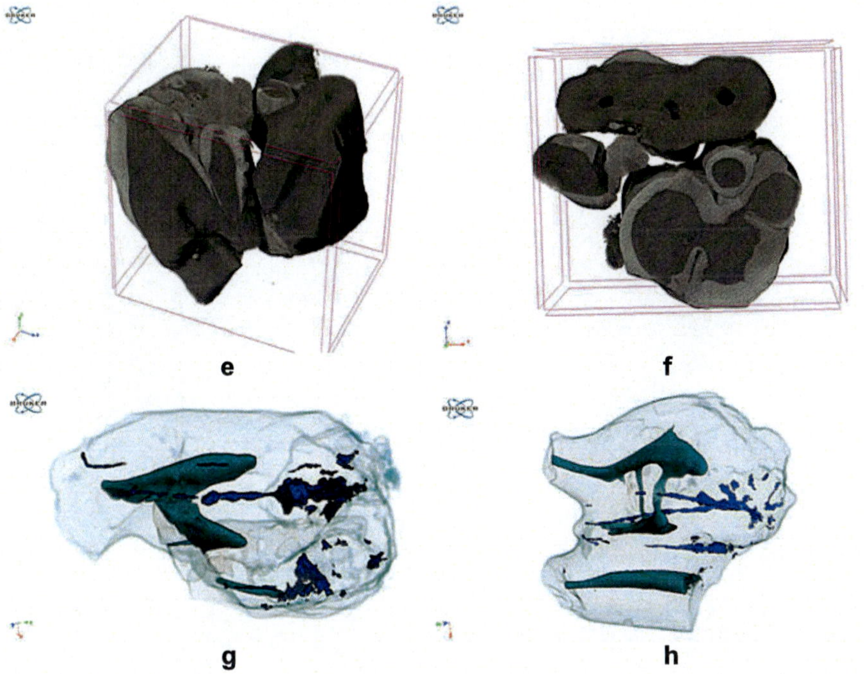

Fig. 6 (continued)

part to the crown. Odontoids are interconnected by connective tissue into a single conglomerate (Plotkina et al. 2018).

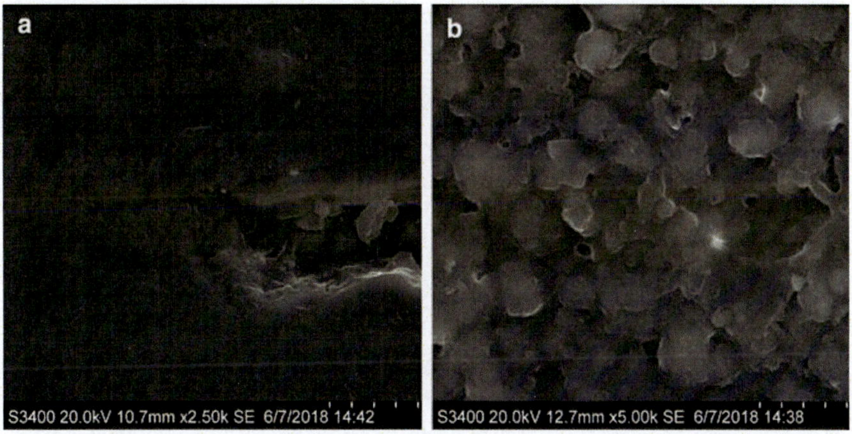

Fig. 7 SEM images of dog's solid complex composite odontoma: **a** enamel surface; **b** dentin surface

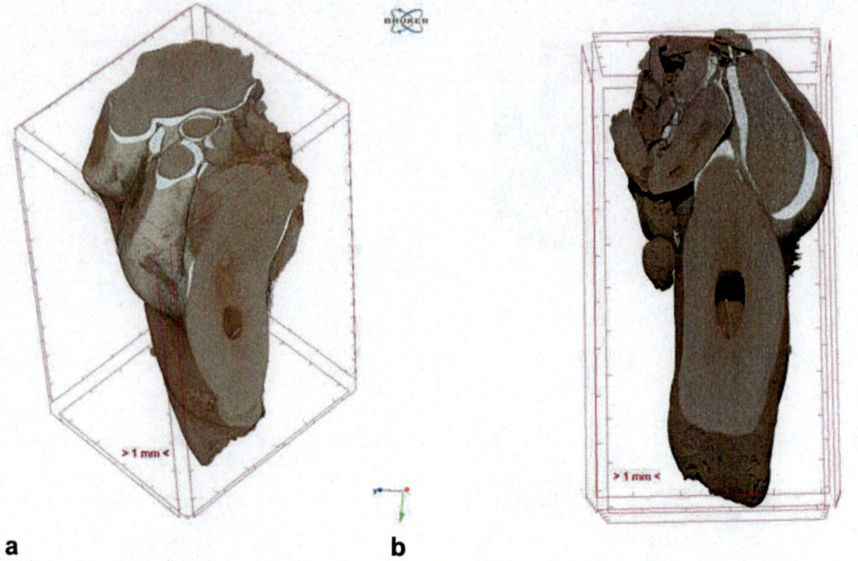

Fig. 8 Micro CT images of dog's solid complex composite odontoma: a the exterior and cross sections in different directions; b the cross section

The SEM study showed that the enamel surface of the dog's solid complex composite odontoma has a heterogeneous morphological structure. There are areas with a smoothed dense surface similar to the enamel of intact teeth and at the same time there are cracks and large potholes (Fig. 7a). The dentin structure is characterized by high porosity and well-defined open dentine tubules (Fig. 7b).

The results of microprobe analysis are shown that the $K_{Ca/P}$ of odontoid for enamel and dentin are 1.56 and 0.80; the $K_{Ca/P}$ of odontoma tissues is 1.58 and 1.03 accordingly. The enamel $K_{Ca/P}$ of odontoid and odontoma totally matches to normal hard tooth tissues, but the dentin $K_{Ca/P}$ is less. The chemical composition of the enamel includes Na (0.60–0.90 wt %), Mg (0.2–0.3 wt %), Cl (0.5–0.6 wt %), K (0.1 wt %).

Hedgehog Odontoma

The odontoma extracted from a hedgehog is classified as a solid complex composite odontoma (Fig. 9). This odontoma consists of 8 soldered odontoids and has a size of 8 × 7 mm. The odontoma contains different types of hard tooth tissues: enamel, dentin, cement (Figs. 10 and 11). The enamel of this odontoma is present on the chewing and one of the lateral surfaces. It has a thickness of 1–2 mm, sometimes thinning to 0 mm. On the side of the chewing surface, the enamel forms deep folds that cut through the dentin 70–80% deep (Fig. 11b). Moreover, atypical enamel formations in the form of granules with a diameter of up to 1 mm are observed on the chewing surface.

The SEM study of a large odontiod, which is part of the solid complex hedgehog's odontoma, showed the presence of several well-formed deep fissures (Fig. 10a, b).

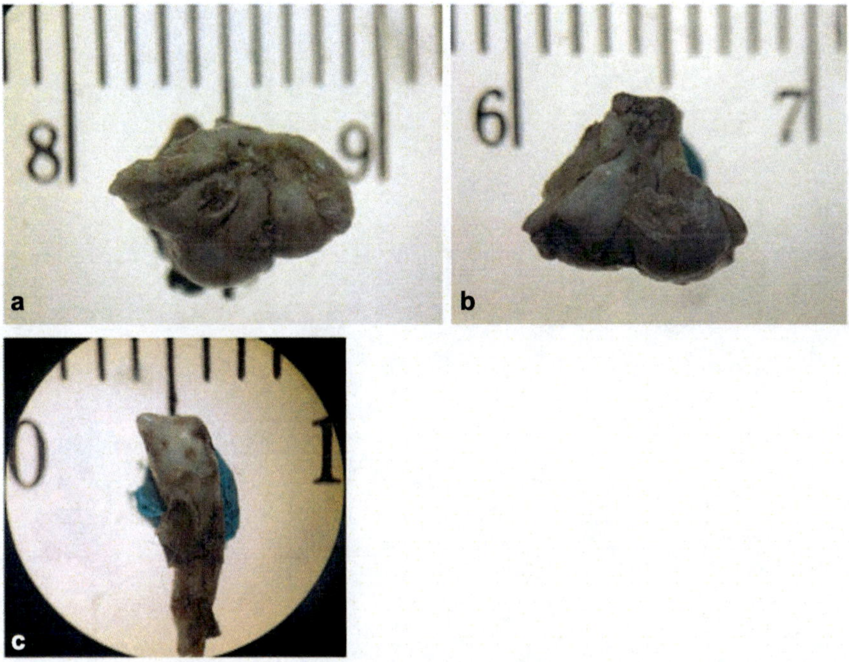

Fig. 9 Optical microscopy images of the hedgehog's solid complex composite odontoma: **a**, **b** the exterior the tumor from different sides; **c** odontoid

The enamel of the chewing surface is well mineralized, smooth with clear contours (Fig. 10a). However, there are defective areas of enamel with a slight exposure of enamel prisms (Fig. 10c). The smaller odontoid also has a mostly smooth, well-formed enamel surface (Fig. 10b–e), however, there are areas with a mesh texture due to the exposure of enamel prisms (Fig. 10a, f). Elements of connective tissue are observed on odontoids, due to which they were joined into a single conglomerate. This is especially noticeable on the root part of the smaller odontoid (Fig. 10d).

According to microprobe analysis, the Ca/P coefficient of enamel is 1.50–1.56. A distinctive feature of the chemical composition of the hedgehog's odontoma is the presence of Na (0.53–1.24 wt %), Mg (0.26–0.61 wt %), S (0.31–0.48 wt %), F (0.68–0.75 wt %).

4 Discussion

The cause of odontoma in the dog's case is probably a genetic predisposition. Black Russian Terrier is a young breed of dog with large amount of original breeds in creating history (more than 17). In this regard there is a high risk of genetic defects

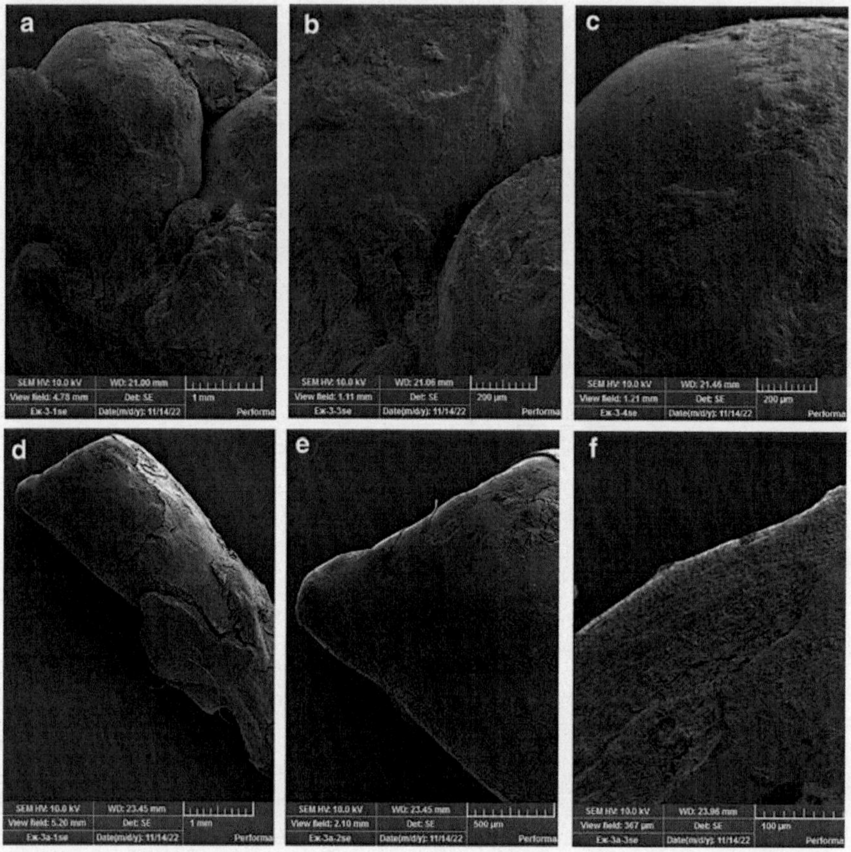

Fig. 10 SEM images of hedghog's solid complex composite odontoma: **a** view of a large odontoid; **b** fissures of a large odontoid; **c** enamel surface; **d** view of a smaller odontoid; **e**, **f** enamel surface

in dog organism. In possible the appearance and progress of odontoma is one of these defects.

It is known that the hedgehog individual had serious nutritional disorders in early childhood, when the rudiments of permanent teeth were formed. This is probably why neoplastic deformation of the tooth rudiment occurred in the area of the first upper molar, resulting in the formation of the odontoma. The process proceeded asymptomatically for a long time. In adulthood (3.5 years old), an inflammatory process developed in the area of the odontoma. Conservative treatment did not bring results, and the odontoma was removed as a result of surgery at the age of the animal 5 years.

Anamnesis data, clinical and radiological examination of patients aged 7 and 20 years did not allow to identify possible causes of the formation of the odontomas in them. For the 7-year-old patient, the odontoma was found in the area of the permanent central upper incisor during the replacement bite. The rudiment of this

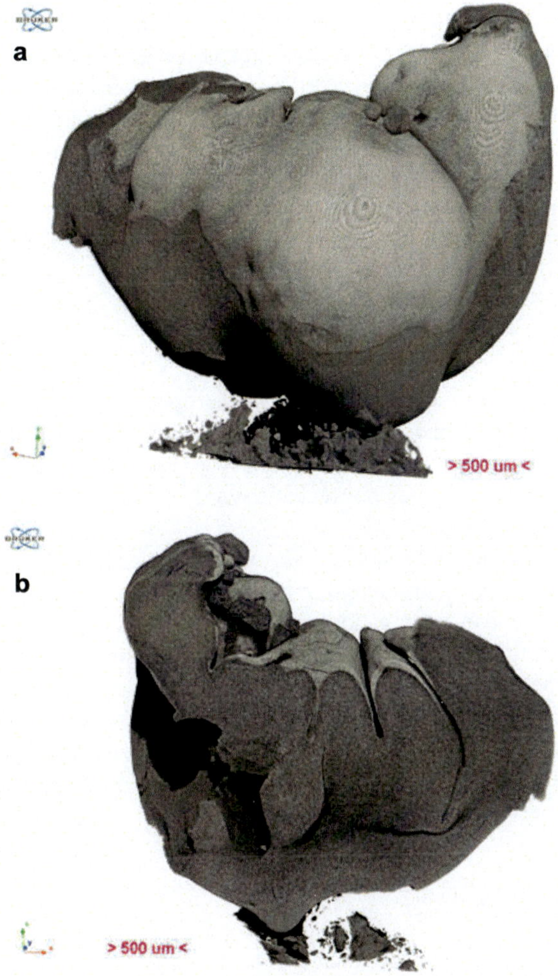

Fig. 11 Micro CT 3D images of the hedgehog's solid complex composite odontoma: **a** the exterior; **b** longitudinal section

tooth was transformed into the odontoma for unknown reasons. For the 20-year-old patient, the odontoma developed asymptomatically and was discovered accidentally during X-ray examination (orthopantomography) for another pathology.

Thus, in childhood, during the period of a replacement bite, odontoma often manifests itself in both humans and animals (in our study—the patient of 7 years and the dog of 7 months). At the same time, an odontoma can exist asymptomatically for a long time and manifest itself at an older age (in our study—the patient of 20 years and the hedgehog of 5 years).

Morphological features of the odontoma of contemporary humans and animals of both gender and different ages are similar and determine the classification type. In our study, solid complex composite odontoma was most often encountered. The studied odontomas belong to the hard ones formed by enamel and dentin, the composition of which is close to the composition of human teeth.

As a result of the use of microprobe analysis by human and animal odontomas, it was found that the Ca/P coefficient can be as in intact human hard tissues (enamel—1.68, dentin—1.44). At the same time, there is a decrease in the mineralization of hard tissues of the odontoma and a decrease in the Ca/P coefficient in enamel (1.50–1.58) and dentin (1.03). However, there are separate areas of enamel with odontoma, where Ca/P coefficient has significant variations (1.37–1.81), which is probably due to the transformation of the odontoma.

5 Conclusions

The use of a set of methods made it possible to study in detail the morphology and composition of human and animal odontomas, considering their age, and to establish belonging to the classification type. Especially important, since each odontoma is unique and does not repeat any other, even if they belong to the same species by classification.

Using the method of X-ray microtomography allowed us to get a representation of the internal structure of odontomas without violating their integrity. Moreover, this method in combination with other methods used by us allows to diagnose "odontoma" accurately, to carry out differential diagnosis with other diseases, as well as to clarify the presence of signs of malignancy.

The data obtained on the basis of these studies allow us to reveal the etiopathogenetic mechanisms of the formation of odontomas and are the basis for the diagnosis and implementation of therapeutic and preventive measures for this pathology.

References

Afanasyev VV (2019) Surgical dentistry. 3rd edn. Geotar-Media. Moscow (in Russian)

Akerzoul N, Chbicheb S, Wady WE (2017) Giant Complex Odontoma of Mandible: A Spectacular Case Report. The Open Dentistry J 11: 413–419

Altay MA, Ozgur B, Cehreli ZC (2016) Management of a Compound Odontoma in the Primary Dentition. J Dent Child 83: 98–101

Andrews C, Gadsden BJ, Carr EA, Kiupel M (2014) Pathology in Practice. Compound Odontoma in a Horse. J Am Vet Med Assoc 244: 417–419

Barros LD, Pedron IG, Utumi ER, Zambon CE, Rocha AC (2010) Complex Odontoma: Report of a Five-Year Follow-Up Case. J Dent Child 77: 183–186

Boyd RC (2002) Ameloblastic Fibro-Odontoma in a German Shepherd Dog. J Vet Dent 19: 148–150

De Clerck NM, Van Dyrk D, Postnov AA (2003) Non-Invasive High-Resolution μCT of the Inner Structure of Living Animals. Microscopy and Analysis (Euro) 81: 13–15

De Oliveira BH, Campos V, Marçal S (2001) Compound Odontoma – Diagnosis and Treatment: Three Case Reports. Pediatr Dent 23: 151–157

De Vasconcellos Machado C, Knop LAH, Barreiros Siquara da Rocha MC, Da Silva Telles PD (2015) Impacted Permanent Incisors Associated With Compound Odontoma. BMJ Case Rep 12: 20–15

El-Naggar AK, Chan JKC, Grandis JR, Takata T, Slootweg PJ (2017) WHO Classification of Head and Neck Tumours. 4th edn. International Agency for Research on Cancer. Lyon
Felizzola CR, Martins MT, Stopiglia A, De Araújo NS, Machado de Sousa SO (2003) Compound Odontoma in Three Dogs. J Vet Dent 20: 79–83
Fletcher CDM, Bridge JA, Hogendoorn PCW, Mertens F et al. (2013) WHO Classification of Tumours of Soft Tissue and Bone. 4th edn. IARC Press. Lyon
Friedrich RE, Fuhrmann A, Scheuer HA, Zustin J (2010) Small Peripheral Developing Odontoma of the Maxilla in a 3-yearold Patient Depicted on Cone-Beam Tomograms. In Vivo 24: 895–898
Gupta A, Vij H, Vij R, Malhotra R (2014) An Erupted Compound Odontoma. BMJ Case Rep 12: 10–14
Hidalgo-Sanchez O, Leco-Berrocal MI, Martinez-Gonzalez JM (2008) Metaanalysis of the epidemiology and clinical manifestations of odontomas. Med Oral Patol Oral Cir Bucal 13: 730–734
Hoyer NK, Bannon KM, Bell CM, Soukup JW (2016) Extensive Maxillary Odontomas in 2 Dogs. J Vet Dent 33: 234–242
Huang P, Bell C, Wallace V, Murphy BG (2019) Mixed Odontogenic Tumors in Four Young Dogs: Ameloblastic Fibroma and Ameloblastic Fibro-Odontoma. J Vet Diagn Invest 31: 98–102
Kalra A, Sheehy EC, Johnson J, McDonald F (2018) Compound Odontoma of the Maxilla. Pediatr Dent 40: 140–142
Kaneko M, Fukuda M, Sano T, Ohnishi T, Hosokawa Y (1998) Microradiographic and microscopic investigation of a rare case of complex odontoma. Oral Surg. 86: 131–134
Klima LJ, Goldstein GS (2007) Surgical Management of Compound Odontoma in a Dog. J Vet Dent 24: 100–106
Knowles S, Blas-Machado U, Butler AM, Gomez-Ibañez SE, Lowder MQ, Fayrer-Hosken RA (2010) Ameloblastic Fibro-Odontoma Associated With a Retained Molar in an Oldenburg Mare. J Vet Diagn Invest 22: 987–990
Pikhur OL, Plotkina YuV, Kulkov AM (2020) Using X-ray Computed Microtomography for Investigation of the Morphology and Composition of the Hard Tooth Tissue. Processes and Phenomena on the Boundary Between Biogenic and Abiogenic Nature. Springer Nature Switzerland AG: 221–232
Plotkina YuV, Pikhur OL, Kulkov AM (2018) A Computed microtomographic study of the dog odontoma. Micro-CT User Meeting. Ghent: 281–284
Sasov A, Van Dyck D (1998) Desktop X-ray Microscopy and Microtomography. Journal of Microscopy 191, 2: 15–158
Savina YuD (2007) Neoplastic diseases of animals / Veterinary doctor 4: 34 (in Russian)
Snetkov AI, Batrakov SY, Morozov AK (2017) Diagnosis and treatment of benign tumors and tumor-like bone diseases in children. Geotar-Media. Moscow (in Russian)
Starchenkov SV (2013) Diseases of the dogs and the cats. Complex diagnostics and therapy. 4th edn. SpecLit, St.Petersburg (in Russian)
Strukov AI, Serov VV (2015) Pathological anatomy. 6th edn. Geotar-Media. Moscow (in Russian)
Sun L, Sun Z (2016) Multiple Complex Odontoma of the Maxilla and Mandible. Reply Oral Surg 121: 443–444
Sviridov EG, Kadykova AI, Redko NA, Drobyshev AY, Deev RV (2019) Genetic heterogeneity of tumor-like lesions of the bones of the maxillofacial region. Genes & Cells 1: 49–53 (in Russian)
Vengal M, Arora H, Ghosh S, Pai KM (2007) Large Erupting Complex Odontoma: A Case Report. J Can Dent Assoc 73: 169–173
Walker KS, Lewis JR, Durham AC, Reiter AM (2009) Diagnostic Imaging in Veterinary Dental Practice. Odontoma and Impacted Premolar. J Am Vet Med Assoc 235: 1279–1281
Willis MB (2000) Genetics of the dog. Centerpolygraph, Moscow (in Russian)

Synthesis and Properties of Hydroxyapatite—Chitosan Biopolymer Composite Materials

Olga A. Golovanova

Abstract The results of a study of the physicochemical properties of composites and first synthesized scaffolds based on hydroxyapatite (HA) and chitosan are presented. It was found that the size of composite crystallites increases with an increase of the chitosan content in the initial solution. For the first time, a technique for obtaining scaffolds based on HA and chitosan gel was proposed. It has been established that the pore sizes of the scaffolds increase with an increase in the polymer content. It was found that the stability of the scaffolds increases with increasing temperature, while the composites remain in the form of powders, regardless the temperature treatment. When studying the dissolution of the synthesized composites and scaffolds in an isotonic solution, it was found that for the HA–chitosan system, an increase in the dissolution rate is observed with an increase in the concentration of chitosan in the sample. Magnesium-substituted hydroxyapatite (Mg-HA) has been synthesized from an aqueous solution of magnesium, calcium, diammonium phosphate, and ammonia salts in the presence of a polymer matrix of chitosan. The results of the study of the physicochemical properties of the synthesized composites are presented. The results of determining the composition, morphological, thermal, and bioactive characteristics of the obtained composites are presented. It has been found that all the samples have a similar phase composition and morphology, which is characteristic of magnesium-substituted hydroxyapatite. It has been shown that the crystallite size of composites with chitosan decreases with an increase in the polymer content. It has been found that upon dissolution of samples in an isotonic solution, the rates of formation of calcium ions in the liquid phase increase with the content of chitosan in the synthesized composites.

Keywords Crystallization · Scaffold · Hydroxyapatite · Chitosan · Magnesium · Composite · Morphology · Thermal properties · Dissolution

---

O. A. Golovanova (✉)
Department of Inorganic Chemistry, Dostoevsky Omsk State University, Mira St., 55a., Omsk 644077, Russia
e-mail: golovanoa2000@mail.ru

O. V. Frank-Kamenetskaya et al. (eds.), *Biogenic—Abiogenic Interactions in Natural and Anthropogenic Systems 2022*, Springer Proceedings in Earth and Environmental Sciences, https://doi.org/10.1007/978-3-031-40470-2_4

1 Introduction

Currently, materials science in the field of biocomposite development sets itself the task of searching for new biopolymers that can decompose in the natural environment, and is considering the possibility of creating safe and environmentally friendly processes for isolating polymers from raw materials and their further processing. to obtain a product that meets the standards for use (Kandyrin 2002).

Thus, the development of new biodegradable compounds for medicine, used for contact with living organisms and important for the development of new dosage forms used to restore damaged tissues, is of great importance (Nuja 2006). It is known that such compounds should be nontoxic, have high porosity with pore sizes up to tens of micrometers, and have channels for the penetration of drugs into the polymer matrix (Homenko et al. 2013, Jayakumar et al. 2011, Volova 2009).

Polysaccharides such as chitosan (Ch) have been studied with the aim of developing new materials for bone implantation (Danilchenko et al. 2009, Kamskaya 2016, Elson 2011, Kurchenko et al. 2016). This compound is an amorphous-crystalline polymer, which is characterized by polymorphism.

It is known that chitosan is a water-insoluble polymer and its chemical property is solubility in dilute acids, such as CH_3COOH acid and HCl acid.

Chitosan, the simplest derivative of chitin, has important biological properties: lowers cholesterol levels, exhibits an immunomodulatory effect, as well as bactericidal and wound healing activity (Muravyov 2017).

In bone engineering, it is best used in the form of a material with one of the most commonly used calcium phosphate, namely hydroxyapatite (HA) (Volova 2009).

Hydroxyapatite (HA) (the least soluble of calcium phosphates) as an inorganic component causes the greatest interest (Bolarinwa 2010, Wahl 2006). Due to the similar chemical composition of hydroxyapatite to bones and teeth, HA-based materials are widely used in medicine (Agrawal and Gurbhinder 2011).

HA-Ch composite materials can demonstrate high biocompatibility and bioactivity (Fedosov, 2014), osteoinductivity (Bakunova et al. 2011), and absorbability. In addition, such materials can be used as a biologically active layer on the surface of metal implants, for example, titanium, titanium and nickel alloys, etc (Gurin 2009).

To improve the characteristics of HA, it is modified with biogenic ions (cations and anions) found in natural bone tissue, which leads to a change in the solubility of new materials (Ling-Hao et al. 2011) and adds a number of important and necessary properties.

An example of such a necessary material is hydroxyapatite doped with magnesium ions (Mg-HA) (Salman et al. 2013; Trisvetova 2013). Magnesium compounds are important for the cells and tissues of a living organism. Since the radius of the Mg^{2+} ion is smaller and the ionization energy is higher, it is able to form stronger bonds than the Ca^{2+} ion. Thus, Mg^{2+} becomes a more active catalyst for a number of enzymatic reactions (Kojima et al. 2018).

The aim of this work is to select the synthesis conditions and establish the physicochemical properties of materials based on HA, magnesium ions, and chitosan.

2 Experimental

2.1 Synthesis

The HA-chitosan composite (Ca/P = 1.67) was synthesized in the presence of chitosan, molecular weight (M = 38,000), the procedure is given in Fadeev and Golovanov (2019).

2.2 Synthesis

The Mg-HA-chitosan composite was synthesized by precipitation from aqueous solutions of initial solutions and chitosan gel at 22–25 °C (Lyasnikova 2017).

2.3 Methods and Approaches

X-ray phase studies of the obtained composites were carried out on a DRON-3M diffractometer (CuKα radiation). The detection limit of the method is 5%. The crystallite sizes were calculated using the Scherrer formula (Fadeeva and Golovanova, 2019).

IR-Fourier spectra were obtained on an FSM 2202 spectrophotometer by pelletizing with KBr, the measurement range was from 400 to 4000 cm^{-1} with a resolution of 4 cm^{-1}. The detection limit is 5%.

Optical microscopy. The type and structure of the composites were revealed using a KhSP-140 microscope (Fadeeva and Golovanova 2019).

For the study of microstructural features, morphology and surface characteristics scanning electron microscopy (TESCAN MIRA 3 LMH microscope), resolution 1.2 nm was used. The limit of permissible relative measurement error is 3%.

Thermogravimetric studies of composites. The synthesized materials m = 0.0002 g were calcined in an oven with temperature programming from 200 °C to 800 °C in increments of 200 °C for 120 min. After the end of the process, the weight loss of the composite after thermal exposure was determined.

BET method The samples were analyzed by the BET method on adsorption device "Sorbtometer" at 77.4 K one point of the isotherm of nitrogen adsorption in the current helium. Specific surface measurement range from 0.5 to 999 m^2/g. Relative limit measurement error of the specific surface in the mode of repeated measurement there are no more than 5%.

Preparation of a gel based on chitosan. To obtain a chitosan gel, its powder was dissolved in 2% HCl, the pH of the system was 7.5–8.0, then anhydrous succinic acid was added in a ratio of 1: 20. The resulting solution was left for 120 min and a chitosan gel was obtained (from 5 to 20% X).

Frame manufacturing. The scaffolds were obtained from synthesized HA and chitosan gel by mixing and keeping for 60 min at a temperature of 473 K.

Study of the bioactivity of synthesized materials in 0.9% NaCl solution. The studies were carried out for 120 min with constant stirring of the solution. The dissolution rate was determined using kinetic curves (Izmailov 2016).

3 Results and Discussion

3.1 HA–Ch composites

X-ray powder diffraction study of the HA–chitosan samples showed that the composition of the dried samples is represented by HA with small inclusions of an amorphous phase (Fig. 1a).

Analysis of the IR spectra of the composites showed the presence of absorption bands assigned to vibrations of phosphate groups and water (Fig. 2a). Characteristic are absorption bands caused by valence and symmetric oscillations (1024 and 1154 cm^{-1}) O–P–O bonds. Peaks 530, 574 correspond to vibrations in tetrahedral. A wide band in the region 3489–3583 cm^{-1} and a peak at 3142 cm^{-1} can be assigned to the valence vibrations of H–O–H and OH^{-}, respectively.

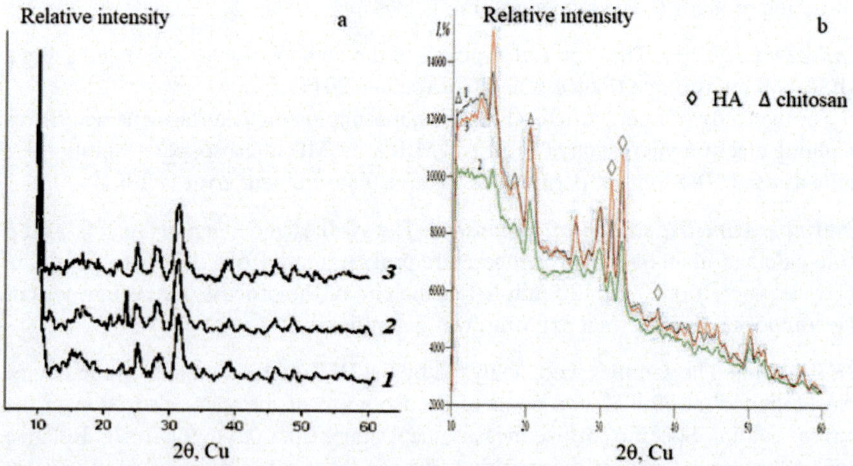

Fig. 1 X-ray diffraction patterns: **a** HA-Ch composites; **b** Mg-HA-Ch composites, $C_{chitosan}$, g/l: 1—0.02; 2—0.08; 3—0.16

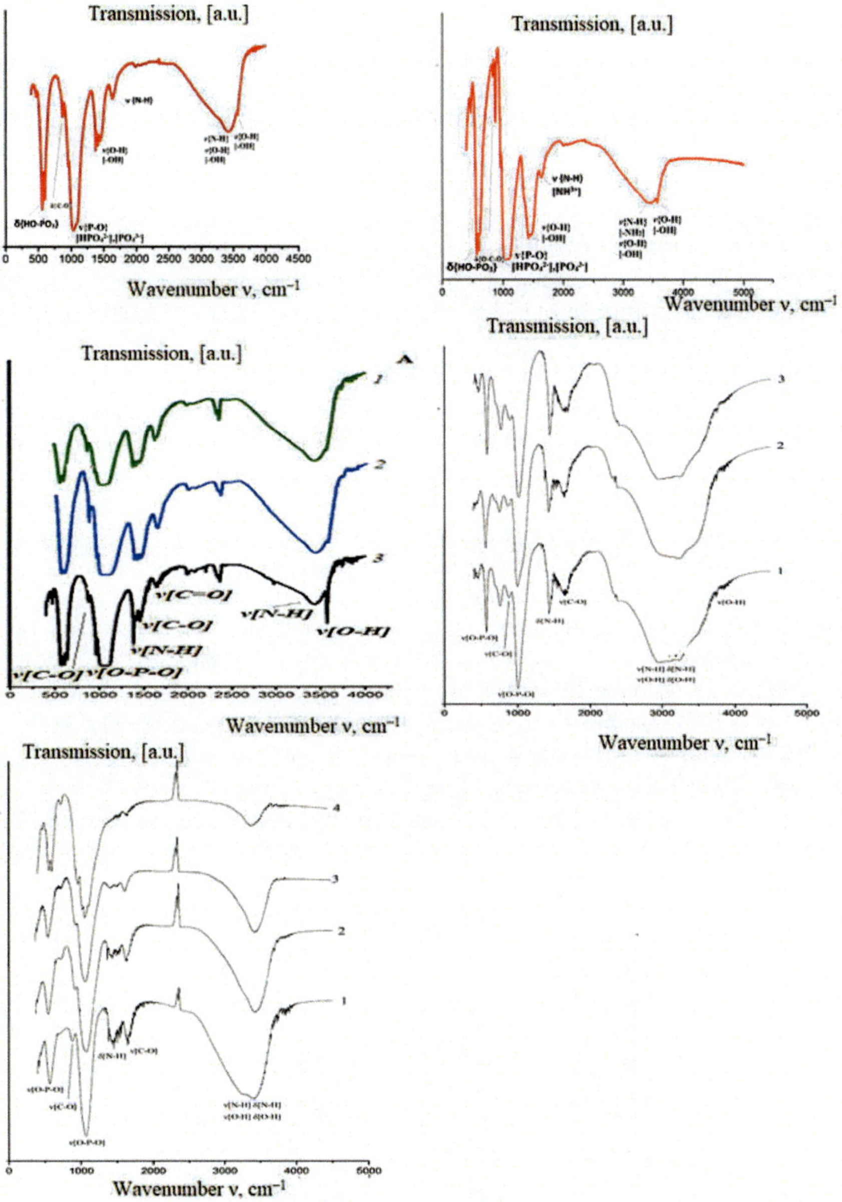

Fig. 2 IR spectra of: HA-Ch composites with $C_{chitosan}$ 0.02 (**a**) and 0.08 (**b**) g/l; **c** HA–chitosan composites with a chitosan content of 0.08 g/l after heat treatment at 1—200 °CC, 2—400 °CC, 3—600 °CC; **d** Mg-HA/chitosan with $C_{chitosan}$, g/l: 1—0.02; 2—0.08; 3—0.16; **e** Mg-HA/chitosan with a chitosan content of 0.08 g/l after heat treatment: 1—200 °C; 2—400 °C; 3—600 °C; 4—800 °C

For chitosan, a wide absorption band at 3290 cm^{-1} can be attributed to valence and deformation vibrations of OH and NH_2 groups participating in the formation of intra- and intermolecular connections. Valence asymmetric vibrations of the C–H bond in the methylene components of Ch are detected at 2876 cm^{-1}. The bending vibration of the associated amino group with NH_3^+ corresponds to a frequency of 1663 cm^{-1}. For the HA–Ch composite of the O–C–O vibrations appear at 877 cm^{-1} (Fig. 2b). The analysis is especially of IR spectra of HA–Ch composites makes the assumption that the most likely process of interaction between HA and Ch in solution is phosphor lionization of Ch with phosphate ions from solution. Besides that, it can be seen from the obtained spectra that the ratio CO_3^{2-} and Ch affects the view of the IR spectra: with growth concentration, in Ch samples (Ch) the intensity of absorption bands characteristic of Ch, increases. The BET method determined the values of the specific surfaces of the resulting composites. For HA–Ch composites, it was obtained that at the Ch content of 0.08 g the specific surface area is 132 m^2/g, and for a composite with the Ch mass Ch of 0.16 g,—124 m^2/g. When comparing specific surface data for different HA–Ch composites, it can be seen that with an increase in the Ch content, the porosity decreases. Thus, varying the concentration of chitosan during synthesis, it is possible to obtain composites of different porosity.

Composites were subjected to thermogravimetric analysis in order to study the thermal properties and phase transformations in crystallization products. In the HA–chitosan samples, the peak weight loss falls on the sample containing 0.16 g (88.8%) of chitosan. A decrease in weight in response to increasing chitosan amount is observed in the HA–chitosan composites after calcinations at 400 °C (Fig. 3).

The alteration of the functional group composition of composites upon calcination was detected by FT-IR spectroscopy (Fig. 2c).

At a temperature of 200 °C, no changes in the functional and group composition were noted for the HA–chitosan composites. And at 600 °C, the absence of bands

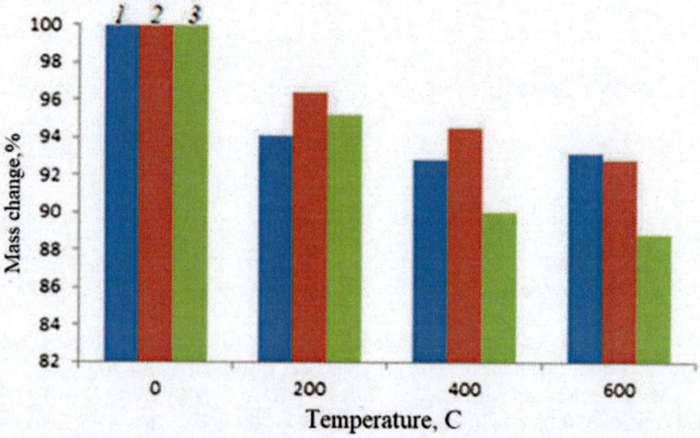

Fig. 3 Change in mass of samples with chitosan content, g/l (1—0.02, 2—0.08, 3—0.16) after heat treatment under different thermal conditions for HA-Ch composites

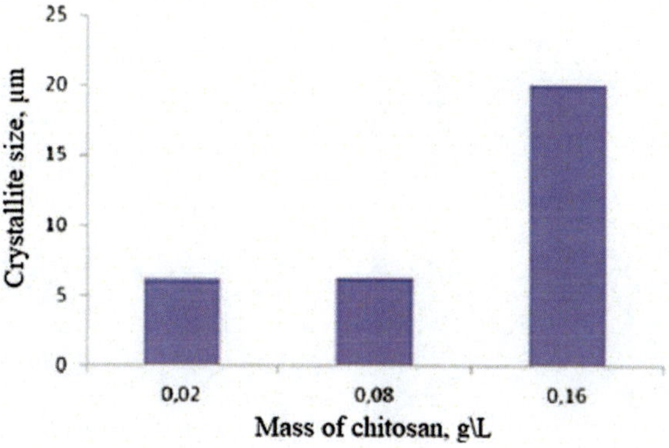

Fig. 4 Dependence of crystal sizes of HA-Ch composites on the $C_{chitosan}$ in the initial solution

of stretching and bending vibrations of the –OH and $-NH_2$ groups, the associated amino group in the chitosan structure, was recorded. It has been found that HA–Ch composites are thermally stable and can be used as surface coatings on metal implants and ceramics.

Microphotographs of HA–chitosan show irregularly shaped HA–Ch crystals. Measurements of the lengths of crystals of hydroxyapatite-chitosan composites showed that with an increase in the content of chitosan in the synthesized solution, the size of the crystals increases (Fig. 4). This pattern is explained by the fact that with an increase in the mass of chitosan, the viscosity of the initial solution increases, which contributes to the diffusion growth of crystals.

For the composite with chitosan content of 0.08 g/L, were obtained SEM images (Fig. 5a). Photomicrograph shows the presence of rounded aggregates characteristic of hydroxyapatite.

Thus, methods for obtaining biomaterials, the kinetics of resorption of which in liquid media is corrected due to the introduction of components into their compounds that accelerating dissolution process.

An analysis of the bioactivity of the synthesized composites showed that with an increase in the concentration of chitosan in the initial solution, the pH value of the solution increases, but the concentration of calcium ions decreases (Fig. 6a).

The absorption of ions at the beginning of the dissolution process leads to a significant increase in pH, which is fixed as a maximum in the kinetic curves pH = f(τ). This indicates a high intensity of resorption of the composites already at the initial stages of interaction with the solvent. Based on the obtained experimental dependences pCa = f(τ), the initial values of the rates of release of calcium ions into the solution were calculated (Fig. 6b). According to the initial dissolution rates composites, it can be seen that the presence of Ch in the solid phase increases passive dissolution for HA (Table 1).

It is known that Ch is a cationic polyelectrolyte with ionized amino groups, which can form a sparingly soluble compound through electrostatic attraction between the protonated amino group Ch and the polyvalent phosphate group of HA. For the

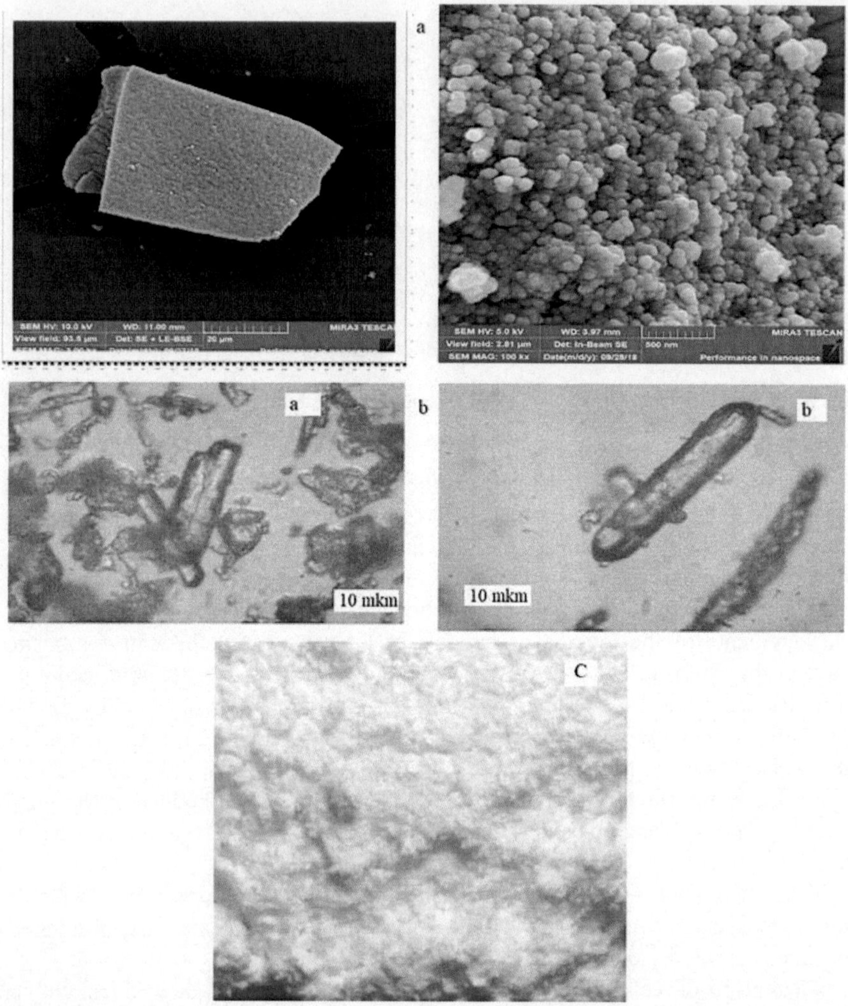

Fig. 5 **a** Morphology of the surface of the HA-Ch composite with $C_{chitosan}$ 0.08 g/l (1—increase 3 thousand times, 2—increase 100 thousand times); **b** micrographs of Mg-HA composites with chitosan **a**, **b** 10 × 100; **c** scaffold surface photo.

HA–Ch composite, the initial dissolution rate increases with increasing chitosan concentration, since the more soluble component, Ch predominates in the HA–Ch composite.

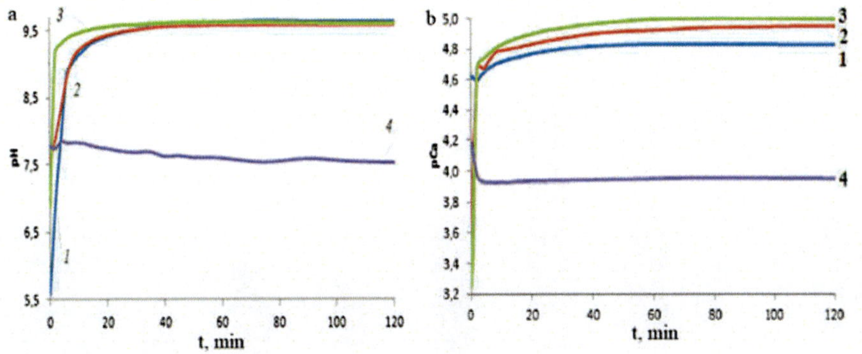

Fig. 6 Dependences of **a** pH and **b** pCa on the dissolution time in 0.9% NaCl of HA-chitosan, with chitosan content, g/l 1—0.02; 2—0.08; 3—0.16; 4—0

Table 1 Initial dissolution rates of HA-chitosan and HA-chitin composites in isotonic solution

Polymer weight during synthesis, g	Kinetic equation	Speed (min^{-1})
	HA-chitosan	HA-chitosan
0	y = –0.0199x + 4.0769	0.0199
0.02	y = 0.0114x + 4.6044	0.0114
0.08	y = 0.0863x + 4.1164	0.0863
0.16	y = 0.1187x + 3.9142	0.1187

3.2 *Caffolds on the base of HA-Ch*

Based on the obtained Ch-gel and mechanical scaffolds were synthesized from mixtures of HA. Analysis of their surface showed that they have interpenetrating pore structure, there are both large and small pores forked, unevenly distributed by the volume of the frame (Fig. 5b). Such a structure, as the analysis of the data (Agrawal and Gurbhinder 2011; Gurin 2009) shows, is suitable for fastening and life activity osteoblasts—cells that form the bone cloth. Based on the micrographs, the sizes of pores are estimated on the surface of the scaffolds. With an increase in the HA concentration in the scaffolds, the pore size is reduced from 11.82 to 2.09 μm 5c).

The thermogravimetric analysis of the scaffolds showed that the mass of the scaffolds decreases with an increase in the calcinations temperature of the samples (Fig. 7). In addition, HA-Ch scaffolds retain their shape best at T = 1273 K. The lower the calcination temperature, the less stable the scaffolds are. Composites, compared to the scaffolds, remain in powder form regardless of the calcination temperature. Thus, the HA-Ch based scaffolds can be used at high temperatures.

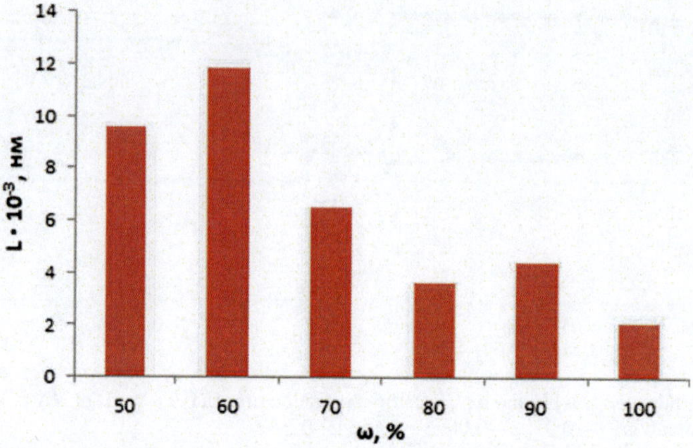

Fig. 7 Dependence of pore sizes ($L*10^{-3}$, nm) on scaffold (ω HA, %)

The study of the phase of passive dissolution of the scaffolds of different compositions in an isotonic solution (Fig. 8a, b) showed that with an increase in the content of Ch in the scaffold, the rate of sample dissolution increases (Table 2).

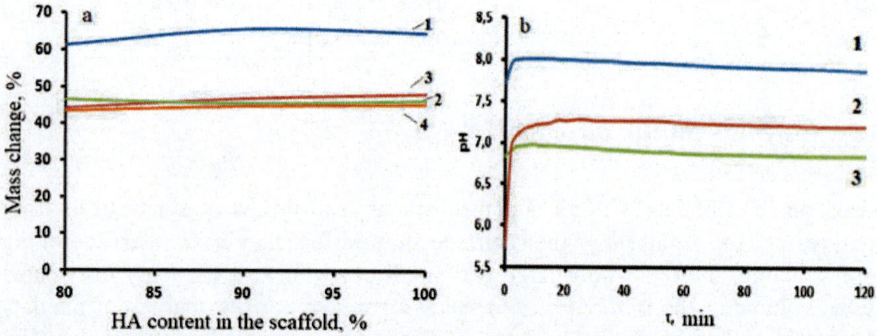

Fig. 8 Dependence of **a** sediment mass change on scaffold (1—T = 473, 2—T = 873, 3—T = 1073, 4—T = 1273 °C); **b** pH on the dissolution time of scaffolds (1—100%:0%; 2—90%:10%; 3—80%:20%) in isotonic solution

Table 2 Initial dissolution rates for scaffolds of different compositions

Frame composition: HA, wt %	Kinetic equation	Speed (min^{-1})
100:0	$y = -0.0141x + 4.3801$	0.0141
90:10	$y = -0.0273x + 4.2642$	0.0273
80:20	$y = -0.0341x + 3.8678$	0.0341

Thus, a method is proposed for obtaining biomaterials, the kinetics of dissolution of which in liquid media is corrected by introducing into their composition components that activate the dissolution process.

3.3 *Mg-HA–Ch composites*

It is known that isomophic structures of Mg-HA in the form of nanoparticles are highly soluble and can be used to deliver drugs to the human body at the cellular level, as well as to obtain nanostructured ceramics. At the next stage, HA-Ch materials were synthesized with the addition of magnesium ions.

X-ray diffraction analysis revealed that Mg-HA-chitosan composites contain hydroxyapatite as the main phase: $2\theta = 31.38°$, $32.73°$, $32.72°$ and $37.68°$

Figure 1b shows the most intense peaks for 2θ angles of 20.30°, 20.38°, 20.28° and a weak peak at 10.49°, characteristic of chitosan (Skryabin et al. 2020), and with an increase in the content of chitosan in the initial solution, the intensity of the peak increases.

The X-ray spectra show a small amount characteristic of the $Mg_3(PO_4)_2$ phase ($2\theta = 26.48°$, $26.58°$) and the $Ca_3(PO_4)_2$ phase ($2\theta = 49.99°$, $43.39°$), their content is not exceeds 5% and they are impurity.

Regularities in the sizes of crystallites showed that with an increase in the content of chitosan in the initial solution (Fig. 9), their geometric characteristics decrease (for 0.02 g/l—1.9 A; for 0.08 g/l—1.69 A; for 0.16 g/l—1.69 A). 1, 1.64 A). This is explained by the accumulation of Mg-HA crystallites in the polymer. At the same time, it is known that with an increase in the content of chitosan in the initial solution, its viscosity increases, which inhibits the growth of large crystals.

The results of FT-IR spectroscopy of the studied materials showed a set of bands corresponding to chitosan and HA (Fig. 4d): O–P–O bonds are characterized by absorption maxima at 1006 and 571 cm^{-1}. A wide absorption band in the region of 2955–3237 cm^{-1} characterizes the stretching and bending vibrations of the –OH

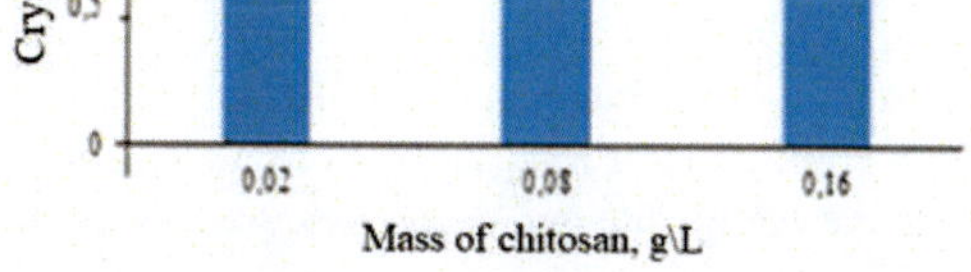

Fig. 9 Dependence of the average crystallite size on the content of chitosan in the initial solution

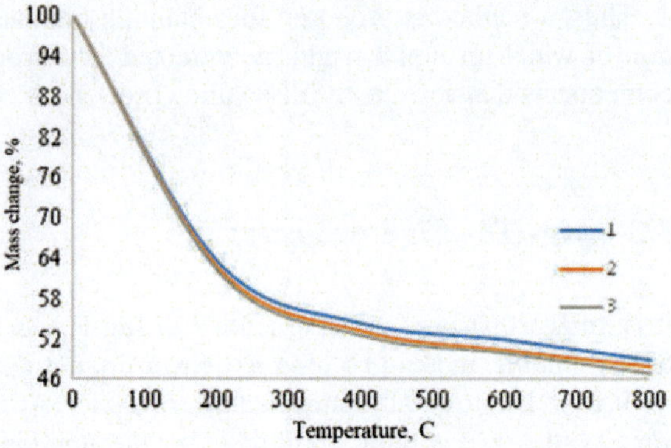

Fig. 10 Change in the mass of samples containing chitosan, g/L (1—0.02, 2—0.08, 3—0.16) after heat treatment Mg-HA/chitosan composites

and NH2 groups, and C–O bond vibrations are observed at 878 cm^{-1} (carbonate-substituted hydroxyapatite belongs to type B) (Frank-Kamenetskaya et al., 2011).

For chitosan, absorption bands at 3737 cm^{-1} are observed in the IR spectrum, which are vibrations of the O–H bond, the stretching vibration of the $-C = O$ bond in the CONHR group is fixed at a frequency of 1648 cm^1.

Thus, it has been proven that the composites obtained by the chitosan transformation in the initial solution have the same phase and group composition.

The difference in the IR spectra of the absorption bands characteristic of chitosan was noted in the different intensities of the peaks of the functional groups.

Analysis of the structure of the synthesized composites showed that Mg-HA-chitosan crystallites are characterized by a lamellar elongated shape, which is typical for isomorphically substituted magnesium-hydroxyapatite compounds (Fig. 5b) (Severin et al. 2020).

Thermal studies have revealed that all the obtained composites are characterized by continuous weight loss up to 800 °C (Fig. 10). And the temperature range of 100–400 °C is characterized by the maximum change in the mass of the sample, which is observed due to the release of volatile impurities and adsorption water.

The maximum change in the mass of the obtained composites was recorded in samples with the highest content of chitosan (0.16 g/l, 46.87%).

Fourier transform IR spectroscopy (Fig. 2e) shows the change in the functional group composition of the synthesized materials after thermal treatment: in the temperature range of 200 and 400 °C, the composition of the composites is constant; after exposure to 600 °C, the absence of peaks characterizing the stretching and bending vibrations of -OH- and -NH2-groups in the composition of chitosan is observed, and their intensity decreases with increasing heating temperature; after heating the samples to 800 °C, the main absorption bands narrow, which confirms the increase in the crystallinity of the Mg–HA composites. The Mg-HA-chitosan composite is characterized by a dependence: the higher the polymer content in the sample, the lower the weight of the precipitate after calcination. This proves the removal of chitosan from the composite and correlates with the data presented (Fadeeva and Golovanova 2019).

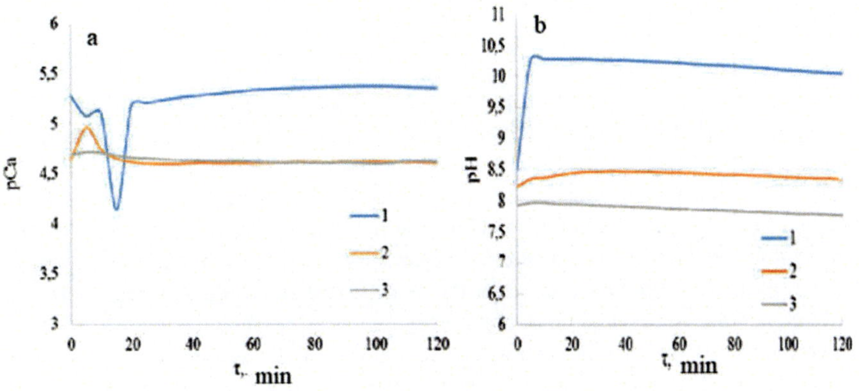

Fig. 11 Dependence of pCa (**a**) and pH (**b**) of the system on the dissolution time of composites with chitosan concentration, g/l: 1—0.02; 2—0.08; 3—0.16

When studying the dissolution rate of the obtained composites (Mg–HA), it was found that at the initial stage of sample dissolution, the concentration of calcium ions increases, and then the kinetic curves reach a plateau and the rate does not change (Fig. 11).

For materials with chitosan, the pH value of the solution decreases with an increase in the polymer content in the composition of the composites (Fig. 11a, b). It is known that chitosan has a reactive functional group (amino group) and this contributes to the binding of metal cations (calcium and magnesium) into chelate complexes (Scriabin et al. 2020) between hydroxyl and aminoacetyl groups.

Based on the kinetic regularities pCa = f(τ), the initial rates of the transition of calcium ions into the solution from the synthesized composites were determined (Table 3). It was found that the initial dissolution rate does not change for materials with a chitosan content of 0.02 and 0.08 g/l, and at the maximum content of the polymer in the initial solution increases by 5 times.

Table 3 The value of the initial dissolution rates of the composites

Polymer weight during synthesis (g)	Kinetic equation Mg-HA/chitosan	Speed (min^{-1}) Mg-HA/chitosan
0.02	y = 5.3657 + 0.0001x	0.0001
0.08	y = 4.6151–0.0001x	0.0001
0.16	y = 4.6508–0.0005x	0.0005

4 Conclusions

Synthesis of HA-Ch composites and matrices based on chitosan gel and hydrokilsapatite was carried out. It was found that the size of crystallites increases with an increase in the concentration of chitosan in the initial mother liquor.

For the first time, a method was developed for the synthesis of matrices from HA and chitosan gel. It has been established that the geometric dimensions of the pores on the surface of the matrices decrease with an increase in the HA content in the composition of the material, while the thermal stability of the matrices increases with an increase in the heating temperature.

The study of the process of passive resorption of the synthesized composites and matrices showed that for HA–chitosan, the dissolution rate increases with the growth of the polymer in the matrices. This makes it possible to change the polymer content in a wide range of concentrations and to control the dissolution kinetics of the synthesized samples.

Materials based on Mg-HA, chitosan have a constant composition, the presence of functional groups of HA and polymer is confirmed by IR spectroscopy data, and the phase composition of XRD.

The synthesized materials have the same morphology and crystal structure. It is shown that their size decreases with increasing polymer content in the initial solution. Thermal analysis showed that with increasing exposure temperature, a loss in mass of all composites is observed. The rate of bioactivity for calcium ions increases with an increase in the concentration of chitosan in the initial solution.

Acknowledgements The work was carried out according to the state order of the Russian Federation on topic No. 1. 075-03-2023-149.

References

Agrawal K, Gurbhinder S (2011) Synthesis and Characterization of hydroxyapatite Powder by Sol-Gel Method for Biomedical Application. J. of Minerals & Materials Characterization & Engineering. 10. 8; 727.

Bakunova NV, Barinov SM, Komlev VS, Smironov VV, Fedotov AYu (2011) Porous chitosan matrices reinforced with bioactive compounds calcium supplementation for bone tissue restoration. Nauch. statements. Series: mathematics, physics. 11 (106); 5. (in Russian).

Bolarinwa AO (2010) The Formulation of a Bioresponsive Ceramic Bone Replacement. School of Chemical Engineering. 220.

Danilchenko SN, Kalinkevich OV, Pogorelov MV (2009) Experimental substantiation of the use of composite materials based on chitosan and calcium phosphates for the replacement of bone defects. J. Orthopedics, traumatology and prosthetics. 1; 66. (in Russian).

Elson Santiago de Alvarenga (2011) Characterization and Properties of Chitosan. Universidade Federal de Viçosa. 91.

Fadeeva TV, Golovanova OA (2019) Physicochemical Properties of Brushite and Hydroxyapatite Prepared in the Presence of Chitin and Chitosan. Russ. J. Inorg. Chem. 64. 7; 690.

Fedosov PA (2014) Chitosan as a polymer of the future and prospects for its use in medicine. APRORI. Ser.: natural and technical sciences. 4;7. (in Russian).

Frank-Kamenetskaya O, Kol'tsov A, Kuz'mina M, Zorin M (2011) Ion substitutions and non-stoichiometry of carbonated apatite-(CaOH) synthesised by precipitation and hydrothermal methods. Mol. Struct. 992. 1–3; 9.

Gurin AN (2009) Comparative evaluation of the effect of various osteoplastic materials based on calcium phosphates on the healing of bone defects: Dis. ... cand. honey. Sciences. 161. (in Russian).

Izmailov RR, Golovanova OA, Tserich Yu, Drozdov VA Leont'eva NN (2016) Crystallization specifics of carbonate-hydroxylapatite in the presence of strontium-containing agents. Russ. J. Inorg. Chem. 61. 7; 1–6.

Jayakumar R, Chennazhi K, Srinivasan S. (2011) Chitin Scaffolds in Tissue Engineering. Int. J. Mol. sci. 12; 1876–1887.

Kamskaya VE (2016) Chitosan: structure, properties and uses. J. Biological sciences. 6;36. (in Russian).

Kandyrin KL (2002) Introduction to materials science of polymers. Uch. allowance. 100. (in Russian).

Khomenko AYu, Popryadukhin PV, Bogomolova TB (2013) Matrices based on chitosan nanofibers for cellular technologies. J. Russian nanotechnologies. 8. 9-10; 41. (in Russian).

Kojima C, Watanabe K, Murata H, Makiura R, Matsunaga K, Nakahira A (2018) Controlled release of DNA from zinc and magnesium ion doped hydroxyapatites. Springer Nature 23–32.

Kurchenko VP, Buga SV, Petrashkevich NV, Butkevich TV (2016) Technological basis for obtaining chitin and chitosan from insects. Proceedings of BSU. 11. 1; 110. (in Russian).

Ling-Hao H, Lu Y, Rui X, Jing S, Rui S (2011) Mineralization of Chitosan/Calcium Phosphate Composite and the Effect of Solvent on the Structure J. Frontier Mater. sci. 5. 3; 282–292.

Lyasnikova AV (2017) Method for obtaining magnesium-substituted hydroxyapatite. Patent. 8. (in Russian).

Muravyov AA (2017) Solutions of mixtures of cellulose and chitin in ionic liquids and composite materials based on them. Diss. cand. chem. Sciences. 111. (in Russian).

Salman SA, Kuroda K, Okido M (2013) Preparation and Characterization of Hydroxyapatite Coating on AZ31 Mg Alloy for Implant Applications. Bioinorg Chem Appl. 175–756.

Severin AV, Rudin VN, Paul ME (2020) Characteristic features of Mg^{2+} behavior and Mg^{2+} effect on the structure and morphology of nanohydroxyapatite in the adsorption method for the fabrication of the ha-mg composite. Russ. J. Inorg. Chem. 65. 9; 1436–1444.

Skryabina KG, Vikhoreva GA, Varlamova VP (2020) Chitin and chitosan: obtaining, properties, application. Ed. M.: Nauka. 368.

Nudga LA Structural and chemical modification of chitin, chitosan and chitin-glucan complexes. / / Diss. doc. chem. Sciences. 2006. 361s.

Trisvetova EL Magnesium in clinical practice // Rational Pharmacotherapy in Cardiology. 2012; 8(4). pp. 545–553. (in Russian).

Volova TG Materials for medicine, cell and tissue engineering. // Uch. allowance. 2009. 262p. (in Russian).

Wahl DA Collagen-hydroxyapatite composites for hard tissue repair. // European Cells and Materials. 2006. No. 11. P.43.

Co-Bearing Hydroxyapatite: Synthesis, Thermal Stability, Crystal Chemistry, Magnetic Properties

Anatolii V. Korneev, Maria A. Kuzmina, and Olga V. Frank-Kamenetskaya

Abstract The numerous ionic substitutions at various sites of apatite crystal structure are giving new properties to apatite-based materials. We studied Co-substituted hydroxyapatites (Co-HAp) precipitated from water solutions with methods of X-ray powder diffraction, IR-spectroscopy, EDX analysis and vibrating-sample magnetometry and showed that cobalt concentration into calcium sites of apatite crystal structure can reach 21 at. % (10.65 wt. %). Cobalt substitution, accompanied by the significant decrease of crystallite sizes, contributes to hydroxyapatite desymmetrization at Co/(Ca+Co) $\geq$ 0.05 and lowers its thermal stability: the beginning of Co-HAp structure decomposition was detected upon annealing at 450 °C, and the complete destruction—at 700 °C, pronounced in the release of cobalt from apatite and formation of Co_3O_4 and β-$Ca_3(PO_4)_2$. The replacement of 5% of calcium with cobalt makes hydroxyapatite paramagnetic, and further incorporation (up to at least 21 at. %) results in linear increase of magnetic susceptibility. Obtained compositions can find an application in various areas of medicine (hyperthermia cancer therapy, medical visualization, drug delivery and others).

Keywords Hydroxyapatite · Cobalt hydroxyapatite · Magnetic nanoparticles · Biomaterials · Co-doped apatite · Co-substituted apatite · Magnetic biomaterials · Biomaterials

1 Introduction

Compositions with apatite structure with formula $M_5(TO_4)_3Z$ (M = Ca^{2+}, Mg^{2+}, Na^+…, T = P^{5+}, S^{6+}, Si^{4+}…, Z = OH^-, F^-, Cl^-, H_2O, $CO_3{}^{2-}$) are widespread in nature and used actively in various areas of material sciences. Hydroxyapatite (HAp, $Ca_5(PO_4)_3OH$), the main mineral component of human bones and teeth, possesses

A. V. Korneev (✉) · M. A. Kuzmina · O. V. Frank-Kamenetskaya
Department of Crystallography, Institute of Earth Sciences, Saint-Petersburg State University, 7/9 Universitetskaya Emb., 199034 St. Petersburg, Russia
e-mail: a_v_korneev@list.ru

O. V. Frank-Kamenetskaya et al. (eds.), *Biogenic—Abiogenic Interactions in Natural and Anthropogenic Systems 2022*, Springer Proceedings in Earth and Environmental Sciences, https://doi.org/10.1007/978-3-031-40470-2_5

a number of properties, making its application attractive especially in tissue engineering, drug delivery and other problems of biomedicine (Vieira et al. 2017). Crystal structure of hydroxyapatite is subject to a wide range of ionic substitutions in various sites, giving new properties to materials based on it (Ptáček 2016). Thus, synthesis and investigation of hydroxyapatite-based materials is one of the foremost problems in a field of biocompatible materials.

Co^{2+} ions are capable of incorporating hydroxyapatite in significant amounts (Veselinović et al. 2010). As Kramer et al. (2014) have shown, Co-doped hydroxyapatites (Co-HAp) possess paramagnetic properties while pure HAp is diamagnetic. Hydroxyapatite biocompatible magnetic nanoparticles are perspective from the point of view of hyperthermia cancer therapy (Fatima et al. 2021), magnetic particle imaging, including cancer visualization (Yang et al. 2022), drug delivery (Price et al. 2018). As was noted by Kramer and co-authors (Kramer et al. 2014), iron oxide nanoparticles, most commonly used in these areas, possess potential toxicity and are non-biodegradable, which limits their potential. On contrary, biocompatibility of hydroxyapatite along with a vast range of synthesis methods, leading to obtainment of various materials, could adapt the use of magnetic hydroxyapatite particles to a wide field of biomedical applications.

As early as 1970, the possibility of cobalt incorporation in apatite lattice was demonstrated by Grisafe and Hummel (Grisafe and Hummel 1970). Cobalt was reported to replace calcium in various amounts depending on Z-anion in the channels of crystal structure: Co/(Ca+Co) ≤ 0.08 for hydroxyapatite (Veselinović et al. 2010), ≤ 0.27 for fluorapatite and ≤ 0.45 for chlorapatite (Grisafe and Hummel 1970). The substitution significantly affects apatite unit cell parameters. The decrease of parameter *c* as Co content grows is common for all three types of apatite, yet the reduction is the most rapid for hydroxyapatite and the least—for chlorapatite. As for parameter *a*, for OH- and F-apatite it tends to decrease with the increase of cobalt content, while for Cl-apatite it stays constant.

Single-crystal XRD study (Anderson and Kostiner 1987) evidences that Ca2 site is more preferable for Co^{2+} in chlorapatite crystal structure. Additionally, Rietveld data of Veselinovic (Veselinović et al. 2010) for hydroxyapatite indicate incorporation of cobalt in Ca2 position along distortion of Ca1 polyhedron, which possibly indicates incorporation of cobalt in this site as well. Generally, previously reported data clearly display that incorporation of cobalt can have different impact on apatite structure depending on Z-site anion. Maximum cobalt content in hydroxyapatite given by Veselinovic is significantly lower than that in fluor- and chlorapatite and probably requires refinement. Additionally, according to Kazin et al. (2017) and Zykin et al. (2019), in apatite structure cobalt can also form dioxocobaltate (II) ions, partially occupying Z-position and resulting in slight increase of parameter *c*. However, such incorporation requires specific synthesis including several high-temperature stages and is not likely to occur in more common syntheses as precipitation or hydrothermal synthesis.

Paramagnetic behavior of Co^{2+}-substituted hydroxyapatite was observed by Kramer et al. (Kramer et al. 2014). It has been noted that magnetic susceptibility of the powders depended significantly on the method of synthesis. More importantly, the powder obtained by wet synthesis possessed lower susceptibility than the powder obtained by ion exchange. The dependencies of magnetic properties on cobalt content haven't been explored, and it is yet to find which cobalt content is optimal to obtain the most remarkable magnetic susceptibility.

In this work we are aiming to:

1. Synthesize hydroxyapatites in wide range of cobalt content.
2. Estimate the effect of cobalt content on thermal stability and crystal structure of hydroxyapatite.
3. Clarify the relations between cobalt content and magnetic properties of Co-bearing hydroxyapatite.

2 Materials and Methods

2.1 *Synthesis*

A series of hydroxyapatites was obtained from Co-containing solutions: 100 ml of 0.2 M $Ca(NO_3)_2 \cdot 4H_2O$ solution was mixed with $Co(NO_3)_2 \cdot 6H_2O$ solution and heated to ~90 °C. The amount of $Co(NO_3)_2 \cdot 6H_2O$ was selected to achieve Co/Ca ratios of 0–0.20 (7 syntheses, see e.g. Table 1). Then 100 ml of 0.1 M $(NH_4)_2HPO_4$ solution were added dropwise into the mixture. The resulting solution was placed into oven at 110 °C for 2 h. The precipitates were washed and dried at 110 °C, then crushed in mortar. The resulting pale purple to purple-grey powders were as well calcined at 200 °C, 450 °C and 700 °C for 24 h.

2.2 *Experimental Techniques*

X-ray powder patterns were obtained using Rigaku Miniflex II diffractometer with $CuK\alpha_{1+2}$ irradiation operating at 30 kV/15 mA, $2\theta = 5$–$80°$ with 0.02° step. A small amount of germanium ($a = 5.658$ Å) was added as internal standard. The data processing was performed by Rietveld refinement by means of Profex 5 software (Doebelin and Kleeberg 2015). The refined variables were: zero shift, scaling factor, phase quantities, unit cell parameters and anisotropic lorentzian broadening caused by size effects. Unit cell parameter of germanium was fixed in order to obtain reliable values of zero shift.

IR spectroscopy was carried out by means of Bruker Vertex 70 FTIR spectrometer. The samples were prepared by pressing of ~0.5 mg of powder with 200 mg of KBr. The resulting spectra of hydroxyapatites were normalized by ν4 PO_4^{3-} band

Table 1 Unit cell parameters* and CSD sizes** of hydroxyapatites in [100] and [001] directions at various cobalt contents and calcination temperatures

Co/Ca in solution	Non-calcine		200 °C		450 °C		700 °C	
	a, Å *c*, Å	CSD_{100}, nm CSD_{001}, nm	*a*, Å *c*, Å	CSD_{100}, nm CSD_{001}, nm	*a*, Å *c*, Å	CSD_{100}, nm CSD_{001}, nm	*a*, Å *c*, Å	CSD_{100}, nm CSD_{001}, nm
0	9.431 6.877	44 19	9.422 6.878	39 16	9.419 6.883	44 17	9.416 6.880	50 35
0.01	9.423 6.880	29 14	9.408 6.879	27 14	9.411 6.881	28 15	9.417 6.881	49 32
0.03	9.419 6.870	22 11	9.403 6.869	25 11	9.408 6.874	22 12	9.418 6.878	49 23
0.05	9.417 6.864	23 10	9.404 6.865	21 10	9.409 6.872	22 11	9.419 6.877	45 22
0.10	9.417 6.835	12 8	9.396 6.842	16 8	9.407 6.860	15 8	9.422 6.878	58 18
0.15	9.432 6.831	10 6	9.405 6.826	11 6	9.421 6.853	11 6	9.426 6.875	81 18
0.20	9.437 6.819	10 5	9.413 6.820	9 6	9.427 6.852	10 6		

Notes * Errors of unit cell parameters determination do not exceed 0.001 Å for samples with Co/Ca (solution) ≤ 0.05 and do not exceed 0.003 Å for the rest samples
**Errors of CSD sizes determination do not exceed 1 nm

for graphical presentation. Areas 450–750 cm^{-1} and 1350–1750 cm^{-1} were deconvoluted into Lorenz functions using MagicPlot software in order to extract integral intensities of PO_4^{3-}, OH^- and CO_3^{2-} bands (see Fig. 3 for details).

SEM images and EDX spectra were obtained using Hitachi TM3000 and Hitachi S-3400N scanning electron microscopes equipped with energy-dispersive spectrometers Oxford Instruments. For measurements of Ca, P and Co content the samples were pressed into pellets to obtain flat surface. Measurements were performed at least in 10 points for each sample and the data were averaged.

Measurements of magnetic properties were carried out by means of Lake Shore 7410 vibrating-sample magnetometer (VSM) for magnetic fields -18 kOe < H < + 18 kOe at 300 K.

3 Results

3.1 Powder X-Ray Diffraction (PXRD)

PXRD has confirmed hydroxyapatite (PDF-2 #01-082-2956) formation in all performed syntheses (Fig. 1a). In non-calcined powders and powders calcined at 200 °C no presence of other phases was detected. In powders calcined at 450 °C Co_3O_4 (PDF-2 #00-042-1467) was detected in amounts of ~1 wt.% at Co/Ca (solution) ≥0.15. In powders calcined at 700 °C β-tricalcium phosphate (β-TCP, PDF-2 #00-055-0898) and Co_3O_4 are present at Co/Ca (solution) ≥0.05, their contents increase with cobalt content in solution (Fig. 1b). Hydroxyapatite peaks become much narrower and there is no significant peak broadening. β-TCP becomes predominant phase at Co/Ca (solution) ≥0.10. Unit cell parameters of β-TCP ($a = 10.347 \pm 0.002$ Å, $c = 37.132 \pm 0.006$ Å) do not change significantly with Co content increase. Content of hydroxyapatite, in turn, reduces with the increase of cobalt content, up to complete disappearance at Co/Ca = 0.20.

Generally, as cobalt content increases, X-ray powder patterns of initial hydroxyapatites and those calcined at 200 and 450 °C demonstrate significant peak broadening, up to complete disappearance of low-intensity peaks (Fig. 1a). In Co/Ca (solution)

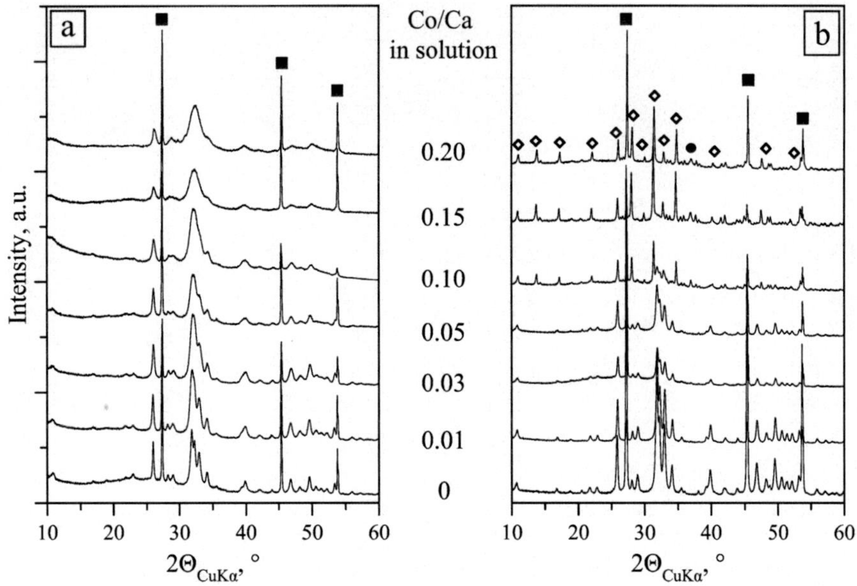

Fig. 1 X-ray powder patterns of synthesized compositions: **a** before calcination; **b** after calcination at 700 °C. ●—peaks of Co_3O_4; ◇—peaks of β-tricalcium phosphate; ■—peaks of germanium (internal standard). Unsigned peaks belong to hydroxyapatite

range of 0–0.03 experimental X-ray powder patterns (Fig. 2a) perfectly match calculated XRD pattern of hydroxyapatite (space group $P6_3/m$, Hughes et al. 1989). At Co/Ca ratio of 0.05–0.10 experimental patterns start to deviate from calculated ones ($P6_3/m$), which is the most well-pronounced in $2\theta_{CuK\alpha} = 31–34°$ range. Reflections $\bar{2}31$ and $\bar{1}22$ are merging into one symmetric peak, rather than an asymmetric one, as calculated pattern presumes. Further increase of cobalt content (Co/Ca $\geq$ 0.15) causes this peak merge with 3 0 0 reflection, forming symmetric peak again. At this point, the difference between experimental and calculated patterns becomes remarkable, though the rest of powder pattern remains apatite-like.

The replacement of hexagonal hydroxyapatite structure with structure of monoclinic hydroxyapatite (space group $P2_1/b$, Ikoma et al. 1999) brings calculated pattern significantly closer to experimental data (Fig. 2b), also resulting in better R_{wp}-factor for all samples with Co/Ca (solution) $\geq$0.05. In refinements of monoclinic unit cell parameters, the initial ratio $b = 2a$, $\gamma = 120°$, typical for monoclinic hydroxyapatite (Elliott 1994; Ikoma et al. 1999) is not met precisely and a, b and γ parameters correlate with each other. One of the possible reasons for this may be poor crystallinity of powders, not allowing to distinguish individual reflections that should be present on X-ray powder pattern of monoclinic HAp. Thus, considering that disparities between X-ray pattern calculated using hexagonal structure and experimental

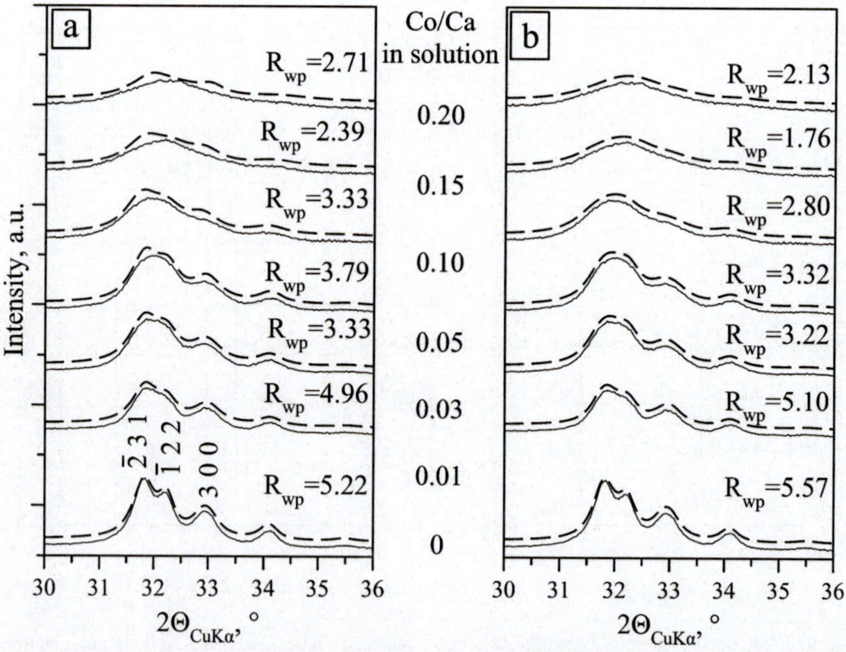

Fig. 2 Fragments of experimental and calculated X-ray powder patterns of hydroxyapatites synthesized at various Co/Ca ratios in solution (calcined at 450 °C). Solid lines—experimental patterns, dashed lines—patterns calculated with hexagonal (**a**) and monoclinic (**b**) hydroxyapatite structures

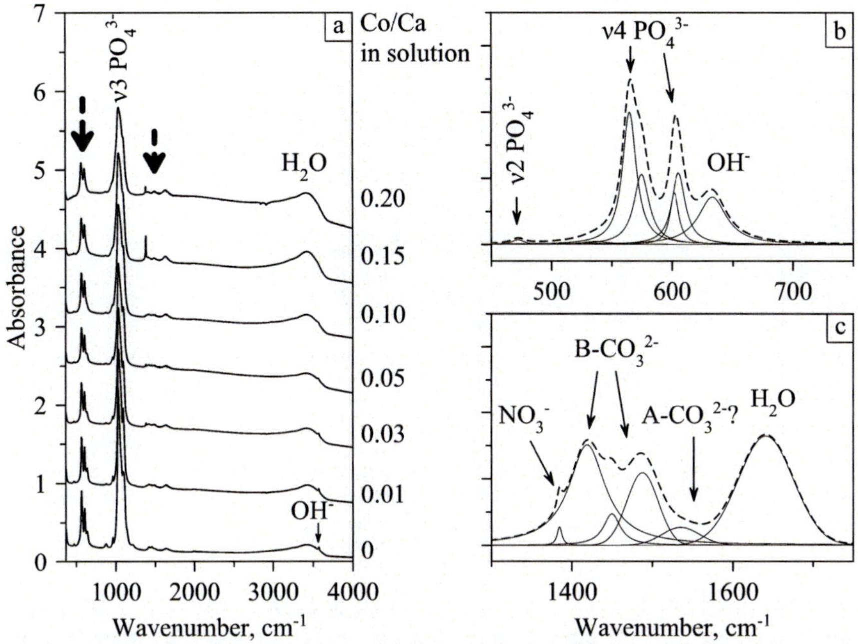

Fig. 3 Examples of IR spectra: **a** synthesized hydroxyapatites (before calcination) at various cobalt contents. Dashed arrows mark areas shown in panels b and c; **b** example of deconvolution of spectra in 450–750 cm^{-1} region; **c** example of deconvolution of spectra in 1350–1750 cm^{-1} region

pattern were minor, we refined the values of unit cell parameters and coherent scattering domains (CSD) using hexagonal variant (Table 1). Surely, using the imperfect fitting model may result in increased errors of determination of parameter *a*.

According to CSD measurements, hydroxyapatite crystallites are elongated along c axis. Average CSD decrease greatly as Co content increases, and the elongation rate (CSD_{001}/CSD_{100}) does not change greatly. Calcination has no effect on CSD sizes, excluding T = 700 °C, when CSD become larger and less elongated.

Excluding T = 700 °C, unit cell parameters of hydroxyapatites vary greatly with cobalt content (solution) increase as well as with calcination temperature. Parameter c tends to decrease as cobalt content increases, while parameter a reduces and reaches minimum at Co/Ca = 0.05–0.10 and then starts to increase. As for T = 700 °C, parameter a tends to increase at very slow rate, while parameter c stays constant.

3.2 IR Spectroscopy

IR spectra of synthesized compositions (Fig. 3a) are typical for hydroxyapatites obtained by wet synthesis (Berzina-Cimdina and Borodajenko 2012). Characteristic

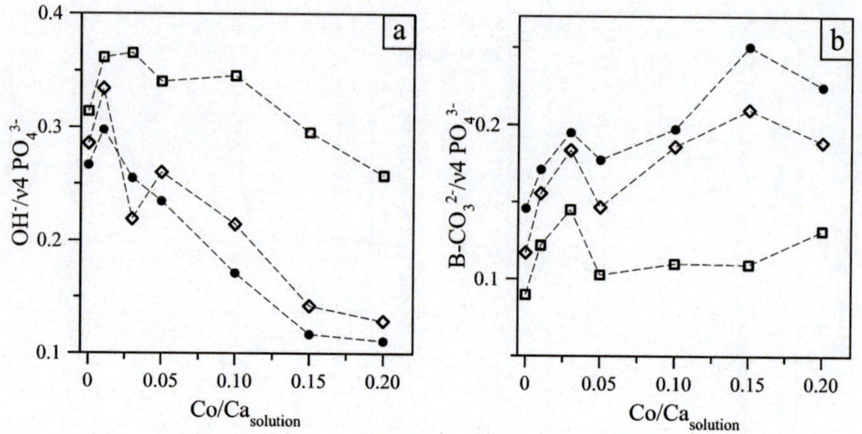

Fig. 4 Variations of relative intensities of OH^- (**a**) and $B\text{-}CO_3^{2-}$(**b**) bands normalized on total intensity of ν4 PO_4^{3-} bands at various cobalt contents and calcination temperatures. Notation: ●—not calcined powders; ◇—calcined at 200 °C; □—calcined at 450 °C

absorption bands of PO_4 (ν2 472 cm^{-1}, ν4 565, 602 cm^{-1}, ν3 1032, 1095 cm^{-1}) and OH (630–632, 3571 cm^{-1}) are present; H_2O (1628, 3424 cm^{-1}) and B-type CO_3 bands (1415–1494 cm^{-1}), which are typical for synthetic hydroxyapatite, are observed as well. Bands of A-type CO_3 (~1550 cm^{-1}) are either not present or on the verge of method sensitivity. Well-pronounced band of NO_3 (1384 cm^{-1}) is observed in samples with high cobalt content (Co/Ca $\geq$ 0.15) and disappears upon calcination at 450 °C. Calcination at 200 and 450 °C does not apply any significant changes to IR spectra, other than variations in intensities of OH, CO_3 and H_2O bands. All bands tend to broaden as cobalt content increases. Besides, intensities of H_2O and CO_3 bands increase with cobalt content while OH bands decrease.

As follows from the results of spectra deconvolution (Fig. 3b, c) and measurements of integral intensities of ν4 PO_4^{3-}, OH- and $B\text{-}CO_3^{2-}$ bands, OH content increases (Fig. 4a) and CO_3^{2-} content decreases (Fig. 4b) upon calcination at 200 °C and even more—upon calcination at 450 °C. Assuming there are no vacancies in structure channels, water content decreases as the calcination temperature rises. Besides, with the increase of cobalt content in solution OH content reduces (H_2O increases) and CO_3 content rises, which is in complete agreement with the pattern described above.

3.3 *Scanning Electron Microscopy (SEM) and Energy-Dispersive X-Ray (EDX) Microanalysis*

In synthesized powders hydroxyapatite tends to form isometric agglomerates up to 30 μm in size. No phases with cobalt as main component were observed in SEM images; the calcination does not bring any significant changes in morphology of agglomerates. Agglomerates of Co-free HAp are formed with intersecting elongated grains (Fig. 5a) up to 500 nm long. Agglomerates of Co-modified hydroxyapatites are smoother, made up of more isometric particles (Fig. 5b).

EDX data clearly show the increase of cobalt content in hydroxyapatite as Co/Ca ratio in solution rises (Table 2). Co/Ca ratio in solid is significantly larger than that in solution. (Co+Ca)/P ratio in solid remains constant at all cobalt contents. Calcination of the samples at 200 °C does not affect cobalt contents in apatite significantly. As for calcination at 450 °C, it does not change cobalt contents in solid as well, but draws Me^{2+}/P rations closer to that value of stochiometric hydroxyapatite (1.67).

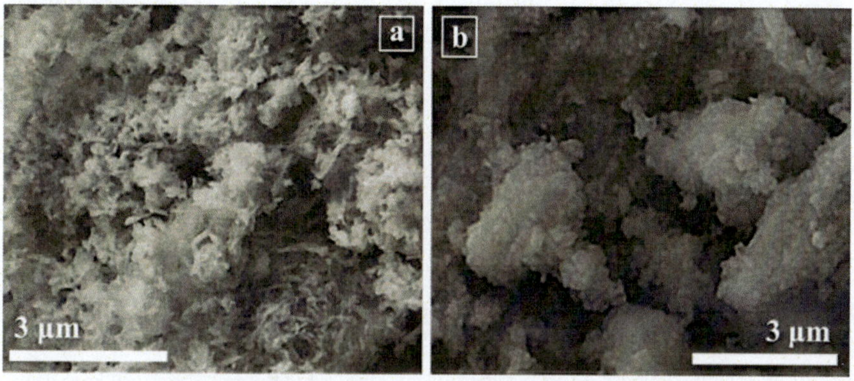

Fig. 5 SEM images of Co-free (**a**) and Co-modified (**b**, Co/Ca = 0.05) hydroxyapatite agglomerates, calcined at 200 °C

Table 2 Chemical composition characteristics of hydroxyapatites synthesized at various Ca/Ca ratios and various calcination temperatures by EDX analysis

Co/Ca in solution	No calcination			Calcined at 200 °C			Calcined at 450 °C		
	$\frac{Co}{Ca}$	$\frac{Co}{Ca+Co}$	$\frac{Co+Ca}{P}$	$\frac{Co}{Ca}$	$\frac{Co}{Ca+Co}$	$\frac{Co+Ca}{P}$	$\frac{Co}{Ca}$	$\frac{Co}{Ca+Co}$	$\frac{Co+Ca}{P}$
0	0.00	0.00	1.49	0.00	0.0	1.49	0.0	0.0	1.67
0.01	0.01	0.01	1.67	0.01	0.01	1.49	0.01	0.01	1.67
0.03	0.04	0.04	1.68	0.04	0.04	1.49	0.05	0.04	1.69
0.05	0.05	0.05	1.61	0.05	0.05	1.51	0.05	0.05	1.56
0.10	0.12	0.11	1.57	0.13	0.12	1.5	0.12	0.11	1.55
0.15	0.20	0.17	1.57	0.20	0.17	1.49	0.21	0.17	1.64
0.20	0.26	0.21	1.57	0.29	0.22	1.49	0.27	0.21	1.56

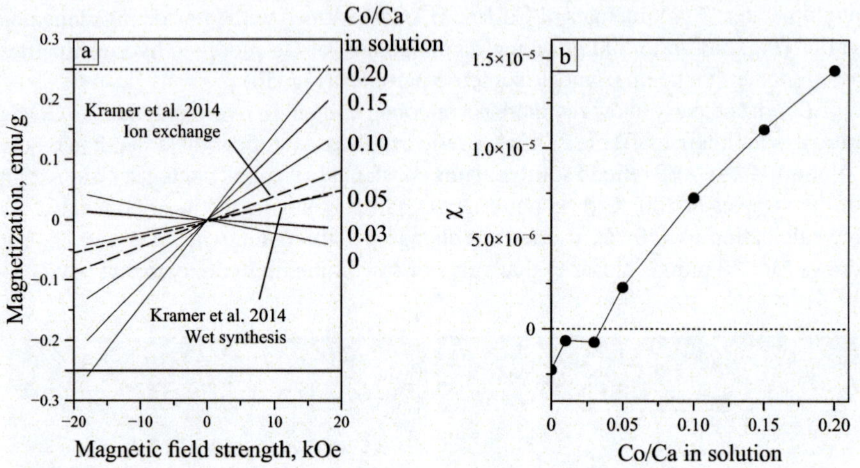

Fig. 6 Magnetic properties of Co-hydroxyapatites. **a** Magnetization curves of hydroxyapatites obtained at various cobalt contents; **b** dependence of magnetic susceptibility of hydroxyapatites on Co/Ca ratio in solution. Data of (Kramer et al. 2014) are given in dashed lines

3.4 Magnetization Measurements.

As follows from magnetization measurements, hydroxyapatites obtained from solutions with Co/Ca ratio $\leq$0.03 possess diamagnetic properties (Fig. 6a). Hydroxyapatites obtained at Co/Ca ratio $\geq$0.05 have positive slope of magnetization curves without any signs of hysteresis loop, therefore, possessing paramagnetic properties. Magnetic susceptibility χ increases linearly with cobalt content in solution, reaching 1.43×10^{-5} at Co/Ca (solution) = 0.20 (Fig. 6b).

4 Discussion

The performed syntheses from Co-containing solutions (Co/Ca = 0–0.20) resulted in formation of hydroxyapatite powders without any impurity phases.

Obtained regularities of variations of hydroxyapatite unit cell parameters evidence cobalt incorporation in hydroxyapatite lattice. Of course, adsorption of cobalt on surface of hydroxyapatite crystallites could be possible as well.

It is known that cobalt is able to replace calcium in hydroxyapatite structure (Veselinović et al. 2010), which should lead to decrease of both unit cell parameters of hydroxyapatite ($R_{Co2+} < R_{Ca2+}$, Shannon 1976). In turn, the incorporation of cobalt into channel as CoO_2^{2-} ion (Kazin et al. 2017) should result in constant parameter *a* and increasing parameter *c*. It is also known that unit cell parameters of hydroxyapatite are affected by replacement of OH by H_2O and PO_4^{3-}—by CO_3^{2-} (Frank-Kamenetskaya et al. 2011). Incorporation of water is leading to the increase of

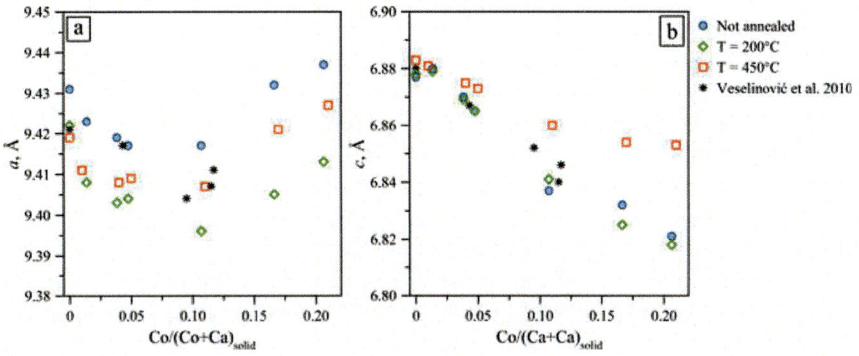

Fig. 7 Unit cell parameters of Co-hydroxyapatites before and after calcination: **a** parameter ***a***, **b** parameter ***c***. Data of (Veselinović et al. 2010) are displayed as well

parameter *a*, while incorporation of CO_3^{2-} is leading to the decrease of parameter *a* and increase of parameter *c*. The extra positive charge caused by these incorporations is often compensated by vacancies in Ca^{2+} position, lowering parameter *c*.

In Co-containing apatites synthesized by us (before calcination) both unit cell parameters decrease at Co/(Co+Ca) ratio in solid = 0–0.05 (Fig. 7), which can be solely explained by the incorporation of Co^{2+} into Ca sites. At higher concentration of cobalt its incorporation is still going on, as follows from continuing decrease of parameter *c*. At the same time, the increase of parameter *a* in range Co/(Ca+Co) ≥ 0.11 cannot be explained by cobalt incorporation alone. Such behavior of parameter *a* can be related to the rising incorporation of water into channel site at Co/(Ca+Co) ≥ 0.11, supported by IR-spectroscopy (Fig. 4).

The Co → Ca replacement is as well accompanied by the significant decrease of crystallite sizes and attributes of hydroxyapatite desymmetrization observed at X-ray powder patterns at Co/(Ca+Co) ≥ 0.05–0.11 (Fig. 2). The potential transition from hexagonal to monoclinic symmetry could be caused by various reasons. As it was described by (Elliott 1994; Ikoma et al. 1999), monoclinic hydroxyapatite is different from hexagonal in ordered alignment of OH-ions in channels. At hypothesis level we can assume that entrance of water in channels, as Co content increases, could cause partial ordering of OH and H_2O. Besides, since relations $b = 2a$, $\gamma = 120°$, typical for monoclinic hydroxyapatite studied before (Ikoma et al. 1999), are not preserved, the structure of Co-containing hydroxyapatite seems to undergo deeper changes. Such changes could be caused, as example, directly by cobalt incorporation in calcium sites. The incorporation of CO_3 needs to be taken into account as well, since it is leading to notable distortions in hydroxyapatite framework (Ivanova et al. 2001). These possible mechanisms are subject for future investigations.

According to EDX data, the maximum cobalt content in hydroxyapatite achieved by us is Co/(Ca+Co) = 0.21 (~10.65 wt.%) at Co/Ca (solution) = 0.20 assuming there was no significant cobalt adsorption on the surface of hydroxyapatite crystallites.

However, the incorporation limit of cobalt into hydroxyapatite should be higher, since we did not observe flattening of curves of unit cell parameters at high cobalt contents.

The revealed regularities of variations of unit cell parameters are in good agreement with data of (Veselinović et al. 2010) and extend them, since wet synthesis performed by us resulted in almost twice higher cobalt incorporation into hydroxyapatite. Veselinovic and co-authors have also experienced the rise of parameter a at Co/(Ca+Co) ≥ 0.11, though it was not interpreted.

Co-substituted hydroxyapatites possess lower thermal stability compared with pure HAp. Calcination of hydroxyapatites at 200 °C affects only unit cell parameters of apatite. Parameter *c* does not change greatly and keeps reducing as Co content increases, indicating that cobalt contents in hydroxyapatites calcined at 200 °C are similar to those in not calcined hydroxyapatites. Parameter *a* lowers at almost same value regardless of the cobalt content, which is well explained by the decrease of water content, observed at IR spectra (Fig. 4).

Calcination of hydroxyapatites at 450 °C leads to the decomposition of Co-HAp. The released cobalt is probably mostly adsorbed on hydroxyapatite particles, explaining the immutability of cobalt contents in solid obtained by EDX. At Co/Ca (solution) ≥ 0.15 the decomposition is as well pronounced in Co_3O_4 formation while the content of Co-HAp is still high. Calcination also resulted in slower decrease of parameter *c* with cobalt content increase (Fig. 7b) and increase of parameter a relatively to T = 200 °C (Fig. 7a). As the result, the values of parameter *a* for T = 450 °C are placed between values of not calcined HAps and calcined at T = 200 °C. These regularities can be explained by removal of CO_3, supported by IR spectroscopy (Fig. 4) along with the release of cobalt from hydroxyapatite crystal structure. Simultaneously, at Co/(Ca+Co) ratio in solid ≥ 0.10 the plot of parameter *c* starts to flatten, and at Co/Ca ≥ 0.15 it becomes almost constant, indicating that all extra cobalt has been removed from hydroxyapatite structure. Basing on the slope of the original plot of parameter *c* (before calcination) we can assume that after thermal treatment at 450 °C maximum cobalt content in HAp (Co/(Ca+Co) ~ 0.10) is approximately twice lower than the maximum reached in hydroxyapatites before calcination.

The calcination at 700 °C causes Co-HAp to decompose almost completely. At Co/Ca (solution) ≤ 0.05 Co-free HAp is preserved, as follows from unit cell parameters of apatite (Table 1). At higher cobalt contents the decomposition manifests itself in formation of β-TCP and Co_3O_4 (Co/Ca in solution ≥ 0.10). The released cobalt does not seem to incorporate β-TCP, as follows from its unit cell parameters (Sect. 3.1).

Incorporation of cobalt into hydroxyapatite is pronounced in paramagnetic properties of the latest at Co $\geq$ 5 at.%. The magnetization curve obtained at Co/(Ca+Co) = 0.05 almost coincides the curve given by (Kramer et al. 2014) for the sample with same cobalt content and synthesized in similar way. Unlike the previous researchers (Kramer et al. 2014) who have reported magnetic properties of Co-HAp at 5 at. % of cobalt we have obtained linear concentration dependency of magnetic susceptibility on cobalt content (Fig. 6b). The maximum magnetic susceptibility achieved is 1.43×10^{-5} at Co/Ca in solution = 0.20 (Co/(Ca+Co) in HAp = 0.21).

As it was shown by (Kramer et al. 2014), Co-HAp does not possess significant toxicity, and thus could be used in biomedicine. Our data have demonstrated that it is possible to vary the magnetic properties of Co-hydroxyapatite by setting the conditions of synthesis and thermal treatment. This opens the way to use "biofriendly" Co-HAp along with iron oxides and other compounds in cancer therapy, medical visualization and drug delivery (Bárcena et al. 2009; Price et al. 2018; Fatima et al. 2021; Yang et al. 2022).

5 Conclusion

In our work Co-substituted hydroxyapatites were synthesized from water solutions in wide range of cobalt concentrations. Obtained maximum Co^{2+} content in apatite (21 at. %, ~10.65 wt. %) is almost twice higher than reported before (Veselinović et al. 2010).

Thermal stability of Co-HAp was for the first time studied in 200–700 °C range of calcination. It was confirmed that Co-HAp has lower thermal stability compared to pure HAp (stable up to 1200 °C, Kramer et al. 2016) and shown that thermal treatment at up to 200 °C doesn't lead to apatite decomposition, and, if necessary, can be used in fabrication of Co-HAp-based nanoparticles.

Analysis of powder diffraction data has allowed us to advance in crystal chemistry of Co-HAp: to confirm the incorporation of Co^{2+} ions into Ca^{2+} sites (at least up to 21 at. %) and show that at Co $\geq$ 11 at.% desymmetrization of hexagonal hydroxyapatite structure occurs.

Magnetic measurements have revealed linear concentration dependence of Co-HAp's magnetic susceptibility on cobalt content and minimum cobalt concentration required for emergence of paramagnetic properties (5 at. %).

The obtained results expand our views on the incorporation of Co^{2+} ions in hydroxyapatite. Obtained compositions can find an application in biomaterials and biotechnologies, where biocompatibility of hydroxyapatite and magnetic properties are required. Further investigations of Co-HAps should include detailed studies on their magnetization and in vivo experiments on their application.

Acknowledgements The research was supported by President School 2022 grant No. 075-15-2022-831. The laboratory experiments were carried out in the Research Park of Saint Petersburg State University: XRD and IR measurements—in Centre for X-ray Diffraction Studies, SEM and EDX measurements—in Centre for Microscopy and Microanalysis and Centre for Geo-Environmental Research and Modelling (GEOMODEL), magnetization measurements—in Centre for Innovative Technologies of Composite Nanomaterials.

References

Anderson JB, Kostiner E (1987) The crystal structure of cobalt-substituted calcium chlorapatite. J Solid State Chem 66:343–349.

Bárcena C, Sra AK, Gao J (2009) Applications of magnetic nanoparticles in biomedicine. Nanoscale Magn Mater Appl 167:591–626.

Berzina-Cimdina L, Borodajenko N (2012) Research of Calcium Phosphates Using Fourier Transform Infrared Spectroscopy. In: Theophanides T (ed) Infrared Spectroscopy - Materials Science, Engineering and Technology. IntechOpen, Rijeka, pp 123–148.

Doebelin N, Kleeberg R (2015) Profex: A graphical user interface for the Rietveld refinement program BGMN. J Appl Crystallogr 48:1573–1580.

Elliott JC (1994) Structure and chemistry of the apatites and other calcium orthophosphates. Elsevier Science B.V., Amsterdam

Fatima H, Charinpanitkul T, Kim KS (2021) Fundamentals to apply magnetic nanoparticles for hyperthermia therapy. Nanomaterials 11:1–20.

Frank-Kamenetskaya O, Kol'tsov A, Kuz'mina M, Zorina M, Poritskaya L (2011) Ion substitutions and non-stoichiometry of carbonated apatite-(CaOH) synthesised by precipitation and hydrothermal methods. J Mol Struct 992:9–18.

Grisafe DA, Hummel FA (1970) Crystal Chemistry and Color in Apatites Containing Cobalt, Nickel, and Rare-earth Ions. Am Mineral 55:1131-1145.

Hughes JM, Cameron M, Crowley KD (1989) Structural variations in natural F, OH, and Cl apatites. Am Mineral 74:870–876.

Ikoma T, Yamazaki A, Nakamura S, Akao M (1999) Preparation and Structure Refinement of Monoclinic Hydroxyapatite. J Solid State Chem 144:272–276.

Ivanova TI, Frank-Kamenetskaya OV, Kol'tsov AB, Ugolkov VL (2001) Crystal structure of calcium-deficient carbonated hydroxyapatite. Thermal decomposition. J Solid State Chem 160:340–349.

Kazin PE, Zykin MA, Schnelle W, Zubavichus YV, Babeshkin KA, Tafeenko VA, Felser C, Jansen M (2017) Cobalt-Based Single-Ion Magnets on an Apatite Lattice: Toward Patterned Arrays for Magnetic Memories. Inorg Chem 56:1232–1240.

Kramer E, Itzkowitz E, Wei M (2014) Synthesis and characterization of cobalt-substituted hydroxyapatite powders. Ceram Int 40:13471–13480.

Kramer E, Conklin M, Zilm M (2016) A Comparative Study of the Sintering and Cell Behavior of Pure and Cobalt Substituted Hydroxyapatite. Bioceram Dev Appl 04:1–8.

Price PM, Mahmoud WE, Al-Ghamdi AA, Bronstein LM (2018) Magnetic drug delivery: Where the field is going. Front Chem 6:1–7.

Ptáček P (2016) Substituents and Dopants in the Structure of Apatite. In: Apatites and their Synthetic Analogues - Synthesis, Structure, Properties and Applications. IntechOpen, Rijeka, pp 289–334.

Shannon RD (1976) Revised Effective Ionic Radii and Systematic Studies of Interatomie Distances in Halides and Chaleogenides. Acta Crystallography A32:751–767.

Veselinović L, Karanović L, Stojanović Z, Bračko I, Marković S, Ignjatović N, Uskoković D (2010) Crystal structure of cobalt-substituted calcium hydroxyapatite nanopowders prepared by hydrothermal processing. J Appl Crystallogr 43:320–327.

Vieira EG, Silva Vieira TW, Silva MP, Santos MVB, Souse Brito CAR, Sousa Bezerra RD, Falho ACV, Osajima JA, Silva Filho EC (2017) Tuned Hydroxyapatite Materials for Biomedical Applications. In: Pignatello R, Musumeci T (eds) Biomaterials - Physics and Chemistry - New Edition. IntechOpen, Rijeka.

Yang X, Shao G, Zhang Y, Wang W, Qi Y, Han S, Li H (2022) Applications of Magnetic Particle Imaging in Biomedicine: Advancements and Prospects. Front Physiol 13:1–17.

Zykin MA, Anokhin EO, Kazin PE (2019) Cobalt-containing calcium fl uorohydroxyapatites with properties of fi eld-induced single-ion magnets. 68:751–756

Effect of Amino Acids on Hydroxyapatite Synthesized from Solutions Modeling Human Joint Synovia Fluid: Case of Glycine and Proline

Svetlana A. Gerk and Olga A. Golovanova

Abstract Carbonate-containing hydroxyapatite has been synthesized from prototype human synovial fluid in the presence of proline and glycine. The resultant powders have revealed to contain 61–71 wt % proline and 75–80 wt % glycine. The presence of amino acids in the model solution has been shown to change the crystallinity and specific surface area of the samples, without influencing the composition or crystallite size of the solid phase. High amino acids concentrations lead to the formation of poorly crystallized composites consisting of smaller nanocrystals. The amino acid-containing samples have lower porosity and are more thermally stable. The dissolution of the samples in a 0.9% NaCl solution and acetate buffer has been occurs to be a two-step process. The highest solubility in weakly acidic solutions has been demonstrated by the precipitates containing the amino acids.

Keywords Carbonate hydroxyapatite · Biocompatible materials · Crystallization · Proline · Glycine · Resorption · Specific surface area

1 Introduction

Human bone tissue is a biochemical system having a multilevel structural organization and multicomponent organic–inorganic composition of the intercellular matrix (Karyakina and Persova 2009; Popov et al. 2014; Epple et al. 2010; Torbenko and Kasavina 1977). Its life activity is known to be based on remodeling, which comprises two mutually connected and interchangeable processes: osteogenesis (the formation of new bone tissue) and resorption (disintegration of the formed bone) (Green et al. 2014). A key element of remodeling is bone mineralization: deposition of ultradisperse crystalline carbonate-containing hydroxyapatite (CHA), on collagen protein microfibers from human body fluids (intercellular, synovial, and other fluids) (Karyakina and Persova 2009).

S. A. Gerk (✉) · O. A. Golovanova
Dostoevsky State University, Pr. Mira 55-A, Omsk 644077, Russia
e-mail: gerksa_11@mail.ru

O. V. Frank-Kamenetskaya et al. (eds.), *Biogenic—Abiogenic Interactions in Natural and Anthropogenic Systems 2022*, Springer Proceedings in Earth and Environmental Sciences, https://doi.org/10.1007/978-3-031-40470-2_6

A great deal of attention is currently paid to designing materials intended for the restoration of bone tissue defects that resulted from changes (osteoporosis, deforming arthrosis, and others), injuries, and surgical interventions. In this context, calcium phosphate-based materials have the greatest promise because they are characterized by high biocompatibility with human bone tissue (Izmailov and Golovanova 2014; Agrawal and Gurbhinder 2011). This is utilized in designing bone implants, which can be either fabricated entirely of calcium phosphates or have a surface coated with these compounds (Fadeeva and Golovanova 2019; Volova 2009; Gerk et al. 2016).

An important principle in designing implant is to reproduce characteristics of natural bone tissue. Hydroxyapatite (HA) is known to be inorganic component of bone tissue, bearing the main mechanical load, whereas collagen imparts to it elasticity and flexibility (Wahl and Czernuszka 2006; Salman et al. 2013; Basu and Basu 2019). Collagen molecule is the polypeptide chain consisting of repeating amino acid triplets. The main amino acids in the composition of collagen are glycine, proline, hydroxyproline, and hydroxylysine. Alternating residues of a proline molecule and hydroxyproline facilitate the formation of a stable triplehelix collagen structure, which imparts strength to the molecule. Moreover, proline prevents aging, maintaining the strength of bone, flexibility of joints, and elasticity of ligaments; preserves vessel wall elasticity, and makes skin smooth.

On the other hand, proline offers the highest cytoprotective activity at increased inorganic ion concentrations in a medium (salt stress) and reduced temperatures (temperature stress) (Fedotova and Dmitrieva 2015). Under stress conditions, proline exhibits an osmoprotective effect as well, controlling cell water balance. However, currently available data on reactions of proline with calcium phosphates are very limited and contradictory (Fedotova and Dmitrieva 2014).

As shown earlier (Gerk and Golovanova 2013; Fleming et al. 2001; DileepKumar et al. 2021), polypeptide collagen chains contain crystalline segments in the form of –Gly–X–Y– triplets consisting largely of neutral amino acids (Gly = glycine, 33 wt %; X = proline or hydroxyproline, 22 wt %; Y = hydroxylysine), and amorphous segments consisting of polar amino acids (lysine, histidine, and others). The mechanism underlying the interaction between the organic and inorganic bone components has not yet been studied in sufficient detail.

This problem can be partially resolved in experiments aimed at investigating calcium phosphate (brushite, hydroxyapatite (HA), octacalcium phosphate, etc.) biomineralization processes in artificial systems (urine, saliva, and others), as well as by computer simulation (Bye et al. 2013; Tavafoghi et al. 2013; Golovanova and Korol'kov 2017). As shown earlier, amino acids and proteins may inhibit or stimulate crystallization of calcium phosphates through adsorption interactions with their surface and complexation with calcium ions, thus influencing the crystallinity, morphology, and particle size of the solid phase. However, since data available in literature are limited mainly to the proposed models, gaining insight into the role of organic substances on HA formation processes is of great current interest.

Table 1 Average molar concentrations of inorganic ions in human synovial fluid (pH 7.40 ± 0.05; ionic strength of the solution, 0.172 mmol/L)

Component	Ca^{2+}	Na^{+}	Mg^{2+}	K^{+}	Cl^{-}	$HCO_3{}^{2-}$	$HPO_4{}^{3-}$	$SO_4{}^{2-}$
C, mmol/L	2.53	140	1.1	4.6	103	27	4.38	11.4

The purpose of this work was to assess the effect of glycine and proline on the morphology, degree of crystallinity, and resorption properties of HA synthesized from model solutions of human joint synovial fluid (synovia).

2 Experimental

2.1 *Synthesis*

CHA was synthesized from a model medium similar in ion-electrolyte composition, pH, and ionic strength to human synovial fluid, as described previously (Golovanova et al. 2013; Table 1). Concentration of proline (Pro, $NH(CH_2)_3CHCOOH$) in the model solutions was: 0 (sample 1), 0.013 (2), 0.026 (3), 0.039 (4), and 0.052 mol/L(5); glycine (Gly, CH_2NH_2COOH) concentration: concentration in the model solutions varied from 0.04 to 0.24 mol/L. Hereafter, the following designations are used: 0 (sample 1), 0.04 (2), 0.08 (3), 0.016 (4), and 0.024 mol/L (5).

2.2 *Methods and Approaches*

Quantitative determination of amino acids in the solid phase was carried out from the difference between the initial and final concentrations of the amino acid and precipitate-forming ions in the model solution. The residual amino acid concentration in the liquid phase was determined spectrophotometrically (KFK-2, Russian Federation State Standard 18309-72).

The phase composition of the resultant powders was determined by X-ray diffraction (Bruker D8 Advance diffractometer, CuKα radiation, $\lambda = 0.15406$ nm). The occurring phases were identified using ICDD Powder Diffraction File data. The average size of coherently scattering domains (CSD) was determined using the Bruker TOPAS 3.0 program.

IR spectra of the precipitates were measured on an FSM-2202 spectrophotometer using dishaped samples prepared by pressing with KBr. Microstructure and microtexture of the powders were examined by scanning electron microscopy (SEM) on a JEOL JSM-6610LV.

Thermal analysis was carried out on a Netzsch STA-449C simultaneous thermoanalytical system.

The specific surface area S of the samples was determined by single-point nitrogen adsorption measurements at 77.4 K on a Sorbtometr adsorption instrument (manufactured by OOO Katakon, Russia). The S values obtained were calculated using the BET method.

To model active and passive resorption stages, the samples were dissolved with constant stirring in an acetate buffer (pH 5.5) and a 0.9% sodium chloride solution (pH $\approx$ 7). The pCa values in the solution were determined by direct potentiometry using an I-160M ion-selective meter. The data thus obtained were processed using regression analysis (SigmaPlot 12.5 software package).

3 Results and Discussion

3.1 *X-ray Powder Diffraction*

Analysis of the X-ray diffraction patterns of the synthesized powders indicated the formation of a crystalline HA phase, no matter whether or not the model solution contained proline and glycine (Fig. 1a, b; Table 2). At the same time, unlike in the case of the powder prepared with no amino acid, the X-ray diffraction patterns of the HA–proline samples showed stronger background signals (in the range 2θ = 5°–20°), and the intensity of the main reflections from HA (002, 121, and 112), was reduced, pointing to a low degree of crystallinity of the synthesized samples and an increased amount of an amorphous phase.

The lattice parameter *a* of the HA–proline powders is smaller than that of the impurity-free phase, whereas the lattice parameter *c* changes insignificantly (Table 2). Because of this, the samples synthesized in the presence of proline have a larger *c/a* axial ratio and a smaller unit-cell volume. The structural characteristics of HA–proline powders, including CSD size, are similar to nonstoichiometric carbonate apatites in human bone tissue (Safronova and Putlyaev 2013). Various data have been reported in literature as to the effect of proline on crystallization processes in solution. Depending on the experimental conditions and the composition of the solid phase, the amino acid can act as an inhibitor of the crystal growth process (Golovanova and Korol'kov 2017) or a catalyst of the nucleation process, primarily due to chelation (Golovanova and Tomashevsky 2019). Also possible is partial substitution of amino acid anions (0.2 molecules) for OH groups in the structure of HA (Gerk and Golovanova 2013). In our case, the presence of proline in the model solution has a negligible effect on the concentration of the apatite component in the precipitates and their crystallite size (Table 2). The crystallite size in the [001] direction is 21–23 nm. At the same time, the weight of the powders increases systematically owing to the increase in the percentage of the amino acid in their composition.

Each precipitate contains 61–71% of (wt) the amino acid present in the solution during the synthesis process.

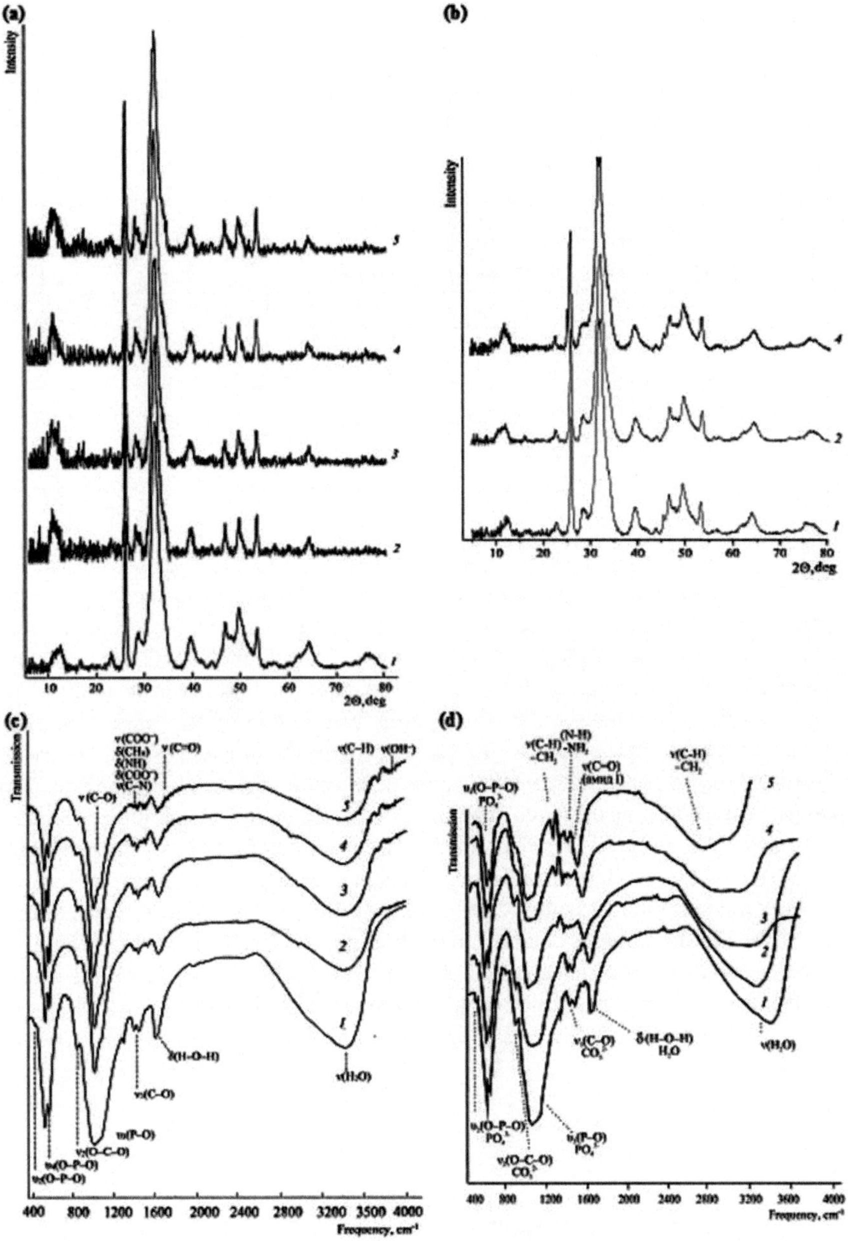

Fig. 1 X-ray diffraction and IR spectra of samples with proline (**a**, **c**) and glycine (**b**, **d**)

Table 2 Lattice parameters, CSD size and specific surface area (S) of the CHA

Sample	*a*, Å	*c*, Å	*c/a*	V, Å	CSD, nm	S, m^2/g
1	9.459 ± 0.002	6.874 ± 0.002	0.727	532.6	21.58	130 ± 7
Proline						
2	9.426 ± 0.004	6.874 ± 0.004	0.729	528.9	22.57	116 ± 6
3	9.420 ± 0.004	6.879 ± 0.004	0.730	528.6	21.07	111 ± 6
4	9.420 ± 0.004	6.882 ± 0.004	0.730	528.7	21.06	114 ± 6
5	9.422 ± 0.004	6.877 ± 0.004	0.730	528.8	22.65	112 ± 6
Glycine						
2	9.474 ± 0.002	6.880 ± 0.001	0.726	534.8 ± 0.717	2.62	70 ± 1
3	9.473 ± 0.004	6.889 ± 0.004	0.727	535.3 ± 0.739	3.54	52 ± 2
4	9.470 ± 0.002	6.887 ± 0.002	0.727	534.9 ± 0.738	2.99	75 ± 3
Bone tissue (Gerk et al. 2016)	9.410	6.891	0.732	528.4	5–10	–

Assessment of the phase composition precipitates obtained in the presence of glycine using the method of corundum numbers showed that in all of the samples the obtained HA was a major phase. The presence of background signals in the X-ray diffraction patterns in the range 2θ = 10°–20° indicates the presence of an amorphous phase. The poorer resolution and considerable broadening of the reflections 002, 121, and 112 in the range 25°–30° in the X-ray diffraction patterns also indicates the formation of a phase with a low degree of crystallinity. As shown by Mendes (Mendes et al. 2012), the presence of amino acids leads to the precipitation of poorly crystallized HA compared to stoichiometric HA. Parameter *a* of HA synthesized in the presence of glycine in the model medium is increased parameter *c*, *c/a* ratio, and unit-cell volume remain essentially unchanged (Table 2). The characteristics of the crystal lattice of the samples obtained in the presence of glycine are similar to those of nonstoichiometric carbonate-containing calcium deficient HAs, including those that constitute a mineral basis of human bone tissue (Neyvis 2010).

3.2 *IR Spectrometry*

The IR spectra of our samples (Fig. 1c) show peaks characteristic of carbonate-containing HA, namely, stretching and bending vibrations in H_2O, $\nu(H_2O) = 3400–3440\ cm^{-1}$ and $\delta(H–O–H) = 1610–1650\ cm^{-1}$; asymmetric stretching and bending vibrations in $\upsilon_3(P–O) = 1030–1090\ cm^{-1}$, $\upsilon4(O–P–O) = 564–605\ cm^{-1}$, and $\upsilon_2(O–P–O) = 470–475\ cm^{-1}$; a doublet due to asymmetric stretching vibrations at $\nu_3(C–O) = 1410–1480\ cm^{-1}$; and $\nu_2(O–C–O)$ bending vibrations at $875–879\ cm^{-1}$ in which suggest that the substitution of carbonate ions for phosphate tetrahedra in the structure of the synthesized CHAs follows a B-type mechanism.

The presence of proline in the composition of the solid phase is evidenced by vibrational modes of organic groups: at 1032–1037 cm^{-1}, due to ν(C–O) stretching vibrations of the pyran ring; at 1422–1440 cm^{-1}, due to ν_s(COO–) symmetric stretching vibrations; at 1460 cm^{-1}, due to δ(CH) and δ(CH_2) bending vibrations of the methyl and methylene groups, respectively; at 1507–1543 cm^{-1}, due to δ(NH) and δ(COO–) bending vibrations and ν(C–N) stretching vibrations (amide II); at 2800–3600 cm^{-1}, due to ν(C–H) stretching vibrations; and at 3420–3600 cm^{-1}, due to ν(OH–) vibrations, associated with intramolecular hydrogen bonds. It is seen that, in the frequency range 900–4000 cm^{-1}, absorption bands of the mineral and organic constituent components overlap with each other. In the frequency range below 900 cm^{-1}, which contains only modal water of HA and characterizes the presence of "water–water" intermolecular bonds. This is accompanied by an increase in the number of vibrational modes of the constituent groups of the amino acid. This can be accounted for by the fact that proline is an imine acid containing a pyrrolidine side radical of intermediate volume and hydrophobicity (with a volume of 113 Å^3, that is, 3.7 times the volume of a water molecule, and a hydrophobicity index of 0.711) and free of N–H groups, which are capable of acting as hydrogen bond donors. As a consequence, proline in solution binds a limited number of water molecules on account of the acceptor properties of COO– (Kirillov et al. 2013). Moreover, during the crystallization process, some of the carboxyl groups are bound to positively charged regions of the HA surface. Low-mobility nonpolar (low-polarity) heterocycles can attract each other owing to intermolecular forces to form associates. Thus the increasing percentage of proline in the composition of the solid phase, on the one hand, reduces the amount of structurally bound water and, on the other, leads to spatial stabilization of its molecules.

The assumed presence of glycine in the solid phase is supported by the IR spectroscopy data (Fig. 1d). Spectra of the samples obtained at glycine concentrations in the model medium above 0.08 mol/L show weak bands in the range 2800–3000 cm^{-1}, corresponding to ν(C–H) bond vibrations in –CHn methylene groups. In the frequency range 1300–1700 cm^{-1}, we observe overlapping modes of water, carbonate ions, and the following peaks of the amino acid: ν(C–H) at 1340 cm^{-1} in –CHn; ν(C = O) asymmetric vibrations (amide I) in the range 1642–1678 cm^{-1}; and ν(N–H) bending vibrations at 1500 cm^{-1} in –NH_2. Thus, glycine concentrations of ≥0.08 mol/L in the model solution lead to the formation of HA–Gly composites. Analysis of residual glycine concentrations in model solutions indicates that, independent of the initial amino acid concentration in solution, the solid phase contains 75–80 wt % glycine. With increasing glycine concentration, the weight of the precipitate increases systematically. One possible reason for this is that, in aqueous solutions, glycine exhibits adsorption activity toward the HA surface because it is present in the form of zwitterions (–CH_2–COO–bipolar ions) and has small molecular dimensions. As a result, there may be electrostatic interaction between the functional groups COO– and Ca^{2+}. Also possible is the formation of hydrogen bonds between a proton of an amino group and OH– and groups in the HA.

3.3 BET and SEM

The results of BET and SEM study demonstrate that the presence of proline in the crystallization medium no effect on the CSD size and causes only a slight change in the specific surface area of the powders (Table 2). Independent content of the amino acid in the composition of the solid phase, its specific surface area is smaller than that of pure HA by just 13–18 m^2/g. It may be, because of its heavy hydrophobic side radical, proline has maximum conformational limitations and reacts with HA mainly through its carboxyl group.

Despite the possible formation of new crystallization centers-chelate complexes and associative proline chains-active crystal growth is spatially limited because of the low-mobility heterocycles and structural stabilization of water, and occurs in a certain direction in accordance with the conformational orientation of the proline chains.

Thus, possible secondary crystallization processes on the one hand and steric hindrances in proline-containing solutions on the other allow obtain a solid phase with good surface characteristics. The above data are consistent with analysis of the morphology of the powder particles. The presence of proline in the model medium was found to have no significant effect on the shape or size of the powder particles. Like particles of a pure HA phase, the HA-proline particles form aggregates up to 200 μm in size, having a flaky shape. The amino acid-containing samples have lower porosity, presumably because of the poorer crystallinity of their mineral component (Fig. 2a).

Fig. 2 Micrographs of samples with the addition of glycine (**a**) and proline (**b**)

The presence of glycine less than 0.08 mol/L in the starting solution was found to have a negligible effect on the CSD size of HA, but tended to reduce their specific surface area by a factor of 1.7–2 (Table 2). Since the samples consisted of rather large particles, it is reasonable to assume that the reduction in S was related to a decrease in the degree of their crystallinity and/or to adsorption interactions of the charged parts of the amino acid with the surface of the nonstoichiometric HA (Cao 2004).

If the model medium contains more than 0.08 mol/L of the amino acid, the CSD size decreases, but the S of the powders rises only slightly. It may be that, at higher concentrations, on the one hand glycine has an inhibiting effect on crystallite growth, but on the other the active centers of the amino acid molecule may act as nuclei for the precipitation of the mineral component, which is responsible for the formation of poorly crystallized samples with a large specific surface area.

The above results correlate with SEM data. Independent of whether the medium contains the amino acid, the particles of the material has the form of spherical aggregates ranging in size from 1 to 200 μm and consisting of HA CSDs according to the X-ray diffraction data Fig. 2b, Table 2). It is seen that the HA obtained from the glycine-free medium has a larger specific surface area. At glycine concentrations under 0.08 mol/L, the aggregates forming in the model solution have lower porosity and their surface was more uniform. Further raising the amino acid concentration in solution (to above 0.16 mol/L) leads to the formation of spheres similar in shape to the particles of the HA synthesized without glycine, which had $S \geq 75\ m^2/g$.

3.4 *Thermal Analysis*

To study the nature of the thermal transformations of our samples, they were characterized by thermal analysis in the temperature range 25–1000 °C. The thermogravimetry (TG), derivative thermogravimetry (DTG), and differential thermal analysis (DTA) curves of the HA–amino acids composites are similar in shape to those of the sample prepared without proline (Fig. 3a, b).

The DTA curves (Fig. 3a) demonstrate four steps of thermal transformations in the temperature range 25–1000 °C (Selifanova et al. 2008; Nurkeev et al. 2005).

The first step (I; 25–280/25–242 (270) °C; samples 1/2, 3 (4, 5); $\Delta H > 0$) comprises the removal of uncombined water and the decomposition of highly volatile impurities.

The second step (II; 280–470/240–410 (464) °C; samples 1/3–5 (2); $\Delta H < 0$) is accompanied by amino acids removal ($t_m = 205–221$ °C) and dehydration of adsorbed water according to the following scheme:

$$[CH \times nH_2O] \times mH_2O_{(s.)} \rightarrow CH \times nH_2O_{(s.)} + mH_2O_{(v.)} + Q \quad (1)$$

The third step (III; 470–750/410 (464)–750(784) °C; samples 1/3–5 (2); $\Delta H < 0$) is the desorption of the water of crystallization and chemically bound water. The water removal processes can be represented by the following scheme:

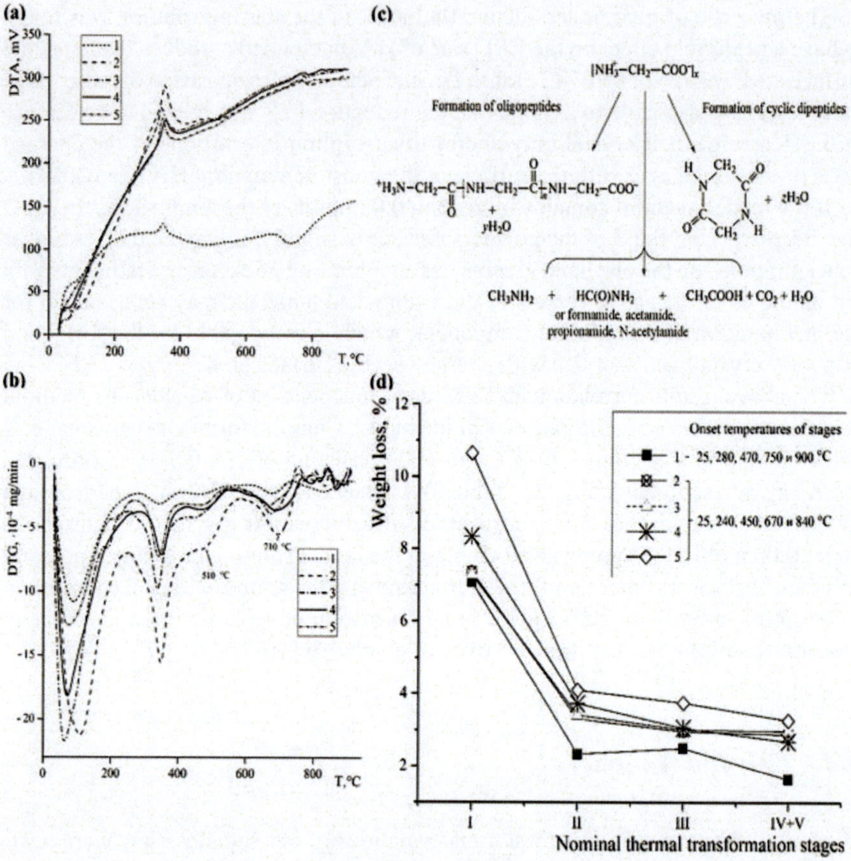

Fig. 3 DTA and DTG curves of samples with proline (**a**, **b**) and schematic of the processes of the fast stage of thermal destruction of glycine (**c**), and weight mass loss stages during annealing of samples (**d**)

$$\mathrm{CHA} \times \mathrm{nH_2O_{(s.)}} \rightarrow \mathrm{CHA_{(s.)}} + \mathrm{mH_2O_{(v.)}} + \mathrm{Q} \quad (2)$$

In the fourth step (IV, 750–1000/750 (784)–1000 °C; samples 1/3–5 (2); $\Delta H > 0$), the CHA transforms into a stoichiometric HA phase (3) or β-$Ca_3(PO_4)_2$ (4):

$$\mathrm{Ca_{10}(PO_4,\ CO_3)_6(OH)_2} \rightarrow \mathrm{Ca_{10}(PO_4)_6(OH)_2} + 6\mathrm{CO_2} \uparrow -\mathrm{Q} \quad (3)$$

$$\begin{aligned}&\mathrm{Ca_9(PO_4)_{6-x-y}(HPO_4)(CO_3)_x(OH)_{2-(s.)}} \xrightarrow{800\,^\circ\mathrm{C}} \\ &3\beta\mathrm{-Ca_3(PO_4)_{2(s.)}} + \mathrm{CO_{2(v.)}} + 2\mathrm{H_2O_{(v.)}} - \mathrm{Q}\end{aligned} \quad (4)$$

Thermoanalytical curves of the HA–amino acids composites were found to be similar in shape to those of the pure HA sample (Fig. 3). During the decomposition of the composites in steps II–IV, three times more heat is released. This is attributable to both the higher content of the components being removed in the composition of the samples and their strongest adhesion to the solid surface. The DTA curves of the composites have a smaller slope (Fig. 3a).

Temperature ranges of the destruction of the components are roughly the same for the HA and HA–amino acids powders. At the same time, the removal of the chemically bound water from the amino acid-containing samples in the third step occurs in two steps (the peaks at 510 and 710 °C).

On the whole, the presence of the amino acids has little effect on the thermal stability of the samples, which have similar extrema. The thermal transformation of the samples synthesized in the presence of proline is accompanied by the largest heat effects in steps II and III of the thermal transformations. This is attributable to the stabilization of the water in the composites prepared in the presence of amino acids and correlates with the IR spectroscopy data. In addition, we observe an increase in the total weight loss due to the removal of the amino acids and water of crystallization (Table 3).

The thermal destruction of glycine in this stage occurs in two steps. In the initial, fast step ($\Delta H > 0$), in the range 200–270 °C, we observe the formation of oligopeptides and diketopiperazine (cyclic dipeptide) and by-products (Fig. 3c). In the range 270–370 °C, we observe the second, slow step, namely, complete oxidation of the amino acid destruction products, which is possibly rate-limiting ($\Delta H < 0$), because it makes a major contribution to the energy of the thermal process in stage II (peak at 321 °C in the DTA and derivative thermogravimetry (DTG) curves.

The present results demonstrate that, even though the thermoanalytical curves of CHA and CHA-*Gly* are very similar, the powders containing the amino acid decompose at lower temperatures (Δt from 20 to 80 °C, depending on the thermal stage), with the largest total weight loss (Fig. 3d). According to the thermal analysis data, the degree of crystallinity of the precipitates is related to the concentrations of glycine (stage II), adsorbed and chemically bound water (stages II and III), and

Table 3 Weight mass (Δm) at the four steps of the thermal transformations of samples with proline

Step	Sample									
	1		2		3		4		5	
	T*, °C	Δm, %	T*, °C	Δm, %	T*, °C	Δm, %	T*, °C	Δm, %	T*, °C	Δm, %
1	25	7.10	25	6.94	25	8.53	25	8.49	25	6.52
2	280	2.34	242	4.85	242	3.31	270	2.43	272	2.28
3.1	470	2.50	464	1.22	410	1.68	410	1.52	408	1.54
3.2			476	1.69	554	1.80	556	2.12	556	1.72
4	750	1.24	784	0.36	750	0.47	750	0.66	756	0.49
1 + 4	25–950	13.18	25–950	15.06	25–950	15.79	25–950	15.22	25–950	12.55

*Transition onset temperature

carbonate ions substituting for phosphate tetrahedra and OH^- ions in the structure of HA (stage IV). It is important to note that, at the highest glycine concentrations in the model medium (0.16 and 0.24 mol/L), the precipitates contain the largest amount of the amino acid, adsorbed water, and highly volatile impurities, including adsorbed carbon dioxide (stages I and II). The IR spectra of these samples show vibrational modes of bonds in organic groups (Fig. 2b). No correlation was detected between the amino acid content of the powders and the concentration of structural carbonate ions. It may be that CHA combines with glycine predominantly through COO^-–Ca^{2+} interactions.

3.5 Dissolution Kinetics Study

The dissolution of the HA-Gly samples in a 0.9% NaCl solution was found to raise its pH from 7.5 to 7.8. In our opinion, the reason for this is that the alkalization of the solution (pH > 7.5, isoelectric point of glycine pI 5.9) leads to Ca^{2+}–COO^- bond breaking and increases the mole fraction of the anion form of glycine: NH_2–CH_2–COO^-. Regression analysis of kinetic curves indicates that the powder dissolution process occurs in two steps. In the initial portion of the curves, the time dependence of the concentration of calcium ions in solution, $C(\tau) = -\log$ is well represented by a linear function (portion of curve I in Fig. 4a; Table 4). It is seen that, in this step, the samples obtained at glycine concentrations in solution above 0.08 mol/L have a higher dissolution rate. As a result, powders 4 and 5 dissolve in a shorter time interval (Table 4). At the highest amino acid concentration, the precipitate dissolves in 240 s, after which the kinetic curve plateaus (dissolution reaches completion). This dissolution behavior of samples 4 and 5 can be accounted for by the high content of adsorbed and constitutional water, volatile components, glycine, and carbonate ions in Ca^{2+}–COO^-, the structure of the HA, in agreement with the thermal analysis data.

The next step of sample dissolution in the 0.9% sodium chloride solution is well represented by an exponential function (region II - final portion in Fig. 4a; Table 4):

$$C(t) = C_0 + C_m\exp(b\tau) \tag{5}$$

where C_0 is the nominal initial concentration, C_m is the saturation concentration, b is a coefficient, and τ is time.

It is seen that, in this case, the calcium concentration in solution increases and the system approaches saturation, so the process can be quantified by the initial dissolution rate at the corresponding values of C_0 and τ_i (Fig. 4a; Table 4). It is shown that the trend in the rate of subsequent dissolution of the precipitates in the exponential step is similar to the linear dependence.

It is important to note that, unlike samples 4 and 5, sample 3, which is readily soluble and has a smaller specific surface area and better crystallinity, has a longer initial dissolution step (I) at the median value of the range of sample dissolution acidities (pH 7.65).

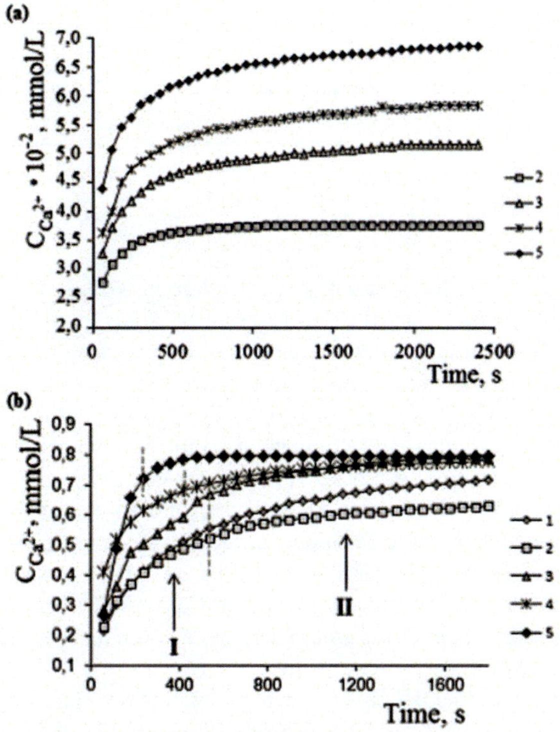

Fig. 4 Ca^{2+} concentration as a function of dissolution time in a NaCl (**a**) and acetate buffer (**b**) fo samples with glycine

Table 4 Dissolution characteristics of the samples with glycine in 0.9% NaCl

Sample	Initial portion		Final portion	
	τ, min	υ, 10^{-4}, min^{-1}	τ, min	υ, 10^{-5}, min^{-1}
1	1–10	6.0 ± 0.6	10–11	5.9 ± 0.3
2	1–10	5.0 ± 0.5	10–11	5.3 ± 0.2
3	1–10	7.0 ± 0.6	10–11	6.7 ± 0.2
4	1–10	6.0 ± 0.9	8–9	7.0 ± 0.2
5	1–4	26.0 ± 4.0	4–6	–

Verifying the adequacy of the regression models chosen above showed a statistically significant difference between the variances for the two dissolution steps in the kinetic curves, at a significance level $\alpha = 0.05$ (Fig. 4a; Table 5).

Unlike in the case of the HA sample, the dissolution rate of the composites drops by a factor of 15–25 if the samples are dissolved in an acetate buffer in weakly acidic solutions (pH 5.40–5.57). The calcium concentration in solution over time rises linearly with increasing content in the composition of the solid phase (Fig. 4b).

For the CHAs obtained at the minimum and maximum glycine concentrations in the model system (0.04 and 0.243 mol/L), we studied the dissolution kinetics in an

Table 5 Fisher's ratio test values for the kinetic dissolution curves of samples 1–5 in a NaCl physiological solution ($\alpha = 0.05$)

F	Kinetic curves				
	1	2	3	4	5
Empirical	9.04	19.77	22.05	20.27	Plateau
Critical	2.42	2.42	2.42	2.49	

acetate buffer solution at temperature of 310 K. The initial dissolution rate of the precipitates was found to increase by a factor of 1.5 with increasing temperature. The highest dissolution rate was observed for sample 5. In particular, in the case of this sample the activation energy for the dissolution process has the lowest value, 43.5 kJ/mol, whereas that for sample 2 is 59.3 kJ/mol.

In the linear and exponential portions of the curve (Table 6), the dissolution rate of the samples synthesized in the presence of proline exceeds that of the prolinefree samples. At the same time, the dissolution rate of the samples is essentially independent of amino acid concentration in the starting solution. The isoelectric point of proline, where the amino acid exists in solution as a neutrally charged zwitterion is 6.3. Solution acidification leads to Ca^{2+}–COO^- bond breaking and the formation of a cation form of the amino acid. As a consequence, the solubility of the precipitates increases.

To model the passive stage of implant resorption in the human body, we studied dissolution of the samples in an isotonic (0.9%) NaCl solution (pH $\approx$ 7.4). As shown by regression analysis, the resorption rate of the proline-containing samples in the 0.9% NaCl solution exceeds that of proline-free CHA (Table 6).

Table 6 Dissolution characteristics of the samples with proline

Sample	Initial portion		Final portion	
	τ, min	υ, 10^{-7}, min^{-1}	τ, min	υ, 10^{-6}, min^{-1}
Acetate buffer				
1	1–15	0.091 ± 0.008	15–26	0.91 ± 0.3
2	1–16	1.9 ± 0.2	16–24	8.1 ± 3.1
3	1–18	1.7 ± 0.2	18–31	7.8 ± 3.9
4	1–18	1.1 ± 0.1	18–31	7.1 ± 2.4
5	1–17	1.5 ± 0.2	17–32	7.1 ± 1.6
0.9% NaCl solution				
1	1–8	2.5 ± 0.4	8–28	0.027 ± 0.043
2	1–7	8.0 ± 1.6	7–20	0.036 ± 0.038
3	1–12	7.2 ± 1.2	12–23	0.036 ± 0.077
4	1–6	12.0 ± 2.5	8–13	0.035 ± 0.014
5	1–6	11.0 ± 2.0	8–13	0.038 ± 0.068

In the initial stage, the samples with a proline concentration above 0.039 mol/L dissolve at a higher rate and in a shorter time. In the exponential region, the disso-lution rate is independent of amino acid concentration.

These findings can be accounted for by the fact that the Ca^{2+}–COO– bonds can become weaker as a result of the reduction in the acceptor properties of the carboxyl group. Raising the solution pH to 7.4 (pH > pI) leads to a decrease in the electron density at the carbon atom of the carboxyl group and reduces the electrostatic attraction between the Ca^{2+} and COO^- ions. As a result, the dissolution rate of the samples prepared in the presence of the amino acid rises.

4 Conclusions

Biomimetic synthesis of calcium-phosphate-organic composites was carried out by varying the content of organic components of the intercellular matrix of bone tissues and synovial joint fluid of a person (amino acids - glycine and proline), studying the effect of synthesizing conditions on the composition and structure, micromorphology, bioactivity of the obtained materials.

The powders thus obtained contain 61–71 wt % proline and 75–80 wt % glycine. The presence of amino acids in the model solution has been shown to cause a change in the degree of crystallinity of the samples and a slight change in their specific surface area, without influencing the crystallite size of the forming HA.

The HA–proline powders are similar in structural characteristics and crystallite size to nonstoichiometric carbonate apatites in human bone tissue.

Increasing the percentage of proline in the composition of the solid phase, on the one hand, reduces the amount of structurally bound water and, on the other, leads to spatial stabilization of its molecules. The amino acid-containing samples have lower porosity and are more thermally stable.

It has been shown that, if the model solution contains less than 0.08 mol/L of glycine, the latter has no effect on the CSD size of the forming HA crystals, but leads to changes their specific surface area. At glycine concentrations in the model solution above 0.08 mol/L, we observe the formation of poorly crystallized HA–Gly composites having a large specific surface area and consisting of smaller nanocrystallites. Such samples contain larger amounts of the amino acid (≈80%), adsorbed and chemically bound water, highly volatile impurities (including adsorbed carbon dioxide), and carbonate ions substituting of phosphate tetrahedra in the structure of HA.

The dissolution of the samples in a 0.9% NaCl solution and acetate buffer solution has been shown to be a two-step process. The samples synthesized in a proline-containing medium dissolve at a higher rate and in a shorter time. The percentage of the amino acids in the samples influences only the initial stage of their dissolution in 0.9% NaCl solution.

The samples synthesized from a medium containing more than 0.08 mol/L of glycine dissolve at a higher rate and in a shorter time. The HA–Gly powders were found to dissolve at a slower rate in an acetate buffer solution.

The highest solubility under weakly acidic conditions has been demonstrated by the precipitates containing the largest amount of the glycine.

Acknowledgements This work was supported by Russian Science Foundation RSF (No. 23-23-00668).

References

Agrawal K, Gurbhinder S (2011) Synthesis and characterization of hydroxyapatite powder by sol-gel method for biomedical application. J of Minerals & Materials Characterization & Engineering 1(8):727–734

Basu S, Basu B (2019) Doped biphasic calcium phosphate: synthesis and structure. Published online: 04 Jul 2019

Bye JW, Meliga S, Ferachou D, Cinque G, Zeitler JA, Falconer RJ (2013) Analysis of the hydration water around bovine serum albumin using terahertz coherent synchrotron radiation. J PhysChem A118(1):83–88

Cao G. (2004) Nanostructures and nanomaterials: synthesis, properties and application. Imperial College Press, London, UK

DileepKumar VG, Sridhar MS, Aramwit P, Valentina K, Krut'ko. (2021) A review on the synthesis and properties of hydroxyapatite for biomedical applications. Published online: 03 Oct 2021

Epple M, Ganesan K, Heumann R, Klesing J, Kovtun A, Neumann S,Sokolova V (2010) Application of calcium phosphate nanoparticles in biomedicine. J Mater Chem 20(1):18–23

Fadeeva TV, Golovanova OA (2019) Physicochemical properties of brushite and hydroxyapatite prepared in the presence of chitin and chitosan. Russ J Inorg Chem 64:847–856

Fedotova MV, Dmitrieva OA (2014) The structure of the hydration shells of the $–NH_2^+$ and $–COO^-$zwitter-ion of L-proline according to the data of the 1D-RISM method of integral equations. ZhFKh. Russ J PhysChem88 (5):801–804

Fedotova MV, Dmitrieva OA (2015) Characterization of selective binding of biologically relevant inorganic ions with the proline zwitterion by 3D-RISM theory. New J Chem 39(11):8594–8601

Fleming D, Bronswijk W, Ryall RL (2001) A comparative study of the adsorption of amino acids on to calcium minerals found in renal calculi. ClinSci 101:159–168

Gerk SA, Golovanova OA (2013) Variations in the amino acid composition of human bone tissues. SustainDev 21(3):299–304

Gerk SA, Golovanova OA, Sharkeev YuP (2016) Synthesis of a two-phase nanopowder from prototype human synovial fluid and the use of the nanopowder for the preparation of coatings on titanium plates. Inorg Mater 52(9):1021–1028

Golovanova OA, Korol'kov VV (2017) Thermodynamics and kinetics of calcium oxalate crystallization in the presence of amino acids. Crystallogr Rep 62(5):787–796

Golovanova OA, Tomashevsky IA (2019) Kinetics and thermodynamics of the formation of compounds of calcium ions and amino acids: their structure and stability. Russ J PhysChem 93(1):7–17

Golovanova OA, Lemesheva (Gerk) SA, Izmailov RR (2013) RF Patent 2 496 150 (in Russian)

Green DW, Goto TK, Kim KS, Jung HS (2014) Calcifying tissue regeneration via biomimetic materials chemistry J R Soc Interface 11:1–11

Izmailov RR, Golovanova OA (2014) Adhesive and morphological characteristics of carbonate hydroxyapatite prepared from a model human synovial fluid on titanium alloys. Inorg Mater 50(6):592–598

Karyakina EV, Persova EA (2009) Features of bone tissue remodeling in inflammatory and degenerative diseases of the hip joint. Saratov J Med Sci Res 5(2):227–230

Kirillov AF, Koz'mik RA, Daskalyuk AP, Kuznetsova NA, Kharchuk OA (2013) Assessment of proline content of plants under drought and salinization conditions. DokEkolPochvoved 1(18):194–202

Mendes LC, Ribeiro GL, Marques RC (2012) In situ hydroxyapatite synthesis: influence of collagen on its structural and morphological characteristic. Mater SciAppl 3:580–586

Neyvis AB (2010) A Computational investigation of the Interaction of the collagen molecule with hydroxyapatite. Dissertation, London

Nurkeev SS, Kazova RA, Akhmetbekova A (2005) Physicochemical study of thermal transformations of phosphorite minerals. VestnKazNTU3(47):82–87

Popov VK, Komlev VS, Chichkov BN (2014) Calcium phosphate blossom for bone tissue engineering. Mater Today 17(2):96–97

Safronova TV, Putlyaev VI (2013) Medical inorganic materials research in Russia: calcium phosphate materials. NanosistFizKhim Mat 4(1):24–27

Salman SA, Kuroda K, Okido M (2013) Preparation and characterization of hydroxyapatite coating on AZ31 Mg alloy for implant applications. BioinorgChem Appl:175–756

Selifanova EI, Chepnova RK, Koblova OE (2008) Thermogravimetric study of L alpha amino acids. IzvSaratovskUniv, SerKhim, Biol, Ekol 8(2):23–28

Tavafoghi JM, Yao G, Cerruti M (2013) The importance of amino acid interactions in the crystallization of hydroxyapatite. J R Soc Interface 10:1–14

Torbenko VP, Kasavina BS (1977) Functional biochemistry of bone tissue. Medicine, Moscow (in Russian)

Volova TG (2009) Materials for medicine, cell and tissue engineering. IPK SFU, Krasnoyarsk (in Russian)

Wahl DA, Czernuszka JT (2006) Collagen-hydroxyapatite composites for hard tissue repair. Eur Cell Mater 11:43–56

Calcium and Magnesium Glutamates: Structure Calculations and IR Spectra by HF and DFT Methods

Dmitry V. Bespalov, Olga A. Golovanova, and Dmitry N. Kugaevskikh

Abstract Structural dynamic models of magnesium glutamate and calcium glutamate complexes were simulated by the methods of Hartree–Fock (HF) and density-functional theory (DFT, v B3LYP) in the 6-31G and 3-21G basis sets. The molecular geometric parameters and frequencies of normal vibrations within the harmonic approximation in the IR spectra of the developed models were calculated. Calcium (II) and magnesium (II) complexes with glutamic acid were synthesized from aqueous solutions of the corresponding salts of calcium chloride and magnesium chloride, and the amino acid. The IR spectra of the synthesized compounds were recorded in the range of 500–4000 cm^{-1}. The calculated and experimental IR spectra of the synthesized calcium (II) and magnesium (II) complexes with glutamic acid were compared in order to validate their structures. Data on the coordination of calcium and magnesium ion complexes with amino acids contributes to understanding the structure of these poorly studied complexes and improving methods for obtaining these complex compounds with a predetermined composition and structure.

Keywords Density-functional theory · Hartree–Fock method · Calcium glutamate · Magnesium glutamate · Molecular modeling · IR spectrum · Synthesis

1 Introduction

Research into the complexation of calcium (II) and magnesium (II) ions with biogenic amino acids, as well as the development of methods to study such compounds, is a promising direction in the field of bioinert interactions. This concerns determination of thermodynamic stability constants between calcium (II) and magnesium (II) ions and amino acids, which is essential for biochemical studies.

Calcium and magnesium in ionized form play an important role in the implementation of a number of cellular and physiological functions of living organisms

D. V. Bespalov (✉) · O. A. Golovanova · D. N. Kugaevskikh
Dostoevsky Omsk State University, Mira St., 55a, Omsk 644077, Russia
e-mail: Dankovskiu@yandex.ru

O. V. Frank-Kamenetskaya et al. (eds.), *Biogenic—Abiogenic Interactions in Natural and Anthropogenic Systems 2022*, Springer Proceedings in Earth and Environmental Sciences, https://doi.org/10.1007/978-3-031-40470-2_7

(Senni et al. 2003; Dobrynina 2007; Gromova 2007; Yudina et al. 2016). Amino acids are building blocks for the formation of all human and animal organs, muscles and ligaments, hormones and enzymes (Levchuk et al. 2017). In these such various settings, calcium and magnesium compounds undergo multiple interactions. The stability of the as-synthesized complexes determines the rate of metabolic processes in the human body. At the same time, solid theoretical foundations explaining the nonequilibrium interaction of inorganic (calcium (II) and magnesium (II) ions) and organic (amino acids) components are yet to developed. In addition, the mechanism of such chemical interactions, as well as the relationship between the reactivity of chemical agents and their spatial structure and implementation conditions, demand elucidation.

The human body is characterized by a multilevel organization of constituent components. The complex nature of interactions between inorganic and organic components requires understanding of the thermodynamical, kinetic, and structural specifics of the synthesized compounds. The absence of reliable knowledge in this field leads to the need to carry out additional tests during biochemical studies, which may result in various side effects (Golovanova 2007). This also explains the relevance of studying the complexation of calcium (II) and magnesium (II) ions with amino acids and developing efficient methods for their investigation.

Metal complexes of amino acids (AA) are biologically active compounds. In such complexes, calcium (II) and magnesium (II) ions frequently act as complexing agents, participating in reactions with AA. The formed complexes are capable of catalyzing reactions and participating in metabolic processes in the human body (Mashina and Shanina 2019). Some human diseases may lead to the formation of stones, i.e., organic mineral aggregates (OMAs) in various living tissues. According to statistical data, the prevalence of such diseases, which are associated with calcium ions and amino acids, is growing annually (Pearce et al. 2002; Larsen and Pearce 2003). OMAs develop in body fluids of complex composition under various factors of exogenous and endogenous nature, which process is insufficiently studied (Golovanova et al. 2004).

Magnesium glycinate is involved in maintaining normal ATP levels. This salt plays an important role in biosynthesis, growth, and thermogenesis, as well as in bone mineralization and intestinal motility. Magnesium glutamate was found to be effective for treating minor epilepsy manifestations and, together with potassium glutamate, cardiovascular diseases (Reyko and Yaroshenko 2010). Calcium glutamate is prescribed for patients suffering from mental disorders due to cerebral atherosclerosis, tuberculous meningitis, epilepsy, polio, and acute period of arachnoencephalitis. AA complexes with metals and their derivatives have important pharmaceutical applications, e.g., as anti-tumor agents or for treating cancer patients with multidrug resistance (Waheed et al. 2019).

Therefore, information about the structure and behavior of calcium and magnesium complexes with AA, as individual compounds or as part of OMAs, can contribute to a better understanding of processes occurring in tissues and fluids of the human body, in both normal conditions and different diseases. Moreover, knowledge about the structure of these complexes can serve as a basis for the development of various drugs and therapeutical approaches. This knowledge can be obtained following a systematic approach to study the complexation of calcium (II) and magnesium (II) ions with AA (Kurochkin 2011; Tomashevsky 2021).

Research into the properties of complex compounds of calcium (II) and magnesium (II) ions with AA, both in the crystalline phase and in the form of a calculated quantum chemical model of an isolated molecule, contributes to the theory of complex compounds, which is relevant for biomedical studies. Data on the coordination of compounds of calcium and magnesium ions with AA elucidates their poorly-studied structure and is helpful in the development of methods for the synthesis of these complex compounds with a predetermined composition and structure (Nakoskin et al. 2012).

A comprehensive study in this area is necessary both from the theoretical and practical point of view, by expanding the pool of thermodynamic and kinetic data on the interaction of calcium (II) and magnesium (II) ions with AA, and contributing to a better understanding of the nature of complexation of AA with calcium (II) and magnesium (II) ions.

Molecular complexes can be simulated using quantum chemical calculations. According to literature data (Babkov et al. 2010; Butyrskaya et al. 2012; Kohn 2002), the density-functional theory (DFT) method and the Hartree–Fock (HF) method can be used to calculate vibrational spectra. The former is considered to be more reliable, while the latter is simpler and faster. For normal IR vibrations, absolute errors of the DFT method does not exceed 5% and 2% in the 2800–3800 cm^{-1} and 2800–500 cm^{-1} regions, respectively.

In this work, we aim to compare the structure and IR spectra of calcium glutamate and magnesium glutamate molecules calculated by DFT and HF methods with those obtained experimentally. Decoding and interpretation of such IR spectra is a complicated issue, which has not been resolved so far. Therefore, we make an assumption that the studied objects belong to complex compounds, where calcium (II) and magnesium (II) ions act as complexing agents, and glutamic acid (Glu) acts as a ligand.

In order to achieve the above aim, the following research objectives were formulated:

1. to calculate the structures of calcium and magnesium glutamate complexes using DTF and HF methods;
2. to validate the obtained data by comparing the calculated and experimental IR spectra of these complexes.

2 Calculations and Experimental

2.1 Calculation by DTF and HF Methods

Molecular structures of the studied compounds were calculated using the standard software package for quantum chemical modeling GAMESS (US), allowing implementation of the HF and DFT (B3LYP) methods in 6-31G and 3-21G basis sets. The 6-31G and 3-21G basis sets are valence-split, have one basis function of the inner shell, which is grouped from six and three, respectively, Gaussian functions, as well as a set of internal functions, grouped from three and two, respectively, Gaussian and external non-grouped Gaussian function for the valence shell of each atom (Gordon and Schmidt 2005). The energy of the calcium and magnesium glutamate models is reduced to their minimum; the structures and frequencies of normal modes for the IR spectroscopy method are calculated.

2.2 Synthesis

Compounds were synthesized under the metal–amino acid molar ratio of 1:1. The method reported in (Bespalov and Golovanova 2021), which had been previously applied for the synthesis of similar compounds, was used. A Glu (0.441 g) sample was dissolved in 10 ml of distilled water (pH ~ 5.46) followed by addition of m($CaCl_2 \cdot 2H_2O$) = 0.441 g and m($MgCl_2 \cdot 6H_2O$) = 0.304 g, respectively. The solution was evaporated at room temperature, during which a chemical reaction of AA with Ca^{2+}, Mg^{2+} ions and crystallization of the resulting compound took place. Following 7–14 days, a crystalline precipitate was obtained. The precipitate was washed with a small amount of water and dried at a temperature of 40–50 °C to remove excess moisture.

2.3 IR Spectroscopy

The samples were characterized by the method of Fourier-transform infrared spectroscopy. IR spectra were obtained using an ФСМ 2202 spectrophotometer (Infraspec, Russia). During the study, the obtained samples of calcium glutamate and magnesium glutamate in powder form were mixed with KBr, placed in a germanium cuvette and pressed. The spectra were recorded with a resolution of 1 cm^{-1}. The spectra of the studied samples were recorded across the range from 500 to 4000 cm^{-1}. Data processing and spectrum construction were performed using the OriginPro 2019b toolsets.

3 Results and Discussion

3.1 Calculation of Structural Models

The energy of calcium glutamate molecules was minimized, comprising −1227.567 Hartree ($-5.35 \cdot 10^{-15}$ J). For magnesium glutamate, it was equal to −750.097 Hartree ($-3.27 \cdot 10^{-15}$ J). A specific structural feature of the studied compounds consists in the complexing agent (Ca^{2+}, Mg^{2+}) interacting with two carboxyl groups located on one amino acid, as shown in Figs. 1b and 2b.

It is worth noting that, in addition to the side group, which contains an additional carboxyl group in this case, the amino group also affects the spatial structure and stability of the compounds under study. Geometrical parameters for the simulated structure of calcium glutamate and magnesium glutamate were calculated by HF and DFT methods (using the B3LYP functional) in the 6-31G and 3-21G basis sets. The bonds and angles between neighboring atoms, as well as close contacts for isolated

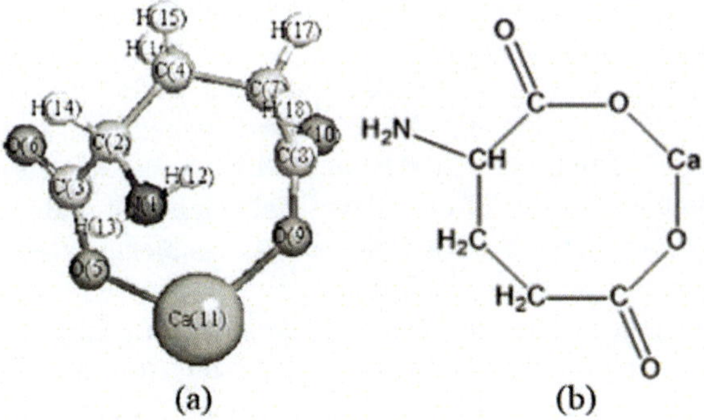

Fig. 1 Calcium glutamate: **a** 3D model; **b** structural formula

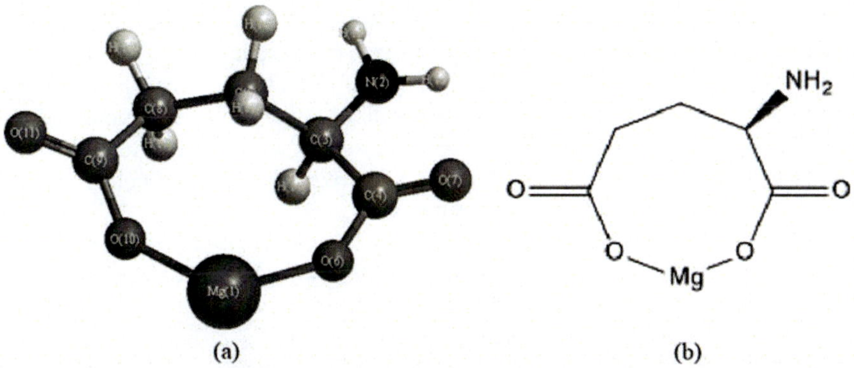

Fig. 2 Magnesium glutamate: **a** 3D model; **b** structural formula

molecules, are shown in Tables 1 and 2. In Figs. 1a and 2a, the atoms are numbered for ease of working with the results given in Tables 1 and 2.

The thermodynamic characteristics of calcium glutamate and magnesium glutamate models calculated at standard temperature showed the possibility of the expected structures of the compounds obtained (Tables 3 and 4). The selection of standard temperature facilitates comparisons with literature data.

3.2 Comparison of Calculated and Experimental IR Spectra

IR spectra for calcium glutamate were obtained by calculations (DFT based on 6-31G), Fig. 3a, and experimentally, Fig. 3b. The analysis of the calculated (using the example of DFT in the 6-31G basis) and experimental IR spectra of calcium glutamate demonstrates strong bands $v_{as}(C=O) = 1685\ cm^{-1}$, $v_{as}(C–O) = 1281\ cm^{-1}$ for the spectrum of the calcium glutamate model and $v_{as}(C=O) = 1627\ cm^{-1}$, $v_{as}(C–O) = 1258\ cm^{-1}$ for the synthesized compound Ca^{2+}:Glu. The error of ~2% indicates that the Ca^{2+} ion in the synthesized compound interacts with the carboxyl group similarly to that in the assumed model (Fig. 1).

In the model spectrum, we observe two pronounced bands $v_{as}(NH_2) = 3482\ cm^{-1}$ and $v_s(NH_2) = 3136\ cm^{-1}$ (Fig. 3), which are absent in the amino acid spectra due to the bipolar structure. The bands $v_{as}(NH_2) = 3392\ cm^{-1}$ and $v_s(NH_2) = 3137\ cm^{-1}$ are also found in the spectrum of the synthesized compound (Table 5), an error of ~2%, which may confirm that the bond between the complexing agent and the ligand forms due to the donor–acceptor mechanism.

The absence of a specific band of the model compound in the region of ~2006 cm^{-1} (Fig. 3) indicates the absence of a strong $N–H^+ \ldots O$ interaction; however, these bands are detected in the spectrum of the synthesized compound (Table 5). It is suggested that the resulting compound has a strong interaction between the amino group of the AA group and structural water, whereas this model does not take into account this influence.

Table 5 presents a detailed analysis of the 500–1500 cm^{-1} fingerprint region in comparison with the IR spectra of the synthesized compound.

An isolated compound model has a number of limitations. Calculations ignore a number of factors that may affect bond vibrations under actual conditions; therefore, the spectra of actual compounds contain some peaks and overtones. These include, e.g., the presence of structural water, hydrogen bonds, neighboring atoms, intermolecular interactions, other impurities, etc.

The presence of the above factors reliably indicates that the synthesized complex compound Ca^{2+}:Glu has the composition similar to that of the model compound. One difference consists in the presence of water in the structure of the synthesized complex. Therefore, the donor–acceptor mechanism of Ca–N bond formation is likely, which is not taken into account in the model compound.

The remaining spectra can be described in a similar manner, including the IR spectra of magnesium glutamate. Thus, the calculated spectrum by the HF method

Table 1 Bond lengths and angles in the structure of calcium glutamate (calculated model)

Chemical bond	Bond length DFT 6-31G, Å	Bond length DFT 3-21G, Å	Bond length HF 6-31G, Å	Bond length HF 3-21G, Å	Angle	DFT 6-31G, °	DFT 3-21G, °	HF 6-31G, °	HF 3-21G, °
Ca11-O9	2.035	2.151	2.060	2.127	O9-Ca11-O5	102.9	105.4	97.7	101.1
C7-H18	1.095	1.096	1.082	1.081	Ca11-O9-C8	132.7	121.7	137.9	123.4
C7-H17	1.097	1.097	1.084	1.086	O10-C8-O9	123.8	125.1	123.9	125.7
O5-Ca11	2.026	2.113	2.051	2.087	O10-C8-C7	121.8	118.7	120.9	119.9
C4-H16	1.098	1.095	1.085	1.081	O9-C8-C7	114.2	116.0	114.9	114.2
C4-H15	1.098	1.096	1.085	1.083	H18-C7-H17	107.2	107.3	107.3	107.4
C2-H14	1.099	1.100	1.082	1.082	H18-C7-C8	109.7	112.7	109.7	110.2
N1-H13	1.014	1.027	0.995	1.008	H18-C7-C4	110.3	110.5	110.5	110.8
N1-H12	1.012	1.024	0.995	1.006	H17-C7-C8	107.8	107.2	107.5	107.3
C8-O10	1.240	1.239	1.222	1.219	H17-C7-C4	108.1	108.0	107.6	107.9
C8-O9	1.353	1.335	1.320	1.312	C8-C7-C4	113.2	110.6	113.6	112.8
C7-C8	1.535	1.568	1.521	1.540	Ca11-O5-C3	152.9	153.6	153.1	159.8
C4-C7	1.547	1.547	1.541	1.539	H16-C4-H15	107.6	108.2	107.5	107.9
C3-O6	1.242	1.241	1.222	1.220	H16-C4-C7	108.0	107.3	108.0	108.0
C3-O5	1.342	1.320	1.315	1.302	H16-C4-C2	106.8	105.8	106.9	106.1
C2-C4	1.565	1.560	1.551	1.551	H15-C4-C7	109.0	110.7	108.5	109.8
C2-C3	1.540	1.567	1.527	1.542	H15-C4-C2	107.5	107.1	107.4	107.0
N1-C2	1.460	1.475	1.449	1.471	C7-C4-C2	117.3	117.1	117.8	117.5
Close contacts					O6-C3-O5	123.7	125.8	124.1	125.1
					O6-C3-C2	121.6	120.3	120.8	120.5
					O5-C3-C2	114.6	113.7	114.9	114.2

(continued)

Table 1 (continued)

Chemical bond	Bond length DFT 6-31G, Å	Bond length DFT 3-21G, Å	Bond length HF 6-31G, Å	Bond length HF 3-21G, Å	Angle	DFT 6-31G, °	DFT 3-21G, °	HF 6-31G, °	HF 3-21G, °
Ca11-H18	3.038	2.784	3.295	2.933	H14-C2-C4	108.1	107.9	108.8	108.1
Ca11-H14	3.478	3.563	3.510	3.731	H14-C2-C3	107.8	108.2	107.6	107.8
O6-H13	–	2.105	–	–	H14-C2-N1	109.5	110.9	108.7	110.5
O10-Ca11	4.142	4.160	4.126	4.027	C4-C2-C3	111.3	112.1	111.8	112.8
C7-Ca11	3.537	3.350	3.700	3.460	C4-C2-N1	112.2	108.3	111.7	107.4
C4-Ca11	4.249	4.160	4.349	4.269	C3-C2-N1	107.5	109.1	107.8	109.9
C2-Ca11	3.847	3.879	3.881	3.979	H13-N1-H12	114.2	111.5	115.0	111.6
N1-O6	2.737	2.710	2.733	2.712	H13-N1-C2	112.7	107.8	114.7	110.1
O6-Ca11	4.283	–	4.248	4.304	H12-N1-C2	115.4	113.1	116.3	113.6

Table 2 Bond lengths and angles in the structure of magnesium glutamate (calculated model)

Chemical bond	Bond length DFT 6-31G, Å	Bond length DFT 3-21G, Å	Bond length HF 6-31G, Å	Bond length HF 3-21G, Å	Angle	DFT 6-31G, °	DFT 3-21G, °	HF 6-31G, °	HF 3-21G, °
Mg(1)-O(6)	1.816	1.853	1.784	1.762	Mg(1)-O(10)-C(9)	139.9	82.4	140.6	108.8
C(8)-H(18)	1.091	1.096	1.083	1.084	O(11)-C(9)-O(10)	122.6	117.0	123.5	122.8
C(8)-H(17)	1.100	1.090	1.088	1.079	O(11)-C(9)-C(8)	122.5	120.0	122.1	121.7
O(10)-Mg(1)	1.802	1.995	1.789	1.812	O(10)-C(9)-C(8)	114.7	121.3	114.2	115.1
C(5)-H(16)	1.092	1.098	1.079	1.084	H(18)-C(8)-H(17)	106.9	108.8	106.6	108.2
C(5)-H(15)	1.097	1.095	1.084	1.083	H(18)-C(8)-C(9)	106.2	109.8	106.9	107.0
C(3)-H(14)	1.108	1.092	1.091	1.082	H(18)-C(8)-C(5)	110.5	108.6	110.1	107.8
N(2)-H(13)	1.004	1.026	0.991	1.002	H(17)-C(8)-C(9)	107.7	107.8	107.1	109.8
N(2)-H(12)	1.004	1.020	0.995	1.001	H(17)-C(8)-C(5)	109.5	112.0	110.3	111.5
C(9)-O(11)	1.235	1.307	1.213	1.214	C(9)-C(8)-C(5)	115.4	109.5	115.1	112.1
C(9)-O(10)	1.355	1.313	1.329	1.342	Mg(1)-O(6)-C(4)	137.5	112.4	147.7	145.9
C(8)-C(9)	1.532	1.506	1.524	1.521	H(16)-C(5)-H(15)	105.6	108.2	104.8	107.0
C(5)-C(8)	1.561	1.559	1.557	1.554	H(16)-C(5)-C(8)	108.5	106.8	108.2	107.2
C(4)-O(7)	1.236	1.237	1.213	1.209	H(16)-C(5)-C(3)	105.3	105.6	105.3	105.3
C(4)-O(6)	1.345	1.355	1.318	1.319	H(15)-C(5)-C(8)	109.8	108.6	110.1	109.2
C(3)-C(5)	1.588	1.577	1.571	1.566	H(15)-C(5)-C(3)	110.2	107.6	110.0	110.1
C(3)-C(4)	1.553	1.549	1.534	1.534	C(8)-C(5)-C(3)	116.6	119.2	117.4	117.2
N(2)-C(3)	1.436	1.472	1.432	1.458	O(7)-C(4)-O(6)	123.2	124.5	123.5	123.8
Close contacts					O(7)-C(4)-C(3)	122.3	120.6	122.4	121.8
					O(6)-C(4)-C(3)	114.3	114.7	114.0	114.2

(continued)

Table 2 (continued)

Chemical bond	Bond length DFT 6-31G, Å	Bond length DFT 3-21G, Å	Bond length HF 6-31G, Å	Bond length HF 3-21G, Å	Angle	DFT 6-31G, °	DFT 3-21G, °	HF 6-31G, °	HF 3-21G, °
					H(14)-C(3)-C(5)	106.6	110.0	107.8	108.8
H(14)-H(17)	2.207	–	2.223	–	H(14)-C(3)-C(4)	105.1	108.6	105.2	107.8
N(2)-H(16)	2.473	–	2.408	–	H(14)-C(3)-N(2)	111.2	110.7	111.4	110.4
C(4)-H(15)	2.581	–	–	–	C(5)-C(3)-C(4)	111.9	111.9	112.0	111.9
Mg(1)-H(15)	2.868	–	2.875	–	C(5)-C(3)-N(2)	113.6	110.4	111.7	111.6
Mg(1)-H(14)	3.196	–	-	–	C(4)-C(3)-N(2)	107.7	104.9	108.4	106.0
Mg(1)-C(8)	3.455	3.529	3.361	3.411	H(13)-N(2)-H(12)	119.4	112.9	119.2	114.5
N(2)-O(7)	2.679	2.705	2.661	2.656	H(13)-N(2)-C(3)	117.1	109.7	118.9	112.6
Mg(1)-C(5)	3.283	–	3.312	–	H(12)-N(2)-C(3)	120.2	113.9	120.3	115.7
Mg(1)-C(3)	3.354	3.227	3.551	3.730	O(6)-Mg(1)-O(10)	132.3	103.2	125.5	117.2

Table 3 Calculated structural characteristics of the calcium glutamate complex at T = 298 K

Characteristic	Value DFT 6-31G	Value DFT 3-21G	Value HF 6-31G	Value HF 3-21G
Molecular mass, g/mol	185.19138			
Enthalpy H, kcal/mol	91.676	92.978	93.364	94.073
Entropy S, cal/(mol K)	98.211	93.746	96.116	92.225
Internal energy U, kcal/mol	91.703	92.067	91.703	93.575
Kinetic energy, kcal/mol	464,856	464,741	465,227	465,242
Potential energy, kcal/mol	−930,963	−930,783	−931,374	−931,301

Table 4 Calculated structural characteristics of the magnesium glutamate complex at T = 298 K

Characteristic	Value DFT 6-31G	Value DFT 3-21G	Value HF 6-31G	Value HF 3-21G
Molecular mass, g/mol	185.19138			
Enthalpy H, kcal/mol	88.796	92.617	92.840	93.784
Entropy S, cal/(mol K)	106.017	101.865	104.529	99.556
Internal energy U, kcal/mol	88.204	92.247	91.959	93.452
Kinetic energy, kcal/mol	763,025	762,879	763,378	763,274
Potential energy, kcal/mol	−1,527,183	−1,527,024	−1,527,459	−1,527,466

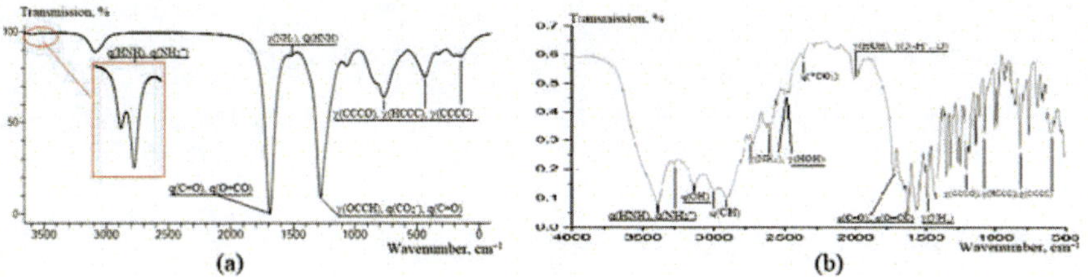

Fig. 3 IR spectra of calcium glutamate: **a** calculated (DFT in 6-31G basis) model; **b** synthesized complex compound Ca^{2+}:Glu.

in the 6–31 basis is shown in Fig. 4: calculated (HF in the 6-31G basis) (a) and experimental (b).

Further, we discuss vibrations in the IR spectra of the calcium glutamate model (DFT in the 6-31G basis) and the complex compound obtained experimentally.

The region of 618–839 cm^{-1}. Complex vibrations, which include the valence C–C bonds of the skeleton and bending vibrations of the carboxyl and amino groups of all forms of amino acids. Scissoring vibrations suggest the presence of a carboxyl group and carbon dioxide. The amino group is characterized by bending vibrations of rocking and wagging modes.

Table 5 Calculated and experimental IR vibration frequencies of the calcium glutamate molecule

$\nu_{experimental}$, cm^{-1}	$\nu_{DFT\ 6\text{-}31G}$, cm^{-1}	$\nu_{DFT\ 3\text{-}21G}$, cm^{-1}	$\nu_{HF\ 6\text{-}31G}$, cm^{-1}	$\nu_{HF\ 3\text{-}21G}$, cm^{-1}	Vibration modes
604	618	580	602	570	γ(OCO), χ(OCCC), χ(OCCH), χ(CCCC), Q(CO)
663	677	658	679	649	χ(CCCO), χ(HCCC), χ(CCCC), γ(CCO), χ(HCCO), χ(*CO_2)
763	728	722	–	731	χ(CCCO), Q(CC), γ(CCO), χ(HCCO), χ(HNH), χ(CCN)
	757	750	759	–	χ(HCCO), γ(NCCO), χ(HCCH), Q(CC), χ(CCCC), χ(CCCO), Q(CO), γ(CCO), χ(HCCC), χ(OCCH), β(CCH), χ(HNH)
	782	–	–	797	Q(CC), χ(CCCO), γ(CCC), χ(CCCO), χ(NCCO), χ(HNH)
821	839	836	842	826	χ(CCCH), β(CCH), χ(HCCH), χ(OCCC), χ(HNH), χ(CCN)
864	877	868	874	904	χ(CCCH), Q(CO), χ(HNH)
919	916	918	905	915	χ(OCCH), χ(HCCO), χ(HCCH), Q(CC), Q(CO), χ(CCCO), β(HCC), γ(CCC), Q(OC), χ(HNH)
983	988	957	946	979	γ(OCC), Q(CO), Q(CC), Q(OC), χ(HCCH), γ(OCC), χ(HNH)
1080	1060	1042	997	1034	χ(CCCC), χ(OCCC), Q(CO), χ(HCCH), χ(CCCH), χ(CCCO), Q(CC), Q(CN)
	1061	1086	1073	–	χ(OCCC), χ(CCCC), Q(CO), χ(HCCH), Q(CC), β(CCH), χ(HCCO), Q(CN)

(continued)

Table 5 (continued)

$\nu_{experimental}$, cm^{-1}	$\nu_{DFT\ 6\text{-}31G}$, cm^{-1}	$\nu_{DFT\ 3\text{-}21G}$, cm^{-1}	$\nu_{HF\ 6\text{-}31G}$, cm^{-1}	$\nu_{HF\ 3\text{-}21G}$, cm^{-1}	Vibration modes
1122	1123	1106	1142	1112	Q(CO), Q(CC), Q(OC), χ(OCCH), Q(CN), χ(CN)
1147	1179	1182	1169	1133	Q(CO), Q(CC), Q(OC), Q(CN), χ(CN), χ(CCN)
1195	1195	–	–	1198	β(CCH)
1214	1229	1232	1226	–	β(OCCH)
1278	1281	1308	1289	1293	β(CCH), χ(HCCO), χ(HCCC), χ(OCCH), q(CO_2^-), q(C = O)
1326	1305	1334	1340	1304	β(HCCO), χ(HCCO), β(CCH), χ(CCCH), χ(NH_2)
1354	1348	1360	1344	1350	χ(HCCH), β(CCH), χ(OCCH), χ(HCCO), β(HCCO), Q(-CO_2^-)
1373	1375	1415	1390	1392	β(HCO), χ(NCCO), χ(HCCH), β(CCH), χ(HCCO), β(HCC), Q(CC), β(HCCO), β(CCH), χ(HCCO), β(CCH), Q(NH_2), χ(NH_2)
	1404	–	–	1421	
1433	1437	1445	1412	1455	Q(OC), Q(CC), β(CCH), χ(HCCH), χ(CCCH), Q(NH_2), χ(NH_2), α(HCH)
1491	1515	1466	1502	1489	α(HCH), β(OCCH), χ(HCCH), χ(HNH), χ(CCN)
1509	1519	1509	1513	1515	α(HCH), β(OCCH), χ(HCCC), χ(NH_2), Q(HNH)
1627	1685	1526	1643	1544	q(C = O), q(O = CO), Q(NH_2), χ(NH_2), χ(HOH)

(continued)

Table 5 (continued)

$\nu_{experimental}$, cm^{-1}	$\nu_{DFT\ 6\text{-}31G}$, cm^{-1}	$\nu_{DFT\ 3\text{-}21G}$, cm^{-1}	$\nu_{HF\ 6\text{-}31G}$, cm^{-1}	$\nu_{HF\ 3\text{-}21G}$, cm^{-1}	Vibration modes
1674	1700	1656	1683	1643	q(C = O), q(O = CO), χ(HOH)
1723	1718	1774	1704	1838	q(C = O), q(O = CO)
2006	–	–	–	–	χ(HOH), χ(N–H⁺…O)
2124–2182	–	–	–	–	χ(HOH), χ(NH_2)
2242–2372	–	–	–	–	q(*CO_2)
2474–2522	–	–	–	–	χ(NH_2)
2603/2626	–	–	–	–	χ(NH_2)
2723/2752	–	–	–	–	q(CH), χ(NH_2)
2903	3034	2901	2961	3176	q(CH)
2960	3060	2933	2961	3229	q(CH)
2995	3090	3035	2990	3261	q(CH)
3136 (Amide B)	3136	3054	3178	3291	q(HNH) q(NH_2^+), q(OH)
3392 (Amide A)	3482	3506	3473	3462	q(HNH), q(NH_2^+), q(OH)
–	3596	3560	3659	3790	q(HNH), q(NH_2^+), q(OH)

Note Conventional coordinate designations are used: stretching vibrations—q, Q, planar bending vibrations—γ, β—with the participation of one atom H, α = α(HCH), non-planar bending vibrations—χ, *CO_2—carbon dioxide

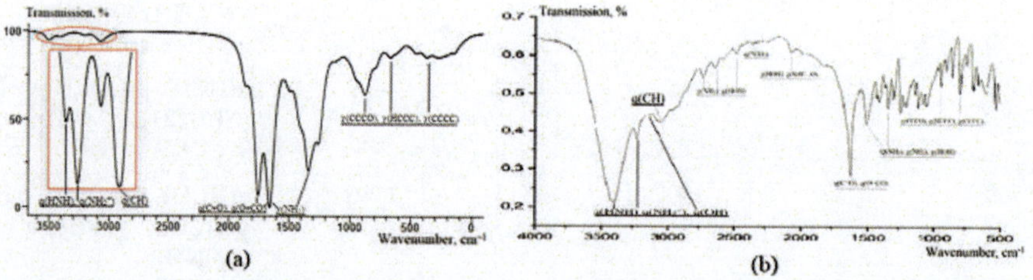

(a) (b)

Fig. 4 IR spectra of magnesium glutamate: **a** calculated (HF in 6-31G basis) model; **b** synthesized complex compound Mg^{2+}:Glu

The region of 877–916 cm^{-1}. In addition to skeletal vibrations, bending (twisting) vibrations of the NH_2 group are observed.

The vibration at 988 cm^{-1}. According to the literature data (Fischer and Evsel 1997), this vibration corresponds to the anionic form of the carboxyl group of an amino acid. This is related to wagging vibrations of the amino group and valence carboxyl groups, which raise the intensity of this band. Skeletal vibrations of C–C also make a large contribution to this band.

The region of 1060–1179 cm^{-1}. The vibration of the C–N bond is stretching and makes a significant contribution. There are bending vibrations of methylene and amino groups (wagging, twisting). In the region of 1179 cm^{-1}, there are stretching and bending vibrations of C–N, CCN.

The region of 1195–1281 cm^{-1}. The bending vibrations of the OCCH and CCH make a contribution. The strong peak of 1281 cm^{-1} is caused by symmetric stretching vibrations of C–O and C = O, as well as wagging vibrations of CH_2 groups.

The 1305 cm^{-1} band is caused by bending vibrations of the methylene and amino groups. Their vibrations are twisting.

The region of 1348–1375 cm^{-1} is characterized by stretching vibrations of the C(=O)O group, as well as by wagging vibrations of the CH_2 groups, as well as various deformations of the OCCH type.

In the region of 1404–1437 cm^{-1}, vibrations of all bonds and groups in the compound under consideration are manifested.

In the region of 1515–1519 cm^{-1}, bending scissoring vibrations of the CH_2 group are observed.

The vibration at 1685 cm^{-1} is partially contributed by bending scissoring vibrations of the amino group, asymmetric. A greater contribution is made by stretching vibrations C = O, as well as by asymmetric stretching vibrations of CO_2^-.

The region of 1700–1718 cm^{-1}. Stretching vibrations of C = O, as well as stretching vibrations of CO_2^-.

The region of 2006 cm^{-1}. The strong N–H$^+$…O interaction contributes to the 2006 cm^{-1} vibrations between the amino group of an amino acid and structural water, as well as bending vibrations of HOH.

In the region of 2124–2182 cm^{-1}, vibrations probably relate to the overtones of bending and libration modes of HOH, overlapping due to low intensity.

The region of 2242–2372 cm^{-1}. Vibrations here refer to asymmetric vibrations of CO_2.

Vibrations in the regions 2474–2522 cm^{-1}, 2603–2626 cm^{-1}, and 2723–2752 cm^{-1} belong to the overtones of bending vibrations of NH_2, their appearance on the IR spectrum may indicate some interaction with the amino group. Both the formation of a hydrogen bond and a Ca–N bond is possible.

Vibrations in the region of 2902–3090 cm^{-1} are attributed to symmetric and asymmetric stretching vibrations of C-H, CH_2.

The region of 3105–3136 cm^{-1}. Vibrations are more likely to be attributed to symmetric stretching vibrations of NH_2, NH_2^+.

The region of 3482–3596 cm^{-1}. Vibrations here are attributed to symmetric and asymmetric stretching vibrations of NH_2, NH_2^+, with OH stretching vibrations also contributing.

A strong shift toward high-frequency vibrations in the calculated spectrum, relative to the experimental data, can be explained by the formation of a hydrogen bond with an amino group in the synthesized compound. Such a bond is not taken into account when calculating the model of an isolated molecule, as well as the possible formation of a Ca–N bond by the donor–acceptor mechanism. As a result, Amide A and Amide B are fixed in a lower frequency region on the actual IR spectrum.

Similar vibrations in the IR spectrum can be observed for the magnesium glutamate molecule, since glutamic acid is also present in this compound. A comparison of the vibrations of the experimental and calculated IR spectra is given in Table 6. The calculated spectrum is shown in Fig. 4.

When comparing the data presented in Tables 5 and 6, which show the IR spectra for calcium and magnesium glutamate, respectively, we observe the similarity of these spectra in terms of functional groups, both calculated (HF, DFT) and obtained experimentally. It can be concluded that the structure of calcium glutamate and magnesium glutamate is similar, namely, the probable ratio of metal ion and AA comprises 1:1. This is evidenced by the similarity of the calculated and experimental spectra. The presence of structural water was also detected in the synthesized compounds.

4 Conclusion

Structural and dynamic models of calcium glutamate and magnesium glutamate were simulated using the Hartree–Fock (HF) and the density-functional theory (DFT) method (B3LYP) in the 6-31G and 3-21G basis sets. The geometric characteristics of isolated molecules were presented. The IR spectra of calcium glutamate and magnesium glutamate models were analyzed and decoded.

According to the conducted quantum chemical calculations of the IR spectra of calcium and magnesium glutamate compounds, these models adequately describe the structure of the synthesized complexes Ca^{2+}:Glu and Mg^{2+}:Glu. The observed differences include the presence of structural water and the probable presence of a covalent bond formed by the donor–acceptor mechanism of an unsettled pair of a nitrogen atom with metal ions. The observed shifts in the low-frequency range of the IR spectra of the synthesized compounds Ca^{2+}:Glu and Mg^{2+}:Glu, compared to those of the models, can be explained by the influence of intermolecular Coulomb and hydrogen interactions that occur in the formed complexes.

Table 6 Calculated and experimental IR vibration frequencies of the magnesium glutamate molecule

$\nu_{experimental}$, cm^{-1}	$\nu_{DFT\ 6\text{-}31G}$, cm^{-1}	$\nu_{DFT\ 3\text{-}21G}$, cm^{-1}	$\nu_{HF\ 6\text{-}31G}$, cm^{-1}	$\nu_{HF\ 3\text{-}21G}$, cm^{-1}	Vibration modes
616	615	612	594	624	γ(OCO), χ(OCCC), χ(OCCH), χ(CCCC), Q(CO)
670	669	641	630	668	χ(CCCO), χ(HCCC), χ(CCCC), γ(CCO), χ(HCCO), χ(*CO_2)
702	700	699	674	–	χ(CCCO), Q(CC), γ(CCO), χ(HCCO), χ(HNH), χ(CCN)
712	721	719	734	739	χ(HCCO), γ(NCCO), χ(HCCH), Q(CC), χ(CCCC), χ(CCCO), Q(CO), γ(CCO), χ(HCCC), χ(OCCH), β(CCH), χ(HNH)
758	748	748	790	792	Q(CC), χ(CCCO), γ(CCC), χ(CCCO), χ(NCCO), χ(HNH)
807	800	793	815	812	χ(CCCH), β(CCH), χ(HCCH), χ(OCCC), χ(HNH), χ(CCN)
866	863	858	859	848	χ(CCCH), Q(CO), χ(HNH)
912	913	917	927	915	χ(OCCH), χ(HCCO), χ(HCCH), Q(CC), Q(CO), χ(CCCO), β(HCC), γ(CCC), Q(OC), χ(HNH)
967	976	992	963	952	γ(OCC), Q(CO), Q(CC), Q(OC), χ(HCCH), γ(OCC), χ(HNH)
1055	1033	1042	1015	1018	χ(CCCC), χ(OCCC), Q(CO), χ(HCCH), χ(CCCH), χ(CCCO), Q(CC), Q(CN)

(continued)

Table 6 (continued)

$\nu_{experimental}$, cm^{-1}	$\nu_{DFT\ 6\text{-}31G}$, cm^{-1}	$\nu_{DFT\ 3\text{-}21G}$, cm^{-1}	$\nu_{HF\ 6\text{-}31G}$, cm^{-1}	$\nu_{HF\ 3\text{-}21G}$, cm^{-1}	Vibration modes
1077	1068	1080	1077	1060	χ(OCCC), χ(CCCC), Q(CO), χ(HCCH), Q(CC), β(CCH), χ(HCCO), Q(CN)
1127	1132	1157	1117	1111	Q(CO), Q(CC), Q(OC), χ(OCCH), Q(CN), χ(CN)
1150	1157	1200	1144	1123	Q(CO), Q(CC), Q(OC), Q(CN), χ(CN), χ(CCN)
1213	1205	1212	1250	1226	β(CCH)
1233	1242	1230	–	1282	β(OCCH)
1260	1265	1263	–	–	β(CCH), χ(HCCO), χ(HCCC), χ(OCCH), $q(CO_2^-)$, q(C = O)
1312	1318	1321	1302	1318	β(HCCO), χ(HCCO), β(CCH), χ(CCCH), $\chi(NH_2)$
1352	1352	1348	1344	1327	χ(HCCH), β(CCH), χ(OCCH), χ(HCCO), β(HCCO), $Q(\text{-}CO_2^-)$
1376	1378	1391	1394	1382	β(HCO), χ(NCCO), χ(HCCH), β(CCH), χ(HCCO), β(HCC), Q(CC), β(HCCO), β(CCH), χ(HCCO), β(CCH), $Q(NH_2)$, $\chi(NH_2)$
1421	1425	1410	1421	1428	
1436	1444	–	1454	1443	Q(OC), Q(CC), β(CCH), χ(HCCH), χ(CCCH), $Q(NH_2)$, $\chi(NH_2)$, α(HCH)
1514	1503	1518	1518	1500	α(HCH), β(OCCH), χ(HCCH), χ(HNH), χ(CCN)
1634	1630	1644	1651	1630	α(HCH), β(OCCH), χ(HCCC), $\chi(NH_2)$, Q(HNH)
1674	1682	1682	1666	1663	q(C = O), q(O = CO), $Q(NH_2)$, $\chi(NH_2)$, χ(HOH)
1716	1701	1701	1753	1687	q(C = O), q(O = CO), χ(HOH)

(continued)

Table 6 (continued)

$\nu_{experimental}$, cm^{-1}	$\nu_{DFT\ 6\text{-}31G}$, cm^{-1}	$\nu_{DFT\ 3\text{-}21G}$, cm^{-1}	$\nu_{HF\ 6\text{-}31G}$, cm^{-1}	$\nu_{HF\ 3\text{-}21G}$, cm^{-1}	Vibration modes
1832	1789	1854	1866	1854	q(C = O), q(O = CO)
1979	–	–	–	–	χ(HOH), χ(N–H$^+$…O)
2079–2256	–	–	–	–	χ(HOH), χ(NH_2)
2257–2363	–	–	–	–	q(*CO_2)
2486–2582	–	–	–	–	χ(NH_2)
2659	–	–	–	–	χ(NH_2)
2742	–	–	–	–	q(CH), χ(NH_2)
2903	–	–	–	–	q(CH)
2963	3008	2993	3023	3010	q(CH)
3069	3034	3022	3066	3129	q(CH)
3238					
(AmideB)	3258	3237	3212	3173	q(HNH) q(NH_2^+), q(OH)
3414 (AmideA)	3451	3406	3411	3382	q(HNH), q(NH_2^+), q(OH)
-	3688	3584	3502	3453	q(HNH), q(NH_2^+), q(OH)

Note Conventional coordinate designations are used: stretching vibrations—q, Q, planar bending vibrations—γ, β—with the participation of one atom H, α = α(HCH), non-planar bending vibrations—χ, *CO_2—carbon dioxide

Future studies should investigate the possibility of considering hydrogen bonds in the framework of the DFT method, as well as carry out quantum chemical calculations and analysis of low-stability complex compounds of calcium and magnesium with other amino acids.

Acknowledgements The Research was Carried Out Within the State Assignment of Ministry of Science and Higher Education of the Russian Federation, Theme No. 075-03-2023-149.

References

Babkov LM, Korolevich MS, Moiseikina OrA (2010) Calculation of the structure and IR spectrum of the methyl-β-D-glucopyranoside molecule by the density functional method. Journal. butt. spectrum. 77(2):179–187

Bespalov DV, Golovanova OA (2021) Synthesis of complex compounds of calcium(II) ions with amino acids. Butlerov Communications. 65(1):15–22

Butyrskaya EV, Nechaeva LS, Shaposhnik VA, Drozdova EI (2012) Assignment of bands in IR spectra of aqueous glycine solutions based on quantum chemical calculation. J Sorption and chromatographic processes. 12(4):501–511

Dobrynina NA (2007) Bio-inorganic chemistry. Moscow State University, Moscow Eremin VV (2007) Systematization of mathematical models of elastic types of Eremin V.V. (2007) Systematization of mathematical models of elastic types of water polarization. J Informatics science and control systems. 14(2):78–88

Fischer WF, Eysel HH (1997) Raman and FTIR spectroscopic study on water structural changes in aqueous solution of aminoacids and related compounds. J. Mol. Struc. 415(3):249–257

Golovanova OA (2007) Pathogenic minerals in the human body. OmSU, Omsk

Golovanova OA, Pyatanova PA, Rosseeva EV (2004) Analysis of the distribution patterns of the protein component of urinary stones. Reports of the Academy of Sciences. 395(5):1–3

Gordon MS, Schmidt MW (2005) Theory and Applications of Computational Chemistry: the first forty years. Elsevier, Amsterdam

Gromova OA (2007) Significance of the calcium deficiency in pediatrics and ways of its correction. Current Pediatrics. 6(2):82–87

Kohn W (2002) The electronic structure of matter: wave functions and density functionals. J Physics-Uspekhi. 172(2): 336–348

Kurochkin VYu (2011) Thermodynamics of the processes of complexation of calcium ions with amino acids in an aqueous solution. Dissertation, State University of Chemical Technology

Larsen MJ, Pearce EIF (2003) Saturation of human saliva with respect to calcium salts. J. Arch. of Oral Bio.48(4):317–322

Levchuk LV, Borodulina TV, Sannikova NE, Danilova IG (2017) Clinical significance of the content of free amino acids for the growth and development of children. Ural Medical Journal. 5:11–15

Mashina EV, Shanina SN (2019) Amino acid composition of gallstones and its connection with the mineral component J. Zapiski RMO. 148(4):95–109

Nakoskin AN, Vorontsov BS, Luneva SN, Vaganova LA (2012) Quantum-chemical modeling of calcium aminoacyl complexes and evaluation of the possibility of their use to compensate for calcium deficiency. J Modern problems of science and education. 3:3–6

Pearce EIF, Dong YM, Gao XJ (2002) Plaque minerals in the prediction of caries activity. J. Comm. Dent. and Oral Epidem. 30(1):61–69

Revko OP, Yaroshenko LA (2010) The physiological role and significance of magnesium in the therapy of internal diseases. J Bulletin of the Pancreatologists' Club. 2(7):62–66

Senni K, Foucault-Bertaud A, Godeau G (2003) Magnesium and connective tissue. Magnes Res. 16(1):70–74

Tomashevsky IA (2021) Thermodynamics and kinetics of formation of complex compounds of calcium (II) and magnesium (II) ions with amino acids. Dissertation, OmSU

Waheed EJ, Obaid SM, Ali-Abbas AAS (2019) Biological Activities of Amino Acid Derivatives and their Complexes a Review. Research Journal of Pharmaceutical, Biological and Chemical Sciences. 10 (2):1624–1641

Yudina NV, Torshin IYu, Gromova OA, Egorova EYu, Bykov AT (2016) Availability of potassium and magnesium ions is a fundamental condition for maintaining normal blood pressure. J Cardiology. 56(10):80-89

Biomineralization in Geosystems

Chronic Maxillary Sinusitis in Ancient Populations: X-Ray Computed Microtomography Data

Alisa V. Zubova, Alexander M. Kulkov, Marianna A. Kulkova, Vyacheslav G. Moiseyev, Maya T. Kashuba, Nikolay N. Potrakhov, Victor B. Bessonov, and Yulia V. Kozhukhovskaya

Abstract A case of chronic maxillary sinusitis (CMS) found in a burial site of a Late Bronze Age settlement in Bai-Kiyat I was studied by X-ray computed microtomography. The aim was to diagnose CMS, analyze the factors in the development of the disease, study its etiopathogenesis, and determine the potential contribution of odontogenic pathology to the overall incidence of CMS in ancient populations. Pathological changes in the left maxillary sinus characteristic of chronic maxillary sinusitis were recorded on a 3D model created using X-ray computed microtomography. The presence of chronic periodontal disease and apical periodontitis of the maxillary molars on the left, which had triggered the development of CMS, was established. The obtained and previously published data confirm the significant contribution of odontogenic pathology to CMS development. The dietary and hygienic behavioral patterns are clearly underestimated in the analysis of the incidence and prevalence of chronic inflammatory diseases of the maxillofacial region among ancient populations. X-ray computed tomography removes most of the limitations in bioarchaeological studies, allowing reliable diagnosis of odontogenic CMS forms.

A. V. Zubova (✉) · V. G. Moiseyev
Peter the Great Museum of Anthropology and Ethnography (Kunstkamera), Russian Academy of Sciences, Saint Petersburg, Russia
e-mail: zubova_al@mail.ru

A. M. Kulkov
Saint Petersburg State University, Saint Petersburg, Russia

M. A. Kulkova
Russian State Pedagogical University Named After A.I. Herzen, Saint Petersburg, Russia

M. T. Kashuba
Institute of the History of Material Culture, Saint Petersburg, Russia

N. N. Potrakhov · V. B. Bessonov
Saint Petersburg Electrotechnical University, Saint Petersburg, Russia

Y. V. Kozhukhovskaya
Crimean Federal University, Simpheropol, Crimea, Russia

O. V. Frank-Kamenetskaya et al. (eds.), *Biogenic—Abiogenic Interactions in Natural and Anthropogenic Systems 2022*, Springer Proceedings in Earth and Environmental Sciences, https://doi.org/10.1007/978-3-031-40470-2_8

Keywords X-ray computed microtomography · Chronic odontogenic maxillary sinusitis · Periodontal diseases · Bai-Kiyat I · Late Bronze Age

1 Introduction

Research into the epidemiology of infectious diseases is of particular importance for the analysis of adaptive models used by ancient populations and the reconstruction of critical biosocial factors in their survival. Infectious diseases that cause high mortality among populations (plague, cholera, etc.) conventionally attract much research attention. However, the spread of chronic diseases that are not characterized by increased mortality is of greater importance for elucidating the adaptation dynamics of human populations. Such diseases not only affect the biological profile of the population, but also require social mechanisms to provide the necessary treatment and care for patients, and, as a result, to mitigate disability-related consequences.

Such diseases include chronic sinusitis—a chronic inflammatory process in one or more paranasal sinuses. Chronic maxillary sinusitis (CMS) is a persistent long-term inflammation of the maxillary sinus mucosa of an infectious, allergic, or traumatic nature (Slavin et al. 2005). At early development stages, CMS is rarely accompanied by severe symptoms that pose a direct threat to human life. CMS manifestations can cause noticeable physical discomfort and decreased performance due to difficulty in nasal breathing, pain in the projection of the paranasal sinuses, general weakness, and such exacerbations, as serous purulent discharge from the nose, increased body temperature, etc. (Arefyeva et al. 2014; Sipkin et al. 2013). However, the spread of inflammation deep into the intracranial structures can lead to bone damage and the development of osteomyelitis. Another common complication of sinusitis is meningitis, a brain abscess.

In the modern population, sinusitis affects children and adults in 5.0% and 5.0–15.0% of cases, respectively. People aged 45–75 years are most susceptible to the disease. Epidemiological studies carried out in more than 30 countries indicate a three-fold increase in the incidence of sinusitis over the past decade (Yaremenko et al. 2015). In the USA, CMS is present in 15.5% of adults. At the same time, odontogenic chronic maxillary sinusitis affects 15.0% of the inhabitants of the Russian Federation and 14.0–20.0% of the world's population (Surin and Pohodenko-Chudakova 2018).

Unlike many other human diseases, CMS is relatively easy to diagnose on ancient materials presented by skeletal remains. Since the mucous membrane of the maxillary sinus is closely connected with the periosteum, representing a whole unit, inflammatory processes may quickly pass to the bone tissue. This results in the development of chronic inflammation of the bone walls of the sinus (osteitis), with the bone tissue becoming heterogeneous and containing foci of osteosclerosis, osteoporosis, and remodeled bone tissue. All these signs can be either detected by visual examination of the maxillary sinus or reconstructed by X-ray computed tomography (Sundman and Kjellström 2013; Boocock et al. 1995; Biedlingmaier et al. 1996; Erdogan et al.

2016; Mafee et al. 2006; Georgalas et al. 2010; Momeni et al. 2007; Snidvongs et al. 2014).

Previous studies discovered a link between the incidence of CMS in different population groups and their living conditions (urban or rural areas), the degree of air pollution, the social status of the buried, the prevalence of dental pathologies, and climatic and geographical factors (Roberts 2007; Zubova et al. 2020a; Zubova et al. 2022b; Lewis et al. 1995; Panhuysen et al. 1997). In bioarchaeological practice, CMS is commonly referred to as a poverty-related disease. Thus, CMS occurred several times more often in the craniological series characterizing the population of the lower strata of European societies than in representatives of the upper classes (Schultz and Schmidt-Schultz 2014). However, due to a lack of data on the epidemiology of CMS in ancient and historical communities, no consensus has so far been reached concerning the factors affecting its incidence.

A statistical correlation showing an increase in the incidence of sinusitis from south to north was found only between CMS and climatic indicators (Zubova et al. 2022b). However, given the general statistical reliability of this correlation, the specific coefficients calculated for the globe as a whole and its individual regions are not high. Hence, the relationship between CMS and particular climate characteristics may differ in different areas. One of the reasons for the observed instability may be the presence of rhinogenic and/or odontogenic etiology in the examined series of CMS. Rhinogenic sinusitis develops with respiratory infections and certain types of allergies. Odontogenic forms develop when microorganisms of the oral cavity penetrate into the sinus cavity, through channels formed as a result of bone resorption of alveolar cells during a long course of chronic periodontal diseases, or osteomyelitis (Buskina and Gerber 2000; Abrahams and Glassberg 1996).

The spread of odontogenic inflammations is theoretically less related to climate than rhinogenic ones. In the former case, the composition of the diet and the peculiarities of oral hygiene, i.e., social factors, are of fundamental importance; in the latter case, the frequency of hypothermia and the composition of the inhaled air gain more significance. Accordingly, when studying the factors affecting the epidemiology of CMS in ancient groups, it is necessary to separate odontogenic and non-odontogenic forms of the disease and compare their incidence rates differentially. This causes some methodological difficulties, since, until recently, the diagnosis of CMS has been performed visually or using an endoscope, based on the presence of porotic changes in the maxillary sinus cavity and newly formed bone tissue in the form of spicules (Sundman and Kjellström 2013). Identification of odontogenic infection foci with this approach is possible only provided that CMS signs are simultaneously present inside the sinus and the oroantral fistula. Non-linear communication channels between the oral cavity and the sinus cavity, formed in chronic apical periodontitis, cannot be detected by visual examination. In addition, infection sources are impossible to determine when bone tissue healing begins after tooth extraction or loss.

X-ray computed microtomography presents a feasible alternative to visual examination. Although this method is yet to receive widespread acceptance in the study of CMS among ancient populations, a number of our investigations have confirmed its advantages (Zubova et al. 2020a, 2022a). Among them are the possibility of a reliable diagnosis of odontogenic CMS and its differentiation from other pathologies. Accordingly, further tomographic studies in this area are required to elucidate the factors affecting the incidence of sinusite. Our work extends these studies by analyzing a new case of odontogenic CMS detected in an individual of the Late Bronze Age buried at the settlement of Bai-Kiyat I.

2 Materials and Methods

2.1 *Description of the Research Object*

The skull of a male individual of mature age (40–60 years old), discovered during excavations in burial *2* of the Bai-Kiyat I settlement (45.6778, 32.9985) in 1990, was analyzed. The state of skull preservation was medium. The bones of the base and the bottom of the alveoli of maxillary molars on the right side were posthumously destroyed; the left zygomatic arch was damaged. Multiple postmortem losses in the dental system were also present. On the maxilla, the central right incisor, the right canine, the right first molar, the right second molar, the left second molar, the left third molar were preserved; on the mandible, the central incisors, the right canine, the first and second premolars, the first and second molars of both sides were present. The rest of the teeth were lost (Fig. 1a).

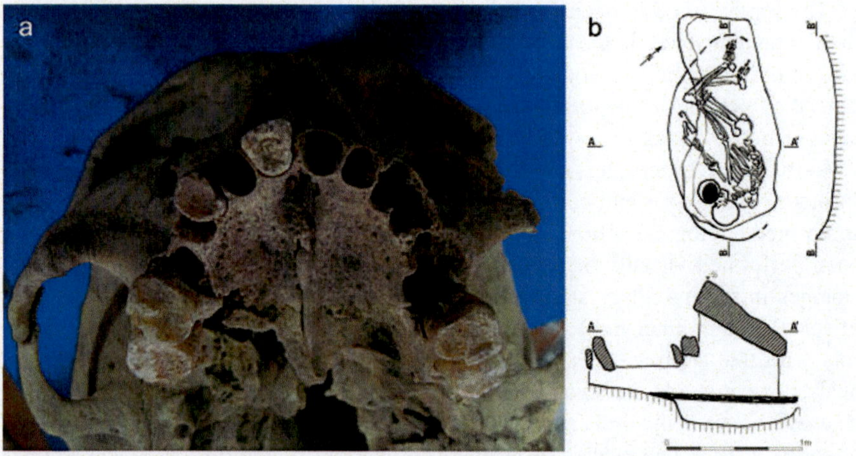

Fig. 1 **a** The skull under study, its maxilla with preserved teeth and lesions. **b** Burial *2* in the settlement of Bai-Kiyat I

2.2 Research Methods

The skull was subjected to X-ray computed microtomography. Scanning was performed on an MSCT-04 tomograph, according to the following protocol: the X-ray tube voltage of 140 kV, the current of 50 μA, no filter, the slice thickness of 0.1 mm. Data reconstruction was performed using an original software package. Postprocessing analysis was performed using CTAn and CTVox (Bruker-microCT). The upper left second molar was scanned by a Skyscan-1172 X-ray microtomography scanner (U = 82 kV, I = 120 μA, Al 0.5 filter, rotation = 0.25°, averaging by 6 fames, image pixel size = 6.92 μm). Reconstruction and post reconstruction processing was performed by the Bruker software package (NRecon, CTAn, CTVox).

The obtained CT scans showed several pathological conditions. First, signs of osteitis in the maxillary sinus walls were recorded. According to the standard approach (Georgalas et al. 2010), we used the maximum thickness of one of the sinus walls of 3 mm and more as the threshold for diagnosing osteitis. The walls were measured at osteosclerotic foci of both left and right maxillary sinuses, avoiding corners of the sinuses.

Sources of odontogenic infections were detected using a number of markers, including the foci of chronic periodontitis and osteomyelitis and the presence of channels connecting the sinus with the alveoli.

Chemical analyses of the bone tissue were carried out by pXRF-technique with an INNOV-X Omega portative scanner.

2.3 Description of the Archaeological Site

The site is located on a high steep bank (cliff height of 4–5 m) and is bounded by the seashore from the north. Across the investigated area of 2100 m^2, about 14 rooms with stone-faced walls for various purposes were excavated. Inside were hearths, household items, and ash pits. The settlement was traced to the end of the Sabatinovka culture and the Belozerka culture, dating back to the end of the 13th–10th centuries BC (Kolotukhin 2003).

Burial *2* (Fig. 1b) was performed in a pit of an irregular oval shape (dimensions 1.50 × 0.95 m, depth 0.15–0.20 m), oriented along the northeast–southwest line. On the southern side, the remains of the overlap of flat limestone slabs were preserved; another large slab was located above the burial. The buried man was laid in a weakly crouched form on his left side, his head oriented to the southeast, the front part—to the west. The left arm was stretched out in front of the trunk, the right hand was placed along the trunk, the left hand was close to the knees. The legs were slightly bent, the right tibia rested on the left tibia, the feet were laid next to each other. In front of the face of the buried, to the west, there was a hand-formed jar-shaped vessel.

3 Results and Discussion

3.1 Results of X-ray Computed Microtomography

The data of X-ray computed microtomography for the skull from the settlement of Bai-Kiyat I showed the absence of frontal sinuses and the presence of chronic inflammation signs in the left maxillary sinus. The posterior wall of the sinus was thickened to 4.79 mm (marked by arrow, Fig. 2a), which corresponds to severe CMS, whereas the maximum thickness of the posterior wall on the right side was only 2.40 mm. At the bottom and walls of the sinus, porosis and small spicules were visually observed, whose etiology and distribution area could not be visually determined due to postmortem damage.

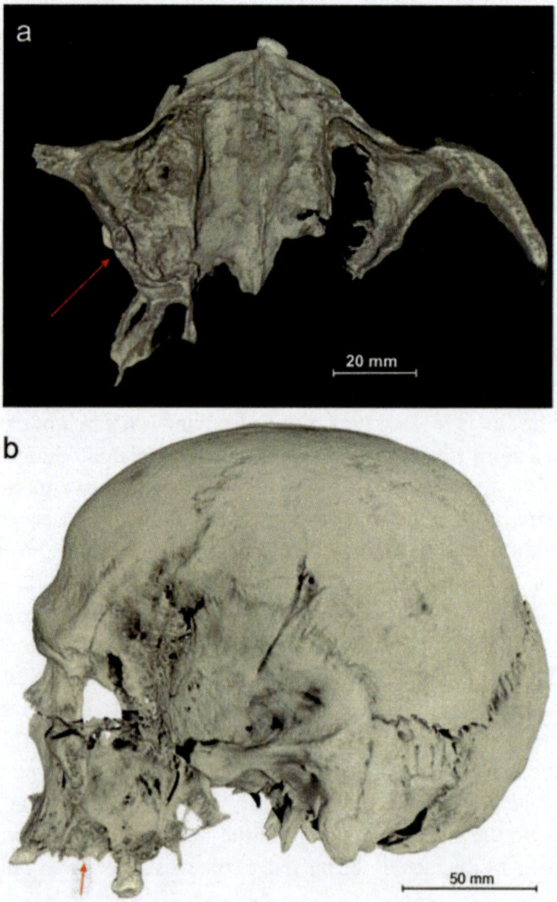

Fig. 2 **a** Micro CT scans of osteosclerosis foci and thickening of the posterior wall of the left maxillary sinus. **b** Micro CT scans of lesions in the area of the left upper molars

The anatomical structure of the maxillary sinuses and dental system of the buried person, accompanied by a number of pathological changes in the dental system, indicates the odontogenic nature of CMS. The bottom of the sinuses is located at the level of the lower edge of the nasal opening, with the roots of the upper premolars and molars identified in the vicinity. Postmortem destruction does not allow the thickness of the bone plate separating the sinus cavity from the alveolar cells to be measured accurately; however, its size of not less than 1.0 mm comprises a predisposing factor for the formation of infection channels from the dental system into the maxillary sinus.

On both jaws, a pronounced resorption of the alveolar margin was observed, indicating the development of chronic generalized periodontal disease of medium degree. The main causes of this disease include genetic predisposition, somatic diseases, and poor oral hygiene. The latter leads to the accumulation of pathogenic microorganisms that destroy the connective tissue and cause bone loss (Clarke and Carey 1985). Due to bone resorption, formation of periodontal pockets, and poor oral hygiene, it obviously led to exacerbations and pain. The individual may have tried to improve oral hygiene by using toothpicks. On the adjacent walls of the lower left first molar and the left second premolar, grooves from toothpicks (3.5 mm long on the premolar and 7.3 mm on the molar) were discovered. A similar groove with a length of 6.6 mm was found on the lower right first molar. The upper molars showed similar signs of damage.

On the preserved teeth of the maxilla and mandible, non-carious lesions of the teeth, having occurred after teething, in the form of increased tooth attrition and wedge-shaped defects were observed (Fig. 3). Increased attrition was typical of all preserved teeth. On the chewing surface of the teeth, the erased enamel led to dentin exposure. A possible cause of wedge-shaped defects was excessive and aggressive use of toothpicks against the background of chronic periodontitis. The increased tooth attrition may have resulted from coarse, thermally and mechanically unprocessed, food (Gerasimova et al. 2012; Pikhur et al. 2020).

The immediate trigger for the development of CMS was most likely to be chronic apical periodontitis of the upper left first and second molars, the signs of which are present on a CT scan (arrow) (Fig. 2b).

In the area of the maxillary bone located under the zygomatic process, an extensive defect of the external wall of the alveolar arch measuring 1.2 × 0.6 cm was discovered. It extends to two teeth—the first and second left molar of the maxilla. The first molar was lost, the second was present. The edge of the defect was sclerosed, and the area of the first molar also showed sclerotic changes in the lingual wall of the alveoli with the formation of a spongy tissue at its bottom, which indicates that the tooth had been lost during life. In modern dental practice, development of apical inflammation alike of Bai-Kiat I would lead to surgical extraction of the tooth. In some ancient samples, cases of such operations also are known (Zubova et al. 2020b; Zubova et al. in press). However, to confirm such treatment was one in this case, a special traceological analysis is required, which has not been performed for the Bai-Kiyat I skull. Thus, there is no positive evidence to assert that the molar was deliberately removed.

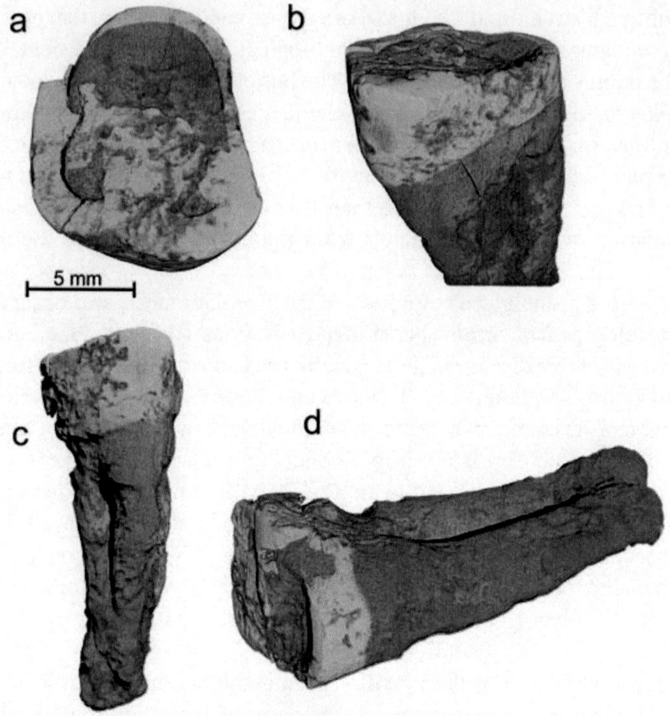

Fig. 3 Micro CT scans of the upper left second molar: **a**, **b**—wear facets on the chewing surface of the tooth; **c**, **d**—wedge-shaped defect in the cervical region of the tooth (root form)

There are no signs of healing in the area near the second molar, which indicates that which indicates that recovery was not completed at the time of the individual's death. At the bottom of the alveoli of both molars, multiple holes are present. Some of them appeared due to bone destruction becouse of periodontitis progression. They became channels for infection to enter the sinus cavity. Thus, the analysis of the obtained X-ray computed microtomography data confirmed the penetration of an odontogenic infection into the cavity of the left maxillary sinus, demonstrating a case of odontogenic CMS.

3.2 Discussion of Maxillary Sinusitis Etiopathogenesis

Periapical inflammatory processes in the area of the roots of the teeth adjacent to the maxillary sinus comprise the main factor in the development of odontogenic lesions of the maxillary sinus.

According to literature data (Yaremenko et al. 2015; Koshel 2017), in 10.0% of cases, an inflammatory process in the maxillary sinus can develop in the presence of a focus of chronic odontogenic infection localized near the sinus.

It should be noted that, in our previous studies into the etiology of CMS using X-ray computed microtomography, odontogenic pathology in all cases occupied the leading position. According to the data published on the Pukara de Tilcara fortress in Northern Argentina, CMS was recorded in more than 20.0% of individuals (Zubova et al. 2020a). In the total number of detected cases, odontogenic forms of maxillary sinusitis caused by chronic periodontitis of maxillary teeth were diagnosed in 80.0%. A fragment of the maxillary sinus floor found at a Neanderthal site of the Middle Paleolithic period in the Chagyrskaya Cave also demonstrated the presence of chronic periodontitis of the maxillary molar, which caused the development of CMS and subsequent infection in the sinus cavity (Zubova et al. 2022a). These findings, along with the present data obtained for the skull from Bai-Kiyat I, confirm the hypothesis that the state of the dental system significantly affected the occurrence and development of CMS in ancient people.

3.3 *Results of pXRF Analysis*

The development of many diseases of the dental system, including those predisposing to CMS, is associated with the diet of a particular individual or population. Data on the chemical composition of the bone for the skull from Bay-Kiyat I were obtained, which allow some assumptions about the diet of the buried person (Table 1). The indicator lg (Sr/Ca) = −0.469 points to the inclusion of a large amount of plant food in the individual's diet (Magee et al. 1994; Fabig and Herrmann 2002). At the same time, the Zn content (109 ppm) and the Zn/Ca ratio ($44 * 10^{-4}$) characterize the inclusion of meat food in the diet as well (Benfer 1995).

The index of diagenetic bone transformations (Ca/P), which characterizes changes in the chemical composition of the bone tissue under the influence of environmental conditions, equals 2.95. This reflects the insignificant contribution of diagenetic changes and the validity of the data obtained.

Table 1 Chemical composition of the skull bone from burial *2* in the Bai-Kiyat I site (pXRF analysis)

Burial *2* in the Bai-Kiyat I site	P, wt %	Ca, wt %	Zn, ppm	Sr, ppm	lg (Sr/Ca)	Zn/Ca	Ca/P
n = 3	8.3	24.5	109	851	−4.469	0.00044	2.95

3.4 *Discussion of the Data Obtained*

The data of X-ray computed microtomography on the skull from the settlement of Bai-Kiyat I suggest that the individual from burial *2* suffered from otolaryngological and dental diseases, the development of which had been influenced by a number of factors. The inflammatory process, which probably developed as a result of chronic periodontitis of the left first and second molars of the maxilla, contributed to the spread of odontogenic infection in the paranasal sinuses, causing odontogenic CMS of the left maxillary sinus. The anatomical structure of the dental system and maxillary sinus were the factors predisposing to the development of CMS.

The probable dominance of plant food in the individual's diet significantly affected his daily life and general health. In order to prevent exacerbations of chronic generalized periodontal disease, the individual tried to improve oral hygiene using toothpicks. However, the frequent and aggressive use of toothpicks in the area of the premolars and molars of the maxilla and mandible led to mechanical trauma to the cervical region of the hard teeth tissues followed by the formation of non-carious lesions in the form of wedge-shaped dental defects. The consumption of coarse unprocessed food increased tooth abrasion, which is clearly manifested on the preserved teeth of the individual.

4 Conclusion

The data obtained by our and previous studies confirm that CMS incidence in ancient populations should be assessed not only based on climatic and geographical contexts, or the social status of the studied community, but also taking the dietary and hygienic behavioral patterns into account. These factors are clearly underestimated in the analysis of the incidence and prevalence of chronic inflammatory diseases of the maxillofacial region among ancient populations. In order to assess the contribution of climatic and environmental (degree of air pollution) factors to the development of CMS, it is necessary to exclude cases of odontogenic CMS from the analysis. This is possible only with an accurate diagnosis of diseases, which can be achieved using X-ray computed microtomography.

Acknowledgements The work was supported by the grant of the Russian National Science Foundation No. 22-18-00065 "Cultural and historical processes and paleocreation in the Late Bronze—Early Iron Age of the Northwestern Black Sea region: interdisciplinary aproach" and MAE RAS research project "The Old and New Worlds: the formation and development of ancient societies and populations". X-ray computed microtomography for the skull under study was performed at the Department of Electronic Devices of the Saint Petersburg Electrotechnical University. The scanning of the upper left second molar and the postprocessing analysis was performed at the Center for X-Ray Diffraction Studies of Saint Petersburg State University. We would like to express our very great appreciation to Prof. O.L. Pikhur for her valuable suggestions and help during preparation of the manuscript.

References

Abrahams JJ, Glassberg RM (1996) Dental disease: A frequently unrecognized cause of maxillary sinus abnormalities. Am J Roentgenol. 166: 1219–1223

Arefyeva NA, Vishnyakov VV, Ivanchenko OA, Karpishchenko SA, Kiselev AB, Kozlov VS, Kozlov RS, Kosyakov SY, Kochetkov PA, Lopatin AS, Nakatis YA, Bravagin IV, Piskunov GZ, Polyakov DP, Turovsky AB (2014) Chronic rhinosinusitis: pathogenesis, diagnosis and principles of treatment (clinical recommendations). Practical Medicine, Moscow (in Russian)

Benfer R (1995) Age of Pregnancies and Age of Weaning from Prehistoric Bone: Are Sr and Zn Indicators? Am. J. Physic. 20: 89–90

Biedlingmaier JF, Whelan P, Zoarski G, Rothman M (1996) Histopathology and CT analysis of partially resected middle turbinates. Laryngoscope. 106: 102–104

Boocock P, Roberts CA, Manchester K (1995) Maxillary sinusitis in Medieval Chichester, England. Am J Phys Anthropol. 98: 483–495

Buskina AB, Gerber VX (2000) On the issue of clinical classification of chronic odontogenic sinusitis. Vestn otorhinolaryngology. 2: 20–22 (in Russian)

Clarke N, Carey S (1985) The aetiology of chronic periodontal disease: an alternative perspective. J Am Dent Assoc. 110: 689–691

Erdogan E, Fidan V, Giritli E (2016) Radiologic imaging in chronic sinusitis. Different Aspects of Rhinosinusitis. 16: 05

Fabig A, Herrmann B (2002) Trace elements in buried human bones: intra – population varability of Sr/Ca and Ba/Ca ratios – diet or diagenesis? Naturwissenschaften. 89(3): 115–119

Georgalas C, Videler W, Freling N, Fokkens W (2010) Global Osteitis Scoring Scale and chronic rhinosinusitis: a marker of revision surgery. Clin Otolaryngol. 35: 455–461

Gerasimova LP, Astakhova MI, Chemikosova TS, Shaidullina HM, Shamsiev MR (2012) Modern aspects of morphology, clinic and treatment of non-carious dental lesions. BSMU, Ufa (in Russian)

Kolotukhin VA (2003) Pozdnii bronzovyi vek Kryma. Stilos Publ, Kiev

Koshel IV (2017) Odontogenic maxillary sinusitis and their pathogenetic treatment. Dissertation, Stavropol (in Russian)

Lewis ME, Roberts CA, Manchester K (1995) Comparative study of the prevalence of maxillary sinusitis in Later Medieval urban and rural populations in Northern England. Am J Phys Anthropol. 98: 497–506

Mafee MF, Tran BH, Chapa AR (2006) Imaging of rhinosinusitis and its complications: plain film, CT, and MRI. Clin Rev Allergy Immunol. 30: 165–186

Magee BJ, Sheridan SG, Van Gerven D, Greene DL (1994) Elemental analysis of Mesolithic bone from Wadi Halfa: A diagenetic and dietary investigation of permineralized human remains. Amer Journal of Physical Anthropology. 18:134

Momeni AK, Roberts CC, Chew FS (2007) Imaging of chronic and exotic sinonasal disease: review. Am J Roentgenol. 189: 35–45

Panhuysen R, Coenen V, Bruintjes T (1997) Chronic maxillary sinusitis in Medieval Maastricht, the Netherlands. Intern J Osteoarchaeol. 7: 610–614

Pikhur OL, Iordanishvili AK, Cherny DA, Tishkov DS (2020) Diagnosis and treatment of wedge-shaped dental defects. Human, Saint Petersburg (in Russian)

Roberts CA (2007) A bioarcheological study of maxillary sinusitis. Am J Phys Anthropol. 133: 792–807

Schultz M, Schmidt-Schultz TH (2014) Paleopathology: Vestiges of Pathological Conditions in Fossil Human Bone. Handbook of Paleoanthropology. 1: 969–981

Sipkin AM, Nikitin AA, Lapshin VP, Nikitin DA, Chukumov RM, Kryazhinova IA (2013) Maxillary sinusitis: a modern view on diagnosis, treatment and rehabilitation. Almanac of Clinical Medicine. 28: 82–87 (in Russian)

Slavin RG, Spector SL, Bernstein IL (2005) The diagnosis and management of sinusitis: A practice parameter update. The Journal of Allergy and Clinical Immunology. 116: 13–47 (in Russian)

Snidvongs K, Earls P, Dalgorf D, Sacks R, Pratt E, Harvey RJ (2014) Osteitis is a misnomer: a histopathology study in primary chronic rhinosinusitis. Intern Forum Allergy Rhinol. 4: 390–396

Sundman EA, Kjellström A (2013) Chronic maxillary sinusitis in Medieval Sigtuna, Sweden: A study of sinus health and effects on bone preservation. Intern J Osteoarchaeol. 4: 447–458

Surin AV, Pohodenko-Chudakova IO (2018) The frequency and the main etiological factors of chronic odontogenic sinusitis of the maxillary sinus at the present stage. International reviews: Clinical practice and health. 4: 54–65

Yaremenko AI, Matina VN, Suslov DN, Lysenko AV (2015) Chronic odontogenic maxillary sinusitis: the current state of the problem. International Journal of Applied and Fundamental Research. 10: 834–837

Zubova AV, Ananyeva NI, Moiseyev VG, Stulov IK, Dmitrenko LM, Obodovskiy AV, Potrakhov NN, Kulkov AM, Andreev EV (2020a) The Use of Computed Tomography for the Study of Chronic Maxillary Sinusitis: Based on Crania from the Pucará De Tilcara Fortress, Argentina. Archeology, Ethnology and anthropology of Eurasia. 3(48): 143–153

Zubova AV, Pikhur OL., Obodovskiy AV, Malyutina AA, Dmitrenko LM., Chugunova KS, Pozdnyakov DV, Bessonov VB. (2020b) A Case of Surgical Extraction of the Lower Third Molarsin a Cranial Series from the Pucará de Tilcara Fortress(Jujuy Province, Argentina). Archeology, Ethnology and anthropology of Eurasia. 2(48): 149–156

Zubova AV, Kulkov AM, Pikhur OL, Moiseyev VG, Kolobova KA, Markin SV (2022a) A case of chronic maxillary sinusitis in a Late Neanderthal Population of the Altai Mountains. Archeology, Ethnology and anthropology of Eurasia. 3(50): 131–139

Zubova AV, Moiseyev VG, Ananyeva NI, Stulov IK, Andreev EV (2022b) Chronic maxillary sinusitis recorded in archaeological series: geographical distribution and predisposing factors. Archeology, Ethnology and anthropology of Eurasia. 1(50): 147–157

Zubova AV, Pikhur OL, Moiseev VG, Malyutina AA, Kalmina OA. (In Press) A case of surgical treatment of the lower molars in the Mesolithic series from the Yuzhny Oleniy Ostrov cemetery (Karelia, Russia). Archeology, Ethnology and anthropology of Eurasia

Magnetic Properties and Composition of Inclusions in Foraminifera Shells at the Mid-Atlantic Ridge

Elena Sergienko, Svetlana Janson, Petr Kharitonskii, Kamil Gareev, Stepan Ilyin, Yaroslav Anoshin, and Andrey Ralin

Abstract Iron-containing magnetic concretions of planktonic foraminifera of sandy and aleurite fractions of sediments obtained from hydrothermal fields of the Mid-Atlantic Ridge were studied. The phase and elemental composition, morphology, and magnetic properties of carbonate foraminifera shells were analyzed. The hysteresis characteristics and microscopy data support the assumption that the magnetic phase is represented by iron oxide similar in composition to non-stoichiometric magnetite with admixtures of manganese and magnesium. The samples were magnetic granulometrically analyzed according to the model of interacting ferrimagnetic particles. The determinations of superparamagnetic particles on the basis of theoretical modeling are in good agreement with the experimental data obtained using frequency-dependent susceptibility and granulometry methods.

Keywords Foraminifera · Biocomposites · Magnetite · Magnetic granulometry · Frequency-dependent susceptibility · Superparamagnetic particles · Magnetic states · Magnetostatic interaction

E. Sergienko (✉) · S. Janson · Y. Anoshin
Saint Petersburg University, Saint Petersburg, Russia
e-mail: e.sergienko@spbu.ru

P. Kharitonskii · K. Gareev · S. Ilyin
Saint Petersburg Electrotechnical University "LETI", Saint Petersburg, Russia

A. Ralin
Far Eastern Federal University, Vladivostok, Russia

P. Kharitonskii
Ioffe Institute, Russian Academy of Sciences, Saint Petersburg, Russia

O. V. Frank-Kamenetskaya et al. (eds.), *Biogenic—Abiogenic Interactions in Natural and Anthropogenic Systems 2022*, Springer Proceedings in Earth and Environmental Sciences, https://doi.org/10.1007/978-3-031-40470-2_9

1 Introduction

Studies of biogenic composite materials based on a calcite matrix (Mishra et al. 2020; Miyauchi et al. 2016; Gupta et al. 2021; Smirnov et al. 2010) are a currently relevant task due to the number of advantages offered by biogenic frameworks over their artificially derived equivalents, including better mechanical properties, biodegradability, biocompatibility, controllability of properties, etc. (Shi et al. 2022; Magnabosco et al. 2019). The most promising ways for obtaining such composites are considered to be methods based on the phenomenon of self-assembly (Magnabosco et al. 2019). Clam shells (Miyauchi et al. 2016) and foraminifera (Magnabosco et al. 2019) are considered as forming a biogenic framework. The term "foraminifera" refers to protist eukaryotic organisms characterized by membranous, agglutinated, or calcareous shells (Pawlowski and Holzmann 2002). Comprising a major source of biogenic calcium carbonate in oceans, these organisms are estimated to contribute 25–50% of the total amount (Langer 2008; Jacob et al. 2017).

Foraminifera shells comprise biogenic structures based on porous microparticles of calcium carbonate. Such shells preserved in marine sediments reflect changes in the composition of microelements and isotopes in the environment over a long period of time. The formation of foraminifera shells proceeds through complex mineralization pathways involving metastable intermediate phases that gradually evolve into the final shell mineral (Jacob et al. 2017). The composition, species diversity, and spatial distribution of benthic foraminifera communities are determined by environmental parameters (Armynot du Châtelet et al. 2013; Ben-Eliahu et al. 2020).

Such porous microparticles of calcium carbonate can be used as a basis for creating composite materials obtained by introducing various fillers into a porous structure, similar to the phenomenon of replacement of calcium atoms in foraminifera shells by heavy metals (Youssef 2015; Titelboim et al. 2021). The main area of application of such biocomposites is medicine, namely targeted drug delivery (Titelboim et al. 2021) and tissue engineering (Dorozhkin 2011). Another option of creating composite materials based on the mineral skeleton of foraminifera is coating the shell surface with nanoparticles of various compositions. Three-dimensional hierarchical structures based on them can be used for water purification by catalytic oxidation (Diab et al. 2018). Calcium carbonate microparticles, which can be controlled by an external magnetic field, are of particular interest for researchers. Magnetic particles are formed in foraminifera during their mineral replacement processes (Pawlowski and Majewski 2011). Magnetic structures on foraminifera skeletons may also appear during the formation of iron-containing biogenic concretions on the surface of mineral skeletons (Yang et al. 2022). However, to date, there are few works specifically aimed at investigating magnetic inclusions in foraminifera.

Magnetometric methods are often used to effectively and informatively reflect the structural and phase features of iron-containing minerals, their genesis, history, and conditions of existence. Thus, the purpose of this work was to perform an experimental and theoretical investigation of the magnetic properties, as well as the phase-

and elemental composition, of iron-containing inclusions in foraminifera shells of the Mid-Atlantic Ridge (MAR).

2 Materials and Methods

2.1 Sampling Location and Geochemical Setting

Samples of Holocene and Upper Pleistocene sediments containing microbiota (mainly benthic and planktonic foraminifera and coccoliths) were collected during studies carried out by the Polar Marine Geosurvey Expedition on the research vessel *Professor Logachev* in the Russian exploration region of MAR (Fig. 1).

This region is characterized by the ubiquitous distribution of relic and active hydrothermal structures. Under the influence of hydrothermal fluids, microbial communities accumulate chemical elements in carbonate shells. There are changes in the composition and structure of foraminifera skeletons up to their complete replacement and disappearance (Gablina et al. 2015, 2021).

The accumulation of metals is affected by various hydrothermal processes. Metals are concentrated on foraminifera shells, forming Fe–Mn crusts containing an atacamite additive, clay minerals with newly formed aragonite, Fe–Mg clay minerals, and bacterial films with copper and iron sulfide crystals. Planktonic foraminifera

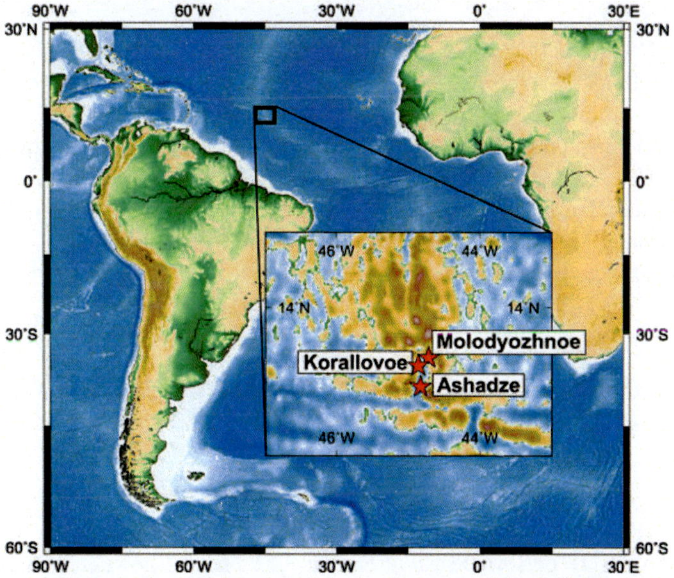

Fig. 1 Mid-Atlantic Ridge. The rectangle shows the Russian exploration region with indication of the hydrothermal fields

change more intensively in comparison with benthic foraminifera, which is explained by the significantly higher porosity of their shells, higher sorption capacity, and solubility (Khusid et al. 2018).

In this work, we investigated iron-containing magnetic inclusions of planktonic foraminifera of sandy and aleurite fractions of sediments from the Ashadze-1, Molodyozhnoye, and Korallovoe MAR hydrothermal fields (Fig. 1).

Among the deepest (4100–4200 m) MAR hydrothermal fields associated with ultrabasites, the Ashadze-1, Molodyozhnoe and Korallovoe fields are located at the foot of the western slope of the rift valley in the highly active area of intersection of a boundary deep rift within a sublatitudinal tectonic deformation zone. The underlying ultrabasites are overlain by modern sediments, which are represented by rock fragments, sometimes ferruginous, and shells of planktonic foraminifera of good preservation. The total number of shells in the obtained samples is high (up to hundreds of thousands of shells per gram of dry sediment). The species composition is dominated by warm-water species: tropical *Globigerinoides ruber* and equatorial-tropical *G. sacculifer*. A comparison of the average concentrations of chemical elements in background planktonic foraminifera with those of foraminifera inhabiting the zone of influence of hydrothermal fluids of the Ashadze-1 field found them to have identical concentrations of Ca (Gablina et al. 2011). However, foraminifera of the Ashadze-1 field are significantly enriched in metals (Fe, Ni, Co and Cr, Ag), as explained by the influence of hydrothermal fluids. A similar pattern is observed in the nearby Molodyozhnoe and Korallovoe fields.

2.2 *Initial Sample Preparation*

The primary sample preparation for our studies consisted in separating sediment samples into magnetic and nonmagnetic fractions with a Sochnev magnet. In this case, the samples not only included shells, but also fragments of various magnetic materials from the sediments. Further, a Leica M205 C stereomicroscope (Leica, Germany) was used to separate only foraminifera shells with iron-containing inclusions from the magnetic fraction.

2.3 *Investigation of the Petrographic Structure and Phase Composition of Samples*

Powder X-ray diffraction (XRD) was used to determine the qualitative and quantitative phase composition of foraminifera shells with iron-containing inclusions. Measurements were performed using Bruker "D2 Phaser" powder diffractometer, Germany) (CuK_α radiation, $\lambda = 1.54178$ Å).

Scanning electron microscopy (SEM) and energy-dispersive X-ray spectroscopy (EDX) were performed to estimate morphological features and element composition of samples. The study was carried out on a QUANTA 200 3D focused electron and ion beams (FIB) system (FEI, Netherlands) with a Pegasus 4000 analytical complex (EDAX, USA) and a TM 3000 scanning electron microscope (HITACHI, Japan).

2.4 Investigation of Magnetic Properties of Samples

A Lake Shore 7410 vibrating sample magnetometer (Cryotronics, Inc., USA) was used to measure hysteresis loops and magnetization curves of the samples. A MFK1-FA magnetic susceptibility meter (AGICO, Czech Republic) was used to measure the frequency-field dependences of the initial magnetic susceptibility of the samples.

Theoretical modeling of the magnetic properties of iron-containing samples was carried out on the basis of the obtained empirical data to establish the most probable ranges of characteristic particle sizes and magnetic characteristics.

3 Results and Discussion

3.1 Morphology, Phase and Elemental Composition of Magnetic Inclusions in Foraminifera Shells

According to the results of quantitative XRD analysis performed by the Rietveld method, it was established that the main mineral of foraminifera was calcite (92.0 wt %). Small amounts of quartz (1.5 wt %), plagioclase (<1 wt %), clinochlor (2.9 wt %), dolomite (< 1 wt %), and kaolinite (2.6 wt %) were also found.

Examination by optical microscopy revealed that the magnetic fraction was represented by black pseudospherules (Fig. 2a, b) with a shiny surface and foraminifera with iron-containing formations on the shell surface in the form of black outgrowths (Fig. 2c, d).

SEM and EDX showed foraminifera shells to have a predominantly carbonate composition. They can be unchanged, well-preserved (Fig. 3a), or include elements of recrystallization (Fig. 3b–d). Morphology is preserved by complete or partial pseudomorphoses occurring in the process of transformation (Fig. 3c). Foraminifera are replaced by opal ($SiO_2 \times nH_2O$), talc ($Mg_3(Si_4O_{10})(OH)_2$), serpentine ($Mg_3(Si_2O_5)(OH)_4$), pyrite (FeS_2), and iron oxides (hydroxides) (Fig. 3e). Sometimes the shells are altered only from the surface, forming pseudomorphoses of draping (Fig. 3 f). EDX spectra obtained from outgrowths and crusts (Fig. 3g, h) showed an iron-manganese (Table 1 analysis 3) and substantially ferruginous (Table 1 analysis 4) composition. These formations can form not only on the surface of whole shells, but also on their fragments (Fig. 3i, j).

Fig. 2 Optical microphotographs of foraminifera of the magnetic fraction

The FIB method was used to determine the characteristics of the internal structure and layer-by-layer changes of the shell's composition. For this purpose, a black foraminifera completely replaced from the surface from the magnetic fraction was selected (Fig. 4a–c). The surface composition corresponds to a mineral of the serpentine group: $(Mg_{2.52}, Fe_{0.52})_{4.04}$ $(Si_{1.98}O_5)$ $(OH)_4$ (Table 1 analysis 1). Skeletal crystallites of magnetite are localized on the shell surface along the layered structure (Fig. 4d, e). The maximum depth of crater was 30 nm (Fig. 4f). In the crater of the ion beam we can see that the layered structure of serpentine with inclusions of small magnetite grains is preserved with depth. Determination of the elemental composition with depth (Table 1 analysis 2) showed that serpentine substitution is preserved. However, the appearance of a weak but distinct calcium peak suggests that the relic

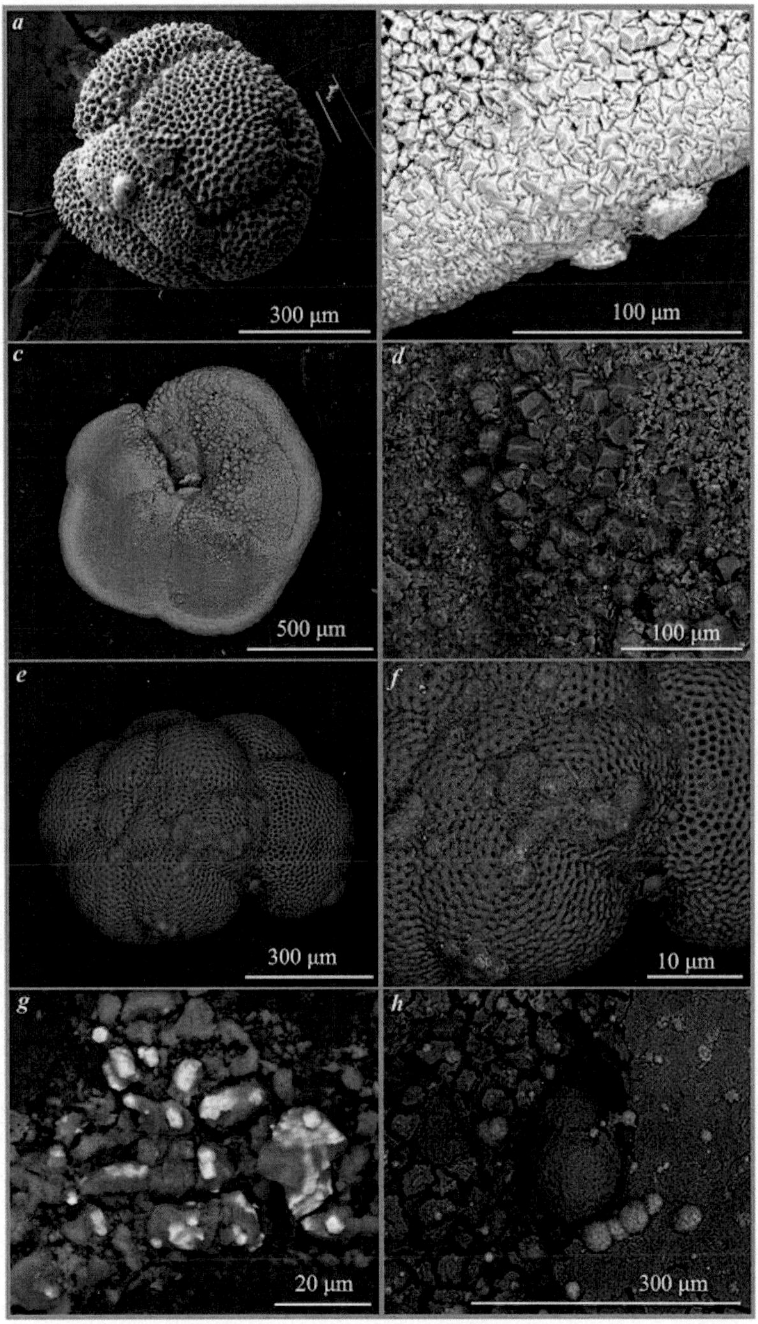

Fig. 3 SEM images of foraminifera shells with iron-containing inclusions

Table 1 Elemental composition of foraminifera shells with iron-containing inclusions*

Chemical element content (wt %)	1	2	3	4
O	43.90	23.63	29.95	9.9
Mg	23.56	20.38	7.33	6.57
Si	21.39	19.70	17.10	6.72
Fe	11.15	29.36	8.76	72.74
Ca	0.00	0.42	17.56	0.00
Mn	0.00	0.00	18.11	4.07
Total	100.00	99.13	98.81	100.00

*1—on surface; 2—in the crater of the ion beam; 3—in outgrowth; 4—crust

composition of the carbonate shell could be preserved in the deeper layers. The presence of a significant amount of magnetite in the form of mechanical inclusions and the high content of isomorphic iron in the serpentine composition explain the high saturation magnetization and the dark color of the foraminifera.

3.2 *Magnetic Properties*

Figure 5a depicts a hysteresis loop of the sample measured at room temperature. Here: saturation magnetization $M_s = 0.973$ A m^2/kg; saturation remanent magnetization $M_{rs} = 0.039$ A m^2/kg, their ratio $M_{rs}/M_s = 0.04$; coercive force $H_c = 3.3$ mT; coercivity of remanence $H_{cr} = 24.3$ mT; their ratio $H_{cr}/H_c = 7.3$. Judging by the non-zero coercive force and remanent magnetization values, there is a ferrimagnetic component in the iron-containing inclusions.

Proceeding from the hysteresis loop parameters and microscopy data, we can assume that the magnetic phase of the samples is represented by iron oxide, which is close in composition to non-stoichiometric magnetite containing admixtures of manganese and magnesium (Wasilewski 1973). The decrease in the magnetization in strong fields observed in the hysteresis loop is explained by the influence of the diamagnetic component associated with the calcite skeleton of the foraminifera.

Measurements of the dependence of the magnetic susceptibility on the excitation field amplitude at three operating frequencies (976, 3904, 15,616 Hz) were performed (Fig. 5b). The frequency dependence of the magnetic susceptibility χ shows the presence of particles in the ferrimagnetic phase that are in a superparamagnetic state (Eyre 1997; Egli 2009; Hrouda 2011; Kharitonskii et al. 2022). The values of χ are small (~10^{-6} m^3/kg), which indicates a very small amount of the magnetic phase. As a consequence, we observe a scattering of values in the graph.

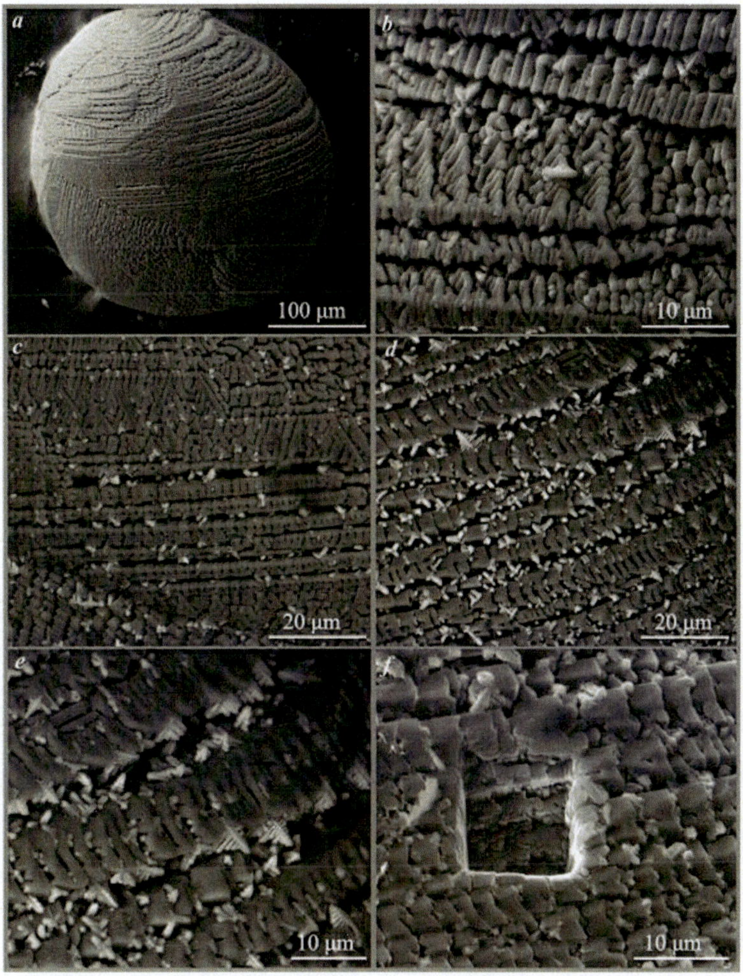

Fig. 4 SEM images of the foraminifera shell before and after FIB

3.3 Theoretical Modeling of Magnetic Characteristics

Using the approximation of lognormal volume distribution of particles and the model of single-domain dipole–dipole interacting particles (Kharitonskii et al. 2018), we have calculated the most probable ranges of characteristic sizes and magnetic characteristics of particles.

The fraction of superparamagnetic particles can be calculated assuming a lognormal particle volume distribution (Olin et al. 2016; Fujihara et al. 2018). The probability density of a lognormal distribution is written as

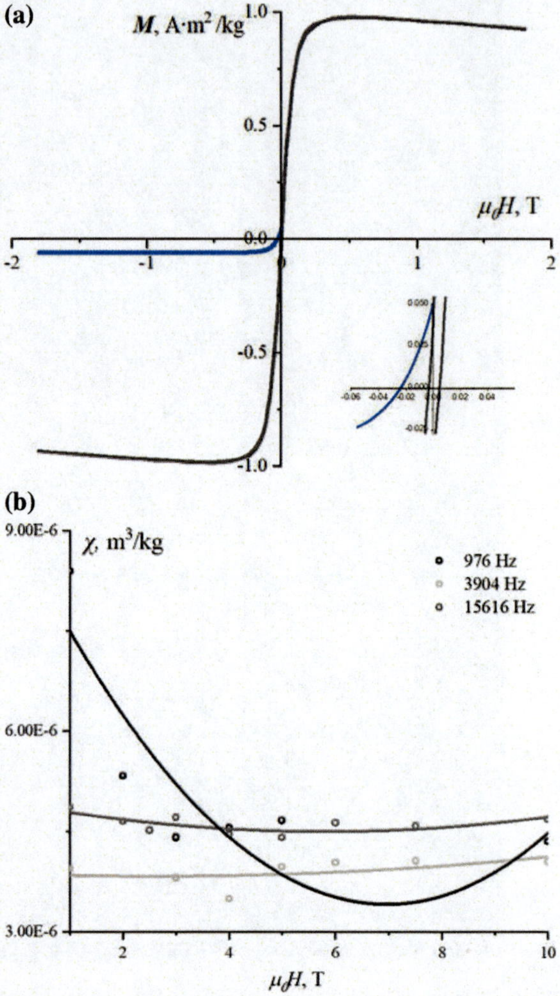

Fig. 5 Magnetic properties of foraminifera shells with iron-containing inclusions: **a** Hysteresis loop and demagnetization curve of the remanent magnetization by the opposite sign field. The inset shows the central part of the loop; **b** the frequency-field dependence of the magnetic susceptibility

$$\varphi(x) = \frac{1}{x\sigma\sqrt{2\pi}} exp\left(-\frac{\left(\ln\left(\frac{x}{\alpha}\right)\right)^2}{2\sigma^2}\right), \quad (1)$$

where $x = v/v_p$ is the volume reduced to the mean, σ is the standard deviation, α is the mean.

Four grainsize ranges corresponding to different magnetic states are considered during the calculation: superparamagnetic (SP), stable single-domain (SD), pseudo-single-domain (PSD), and multi-domain (MD) particles with the respective particle diameters d of 0–25, 25–40, 40–100 and 100–500 nm. We introduce the relative fractions of particles corresponding to the above magnetic states: n_{sp}, n_{sd}, n_{psd} and n_{md}. The relative proportion of each group is written as:

$$n = \int_{x_1}^{x_2} \varphi(x)dx / \int_{x_{min}}^{x_{max}} \varphi(x)dx, \quad (2)$$

where x_1 and x_2 are the lower and upper limits of the volume range of a given particles group, respectively, x_{min} ($d = 0$) и x_{max} ($d = 500$ nm) are the minimum and maximum relative particle volumes, respectively.

The paper (Kharitonskii et al. 2022) demonstrated the possibility of applying the model to samples containing superparamagnetic particles blocked by magnetostatic interaction, which contribute, not only to the saturation magnetization, but also to the remanent magnetization of the sample. Since the frequency dependence of magnetic susceptibility is related to unblocked (true) superparamagnetic particles, two groups of particles should be distinguished: SP—unblocked superparamagnetic (0–18 nm); bSP—blocked superparamagnetic (18–25 nm).

For further calculations the following volumetric concentrations of SP particles were used: $C_{sp} = 0.1$; 0.2; 0.5; 1.0%. The best results for different C_{sp} were obtained with a characteristic size of all ferrimagnetic particles d_p = 50 nm and $v_p = (\pi/6)\, d_p{}^3$. The average sizes of each particles group (SP, bSP, SD, PSD, and MD) were conformed with the average value of the lognormal distribution.

Using the model of interacting single-domain particles (Kharitonskii et al. 2018) and experimental values of magnetization M_s and M_{rs}, we calculate the effective spontaneous magnetization I_{eff} of particles considering their different magnetic states. It is known that in real grains with an inhomogeneous distribution of the magnetic moment over the grain volume that can be related both to grain size and to chemical inhomogeneity, closure domains and various vortex-like structures can occur (Roberts et al. 2017, 2018; Starowicz et al. 2011). According to Figs. 3 and 4, the ferrimagnetic particles are distributed inhomogeneously and form clusters with characteristic sizes of the order of one micrometer, which was taken into consideration in the calculations. All particles were additionally assumed to contribute to the saturation magnetization M_s, whereas only particles in the bSP, SD, and PSD states contribute to the saturation remanence M_{rs}.

We now solve an inverse problem of determining I_{eff} by comparing the experimental and theoretical values of the magnetization M. Since the magnetic phase is non-stoichiometric magnetite, it was assumed that the spontaneous saturation magnetization I_{seff} = 450 kA/m is close to the value for magnetite. Then, from the relation $M_s = C_f \cdot I_{seff}$, we can estimate the value of the volume concentration of the ferrimagnet in the sample: $C_f = 5.1 \cdot 10^{-3}$. Values of relative volumetric concentrations C of particles in different magnetic states and effective spontaneous magnetizations I_{rseff} calculated from saturation remanence are listed in Table 2. For the values obtained, the average calculated particle size, assuming a lognormal distribution, is close to 44 nm, which agrees with the chosen characteristic size of ferrimagnetic particles d_p.

We further assume that easy magnetization axes of the particles are randomly distributed on a sphere. Therefore, H_{cr} can be considered as an average magnetization reversal field of a single particle, and H_c characterizes the magnetization reversal field of the sample as a whole. Indeed, as the external field H_e decreases from its

Table 2 Volume concentrations C (subscript for superparamagnetic (SP), single-domain (SD), pseudo-single-domain (PSD), and multi-domain (MD) particles, correspondingly) and effective magnetizations I_{rseff} (kA/m)

C_{sp}	C_{bsp}	C_{sd}	C_{psd}	C_{md}	I_{rseff}
0.001	0.015	0.202	0.751	0.030	138
0.002	0.019	0.187	0.732	0.060	142
0.005	0.025	0.153	0.661	0.156	158
0.010	0.023	0.096	0.469	0.403	226

maximum value to zero, particles magnetic moments approach easy axis directions, and magnetic inhomogeneities, e.g., vortices, domains, and domain walls, arise. Thus, to estimate the value of M_{rs}/M_s ratio, the theoretical ratio (I_{rseff}/I_{seff}) has to be multiplied by the coercivity ratio H_c/H_{cr} determined from experiment (Kharitonskii et al. 2022):

$$\frac{M_{rs}}{M_s} = \frac{H_c}{H_{cr}} \cdot \frac{C_{rs} I_{rseff}}{C_s I_{seff}}. \tag{3}$$

Here, since unblocked superparamagnetic and multidomain particles in the used approximation do not contribute to the remanent magnetization, $C_s = C_f$, while $C_{rs} = C_f\,(1 - C_{sp} - C_{md})$. The I_{rseff} estimates using Eq. (3) practically coincide with those given in Table 2.

4 Conclusions

- The Holocene and Upper Pleistocene sediments of the ore fields of the Mid-Atlantic Ridge, which are influenced by hydrothermal fluids, contain unsubstituted, well-preserved, and partially or completely recrystallized foraminifera with iron-containing minerals at various growth stages observed on their surfaces.
- Judging by the non-zero coercive force and saturation remanence, there is a ferrimagnetic component in the iron-containing inclusions.
- The frequency dependence of magnetic susceptibility shows the presence of particles in the ferrimagnetic phase that are in a superparamagnetic state. The values of χ are small (~10^{-6} m^3/kg), indicating a very small amount of magnetic phase.
- It can be assumed that the inclusions contain minerals with a spinel structure similar in composition to magnetite (Fe_3O_4).
- The theoretical simulations showed good agreement with the experimental data under the assumption that the sample contains particles in different magnetic states (predominantly in the pseudo-single-domain state) with an average size of 44 nm and a volume concentration of the ferrimagnetic component of about 0.5%.

Acknowledgements The authors are grateful to I.G. Dobretsova, mineralogist of the Ocean Exploration and Prospecting Party "PMGE" for the samples provided for research and V.V. Karpinskiy for the graphical representation of MAR. Scientific research was performed at the Research Park of St. Petersburg State University: "Center for Microscopy and Microanalysis", "Centre for X-ray Diffraction Studies", "Center for Innovative Technologies of Composite Nanomaterials", "Center for Geo-Environmental Research and Modeling (GEOMODEL)".

References

Armynot du Châtelet E, Bout-Roumazeilles V, Coccioni R, Frontalini F, Guillot F, Kaminski MA, Recourt P, Riboulleau A, Trentesaux A, Tribovillard N, Ventalon S (2013) Environmental control on shell structure and composition of agglutinated foraminifera along a proximal-distal transect in the Marmara Sea. Mar Geol 335:114–128

Ben-Eliahu N, Herut B, Rahav E, Abramovich S (2020) Shell growth of large benthic foraminifera under heavy metals pollution: Implications for geochemical monitoring of coastal environments. Int J Environ Res Public Health 17:3741

Diab M, Shreteh K, Afik N, Volokh M, Abramovich S, Abdu U, Mokari T (2018) Design of Hierarchal 3D Metal Oxide Structures for Water Oxidation and Purification. Adv Sustain Syst 2:1800001

Dorozhkin SV (2011) Biocomposites and hybrid biomaterials based on calcium orthophosphates. Biomatter 1:3–56

Egli R (2009) Magnetic susceptibility measurements as a function of temperature and frequency I: inversion theory. Geophys. J. Int. 177: 395–420

Eyre JK (1997) Frequency dependence of magnetic susceptibility for populations of single-domain grains. Geophys. J. Int. 129:209–211

Fujihara A, Tanimoto S, Yamamoto H, Ohtsuki T (2018) Log-normal distribution in a growing system with weighted and multiplicatively interacting particles. J. Phys. Soc. Japan 87:034001

Gablina IF, Demina LL, Dmitrenko OB, Os'kina NS, Khusid TA, Popova EA, Shilov VV (2011) Composition and secondary alterations of microfossils in sediments of the Ashadze-1 hydrothermal field (tropical Mid-Atlantic ridge). Oceanology 51(3):476–490

Gablina IF, Dmitrenko OB, Os'kina NS, Khusid TA, Popova EA (2015) Impact of hydrothermal activity on carbonate fossils in bottom sediments of the tropical Atlantic. Oceanology 55(1):100–111

Gablina IF, Sadchikova TA, Gor'kova NV, Mikheev VV, Dobretsova IG, Popova EA, Dara OM (2021) Mineral composition and geochemical zoning of bottom sediments in the Pobeda hydrothermal cluster (17°07.45′N – 17°08.7′N Mid-Atlantic Ridge). Lithology and Mineral Resources 56(2):13–131

Gupta B, Mishra A, Singh R, Thakur IS (2021) Fabrication of calcite based biocomposites for catalytic removal of heavy metals from electroplating industrial effluent. Environ Technol Innov 21:101278

Hrouda F (2011) Models of frequency-dependent susceptibility of rocks and soils revisited and broadened, Geophys. J. Int. 187:1259–1269

Jacob DE, Wirth R, Agbaje OBA, Branson O, Eggins SM (2017) Planktic foraminifera form their shells via metastable carbonate phases. Nat Commun 8:1265

Kharitonskii P, Kamzin A, Gareev K, Valiullin A, Vezo O, Sergienko E, Korolev D, Kosterov A, Lebedev S, Gurylev A, Reinyuk A (2018) Magnetic granulometry and Mössbauer spectroscopy of Fe_mO_n–SiO_2 colloidal nanoparticles. Journal of Magnetism and Magnetic Materials 461:30–36

Kharitonskii P, Bobrov N, Gareev K, Kosterov A, Nikitin A, Ralin A, Sergienko E, Testov O, Ustinov A, Zolotov N (2022) Magnetic granulometry, frequency-dependent susceptibility and magnetic states of particles of magnetite ore from the Kovdor deposit. Journal of Magnetism and Magnetic Materials 553:169279

Khusid TA, Os'kina NS, Libina NV, Matul AG, Lukashina NP, Gablina IF (2018) Benthic and plankton foraminifers in hydrothermally active zones of the Mid-Atlantic Ridge (MAR). Stratigraphy and Geological Correlation 26(1):109–120

Kulsoom R, Sarfraz M, Afzal A, Farooq M, Adnan S, Ashraf MU, Khan SA (2022) Synthesis of calcium carbonate-quince bio-composite for programmed and on-demand drug release of paracetamol at target site: a green chemistry approach. Polym. Bull 1–22

Langer MR (2008) Assessing the contribution of foraminiferan protists to global ocean carbonate production. Journal of Eukaryotic Microbiology 55:163–169

Magnabosco G, Hauzer H, Fermani S, Calvaresi M, Corticelli F, Christian M, Albonetti C, Morandi V, Erez J, Falini G (2019) Bionic synthesis of a magnetic calcite skeletal structure through living foraminifera. Mater Horizons 6:1862–1867

Mishra A, Gupta B, Kumar N, Singh R, Varma A, Thakur IS (2020) Synthesis of calcite-based bio-composite biochar for enhanced biosorption and detoxification of chromium Cr (VI) by Zhihengliuella sp. ISTPL4. Bioresour Technol 307:123262

Miyauchi S, Imoto H, Naka K (2016) Fabrication of polymer-calcite composite thin films by phase transition of vaterite composite particles with octacarboxy-terminated T8-caged silsesquioxane. Polym J 48:1019–1027

Olin M, Anttila T, Dal Maso M (2016) Using a combined power law and log-normal distribution model to simulate particle formation and growth in a mobile aerosol chamber. Atmos. Chem. Phys. 16:7067–7090

Pawlowski J, Majewski W (2011) Magnetite-bearing foraminifera from admiralty bay, west antarctica, with description of psammophaga magnetica, sp. nov. J Foraminifer Res 41:3–13

Pawlowski J, Holzmann M (2002) Molecular phylogeny of Foraminifera – A review. Eur. J. Protistol. 38:1–10

Roberts AP, Almeida TP, Church NS, Harrison RJ, Heslop D, Li Y, Li J, Muxworthy AR, Williams W, Zhao X (2017) Resolving the Origin of Pseudo-Single Domain Magnetic Behavior. J. Geophys. Res. Solid Earth 122:9534–9558

Roberts AP, Tauxe L, Heslop D, Zhao X, Jiang Z (2018) A Critical Appraisal of the "Day" Diagram. J. Geophys. Res. Solid Earth 123:2618–2644

Shi R-J, Lang J-Q, Wang T, Zhou N, Ma M-G (2022) Fabrication, Properties, and Biomedical Applications of Calcium-Containing Cellulose-Based Composites. Front. Bioeng. Biotechnol. 10:937266

Smirnov VV, Goldberg MA, Shvorneva LI, Fadeeva IV, Shibaeva TV, Barinov SM (2010) Synthesis of composite biomaterials in the hydroxyapatite-calcite system. Dokl Chem 432:151–154

Starowicz M, Starowicz P, Żukrowski J, Przewoźnik J, Lemański A, Kapusta C, Banaś J (2011) Electrochemical synthesis of magnetic iron oxide nanoparticles with controlled size. J. Nanoparticle Res. 13:7167–7176

Titelboim D, Sadekov A, Blumenfeld M, Almogi-Labin A, Herut B, Halicz L, Benaltabet T, Torfstein A, Kucera M, Abramovich S (2021) Monitoring of heavy metals in seawater using single chamber foraminiferal sclerochronology. Ecol Indic 120:106931

Wasilewski PJ (1973) Magnetic hysteresis in natural materials, Earth and Planetary Science Letters 20(1):67–72

Yang H, Peng X, Gooday A.J, Jones B, Li J, Liu S, Huang W, Sun Z, Chen S, Dasgupta S, Xu H, Liu S, Xu W, Ta K. (2022) Magnetic foraminifera thrive in the Mariana Trench. Geochem. Persp. Let. 21:23–27

Youssef M (2015) Heavy metals contamination and distribution of benthic foraminifera from the Red Sea coastal area, Jeddah, Saudi Arabia. Oceanologia 57:236–250

Calcite Microspherulites as a Reflection of the Relationship Between Abiotic Processes and Biological Mechanisms

Anna I. Antoshkina

Abstract Identified clusters of calcite microspherulites in carbonate fine-clastic breccia comprising inter-bioherm carbonates of the Upper Miocene serve as an important indicator of paleoenvironments. The formation of such structures is part of a long-studied sedimentological problem concerning the interaction between abiotic processes and biological mechanisms in the formation and early diagenesis of non-skeletal carbonates. Calcite microspherulites were studied along with the host rocks from a section of the carbonate bioherm sequence on the territory of the Kazantip Nature Reserve. Lithological, geochemical and physicochemical research methods were used. Syngenetic authigenic minerals barite, halite, celestine, high-Mg calcite and dolomite intergrown with calcite and framboidal pyrite were identified along with fossilized biofilms having a complex elemental composition. It is suggested that the formation of the microspherulites could have occurred upon contact of carbonate silt with microbial matter in a very small lagoon under conditions of falling sea level with the presence of elevated concentrations of Na, Ba, Cl^- and SO_4^- ions relative to normal sea water, which are explained by local gas-fluid seeps due to the activation of an ancient mud volcano.

Keywords Calcite microspherulites · Pelitomorphic matrix · Authigenic minerals · Gas-fluid seep · Late Miocene

1 Introduction

One of the longstanding conundrums in sedimentology concerns how to demarcate the influences exerted by abiotic processes and biologically influenced mechanisms in the formation and early diagenesis of non-skeletal carbonates. Among the first researchers who described spherulitic calcites is Godovikov (1961), who proposed

A. I. Antoshkina (✉)
Institute of Geology, FRC Komi SC UB RAS, 54, Pervomayskaya St., Syktyvkar 167982, Russian Federation
e-mail: Antoshkina@geo.komisc.ru

O. V. Frank-Kamenetskaya et al. (eds.), *Biogenic—Abiogenic Interactions in Natural and Anthropogenic Systems 2022*, Springer Proceedings in Earth and Environmental Sciences, https://doi.org/10.1007/978-3-031-40470-2_10

a mechanism whereby noticeable magnesium impurities and specific growth conditions in dolomite rocks at low temperatures could contribute to the appearance of spherulites with sizes from 3 to 5 mm. One of the first researchers to raise the problem of the relationship between biotic and abiotic factors affecting the morphology of carbonate bodies was Serebryakov (1976), who studied the Riphean stromatolites of Siberia. Subsequently publications by Wilkinson and Given (1986), as well as Kelts and Talbot (1990), appeared. With the turn of the century, the problem began to be covered in an increasing number of works by foreign sedimentologists. However, an outstanding problem concerns the unification of criteria for assessing whether non-enzymatic carbonates are due to a biogenic effect, or whether they can be formed completely abiotically. This fundamental sedimentological problem remains critical in terms of defining models of how life has influenced substrate formation, lithification, and diagenesis (Hodgson et al. 2018). Recent discussions have focused on the origin of some of the abundant puzzling carbonate structures that form pre-salt Cretaceous hydrocarbon reservoirs at the South Atlantic margin. These carbonate structures have been described as being formed by a repeating combination of fibrous-radial polycrystalline calcite spherulites with forked and densely divergent, fibrous to lamellar, beam-optic to radial-fibrous, polycrystalline clusters (Farias et al. 2021; references therein). There are many publications on modern salt lakes that describe the formation of microspherulitic calcites (Mercedes-Martín et al. 2017, 2021; Bischoff et al. 2020 references therein). It has been shown that in contemporary microbial mats of salt lakes, as well as under laboratory experimental conditions, spherulitic calcites are formed in waters with moderate and high $[Ca^{2+}]/[CO_3^{2-}]$ ratios. Unusual microspherulitic calcites combined into various clusters in a microbial-pelitomorphic matrix of the fine-grained breccia were also identified during our study of the Upper Miocene bryozoan limestones of the Kazantip Natural Reserve.

This paper presents the results of complex analytical investigations of calcite microspherulites as evidence of the relationship between biological mechanisms and abiotic processes of their origin.

2 Materials and Methods

When examining the samples, a consistent set of methods was applied. Polished slabs, thin sections, and chips were prepared from the breccias and surrounding carbonate rocks for various studies. Specimens for analysis were drilled from rock sections least affected by secondary processes (recrystallization, dolomitization, and leaching) using thin dental drills. First, chemical 8-component carbonate analysis was used to determine the chemical composition of the samples. The gross C_{org} content was determined in concentrated hydrochloric acid in the insoluble rock residue using an AN-7529 express analyzer for carbon; glucose was used as a standard. To determine the structure of organic matter (OM), the samples were studied by Raman spectroscopy on a high resolution LabRam HR800 spectrometer (Horiba

Jobin Yvon). Spectrum registration parameters: He–Ne laser (power 2 mW, λ = 632.8 nm); spectrometer grating—600 f/mm; confocal hole size—300 μm; slit—100 μm; objective—x50; signal accumulation time—10 s; number of measurements in one part of the spectral range—3. The spectra were recorded at room temperature. The optical properties of microspherulites and the pelitomorphic matrix of breccias were studied using a Nikon Eclipse LV100ND universal microscope. SEM images of the elemental composition were obtained using a Tescan Vega3 LMH scanning electron microscope with an accelerating voltage of 20 kV and a beam current of 15 nA. The chemical composition of the carbonate was analyzed in the EDS mode using an INCA X-MAX 50 mm from Oxford Instruments with the following parameters: voltage—20 kV; current—15 nA; vacuum—0.05 Pa; beam diameter—2 μm; exposure time—500,000 pulses. Standards, characteristic lines, lower detection limits (wt %) as follows: Pd (Pd, Lα, 0.10 and 0.16 for analyses 7 and 8 in the table); Te (PbTe, TeLα, 0.26–0.35); Bi (Bi, Mα, 0.28); Sb (Sb, Lα, 0.21–0.30); Ni (Ni, Kα, 0.19–0.25); Co (Co, Kα, 0.11). The isotope composition of carbon and oxygen in carbonates was determined using a Delta V Advantage mass spectrometer. The $\delta^{13}C$ values are given in ppm relative to the VPDB standard, while the $\delta^{18}O$ values are given according to the SMOW standard calibrated using the NBS 19 international standard (TS-limestone). The error in the determination of $\delta^{13}C$ and $\delta^{18}O$ did not exceed $\pm$ 0.15‰ (1σ). Analytical studies were carried out at the "Geonauka" Center for collective use (IG FRC Komi SC UB RAS, Syktyvkar).

3 Brief Geological and Sedimentological Background

The Kazantip Nature Reserve is located in the northern Kerch Peninsula, Crimea. Here, Miocene deposits make up an elliptical cryptodiapir fold with an injection core of strongly folded, crushed, and powdered clays of the Maikop Series. Based on drilling materials, Klyukin (2006) characterized the stratification of the Kazantip sedimentary succession in the following way. The base comprises thick (4–5 km) Oligocene–Lower Miocene crushed clays of the Maikop Series. These are overlain by Middle Miocene clays (0.2 km) with layers of siltstones, sandstones, marls and limestones. The section is completed by rocks of the Keza Formation consisting of schistose gypsified clays with thin and sparse interlayers of the Upper Sarmatian marls, limestones, sandstones, and conglomerates, as well as a thick layer of Lower Maeotian bryozoan limestones, partly shelly limestones, marls, and clays (0.3–0.5 km). Clays of the Sarmatian Stage make up the inner basin, whereas the Cape Kazantip outer ridge is composed of Maeotian limestones. Thin Quaternary marine and unconformable continental sediments overlie the bedrock.

Peculiar fine-clastic breccias with calcite microspherulitic were marked in the bryozoan limestones. In general, the rocks are porous, often strongly weathered. Frameworks of biohermal limestones are composed of skeletons of bryozoans growing on top of each other, between which there is small lithoclastic and bioclastic material (Fig. 1).

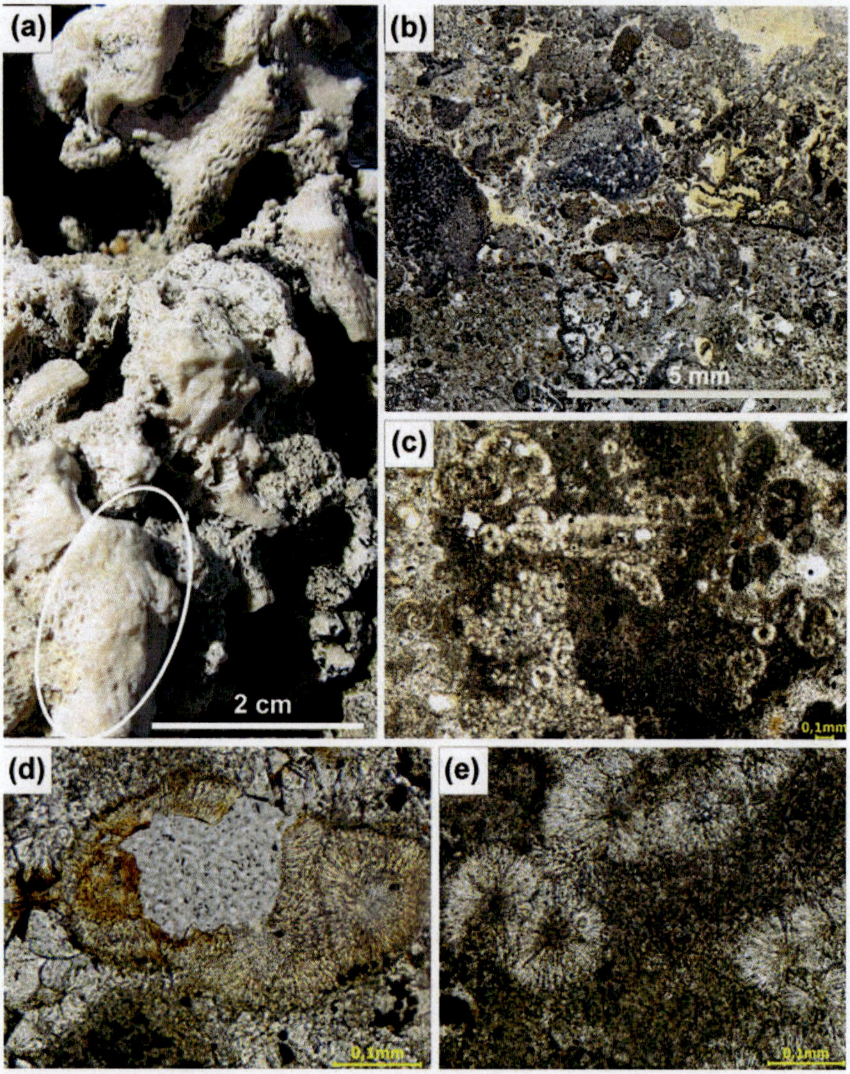

Fig. 1 Lithology of host rocks and petrography details of breccia: **a** frameworks of biohermal limestones are composed of skeletons of bryozoans as well as mucoid-like fossil biofilms (in a white oval) on their surface and intraframe filling; **b** thin-section showing the complex in composition and structure the studied fine-grained breccia; **c** clasts of calcite microspherulites in the pelitomorphic matrix; **d** vadose leaching of a calcite spherulite; **e** details (c) of well-marked fibrous structures of calcite microspherulites with pelitic matter in the center

On the surface of the bryozoan and intraframe filling, mucoid-like fossil biofilms are clearly visible (Fig. 1a). Fine-clastic breccias include pelitomorphic matrix, vadose cement, and clastic material ranging in size from 1–2 mm to 1 cm (Fig. 1b). The pelitomorphic matrix includes microbial and filamentous cyanobacteria-like accumulations, calcite microspherulites, terrigenous admixtures, and authigenic mineralization. Calcite microspherulites of 0.05–0.15 mm in size form clusters up to 1 mm (Fig. 1c, e). Vadose cement is associated with cavities in which crystals of newly formed calcite are developed. There are leaching cavities in microspherulites (Fig. 1d). The largest fragments of breccia are represented by bryozoan bioherm and bioclastic limestones, as well as microbial accumulations with terrigenous minerals and vadose pisolites. Fragments of biohermal limestone with a surface crust of fibrous calcite were also found.

4 Results

4.1 Chemical and Isotopic Composition of Host Rocks

Limestone variations from weakly dolomitic to strongly dolomitic were determined according to the normative mineral composition of the rocks hosting the breccia as established by the calculations of the 8-component carbonate analysis. Calcites and dolomites are characterized by an insignificant admixture of Mn and Fe. The insoluble part of carbonate rocks is represented by clastic and authigenic quartz, chalcedony (in cavities of vadose leaching), bone detritus, feldspars (most often orthoclase), muscovite, chlorite, gypsum, iron hydroxides, and apatite.

The microbial organisms represent a mineral record of carbonate chemistry, including microbial influences on local dissolved inorganic carbon (DIC). All microbial communities cycle CO_2, changing local carbonate mineral stability and influencing the $\delta^{13}C$ of DIC. When carbonates are precipitated, they record the $\delta^{13}C$ signature of local DIC, thus capturing microbial influences on isotopic ratios. The isotope analysis of the studied carbonates showed a very variable carbon composition from –0.32 ‰ to 4.53 ‰, which may depend on water productivity. At the same time, significant variations in the oxygen isotopic composition from normal sea waters, both downward (23.06–26.92 ‰) and upward (32.51–39.74 ‰), may indicate both salinity fluctuations, as well as variations in water temperature. The isotopic composition of organic carbon, ranging from –22.28‰ to –24.62‰, indicates microbially induced abiotic methane (e.g., Stahl 1984; Temescu et al. 2009; Silaev et al. 2021).

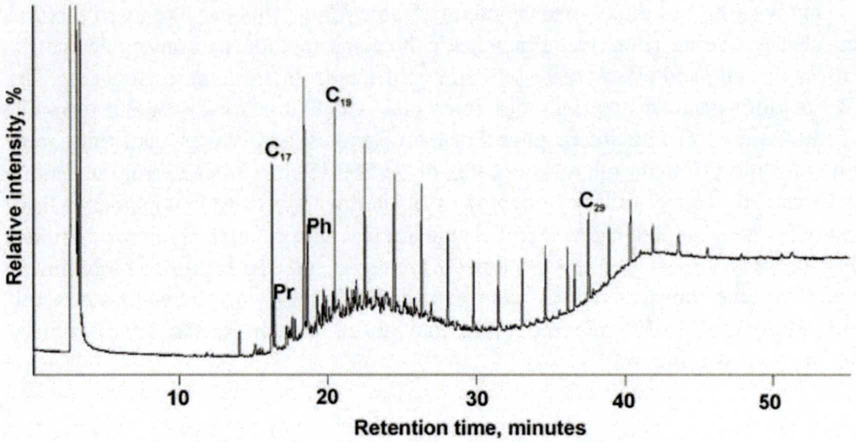

Fig. 2 Chromatogram of the aliphatic fraction of bitumoids, Shir-8–19 sample. *Note*: Pr—pristane, Ph—phytane

5 Geochemical Composition of Organic Matter

In geochemical studies using gas–liquid chromatography to determine the composition of the initial organic matter (OM) of the rocks, the conditions of its sedimentation, as well as the degree of catagenetic transformation, data on the distribution of such acyclic hydrocarbons as normal alkanes (n-alkanes) were used. These compounds are relatively stable over time and under changing geological conditions. Each genetic type of OM is characterized by a specific distribution of n-alkanes depending on the source. According to gas chromatography data (Fig. 2), low bituminous values equal to 2.31% in pelitomorphic limestones indicate the autochthonous and residual nature of bitumoids (Vassoevich 1986). The chromatogram clearly shows the bimodal distribution of n-alkanes with a maximum at n-C_{18} and n-C_{29}.

According to (Hunt 1982; Dorogochinskaya et al. 1993), the predominance of n-C_{18} indicates an increase in bacterial activity, while the low pristane-phytane ratio of 0.30 indicates reducing conditions. Under such conditions, biofilm and microbial mat microorganisms can create an anoxic environment in millimeter-thick layers (Lin et al. 2020).

6 Structure of OM in the Microspherulites

Carbonaceous substances are an important object of research from the standpoint of informativeness, both in terms of geological prevalence and the extraordinary variety of modifications. At the same time, the predominantly non-crystalline nature of such substances presents serious problems for mineralogical and geochemical study, requiring the use of special research methods and techniques. Raman spectroscopy, which offers specific sensitivity to carbon–carbon covalent chemical bonds, is one such method (Tikhomirov and Kimstach 2011). Studies of the calcite microspherulites by Raman spectroscopy demonstrated intense luminescence from OM with poorly pronounced maxima in the complete absence of Raman lines (Fig. 3). In this case, OM in the microspherulites is characterized by a low degree of transformation due to the lack of pronounced peaks in the Raman-scattered light spectra, which indicates a lack of primary OM structure in them (Wopenka and Pasteris 1993; Silaev et al. 2013).

The exceptions here are authigenic calcite crystals, which have standard but somewhat less pronounced peaks. According to the character of the Raman spectra, a number of studied substances include primary and slightly altered organic substances characterized by luminescence spectra with minor form of Raman lines. In addition, it was found that oval-shaped nanograins having a dark rim, which are comprised of pyrite, hematite, and magnetite, are widespread in the pelitomorphic matrix. The extended peaks in the Raman spectra indicate a weak structure ordering in these minerals.

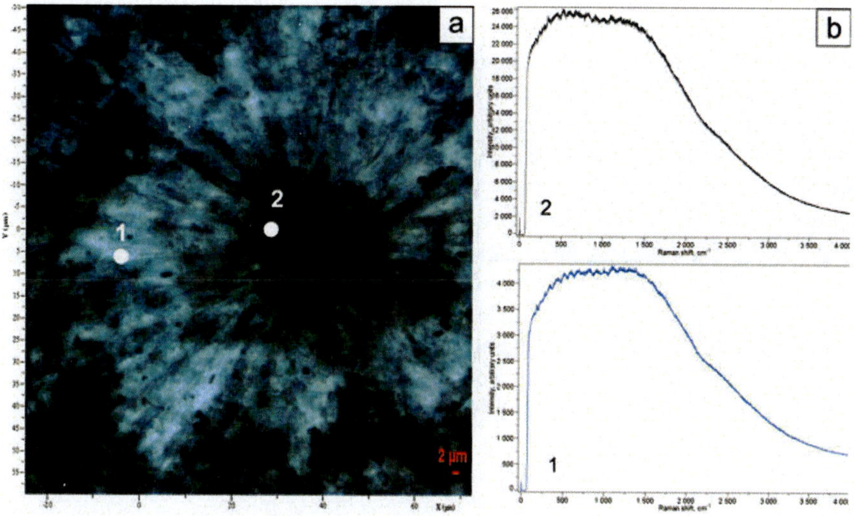

Fig. 3 Raman image of a microspherulite (**a**) and luminescence spectra of unstructured primary organic substance at points 1, 2 (**b**)

7 Mineral Composition and Structures of the Microspherulites

Electron microscope studies of fresh cleavages of the breccia sample provided important information for genetic interpretation in the mineral composition and structures of the microspherulites. Microspherulites are shown to form a dense packing (Fig. 4). According to (Mercedes-Martin et al. 2021), the dense-packing of spherulites is a response to the presence of a liquid mineral polymer temporarily stabilized by an exopolymer substance (EPS) performing the role of a precursor of amorphous calcium carbonate, which is then transformed into calcite or aragonite.

As can be seen from the X-ray spectral microprobe analysis, the elements Sr and Mg, as well as S and Fe, are consistently included in the calcite composing the microspherulite aggregates. At the same time, magnesium content characterizes low-Mg (a) points 1, 2; (b) points 3, 4 and high-Mg (a) point 3 calcites, up to dolomite (b) point 1. Sr is typically contained in aggregates from the low-Mg calcites. High-Mg calcite is known to be an unstable mineral; due to its transformation into low-Mg calcite during diagenesis, it is poorly preserved in ancient rocks. However, it may still be present in sediments formed under high salinity conditions (Stanienda-Pilecki 2018). In the aeration zone, the less stable aragonite and low-Mg calcite break down to form stable high-Mg calcite (Boggs 2010).

The close-packed microspherulites in another area of the same sample show the presence of mineralized biofilms, which persists as a cloudy-gray coating clearly visible on some aggregates in the SEM image (Fig. 5). Simultaneously with the non-permanent elements Al, Si, Fe, Sr, the EDS spectra demonstrated the presence of halite elements (Na, Cl) in the biofilm at all studied points (Table 1).

In aggregates of the studied microspherulites, magnesian calcites appear. In some sections, the relationship between individuals of crystal-grained and fibrous structures can be observed. On the Fig. 6a an oval-shaped "core" of a microspherulite is a crystal-grained calcite with the presence of the elements Mg, Si, P, Fe (wt. %), as shown in the Table 2. The data from points 2–4 demonstrate significant fluctuations in the permanent presence of Mg in fibrous calcite individuals developed in the crust zone; in other words, the radial zone is formed by magnesian calcite. Among the mineral phases calculated by X-ray spectral microprobe analysis, the data of an authigenic crystal (point 5) located at the junction of microspherulites stand out sharply. This phase characterizes a complex mineral mixture of dolomite, muscovite, apatite, and quartz. According to many published papers, certain elements such as magnesium can not only affect the stability of the crystalline $CaCO_3$, but also play important roles in the formation of spherulitic morphology (Wu et al. 2017). Similar mineral relationships were revealed in the study of the bacterial habitat represented by the Khoito-Gol spring, Eastern Sayan. Among the normative nanominerals calculated from the stoichiometric ratios of chemical elements in a dry biofilm, halite and calcite are not always present together, while magnesian calcite is present only with dolomite (Tatarinov et al. 2017). The authors found that the behavior, distribution

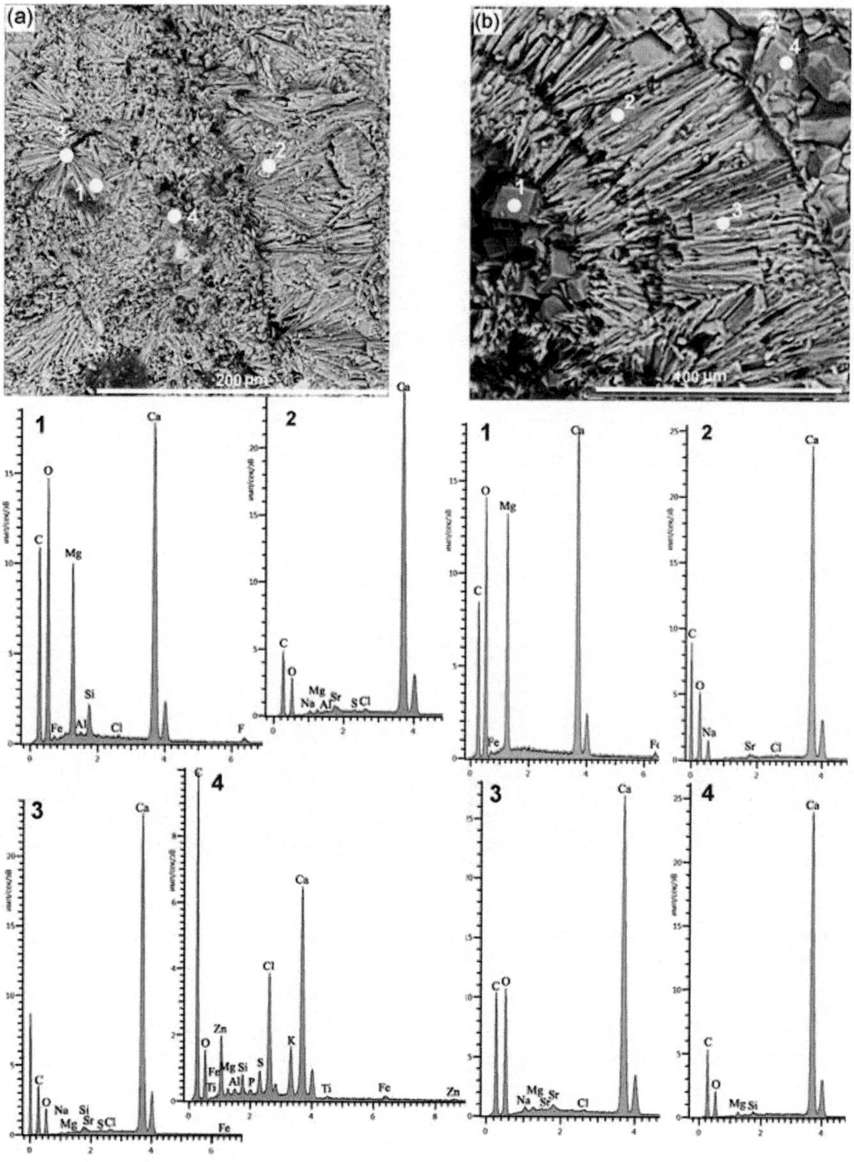

Fig. 4 SEM images in secondary (**a**) and reflected (**b**) electrons and X-ray spectral microprobe analysis for research points: **a**—close-packed microspherulites with disparate aggregates of microspherulites; **b**—fragment of a bizonal microspherulite with a parallel columnar structure and authigenic dolomite crystals in the central part

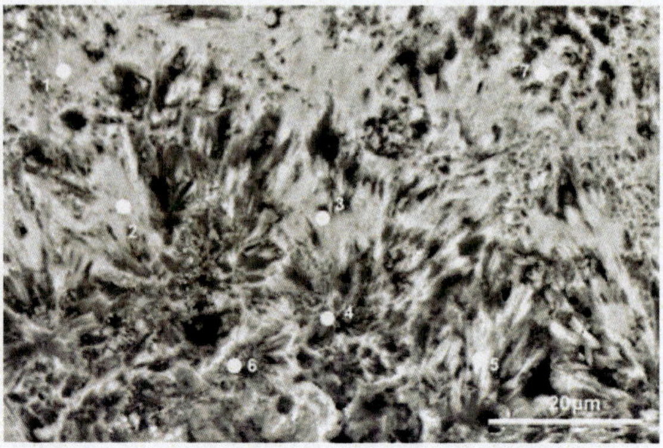

Fig. 5 SEM image of the close-packed microspherulites structure (in reflected electrons) sectorally radiant aggregates are visible

Table 1 The elemental composition according to X-ray spectral microprobe analysis for seven points in Fig. 5

Chemical composition, weight %							
Element	1	2	3	4	5	6	7
O	51.5	50.7	53.8	53.3	52.5	46.0	51.2
Na	0.5	0.6	0.7	0.8	0.9	0.8	0.5
Mg	0.3	–	0.2	0.6	0.2	–	–
Si	0.7	–	0.5	1.2	0.4	1.1	–
Cl	0.3	0.3	1.1	1.3	0.6	1.2	0.6
Ca	46.7	47.2	43.9	42.9	44.7	45.0	46.5
Sr	–	1.2	–	–	–	–	1.2
Al	–	–	–	–	0.7	5.9	–

and accumulation of trace elements in various biofilms are mainly determined by geochemical barriers of geological and biological origin.

The presence of crystal-grained individuals among fibrous individuals with a fossilized biofilm is clearly observed in Fig. 6b. According to (Zhang et al. 2017), data from experiments with bacteria *Curvibacter lanceolatus* suggest the following possible mechanisms of biomineralization of Mg-calcite spherulites: (1) the precipitation and association of amorphous calcium carbonate particles into a spherical core in combination with organic molecules; (2) the growth of an organic film on the surface of Mg-calcite spherulites as a result of bacterial metabolism and calcification.

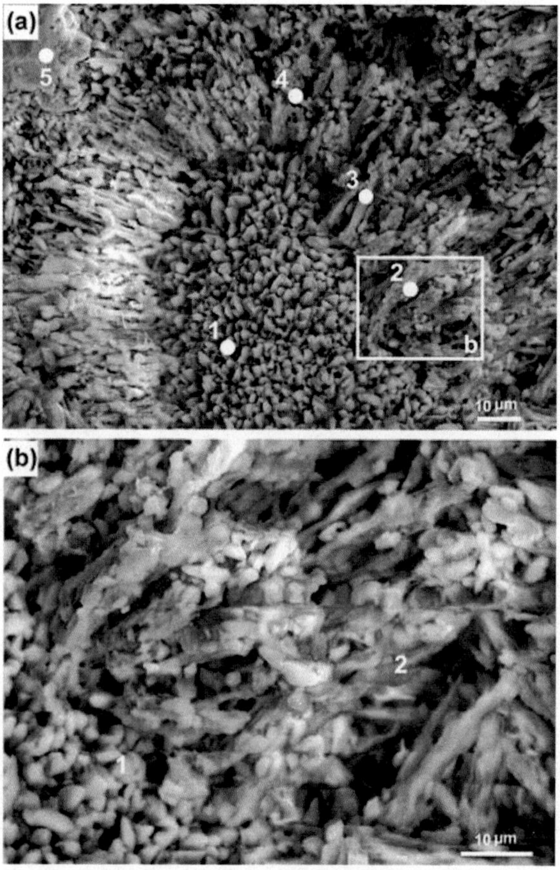

Fig. 6 SEM images in secondary electrons: **a**—oval-shaped "core" of individuals of a crystal-grained structure with an increase in a crust of aggregates of a parallel columnar structure; **b**—intergrowth of crystal-grained (1) and fibrous (2) individuals

Table 2 The results of X-ray spectral microprobe analysis of points in Fig. 6 normalized to 100%

Chemical composition, weight %					
Component	1	2	3	4	5
MgO	5.16	7.21	15.88	20.08	26.55
Fe_2O_3	0.49	1.01	0.81	–	2.89
K_2O	–	–	–	–	0.29
P_2O_5	0.77	–	–	1.02	1.06
SiO_2	1.65	3.21	3.93	1.16	20.78
Al_2O_3	–	–	–	–	5.90

8 Atypical Mineral Associations in the Pelitomorphic Matrix

The presence of syngenetic authigenic crystals of framboidal pyrite, halite, barite, celestine, dolomite-calcite (Fig. 7) in the pelitomorphic matrix with microspherulites is atypical for normal sedimentary deposits. Scanning microscopic and X-ray spectral studies have found accumulations of halite (Fig. 7a) and pyrite crystals (Fig. 7b) containing impurity elements Si, Cl, Na, Mg, Fe, and Al in the biofilm; however, Ti is rarely observed. Ca, Ba, and Sr carbonates are known to be difficult to dissolve in water. Carbonic acid is a weak acid; solutions of its salts, due to hydrolysis, have an alkaline reaction, which is stronger for carbonates. The precipitation of barite (Fig. 7c) and celestine (Fig. 7d) in such amounts requires additional triggers, which, according to some researchers, can include microbial organic polymers and ions of light and heavy metals.

For example, barite occurs in sedimentary rocks in a redeposited pre-hydrothermal-sedimentary genesis, which mark the zones of discharge of gas-fluid seeps. According to (Aloisi et al. 2004), barite sedimentation occurs in the upper few meters of sediments at a slow flow rate of gas-fluid solutions to the bottom (<5 cm/yr), resulting in the formation of microcrystalline phases and nodules. Gonzalez-Muñoz et al. (2012) demonstrated the precipitation of barite by marine bacteria through a metabolically mediated biomineralization process. An experimental study

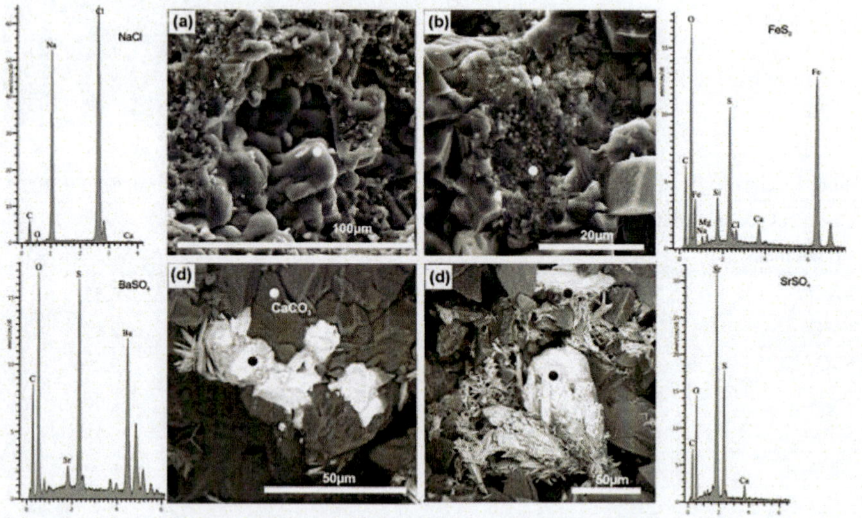

Fig. 7 SEM images of destroyed framboidal pyrite nanograins in a fossilized biofilm (**a**) of mineralized biofilm with halite on fibrous individuals of magnesian calcite (**b**) of barite crystals (**c**), and celestine crystals (**d**)

by Martinez-Ruiz et al. (2018) showed that marine bacterial biofilms such as extracellular polymeric substances (EPS) can bioaccumulate Ba to provide adequate conditions for barite precipitation. In the absence of bioturbation and active hydrodynamics, microbial matter and biofilms enhance the binding properties of bottom sediments to facilitate the burial of organic matter (Krumbein et al. 1994). When a sediment is isolated from the water environment by a microbial mat, it becomes a closed dynamic system of physical, chemical, and biological processes, allowing the microbial component buried in this sediment to serve as a matrix for early diagenetic mineralization. The excess of calcium ions created during sulfate reduction under brackish water conditions due to the consumption of the SO_4^- ion leads to an alkalization of the medium. Both of these processes shift the carbonate equilibrium towards the precipitation of early diagenetic calcium carbonates (Lein and Ivanov 2009). Halite is a characteristic crystalline admixture to inter-bioherms breccias in almost all the studied carbonates of the Kazantip sections (Leonova et al. 2020). However, mineral halite is not found in sedimentary rocks in the same quantity due to its water solubility. While chlorides form an integral component of sea water, their presence in carbonates in the solid-solution polycomponent form with the formation of authigenic minerals indicates an additional active replenishment source. With modern gas-fluid seepage in the seas and oceans, this process causes supersaturation and precipitation of crystals (Lein and Ivanov 2009).

9 Discussion

The complex analytical studies of the calcite microspherulites in fine-clastic breccias have revealed the polycomponent composition of the microspherulites themselves, as well as that of the host pelitomorphic matrix. Fossilized biofilms typically occurring in the pelitomorphic matrix and close-packed microspherulites are associated with authigenic crystals of framboid pyrite, halite, and magnesian calcite. The structure and construction of the EPS is formed by the glycocalyx as a natural process of the life of the microbial community, including bacteria, whose main functions were to mediate the initial attachment of cells to various substrates and protect against environmental stress and dehydration (Vu et al. 2009). The authigenic crystals of framboidal pyrite, barite, celestine, and halite indicate gas-fluid infiltration and active participation of sulfate-reducing bacteria in the processes of early diagenetic mineral formation.

The obtained results identify the presence of calcite microspherulites in the zone of contact of carbonate silt with microbial matter in the inter-biotherm space, which form clasts in carbonate fine clastic breccias. It is likely that the very shallow lagoons with sediments subjected to subaerial leaching and diagenesis processes, to which our samples bear testament, could have formed during a strong lowering of the water level in the marine basin. The manifestation of gas-fluid filtration introduced additional chlorides and Ba, Sr, and Fe ions. Interestingly, calcite microspherulites were found near the Dvurechensky mud volcano in the Black Sea under a bacterial layer covering, on the one hand, newly formed carbonate crusts and small tubular formations (Shnyukov et al. 2006). It should be noted that calcite microspherulites that form accumulations in microbial mats of modern salt lakes (Bischoff et al. 2020; Mercedes-Martín et al. 2021) show morphological similarity with the accumulations of the studied calcite microspherulites in fine clastic breccias. Hydrothermal seeps were previously established to constitute an additional process during the formation of the Kazantip bryozoan bioherm complex (Antoshkina et al. 2020).

10 Conclusion

On the basis of the available data, it is possible to state with some certainty that the biogenic and authigenic formation of calcite microspherulites, sulfides, sulfates, and high-magnesian calcite began with an abiogenic mechanism involving the activation of gas-fluid seepages. Although this abiogenic mechanism contributed to an increase in water salinity in local areas, the activation of bacteria led to the bioinduction of abiogenic methane and synsedimentary formation of minerals in the pelitomorphic matrix of carbonate fine clastic breccias. On the territory of the Kazantip Nature Reserve, such a specific near-bottom situation could be associated with a discharge zone of ascending gas-fluid flows along faults or the activation of gryphons of an ancient mud volcano. At this time, the most active phase of the Alpine Orogeny had already appeared in the Mediterranean Sea.

Acknowledgements The authors express gratitude to I.V. Smoleva, S.I. Isaenko, V.N. Filippov, and E.M. Tropnikov for conducting analytical studies within the framework of this research. The investigations were supported the State Assignment of the IG FRC Komi SC UB RAS (No. 122040600013-9) and the Agreement between the IG FRC Komi SC UB RAS and the FSBI "Joint Directorate of Protected Areas "Reserved Crimea".

References

Aloisi G, Wallmann K, Drews M, Bohrmann G (2004) Evidence for the submarine weathering of silicate minerals in Black Sea sediments: Possible implications for the marine Li and B cycles. Geochem. Geophys. Geosyst. 5 Q04007. https://doi.org/10.1029/2003GC000639

Antoshkina AI, Leonova LV, Simakova YuS (2020) The Development of Miocene biohermal bryozoan limestones of Kazantip Cape (Crimea): A New Insight. Doklady Earth Sciences 491 (2):195–198

Bischoff K, Sirantoine E, Wilson M E J, George AD, Monteiro J M, Saunders M (2020) Spherulitic microbialites from modern hypersaline lakes, Rottnest Island, Western Australia. Geobiology 00:1–17

Boggs SJr (2010) Petrology of sedimentary rocks, 2nd edn. Cambridge University Press, Cambridge

Dorogochinskaya VA, Stepanov AN, Fadeev VS (1993) Geochemical factors influencing the formation of the composition of relic C_{17}-C_{20} alkanes in caustobiolites. Petrochemistry 33(1):7–16 (In Russian)

Godovikov AA (1961) About a calcite from the Amerovo quarry in the Moscow region. Tr. Mineralogical Museum 12:177–181 (in Russian)

Gonzalez-Muñoz MT, Martinez-Ruiz F, Morcillo F, Martin-Ramos JD, Paytan A (2012) Precipitation of barite by marine bacteria: A possible mechanism for marine barite formation. *Geology* 40(8):675–678

Farias F, Szatmari P, Bahniuk A, França AB (2021) Evaporitic carbonates in the pre salt of Santos Basin: Genesis and tectonic implications – A reply. Marine and Petroleum Geology 133:105–203

Hunt J (1982) Geochemistry and geology of oil and gas. Moscow Mir (In Russian)

Hodgson DM, Bernhardt A, Clare MA, Da Silva A-C, Fosdick JC, Mauz B, Midtkandal I, Owen A, Romans BW (2018) Grand challenges (and great opportunities) in sedimentology, stratigraphy, and diagenesis research. Frontiers in Earth Science 6:173

Kelts K, Talbot M (1990) Lacustrine carbonates as geochemical archives of environmental change and biotic/abiotic interactions. In: Tilzer MM, Serruya C (eds). Large Lakes. Brock/Springer Series in Contemporary Bioscience. Springer, Berlin, Heidelberg

Klyukin AA (2006) Factors determining the biodiversity of the Kazantip nature reserve. Proceedings of the Nikiski botanic garden. National scientific center 126:133–148 (In Russian)

Krumbein WE, Paterson DM, Stal LJ (Eds.) (1994) Biostabilization of sediments. BIS Univ. of Oldenburg

Lein AYu, Ivanov MV (2009) Biogeochemical cycle of methane in the ocean. Publisher: Nauka (in Russian)

Leonova LV, Antoshkina AI, Simakova YuS (2020) Specific minerals in bryozoan bioherms and polychaete limestones. Proceedings of the All-Russian lithol. conf. "Geology of Reefs". Syktyvkar Geoprint 71–73

Lin I, Wiles L, Waller R, Goucke R, Nagree Yu, Gibberd M, Straker L, Maher CG, O'Sullivan PPB (2020) What does best practice care for musculoskeletal pain look like? Eleven consistent recommendations from high-quality clinical practice guidelines: systematic review. Br J Sports Med 54:79–86

Martinez-Ruiz F, Jroundi F, Paytan A, Guerra-Tschuschke I, del Mar Abad M, González-Muñoz MT (2018) Barium bioaccumulation by bacterial biofilms and implications for Ba cycling and use of Ba proxies. Nat Commun **9**:16–19

Mercedes-Martín R, Brasier A T, Rogerson M, Reijmer JJ, Vonhof H, Pedley M. (2017). A depositional model for spherulitic carbonates associated with alkaline, volcanic lakes. Marine and Petroleum Geology 86:168–191

Mercedes-Martín R, Rao A, Rogerson M, Sánchez-Román M. (2021) Effects of salinity, organic acids and alkalinity on the growth of calcite spherulites: Implications for evaporitic lacustrine sedimentation. Depositional Rec. P. 00:1–22

Serebryakov S N (1976) Biotic and abiotic factors controlling the morphology of Riphean stromatolites. Developments in Sedimentology 20:321–336

Silaev VI, Kokin AV, Pavlovich NV, Shanina SN, Kiseleva DV, Vasiliev EA, Martirosyan OV,. Smoleva IV, Filippov VN, Khazov AF, Shuisky AS, Schemelinina TN, Ignatiev GV, Slyusar AV (2021) The first results of complex studies of modern microorganisms by physical-chemical and mineralogical-geochemical methods. Vestnik of Geonauk 9:3–33 (In Russian)

Silaev VI, Lyutoev VP, Petrovsky VA, Khazov AF (2013) Experience in studying natural carbonaceous substances and some synthetic analogs by Raman spectroscopy. Mineral. Journ. (Ukraine) 35(3):33–47 (In Russian)

Shnyukov EF, Kutniy VA, Maslakov NA, Shnyukova EE (2006) On the mineralogy of carbonate formations of gas sources in the Black Sea. Geology and Minerals of the World Ocean 69–81 (In Russian)

Stahl W (1984) Carbon isotopes in oil geochemistry. In: Jager E and Hunziker JC (eds) Isotope Geology Moscow Nedra 289–297 (In Russian)

Stanienda-Pilecki KJ (2018) Magnesium calcite in Muschelkalk limestones of the Polish part of the Germanic Basin. Carbonates and Evaporites 33:801–821

Tatarinov AV, Yalovik LI, Kashkak ES, Danilova EV, Khromova EA, Khakhinov VV, Namsaraev BB (2017) Mineralogical and geochemical features of bacterial mats and travertines of the Khoito-Gol termal spring (East Sayan). Geology and Geophisics 58(1):60–73 (In Russian)

Temescu AMF, Pratt LM, Rothwell GW, Strother PK, Nadon GC (2009) Carbon isotopes Support the presence of extensive land flores pre-dating the origin of vascular plants. Palaeoclimatology, Palaeoecology 283:46–59

Tikhomirov S, Kimstach T (2011) Raman spectroscopy is a promising method for studying carbon nanomaterials. Analytics 1:28–32

Vassoevich NB (1986) Geochemistry of organic matter and the origin of oil: selected works. M.: Nauka (In Russian)

Vu B, Chen M, Crawford RJ, Ivanova EP (2009) Bacterial extracellular polysaccharides involved in biofilm formation. Molecules 2009 14(7):2535–2554

Wilkinson BH, Given R K (1986) Secular variation in abiotic marine carbonates: constraints on Phanerozoic atmospheric carbon dioxide contents and oceanic Mg/Ca ratios. The Journal of Geology 94(3): 321–333

Wopenka B, Pasteris JD (1993) Structural characterization of kerogens to granulite-facies graphite: Applicability of Raman microprobe spectroscopy. American Mineralogist 78:533–577

Wu S, Chiang C-Y, Zhou W (2017) Formation mechanism of $CaCO_3$ spherulites in the Myostracum layer of Limpet Shells Crystals 7:319. https://doi.org/10.3390/cryst7100319

Zhang C, Lv J, Li F, Li X (2017) Nucleation and growth of Mg-calcite spherulites induced by the bacterium Curvibacter lanceolatus strain HJ-1. Microsc. Microanal. 23:1189–1196

Morphological Features of Framboidal Pyrites in Various Geological Settings: The Contribution of Microorganisms to Their Formation

Lyubov V. Leonova, Oxana B. Azovskova, Elena I. Soroka, and Yulia S. Simakova

Abstract The objects of our research are framboidal pyrites. They are named for their resemblance to raspberries («la framboise» in French) (Rust 1935). The term «framboid» characterizes only micromorphology, and the mineral composition can be represented by pyrite, marcasite, troilite, gel-pyrite, greigite, magnetite, makinavite. But more often—pyrite. At present there are no exact criteria to establish the biogenic or abiotic origin of framboids. The aim of our work is to study and compare the morphology of framboids from the sediment of a modern mud volcano and rocks of different ages and genesis; identification of direct and indirect signs indicating their microbial origin. The following methods have been applied: X-ray diffractometry, scanning electron microscopy and energy-dispersive analysis. Thermal methods have been used to study bituminized lenses with framboids in some rocks of the Urals. We have studied the samples from the sediment of an active mud volcano (Kuchugury, Krasnodar Territory) and from the rocks of some deposits of the Urals (Safyanovskoe, Mikheevskoe, Moscow placer). All framboidal pyrites characterizes by a great morphological diversity: in one cluster there can be nearby well-shaped, and those without faceting; calibrated and very different microcrystals in size. This is difficult to explain only by a chemogenic origin. However, a number of features point to their bacterial genesis: framboids are confined to macrobiota remains or bitumen lenses in the rock. On the surface of some aggregates, sheaths are observed, often replaced by silicon dioxide. Biofilms substituted S, Pb Cu, Zn, Sb, As have also been found.

Keyword Framboidal pyrites Mineralized biofilms Mud volcano Ural Copper ore deposits

L. V. Leonova (✉) · O. B. Azovskova · E. I. Soroka
Institute of Geology and Geochemistry UB RAS, Ekaterinburg, Russia
e-mail: lvleonova@yandex.ru

Y. S. Simakova
Institute of Geology, FRC Komi SC UB RAS, Syktyvkar, Russia
e-mail: yulia5-07@mail.ru

O. V. Frank-Kamenetskaya et al. (eds.), *Biogenic—Abiogenic Interactions in Natural and Anthropogenic Systems 2022*, Springer Proceedings in Earth and Environmental Sciences, https://doi.org/10.1007/978-3-031-40470-2_11

1 Introduction

Framboid pyrites are microscopic aggregates with a diameter of 2–50 microns, which are spherical accumulations of crystallites (microcrystals) of predominantly octahedral habit with a size about 0.25–2 microns. They are named for their resemblance to raspberries (French «la framboise») (Rust 1935). The term of «framboid» characterizes only micromorphology of objects, according to mineral composition they are often represented by pyrite and marcasite (FeS_2), troilite (FeS), gel-pyrite, as well as sedimentary minerals: greigite (Fe_3S_4), magnetite (Fe_3O_4), hematite (Fe_2O_3), makinavite $(Fe, Ni)_1 + x\, S$ ($x = 0$–0,11).

Usually framboidal pyrite is formed in reducing environment. Under conditions of a well-aerated environment, framboids are also formed, but they have iron oxide composition of crystallites. A number of the researchers believe that such framboidal pyrites have been formed as a result of the vital activity of bacteria (Maslennikov et al. 2016; Antoshkina et al. 2017; Butler and Rickard 2000; Astafieva et al. 2005; Gong et al. 2008).

However, despite many years of study and discussion there are still no exact criteria to establish the biogenic or abiotic origin of framboids.

The purpose of our work is to study the framboidal pyrites of different ages from different settings and identify direct and indirect signs indicating microbial or chemogenic origin.

In the process of investigation, the following tasks were solved. To collect the samples with framboidal pyrites of rocks of various geological ages; determination of a relative time of framboids' formation—synsedimentary ones or associated with superimposed hydrothermal processes; study of the geological settings and facies conditions that caused the formation of framboids.

2 Materials and Methods

2.1 *Objects of Study*

We found framboids in sediments of different ages and in a wide range of primary sedimentary, hydrothermal-sedimentary settings, often with traces of bituminization.

We have made a working collection (more than 50) for a detailed study of framboidal pyrites. Framboids of typical morphology of our collection are presented in Fig. 1.

These framboids (Fig. 1) are interesting, but pyrites from the deposits of the Urals (Safyanovskoe Cu–Zn-sulfide, Mikheevskoe porphyry copper, Moscow Au—placer) turned out to be the most informative in our collection. The study of pyrite and associated ores revealed framboids with mineralized crusts and biofilms. In this case it is natural that the question concerning the appearance of modern framboids and whether similar traces of microbial activity will be visible on their surface or

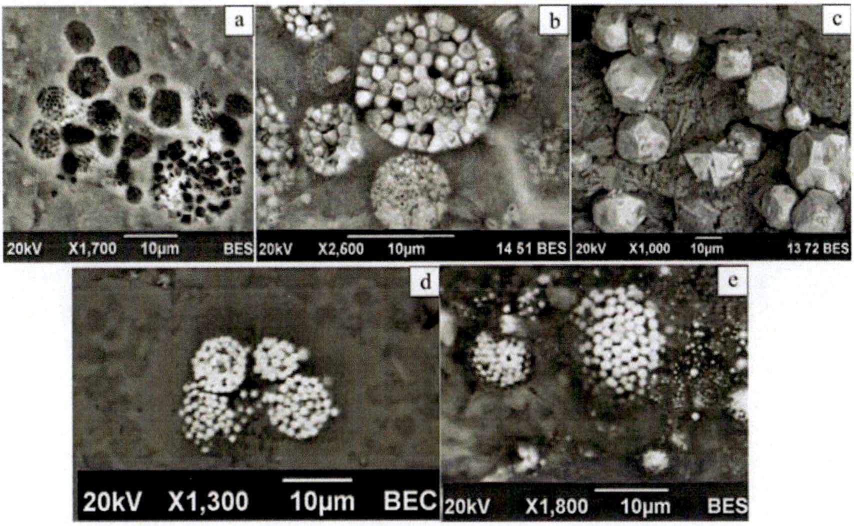

Fig. 1 Framboids of a typical habitus of different environments (author's collections): **a** pyrites in cristobalite-chalcedony rock of hydrothermal-sedimentary genesis. Marginal zone of outpouring of andesites on the seabed (D/C). Bichur village, Artemovsky district, Sverdlovsk oblast; **b** framboids of iron oxide composition in goethite-limonite crusts. Clay-gypsum-carbonate deposits of a mud volcano (Miocene N_1). Kazantip Reserve, Leninsky District, Republic of Crimea; **c** accumulations of pyrite grains on the surface of glauconite grains. Bituminized and clay shales (J_3). Oil shale deposit Chim-Loptyug. Udorsky district of the Komi Republic; **d** pyrites from sandstones: the clastogenic component (about 90%) is represented by siliceous organogenic rock (gaize?), fragments partially replaced by glauconite. Chalcedony cement (P_2l). Aktai area, Novolyalinsky district, Sverdlovsk oblast; **e** framboidal pyrite in the zone of bituminization of colonial coral (Favosites). Limestone (D_1 l). Aramashevo village, Sverdlovsk oblast

not arises. We have resolved the question when studying the sediments of a modern mud volcano located near the village of Kuchugury (Krasnodar Krai).

In this article, we present the results of studying the framboidal pyrites from the sediments of a modern mud volcano «Plevak» located near the village of Kuchugury (Krasnodar Krai) (Fig. 2a) and from the number of Ural deposits (Fig. 2b, c) of different ages and genesis (Safyanovskoe Cu–Zn-sulfide, Mikheevskoe porphyry copper, Moscow Au—placer).

Special interest to the occurrences of mud volcanism is caused by the accompanying emissions of water, oil and gases, such as methane (CH_4), hydrogen sulfide (H_2S), CO_2, N_2 and inert gases (helium, argon), and heavy hydrocarbons may also present in small quantities (Kropotkin and Valyaev 1980; Kikvadze et al. 2014; Kholodov 2002a, b). When investigating the mud volcanoes including those of the Kerch-Taman region the great attention was paid to accessory minerals of breccia rocks (Avdusin 1948) erupted together with the mud volcanic substrate. Priority in the study was given to minerals—indicators of the depth of this process (Kalyuzhny

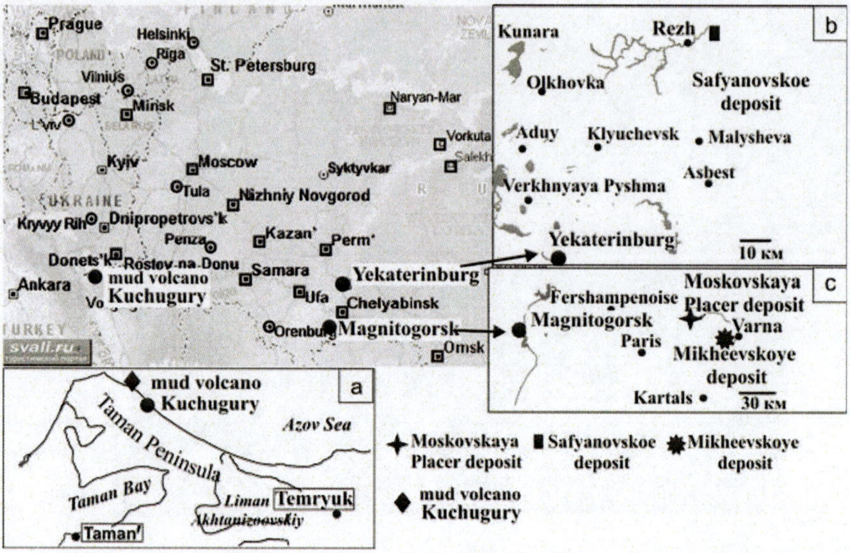

Fig. 2 Geographical location geological objects of study: **a** of the mud volcano Kuchugury («Plevak»); **b** Safyanovskoye Cu–Zn pyrite deposit; **c** Mikheevskoe (Mo, Au)-Cu-porphyry ores deposit and Moskovskaya Au—placer deposit

et al. 1984; Shnyukov and Kobolev 2018; Kropotkin and Valyaev 1980; Sobisevich and Sobisevich 2018) since the problem of depth formation and exhalation mechanism of mud volcano deposits was unsolved. But of great interest to us is the mineral composition of mud and the new formed exogenous «here and in a short time» minerals (especially pyrite framboids). Although the mud volcanism of the Taman-Kerch region has been actively studied and continued to be studied (Ovsyuchenko 2005; Ismagilov et al. 2006; Ershov et al. 2019; Kulakovskiy et al. 2022; Beloborodov 2022), there is very little scientific information especially about studying, in particular about «Plevak» volcano (Tveritinova et al. 2015; Sobisevich et al. 2015). Therefore our research is relevant at the present time.

Currently the mud volcano is popular as an object of therapeutic mud and is a small lake between the hills (Fig. 3) with periodic mud volcanic eruptions of a liquid mud substrate.

The photo (Fig. 3a) shows a cover of mud breccia flowing down to the sea and creating a peculiar landscape. During sampling the sediment was viscous (Fig. 3b), it was a hot summer. But depending on the dryness of the season, the viscosity changes from liquid silt to solidified substrate. Objects of the study: modern framboidal pyrite crusts, biomorphoses, and individual pyrite aggregates were formed under such lacustrine-mud conditions.

Framboidal pyrites are preserved in ancient rocks as fossil analogues of synsedimentation environments, but can also be formed in late diagenesis, with regional metamorphism and hydrothermal processes. In our collection framboidal pyrites are

Fig. 3 Mud volcano «Plevak» in 2021: **a** mud volcano with a flow of hilly breccia flowing down to the sea; **b** viscous substrate in the volcanic crater, sampling for study

represented by findings in the rocks of different ages from some deposits of the Urals (Safyanovskoye Cu–Zn pyrite, Mikheevskoe (Mo, Au)–Cu-porphyry ores and Moskovskaya Au—placer) (Fig. 2b, c).

There are numerous findings of synsedimentary framboids in sedimentary-hydrothermal and massive sulfide rocks formed in the discharge zones of the ancient gas-fluid seepage «black smokers» (Maslennikov et al. 2016; Leonova et al. 2022).

2.2 *Geological Setting*

2.2.1 «Plevak» Mud Volcano

An active mud volcano «Plevak» (Fig. 2a). is associated with a diapiric anticline which strikes northeast and is composed mainly of clay strata.

The sediments of the oil-bearing Maikop suite are overlain by clays interbedded with marls of the undivided North Shirvan and Matrosskaya suites (N_1ss–mt) (Tarkhan–Chokrak–Karagan–Konka) which are replaced upsection by clays with interlayers of marls, limestones, sandstones and siltstones of the Zelenskaya (N_1zl) and Panagian (N_1pn) suites (Sarmatian) (Tveritinova et al. 2015). This rocks are unconformably overlain by a sequence of interbedded clays, sands, diatomites with lenses of hill breccia and bryozoan bioherms in some places at the base of the Kholodnodolinovskaya suite (N_1hd) (Meotis) replaced by greenish-gray clays with interlayers of siltstones, sands, shelly detritus, and brown ironstones of the Kruglovskaya ($N_{1\text{-}2}$kr) and Zheleznorogovskaya (N_2zr) suites (Pontian–Cimmerian) and higher with light quartz sands with interlayers of clays and lenses of conglomerates of the Sennovskaya sequence (N_2sn) (Lower Akchagylian). The deposits near the mud volcano are composed mainly of sandy and loess-like rocks of the Akchagyl age (Tveritinova et al. 2015).

As noted by researchers (Tveritinova et al. 2015) mud volcano «Plevak» is confined to a zone of increased permeability and degassing. There are a modern mud griffin and a barrier lake within the complex structure of the volcano. Geologists suggest that the lake is not only barrier but exists due to modern fluid recharge through the degassing channel (Tveritinova et al. 2015).

2.2.2 Ural Deposits

Safyanovskoye Cu–Zn pyrite deposit. This deposit is located within the East Ural megazone of the Middle Urals, 9 km northeast of the city of Rezh (Fig. 2b) and is confined to the southern part of the Rezhevskaya structural-formational zone, the internal structure of which is characterized by a complex of thrust dislocations, which caused the spatial overlap of complexes of different ages, formations and facies (Yazeva et al. 1992). The ore-bearing volcanogenic-sedimentary series of the deposit are about 500 m thick and belongs to the rocks of the Safyanovsky complex.

Some researchers suggest the deposit is a typical representative of the dacite-rhyolite-black shale association (Maslennikov 2006). According to (Korovko et al. 1999), the volcanogenic rocks of the ore-bearing complex are assigned to the Middle Devonian basalt-rhyolite formation or the Lower-Middle Devonian basalt-andesite-dacite-rhyolitic formation (Smirnov 2012). The faunal findings show the Early-Middle Devonian age of volcanogenic-sedimentary rocks and limestones (Puchkov et al. 1990; Amon and Korovko 1992; Korovko et al. 1999; Chuvashov et al. 2011; Anfimov et al. 2015). By the U–Pb dating of zircon (422.8 ± 3.7 and 422.8 ± 2.0 Ma) of the volcanic rocks (andesite, rhyolite), the volcano-sedimentary ore-bearing series of the Safyanovskoye deposit include volcanic rocks of the Przhidol Stage of the Upper Silurian (Pritchin et al. 2021). The volcano-sedimentary series of a deposit were formed under back-arc paleobasin conditions (Koroteev et al. 1997). But the fragments of island-arc formation rocks of the Silurian Trans-Ural paleoarc also present at the deposit (Pritchin et al. 2021).

Framboidal pyrites were also formed in a later hydrothermal process (argillization).

Mikheevskoe (Mo, Au)-Cu-porphyry ores deposit. The deposit is located near v. Varna of the Chelyabinsk region, 250 km south from Chelyabinsk (Fig. 2c) and is one of the most significant (Mo, Au)–Cu-porphyry type deposit in Russia.

In the regional geotectonic plan, the deposit is confined to the border of large structures—the East Ural and Trans-Ural megazones. Porphyry mineralization is of Early Carboniferous (Tournaisian) age (Tessalina and Plotinskaya 2017). It is spatially and genetically associated with small intrusions of the magmatic Mikheevka complex D3–C1 (Grabezhev 2014).

One of the features of the deposit is the wide manifestation of low-temperature metasomatism (argillization) and the industrial ore content of this type of metasomatites. Argillisites are often steeply dipping zones confined to tectonic faults, dikes, and their exocontacts. In the upper part of the section, they turn into uneven areal formations up to 46 m thick. Two or three stages of argillization are assumed. The last of them is probably associated with the Late Mesozoic tectonic–magmatic activation of the region and with the Late Cretaceous marine transgression (Azovskova et al. 2019; Azovskova and Rovnushkin 2021).

The framboidal pyrites of this deposit are associated with areas of rock impregnation with condensed hydrocarbons.

Moskovskaya Au—placer deposit. The Moskovskaya Au—placer deposit is located in the Chesmensky district, Chelyabinsk region (Southern Urals) (Fig. 2c) near the village of Moskovsky, about 30 km from the village of Varna. This placer deposit is confined to the Redutovskaya erosion-structural depression, the Paleozoic basement of which in this part is composed mainly of the Carboniferous formations of the Birgilda (C_1br) and Kuzeya (C_2ks) formations—metamorphosed volcanogenic-carbonaceous-carbonate, terrigenous-carbonate complexes.

2.3 Methods

Panning method. Viscous sediment of a mud volcano (1 kg) was washed with distilled water under laboratory conditions. The resulting granular material was divided into fractions: magnetic and non-magnetic; heavy and light. Next, the samples were studied by scanning electron microscopy and energy-dispersive analysis. A sample of bottom sediments was taken in the amount of 5 kg from a depth of 1 m (Fig. 3b) in 2021.

The phase composition of the samples of volcanic mud was determined according the diffraction patterns of non-oriented samples, the clay components—according the diffraction patterns of oriented samples under standard diagnostic treatments (air-dried and saturated with ethylene–glycol).

The XRD characterizations were carried out with an X-ray diffractometer Shimadzu XRD-6000, CuKα radiation, Ni filter, 30kV, 30mA. Scan areas 2θ°: 2–52°—for oriented samples, 18–45° for non-oriented and 55–65°—for 060 area. Semi-quantitative X-ray diffraction analysis was performed using the Sybilla© program which allows modeling diffraction profiles for mixtures of clay minerals and mixed-layer phases in oriented samples.

The elemental composition and microfossils from the granular component of the mud volcanic substrate were determined by the method of scanning electron microscopy (SEM) (JSM-6390LV JEOL) and (TESCAN MIRA LMS (S6123), carbon sputtering) and energy dispersive spectrometry (EDS) (INCA Energy 450 X-max 80). Also for a detailed study of framboidal pyrites, polished sections were prepared from the following samples of the Ural rocks: carbonaceous-siliceous rock, pyrite ore, argillisites with bitumen lenses.

The bituminized lenses with framboids of the rocks of the Urals were studied by the thermal methods. Thermal effects were investigated by the methods of differential thermal analysis (DTA) and synchronous thermal analysis (STA) by 449 F5 Jupiter. TG–DTA measurements were performed in platinum bowls in air with a heating rate of 10 °C/min in the temperature interval of 30–1100 °C using the Q-1500 D derivatograph.

3 Results

3.1 *Framboidal Pyrites of the «Plevak» Mud Volcano (Kuchugury)*

The X-ray diffraction pattern of the clay fraction is shown in Fig. 4. The dark gray clay of the mud volcano «Plevak» is rich and plastic. Due to the salinity of the stratum a significant amount of halite was determined in the samples. Smectite is practically absent in the clay component of the samples.

The clay fraction consists of dioctahedral mica (illite), kaolinite, chlorite, quartz; the swelling phase is represented by mixed-layer illite/smectite. According to the results of modeling diffraction profiles the illite/smectite structure has a short-range order factor R = 2 (i.e., some degree of segregation), which corresponds to the zone of the appearance of mixed-layer structures and beginning of illitization according to Kholodov (Kholodov 2013). Disordered mica structure 1Md was identified.

Approximate semi-quantitative content of clay minerals is: illite—30–35%, chlorite—7–10%, kaolinite—25–34%, illite/smectite (R = 2)—20–30%. In the region of 060 reflections in addition to illite, kaolinite, and chlorite peaks there are also reflections of interstratified phases with spacings d/n ≈ 1.48–1.52 Å.

The presence of quartz, halite, feldspar and a small amount of pyrite is also noted in the mud (Fig. 4).

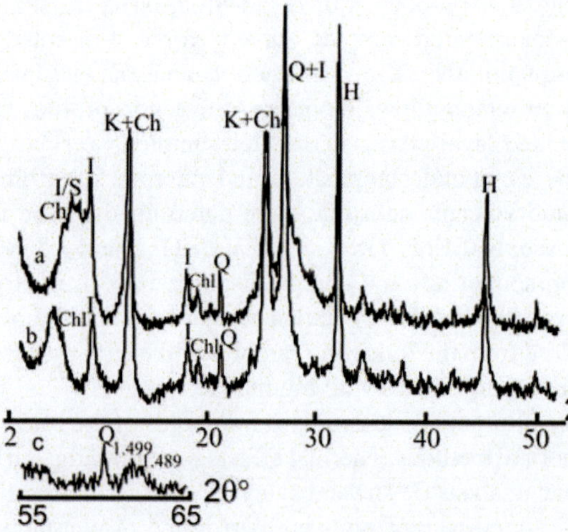

Fig. 4 XRD pattern of clay fraction sample: **a** oriented air-dried, **b** oriented glycolated, **c** 060 area. K—kaolinite, I—illite, Chl—chlorite, I/S—illite/smectite, Q—quartz, H—halite. Spacing in angstroms

The results of an electron microscopic study showed the presence of residues of non-mineralized biota in granular fractions: a bone fragment (Fig. 5a) and a fish vertebra (Fig. 5b). It is interesting that the EDS spectrum of bone remains (Fig. 5a) shows peaks of Sr and Ce which indicates the lifetime accumulation of these elements by organisms from the environment. Whole shells (Fig. 5c), fragments of diatoms and spicules of siliceous sponges were found (Fig. 5d). Accumulation of framboidal pyrites and marcasite was found on a modern algae bundle (Fig. 5e). The EDS spectrum of algae shows peaks of such biogenic elements as C, S, P, Ca, K, Cl and even Sr.

Granular component of mud volcanic substrate contains sulfidized plant remains (Fig. 5f) and fragments of elongated cylindrical biomorphoses (Fig. 5g) replaced by framboidal pyrites and marcasite. The most informative fossil is a foraminifera shell completely replaced by framboidal pyrite (Fig. 5h). This finding indicates that pyrite framboids were formed during sediment accumulation and the shell of the dead organism was empty.

Among the granular component of the mud volcanic substrate, individual framboids (Fig. 6a) and accumulations in the form of fragile thin crusts (Fig. 6b) are often encountered. The presence of such pyrite forms clearly indicates their formation in situ i.e. in the surface conditions of a mud volcanic reservoir. Detailed SEM investigations revealed sheaths (Fig. 6c, d) that look like a translucent cover on the surface of some framboidal pyrites. The SEM image (Fig. 6d) shows that surfaces of the covers have a clumpy structure and octahedral crystallites are visible in the area without the cover. Such covers are usually formed on colonies of bacteria (as well as cyanobacteria, actinomycetes, micromycetes, etc.) in the form of mucus from extracellular biopolymers as a protective reaction to adverse environment (Mudretsova-Viss et al. 2014). Similar formations observed on pyrite framboids are interpreted as the remains of microbial colonies (Gong et al. 2008).

Also of great interest are framboidal formations with a «sunflower» structure (Fig. 6e, f), where the central part is composed of ordered layers of crystallites and the peripheral part is composed of prismatic microcrystals (Fig. 6e). However, there may be formations with a central part of randomly arranged octahedra of different sizes and the edge zone on the contrary is composed of microlayers of calibrated crystallites (Fig. 6f). Such an unusual variability of properties is characteristic of microbial formations.

In biofilms or under conditions controlled by microorganisms the growing minerals are affected by the structures of various biopolymers and metabolites secreted by bacteria, pH microfluctuations and electrical and electrochemical potentials of living and non-living cells. Therefore, organomineral formations often differ from their chemogenic counterparts.

Framboids are usually composed of crystallites of the same size and framboid size correlates with the size of crystallites (Fig. 6g). However, sometimes one can see framboids consisting of octahedral crystallites with different sizes (Fig. 6h). Also large octahedrons can be located between framboids (Fig. 6g) or form separate microdruses.

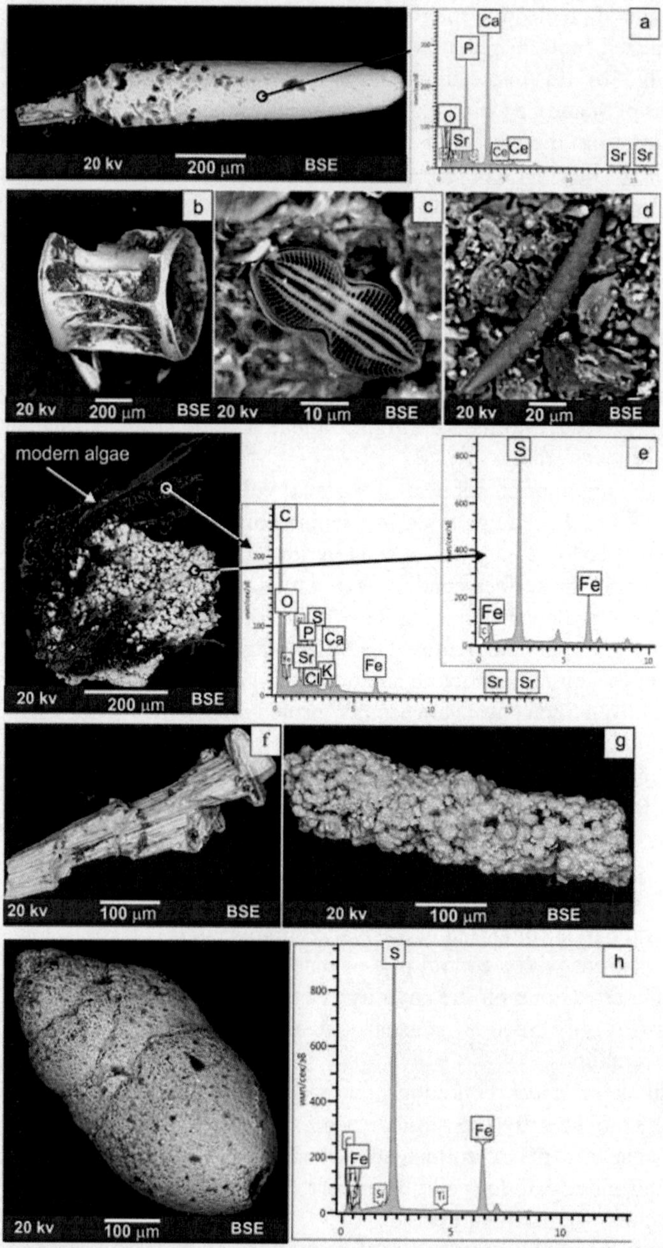

Fig. 5 Remains of biota from the granular component of the studied mud volcanic sediment: **a** SEM photo of a bone fragment, peaks of Ce and Sr are noted on the EMF spectrum; **b** fish vertebra; **c** diatom; **d** spicule of flint sponge; **e** recent algae with overgrown accumulations of framboidal pyrite and marcasite; **f** sulfidized plant residue; **g** fossil composed of pyrite framboids; **h** SEM image and EDS spectrum of foraminifera shell replaced by framboidal pyrite

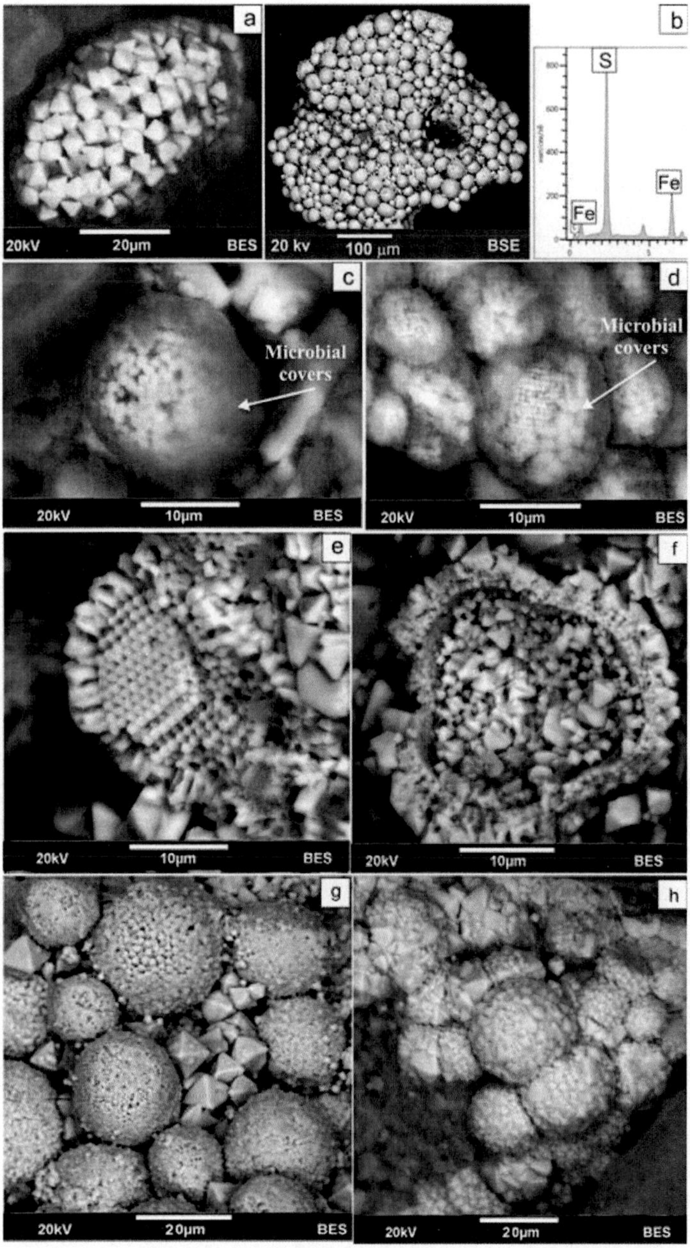

Fig. 6 SEM photo of the morphological diversity of framboid pyrites from the mud volcanic sediment: **a** single framboid; **b** framboidal pyrite crust; **c** covers on the framboid surface in the form of a translucent film; **d** microbial covers with a clumpy structure on framboids. Pyrite microcrystals are visible in the area without a cover; **e** framboidal formations with a «sunflower» structure in the center and octahedrons of different sizes in the marginal part; **f** «sunflower» with ordered crystals in the border and octahedrons of different sizes in the center; **g** framboidal pyrites with small crystals of the same size and large pyrite octahedrons between segregations; **h** framboidal pyrites with octahedral crystals of various sizes in segregations

Plates composed of small framboids (Fig. 7a, b) may contain cubic pyrites. At high magnification (Fig. 7b, c), it was found that the cubes are composed of small fragments like puzzles (Fig. 7c). Rather large (1 mm) intergrowths of pyrite cubes are found sometimes (Fig. 7d). On closer examination, they look like skeletal crystals (Fig. 7e) with framboidal pyrites embedded between the sheets (Fig. 7e). In addition to morphologically diverse pyrites, quite a lot of marcasite is noted among the granular component of the sediment. It forms aggregates of acicular microcrystals or lamellar intergrowths—the so-called «cockscombs». It is interesting that these structural varieties of FeS_2 occur together in the sediment. An intergrowth of a truncated pyrite octahedron with lamellar marcasite is presented on Fig. 7f.

A spherical formation in fact is a large framboid composed of pyrite crystals and includes small isolated framboids (Fig. 7g) and «cockscombs» of marcasite (Fig. 7h).

3.2 *Framboidal Pyrites of the Ural Deposits*

Safyanovsky Cu–Zn-sulfide deposit. The ore-bearing complex of the deposit includes massive ores (copper and copper-zinc), vein-stockwork copper (ore pillars), disseminated copper and copper-zinc ores. The ore-bearing rocks also include carbonaceous-siliceous formations (Fig. 8a) (Yaroslavtseva et al. 2012). They are promising for micropaleontological study, since they are less altered by hydrothermal processes compared to pyrites. In accordance with microlayering, they contain fossils of presumably foraminifers with apatite shell (Anfimov et al. 2015), within which framboidal pyrites occur (Fig. 8b). It should be noted that spherical framboidal segregations can be composed of both crystallites significantly different in size and habit (Fig. 8c, d), and specific massive iron sulfide, perhaps initially gel-pyrite (Fig. 8d). Before studying the framboidal pyrites in the sediment of the modern Kuchugury mud volcano (Azovskova et al. 2022), such formations were thought to be a result of late diagenesis recrystallization or an effect of the secondary hydrothermal processes.

The findings of fossilized sheaths enveloping the surfaces of some spherical segregations give important information. According to EDS analysises, they are composed of silicon dioxide. At the SEM photos, they look like a translucent film, which negatively affects the image quality (Fig. 8e). It is possible that some researchers observed such images, but interpreted the haziness of an image as a defect of the preparation or as a result of a poorly tuned instrument.

We consider such sheaths on framboids as originally mineralized slimy films that covered fossil bacterial colonies. It is known that under adverse external conditions, many microbial organisms (bacteria, cyanobacteria, actinomycetes, et al.) and micromycetes secrete protective mucus—extracellular biopolymer substances (Mudretsova-Viss et al. 2014) that can be mineralized, including by silicon dioxide (Toporski et al. 2002; Orleansky et al. 2007).

It is interesting, that similar sheaths, also represented by SiO_2, occur at the surfaces of framboids of the Safyanovskoye deposit sulfide ores (Fig. 9a) which were formed under reducing conditions. Framboids were found in the pyrite rock sample (Saf 7/

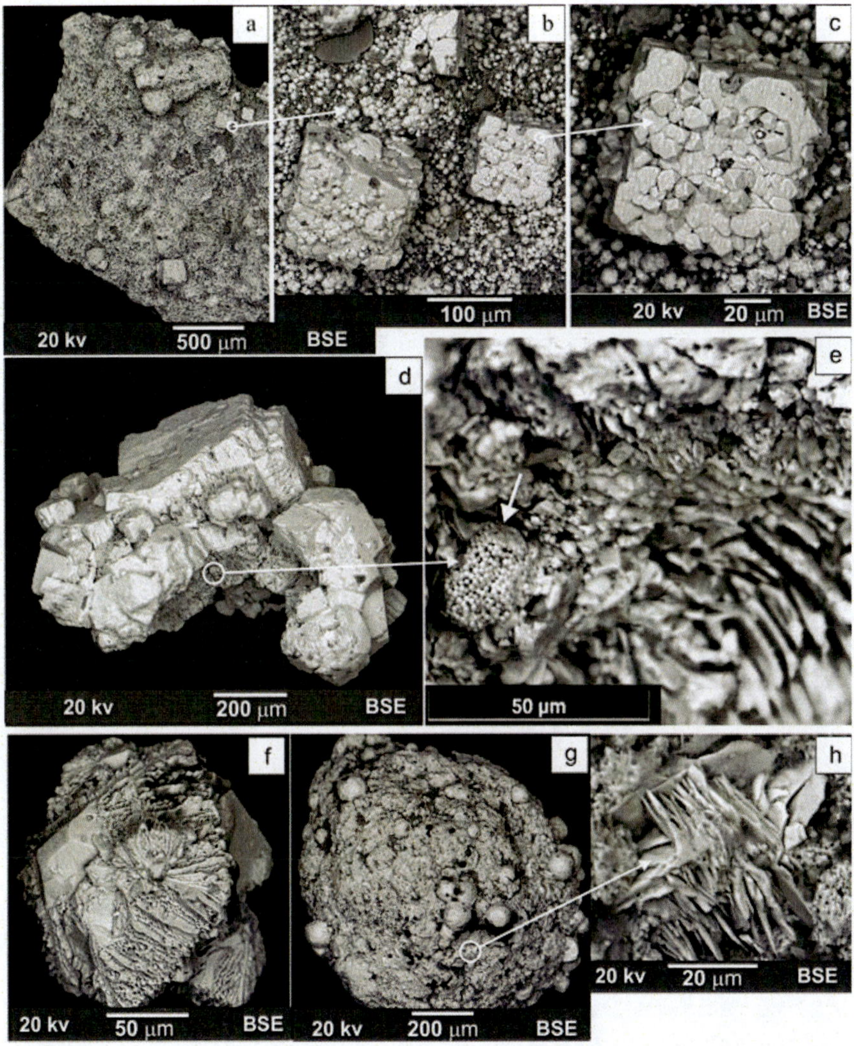

Fig. 7 Pyrite aggregates and intergrowths with marcasite: **a**, **b**, **c** framboidal pyrite plate with cubic pyrite «puzzles»; **d** large (1 mm) intergrowth of cubic pyrites; **e** enlarged fragment of the central part of this pyrite with a skeletal structure and an embedded framboid; **f** intergrowth of a truncated pyrite octahedron with lamellar marcasite of the «cockscomb» type; **g** large pyrite framboid with small framboids and marcasite; **h** enlarged fragment of marcasite

18) in the mine at the depth of 232 m together with mineralized pipes of vestimentifer and polychaetes (Fig. 9b, c) (Leonova et al. 2022).

In polished thin sections of this sample in the course of electron microscope research, clusters of framboidal pyrites were determined (Fig. 9d), which encrust in the inner side of the wall of polychaete tubes (Fig. 9e). It should be noted that earlier

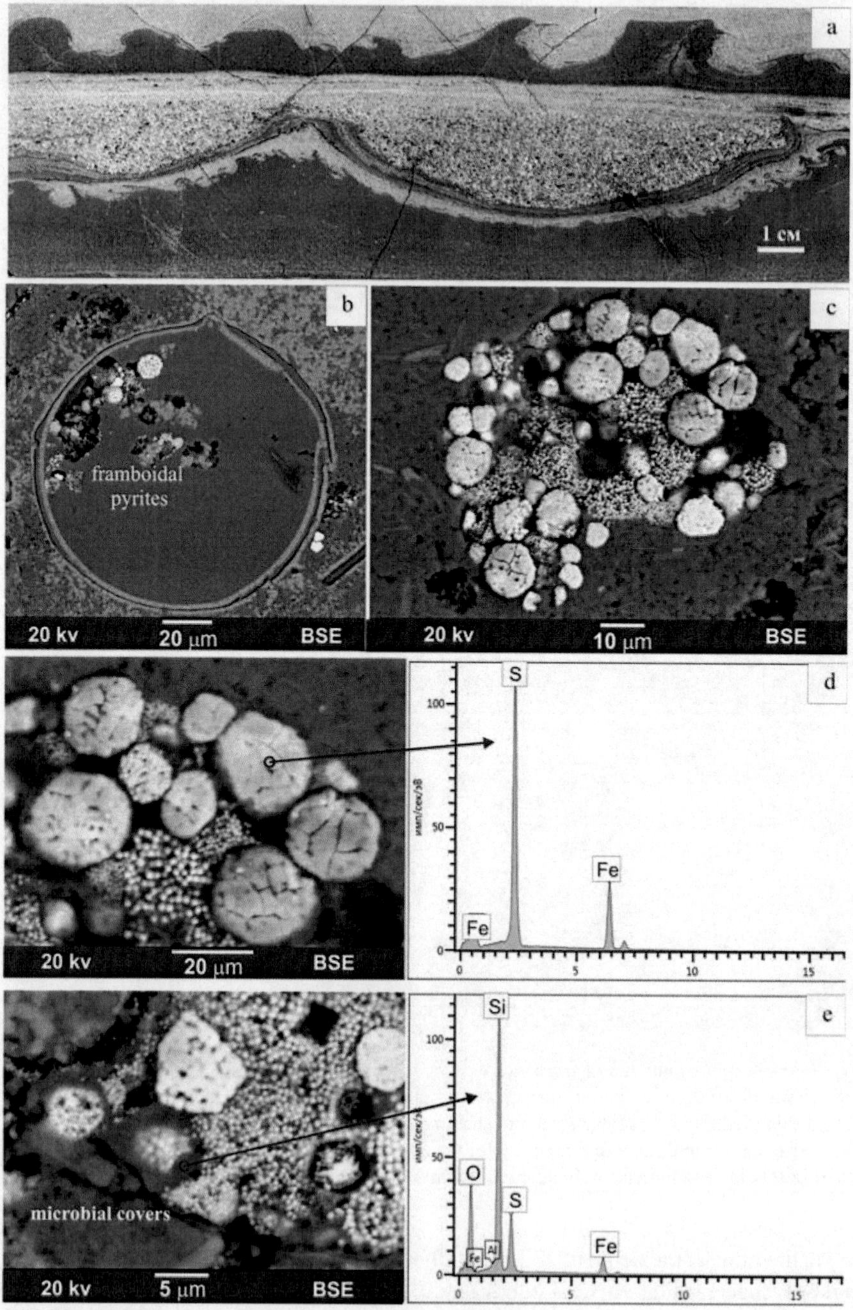

Fig. 8 Safyanovskoye deposit, carbonaceous-siliceous rocks: **a** photo of the rock sample in a cross section; **b** SEM photo of a foraminifera with framboidal pyrites; **c** accumulation of framboids with different structures; **d** enlarged SEM photo and EDS spectrum of a framboid with massive iron sulfide (possibly originally gel-pyrite); **e** framboidal pyrites in a microbial covers: SEM photo and EDS spectrum

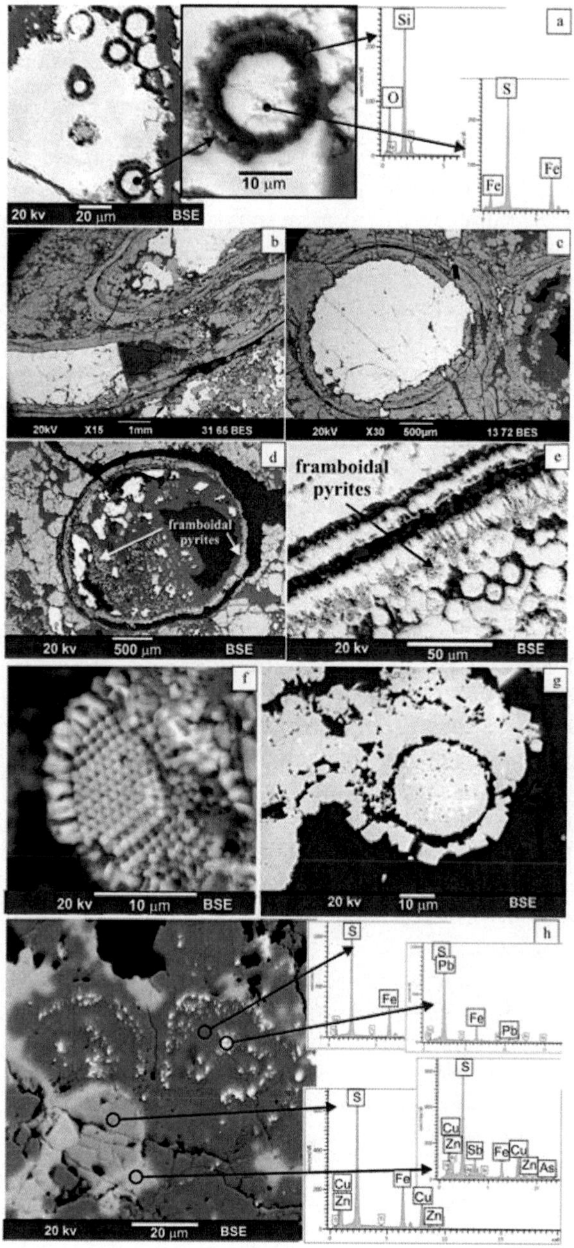

Fig. 9 Safyanovskoye deposit, pyrite ores (sample Saf 7/18): **a** SEM photo and EDS spectra of framboidal pyrites with SiO_2 covers; **b** tubes of polychaete worms in longitudinal section; **c** polychaete tubes in cross section; **d** tubes of polychaete worms in cross section with numerous framboids in the cavity; **e** incrustation of the walls of the polychaete pipe with framboidal pyrites; **f** framboid pyrite with a «sunflower» structure, a modern framboid from the deposits of the Kuchugur mud volcano; **g** SEM photo of framboidal pyrite with a «sunflower» structure from pyrite rock of the Sayanovsky Cu–Zn deposit; **h** SEM photo and EDS spectra of framboidal pyrites with polysulfide microimpurities

a similar association of framboidal pyrites with polychaete pipes V. V. Maslennikov has been observed while studying the of the Safyanovskoye deposit (Maslennikov et al. 2016).

Among modern framboid pyrites, there are segregations with an ordered arrangement of crystallites, for example, framboids with a «sunflower» structure from the sediment of the Kuchugury mud volcano (Fig. 9f), (Azovskova et al. 2022). In polished sections of massive sulfide rocks of the Safyanovskoye deposit, such framboids look like microconcretions with small crystallites in the center and large crystallites along the periphery (Fig. 9g). Previously we considered them as a result of late diagenetic recrystallization.

Microinclusions of PbS also associated with the processes of later hydrothermal alteration. However, the detailed study showed (Fig. 9h) that microlayers with PbS mark the periodic entry of Pb2 + into the environment and its fixation as a compound with sulfur. It's interesting, that the other chalcophile elements Cu, Zn, Sb, and As are noted as microimpurities in the substrate hosting the framboids (Fig. 9h).

Mikheevskoe (Mo, Au)-Cu-porphyry ores deposit. The argillisites of this deposit in some parts are weakly bituminized (Fig. 10a). According to the results of analysis (Fig. 10b), the values of organic carbon correspond to kerite, more precisely, slightly altered organic substances (Fig. 10c).

On the surface of some framboidal pyrites, SEM images show the sheaths substituted with SiO_2 (Fig. 10d), similar to those described earlier on the framboids from carbonaceous-siliceous and sulfide rocks of the Safyanovskoye deposit. Detailed study of framboids showed that, due to rapid mineralization, film covers were preserved on the crystallites composing these framboids (Fig. 10e).

Moskovskaya Au—placer deposit. The placer has a complex structure, with several gold-bearing horizons, including those in lacustrine-alluvial and coastal-marine (?) sediments of the Lower Cretaceous age (Azovskova and Glavatskikh 2013). The investigation of polished Section (251,086/25–26) with a quartz-sulfide «nodule» among sulfidized loose sediments shows presumably of their coastal marine origin.

Interesting mineralization was found in the weakly lithified sulfidized sediments (lacustrine, coastal-marine) on the southern flank of the placer. In some parts, they contain native gold of a distinctly ore appearance, which is characterized by variations in the composition of Au–Ag and often by an admixture of mercury. Relatively strong separate quartz-sulfide «nodules» are also found here.

A large number of microinclusions of various mineral phases were revealed in the polished sections of such formations. Among them, globular segregations of microaggregates of Ni-bearing pyrite are the most widespread. These pyrites are more or less replaced by As–Co–Fe–Ni–S phases (arsenic Co–Ni-pyrite, possibly a mixture of phases close in composition to the cobalt-gersdorfite series) (Azovskova and Glavatskikh 2013).

According to the results of the SEM study in the polished Section (251,086/25–26), framboidal pyrites were identified, in which both the shells around the framboids (Fig. 11a) and the films covering the crystallites (Fig. 11b) were replaced by polysulfides. On the SEM photo (Fig. 11c, d), in the clusters of framboidal segregations

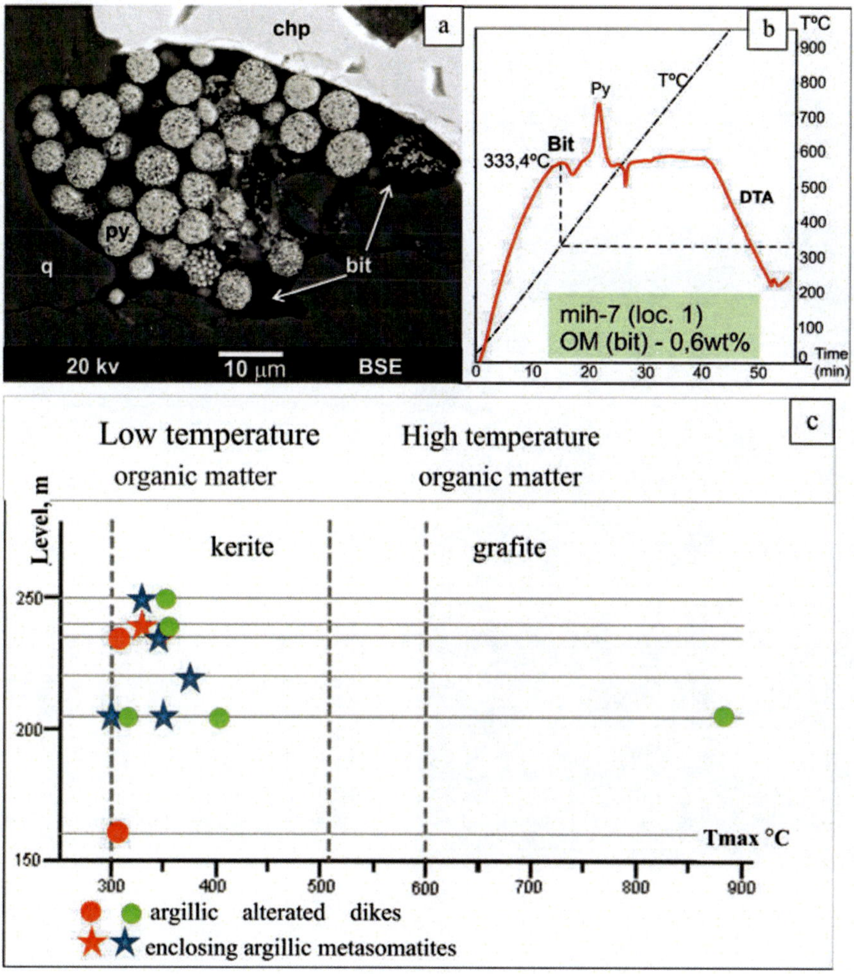

Fig. 10 The Mikheevskoe deposit, framboidal pyrites of argillisites: **a** SEM photo of a lens with bitumen and framboidal pyrites; **b** thermogram of analysis of bitumen and framboidal pyrites; **c** graph showing a degree of organic matter (bitumen) alteration. Designations: Py—pyrite, Bit—bitumen, **g** quartz, chp—chalcopyrite; SEM photo and EDS spectra: **d** framboids with crystallites covered by cover (films) substituted by SiO_2; **e** the film covers with crystallites in the cut

in the BSE mode, a bright rim is observed in the peripheral zone and the substrate, in which the crystallites are located, is also lighter than the pyrites themselves. The EDS spectra (Fig. 11e) contains: S, Fe, Ni, Co, As.

The mapping by elements showed a predominance of As, Co, and Ni sulfide in the rim and in peripheral zone of the substrate between crystallites (Fig. 12). Whereas the central parts of the framboids and substrate between them in clusters are represented by Zn and Fe sulfides (Fig. 12).

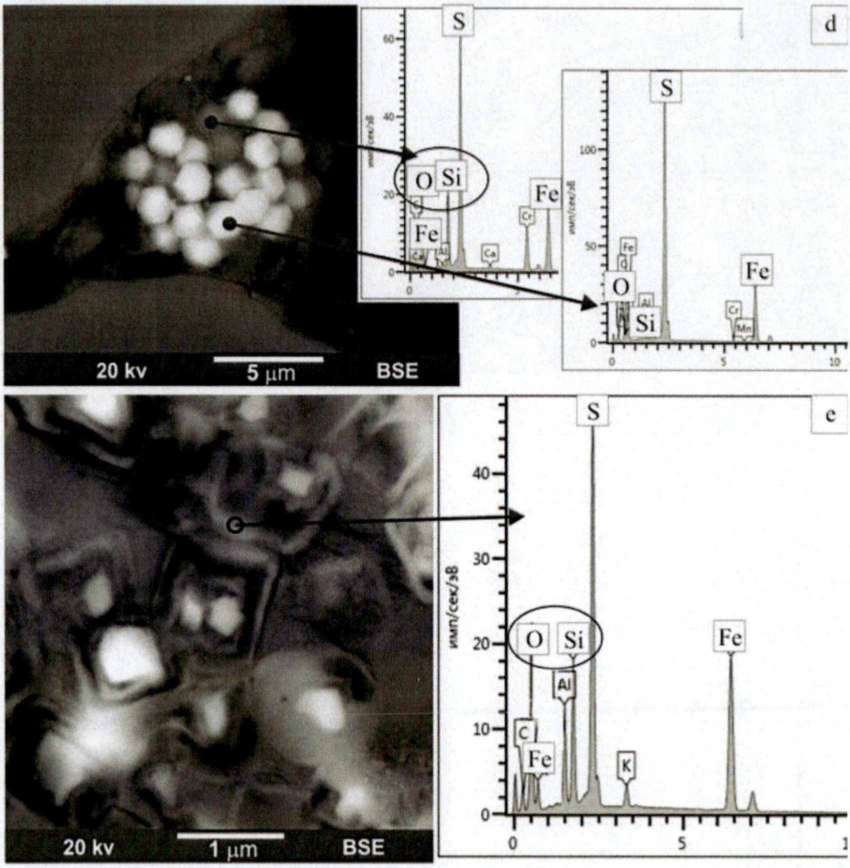

Fig. 10 (continued)

4 Discussion

All samples are characterized by a large morphological diversity of framboidal pyrite.

However, the mechanism of formation of some morphotypes and their combinations does not have an unambiguous explanation. For example there are small framboids (diameter 5–10 μm) consisting of sub-micron cubic pyrite crystals 0.5–1 μm in size. And neighboring framboids in the cluster or crust can have crystallite sizes ten times larger. Although, they were clearly formed at the same time and under the same conditions. It is also not clear how pyrite puzzle cubes located within of massive of polygonal crystallites (Fig. 7a) or spherical framboids located in a massive of scattered crystallites of a large framboid (Fig. 7g) were formed.

It is very difficult to explain such complex morphologycal combinations of crystals only by the chemogenic genesis. But if we consider them as a product of the vital activity of bacteria, then it can be assumed that the community includes various strains

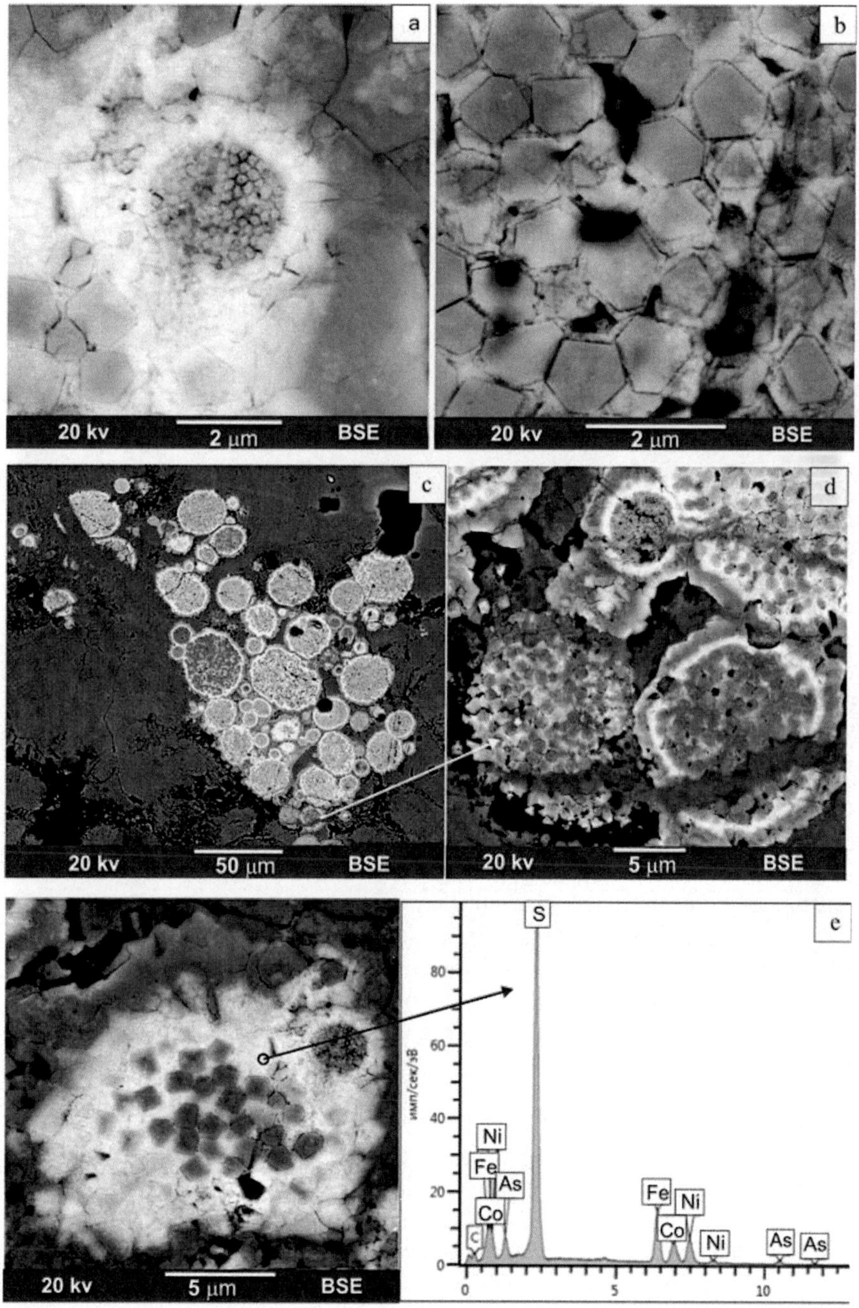

Fig. 11 The Moskovskaya Placer deposit, framboidal pyrites of a quartz-sulfide formation, polished Sect. (251,086/25–26). SEM photo: **a** shell of framboidal pyrite replaced by polysulfides; **b** films covering crystallites, mineralized with sulfides; **c** accumulations of framboidal pyrites with polysulfide mineralization of covers and substrate between crystallites; **d** an enlarged fragment of this cluster; **e** the element composition of the substrate between crystallites (by EDS spectrum)

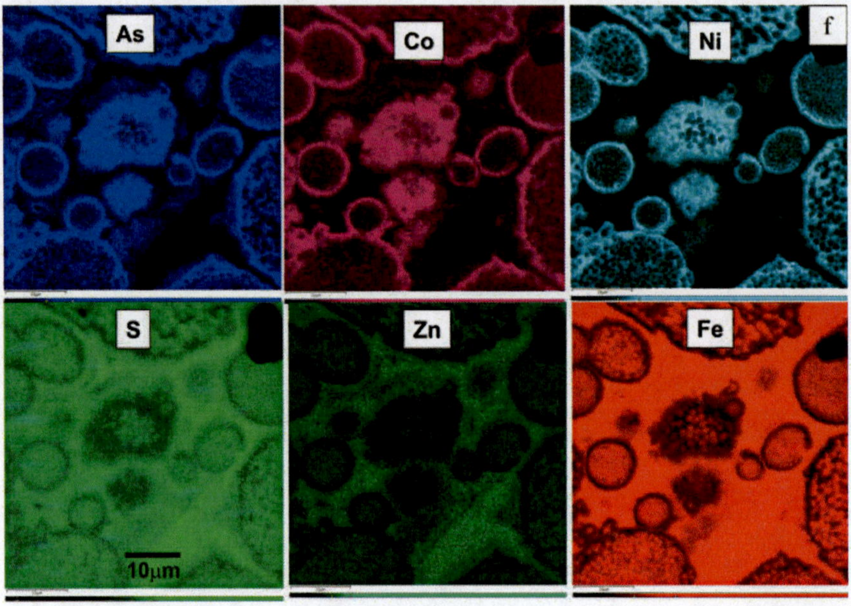

Fig. 12 The Moskovskaya Placer deposit, mapping by elements of a framboidal pyrite accumulation, polished Section (251,086/25–26). In the upper row, As, Co, Ni sulfide is located in the rim and in peripheral zone of the substrate between crystallites; in the lower one, Zn and Fe sulfide dominates in the central parts of framboids and in the substrate between them

of element-precipitating bacteria, which have their own specifics of precipitating the substances, but this mechanism has not yet been evident.

There are known the prokaryotic organisms that can extract energy from the oxidation of sulfur from hydrogen sulfide, thiosulfate, native sulfur (Lengeler et al. 2005; Sokolova and Karavaiko 1964). It is assumed that the activity of microorganisms including nucleation on or within the biomass contributes to the formation of conditions favorable for the formation of framboidal pyrite (Berberyan 1983; Schieber 2002; Astafieva et al. 2005; Maclean et al. 2008). Modern studies support the conclusion that natural framboids usually form under weakly reducing conditions during early diagenesis in organic-rich systems which suggest a significant role of biomineral formation (Weiner and Dove 2003; Popa et al. 2004; Cavalazzi et al. 2012; Duverger et al. 2021). Also, the formation of framboids with the «sunflower» structure (Fig. 9f, g) can be explained by bacterial origin: larger prismatic crystals grew in a thicker sheath of bacterial metabolites surrounding the colony.

It was noted that not only in a siliceous-carbonaceous facies, but even in a sulfide-forming environment, covers on framboids are replaced by SiO_2 (Fig. 9a). This is probably due to the specifics of the processes of replacement of released by the microbial community extracellular biopolymers, which can change the pH of the medium and promote selective mineralization.

Bacterial cells are not directly preserved as fossils, but the products of microbial life activity (extracellular biopolymers, biofilms, some metabolites) could be mineralized in sediment and preserved in rocks for a long time (Zavarzin 1993; Orleansky et al. 2007).

The covers on framboids and biofilms mineralized with polysulfides (Fig. 11) are also of particular interest. Such specific mineralization (Co, Ni, As, Pb) associated with framboids indicates microbial involvement in the accumulation of these elements from the environment. This is the first finding of such interesting mineralization of framboids (Fig. 12).

The following questions also remain debatable. Framboidal pyrites are associated with condensed hydrocarbons.

Pyrite and especially pyrite framboids are considered a common authigenic mineral in hydrocarbon seep-related deposits. Hydrogen sulfide produced as a result of anaerobic oxidation of higher hydrocarbons and methane through the contribution of sulfate-reducing bacteria in the anoxic marine sediment zone may combine with reduced iron to form sulfide (Sawlowicz 1993; Cavalazzi et al. 2014). It should be noted that the confinement of prokaryotic communities to oils is currently a hot theme (Rozanova and Nazina 1982; Nazina and Rozanova 1978; Nazina 2000). However, this theme requires further research, since the reason for the formation of framboids in lakes remains unclear.

5 Conclusions

At present time, there is an opinion that aggregates of pyrite framboids are fossilized microorganisms or cell colonies (Popa et al. 2004; Gong et al. 2008; Wacey et al. 2015).

- We found the presence of biofilms (organic structures) in modern pyrite framboids and thus confirm the role of the biogenic factor in their formation.
- In framboidal pyrites from ancient rocks these covers are more noticeable than in modern ones due to the mineralization of originally biopolymers mostly by silicon dioxide. Of the greatest interest in our collection are covers mineralized with polysulfides indicating the primary microbial deposition of Co, Ni, As, Pb.
- Framboids with «sunflower» structures are found both in the modern deposits that are still being formed, and in ancient rocks, which indicate the primacy of this appearance.
- The following fact can serve as an indirect sign of the presence of biofilms in framboids: isolated framboids do not disintegrate into individual crystals, although no continuous cement is seen between the crystallites (Fig. 6a). In modern framboids the bacterial films are not always visible due to weak mineralization.
- Indication of the bacterial genesis of framboids is their association with macrofaunal remains.

- In environments with organic carbon deficiency, heterotrophic bacteria use the substrate of dead macrobiota, because its protein is an available source of substances for the vital activity of bacteria.
- Another indirect sign of the participation of bacteria in the formation of framboids may be the identity of morphology with different mineral compositions. So, (for example, the crystallites in framboids in Fig. 1b, c) have the same polygonal habit. But the framboids shown in Fig. 1b are composed of iron oxides, while the rest are iron disulfide (pyrite or marcasite). But the chemogenic analogues of these minerals differ in their structural and crystalline characteristics.
- Previously, it was believed that enlarged crystallites in the «sunflower» shell are the result of recrystallization of the marginal zone of aggregates. In addition, the host rocks bear traces of secondary hydrothermal alteration. However, modern framboids from the sediment of a mud volcano, which have not undergone any changes, have the same structure.

Thus, the study of modern framboids contributes to the correct interpretation of the morphological features of fossil framboid pyrites.

The studies are carried out as a part of the IGG UB RAS State assignment (state registration No. 123011800012-9; No. 123011800011-2; No. 122040600009-2) using the «Geoanalitik» shared research facilities of the IGG UB RAS and «Geonauka» shared research facilities of the Institute of Geology FRC Komi SC UB RAS. The re-equipment and comprehensive development of the «Geoanalitik» shared research facilities of the IGG UB RAS is financially supported by the grant of the Ministry of Science and Higher Education of the Russian Federation (Agreement No. 075-15-2021-680).

References

Amon EO, Korovko AV (1992) New data on stratigraphy and lithology of the paleozones of the Urals and Central Asia. Ural Branch of the Russian Academy of Sciences Yekaterinburg: 69–77 (in Russian)

Anfimov AL, Soroka EI, Leshchev NV (2015) New data on foraminifers in the ore-bearing strata of the Safyanovsky copper pyrite deposit (Middle Urals). Bull. MOIP. Dep. geol. 90-3:23-26 (in Russian)

Antoshkina AI, Ryabinkina NN, Valyaeva OV (2017) Genesis of siderite nodules from the Lower Carboniferous terrigenous sequence in the Subpolar Urals. Lithology and Mineral Resources 2:130-144

Astafieva MM, Rozanov AYu, Hoover R (2005) Framboids: their structure and origin. Paleontological Journal, 5:1–7

Avdusin PP (1948) Mud volcanoes of the Crimean-Caucasian geological province. Ed. USSR Academy of Sciences Moscow, 192 (in Russian)

Azovskova OB, Glavatskikh SP (2013) Rare and exotic minerals from the plotic part and deep horizons of the Moscow placer (Chelyabinsk region). Materials of the V Russian. Youth Scientific Conference. IGG UB RAS Yekaterinburg: 14–17 (in Russian)

Azovskova OB, Rovnushkin MYu (2021) Features of mineralogy and a probable model of the formation of argillisites («loose sulfide ores») of the Mikheevskoe Cu-porphyry deposit, Southern

Urals. Materials of the XVI International. Meeting on the geology of placers and weathering crust deposits. Voronezh: 12–14 (in Russian)

Azovskova OB, Plotinskaya OY, Rovnushkin MY, Gemel VA (2019) Argillic alteration of the Mikheevskoe porphyry copper deposit (South Urals, Russia). 15th SGA Biennial Meeting 2:1038–1041

Azovskova OB, Leonova LV, Simakova YuS (2022) Morphological diversity of framboidal pyrites from the Kuchugury mud volcano The VII Intern. Sympos. Biogenic - abiogenic interactions in natural and anthropogenic systems. St. Petersburg State University:57–58 (in Russian)

Beloborodov DE (2022) Geodynamic conditions for the manifestation of mud volcanism in the Kerch-Taman region. Dissertation, Institute of Physics of the Earth RAS (in Russian)

Berberyan TK (1983) Framboid-pyrite aggregates in the ores of pyrite deposits and their genetic and prospecting significance. Dissertation, Tbilisi (in Russian)

Butler IB, Rickard D (2000) Framboidal pyrite formation via the oxidation of iron (II) monosulfide by hydrogen sulfide. Geochimica et Cosmochimica Acta. 64-15:2665-2672

Cavalazzi B, Agangi A, Barbieri R, Franchi F, Gasparotto G (2014) The formation of low-temperature sedimentary pyrite and its relationship with biologically-induced processes. Geology of ore deposits, 56-5:395-408

Cavalazzi B, Barbieri R, Cady SL, George AD, Gennaro S, Westall F, Lui A, Canteri R, Rossi AP, Ori GG, Taj Heddine K (2012) Iron-framboids in the hydrocarbon-related Middle Devonian Hollard Mound of the Anti-Atlas mountain range in Morocco: Evidence of potential microbial biosignatures. Sedimentary geology, 263-264:183-193

Chuvashov BI, Anfimov AL, Soroka EI, Yaroslavtseva NS (2011) New data on the age of the ore-bearing strata of the Safyanovskoye deposit (Middle Urals) based on foraminifers. Doklady Earth Sciences 439-5:648-650

Duverger A, Bernard S, Viennet JC, Miot J, Busigny V (2021) Formation of pyrite spherules from mixtures of biogenic FeS and organic compounds during experimental diagenesis. Geochemistry, Geophysics, Geosystems. 22 10:1-16 https://doi.org/10.1029/2021GC010056

Ershov VV, Elovsky EV, Puzich IN (2019) Distribution patterns of rare earth elements in mud volcanic waters. Reports of the Academy of Sciences, 488-1:71-73

Gong Y-M, Shi GR, Weldon EA, Du Y-S, Xu R (2008). Pyrite framboids interpreted as microbial colonies within the Permian Zoophycos spreiten from southeastern Australia. Geological Magazine, 145 1:95–103 https://doi.org/10.1017/S0016756807003974

Grabezhev AI (2014). Novonikolaevsky (Mo, Au)-Cu-porphyry ore cluster (Southern Urals): petrogeochemistry of ore-bearing granitoids and metasomatites. Lithosphere 2:60-76 (in Russian)

Ismagilov AF, Kozlov VN, Terekhov AA, Hortov AV (2006) Clay diapirism and mud volcanism during the formation of local structures in the Russian part of the Black Sea. Geology, geophysics and development of oil and gas fields, 2:4-10 (in Russian)

Kalyuzhny VA, Telepko LF, Shiritsa AS (1984) Composition and P/T1 parameters of deep fluids according to the study of inclusions in calcites from ejecta of mud volcanoes. Mineralogical collection, 2: 32-38 (in Russian)

Kholodov VN (2013) Patterns of distribution and conditions for the formation of salt diapirs and mud volcanoes. Lithology and minerals, 5:441-460

Kholodov VN (2002a) Mud volcanoes: distribution patterns and genesis. Lithology and minerals, 3:227–241 (in Russian)

Kholodov VN (2002b) Mud volcanoes: distribution patterns and genesis. Lithology and minerals, 4:339–358 (in Russian)

Kikvadze OE, Lavrushin VYu, Pokrovsky BG, Polyak BG (2014) Isotope-chemical composition of mud volcanic gases of the Taman Peninsula and the problem of their genesis. Lithology and minerals, 6:525–538

Koroteev VA, Yaseva RG, Bochkaryov VV, Moloshag VP, Korovko AV, Sheremetev YuS (1997) Geological setting and composition of the sulfide-ore Safyanovka deposit in the Middle Urals. IGG Ekaterinburg (in Russian)

Korovko AV, Postoyalko MV, Stepanova TI, Petrova LG, Shurygin MV, Ivanov KS, Amon EO (1999) Problems of stratigraphy and paleontology of the Urals. Committee of Nature for the Sverdlovsk region. Yekaterinburg: 141–144 (in Russian)

Kropotkin PN, Valyaev BM (1980) Geodynamics of mud volcanic activity (due to non-gas content). Geological and geochemical foundations of oil and gas prospecting. «Naukova Dumka» Kyiv: 148–178 (in Russian)

Kulakovskiy AL, Sysolin AI, Sysolin MI (2022) Structural criteria for the localization of mud volcanoes in the Kerch-Taman province (according to interpretation of aerial and satellite images). The scientific heritage, 83-1(83):8-18 (in Russian)

Lengeler J, Drews G, Schlegel G (2005) Modern microbiology. Prokaryotes: in 2 volumes. «Mir» Moscow (in Russian)

Leonova LV, Soroka EI, Galeev AA (2022) Microfossils of hydrothermal-sedimentary deposits of the Safyanovsk pyrite deposit (Middle Urals). Lithosphere 22-3:376–390

Maclean LCW, Tyliszczak T, Gilbert P, Zhou D, Pray TJ, Onstott TC, Southam G (2008) A high-resolution chemical and structural study of framboidal pyrite formed within a low-temperature bacterial biofilm. Geobiology, 6-5:471-480

Maslennikov VV (2006) Lithogenesis and pyrite formation. Institute of Mining Ural Branch of the Russian Academy of Sciences, Miass (in Russian)

Maslennikov VV, Ayupova NR, Maslennikova SP, Tseluiko AS (2016) Hydrothermal biomorphoses of sulfide deposits: microtextures, trace elements and detection criteria. RIO UB RAS Yekaterinburg (in Russian)

Mudretsova-Viss KA, Dedyukhina VP, Maslennikova EV (2014) Fundamentals of microbiology: textbook. Vladivostok University of Economics and Service. 5th ed., corrected, revised and enlarged. «INFRA-M» Moscow: 354 (in Russian)

Nazina TN, Rozanova EP (1978) Thermophilic sulfate reducing bacteria from oil reservoirs. Microbiology 47:142-148

Nazina TN (2000) Microorganisms of oil reservoirs and their use in biotechnology for enhanced oil recovery. Dissertation, Institute of Microbiology RAS, Moscow (in Russian)

Orleansky VK, Karpov GA, Zhegallo EA, Gerasimenko LM (2007) Biogenic-siliceous structures of thermal fields and their laboratory modeling. Mineralogy and life: the origin of the biosphere and the co-evolution of the mineral and biological worlds, biomineralogy. IG Komi Science Center Ural Branch of the Russian Academy of Sciences Syktyvkar:127–128 (in Russian)

Ovsyuchenko AN (2005) Manifestations of young tectonic activity in the structure of mud volcanoes in Taman and the Northwestern Caucasus. Modern methods of geological and geophysical monitoring of natural processes on the territory of the North Caucasus IPE RAS, Moscow: 235–248 (in Russian)

Popa R, Badescu A, Kinkle BK (2004) Pyrite framboids as biomarkers for iron sulfur systems. Geobiology, 1-21:1–14

Pritchin ME, Soroka EI, Puchkov VN (2021) New U-Pb isotope data of zircon of a rhyolite from the Safyanovsky copper pyrite deposit (Middle Urals). Lithosphere 21-6: 884–893

Puchkov VN, Ivanov KS, Korovko AV (1990) About age of volcanogenic formations and the time of initiation of the island arc in the east of the Middle Urals. Doklady Earth Sciences 315-5:1203-1205

Rozanova EP, Nazina TN (1982) Hydrocarbon-oxidizing bacteria and their activity in oil reservoirs. Microbiology 51: 342-348

Rust GW (1935) Colloidal primary copper ores in the mines of Cornwall, southeastern Missouri. Journal of Geology 43:398-426

Sawlowicz Z (1993) Pyrite framboids and their development: A new conceptual mechanism. Geologische Rundschau, 82-1:148–156 https://doi.org/10.1007/BF00563277

Schieber J (2002) Sedimentary pyrite: A window into the microbial past. Geology, 30-6:531 https://doi.org/10.1130/0091-7613(2002)030<0531:spawit>2.0.co;2

Shnyukov EF, Kobolev VP (2018) Mud volcanic deposits of methane gas hydrates in the Black Sea. Geology and Minerals of the World Ocean, 1-51:5-34 (in Russian)

Smirnov VN (2012) Main stages of magmatism of the eastern slope of the Middle Urals. Lithosphere 5:4-15 (in Russian)

Sobisevich AL, Sobisevich LE (2018) Study of the deep structure of zones of concentrated fluid activity in the Kerch-Taman mud volcanic province. Actual problems of oil and gas, 4-23: 1–11 (in Russian)

Sobisevich AL, Dudarov ZI, Likhodeev DV, Dolov SM, Beloborodov DE, Presnov DA (2015) New data on the deep structure of the Hell of Azov mud volcano (Russia, Krasnodar Territory, Taman Peninsula). Materials of the 10th Intern. Seismological School Modern methods of processing and interpretation of seismological data GS RAS, Republican Center for Seismological Service under the National Academy of Sciences of Azerbaijan: 308–311 (in Russian)

Sokolova GA, Karavaiko GI (1964) Physiology and geochemical activity of thionic bacteria. Institute of Microbiology, Academy of Sciences of the USSR. Nauka Moscow (in Russian)

Suits NS, Wilkin RT (1998) Pyrite formation in the water column and sediments of meromictic lake. Geology 26-12:1099-1102

Tessalina S, Plotinskaya OY (2017) Silurian to Carboniferous Re-Os molybdenite ages of the Kalinovskoe, Mikheevskoe and Talitsa Cu- and Mo porphyry deposits in the Urals: implications for geodynamic setting. Ore Geology Reviews. 85:174-180

Toporski JKW, Steele A, Westall F, Kathie L, Thomas-Keptra, McKay DS (2002) The Simulated Silicification of Bacteria – New Clues to the Modes and Timing of Bacterial. Preservation and Implications for the Search for Extraterrestrial Microfossils. Astrobiology. 2-1:1-26

Tveritinova TYu, Dolov SM, Dudarov ZI, Beloborodov DE (2015) Peklo Azovskoye mud volcano: position in the regional structure, surface manifestation, structural features of the degassing channel. Collection of the III school-seminar «Gordinsky Readings» IPE RAS. Moscow: 119–123 (in Russian)

Wacey D, Kilburn MR, Saunders M, Cliff JB, Kong C, Liu AG, Matthews JJ, Brasier MD (2015) Uncovering framboidal pyrite biogenicity using nano-scale CNorg mapping. Geology, 43-1:27-30 https://doi.org/10.1130/G36048.1

Weiner S, Dove PM (2003) An Overview ofbiomineralization processes and the problem of the vital effect, Rev. Mineral. Geochem. 54-1:1-29

Yaroslavtseva NS, Maslennikov VV, Safina NP, Leshchev NV, Soroka EI (2012) Carbon-bearing aleuropelites of the Safyanovsky copper-zinc deposit (Middle Urals). Lithosphere 2:106-124 (in Russian)

Yazeva RG, Moloshag VP, Bochkarev VV (1992) Geology of the Safyanovsky pyrite deposit (Middle Urals). Nauka Yekaterinburg: 70 (in Russian)

Zavarzin GA (1993) Development of microbial communities in the history of the earth. Problems of pre-anthropogenic evolution of the biosphere. Nauka Moscow: 212–222 (in Russian)

Biogenic Processes and Their Role in the Formation of Vendian Oncolites

Tatiana V. Litvinova and Petr N. Kolosov

Abstract Among biogenic structures, oncolites are notoriously difficult objects to study. External similarities and differences of these structures hamper an unambiguous conclusion regarding their origin. This complicates the establishment of a reasonable systematization of oncolites and, therefore, the possibility of utilizing them in the stratigraphic breakdown of Precambrian rocks. The nature of oncolite formations and their specific features can be established solely through the detection of organic residues contained therein, which practice has appeared relatively recently. This paper presents the results of a study of oncolites found in two suites of Vendian deposits of the Baikal-Patom Highlands. These are black oncolites of the Kalancha formation and microoncolites of the Tinnaya formation. The conducted comparison of the obtained results with those for the host rocks revealed another, more complex organic world of these formations. The participation of microorganisms and fungi and their role in the occurrence and development of oncolites are discussed. Questions related to the interaction of biogenic and abiogenic factors in their formation are considered. Although both studied types are of organogenic sedimentary origin, the composition of microorganisms and their participation in the formation of oncolites differs significantly.

Keywords Oncolites · Microorganisms · Cyanobacteria · Microfossils · Fungi

T. V. Litvinova
Geological Institute of the Russian Academy of Sciences, Moscow, Russia
e-mail: litvinova-geo@rambler.ru

P. N. Kolosov (✉)
Diamond and Precious Metal Geology Institute SB RAS, Yakutsk, Russia
e-mail: petrkolosov36@mail.ru

O. V. Frank-Kamenetskaya et al. (eds.), *Biogenic—Abiogenic Interactions in Natural and Anthropogenic Systems 2022*, Springer Proceedings in Earth and Environmental Sciences, https://doi.org/10.1007/978-3-031-40470-2_12

1 Introduction

The term "oncolithi" was first coined by Pia to describe loose concentrically layered carbonate grains formed with the participation of microorganisms (Pia 1927). Oncolites are often difficult to distinguish from oolites (Maslov 1973). Some of the oncolites present in the Neoproterozoic deposits of the Ura Uplift transformed into oolites as a result of recrystallization (Rejtlinger 1959; Voronova and Radionova 1976; Kuznetsov 2003). Some forms (e.g., osagia tenuilamellata), which cannot be distinguished from oolites without the evidence of microorganism participation, were sometimes mistaken for oncolites of the osagia group. The term "osagia" was introduced by U. Twenhofel to refer to rounded nodules of various sizes with concentric layering, formed with microorganism participation (Twenhofel 1919).

The most acceptable definition seems to refer to oncolites as grains composed of concentric layers of various thicknesses, more often with wavy outgrowths in places favorable for photosynthesis, less often (in Paleozoic sediments) with remains of microorganisms (Maslov 1973; Vologdin 1962; Selley 1977; Odum 1957; Davydov 1975). Other researchers, when investigating spherical carbonate grains, associated the genesis of some grains with the vital activity of microorganisms. Morphologically indistinguishable forms of oncolites were found in sediments of different ages, obviously having formed under similar sedimentary conditions (Vologdin 1962; Logan et al. 1964; Kolosov 1970; Butakov et al. 1972; Chuvashov 1988).

The carbonate type of the Neoproterozoic section is widespread on the Siberian platform. It is associated with the prime of calcareous cyanobacteria and algae precipitating carbonates in the sedimentation stage (Kolosov 1970, 1977, 1982). In addition to stromatolites, the abundance of microorganism activity traces in the form of oncolites in the Neoproterozoic section is well known from numerous works of Russian researchers published in 1960–1980s (I. K. Korolyuk, Ye. A. Rejtlinger, Z. A. Zhuravleva, L. I. Narozhnykh, V. Ye. Milshteyn, G. A. Vorontsova, M. S. Yakshin, etc.). The research results obtained during that period were used in the Neoproterozoic stratigraphy of the region as markers for the correlation of deposits within the structural-facies zones. In the Baikal-Patom Highlands of Eastern Siberia, the organogenic sedimentation nature of Neoproterozoic oncolites remains insufficiently substantiated by factual data. Thus, without detection of either cells or filaments, oncolites were claimed (Zhuravleva 1964) to be deposits of carbonate around algae filaments.

The participation of lower plants in the origin of oncolites can only be asserted based on the evident signs of cyanobacteria and algae residues in their composition. The difficulty in detecting the residues of life in Neoproterozoic oncolites is related to the microscopic scale of the life forms involved in those formations, which rarely remain in a definable state. In addition, due to the initial (associated with the presence of organic residues) heterogeneity of the sediment composition, carbonate rocks and oncolites are often recrystallized (Kolosov et al. 2021).

In recent years, interest in the study of oncolites has almost disappeared. All the numerous attempts to systematize oncolites according to their formal features and to employ them in some way for the purpose of stratigraphic breakdown of the Precambrian have so far been unsuccessful. These attempts are hampered by lithological similarities and differences in the structure of oncolites. At the same time, their origin, as well as reasons for their similarity and diversity, continue to raise many research questions. Since recently, it has become possible to study oncolite microcosms in comparison with the organisms of the host rocks utilizing a scanning microscope (hereinafter referred to as SEM). This methodology enables observation of the visual features of oncolites, thus providing factual material for theoretical discussion of the causes of their occurrence.

To the best our knowledge, this paper is the first to discuss the structure of two different types of oncolites from different—Kalancha and Tinnaya—Vendian formations, the Baikal-Patom Highlands (Eastern Siberia, Russia). Some of the studied samples exhibit a radial laminated structure and sizes typical of oncolites, while others are quite different in internal structure, which can be seen only with the use of SEM. Both types were established to contain a variety of microorganisms, whose vital activity was studied. This work is the first attempt to elucidate both the issues of microorganism participation in the development of these formations and the destructive action (parasitism) of aquatic fungi. During the study of oncolite origin, we analyze and consider the paleofacies setting at the time of their formation, including the impact of hydrodynamic factors that had a tremendous effect on the described events.

2 Geological Structure of the Area Under Study

The Ura transverse uplift is located on the northeastern margin of the Patom folded arc (Chumakov et al. 2013). The most complete section of the Riphean and Vendian deposits in the south of Central Siberia is exposed near the mouth of the Bolshoi Patom River and stretches to the north, crossing the Lena River between the mouths of the Zherba and the Malyj Patom rivers (Fig. 1, Litvinova and Sergeev 2018), the Upper-Riphean Ballaganakh group and the Vendian Dalnyaya Taiga, Zhuya and Trekhverstnaya groups are exposed (Dol'nik 2000; Chumakov et al. 2013).

The Dalnyaya Taiga group comprises four conformable formations from bottom to top: Bolshoi Patom, Barakun, Ura, and Kalancha. The Bolshoi Patom formation is composed primarily of massive and laminated tillites with a total thickness of 900–1100 m (Chumakov et al. 2013). The basal horizon of the Barakun formation is comprised of thin-laminated carbonates. It is considered to be the final cycle of the tillite strata, and is referred to as the "crowning dolomite" or cap-dolomites (Chumakov et al. 2013). The main part of this formation is presented by layers of dark limestone and black slate with a total thickness of 1100–1200 m. The Ura formation is composed primarily of siltstones and mudstones, with subordinate interlayers of dolomites and limestones, the total thickness is about 300–350 m (Kolosov 1975). The Dalnyaya Taiga group ends with the Kalancha formation that was accumulated in the littoral settings of the shallow-water basin. It is represented by strata

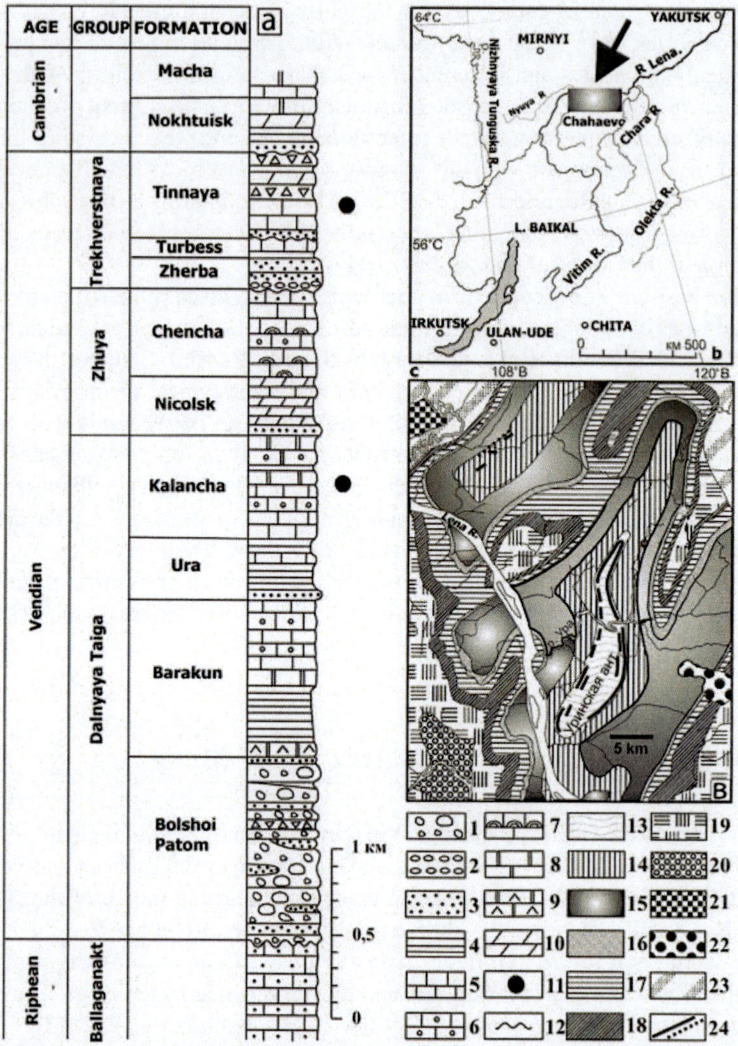

Fig. 1 Location of the studied objects (oncolites) in space and time. **a** Summary Vendian section in the Ura Uplift (Chumakov 1959; Chumakov et al. 2013; Sergeev et al. 2011): 1—diamictites; 2—conglomerates; 3—sandstones; 4—siltstones and mudstones; 5—limestones; 6—oolitic limestones and dolomites; 7—stromatolitic limestones and dolomites; 8—dolomites; 9—cap dolomites; 10—marls; 11—position of the studied stromatolite samples in the section and area; 12—stratigraphic unconformities. **b** geographical location of the research area (Ura anticline). Sampling sites are shown by a gray rectangle and designated by an arrow. **c** geological map of the Ura anticline: 13—Bol'shoi Patom Formation; 14—Barakun Formation; 15—Kalanchev and Ura formations; 16—Nikola Formation; 17—Chencha Formation; 18—Zherba Formation; 19—Tirbess and Tinnov Formations; 20—Nokhtuisk Formation; 21—Macha Formation; 22—Jurassic rocks; 23—Quaternary sediments; 24—sills, dikes, and major faults

of carbonate and argillaceous carbonate rocks, with a thickness not exceeding 500 m. The Kalancha formation is characterized by relict stromatolite texture, and a fairly wide spread of various oncolites (Kolosov 1975). Most of them have a benthic origin, although some were formed in suspension (Kolosov 2016). The former were rolled evenly or intermittently along the bed, gradually becoming overgrown with concentric layers of the same or varying thickness; their formation involved cyanobacteria. The latter underwent two stages of construction: the emergence of the nucleus and its overgrowth with cyanobacteria. The nucleus is represented by a tightly packed carbonate material due to over-rolling, with a lumpy coating around it (Kolosov 2016). This formation also hosts unique black oncolites that are discussed in the present paper.

The deposits of the Dalnyaya Taiga group are concordantly overlapped by the Zhuya group that includes the Nikolsk and Chencha formations. The former consists of a sequence of variegated thin-laminated sandstones, siltstones, and marlstones with rare limestone interlayers, with the thickness ranging from 120 to 360 m (Chumakov et al. 2013). Upwards along the section, it is gradually replaced by stromatolite limestones of the Chencha formation that hosts relics of Chroococcales and Hormogonales stromatolite-forming communities of cyanobacteria, as well as problematical remains of eucaryotic microorganisms of the Pertataka type (Litvinova and Sergeev 2019). The total thickness of the Chencha formation on the Ura uplift was estimated to vary from 400 to 600 m on the Lena River (Kontorovich 2005) to 800–1200 m in further southern areas (Melezhik et al. 2009).

Biostratigraphic, lithological, and isotopic-geochronological studies conducted in combination indicate the Lower Vendian age of the Dalnyaya Taiga and Zhuya groups of the Patom complex (Chumakov et al. 2013). This agrees with the paleontological data on the microbiota findings in the Ura formation (Sergeev et al. 2011; Moczydlowska and Nagovitsin 2012).

The Vendian section is completed by a Trekhverstnaya group composed of unmetamorphosed facies of the ocean shelf and the upper part of the continental slope (Dol'nik 2000; Chumakov et al 2013). It is divided into the Zherba, Tirbess and Tinnaya formations. During Zherba (Djerba) time, a drastic change in the nature of sedimentation occurred in the marginal part of the Patom basin; carbonate sediments were replaced by sandstones and gritstones of quartz composition, gradually turning into siltstones and mudstones. In this process, the Zhuya group of the Vendian was overlapped with the erosion of the Trekhverstnaya ("Yudoma") group. The latter includes the Zherba formation, which is divided by a number of researchers into the Zherba, Tirbess, and Tinnaya formations. The overlying conformable Tirbess (or the upper part of the Djerba) and Tinnaya formations were re-composed by units of carbonate sediments with sparse interlayers of clastic rocks.

The Tinnaya formation is divided into two unequal sub-formations: the lower (112 m) and the upper (363 m). The former is represented by bituminous, sometimes strongly fractured, brecciated, or organogenic dolomites, and interbedded units of dark gray siltstones and mudstones with interlayers of limestones, dolomites, and occasional sandstones. At 25 m from the bottom of the unit, traces of gypsum leaching in the gray dolomites are observed. The latter, i.e., the upper sub-formation

(20 m), is composed of a unit of gritstone-sandstones and inequigranular gray sandstones, including interlayers of dolomites, dark gray siltstones, and maroon-gray sandstones. Above, there are locally brecciated limestones, black and dark gray bituminous dolomites with interlayers of dolomitic limestones, gray dolomites, and dark gray mudstones. The studies of the Tinnov formation identified the remains of small shelly fauna of the Nemakit-Daldyn tier (Khomentovsky et al. 2004; Kochnev and Karlova 2010), and established cyanobacteria *Shanganella* sp., algae *Proaulopora* sp. (Kolosov 1977), as well as problematic organic (?) formations *Problematicum* (Rejtlinger 1959). On the Lena River, its thickness reaches 470–500 m.

3 Research Methods

The microorganisms of oncolites (Fig. 2, 3) and their elemental composition were studied with a TesScanMV-2300 SEM equipped with EDS (INCA 200, Oxford Instruments) under the following conditions: a scan diameter of 1 μm and the sensitivity of polarized light scattering measurements of 0.001%. Gold sputtering of samples was conducted to judge whether the carbon content in the allegedly biogenic structures is increased (the standard rock fragment for SEM analysis has a base of about 20 × 20 mm). All specimens along with their paleontological smear slides and samples are preserved at the Geological Institute, Russian Academy of Sciences (collection No. 1005/16).

The materials from the Tinnov formation were studied according to the method described below (Fig. 4). The search, study, and photographing of oncolite-like formations were performed in the following sequence:

1. Under transmitted light using a microscopy Axioskop 4 at magnifications of 5, 10, 20, and 50 times; stereoscopic microscopy Olympus SZX 12 at magnifications from 10 to 90 times; photographing by a digital camera Olympus C5060 (5.1 megapixel matrix) under combined light.
2. To search and study microscopic organic remains of the Tinnov formation by SEM, several thin sections oriented vertically of the stratification were cut out of partially silicified black limestones samples (Fig. 4). These sections were sanded with a fine-grained powder, washed with distilled water, dried, and gold-coated under sterile conditions. This method of obtaining polished sections is used by many researchers. The JEOL JSM-6480LV SEM was operated in a high vacuum mode, an accelerating voltage of 10 kV, and a diaphragm diameter of 10 mm (twice the SEI signal, operating distance).

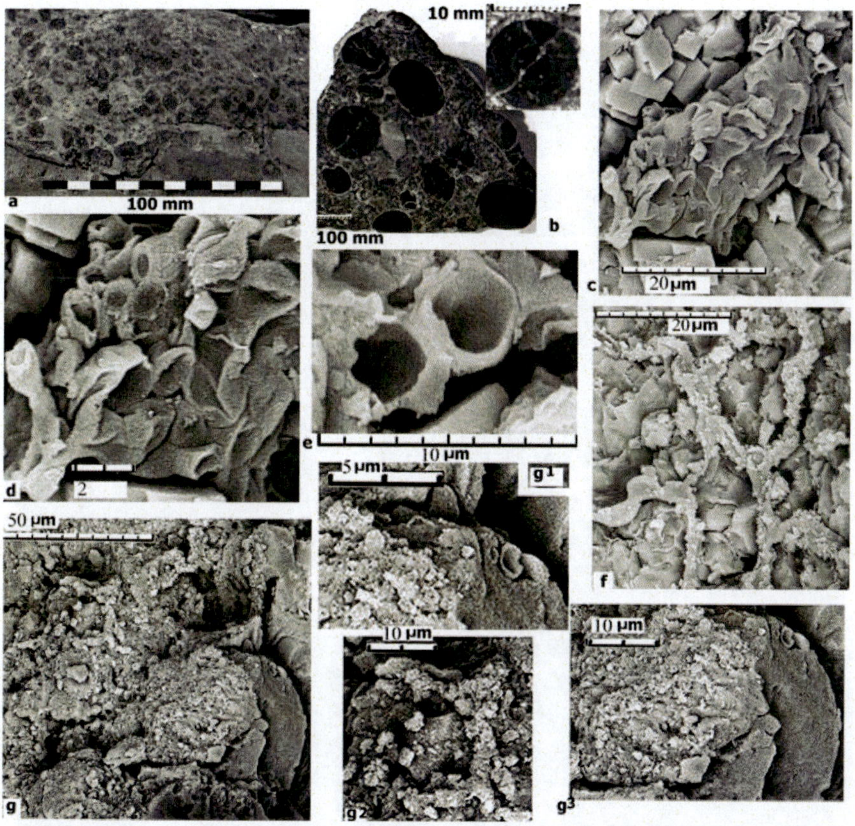

Fig. 2 Black oncolites of the Kalancha formation and microorganisms contained therein. **a** layer containing black oncolites; **b** radial form of oncolites; **c** phenocrysts isolated from the host rock by a hollow space; **d** remains of tubular microfossils that form phenocrysts; **e** cross-section of recrystallized tubular microfossils; **f** remains of recrystallized filamentous microfossils; **g** spherical ultramicrostructure of inorganic components of oncolite: $\mathbf{g}^1$ tubular microfossils on the surface of ultramicrooncolite, $\mathbf{g}^2$ recrystallized filamentous microfossils encircling a spherical inorganic formation; $\mathbf{g}^3$ ultramicrooncolite within oncolite

4 Microorganisms in Oncolites of the Kalancha Formation

In the Kalancha formation of the Vendian of the Baikal-Patom Upland, there are oncolites of various sizes and preservation. This paper presents uncommon black oncolites (Fig. 2a) with a rounded shape (Fig. 2 b, b_1), concentric structure, and sizes ranging from 4–5 to 12 mm, rarely up to 23 mm in diameter.

In the central part of an oncolite sample, irregularly shaped phenocrysts were found (Fig. 2c), folded by closely touching hollow non-segmented tubular formations with a diameter from 0.8 to 2.1 μm, a length of up to 3.5–4 μm, and circular cross section (Fig. 2d, e). All formations are inclined in one direction, partially flattened,

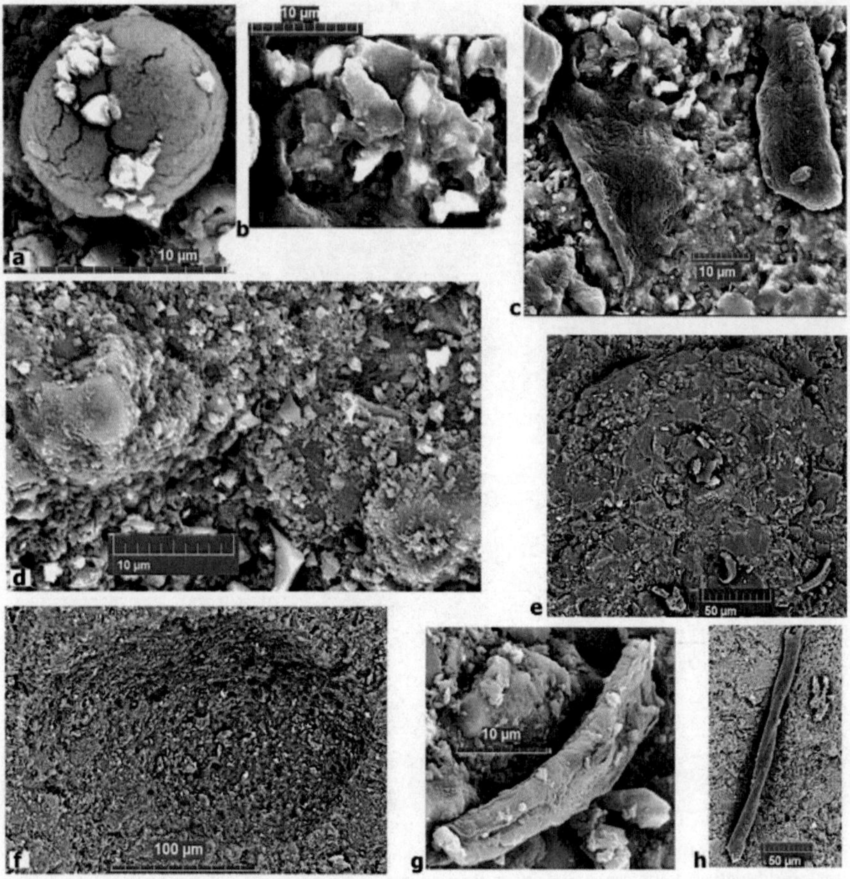

Fig. 3 Microorganisms and biogenic ultramicrostructures in the oncolites of the Tinnaya formation. **a** coccoid microfossil; **b** cluster of deformed coccoid microfossils; **c** bacterial films in areas of clasters of deformed coccoid microfossils (top right); **d** cluster of coccoid microfossils overlain by sedimentary cover; **e** microoncolite and microfossils; **f** notch left in the rock after microoncolite deposition; **g**, **h** rod-like microfossils

distinctly expanding toward the open end. The remains of relatively large ($2 \times 30\ \mu$m) recrystallized filamentous cyanobacteria (Fig. 2 f^1) in oncolites were also found, with the degree of their preservation increasing toward the peripheral part of the oncolite.

A comparison of the remains of fossilized microorganisms in the host rocks and in oncolites (SEM) showed that both contain bacterial films and coccoid cyanobacteria of 10–15 μm in size. In oncolites, they are flattened, crumpled, and severely deformed. Bacterial films were identified within the limits of their location. The area-limited distribution of coccoid cyanobacteria residues in oncolites also differs significantly from their scattered placement in the host rocks: they form rare scattered clusters across a small area ($100 \times 100\ \mu$m). In oncolites, these residues are partially

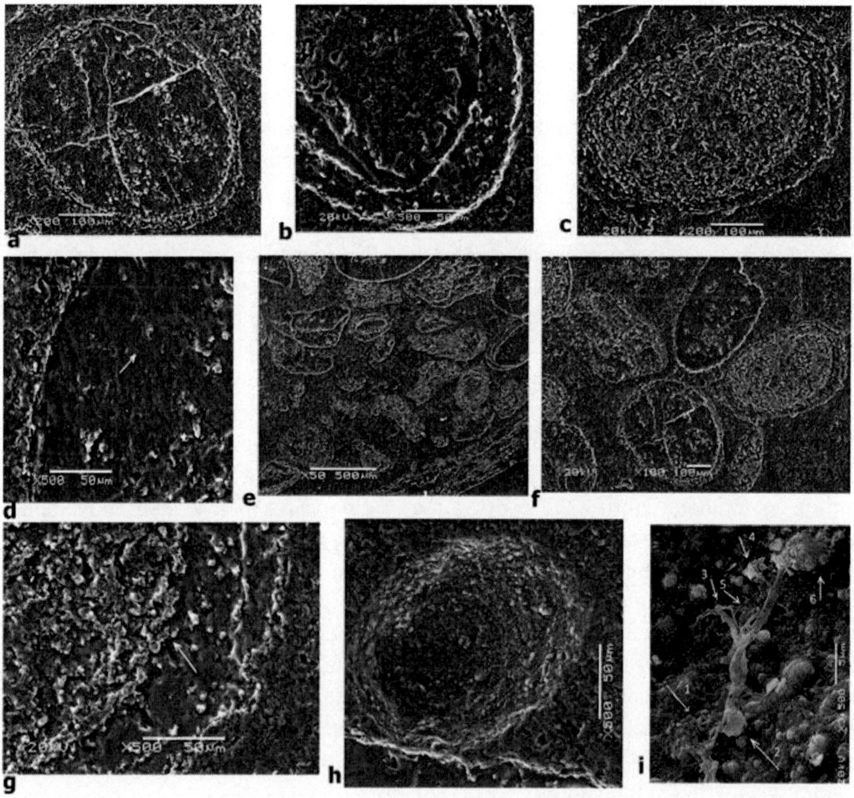

Fig. 4 Structure of microoncolites of the Tinnaya formation and mycelial fungi. **a–h**—microoncolites and their types: **a**light coating encircling a dark, fine-crystalline rock with a lumpy structure; **b** multilayered coating, with central part composed of dark microgranular material; **c** dark amorphous coating, with central part composed of light microgranular material; **d** mucous coating with single filamentous fungi; **e f** microoncolites composed of oncolite-like formations with a common coating; **g** microoncolite, arrow shows the remains of coccoid microfossils; **h** microoncolite with remains of coccoid microfossils and fungi; **i** mycelial fungi Matchi sitnikovi gen. et sp. nov.: 1—mycelium base, 2—rhizoid, 3—reproduction organ, 4—indistinct formation from hyphae, 5—hypha, 6—terminal end of the mycelium

surrounded by a sedimentary cover. This indicates that the remains of coccoid microfossils were accumulated during the oncolite movement, rather than during its lifetime. Coccoid microfossils are likely to have been collected from the bottom of the basin by a rolling mineral nodule or stuck to it while being in a suspended state. As a result of the oncolite movement, these remains were either damaged or acquired a sedimentary cover.

However, any filamentous forms of biogenic formations (tubular, filamentous) were found only in oncolites, although being completely absent in the host rocks. While tubular microfossils form clusters (phenocrysts) or occupy notches within the oncolite, filamentous forms act differently. They encircle the surface of the oncolite on each subsequent layer. Fragments of the bacterial films located next to filamentous microfossils are significantly thinner than their usual size, have a smooth surface, and are partially or completely recrystallized. Each successive layer, trapped during the oncolite movement, compressed their core. During lithification, filamentous cyanobacteria and their mucous substance (glycocalyx) formed black interlayers in oncolites.

Colonies of tubular organisms could act as the core of an oncolite; such formations had a carbonate cover of the smallest particles of silt around them, which appeared while moving along the bottom or in suspension. Filamentous cyanobacteria developed on the surface of such an organogenic sedimentary nodule, with their mucous mass (glycocalyx) partially or completely enveloping the oncolite. The notches occurring as a result of the fallout of sand grains were re-assimilated by colonies of tubular cyanobacteria. Large filamentous microfossils, which were surrounded by mucus, formed a black concentric layer with a high content of organic carbon accumulated during the dehydration of glycocalyx. Subsequent movements of the mucus-covered oncolite were accompanied by a build-up of carbonate material on its surface, on which the most sustainable trichomes resurfaced to create a new colony. As a result of dehydration and material compaction, a concentric lamination was formed during the lithification of the sediment. Organogenic layers of a black color were formed due to the vital activity of filamentous cyanobacteria and acquired a monolithic ultramicrostructure. Mineral carbonate layers were irregularly diluted by organogenic material, which colored them dark gray with black spotting resulting from the presence of other organisms (clusters of remaining coccoid microfossils, phenocrysts with colonies of tubular organisms).

The lamination of oncolites is more complex and less differentiated in comparison with the alternating layers in stromatolites (Litvinova 2014 2018). The presence of rounded formations and alignment of their surface through the bios activity is predominant at each stage of oncolite growth (Fig. 2g, g^2); this indicates the unity of biotic and abiotic processes during its formation. Some mineral oncolite components have a spherical shape (Fig. 2g), i.e., they represent an ultramicrooncolite (Fig. 2g^3). Black oncolites are to be considered as complex sedimentary-biogenic formations occurred as a result of hydrodynamic bottom processes.

A small amount of relatively evenly dispersed biogenic material (coccoid microfossils and bacterial films) imparts a dark gray color to the host rocks.

5 Organisms in the Minioncolites of the Tinnaya Formation

A microscopic study of the Tinnaya formation deposits revealed unique structures that represent formations of an irregular spherical shape, with a size from 200 to 500 μm in diameter. The increased amount of carbon in comparison with the host rocks, as well as the presence of some biophile elements specific only to organisms, indicates their possible biogenic nature. Further in the work, these formations of undetermined genesis are referred to as microoncolites. A SEM study of their host rocks revealed round coccoid microfossils measuring slightly more than 10 μm in diameter (Fig. 3a). These formations have a smooth surface and a regular spherical shape; no indentations and creases were observed on the surface. Microfossils are often accompanied by fragments of large bacterial films up to 40 μm in length (Fig. 3c). As a result of dehydration of sedimentary material in the course of lithification, bacterial films acquired a distinct wrinkled surface. The samples also revealed an accumulation of peculiar organogenic sedimentary formations comparable in size to the above-described coccoid microfossils (Fig. 3c, d). These might have provided the core for the formation of the currently observed microoncolites, given the absence of another theory that could explain their origin.

In highly silicified rocks, it is essentially possible to observe all stages in the formation of microoncolites (Fig. 3b–f), including traces remaining after their fallout from the rock (Fig. 3f). This interpretation of events appears to be logical and consistent, especially since the available factual material complicates finding a different reason for their occurrence. Only coccoid cyanobacteria and their modified microstructures were established in the rocks in large amounts. Based on the results of a SEM study and interpretation of the obtained material, it is possible to present the following stages in the formation of microoncolites.

1. Hydrodynamic processes lead to water resuspension. 2. Covering of coccoid microfossils or their remains with a carbonate sedimentary material, leading to their overgrowth with silt during their movement in suspension (Fig. 3d). 3. Accumulation and destruction of the accumulated remains of organisms. 4. Formation of a single rounded nodule (Fig. 3e). 5. Rounding and compression of the organogenic sedimentary material. 6. Submerging of the resulting microoncolite in carbonate silt (Fig. 3e). 7. Deposition or roll-out of the new formation due to certain reasons (movement of microoncolites under active hydrodynamic conditions, or uneven rate of dehydration of organic-biogenic material and mineral sediment containing it, etc.). Occurrence of a spherical notch in the host rock (Fig. 3f) that corresponds to the size of the microoncolite. 8. Silicification of the rock that records all the above-mentioned processes.

Within these forms of spheroidal nodules (microoncolites), the presence of tubular (rod-like) formations (Fig. 3g, h), which can be nominally distinguished into two types, was established. The first type ranges 5–6 μm in width and 50–60 μm in length (Fig. 3g), has an uneven surface, and represents a layering of several thin coatings, similar to a cabbage head. In the past, these coatings may have consisted of the same bacterial films that surrounded the rod-like formation and were dehydrated after lithification of the rock. The second type of remains of tubular microfossils is represented by formations with a width of not exceeding 15 μm and a length of up to 60 μm (Fig. 3h). These formations have a smooth surface, can be split into roughly equal, partially intertwined parts, and might have been represented polytrichomous filaments enclosed in a common coating in the past.

Despite the external similarity, microoncolites are rather diverse in structure. Some of them are characterized by a thin light coating that encircles a dark, lumpy-structured fine-crystalline rock (Fig. 4a). Other microoncolites have a double

(Fig. 4b), and sometimes a multilayered coating, with their central part composed of light microgranular carbonate rocks. The third oncolite-like formations are surrounded by a dark amorphous coating with a thickness of 45–55 μm (Fig. 4c, g), which allows us to assume its original mucous consistency. In this coating, single filamentous fungi (Fig. 4d) and their traces were detected. Fungi were parasitizing on these structures, feeding on their organic matter. Some microoncolites are composed of oval and rounded oncolite-like formations (Fig. 4f), combined by a single multilayered coating (Fig. 4e, f). As indicated above, it can be assumed that these complex formations of unidentified genesis could have occurred with the active participation of coccoid cyanobacteria (Fig. 3a–c, h).

The formation of oncolites involved both organic and mineral matter. Their inner part, in addition to microfossils, revealed an organic residue of a branched filamentous form, which we identified as a mycelial fungus (Fig. 4d, h, i). The latter parasitized on the organic matter of coccoid cyanobacteria, so-called constructors of microoncolites. The main long branch consists of quite thin parallel filaments identical to the hyphae of fungi (Fig. 4i). This and the presence of reproduction organs at the ends of the branches, as well as the shape, size, and ecological characteristic (parasitizing) allowed us to most certainly identify the established microfossilia as belonging to the lower aquatic mycelial fungi.

The following is a paleontological description of fungi.

Collection No. 87 is stored at the Geological Museum of the Diamond and Precious Metal Geology Institute of the Siberian Branch of the RAS (Yakutsk). The research was conducted within the framework of the state assignment.

6 Fungi Incertae Sedis

6.1 Genus Matchi Kolosov Et Litvinova, Gen. Nov.

The genus name derives from the village of Macha, located near the location of fungi.

Type species—Matchi sitnikovi sp. nov.

Diagnosis. Microscopic aquatic mycelial bush-like branched formations. They are attached to the ground by a rhizoid base that consists of numerous hyphae. The branching of the mycelium is located at some distance from its base. The branches are of two orders and of different length and thickness. The main (central) branch is an extension of the lower part of the mycelium and has a flat surface. It consists of distinctly visible hyphae. From this branch, many of the thinnest hyphae are directed in different sides in order to search for organic material. Some branches of the second order have an oval-shaped reproduction organ at their end.

At the terminal end of the main branch of the mycelium there is a lumpy tangle consisting of an interlacing of hyphae. The fungi fed on the organic material of microorganisms, which led to the development of organic sedimentary formations.

Species composition. Type species.

Notes. The selected genus is similar to the mycelial fungus Surninia Kolosov, 2016, on the basis of its ecological features. The fungi of both these genera fed on the organic matter of microorganisms that participated in the development of organic sedimentary formations of the Vendian (Ediacaran): columnar stromatolites were inhabited by Surninia, while oncolite-like formations were inhabited by Matchi. However, they possess different mycelium structures: Surninia has a set of intertwining rod-like hyphae, while Matchi has a presumably simpler, bush-like mycelium; location of the reproduction organs and branches also varies.

6.2 *Matchi Sitnikovi Kolosov Et Litvinova, Sp. Nov.*

The species is named after V.S. Sitnikov, a researcher in the field of geology.

Holotype—DPMGI SB RAS, No. 87/183, Eastern Siberia, Western Yakutia, Lena River, middle course, Ura Uplift, Macha settlement; Neoproterozoic, Upper Vendian, Tinnaya formation.

Diagnosis. The same as for the genus.

Description (Fig. 4i) (the dimensions of the organism are given in μm, according to the holotype).

The holotype is located inside a nest-shaped oncolite-like formation (Fig. 4h).

The lower half of the mycelium has a bumpy surface, which is presumably due to the location of hyphae; in some places they are interlaced in large quantities. This part of the mycelium ends with branches. One of them has a reproduction organ at the end (Fig. 3i:3). Oncolite-like formations around the mycelium contain many coccoid cyanobacteria, with their mucus being used as nutriment for fungi.

The mycelium is attached to the microoncolite by its basal part that looks like a tangle of the thinnest hyphae (Fig. 4i: 1). A rhizoid-like formation (Fig. 4i: 2) is observed, usually it attaches fungi to the ground, with a size of 2.4×3.4. The branching of the mycelium is observed at 10.8 from its base. Branches are characterized by different lengths and widths (thicknesses). The total length of the mycelium is 22.4, with the longest being 11.6, the width ranging from 0.59 to 1.03. This branch consists of very thin hyphae. The sizes of the other two branches: curved—length 10.3, width 0.9–1.0; thin, located on the left side—length 6.0, width 0.3–0.5. Branches of the second order of the following sizes (length and width, respectively): 2.27 and 0.27; 3.65 and 0.41; 3.13 and 0.34; 4.34 and 0.27. The most prominent branch is 3.17 in length and 0.06 in thickness. Reproduction organs (Fig. 4i: 3) are 1.03×0.38 in size, having an oblong shape. An indistinct distorted (?) formation consisting of hyphae is observed (Fig. 4i: 4). Individual hyphae (Fig. 4i: 5) are clearly distinguishable as well as the terminal end of the mycelium (Fig. 3i: 6).

Comparison. The sole species of the genus.

Material. One complete (holotype) specimen of fine preservation and fragments of three mycelia.

Therefore, there is every reason to confirm the organogenic sedimentary nature of these spherical and oval oncolite-like formations occurred with the participation of coccoid cyanobacteria. These formations are the size of dust particles and were formed in an aquatic environment in suspension, thus differing from the majority of oncolites developed in the process of rolling along the basin bottom. At the same time, both microoncolites and notches formed after their fallout were actively inhabited by numerous fungi that parasitized on these organogenic sedimentary formations.

7 Conclusions

The conducted study of oncolites and microoncolites of various Vendian formations, which seem to differ so much in size, appearance, and microorganism composition that actively participated in their occurrence, allows us to postulate their common characteristics:

1. The fundamental role in the formation of the oncolites discussed in the work belongs to microorganisms. Microorganisms served as the basis for oncolite and microoncolite formation and influenced the development of their morphological features (microstructure).
2. A favorable paleofacies setting, manifested in hydrodynamic processes that are quite active in shallow waters, was an essential condition for oncolite and microoncolite occurrence, as well as for the development of their spherical shape.
3. Parasitism of mycelial aquatic fungi in the late stages of oncolite and microoncolite formation did not have a significant impact on oncolites.
4. The structure of oncolites underwent some changes in the process of secondary transformations.

Hence, the Vendian formations (oncolites and microoncolites) discussed in this work developed due to a complex interaction of biogenic and abiogenic processes that defined their size, shape, structure, and distribution in the rock. The presence of fungi in oncolite-like formations was established, which extends our knowledge of the biodiversity and paleoecology of Vendian sediments.

Acknowledgements The research was funded within the framework of the state assignment of the Geological Institute of the Russian Academy of Sciences (Litvinova) and Diamond and Precious Metal Geology Institute SB RAS (Kolosov).

References

Butakov EP, Khomentovsky VV, Shenfil VY, Yakshin MS (1972) Base sections of Upper Precambrian and Lower Cambrian deposits the Siberian Platform. Nauka, Moscow (in Russian)
Chumakov NM, Semikhatov MA, Sergeev VN (2013) Vendian reference section of southern Middle Siberia. Stratigr. Geol. Correlation 21. 4: 359–382 (in Russian)
Chumakov NM (1959) Stratigraphy and tectonics of the southwestern Vilyui depression. Tektonika SSSR (Tectonics of the Soviet Union).: AN SSSR Moscow 4: 345–460 (in Russian)

Chuvashov BI (1988) Stromatolites and oncolites of Devonian, Carboniferous and Permian. In: Dubatolov VN (ed) Calcareous algae and stromatolites. Nauka, Novosibirsk (in Russian)
Davydov YV (1975) Riphean carbonate deposits of the south-east of the Siberian Platform and its framing (composition and origin). Nauka, Novosibirsk (in Russian)
Dol'nik TA (2000) Stromatolites and Microphytoliths in the South Siberian Platform. SO RAN, Novosibirsk (in Russian)
Khomentovsky VV, Postnikov AA, Karlova GA et al. (2004) Vendian in the Baikal–Patom Uplift, Siberia. Geol. Geofiz 45: 465–484 (in Russian)
Kochnev BB and Karlova GA (2010) New data on biostratigraphy of the Vendian Nemakit–Daldynian Stage in the southern Siberian Platform. Stratigr. Geol. Correlation vol. 18 5: 492–504
Kolosov PN (1975) Upper Precambrian Stratigraphy in southern Yakutia. Nauka, Novosibirsk (in Russian)
Kolosov PN (1977) Ancient oil and gas bearing strata of the Southeast Siberian platform. Nauka, Novosibirsk (in Russian)
Kolosov PN (1982) Upper Precambrian paleoalgological remains on the Siberian Platform. Nauka, Moscow (in Russian)
Kolosov PN (2016) New microorganisms from the Vendian (Ediacaran) of the Berezovsky Trough, southern Siberian Platform Paleontological Journal 50 (6): 549–556
Kolosov PN (1970) Organic remains of the Upper Precambrian of the south of Yakutia. Stratigraphy and paleontology of the Proterozoic and Cambrian of the East Siberian Platform. Nauka, Yakutsk (in Russian)
Kontorovich AE (2005) Stratigraphy of Petroliferous Basins in Siberia, Riphean and Vendian in the Siberian Platform, and Its Orogenic Framing Geo, Novosibirsk (in Russian)
Kuznetsov VG (2003) Evolution of the carbonate formation in the Earth's history. Geos, Moscow (in Russian)
Litvinova TV (2014) Ultramicrostructures of the stromatolite reef in the Northern Anabar Region and their origin. Lithol Miner Resour Soc Notes 5: 416–426 (in Russian)
Litvinova TV (2018) Possible role of microorganisms in the formation of structural and textural features of stromatolites Lethaea rossica. 16: 1-16 (in Russian)
Litvinova TV, Sergeev VN (2019) Integrated Study of Microfossils in Stromatolites of the Vendian Chencha Formation, Central Siberia. Paleontological Journal Soc Notes, 53(9): 98–103.
Litvinova TV, Sergeev VN (2018) Biogenic Microstructures in Stromatolites of the Baikal – Patom Highland: Results of Complex Study Lithol Miner Resour Soc Notes 53(2): 159–169
Logan BW, Rezak R, Ginsburg RN (1964) Classification and environmental significance of algal stromatolites. The Journal of Geology 72 (1): 68–83 (in Russian)
Maslov VP (1973) Atlas of rock-forming organisms. Science, Moscow (in Russian)
Melezhik VA, Pokrovsky BG, Fallick AE et al. (2009) Constraints on 87Sr/86Sr of Late Ediacaran seawater: insight from Siberian high-Sr limestones. J Geol Soc London 166: 183–191
Moczydlowska M, Nagovitsin KE (2012) Ediacaran radiation of organic-walled microbiota recorded in the Ura Formation, Patom Uplift, East Siberia, Precambrian Res (198)199: 1–24
Odum HT (1957) Biochemical deposition of strontium. Marine Science 4(2): 38–114
Pia J (1927) The conservation of fossil plants. Munich, Berlin
Rejtlinger E A (1959) Atlas microscopic of organic residues and perspective ancient strata of Siberia. Academy of Sciences USSR, Moscow (in Russian)
Selley RC (1977) An introduction to sedimentology. Academic Press, London
Sergeev VN, Knoll AH, Vorob'eva NG (2011) Ediacaran microfossils from the Ura Formation, Baikal–Patom Uplift, Siberia: taxonomy and biostratigraphic significance. J. Paleontol 85 (5): 987–1011
Twenhofel WH (1919) Pre-Cambrian and Carboniferous algal deposits. American Journal of Science 48(287): 339-352
Vologdin AG (1962) The oldest algae of the USSR. Academy of Science of the USSR, Moscow (in Russian)

Voronova LG, Radionova EP (1976) Paleozoic algae and microphytolites. Nauka, Moscow (in Russian)

Zhuravleva ZA (1964) Riphean and Lower Cambrian oncolithes and catagraphes of Siberia and their stratigraphic importance. Nauka, Moscow (in Russian)

Geochemistry of Biogenic–Abiogenic Systems

The Carbon Isotopic Composition of the Pai-Khoi Amber-Like Fossil Resin (NW Asia)

Andrey V. Zhuravlev and Irina S. Astakhova

Abstract Amber-like fossil resin localities are known in the Cretaceous—Quaternary deposits of the Eurasian Arctic. The fossil resins at these localities have been less studied than those from the European, SE Asian, American, and African localities. In this article, the properties and carbon isotopic compositions of the amber-like fossil resins from the Yugor Peninsula (Pai-Khoi Ridge, NW Asia) were considered and compared with those of Baltic amber (Sambian Peninsula, eastern Europe) and North Siberian amber-like fossil resins (Maimecha River, Khatanga Depression, North Asia). The carbon isotope compositions of the Arctic ambers represented by gedanite were investigated for the first time. The North Siberian Late Cretaceous (Santonian) gedanite has a $\delta^{13}C_{org}$ value of −21.7‰. The Pai-Khoi gedanite, which probably was produced by the Late Cretaceous Pinnacea plants, possesses $\delta^{13}C$ values ranging from −27.6 to −22.1‰ (mean −25.2‰). The $\delta^{13}C$ values of the Pai-Khoi gedanite are similar to those of the late Campanian amber-like fossil resins. The wide range of variations and bimodal distribution of the $\delta^{13}C$ values of gedanite of the Peschanaya River locality suggest diverse sources of the gedanite in this placer. Probably the placer had been formed by the material reworked from different parts of the sequence of the Sayakha Formation corresponding to diverse Late Cretaceous landscapes.

Keywords Gedanite · Succinite · C-isotopes · Arctic · Late cretaceous

A. V. Zhuravlev (✉) · I. S. Astakhova
N.P. Yushkin Institute of Geology FIC Komi SC UrB RAS, Syktyvkar 167000, Russia
e-mail: micropalaeontology@gmail.com

I. S. Astakhova
e-mail: astakhova@geo.komisc.ru

O. V. Frank-Kamenetskaya et al. (eds.), *Biogenic—Abiogenic Interactions in Natural and Anthropogenic Systems 2022*, Springer Proceedings in Earth and Environmental Sciences, https://doi.org/10.1007/978-3-031-40470-2_13

1 Introduction

Amber (sensu lato) is a fossil tree resin that has achieved a stable state through "amberisation processes" (Poinar 1992): the loss of volatile constituents and its chemical alteration after burial in the sediment (Kosmowska-Ceranowicz 2006; Gaigalas and Halas 2009; Vavra 2009). The known amber-like fossil resins range from the Carboniferous through to the Quaternary, and their occurrences are widely distributed over the world (for reviews see Yushkin 1973; Zherikhin and Eskov 1999; Bray and Anderson 2009; Tappert et al. 2013; Bogdasarov 2017). Fossil resins commonly form secondary placers in costal and alluvial deposits; examples include Baltic amber of Eocene (Bartonian) age (Bukejs et al. 2019), which occurs in late Eocene Prussian Formations and Pleistocene and Holocene sediments (Kosmowska-Ceranowicz 2006). The amber-like fossil resin localities are known in the Cretaceous—Quaternary deposits of the Eurasian Arctic (Yushkin 1973; Bogdasarov 2007, 2017). The resins of these localities have received less research attention than their European, SE Asian, American, Australian, and African counterparts.

The fossil resins represent a valuable geochemical archive of ancient floras and the Earth's atmosphere. It is thought that the fractionation of ^{13}C in resin-producing plants is primarily influenced by mean annual precipitation and atmospheric oxygen concentrations (pO_2) (Tappert et al. 2013; Corso et al. 2017). Increasing mean $\delta^{13}C$ resin values correspond to decreasing atmospheric pO_2 and vice versa (Tappert et al. 2013). The $\delta^{13}C$ values of fossil resins range from −28.4‰ to −18.2‰ (Tappert et al. 2013). The lowest values are known for the Pleistocene Columbian copal (down to −28.4‰), and the highest ones are characteristic of the Early Cretaceous (Spanish amber) and Eocene (Ellesmere Island amber) fossil resins (Tappert et al. 2013). The carbon isotope compositions of Baltic amber have been well-studied and demonstrate nearly constant values (c.a. −23‰) (Gaigalas and Halas 2009; Tappert et al. 2013).

This article aims to study the properties and carbon isotopic compositions of amber-like fossil resins from the Yugor Peninsula (Pai-Khoi Ridge, NW Asia), provide a comparison of these with Baltic amber and North Siberian amber-like fossil resin (Khatanga Depression), and decipher the origin of the Pai-Khoi fossil resin.

2 Material

This study is based on the collection of amber-like fossil resins stored in the Geological Museum (Syktyvkar, Russia; collection #160) and the authors' collection. A total of two samples of Baltic amber (succinite), 19 samples of Pai-Khoi amber-like fossil resin, and one sample of North Siberian amber-like fossil resin were studied. The Baltic amber samples came from the Baltic Sea shoreline placer near the town of Zelenogradsk (Kaliningrad district, the Sambian Peninsula). The late Eocene Prussian Formation is considered to be the most probable source of the amber in this placer.

The Pai-Khoi samples were collected from the Quarternary placer cropping out on the side of the Peschanaya River valley (N 69°36′44.91 "E 62° 7′17.24") (Fig. 1). The sequence of the alluvial Quarternary sediments in this locality is represented by the following units (from bottom to top) (Yushkin 1973):

Unit 1. Gray coarse-grained sand with wavy bedding. Unit is 0.3 m thick.

Unit 2. Gray coarse-grained sand containing numerous lignitized wood fragments. Unit is 0.25 m thick.

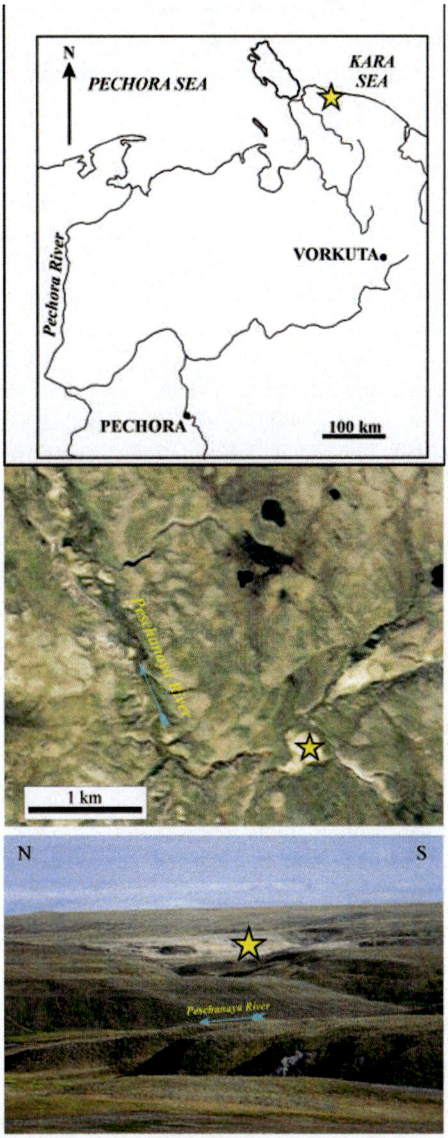

Fig. 1 Locality of the Pai-Khoi amber-like resin placer (valley of the Peschanaya River) (marked with yellow star)

Unit 3. Bluish-gray sandy clay containing lignitized wood fragments. Unit is 0.5 m thick.

Unit 4. Bluish-gray clayey fine-grained sand with lens-like and wavy bedding. Unit is 0.4 m thick.

Unit 5. Yellow fine-grained sand with lens-like and wavy bedding. Unit is 0.2 m thick.

Unit 6. Gray fine-grained sand with lens-like bedding and lenses enriched by plant remains and grains of the amber-like fossil resin. Unit is 0.8 m thick.

Unit 7. Yellow fine-grained sand with lens-like and wavy bedding. Unit is 0.4 m thick.

Unit 8. Gray coarse-grained sand containing lignitized wood fragments. Unit is 0.2 m thick.

Unit 9. Yellow fine-grained sand with lens-like and wavy bedding. Unit is 0.4 m thick.

Unit 10. Yellow fine-grained sand with wavy bedding. Unit is 0.2 m thick.

The lithological characteristics of the sequence suggest its deposition in the flood-plain or oxbow of the paleo-river delta. The grains of fossil resin were deposited along with plant fragments, allohtonous coal, and fine-grained sediments in the calm water conditions. The supposed age of the deposits based on the spore and pollen associations is the Holocene (Yushkin 1973).

The material of the placer is supposed to consist of reworked deposits of the Late Cretaceous—Middle Paleogene Sayakha Formation (Yushkin 1973; Zarkhidze et al. 2017). The Sayakha Formation is composed of three units (Zarkhidze et al. 2017). The lower unit 40–45 m thick is represented by gray siltstones with scattered gravel, and layer of clay in the lower part. The middle unit is composed of alternation of sandstone, siltstone and sandy clay with thin beds of gaize in the upper part. This unit is about 130 m thick. The upper unit comprises dark-gay siltstones with beds of clay and sandstone. The upper unit is about 100 m thick. The lower and middle units of the Sayakha Formation contain Late Cretaceous inoceramid mollusks (*Inoceramus* (*Sphenoceramus*) *cardissoides* (Goldfuss), *Inoceramus* (*Sphenoceramus*) aff. *pachti* (Archanguelsky), *Inoceramus* aff. *lobatus* Goldfuss), foraminifers, radiolarians, and ostracods (Zarkhidze et al. 2017). The age of this part of the formation is supposed as Campanian—Maastrichtian. The upper unit of the Sayakha Formation yields spores and pollen of the middle Eocene tropical flora (Zarkhidze et al. 2017). The spore association contains representatives of *Picea* and *Pinus* among gymnosperms. The amber-like fossil resins are known in sandstones of the Upper Cretaceous part of the formation.

The North Siberian amber-like resin sample is from the Khatanga Depression, at the Yantardakh locality on the bank of the Maimecha River (N 71°18′38.41″ E 99°29′14.73″), within the Kheta Formation of Santonian (Late Cretaceous) age (collection of M. A. Bogdasarov; collection #160/21). The Kheta Formation is composed of cross-bedded sandstone with lenses containing lignitized wood (Rasnitsyn et al. 2016). The Santonian age of the formation is supported by plant remains and inoceramid mollusks *Inoceramus patootensis* Loriol, *I. sibiricus* Dobr., and *I.* cf. *patootensis angustus* Beyenburg (Rasnitsyn et al. 2016). Rasnitsyn et al.

(2016) suggested *Cedrus* or other coniferous trees as probable resin-producing plants for the Yantardakh locality.

3 Methods

The fossil resins were studied with a number of methods. The investigations were focused on determination of the fossil resin type prior to the carbon isotope analyses. Fourier-transform infrared spectroscopy (FTIR) and differential thermal analyses were used to determine the type of fossil resins examined.

The absorption FTIR spectra in the range of 4000–400 cm^{-1} with a resolution of 4 cm^{-1} were taken for six samples from the study localities. The material was studied using an Infralum FT-801 Fourier-spectrometer. The samples were prepared by pressing 1.0–1.7 mg of resin with 800 mg of optically pure potassium bromide in a 20 mm diameter cage. The spectra were analyzed with SpectraLUM software (Lumex Ltd.) and Spectragryph software (Menges 2022).

Differential thermal and thermogravimetric analyses of three samples of the Pai-Khoi amber-like resin and one sample of Baltic amber were performed with a TGA/DSC 3 + derivatograph (MettlerToledo). Maximum heating reached 400 °C (three samples) and 600 °C (one sample), with a heating rate of 10 °C/min and a sample weight of 25–30 mg.

Some other physical properties of the Pai-Khoi amber-like resin were studies as well. The Vickers' microhardness of the fossil resin was measured by PMT-3 hardness testing machine at 20 g load and the exposition time of 5 s. The refractive index and luminescence were studied by Modul 1 gemological equipment (System Eikhorst).

The organic carbon isotopic composition of 22 samples was analyzed (two of Baltic amber, 19 of the Pai-Khoi amber-like resin, and one of the North Siberian amber-like resin). The samples were placed in a tin capsules and combusted in the presence of oxygen at 900 °C. The carbon isotope values were measured with a DELTA V Advantage mass spectrometer equipped with a Thermo Electron Continuous Flow Interface (ConFlo III) and Element Analyser (Flash EA 1112). An international standard USGS-40 (L-Glutamic acid) and internal standard Acetanilide (C_8H_9NO) were used. The precision of the $\delta^{13}C_{org}$ value is ±0.15‰, and $\delta^{13}C_{org}$ values are reported relative to the VPDB (Vienna Pee Dee Belemnite) standard.

4 Results

The examined Pai-Khoi amber-like fossil resin is honey-yellow or lemon-yellow in color, brittle, and almost transparent. It is almost free of organic debris and other contaminants that are critical for the correct carbon isotope analyses (McKellar et al. 2011). The refractive index of the studied samples fluctuates within a small range of

1.520–1.535. The luminescence of the resin is blue at an excitation wavelength of 354 nm, and is dark blue at an excitation wavelength of 254 nm. Some of the fossil resin pieces display an abundance of vacuoles occupied by air. The vacuoles are nearly spherical in shape and arranged along the flow lines. Diameter of the vacuoles varies from <1 μm up to 40 μm (mean 11.6 μm, standard deviation 8.9, N = 93). The small vacuoles (up to 10 μm) dominate. The spherical form of the vacuoles and their small size suggests quite slow hardening and high viscosity of the resin. The fossil resin grains are quite small; their size varies from 3 mm up to 10 mm. The mean hardness value of the Pai-Khoi fossil resin is of 0.207 GPa (standard deviation 0.02, N = 9). The brittle fracture was detected at load of 30 g. Due to high fragility, the grains bear numerous cracks. The studied Baltic amber is honey-yellow in color and almost transparent. The luminescence of the amber is whitish blue at an excitation wavelength of 354 nm, and is blue at an excitation wavelength of 254 nm. The mean hardness value of the amber is of 0.222 GPa (standard deviation 0.002, N = 4). The grains without of cracks and inclusions were used in this study. The examined North Siberia (Khatanga Depression, the Maimecha River) amber-like fossil resin is yellowish-brown in color and transparent. The studied grain contains no cracks and inclusions.

4.1 *Fourier-Transform Infrared Spectroscopy (FTIR)*

The FTIR spectra of the studied samples contain absorption bands of various functional groups characteristic of fossil resins (Fig. 2 and Table 1). The wide absorption band near 3420 cm^{-1} refers to water molecules adsorbed on KBr matrix. The bands at 2925 cm^{-1} and 2867 cm^{-1} are characteristic of symmetrical and asymmetrical stretching vibrations of aliphatic $–CH_2$ and $–CH_3$ groups. The intense band near 1725 cm^{-1} indicates the presence of carboxyl groups. The appearance of this band is attributed to vibrations of the C–O group. The weakly expressed absorption band in the region of 1620–1650 cm^{-1} is assigned to unsaturated bonds of the $CR_1R_2 = CH_2$ type.

The examined Pai-Khoi amber-like fossil resin shows a number of absorption peaks in the "fingerprints area" of FTIR spectrum (interval from 1500 cm^{-1} to 700 cm^{-1}) (Fig. 3). The band at 1458 cm^{-1} with shoulder at 1448 cm^{-1} is of high intensity. The next series of bands of high intensity corresponds to the interval from 1390 cm^{-1} to 1365 cm^{-1}. In the spectrum interval from 1248 cm^{-1} to 700 cm^{-1} there are a number of absorption bands of medium and low intensity (1242 cm^{-1}; 1175 cm^{-1}; 1125 cm^{-1}; 1083 cm^{-1}; 1037 cm^{-1}; 972 cm^{-1}; 938 cm^{-1}; 910 cm^{-1}; 887 cm^{-1}; 823 cm^{-1}; and 710 cm^{-1}) (Fig. 3).

The bands located in the short-wavelength part of the spectrum between 1250 and 1120 cm^{-1} are diagnostic for succinite. The FTIR spectrum of the Baltic amber is characterized by the presence of a "Baltic shoulder" at interval from 1234 cm^{-1} to 1167 cm^{-1} (Fig. 2).

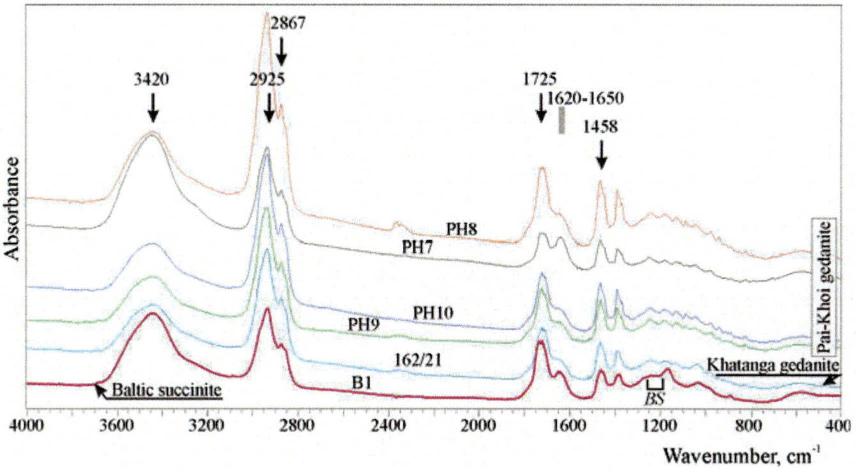

Fig. 2 Fourier-Transform Infrared (FTIR) spectra of the Pai-Khoi (Peschanaya River) and North Siberian (Khatanga Depression, Maimecha River) amber-like resins (gedanite) and the Baltic amber (succinite). *BS*—"Baltic shoulder"

The FTIR spectra of the Pai-Khoi and North Siberia amber-like resins show a lack of a "Baltic shoulder" at 1235–11175 cm^{-1} in association with a series of bands of low intensity in the interval from 1247 cm^{-1} to 972 cm^{-1} (Figs. 2 and 3). This FTIR pattern differs from that of Baltic amber (succinite) and is similar to that of gedanite (Yushkin 1973; Martirosyan and Bogdasarov 2014).

4.2 *Differential Thermal Analyses*

The differential thermal patterns for the Pai-Khoi amber-like resins demonstrate an exothermic peak at 150–160° and weak endothermic peaks near 120 and 170° (Fig. 4 and Table 2). The Baltic amber demonstrates a lack of prominent thermal effects in the temperature range from 100° to 200 °C. The differential thermal patterns of the Pai-Khoi amber-like resins correspond to those of gedanite (Savkevich 1970).

4.3 *Differential Thermogravimetric (DTG) Analyses*

The differential thermogravimetric analysis of the Pai-Khoi amber-like resin (sample PH14) shows the Main Thermal Point (MATHEP), corresponding to the maximal rate of weight loss, at 436 °C (Fig. 5 and Table 2). The MATEP represents the loss of more than 75% of the sample. The derivative weight loss at MATHEP is about

Table 1 Absorption bands and their assignment for studied amber-like fossil resins

Amber-like resin	Absorption bands (Wavenumber cm^{-1})							
	C = O acid/C = O ester (1700/1730)	C = O acid/C = O ester (1250/1160)	C–O primary alcohols (1000–1075)	C–O secondary alcohols (1030–1125)	O–H (3400–2550)	C = C aromatic ring (1600)	$R_2R_2C = CH_2$ (885–895)	$HR_1C = CR_2H$ (960–980)
Baltic succinate	◯ 1725	◯ 1250 < Δ 1160	Δ 1030	–	*3400–3500	Δ 1633	–	Δ 972
Pai-Khoi gedanite	◯ 1725	1250 > 1175	◯ 1037	Δ 1125	*3400–3500	Δ 1633	Δ 880	Δ 972
Khatanga gedanite	◯ 1725	1250 > 1175	◯ 1037	Δ 1125	*3400–3500	Δ 1633	Δ 880	Δ 972

Band intensity: *strong, ◯ medium, Δ weak, – absent

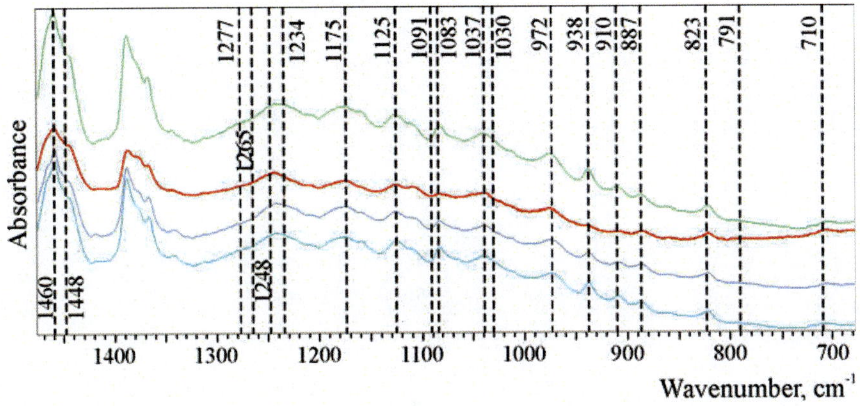

Fig. 3 Close-up of the 1470–680 cm^{-1} spectral region of the Fourier-Transform Infrared (FTIR) spectra of the Pai-Khoi amber-like resin

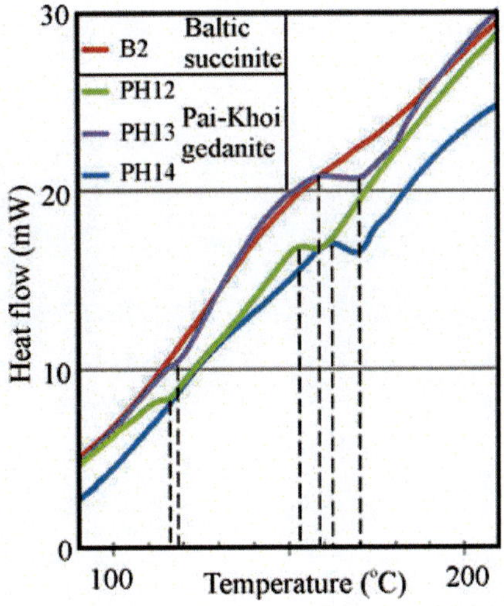

Fig. 4 Results of differential thermal analysis of the Baltic amber (succinite) and the Pai-Khoi amber-like fossil resin (gedanite)

4.5%/min. The minor peaks of rate of weight loss at 310 and 350 °C were detected for one sample (PH13) of the Pai-Khoi gedanite (Fig. 5).

4.4 *Carbon Isotope Analyses*

Table 3 presents the results of the carbon isotope study of the amber-like resins. The $\delta^{13}C$ value ranges from −27.6 to −22.1‰ in the gedanite samples from the Pai-Khoi Ridge. The mean value of $\delta^{13}C_{org}$ for the measured samples is about −25.2‰

Table 2 Data of thermic analysis of the Pai-Khoi amber-like resin (sample PH14)

Ts (°C)	t (s)	HF (mW)	Weight (mg)
30.05	15.00	−8.68	25.07
40.07	72.00	−7.47	25.06
50.85	132.00	−3.68	25.06
60.05	185.00	−1.70	25.05
70.15	244.00	−0.06	25.03
80.05	302.00	1.36	25.02
90.07	361.00	2.68	25.00
100.14	420.00	4.43	24.98
110.10	478.00	6.63	24.94
120.08	536.00	9.02	24.88
130.02	594.00	11.39	24.80
140.05	653.00	13.16	24.69
150.07	712.00	14.94	24.58
160.09	771.00	16.86	24.50
170.82	836.00	16.67	24.37
180.10	890.00	19.13	24.26
191.01	954.00	21.68	24.13
200.16	1008.00	23.38	24.02
210.10	1067.00	24.79	23.88
220.01	1126.00	25.96	23.71
230.05	1186.00	27.00	23.52
240.12	1246.00	28.15	23.30
250.02	1305.00	29.21	23.07
260.08	1365.00	30.48	22.79
270.17	1425.00	31.84	22.49
280.07	1484.00	33.09	22.16
290.08	1543.00	35.41	21.79
300.01	1603.00	35.33	21.40
310.12	1664.00	35.53	20.98
320.11	1724.00	35.98	20.54
330.06	1784.00	36.05	20.05
340.15	1845.00	35.86	19.48
350.06	1905.00	35.61	18.82
360.64	1969.00	35.18	18.00
370.01	2026.00	34.41	17.12
380.04	2087.00	33.50	16.01
390.03	2148.00	32.10	14.69
400.00	2208.00	32.23	13.16

(continued)

Table 2 (continued)

Ts (°C)	t (s)	HF (mW)	Weight (mg)
410.02	2269.00	31.20	11.31
420.05	2330.00	29.97	9.31
430.01	2391.00	28.14	7.07
440.06	2452.00	27.20	4.76
450.04	2512.00	27.29	2.82
460.05	2572.00	27.96	1.38
470.06	2632.00	28.65	0.65
480.16	2693.00	28.10	0.39
490.14	2754.00	26.27	0.32
500.11	2815.00	24.08	0.29
510.04	2876.00	21.50	0.28
520.00	2937.00	19.07	0.28
530.09	2999.00	16.22	0.27
540.03	3060.00	13.54	0.27
550.13	3122.00	10.58	0.27
560.09	3183.00	7.88	0.26
570.04	3244.00	5.11	0.26
580.14	3306.00	1.93	0.26
590.03	3367.00	−1.86	0.26
598.59	3420.00	−5.62	0.26

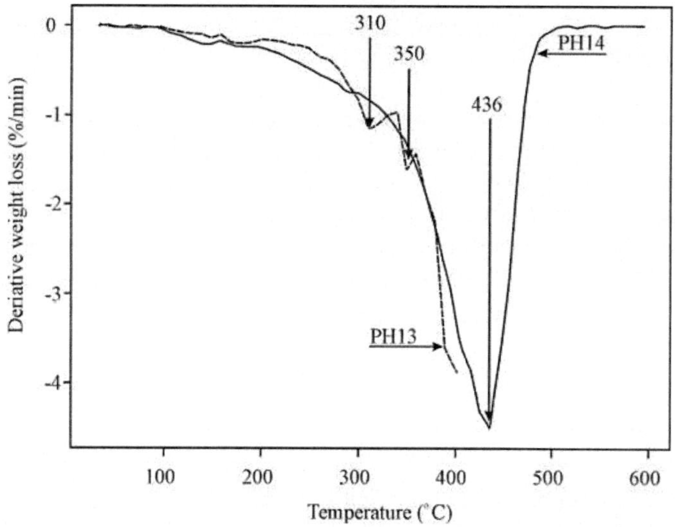

Fig. 5 Differential thermogravimetric (DTG) profile of specimens (PH13 and PH14) of the Pai-Khoi amber-like fossil resin

Table 3 Carbon isotope composition (VPDB, ‰) of the studied amber-like fossil resins

Sample	Locality	$\delta^{13}C_{org}$
B1	Sambian Peninsula	−23.4
B2	Sambian Peninsula	−23.3
4	Pai-Khoi, the Peschanaya River	−24.2
5	Pai-Khoi, the Peschanaya River	−26.4
6	Pai-Khoi, the Peschanaya River	−25.0
PH1	Pai-Khoi, the Peschanaya River	−27.6
PH2	Pai-Khoi, the Peschanaya River	−24.6
PH3	Pai-Khoi, the Peschanaya River	−26.1
PH4	Pai-Khoi, the Peschanaya River	−26.4
PH7	Pai-Khoi, the Peschanaya River	−22.2
PH8	Pai-Khoi, the Peschanaya River	−22.6
PH9	Pai-Khoi, the Peschanaya River	−27.0
PH10	Pai-Khoi, the Peschanaya River	−23.1
PH11	Pai-Khoi, the Peschanaya River	−26.2
PH12	Pai-Khoi, the Peschanaya River	−26.4
PH13	Pai-Khoi, the Peschanaya River	−22,7
PH14	Pai-Khoi, the Peschanaya River	−27.1
PH15	Pai-Khoi, the Peschanaya River	−26.6
PH16	Pai-Khoi, the Peschanaya River	−22.1
PH17	Pai-Khoi, the Peschanaya River	−26.3
PH18	Pai-Khoi, the Peschanaya River	−26.6
162/21	Khatanga Depression, the Maimecha River	−21.7

(standard deviation 1.85, $n = 19$). It is notable that distribution of the $\delta^{13}C$ values of the studied samples of the Pai-Khoi gedanite is bimodal (Fig. 6). The mixture analysis reveals two populations having mean values of −26.6‰ and −23.3‰, and standard deviations of 0.44 and 1.11 respectively.

The North Siberia (Khatanga Depression, the Maimecha River) gedanite shows $\delta^{13}C$ value of −21.7‰; and studied Baltic amber reveals $\delta^{13}C$ values of about −23.3‰.

5 Discussion

Results of FTIR spectroscopy and differential thermal analyses suggest that the study samples of the Pai-Khoi and North Siberia amber-like resins correspond to gedanite. The FTIR spectra show that absorption bands associated with oxygen compounds are strongly manifested. These include bands of hydroxyl group (3400–3480 cm^{-1}), C = O groups of aliphatically saturated (1175 cm^{-1}) and unsaturated esters (1725 cm^{-1}).

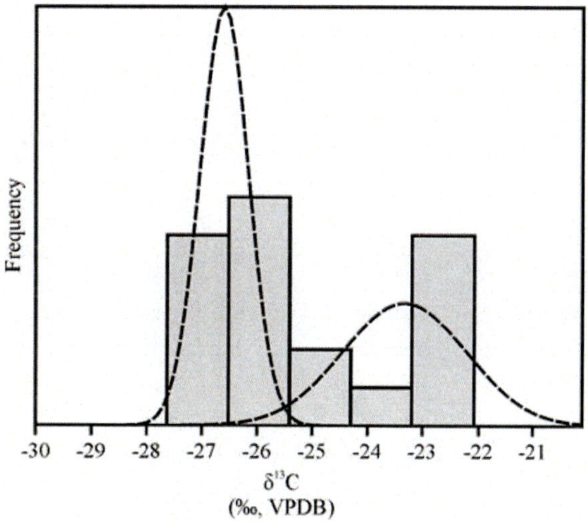

Fig. 6 Distribution of the $\delta^{13}C$ values of the Pai-Khoi amber-like fossil resin with the kernel densities of two populations

The revealed features indicate a high degree of oxidation of the resin samples, and their fossilization in anaerobic reducing conditions of waterlogged soils and peatlands. It can be assumed that under the conditions of amber formation, autoreduction dominated—primary resin acids were subjected to decarboxylation and hydration with the formation of brittle fossil gedanite resins.

The upper part of the infrared spectrum provides information on probable origin of the resin (Langenheim and Beck 1968; Tappert et al. 2011). The spectrum part from 1500 cm^{-1} to 700 cm^{-1} is useful as a "fingerprint" of amber-like resins (Martirosyan and Bogdasarov 2014).

Moreover, this part of the spectrum allows chemotaxonomical interpretation (Tappert et al. 2011) (Table 4). The infrared spectra permit distinguishing two main types of resin. According to Tappert et al. (2011) the first type of resin produced by Pinaceae (pinaceous type) consists mainly of diterpenes that are based on abietane/pimarane skeletal structures. The second type of resin associated with Cupressaceae, Sciadopityaceae, Araucariaceae, and Podocarpaceae (cupressaceous resin) consists mainly of diterpenes that are based on the labdanoid skeletal structures. The differences in composition of these types of resin are reflected in their infrared spectra, especially in the spectrum part from 2848 cm^{-1} to 700 cm^{-1} (Tappert et al. 2011, Table 2).

The examined Pai-Khoi fossil resin shows presence the distinct band at 1460 cm^{-1} in association of characteristic absorption spectrum structure in the interval from 1450 cm^{-1} to 700 cm^{-1}, which are similar to the spectrum of pinaceous type of resin (Tappert et al. 2011) (Table 4 and Fig. 3).

Some difference between spectra of the Pai-Khoi fossil resin and the pinaceous type of resin may be attributed to diagenetic processes and/or specific type of the resin-producing plants. In any case, it is reasonable that the Pai-Khoi resin-producing plants had close relation to Pinaceae.

Table 4 Relative intensity of the characteristic infrared absorption peaks in spectra of different types of resin (based on Lambert et al. 2008; Tappert et al. 2011) and Pai-Khoi fossil resin

Wavenumber (cm^{-1})	Pinaceous type of resin	Cupressaceous type of resin	Pai-Khoi fossil resin
1460	Medium	Absent	Strong
1448	Weak or shoulder	Medium	Shoulder
1277	Variable	Absent	Absent
1265	Absent	Variable	Absent
1248	Variable	Absent	Absent
1234	Absent	Variable	Absent
1091	Absent	Weak	Absent
1030	Absent	Medium	Absent
887	Absent	Strong	Weak
823	Medium	Weak	Weak
791	Absent	Medium	Absent
710	Weak–medium	Absent	Weak

The age of the Pai-Khoi gedanite is not yet clear. Yushkin (1973) supposed Mesozoic age of the gedanite from the Peschanaya River locality. This supposition was based on the mixed microphytofossil association discovered from the gedanite-bearing Quarternary placer. The association contains Quarternary, Cretaceous, and Jurassic microphytofossils, with predominance of the Cretaceous forms (Yushkin 1973). The exotic sandstone boulders and blocks occurring in the Quarternary deposits of the northern Pai-Khoi contain the amber-like fossil resins in association with the Cretaceous microphytofossils and fauna (Romanov et al. 2018). The middle part of the Sayakha Formation of the Campanian–Maastrichtian age is probable source of these boulders and blocks. It may be reasonable to suggest, therefore, that the age of the Pai-Khoi amber-like fossil resin is Campanian–Maastrichtian. This assumption is supported by the results of the differential thermogravimetric analysis, which indicates MATHEP at 436 °C. The high value of MATHEP suggests at least Mesozoic age of the resin (Ragazzi et al. 2003).

The mean value of $\delta^{13}C_{org}$ of the Pai-Khoi gedanite differs significantly from those of Baltic succinite (−23.4‰, see also Gaigalas and Halas 2009) and North Siberian (Khatanga Depression, the Maimecha River) gedanite (−21.7‰) (Fig. 7). The mean value and range of variation of $\delta^{13}C$ in the Pai-Khoi gedanite span those of the late Campanian (Grassy Lake, Cedar Lake, Albertian, and Genesee ambers), early Eocene (Tadkeshwar and Panda kimberlite ambers), and Miocene (Malysian, Mexican, and Dominican ambers) fossil resins according to data presented by Tappert et al. (2013) (Fig. 7). The $\delta^{13}C$ value of the North Siberian (Khatanga Depression, the Maimecha River) gedanite corresponds to that of the Late Cretaceous amber-like fossil resins (Tappert et al. 2013).

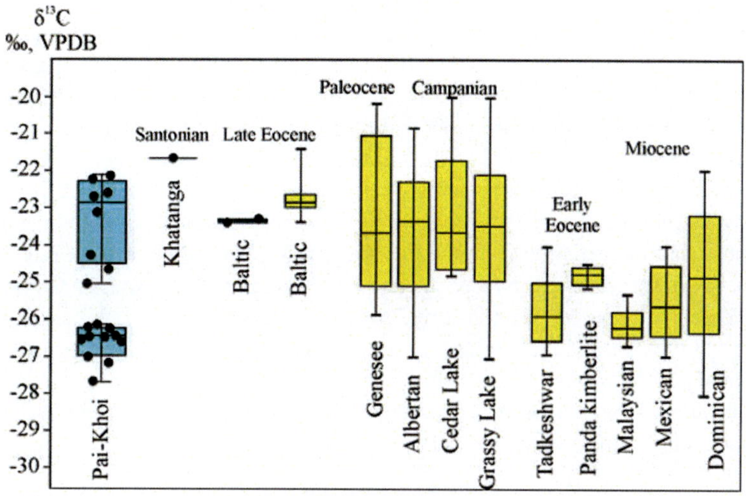

Fig. 7 Compiled record of carbon isotopic compositions of the studied samples (blue) and amber-like fossil resins based on data of Tappert et al. (2013), Gaigalas and Halas (2009) (yellow). The 25–75% quartiles are drawn using a color box. The median is shown with a horizontal line inside the box. The minimal and maximal values are shown with whiskers

The carbon isotope values of the studied fossil resins fall in the range expected for the bulk of organic matter in C3 plants (Nissenbaum and Yakir 1995). According to the data of Tappert et al. (2013) resin-producing plants of different taxonomic groups generally have a similar fractionation behavior during resin biosynthesis, and the fractionation observed in resins is similar to that of bulk plant matter. Moreover, diagenetic processes, polymerization, and weathering affect the carbon isotopic composition of fossil resins in a small degree (Stout 1995; Tappert et al. 2013; Corso et al. 2013). In contrary, Corso et al. (2017) reported that carbon isotope fractionation during resin biosynthesis results in more ^{13}C enriched $\delta^{13}C$ signature of resin (by approximately 2–4‰) than the $\delta^{13}C$ of other plant parts. Furthermore, $\delta^{13}C$ values of resin of different tree genera growing in the same locality shows differences of about 2–5‰ (Corso et al. 2017).

Thus, the carbon isotopic signal preserved in the fossil resin reflects the primary isotopic composition of the ancient resin producing plants with ^{13}C enrichment of about 2–4‰. The primary carbon isotope composition of resin-producing plants, in turn, was controlled by the global and local environment, such as atmospheric pO_2, isotopic composition of atmospheric CO_2, mean annual precipitation, water and nutrient availability, light, and temperature (Tappert et al. 2013; Corso et al. 2013, 2017).

It is notable that range of $\delta^{13}C$ variations in the Pai-Khoi gedanite is more than 4‰, and distribution of $\delta^{13}C$ is bimodal (Figs. 6 and 7). Probably these range and distribution are caused by diverse sources of the gedanite (different parts of the Sayakha Formation) in the study placer, or by high environmental variability/differentiation during the formation of this fossil resin or during its preservation. High consistency in FTIR parameters of the investigated Pai-Khoi samples suggests lack of significant variations in the resin composition and preservation. Thus, variations in preservation may affect the carbon isotope composition in a small degree; and environment diversity seems to be a most probable factor caused differentiation of $\delta^{13}C$ values of the studied samples of the Pai-Khoi fossil resin. According to Tappert et al. (2013) the lower $\delta^{13}C$ values may correspond to cooler and wet environment, but the higher $\delta^{13}C$ values may be caused by warmer and dry environment. Thus it is reasonable to suppose that the gedanite from the Peschanaya River locality corresponds to two groups of resin-producing Pinnacea plants which inhabited different Late Cretaceous landscapes. The $\delta^{13}C$ values of the first group (mean value of −23.3‰) fall into the interval typical for Late Cretaceous amber-like resins. The second group shows significantly lower $\delta^{13}C$ values (mean value of −26.6‰). It is possible that cool and wet environment corresponding to the second group of the Pai-Khoi gedanite existed locally, or reflects the seasonal variations.

6 Concluding Remarks

The Pai-Khoi amber-like fossil resin from the Peschanaya River locality is represented by gedanite that probably was produced by the Late Cretaceous Pinnacea plants. This gedanite possesses $\delta^{13}C$ values ranging from −27.6‰ to −22.1‰ and having bimodal distribution. The $\delta^{13}C$ values of the Pai-Khoi gedanite are similar to those of the late Campanian (Grassy Lake, Cedar Lake, Albertian, and Genesee) amber-like fossil resins. The wide range of variations and bimodal distribution of the $\delta^{13}C$ values of gedanite of the Peschanaya River locality suggest diverse sources of the gedanite in this placer. Probably the placer had been formed by the material reworked from different parts of the sequence of the Sayakha Formation corresponding to diverse Late Cretaceous landscapes. Another possibility is that the fossil resins were formed in the highly variable or differentiated environment.

7 Declaration of Competing Interest

The authors declare that they have no known competing financial interests or personal relationships that could have appeared to influence the work reported in this paper.

Acknowledgements Irina V. Smoleva, Vladimir P. Lutoev, and Eugeniy M. Tropnikov (CKP "Geonauka" of the Institute of Geology, Syktyvkar, Russia) are acknowledged for their help in the analytical study of the amber samples. The authors thank the Proof-Reading-Service.com for professional proof-reading of the early version of the manuscript.

References

Bogdasarov MA (2007) Mineralogy of Fossil Resins in Northern Eurasia. Geology of Ore Deposits 49: 630-637. https://doi.org/10.1134/S1075701507070215

Bogdasarov MA (2017) Mesozoic fossil resins of Northern Eurasia. Vestnik IG Komi SC UB RAS 7: 4–11. https://doi.org/10.19110/2221-1381-2017-7-3-11 (in Russian)

Bray PS, Anderson KB (2009) Identification of Carboniferous (320 million years old) class Ic amber. Science 326(5949): 132–134. https://doi.org/10.1126/science.1177539.

Bukejs A, Alekseev VI, Pollock DA (2019) Waidelotinae, a new subfamily of Pyrochroidae (Coleoptera: Tenebrionoidea) from Baltic amber of the Sambian peninsula and the interpretation of Sambian amber stratigraphy, age and location. Zootaxa 4664(2): 261–273. https://doi.org/10.11646/zootaxa.4664.2.8

Dal Corso J, Roghi G, Ragazzi E, Angelini I, Giaretta A, Soriano C, Delclòs X, Jenkyns HC (2013) Physico-chemical analysis of Albian (Lower Cretaceous) amber from San Just (Spain): implications for palaeoenvironmental and palaeoecological studies. Geologica Acta 11(3): 359-370. https://doi.org/10.1344/105.000001871

Dal Corso J, Schmidt A, Seyfullah L, Preto N, Ragazzi E, Jenkyns H, Delclòs X, Néraudeau D, Roghi G (2017) Evaluating the use of amber in palaeoatmospheric reconstructions: The carbon-isotope variability of modern and Cretaceous conifer resins. Geochimica et Cosmochimica Acta 199: 351-369. https://doi.org/10.1016/j.gca.2016.11.025

Gaigalas A and Halas S (2009) Stable isotopes (H, C, S) and the origin of Baltic Amber. Geochronometria 33: 33-36. https://doi.org/10.2478/v10003-009-0001-9

Kosmowska-Ceranowicz B (2006) Polska, opowieść o bursztynie. [Poland, The story of Amber]. Sport i Turystyka, Muza SA

Lambert JB, Santiago-Blay JA, Anderson KA (2008) Chemical signatures of fossilized resins and recent plant exudates. Angewandte Chemie 47: 9608–9616. https://doi.org/10.1002/anie.200705973

Langenheim JH, Beck CW (1968) Catalogue of infrared spectra of fossil resins (ambers) 1. North and South America. Botanical Museum Leaflets, Harvard University 22(3): 65–120

Martirosyan OV, Bogdasarov MA (2014) Fossil resins: diagnosis, classification and structural transformation under conditions of thermal impact. Vestnik IG Komi SC UB RAS 4: 10-15 (in Russian)

McKellar RC, Wolfe AP, Muehlenbachs K, Tappert R, Engel MS, Cheng T, and Sanchez-Azofeifa GA (2011) Insect outbreaks produce distinctive carbon isotope signatures in defensive resins and fossiliferous ambers. Proceedings of the Royal Society B: Biological Sciences 278(1722): 3219–3224. https://doi.org/10.1098/rspb.2011.0276

Menges F (2022) Spectragryph - optical spectroscopy software, Version 1.2.16, 2022. http://www.effemm2.de/spectragryph/

Nissenbaum A and Yakir D (1995) Stable isotope composition of amber. In: Anderson KB and Crelling JC (eds) Amber, resinite, and fossil resins. ACS Symposium series, American Chemical Society, Washington: 32–42

Poinar GO (1992) Life in Amber. Stanford University Press, Stanford

Ragazzi E, Roghi G, Giaretta A, Gianolla P (2003) Classification of amber based on thermal analysis. Thermochimica Acta 404: 43–54. https://doi.org/10.1016/S0040-6031(03)00062-5

Rasnitsyn AP, Bashkuev AS, Kopylov DS, Lukashevich ED, Ponomarenko AG, Popov YA, Rasnitsyn DA, Ryzhkova OV, Sidorchuk EA, Sukatsheva ID, Vorontsov DD (2016) Sequence and scale of changes in the terrestrial biota during the Cretaceous (based on materials from fossil resins). Cretaceous Research 61: 234-255. https://doi.org/10.1016/j.cretres.2015.12.025

Romanov AA., Romanova EA., Repin YS (2018) Cretaceous deposits of the Ugra Peninsula: new finds and generalization of materials. Vestnik IG Komi SC UB RAS 2: 3–9. https://doi.org/10.19110/2221-1381-2018-2-3-9 (in Russian)

Savkevich SS (1970) Amber. Nedra Publ., Leningrad (in Russian)

Stout S (1995) Resin-derived hydrocarbons in fresh and fossil Dammar resins and Miocene rocks and oils in the Mahakam delta, Indonesia. In: Anderson KB and Crelling JC (eds) Amber, resinite, and fossil resins. ACS Symposium series, American Chemical Society, Washington: 43–75

Tappert R, McKellar RC, Wolfe AP, Tappert MC, Ortega-Blanco J, Muehlenbachs K (2013) Stable carbon isotopes of C3 plant resins and ambers record changes in atmospheric oxygen since the Triassic. Geochimica et Cosmochimica Acta 121: 240–262.

Tappert R, Wolfe AP, McKellar RC, Tappert M, Muehlenbachs K (2011) Characterizing modern and fossil exudates of gymnosperms using micro-Fourier transform infrared spectroscopy. International Journal of Plant Sciences 172(1): 120–38. https://doi.org/10.1086/657277

Vavra N (2009) Amber, fossil resins, and copal: Contributions to the terminology of fossil plant resins. Denisia 26: 213–22

Yushkin NP (1973) Amber of Arctic regions. Preprints, Komi branch of USSR AS 7 (in Russian)

Zarhidze DV, Piskun PP, Krasnozhen AS. Devyatuha YA, Starikova EV, Bartova AV, Klevcov AS, Vojtovich ZN, Cybul'skaya AE, Alekseeva NA, Bogatyreva EV, Uhach TN, Romanov AA (2017) Gosudarstvennaya geologicheskaya karta Rossijskoj Federacii. Masshtab 1 : 200 000. Izdanie vtoroe. Seriya Vajgachsko-Pajhojskaya. Listy R-41-XXVIII, XXIX (Ust'-Kara). Ob"yasnitel'naya zapiska. Moskovskij filial FGBU «VSEGEI», Moscow (in Russian)

Zherikhin VV, Eskov KY (1999) Mesozoic and lower Tertiary resins in former USSR. Est. Mus. Estudios del Museo de Ciencias Naturales de Álava 14(2): 119–131.

Bioavailable Sr Isotope Ratio in the Caspian Catchment Basin: Insight from Mollusk Shells and Model Calculation

Anton B. Kuznetsov and Anastasia A. Gavrilova

Abstract The Sr isotope composition in *natural biomineral system* such as mollusk shells *represents the composition of* terrestrial *water which flow down the catchment area.* The bioavailable $^{87}Sr/^{86}Sr$ ratio depends on solubility and Sr isotope composition of different bedrock types. The article represents a map of the litho-formational complexes of the Caspian catchment basin. Based on geochemical data, the average $^{87}Sr/^{86}Sr$ ratios have been calculated for major river basins of Caspian *catchment area.* The direct measurement of the $^{87}Sr/^{86}Sr$ ratio in shells of modern freshwater mollusk from the Lower Volga, Ural and Belaya rivers are 0.7081–0.7082, 0.7086 and 0.7079–0.7125 respectively. The obtained values correspond to the calculated $^{87}Sr/^{86}Sr$ ratios. The low $^{87}Sr/^{86}Sr$ ratios in plain stream of the Volga, Ural and Belaya rivers due to predominance of carbonate and salt-sulfate complexes in *catchment* areas. The importance of the bioavailable $^{87}Sr/^{86}Sr$ ratio is to provide the paleogeographic, archaeologic and ecologic reconstructions in region.

Keywords Geochemistry · Mollusk · Unio · Strontium isotopes · Catchment · Caspian basin

1 Introduction

The chemical and isotopic composition of terrestrial water depends on drained areas through the destruction, dissolution and weathering of surface rocks. Among all the variety of dissolved chemical elements transported, the strontium holds a special

---

A. B. Kuznetsov (✉) · A. A. Gavrilova
Institute of Precambrian Geology and Geochronology, Russian Academy of Sciences, St. Petersburg, Russia
e-mail: antonbor9@mail.ru

A. A. Gavrilova
St. Petersburg State University, St. Petersburg, Russia

O. V. Frank-Kamenetskaya et al. (eds.), *Biogenic—Abiogenic Interactions in Natural and Anthropogenic Systems 2022*, Springer Proceedings in Earth and Environmental Sciences, https://doi.org/10.1007/978-3-031-40470-2_14

importance. The Sr isotope composition in river waters strictly depends on the petrographic and mineralogical composition of eroded rocks (Wadleigh et al. 1985; Goldstein and Jacobsen 1988; Brennan et al. 2014; Pearce et al. 2015). The $^{87}Sr/^{86}Sr$ ratio is an indicator in reconstructing the chemical evolution of the hydrological system (Ingram and Sloan 1992; Hodell et al. 1991, 2004; Evans et al. 2010; Slovak and Paytan 2012; Pestle et al. 2013; Kuznetsov et al. 2014, 2018; Willmas et al. 2018). Knowledge of the Sr isotopic composition in rivers allows us to reconstruct the geological environment background and the changes into drainage basin in the past. This is the key to palaeogeographic, archaeologic and ecologic reconstructions.

Despite the very low concentrations in river water, the dissolved strontium is bioavailable for different consumers such as mollusks, fishes, plants and animals inhabitant in terrestrial waters. This element incorporates into crystal lattices of calcium in Ca-bearing minerals as an isomorphic impurity. Thus, the Sr isotope composition in shells, bones and skeletal tissues have been chosen to determine the $^{87}Sr/^{86}Sr$ ratio in the aquatic environment where the organisms developed (Brand et al. 1987; Prohaska et al. 2001; Kuznetsov et al. 2012, 2014; Slovak and Paytan 2012).

Mollusk is a natural biomineral system that produces solid minerals during the life of organisms. The mollusk takes the bioavailable material to build up the external shell by feeding from the water during the filtration process. Simultaneously with calcium, the mollusk captures dissolved strontium and accumulates it in the shell in the form of isomorphic impurity. Therefore, the isotope-geochemical characteristics of carbonate shells reflect the composition of water, which in turn contains elements obtained during the drainage of rocks of the catchment area.

The history of the Caspian Sea began 3.4 million years ago, when tectonic uplift was divided Pontic Sea into the Black, Aral and Caspian seas. The development of the Caspian Sea in Pliocene–Pleistocene evolved multiple transgressions and regressions. The Sea finally lost the connection with the world ocean after early Khvalyn transgression about 40 thousand years ago (Svitoch 2014). The recent catchment area of the Caspian great lake covers the East European Plain, the Caucasus Range, the Ural Mountains and the Elbursk Mountain system. The catchment of the Caspian Sea includes Volga, Ural, Terek, Kura, Sulak and Sefid-Rud rivers which are covered nearly 60% of the total area. These rivers are the most abundant and supply about 80–90% of the sediment and dissolved substances to the lake (Svitoch 2014).

This article summarizes the first Sr isotope data obtained for carbonate shells of modern freshwater mollusk Unio within the catchment of the Caspian Sea. *Simultaneously with direct measuring of* the $^{87}Sr/^{86}Sr$ ratio, we propose mathematical model to predict the bioavailable $^{87}Sr/^{86}Sr$ ratio and demonstrate the potential of this approach to produce a detailed map of dissolved strontium isotope distribution on regional scale for the East European Platform.

2 Material and Methods

This work took place in 3 stages: mapping, model calculation and Sr isotope analysis of mollusk shells from different parts of the studied region.

Carbonate shells of modern freshwater mollusk Unio have been taken along lower stream of the largest rivers of the studied region: the Volga (5 samples) and the Ural (3 samples). The additional collection represents carbonate shells from upstream to downstream (5 samples) of the Belaya river. The Belaya river is drained two different geological bedrocks such as Precambrian crystalline basement and Permian sedimentary cover. The Belaya river water follows to the Kama river which is large tributary of the Volga river.

The chitinous cover of each shells was scraped off to leave only the carbonate component. The pounded shells of about 0.02–0.03 g were dissolved in 0.5N hydrochloric acid. Through centrifugation, additional precipitation of silicate impurities not removed earlier was carried out. The isotope work was carried out at the Institute of Precambrian Geology and Geochronology of RAS, St. Petersburg. The Sr was separated by the ion-exchange method with the Dowex AG50Wx8 (200–400 mesh) resin and with 2.5 N HCl as eluent (Kuznetsov et al. 2012). Direct isotope measurements were carried out on a Triton TI thermal ionization mass spectrometer in static mode. The $^{87}Sr/^{86}Sr$ ratio in the samples was controlled by measuring NIST SRM-987 and USGS EN-1 standard samples. During the period of data acquisition, the average $^{87}Sr/^{86}Sr$ were 0.71027 ± 0.00001 ($2\sigma, n = 26$) for NIST SRM 987 and 0.70920 ± 0.00001 (2σ, $n = 14$) and USGS EN-1 standard samples. The average discrepancy of the measured $^{87}Sr/^{86}Sr$ values in repeat samples was lower 0.000005.

3 Results

3.1 Sr Isotope Composition in Freshwater Shells

Among 12 studied samples of Unio shell, five were from lower and middle stream of the Volga river, and two—from the middle stream and mouth of the Ural river (Fig. 1). The five shells represent upper, middle and lower stream of the Belaya river (Table 1).

The $^{87}Sr/^{86}Sr$ ratios in shells from the lower Volga basin (from Tolyatti to Astrakhan) vary within a narrow range of 0.7081–0.7082, this ratio in the shell from terminal part of the middle Volga basin (above Kazan city) is slightly higher 0.7085 (Table 1). The $^{87}Sr/^{86}Sr$ ratio in shell from Orsk in the middle part of the Ural river is 0.7094, whereas it decreases in the mouth of the Ural river down to 0.7086. The $^{87}Sr/^{86}Sr$ ratios in shells from mountains part of the Belaya river are higher than those at plain stream, 0.7110–0.7125 against 0.7079–0.7083 respectively. The latter corresponds to environmental setting within watershed such as change of Precambrian bedrocks to Permian sediments.

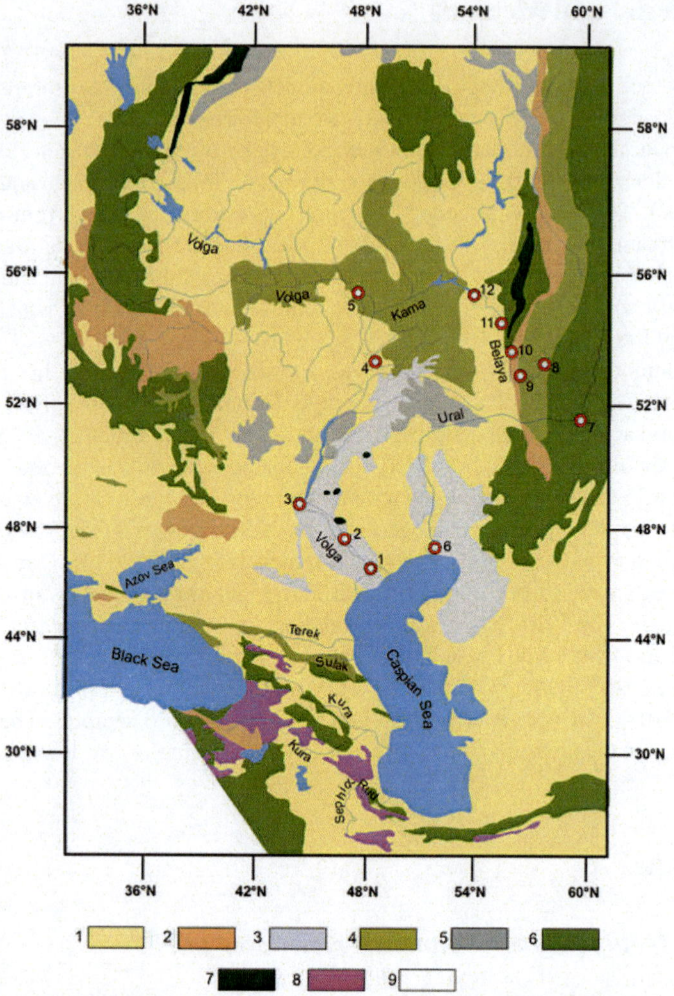

Fig. 1 Map of the formational complexes around Caspian Sea catchment basin, and location of Unio shell samples. Formational complexes: 1—terrigenious and carbonate-terrigenious, poorly litified, sandy-clayey; 2—terrigenious; 3—terrigenious and carbonate-terrigenious, poorly litified, clayey; 4—clayey-carbonate; 5—clayey-sandstone carbonate; 6—carbonate; 7—halogenic-sulfate-carbonate; 8—volcanogenic basalt. Sample shell location: 1—Astrakhan, 2—Kosika, 3—Volgograd, 4—Tolyatti, 5—Kazan, 6—Atyrtau, 7—Orsk, 8—Muradymovo, 9—Yumaguzovo, 10—Sterlitamak, 11—Ufa, 12—Dyurtyuli

Comparison of the Sr isotope data for shells from the Volga, Ural and Belaya rivers reveals significant differences. The highest $^{87}Sr/^{86}Sr$ ratios have shells in stream at Urals mountains, whereas these ratios in the Pre-Uralian and plain streams are significantly lower. In general, the bioavailable $^{87}Sr/^{86}Sr$ ratio strictly decreases in rivers following from the Urals mountains to the Caspian Sea.

Table 1 Sr isotope data for shells of freshwater mollusk Unio from different locations

Sample	River	Location	$^{87}Sr/^{86}Sr$ measured
1	Volga	Astrakhan	0.7081
2	Volga	Kosika	0.7082
3	Volga	Volgograd	0.7082
4	Volga	Tolyatti	0.7081
5	Volga	Kazan	0.7085
6	Ural	Atyrtau	0.7086
7	Ural	Orsk	0.7094
8	Belaya	Muradymovo	0.7125
9	Belaya	Yumaguzovo	0.7110
10	Belaya	Sterlitamak	0.7083
11	Belaya	Ufa	0.7079
12	Belaya	Dyurtyuli	0.7079

The observed $^{87}Sr/^{86}Sr$ ratios in plain streams of Volga, Ural and Belaya rivers are much lower than those in large continental rivers such as Amazon, Congo, Yangtze, Brahmaputra, Ganges, Mississippi, Mekong, St. Lawrence, Yukon, Neva etc., 0.7081–0.7082 against to 0.7095–0.7210 (Wadleigh et al. 1985; Goldstein and Jacobsen 1988; Kuznetsov et al. 2014). The "unusually" low $^{87}Sr/^{86}Sr$ ratios in the lower stream of rivers around the Caspian Sea could be due to a large volume of ^{87}Sr-depleted discharge. The source of low-radiogenic ^{87}Sr in surface waters would be provide by the weathering of bedrocks in drainage basin. In order to understand the reason of the changes in the $^{87}Sr/^{86}Sr$ ratio of peri-Caspian riverine it is necessary to determine the mass balance in drainage bedrock basin.

3.2 *Model Calculation of the $^{87}Sr/^{86}Sr$ Ratios in Peri-Caspian Riverine*

The model $^{87}Sr/^{86}Sr$ values might be obtained by calculation using empirical data such as the lithological composition of bedrocks, the volume of different rock types in each basin, the volume of dissolved Sr which each rock able to supply to surface water, and Sr isotope composition of each lithotypes. As the result, the calculated Sr isotope composition of each environmental area predicts the one of detached Sr input to total riverine system. Finally, the calculated $^{87}Sr/^{86}Sr$ values could be compared to measured bioavailable $^{87}Sr/^{86}Sr$ ratio in river shells.

Presently, the catchment area of the Caspian Sea covers the East European Plain, the Caucasus Range, the Ural Mountains and the Elbursk Mountain System. The structural-formational Map of Russia (Smyslov 1995) and Geological Map of Russia and adjacent territories (Choubert et al. 1976; Rad and Afzal 2008) have been used to identify lithological-formational complexes or zones. The basis of the formational zones adopted sets of lithological formations and unstratified magmatic, metamorphic and metasomatic (including weathered) complexes that are associated with specific structural elements of the earth's crust.

The formational complexes have been divided into eight main rock types (Fig. 1), which represent 3 groups.

terrigenious:

- terrigenious and carbonate-terrigenious, poorly litified, sandy-clayey
- terrigenious
- terrigenious and carbonate-terrigenious, poorly litified, clayey

carbonate:

- clayey-carbonate
- clayey-sandstone carbonate
- carbonate
- salt-sulfate-carbonate

volcanogenic:

- basalt

For the model calculation of the strontium isotope composition for the individual formational complexes, the following geochemical data was used for the different rock types:

- Average Sr concentration (ppm)
- Average $^{87}Sr/^{86}Sr$ ratio
- Average volume of mobile Sr, %.

The principle geochemical data are based on published Sr isotope characteristics for carbonate and clay rocks of the South Urals and East European Platform (Gorokhov et al. 1994, 2019; Kuznetsov et al. 2003, 2006, 2008, 2017; Korte et al. 2006; Zaitseva et al. 2008, 2019; Semikhatov et al. 2009; Maslov et al. 2022; Zakharov et al. 2022) and leaching experiments (Ohde and Elderfield 1992; Chaudhuri and Clauer 1993; Wada and Furumura 1994; Bailey et al. 2000; Gorokhov et al. 1994, 2007, 2019; Kuznetsov et al. 2003, 2006; Gryschko et al. 2005; Li et al. 2011). The geochemical data for six formational complexes are summarized in Table 2. In order to simplify the calculation, terrigenious and carbonate-terrigenious, poorly lithified (sandy-clayey and clayey) rocks were combined into one group* and the same characteristics were used for them as they have similar values for the average Sr concentration, the fraction of mobile strontium in the rock reservoir, and the volume of dissolved strontium.

The most soluble rocks such as salt and sulfate supply the most amount of dissolved Sr. Compared to gypsum, the carbonates are less soluble, and terrigenous rocks almost do not supply Sr because of their poor solubility. The salt-sulphate complex (halite and gypsum) is characterized by the highest Sr concentration, while the terrigenous complex—the lowest Sr concentration, in average 1500 ppm against 20 ppm respectively. The average Sr concentration in carbonate

Table 2 Geochemical parameters for calculation the average $^{87}Sr/^{86}Sr$ ratio in individual formational rock complex

Formational (rock type) complexes	Average Sr concentration in rock, ppm	Average $^{87}Sr/^{86}Sr$ ratio in rock	Average fraction of mobile Sr, %	Average volume of dissolved Sr, ppm
Carbonate	800	0.7085	4	32
Clayey-carbonate	150	0.7120	2	3
Clayey-sandstone carbonate*	50	0.7180	1	0.5
Terrigenious	20	0.7300	0.1	0.02
Salt-sulfate-carbonate	1500	0.7070	10	150
Volcanogenic basalt	200	0.7050	2	4

complex is about 800 ppm, but its decreased with increasing of clay amounts down to 150 ppm. The volcanogenic basalt complex contains 200 ppm of Sr in average. The average $^{87}Sr/^{86}Sr$ ratios in terrigenous, clayey-carbonate, carbonate, salt-sulfate-carbonate, basalts complexes decrease according to the consistent row $0.7300 \rightarrow 0.7120 \rightarrow 0.7085 \rightarrow 0.7070 \rightarrow 0.7050$ respectively.

Based on the geochemical data on fraction of mobile Sr in typical rock and their average $^{87}Sr/^{86}Sr$ ratio (Table 2), we calculate the average amount of mobile Sr and average $^{87}Sr/^{86}Sr$ ratio in surface waters that flush rocks of individual geological formations within the study catchment basins.

3.3 *Evaluation of Volume of Formational Complexes into Catchment Area*

The volume of individual formational complex into catchment area of each river basins are different. To evaluate the volume of each formational complex we use the formational maps (Fig. 1).

The largest rivers such as Volga, Ural, Terek, Kura, Sulak and Sefid-Rud supply about 80–90% of the sediment and dissolved substances to the Caspian Sea (Svitoch 2014). The other rivers are located in the eastern part of the catchment area, where its are dried up or extremely regulated. The terrigenous rocks predominate in eastern part of the catchment area, suggesting that these rivers bring no dissolved strontium. The implementation of formational maps provides the calculated square of catchment areas for six rivers (Table 3). As a result of the mapping calculation, the theoretical and calculated squares of catchment area differ less than 0.4%.

The Volga basin is divided into three parts: from the source to the Oka, then to the Kama and finally to the Caspian Sea (Fig. 1). The river has its origin in the Valdai Upland, which is consisted of the Paleozoic carbonate complex (limestone, dolomites) with clay admixture. Then Volga flows out into the Russian Plain, which is

Table 3 The square of catchment area of the rivers of the Caspian basins

River	Catchment area (reference source), km^2	Catchment area (calculated), km^2	Catchment area, %
Volga	1,360,000	1,283,191	43.9
Ural	231,000	226,701	7.5
Kura	188,000	179,522	6.1
Terek	43,200	43,221	1.4
Sulak	15,200	15,052	0.5
Sephid-Rud	56,200	56,263	1.8
Others	1,206,400	1,202,633	38.9

mostly consisted of terrigenous sandstone rocks. In the middle stream, the Volga flows through the Paleozoic clay-sand-carbonate complex and the Mesozoic limestones. After the confluence of the Kama River, the Volga River flows through terrigenous complexes of the Mesozoic-Cenozoic age. The lower Volga basin is represented by sands and clays of the Neogene-Quaternary age as well as salt rocks.

The Ural River drains mostly the terrigenous and carbonate complexes of the Paleozoic age.

The headwater of the Kura River is located in the Armenian highlands, which is consisted by mainly the volcanogenic basaltic complex. The river erodes Jurassic and Cretaceous limestones in the middle stream and clayey-carbonate rocks in the lower stream.

The catchments of the Terek and Sulak rivers are similar: both flow predominantly through terrigenous, clayey-sandstone carbonate and carbonate rocks of the Mesozoic-Cenozoic age.

The Mesozoic terrigenous complex consists of 70% of the Sefid-Rud river basin. The other area of Sefid-Rud catchment area is occupied by volcanogenic basaltic complexes of Mesozoic and Cenozoic age.

The square of the above-mentioned formational complexes has been evaluated after the formational map (see Fig. 1). Thus, the evaluated squares has been used to calculate the proportion of the individual formational complex into catchment area of each river (Table 4).

3.4 Calculation of Volume of the Dissolved Sr for Formational Complexes into Catchment Area

The terrigenous rocks are predominant into catchment area of the Caspian Sea. The Volga basin being the largest catchment area consists of 56% of the terrigenous rocks. The proportion terrigenous rocks vary from 37% in Ural river basin up to 70–85% in Terek, Sulak and Sefid-Rud river basins. The carbonate complex covers

Table 4 The proportion of formational complexes (in percentage) of each catchment area into the Caspian basin

River	Volga	Ural	Kura	Terek	Sulak	Sefid-Rud
Formational (rock type) complex						
Carbonate	14	60	17	18	4	9
Clayey-carbonate	7	–	–	–	–	–
Clayey-sandstone carbonate	21	3	–	10	11	–
Terrigenious	56	37	44	70	85	70
Salt-sulfate-carbonate	2	–	–	–	–	–
Volcanogenic basalt	–	–	39	1	–	21

the second volume into studied catchment area. The large proportion (21–39%) of the volcanogenic basalts are situated into Kura and Sefid-Rud river basins.

The compilation of the geochemical data on different rock types from the Table 2 and the volume of formational complexes from the Table 4 provide the calculation the volume of dissolved Sr for individual formational complex into each catchment area within Caspian basin. The calculated data are summarized on Fig. 2.

Despite the fact that terrigenous rocks predominate within catchment areas of all rivers of the Caspian drainage basin, the main contribution of the dissolved Sr is carbonate complex. This fact is supported by the geochemical data which demonstrate

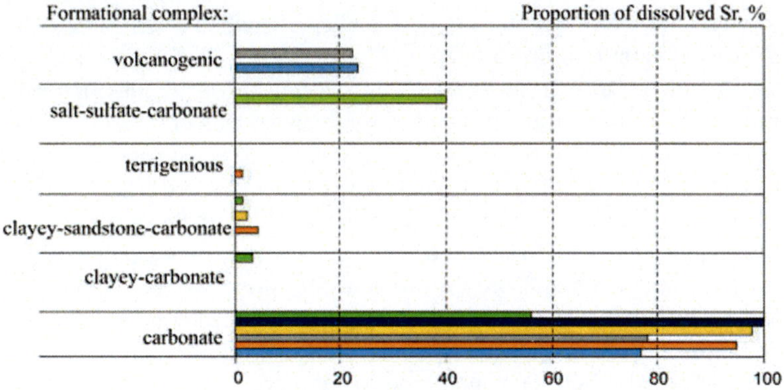

	carbonate	clayey-carbonate	clayey-sandstone-carbonate	terrigenious	salt-sulfate-carbonate	volcanogenic
Volga	56	3	1	-	40	-
Ural	99	-	1	-	-	-
Terek	98	-	2	-	-	-
Kura	78	-	-	-	-	22
Sulak	95	-	4	1	-	-
Sefid-Rud	77	-	-	-	-	23

Fig. 2 Proportion of dissolved Sr in different rock types for each catchment area, %

Table 5 The calculated and measured $^{87}Sr/^{86}Sr$ ratios in major rivers within the Caspian catchment area

Water object	Calculated $^{87}Sr/^{86}Sr$	Measured $^{87}Sr/^{86}Sr$
Volga	0.7083	0.7081
Ural	0.7085	0.7086
Sulak	0.7079	0.7080
Kura	0.7078	0.7075
Terek	0.7086	no data
Sefid-Rud	0.7078	no data
Caspian Sea	0.7083	0.7082

the low solubility of siliciclastic rocks and small volume of dissolved Sr (Table 2). The proportion of carbonate Sr in rivers reveal from 56 up to 99% (Fig. 2). The salt-sulfate formation of the Volga area perform the large portion of dissolved Sr into Caspian Sea. The Sefid-Rud and Kura rivers supply large (about 22–23%) portion of dissolved Sr from the volcanogenic plateaus of the Caucasus and the Elburz Ridges.

3.5 *Calculation of the $^{87}Sr/^{86}Sr$ Ratio for Catchment Area*

Taking into account the average of the $^{87}Sr/^{86}Sr$ ratio in individual rock type (Table 2) and volume of the mobile Sr supplied by different formational complexes (Fig. 2), we calculated the total $^{87}Sr/^{86}Sr$ ratio in end-member of studied river basins. The calculated data are summarized in Table 5.

Model calculation of $^{87}Sr/^{86}Sr$ ratio in the Volga basin is 0.7083 (Table 5), which is corresponds to the Sr isotope data obtained by direct measuring of $^{87}Sr/^{86}Sr$ ratios in shells from the Lower Volga—0.7081–0.7082 (Table 1). The calculated $^{87}Sr/^{86}Sr$ ratio for the Ural river is of 0.7085, that is very close to the measured $^{87}Sr/^{86}Sr$ ratio in the shell from the mouths of this river. The calculated $^{87}Sr/^{86}Sr$ values for the Kura and Sulak rivers are of 0.7076 and 0.7079 respectively. These Sr isotope values also coincide with the measured $^{87}Sr/^{86}Sr$ ratios in the spring waters of the same rivers—0.7075 and 0.7080 respectively (Clauer et al. 2009; Shishlina et al. 2016). The measured $^{87}Sr/^{86}Sr$ ratios in shells from the Belaya river vary from 0.7110–0.7125 to 0.7079–0.7083 (Table 1), which represent the average Sr isotope ratio in different litho-formational complexes of the Uralian region: terrigenous (0.7097–0.7189), carbonate (0.7075–0.7083) and salt-sulfate (0.7075–0.7078).

Based on this model and input of major rivers, the $^{87}Sr/^{86}Sr$ ratio has been calculated for the Caspian seawater. The calculated $^{87}Sr/^{86}Sr$ ratio is about 0.7083. This value corresponds to the measured $^{87}Sr/^{86}Sr$ ratios in the Caspian Sea and Kara-Bogaz Gol water-samples varying about 0.7082 ± 0.0001 (Clauer et al. 2009).

Thus, the obtained $^{87}Sr/^{86}Sr$ values correspond to the empirical Sr isotope data in the shells within catchment areas (Table 5). This fact reveals that the bioavailable $^{87}Sr/^{86}Sr$ ratio in continental river basins in other regions could be calculated on the base of model parameters. The using of the model calculation is simplified the work. However, the each model requires comprehensive assessment of the geological structure of the area and its topography. Moreover, the Quaternary sediments (loess, glacial till, cover sands, peat) might affect the mobile Sr volume and their isotope composition (Bentley 2006; Evans et al. 2010; Slovak and Paytan 2012; Frei and Frei 2013; Willmes et al. 2018). Therefore the most precise information on bioavailable $^{87}Sr/^{86}Sr$ ratio provides the direct measurement of Sr isotope composition in shells of mollusk being the *natural biomineral system.*

4 Conclusions

The article represents a formational map of the catchment area of the Caspian Sea. The formational zones are based on assemblages of lithological formations and unstratified magmatic complexes. The produced map provides the evaluation of the $^{87}Sr/^{86}Sr$ ratio in the catchment areas through model calculations.

The analysis of Sr isotope composition in samples of freshwater mollusk shells from the Volga, Ural and Belaya rivers has shown the bioavailable $^{87}Sr/^{86}Sr$ ratios. The $^{87}Sr/^{86}Sr$ ratios are depend on lithotypes of drainage bedrocks. The observed $^{87}Sr/^{86}Sr$ ratios in carbonate and salt-sulfate rocks is much lower than it is in terrigenous, clay-carbonate and clay-sandy-carbonate rocks.

Model calculation of bioavailable $^{87}Sr/^{86}Sr$ ratio for major rivers within the Caspian Sea catchment area coincide with Sr isotope data obtained by direct measurement. The $^{87}Sr/^{86}Sr$ ratios in water of Lower Volga are 0.7081–0.7082, the one in downstream of the Ural river is 0.7086. The $^{87}Sr/^{86}Sr$ ratios in shells from mountains part of the Belaya river are higher than those at plain stream, 0.7110–0.7125 against 0.7079–0.7083 respectively. The “unusual” low $^{87}Sr/^{86}Sr$ ratios in plain streams of Volga, Ural and Belaya rivers are explained by predominance of carbonate and salt-sulfate complexes in drainage areas.

The calculated and measured $^{87}Sr/^{86}Sr$ ratios in the Caspian Sea are coincident, 0.7083 and 0.7081–0.7082 respectively. The low $^{87}Sr/^{86}Sr$ ratio in the Caspian Sea due to the abundant of carbonate and salt-sulfate rocks in catchment area.

The revealed features of bioavailable Sr in the shells taken from various rivers assist to distinguish the different lithotype rocks into catchment area. These Sr isotope data might be used in palaeogeographic, archaeologic and ecologic reconstructions.

Acknowledgements This field work and Sr-isotope work were supported by RSF (project no. 18-17-00247, ABK). The *scientific research were performed at the Research park of St.Petersburg State University "Center for Chemical Analysis and Materials Research" and* using the equipment of the AIRES Shared Equipment Center (Institute of Precambrian Geology and Geochronology, Russian Academy of Sciences, St. Petersburg).

References

Bailey TR, McArthur JM, Prince H, Thirwall MF (2000) Dissolution methods for strontium isotope stratigraphy: wholerock analysis. Chemical Geology, 167: 313–319

Bentley RA (2006) Strontium isotopes from the earth to the archaeological skeleton: a review. J. archaeological method and theory: 135–187.

Brand U, Morrison JO, Brand N, Brand E (1987) Isotopic variation in the shells of recent marine invertebrates from the Canadian Pacific coast. Chemical Geology, 65: 137-145.

Brennan SR, Fernandez DP, Mackey G, Cerling TE, Bataille CP, Bowen GJ, Wooller MJ (2014) Strontium isotope variation and carbonate versus silicate weathering in rivers from across Alaska: implications for provenance studies. Chemical Geology, 389: 167–181.

Chaudhuri S, Clauer N (1993) Strontium isotope compositions and potassium and rubidium contents of formation waters in sedimentary basins: Clues to the origin of the solutes. Geochimica Cosmochimica Acta, 57: 429-437.

Choubert G, Faure-Muret A, Chanteux P (1976) Atlas géologique du monde: 1:10000000. Unesco.

Clauer N, Pierret MC, Chaudhuri S (2009) Role of subsurface brines in salt balance: the case study of the Caspian Sea and Kara Bogaz Bay. Aquatic geochemistry, 15(1): 237-261.

Evans JA, Montgomery J, Wildman G, Boulton N (2010) Spatial variations in biosphere $^{87}Sr/^{86}Sr$ in Britain. Journal of the Geological Society, 167 (1): 1-4.

Frei R, Frei KM (2013) The geographic distribution of Sr isotopes from surface waters and soil extracts over the island of Bornholm (Denmark). Applied Geochemistry, 38: 147-160.

Goldstein SJ, Jacobsen SB (1988) Nd and Sr isotopic systematics of river water suspended material: implications for crustal evolution. Earth and Planetary Science Letters, 87(3): 249-265.

Gorokhov IM, Clauer N, Turchenko TL, Melnikov NN, Kutyavin EP, Pirrus E, Baskakov AV (1994) Rb-Sr systematics of Vendian-Cambrian claystones from the east European Platform: Implications for a multi-stage illite evolution. Chemical Geology, 112(1/2): 71-89.

Gorokhov IM, Mel'nikov NN, Kuznetsov AB, Konstantinova GV, Turchenko TL (2007) Sm-Nd systematics of fine-grained fractions of the Lower Cambrian blue clay in northern Estonia. Lithology Mineral Resources, 42(5): 482–495

Gorokhov IM, Zaitseva TS, Kuznetsov AB, Ovchinnikova GV, Arakelyants MM, Kovach VP, Konstantinova GV, Turchenko TL, Vasil'eva IM (2019) Isotope systematics and age of authigenic minerals in shales of the Upper Riphean Inzer Formation, South Urals. Stratigraphy and Geological Correlation, 27(2): 133–158

Gryschko R, Kuhnle R, Terytze K, Breuer J, Stahr K (2005) Research articles soil extraction of readily soluble heavy metals and as with 1 M NH4NO3-solution. J. Soils Sediments 5: 101–106.

Hodell DA, Mueller PA, Garrido JR (1991) Variations in the strontium isotopic composition of seawater during the Neogene. Geology, 19(1): 24-27.

Hodell DA, Quinn RL, Brenner M, Kamenov G (2004) Spatial variation of strontium isotopes (87Sr/86Sr) in the Maya region: a tool for tracking ancient human migration. J. Archaeol. Sci. 31: 585–601. http://dx.doi.org/https://doi.org/10.1016/j.jas.2003.10.009.

Ingram BL, Sloan D (1992) Strontium isotopic composition of estuarine sediments as paleosalinity-paleoclimate indicator. Science, 255(5040): 68-72.

Korte C, Jasper T, Kozur HW, Veizer J (2006) $^{87}Sr/^{86}Sr$ record of Permian seawater. Palaeogeography, Palaeoclimatology, Palaeoecology, 240: 89–107.

Kuznetsov AB, Bekker A, Ovchinnikova GV, Gorokhov IM, Vasilyeva IM (2017) Unradiogenic strontium and moderate-amplitude carbon isotope variations in early Tonian seawater after the assembly of Rodinia and before the Bitter Springs Excursion, Precambrian Research, 298: 157–173.

Kuznetsov AB, Makarikhin VV, Pokrovskii BG, Konstantinova GV (2014) $^{87}Sr/^{86}Sr$ and $\delta^{18}O$ chemostratigraphy and facies conditions of Pleistocene mollusks in Karelia (Gridino). Doklady Earth Sciences, 459(1): 1413–1417.

Kuznetsov AB, Semikhatov MA, Gorokhov IM (2012) The Sr isotope composition of the World Ocean, marginal and inland seas: implications for the Sr isotope stratigraphy. Stratigraphy and Geological Correlation, 20(6): 501–515.

Kuznetsov AB, Semikhatov MA, Gorokhov IM (2018) Strontium isotope chemostratigraphy: Principles and State of the Art. Stratigraphy and Geological Correlation, 26 (4): 367–386.

Kuznetsov AB, Semikhatov MA, Gorokhov IM, Melnikov NN, Konstantinova GV, Kutyavin EP (2003) Sr isotope composition in carbonates of the Karatau Group, Southern Urals, and standard curve of $^{87}Sr/^{86}Sr$ variations in the Late Riphean ocean. Stratigraphy and Geological Correlation, 11: 415-449.

Kuznetsov AB, Semikhatov MA, Maslov AV, Gorokhov IM, Prasolov EM, Krupenin MT, Kislova IV (2006) New data on Sr- and C-isotopic chemostratigraphy of the Upper Riphean type section (Southern Urals). Stratigraphy and Geological Correlation, 14: 602-628.

Kuznetsov AB, Ovchinnikova GV, Semikhatov MA, Gorokhov IM, Kaurova OK, Krupenin MT, Vasil'eva IM, Gorokhovskii BM, Maslov AV (2008) The Sr isotopic characterization and Pb–Pb age of carbonate rocks from the Satka formation, the Lower Riphean Burzyan Group of the Southern Urals, Stratigraphy and Geological Correlation, 16(2): 120–137

Li D, Shields-Zhou GA, Ling H-F, Thirlwall M (2011) Dissolution methods for strontium isotope stratigraphy: guidelines for the use of bulk carbonate and phosphorite rocks. Chemical Geology, 290 (3-4): 133-144.

Maslov AV, Kuznetsov AB, Kramchaninov AYu, Shpakovich LV, Gareev EZ, Podkovyrov VN, Kovalev SG (2022) Provenances of the Upper Precambrian clay rocks in the Southern Urals: results of geochemical and Sm–Nd isotope geochemical investigations. Stratigraphy and Geological Correlation, 30 (1): 30–51

Ohde S, Elderfield H (1992) Strontium isotope stratigraphy of Kita-daito-jima Atoll, North Phillipines Sea: implications for Neogene sea level change and tectonic history. Earth Planet. Sci. Lett. 113: 473-486.

Pearce CR, Parkinson IJ, Gaillardet J, Charlier BL, Mokadem F, Burton KW (2015) Reassessing the stable ($\delta^{88}Sr/^{86}Sr$) and radiogenic ($^{87}Sr/^{86}Sr$) strontium isotopic composition of marine inputs. Geochimica et Cosmochimica Acta, 157: 125-146.

Pestle WJ, Simonetti A, Curet LA (2013) $^{87}Sr/^{86}Sr$ variability in Puerto Rico: geological complexity and the study of paleomobility. J. Archaeol. Sci., 40: 2561–2569.

Prohaska T, Latkoczy C, Schultheis G, Teschler-Nicola M, Stingeder G (2001) Investigation of Sr isotope ratios in prehistoric human bones and teeth using laser ablation ICP-MS after Rb/Sr separation. J. Analytical Atomic Spectrometry, 16: 806–811.

Rad AJ, Afzal P (2008) Iran Geological Maps 1:250 000. Geological Survey of Iran, Teheran.

Semikhatov MA, Kuznetsov AB, Maslov AV, Gorokhov IM, Ovchinnikova GV (2009) Stratotype of the Lower Riphean, the Burzyan Group of the Southern Urals: Lithostratigraphy, paleontology, geochronology, Sr and C-isotopic characteristics of its carbonate rocks. Stratigraphy and Geological Correlation, 17 (6): 574–601.

Shishlina NI, Larionova YO, Idrisov IA, Azarov ES (2016) Variations in $^{87}Sr/^{86}Sr$ ratios in contemporary snail samples obtained from the eastern Caucasus. Arid Ecosystems, 6(2): 100-106. (in Russian)

Slovak NM, Paytan A (2012) Applications of Sr isotopes in archaeology. In Handbook of environmental isotope geochemistry Springer, Berlin, Heidelberg: 743–768.

Smyslov AA (1995) Map of structural and formational complexes of Russia.1: 10000000. Geological Atlas of Russia. VSEGEI, Saint Petersburg. (in Russian)

Svitoch AA (2014) The Great Caspian Region: its structure and history of development. MSU, Moscow, 272p (in Russian)

Wada S-I, Furumura S (1994) Solubility of $CaCO_3$ in mol L^{-1} ammonium acetate for extracting exchangeable bases. Soil Sci. Plant. Nutr, 40(2): 361-364.

Wadleigh MA, Veizer J, Brooks C (1985) Strontium and its isotopes in Canadian rivers: Fluxes and global implications. Geochimica et Cosmochimica Acta, 49(8): 1727-1736.

Willmes M, Bataille CP, James HF, Moffat I, McMorrow L, Kinsley L, Armstrong, RA, Eggins S, Grün R (2018) Mapping of bioavailable strontium isotope ratios in France for archaeological provenance studies. Applied Geochemistry, 90: 75-86.

Zaitseva TS, Kuznetsov AB, Gorozhanin VM, Gorokhov IM, Ivanovskaya TA, Konstantinova GV (2019) The lower boundary of the Vendian in the Southern Urals as evidenced by the Rb–Sr age of glauconites of the Bakeevo Formation. Stratigraphy and Geological Correlation, 27(5): 573–587.

Zaitseva TS, Gorokhov IM, Ivanovskaya TA, Semikhatov MA, Kuznetsov AB, Mel'nikov NN, Arakelyants MM, Yakovleva OV (2008) Mössbauer characteristics, mineralogy and isotopic Age (Rb–Sr, K–Ar) of Upper Riphean glauconites from the Uk Formation, the Southern Urals. Stratigraphy and Geological Correlation, 16(3): 227–247

Zakharov YD, Kuznetsov AB, Seltser VB, Gavrilova AA, Stativko VS, Smyshlyaeva OP, Kirienko AP, Grigorev VE (2022) A $^{87}Sr/^{86}Sr$, $\delta^{18}O$ and $\delta^{13}C$ record of Turonian-Santonian belemnites from Lower Volga region of the East European Platform: stratigraphic significance and palaeoenvironmental reconstructions. Geobios, 74: 77-94.

Chemical and Biological Weathering of Black Shales

Elena G. Panova, Dmitriy O. Voronin, and Arshavir E. Hovhannisyan

Abstract Black shales containing 10–25% organic matter are representatives of rocks that formed as the result of bioinert interactions. Alum has been extracted from black shales since the fifteenth century for bleaching paper, fabrics and leather dressing, along with kerogen, radium, uranium and rhenium. A number of black shale elements have a high Clarke concentration: Mo—12.5; U—8.3; V—3.2. Mobile forms of elements washed out of shale during chemical weathering form secondary minerals (gypsum, copiapite, pickeringite and tamarugite), which inherit the geochemical characterization of parent rocks. As a result of biological weathering, 29 species of micromycetes were found on the surface of black shales. In the primitive soil formed on shales, the following accumulate (number of times): Ni—3.4; Co—3.4; Zn—3.1. Grass growing on such substrates is characterized by accumulations relative to black shales: Zn—43.6; Ni—15.3; Mo—1.2.

Keywords Black shales · Biological weathering · Chemical weathering · Secondary minerals · Soil and grass geochemistry

1 Introduction

Black shales (BS) are widespread in paleobasins of different ages around the world. A feature of the rocks is their enrichment in carbonaceous matter and significant amounts of U, V, Mo, Cu, Ni, Co, Zn, and Pb.

During the weathering process, some elements can move into a mobile state and accumulate in soils and water (Zheng et al. 1999; Islam et al. 2000; Peng et al. 2004; Peters and Burkert 2008; Baruah and Khare 2010; Paikaray 2012; Fontenot et al.

E. G. Panova (✉) · D. O. Voronin
Saint Petersburg State University, Saint Petersburg, Russia
e-mail: e.panova@spbu.ru

A. E. Hovhannisyan
Institute of Geological Sciences of National Academy of Sciences of the Republic of Armenia, Yerevan, Armenia

O. V. Frank-Kamenetskaya et al. (eds.), *Biogenic—Abiogenic Interactions in Natural and Anthropogenic Systems 2022*, Springer Proceedings in Earth and Environmental Sciences, https://doi.org/10.1007/978-3-031-40470-2_15

2013; Cary et al. 2014; Tuttle et al. 2014a, b; Cappuyns et al. 2019; Huyen et al. 2019).

Carbon-containing rocks rich in toxic elements in natural conditions can have a negative impact on the environment. With their active mining, an increase in the removal of toxicants from bedrock and accumulation in various biotopes is recorded everywhere (He et al. 2002; Changxun et al. 2013; Tabelin et al. 2014; Tuttle et al. 2014a, b; Duan et al. 2020).

The current study aims to identify the behavior of chemical elements in the process of chemical and biological weathering of black shales.

2 Materials and Methods

The object of the study was the black shales of the Lower Paleozoic age (E3–O1), which are widely developed in the north-west of the East European Platform on the territory of Southern Sweden, which is framed by the Baltic Shield. Their intensive mining has been carried out in Sweden since the fifteenth century. Initially, they were used to obtain alum for bleaching paper, fabrics and leather dressing; from the twentieth century, kerogen, radium, uranium and rhenium were also extracted (Andersson et al. 1985; Falk et al. 2006).

Under surface conditions, processes of physical, chemical and biological weathering are actively taking place in the many quarries been formed over long years of mining. Chemical weathering is manifested in the development of secondary minerals that form crusts, incrustations, and individual crystals in spots of water seepage. Biofilms attached to the bedrock contribute to the biodegradation of parent rocks. A soil layer is formed on the slates, which is covered with herbal vegetation.

The material for the study was selected during many years of field work by a detachment of the Department of Geochemistry of St. Petersburg State University in southern Sweden at the Kvanthorp, Latorp, Billingen, Kinnekule, Hallberg, Hanneberg and Öland deposits (Fig. 1).

Samples of black shales were taken layer by layer along the section of the strata from the most organic part. Secondary minerals (SM) were selected separately by color to identify their mineral and chemical composition with reference to the BS substrate. Samples were taken of the primitive soil formed directly on the BS and the grass growing on it (meadow grass (Poa praténsis)). A total of 53 samples of BS, 15 samples of SM, and 10 samples of soil and grass growing on it (10 samples) were analyzed.

Black shales and secondary minerals were studied using petrographic analysis. To determine the mineral composition of rocks and minerals, X-ray phase analysis (Rigaku MiniFlex II device and PDXL software package, Cu-cathode), Raman spectroscopy using the ruff.info database (Horiba LabRam HR800), scanning electron microscopy and electron probe studies were applied (TM 3000 Hitachi).

The chemical composition of BS, HM, soil, and grass was studied using ICP-MS (ELAN-6100 DRC, PERKIN ELMER). The content of petrogenic oxides was

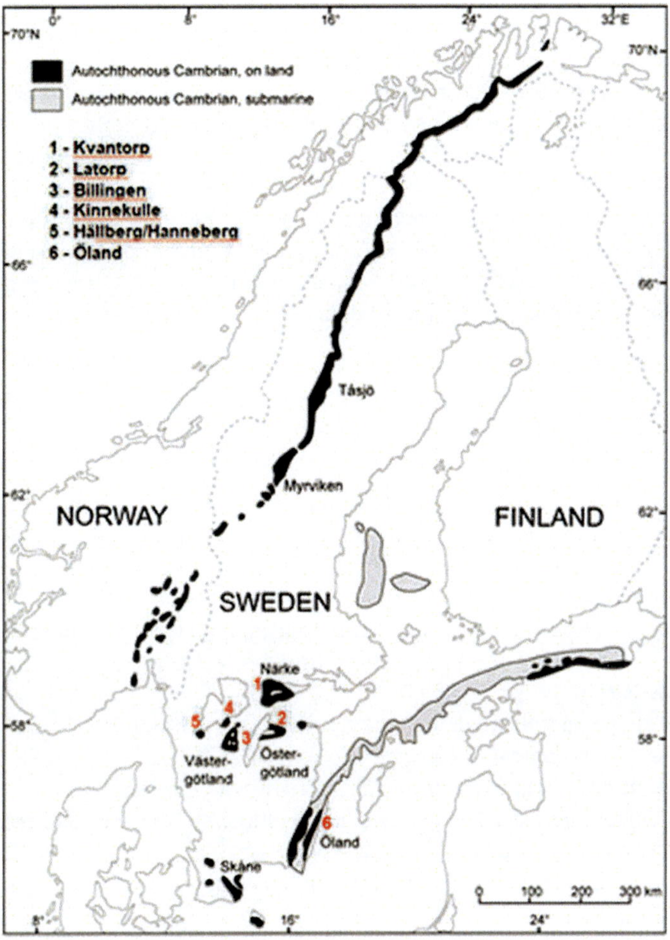

Fig. 1 Scheme of the studied deposits location (Andersson et al. 1985 with additions)

determined on an ARL 9000 instrument. Organic carbon was determined by the method of loss on ignition at 550 °C.

Samples with biofilms were selected using a laser confocal microscope (Leica TCS SPE). For their identification in biofilms on the shale surface, traditional methods of mycology and microbiology were used. When conducting the bacteriological study, the isolation of microorganisms was carried out on solid nutrient media using enzymatic meat hydrolysate to detect heterotrophic bacteria and determine the total microbial number, Aleksandrov medium with sand to isolate silicate bacteria, potato-ammonia agar to isolate actinomycetes, and liquid nutrient media to isolate chemolithotrophic bacteria. Quantification of bacteria was carried out using the dilution method (Tepper et al. 2004).

Nutrient media were used for primary isolation, maintenance in culture and identification of micromycetes: classical Czapek-Dox agar and in various modifications in terms of the content of glucose and some salts, agarized oatmeal broth with the addition of glucose, Sabouraud agar, potato-glucose agar, DRBC medium (King et al. 1979), water agar, 2% Malt Extract Agar, Wort Agar. For isolating fungi in culture from black shale samples, the following methods were used: scattering crumbs and small fragments of the substrate onto the surface of the nutrient medium; flushing from the substrate surface with subsequent dilution of the resulting suspension and sowing on a nutrient medium; selective isolation of fungi from the substrate surface to the nutrient medium using an injection needle.

3 Results and Discussion

Black shales comprise a black thin-lamellar mudstone (Figs. 2 and 3), consisting of organic (9–26 rel. %), clayey (40–60 rel. %), and silty–sandy (25–50 rel. %) components along with a nodule complex (2–5 rel. %). Organic matter is represented by graptolites and lower algae (Hints et al. 2014; Lecomte et al. 2017).

Inclusions of gelified humus, coal detritus and cutinite were identified, along with two types of macerals (amorphous and vitrinite-like). The average reflectance is 0.32–0.52% Ro.

Table 1 shows the macroelement composition according to XRF analysis. The comparison of the contents of petrogenic oxides in BS with the sialite standard (Ketris and Yudovich 2009) confirms the similarity of compositions.

Trace elements were determined by ICP MS (Table 2). BS are characterized by an increased content of a wide range of trace elements. The concentration Clarke was calculated by comparison with the corresponding Clarke for carbonaceous rocks.

Fig. 2 Wall of a quarry for the BS mining

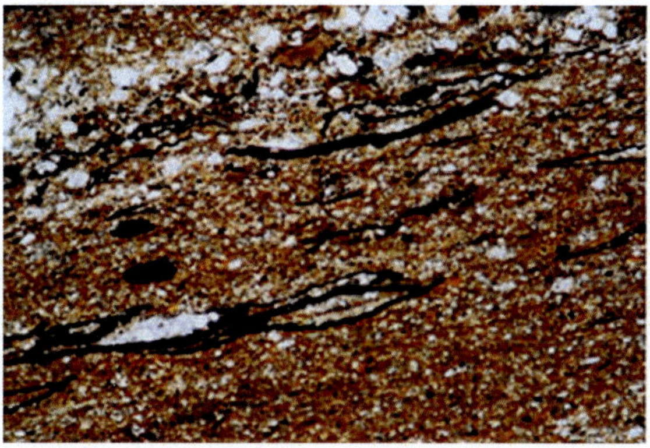

Fig. 3 Layered texture of BS in thin section. Graptolite thecas are black (60 × zoom, crossed nicols)

Table 1 Average content of petrogenic oxides in black shales, wt.%

Oxide	Mean (n = 53)	Min	Max	σ	Carbonaceous sialite
SiO_2	52.3	50.23	58.1	2.0	54.02
TiO_2	0.7	0.28	1.1	0.25	0.87
Al_2O_3	10.6	7.4	15.9	2.4	9.11
Fe_2O_3	5.0	1.45	12.7	2.4	2.68
MnO	0.01	0.004	0.05	0.01	0.01
MgO	0.8	0.45	1.3	0.2	1.19
CaO	0.3	0.066	1.1	0.2	0.56
Na_2O	0.2	0.081	0.3	0.05	0.17
K_2O	3.5	2.21	4.7	0.5	4.31
P_2O_5	0.1	0.032	0.2	0.04	0.11
LOI	30.1	24.9	33.2	2.6	26.97

The highest concentration Clarkes are typical for: Mo—12.5; U—8.3; Tl—5.2; V—3.2.

3.1 *Chemical Weathering of Black Shale*

Black shales are subject to mechanical, chemical and biological weathering, which are closely related. Under the influence of temperature and frost weathering, in the places where the black shale horizon comes to the surface, stratification and destruction of thin-lamellar rock occurs (Fig. 4).

This increases the area of rock interaction with the environment, enhancing chemical and biological weathering, as well as creating favorable conditions for the colonization of microorganisms and higher plants.

Table 2 Average contents (ppm) and concentration Clarkes of BS trace elements

Elements	Mean (n = 53)	Clarke (Ketris, Yudovich 2009)	Concentration Clark
U	82.6	9.9	8.3
Th	8.5	7.8	1.1
V	578.5	180	3.2
Mo	175.2	14	12.5
Cu	61.2	140	0.4
Co	8.5	18	0.5
Ni	49.2	67	0.7
Pb	35.8	25	1.4
Bi	0.4	1.8	0.2
Zn	54.3	160	0.3
Sb	6.3	5.3	1.2
Cd	0.7	6.2	0.1
Tl	11.5	2.2	5.2
Ga	20.8	17	1.2
Ge	1.2	2.5	0.5
Sc	9.7	5.7	1.7
Cr	85.8	93	0.9
Rb	153.8	76	2.0
Cs	12.3	4.2	2.9
Sr	74.0	230	0.3
Ba	722.5	630	1.1
Be	3.2	2.6	1.2
Ce	65.0	55	1.2
Y	15.9	23	0.7
Zr	118.7	200	0.6
Nb	16.1	15	1.1
Sn	3.3	5.7	0.6
Hf	3.1	4.5	0.7
Ta	1.5	0.8	1.9
W	2.8	2.7	1.0
Ag	1.1	1.7	0.7

Chemical weathering involves a process of transformation of minerals and rocks under the influence of water, oxygen, carbon dioxide, hydrogen sulfide, methane, ammonia and organic acids, as well as due to biochemical processes. The leading processes of chemical weathering are dissolution, leaching, oxidation, hydration, carbonatization, and hydrolysis. According to a number of authors, the oxidation of organic matter and pyrite is the main driving force for the weathering of alum

Fig. 4 Mechanical weathering of black shale

black shales. The appearance of sulfuric acid contributes to the further destruction of minerals, which leads to the release of many elements from them (Dang et al. 2002; Hammarstrom et al. 2003; Descostes et al. 2004; Sracek et al. 2004; Fernández-Remolar and Knoll 2008; Joeckel et al. 2005; Fischer et al. 2007 2009; Lavergren 2009; Tuttle and Breit 2009; Zhu 2010; Armstrong et al. 2019).

In the weathering zone on the surface of black shales, secondary minerals (SM) are formed; comprising aggregates of small crystals of predominantly white and yellow color, these sometimes form sinter forms in water seepage zones (Fig. 5).

According to the results of X-ray phase analysis and Raman spectroscopy, gypsum ($CaSO_4 \times 2H_2O$), copiapite ($Fe^{+2}Fe_4^{+3}(SO_4)6(OH)^2 \times 20H_2O$), pickeringite ($MgAl_2(SO_4)4 \times 22H_2O$), tamarugite ($NaAl\ (SO_4)2 \times 6H_2O$). The most widespread are gypsum (rel. 70%) and pickeringite (25 rel. %). Typically, samples are either practically pure gypsum or pickeringite. Copiapite and tamarugite act as addition. Copiapite usually occurs with gypsum; tamarugite—with pickeringite.

Fig. 5 Secondary minerals on the surface of black shales

The chemical composition of SM determined by ICP MS is characterized by the content of a wide range of trace elements (Table 3).

From the data presented in the table, it follows that secondary elements accumulate chemical elements characteristic of black shales. The highest contents differ (ppm): U—229; V—509; Mo—165; Zn—411; Ni, 214; Cu—112; Sr—968; Ba-337.

Table 3 Content of impurity elements in secondary minerals, ppm

Elements	Average (n = 15)	Min	Max	σ
U	49.4	4.5	229.0	67.2
Th	2.8	0.3	8.9	2.7
V	63.0	1.5	509.0	126.8
Mo	39.0	2.9	165.0	39.5
Cu	38.1	1.1	112.0	38.5
Co	9.0	0.3	39.6	11.7
Ni	50.6	2.3	214.0	63.8
Pb	2.7	0.6	9.9	2.7
Bi	0.1	0.1	0.1	0.01
Zn	66.7	0.6	411.0	121.3
Sb	0.6	0.1	2.5	0.6
Cd	1.1	0.1	4.4	1.3
Tl	3.5	0.3	9.7	2.4
Ga	4.0	0.3	15.6	4.5
Ge	0.2	0.1	0.4	0.1
Sc	1.6	0.1	7.6	2.1
Cr	10.9	0.6	44.4	14.5
Rb	15.8	1.3	109.0	26.2
Cs	1.2	0.1	10.5	2.6
Sr	280.7	8.6	968.0	275.1
Ba	46.6	5.2	337.0	82.0
Be	0.9	0.5	2.9	0.7
Ce	54.9	2.6	108.0	33.0
Y	20.0	2.2	67.9	15.2
Zr	12.0	1.9	69.5	16.5
Nb	1.3	0.3	11.7	2.9
Sn	0.4	0.1	0.9	0.2
Hf	0.3	0.0	2.2	0.5
Ta	0.2	0.1	1.4	0.4
W	0.5	0.0	1.9	0.4
Ag	0.1	0.0	0.9	0.2

Thus, during chemical weathering, some of the elements pass into a mobile state and crystallize on the surface of shale in the form of secondary minerals. The latter inherit a wide range of BS trace elements (U, V and chalcophile elements).

3.2 *Biological Weathering of Black Shale*

Microorganisms. Biofilms, various groups of microorganisms, and lichen colonies develop on the surface of BS (Fig. 6).

A microbiological experiment carried out to identify species diversity was based on differences in growth rate and in relation to micromycete food sources. Based on the results of the experiment, the following types of micromycetes were identified (Table 4).

Mycelial fungi tend to accumulate in places where lichens and mosses grow, which enhances the destruction of the rock surface, contributing to the formation of primitive soils and aggravating the impact of other weathering agents.

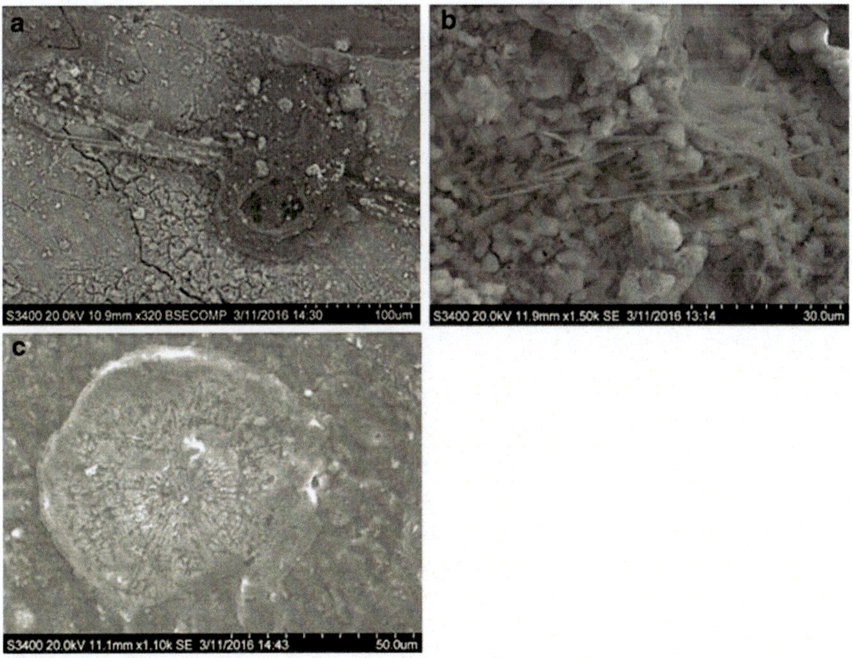

Fig. 6 Biofilms and lichens on the surface of black shales. **a** biofilm; **b** fungal hyphae; **c** lichen colony

Table 4 Micromycetes formed on the surface of Black Shales

1. Acremonium hyalinulum	11. Epicoccum nigrum	21. Penicillium oxalicum
2. Alternaria alternata	12. Fusarium oxysporum	22. Penicillium purpurogenum
3. Arthrinium phaeospermum	13. Melanospora sp.	23. Penicillium waksmanii
4. Aspergillus niger	14. Mortierella lignicola	24. Phoma sp.
5. Aureobasidium pullulans	15. Mucor racemosus	25. Rhizopus stolonifer
6. Botrytis cinerea	16. Paecilomyces lilacinus	26. Scytalidium lignicola
7. Chaetomium globosum	17. Paecilomyces marquandii	27. Trichoderma viride
8. Cladosporium cladosporioides	18. Penicillium brevicompactum	28. Non-stemming, light-colored fungus
9. Cladosporium herbarum	19. Penicillium citrinum	29. Non-stemming, dark-colored fungus
10. Coniosporium sp.	20. Penicillium decumbens	

3.3 Soil and Herbage

On black shales, in the process of erosion under surface conditions, a soil layer is actively formed, consisting of a mixture of humus and fragments of black shales. The most common grass species growing on these primitive soils is meadow bluegrass (Poa praténsis) (Fig. 7).

Black shale–soil–grass sample triples were taken for analysis. The contents of the elements were determined by the ICP-MS method; grass decomposition was carried out in nitric acid (Figs. 8 and 9).

As can be seen from Fig. 8, chalcophile elements and gallium are actively accumulated in soils. As compared to soil, shale is enriched in uranium, vanadium, and molybdenum.

Fig. 7 Primitive soil and bluegrass meadow on black shales

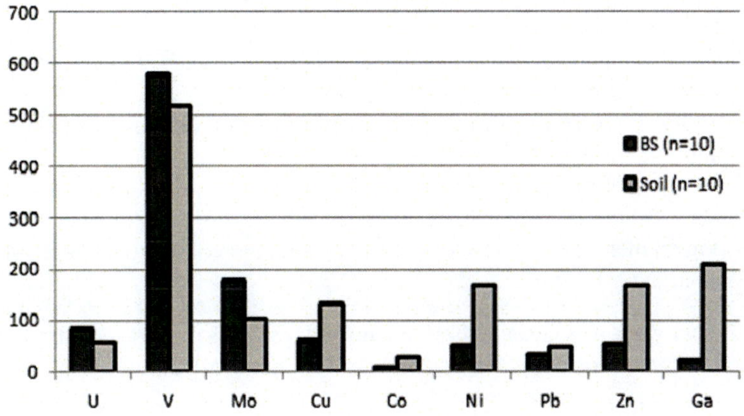

Fig. 8 Average content of trace elements in soil and black shale, ppm

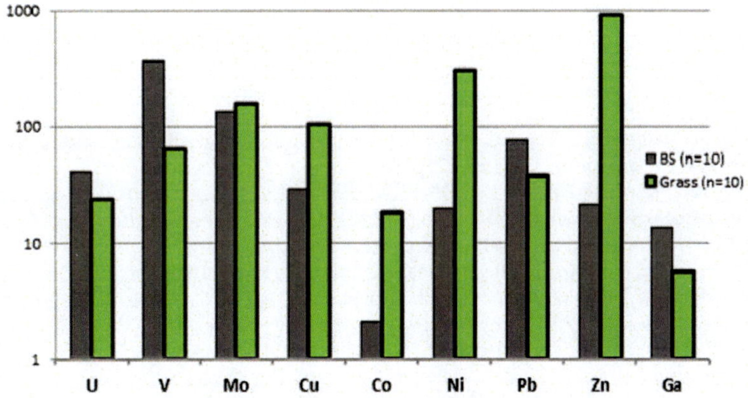

Fig. 9 Average content of trace elements in BS and grass growing on them, ppm

Grass accumulates chalcophilic elements. The content of uranium, vanadium, gallium and lead in grass is lower than in shale. The highest values in the grass are noted for (ppm): Zn—922, Ni—301 and Mo—158. Zn, Ni, Co and Cu accumulate most actively both in the soil and in the grass.

4 Conclusions

1. There is a wide range of trace elements in black shales, whose content exceeds Clarke values (Clarke concentration): Mo—12.5; U—8.3; V—3.2.

2. Secondary minerals formed on the surface of black shale during chemical weathering borrow chemical elements from BS (ppm): U—229; V—509; Mo—165; Zn—411; Ni, 214; Cu—112; Sr—968; Ba-337.
3. According to the results of the experiment, 29 species of micromycetes were found on the surface of BS.
4. In the course of biological weathering in the primitive soil formed on BS, the following accumulate (number of times): Ni—3.4; Co—3.4; Zn—3.1. The grass growing on them is characterized by the accumulation of relatively black shales: Zn—43.6; Ni—15.3; Mo—1.2.

Analytical work was carried out in the resource centers of St Petersburg University (Centre for Microscopy and Microanalysis, Chemical Analysis and Materials Research Centre, Centre for X-ray Diffraction Studies) and Central laboratory of Russian Geological Research Institute (VSEGEI). The authors are grateful to M. S. Zelenskaya for conducting microbiological studies.

References

Andersson A, Dahlman B, Gee DG, Snäll S (1985) The Scandinavian Alum Shales. Sveriges Geologiska Undersökning, Uppsala, 125p

Armstrong JGT Parnell J, Bullock LA, Boyce AJ, Perez M, Feldmann J (2019) Mobilisation of arsenic, selenium and uranium from Carboniferous black shales in west Ireland. Appl Geochem 109: 13 1044014. https://doi.org/10.1016/j.apgeochem.2019.104401

Baruah, B.P., Khare, P., 2010. Mobility of trace and potentially harmful elements in the environment from high sulfur Indian coal mines. Appl. Geochem. 25, 1621–1631 https://doi.org/10.1016/j.apgeochem.2010.08.010

Cappuyns V, Claes, H., Geris, P., Maquil, R., Swennen, R., 2019. Mineralogical and geochemical characterization of Mo-rich black shales in the Grand Duchy of Luxembourg. J. Geochem. Explor. 206, 106351. https://doi.org/10.1016/j.gexplo.2019.106351

Cary, L., Benabderraziq, H., Elkhattabi, J., Gourcy, L., Parmentier, M., Picot, J., Khaska, M., Laurent, A., Négrel, P., 2014. Tracking selenium in the Chalk aquifer of northern France: Sr isotope constraints. Appl. Geochem. 48, 70–82. https://doi.org/10.1016/j.apgeochem.2014.07.014

Changxun Y., Lavergren U., Peltola P., Drake H., Bergback B., Astrom M.E. Retention and transport of arsenic, uranium and nickel in a black shale setting revealed by a long-term humidity cell test and sequential chemical extractions // Chemical Geology. 2013. Vol.363. P.134–144

Dang, Z., Liu, C., Haigh, M.J., 2002. Mobility of heavy metals associated with the natural weathering of coal mine spoils. Environ. Pollut. 118, 419–426

Descostes, M., Vitorge, P., Beaucaire, C., 2004. Pyrite dissolution in acidic media. Geochem. Cosmochim. Acta 68, 4559–4569. https://doi.org/10.1016/j.gca.2004.04.012

Duan Y, Yang Z, Yu T, Yang Q, Liu X, Ji W, Jiang H, Zhuo X, Wu T, Qin J, Wang L (2020) Geogenic cadmium pollution in multi-medians caused by black shales in Luzhai, Guangxi. Environ Pollut 260. https://doi.org/10.1016/j.envpol.2019.113905

Falk H., Lavergren U., Bergback B. Metal mobility in alum shale from Oland, Sweden // Journal of Geochemical Exploration. 2006. Vol. 90. P.157–165

Fernández-Remolar, D.C., Knoll, A.H., 2008. Fossilization potential of iron-bearing minerals in acidic environments of Rio Tinto, Spain: Implications for Mars exploration. Icarus 194, 72–85

Fischer, C., Karius, V., Thiel, V., 2007. Organic matter in black slates shows oxidative degradation within only a few decades. J. Sedimentol. Res. 77, 355–365

Fischer, C., Scmidt, C., Bauer, A., Gauppb, R., Heide, K., 2009. Mineralogy and geochemical alteration of low-grade metamorphic black shale slates due to oxidative weathering. Chemie der Erde – Geochem. 69, 127–142

Fontenot, B.E., Hunt, L.R., Hildenbrand, Z.L., Carlton Jr., D.D., Oka, H., Walton, J.L., Hopkins, D., Osorio, A., Bjorndal, B., Hu, Q.H., Schug, K.A., 2013. An evaluation of water quality in private drinking water wells near natural gas extraction sites in the Barnett shale formation. Environ. Sci. Technol. 47, 10032–10040. https://doi.org/10.1021/es4011724

Hammarstrom, J.M., Seal II, R.R., Meier, A.L., Jackson, J.C., 2003. Weathering of sulfidic shale and copper mine waste: Secondary minerals and metal cycling Great Smoky Mountains National Park, Tennessee, and North Carolina, USA. Environ. Geol. 45, 35–57

He, B., Liang, L., Jiang, G., 2002. Distributions of arsenic and selenium in selected Chinese coal mines. Sci. Total Environ. 296, 19–26

Hints R., Hade, S., Soesoo, A., Voolma M. Depositional framework of the East Baltic Tremadocian black shale revisited // Geologiska Föreningen. 2014. Vol. 4. P. 1-19

Huyen, D.T., Tabelin, C.B., Thuan, H.M., Dang, D.H., Truong, P.T., Vongphuthone, B., Kobayashi, M., Igarashi, T., 2019. The solid-phase partitioning of arsenic in unconsolidated sediments of the Mekong Delta, Vietnam and its modes of release under various conditions. Chemosphere 233, 512–523. https://doi.org/10.1016/j.chemosphere.2019.05.235

Islam, M.R., Salminen, R., Lahermo, P.W., 2000. Arsenic and other toxic elemental contamination of groundwater, surface water and soil in Bangladesh and its possible effects on human health. Environ. Geochem. Health 22, 33–53.https://doi.org/10.1023/A:1006787405626

Joeckel, R.M., Ang Clement, B.J., VanFleet Bates, L. R., 2005. Sulfate-mineral crusts from pyrite weathering and acid rock drainage in the Dakota Formation and Graneros Shale, Jefferson County, Nebraska. Chem. Geol. 215, 433–452

Ketris M.P., Yudovich Ya.E. Estimation of Clarkes for carbonaceous biolithes: World averages for trace elements contents in black shales and coals // Intern. J. Coal Geol. 2009. V. 78. P. 135–148

King DA, Hocking AD, Pitt JJ (1979) Dichloran-Rose Bengal medium for enumeration and isolation of molds from foods. Appl. Environm. Microbiol. 37:959-964

Lavergren U., Mats E.Åström, Falk H., Bergbäck B. Metal dispersion in groundwater in an area with natural and processed black shale – nationwide perspective and comparison with acid sulfate soils. Applied Geochemistry, 2009. Vol. 24. Issue 3. P.359–369

Lecomte A, Cathelineau M, Michels R, Peiffert C, Brouand M (2017) Uranium mineralization in the Alum Shale Formation (Sweden): evolution of a U-rich marine black shale from sedimentation to metamorphism. Ore Geol Rev 88:71–98

Paikaray, S., 2012. Environmental hazards of arsenic associated with black shales: a review on geochemistry, enrichment and leaching mechanism. Rev. Environ. Sci. Bio.Technol. 11, 289–303. https://doi.org/10.1007/s11157-012-9281-z

Peng, B., Song, Z., Tu, X., Xiao, M., Wu, F., Lv, H., 2004. Release of heavy metals during weathering of the Lower Cambrian Black Shales in western Hunan, China. Environ. Geol. 45, 1137–1147. https://doi.org/10.1007/s00254-004-0974-7

Peters, S.C., Burkert, L., 2008. The occurrence and geochemistry of arsenic in groundwaters of the Newark basin of Pennsylvania. Appl. Geochem. 23, 85–98 https://doi.org/10.1016/j.apgeochem.2007.10.008

Sracek, O., Choquette, M., Gélinas, P., Lefebvre, R., Nicholson, R.V., 2004. Geochemical characterization of acid mine drainage from a waste rock pile, Mine Doyon, Québec, Canada. J. Cont. Hydrol. 69, 45–71. https://doi.org/10.1016/S0169-7722(03)00150-5

Tabelin, C.B., Hashimoto, A., Igarashi, T., Yoneda, T., 2014. Leaching of boron, arsenic and selenium from sedimentary rocks: I. Effects of contact time, mixing speed and liquid-to-solid ratio. Sci. Total Environ. 472, 620–629. https://doi.org/10.1016/j.scitotenv.2013.11.006

Tepper E Z, Shilnikova V K, Pereverzeva G I (2004) Microbiology practicum, Drofa, Moskow (in Russian), 154p

Tuttle, M.L.W., Breit, G.N., 2009. Weathering of the New Albany Shale, Kentucky, USA: I. Weathering zones defined by mineralogy and major-element composition. Appl. Geochem. 24, 1549–1564

Tuttle, M.L.W., Fahy, J.W., Elliott, J.G., Grauch, R.I., Stillings, L.L., 2014a. Contaminants from Cretaceous black shale: I. Natural weathering processes controlling contaminant cycling in Mancos Shale, southwestern United States, with emphasis on salinity and selenium. Appl. Geochem. 46, 57–71. https://doi.org/10.1016/j.apgeochem.2013.12.010

Tuttle, M.L.W., Fahy, J.W., Elliott, J.G., Grauch, R.I., Stillings, L.L., 2014b. Contaminants from cretaceous black shale: II. Effect of geology, weathering, climate, and land use on salinity and selenium cycling, Mancos Shale landscapes, southwestern United States. Appl. Geochem. 46, 72–84. https://doi.org/10.1016/j.apgeochem.2013.12.011

Zheng, B., Ding, Z., Huang, R., Zhu, J., Yu, X., Wang, A., Zhou, D., Mao, D., Su, H., 1999. Issues of health and disease relating to coal use in southwestern China. Int. J. Coal Geol. 40, 119–132. https://doi.org/10.1016/S0166-5162(98)00064-0

Zhu W.Y., 2010. Chemical and microbial control of pyrite weathering and its implications to arsenic mobility and sulfur and iron geochemistry. Dissertation, New Brunswick, New Jersey. The State University of New Jersey

Effects of Bryophyte Vegetation on the Chemical Composition of Sandstone

Evgeniia V. Smirnova, Helena V. Kushnevskaya, Elena G. Panova, and Elena E. Orlova

Abstract The study of plant-substrate interaction in biocrusts has focused less on bryophytes than on other phototrophs for a long time. Many authors, however, consider it as part of early stage of primary succession. The role of various bryophyte species in redeposition of chemical elements and organic deposition is poorly understood. The aim of this study is to estimate the influence of bryophytes on sandstone weathering. The data were collected on outcrops of Devonian sandstones in three locations in Northwest Russia. Devonian sandstones are inhabited by plenty of bryophyte species, and the composition of their moss-liverwort communities depends on the mineral composition of the sandstone. Red sandstones rich in metals (aluminium, iron, potassium) are occupied mainly by mosses. 0.2% calcium oxide in the substrate composition is already sufficient for the appearance of calciphilous species in moss communities. Liverworts prefer white, generally element-poor sandstones, but some calcicolous liverwort species may inhabit red ferruginous sandstones containing more than 0.2% calcium. The analyses reveal different changes in the contain of some elements under various bryophyte-dominated crusts. The four categories of communities according to the ability to deposit calcium and other metals are given. The composition of the organic matter under different bryophyte communities was analyzed to evaluate the degree of soil formation. The content of total organic matter does not exceed 1% and does not differ between categories. However, the average content of humic acids differs significantly. The results show the lowest rates of pedogenesis under communities of calcicoles.

E. V. Smirnova (✉) · H. V. Kushnevskaya · E. G. Panova · E. E. Orlova
Saint Petersburg State University, Universitetskaya Embankment, 7/9, 199034 St. Petersburg, Russia
e-mail: esmir98@ya.ru

H. V. Kushnevskaya
Komarov Botanical Institute, Russian Academy of Sciences, Professor Popov Street, 2, 197376 St. Petersburg, Russia

Saint Petersburg State Forest Technical University, Institutskaya Street, 5, 194021 St. Petersburg, Russia

O. V. Frank-Kamenetskaya et al. (eds.), *Biogenic—Abiogenic Interactions in Natural and Anthropogenic Systems 2022*, Springer Proceedings in Earth and Environmental Sciences, https://doi.org/10.1007/978-3-031-40470-2_16

Keywords Mosses · Liverworts · Plant communities · Weathering · Biodeterioration · Pedogenesis · Soil formation · Primary soil

1 Introduction

Bryophytes (mosses and liverworts) are renowned as plants that appear on the early stages of primary succession. Due to the peculiarities of morphology, anatomy and physiology, these plants are adapted to life on solid substrates in conditions of poor nutrition, significant changes of temperature and moisture. The rhizoids allow bryophytes to gain a foothold even on very solid rocks, using microscopic cracks and pits. The absence of covering tissues ensures the absorption of water and nutrients through the entire surface of the shoots and the ability to quickly rehydrate when the air humidity changes. Also mosses and liverworts can maintain a dormant state for a long time escaping unfavorable environmental conditions. All these features give the bryophytes a significant competitive advantage over tracheophytes in the colonizing of rocky habitats.

During their life epilithic bryophytes extract the necessary elements from the rocky substrate (Bates 1978, 1982; Altieri and Ricci 1997). Moss cushions accumulate soil particles and dust carried by wind and rain (Danin and Ganor 1991; Pérez 2010, 2020). If plant residues are collected on the rock surface, the formation of primary soil is possible (Giordano et al. 2012; Popova and Vlasov 2015; Panova et al. 2015). Thus, the composition of the entire surface layer changes in contact with vegetation.

Bryophytes are regularly mentioned in literature as agents of biological weathering. At the same time, their role in the transformation of the substrate remains controversial. For example, some authors believe that bryophytes can only accumulate the dust and organic matter (Lisci et al. 2003) or affect a little and do not cause serious damage to rocky substrate (Tiano 2002). Even in large, widely cited reviews authors mentioned mosses only among other photoautotrophs (Warscheid and Braams 2000) and did not cite real studies of biogenic-abiogenic interactions between mosses and rocks for a long time due to their scarcity.

Today many researchers are interested in bryophyte-rock interaction. Recent studies show that bryophytes can directly and indirectly affect the substrate by excreting organic compound (Galloway et al. 2018), absorbing elements (Giordano et al. 2012), protecting the surface from wind erosion (Langston and Neuman 2005; Wang et al. 2011), changing the temperature and humidity near the surface (Jang 2022). The character and ratio of weathering depend on the type of rock (Chen et al. 2021). The species of bryophyte determines the dust collection depending on the form of growth and cushion size (Mitchell et al. 2021). Also, it is important that algae and lichens often occur in bryophyte patches in natural conditions. Microorganisms, including symbiotic bacteria, are also constantly present (Cao et al. 2020; Bai et al. 2021). Thus, we can work only with bryophyte-dominated communities.

The effect of bryophytes in the processes of bioweathering and pedogenesis are still poorly researched, but the new data help to reveal the different aspects of this complex interaction. The aim of present paper is to show the role of epilithic bryophyte vegetation in the transformation of the sandstone located in the Leningrad Region (Russian Federation).

2 Materials and Methods

2.1 *Materials*

The biological weathering by bryophytes was studied on sandstone plant communities. Devonian sandstones are widespread in the southwest of the Leningrad Region (Russian Federation), which is a part of the Main Devonian Field. The sandstone plant communities are a great object for studying the bryophyte-rock interaction for few reasons. Firstly, the sandstone outcrops are numerous on the riverbanks and differ in chemical and mineral composition. A wide range of plant communities can be considered both in geographic and substrate-dependent aspects. Secondly, the vertical exposure of the outcrops makes it difficult to collect dust and soil particles from other substrates, prevents vascular plants from establishing and germinating. This reduces competition and helps to determine the effect of mosses on rock composition more carefully.

The material on moss-liverwort dominated communities was collected on natural Devonian sandstone outcrops located in the Luzhsky District in the Leningrad Region (Fig. 1a). The outcrops were observed on the banks of big lowland Luga River and its tributaries in three geographic locations: “Saba” (Saba River, 58°58′31″N, 29°05′00″E), “Yashchera” (Yashchera River, from 58°53′47″N, 29°50′25″E to 58°53′49″N, 29°51′01″E), and “Luga” (Luga River, 59°00′30″N, 29°37′35″E, and Rastishenskii Stream, 59°00′40″N, 29°37′17″E). The data were collected during the field seasons in 2019–2021.

2.2 *Vegetation Survey*

We studied vegetation only on vertical surfaces of sandstone (70–110°) (Fig. 1b–d). Bryophyte-dominated plant communities were examined using sample plot method. In every sample plot the projective cover (%) of all bryophyte species was determined. The species occurrence in sample plots was calculated for each separate outcrop. The plots ($S = 60$ cm^2) were arranged regularly with an interval of 60–70 cm within the natural boundaries of the vegetation patches. In patches less than $S = 1500$ cm^2, three plots were placed at same distances to each other.

Fig. 1 The studied sandstone outcrops: **a** geographic locations on a map (1—"Saba", 2—"Luga", 3—"Yashchera"). Vegetation on the surface of outcrop in **b** "Luga" location, **c** "Saba" location, **d** "Yashchera" location

2.3 Rock Sampling

After the vegetation survey, the samples of the substrate under several vegetation patches were taken. We selected different patches based on differences in vegetation, substrate color and substrate density. For each patch two types of samples were taken. The first—the upper layer of the substrate with bryophytes (hereinafter: crust)—was taken from the surface of the outcrop (thickness <5 mm; transported in kraft paper packages). Then a second sample (hereinafter: intact substrate) was taken from a depth of >3 cm (weight 500 g; transported in cloth bags). All samples were air dried at room temperature in the laboratory. A total of 16 pairs of samples were collected and analyzed.

2.4 *Identification of Bryophytes*

Bryophyte species were identified by standard anatomical and morphological method using light microscopy—MSP-2 var. 2 (LOMO, Russia), Leica DM 500 (LeicaMicrosystems, Germany) and Axio Scope.A1 (Zeiss, Germany) microscopes. Identification was carried out according to several keys (Schljakov 1976, 1979, 1980, 1981, 1982; Paton 1999; Damsholt 2002; Ignatov and Ignatova 2003, 2004; Szweykowski et al. 2005; Bakalin 2005; Potemkin and Sofronova 2009; Ignatov et al. 2017, 2018, 2020). The species names follow Hodgetts et al. (2020).

2.5 *Substrate Analyzes*

The major oxide composition of intact sandstone was determined using X-ray fluorescence (XRF) silicate analysis. 16 samples of intact rock were powdered on a vibration cup grinder IV-3 (Vibrotekhnik, Russia). XRF silicate analysis was carried out in the Resource Center "Chemical Analysis and Materials Research Centre" of St. Petersburg State University on an energy-dispersive X-ray fluorescence spectrometer EDX-8100P (Shimadzu, Japan) with the same parameters for all samples (range of determined elements from carbon to uranium; diameter of the irradiated area 10 mm; polypropylene cuvette) except the atmosphere parameter (air and helium). The difference between the results obtained in different atmospheres was 0.5%, which can be considered insignificant.

The chemical composition of the intact fresh rock and the crusts under moss cushions were compared using XRF spectroscopy. The major and trace elements were determined for 16 pairs of samples (crust and intact substrate). The analysis was carried out using a portable X-ray fluorescence spectrometer OMEGA (INNOV-X SYSTEMS, USA) in the range of determined elements from phosphorus to uranium.

The organic matter composition was studied using Tyurin dichromate-oxidation method (total carbon content) and Kononova-Belchikova method (humus fraction composition and the content of humic acid carbon). To characterize humus substances, the humus acid optical density was determined at 465 nm on a photoelectric colorimeter using Ponomareva-Plotnikova method. The analyzes included 16 samples of crusts (hypothesis—the presence of primary soil) and 4 samples of intact sandstone as controls (hypothesis—the absence of primary soil). The sample preparation was modified to improve the accuracy of the analyzes: (1) weight of the dry sample was increased to 5 g (in accordance to the recommendations of V. V. Ponomareva and T. A. Plotnikova for sandy soils with a low content of humus); (2) unfragmented organic matter and rhizoids were removed using tweezers and microscope, and the samples were powdered with porcelain and agate mortars and pestle; (3) for all analyzes, extracts from soil samples in a highly alkaline solution of sodium pyrophosphate were used (modified from Kononova-Belchikova method). Each sample was analyzed in 2 replications.

2.6 Statistics

For comparison of average values, the nonparametric Mann–Whitney U test was used to identify significant differences. Kendall's tau correlation statistic was used to assess correlation between content of different elements. These statistical analyzes were conducted using PAST version 4.06.

For analyzing trends in substrate transformation, a classification of substrate samples was created and tested using discriminant analysis with the Mahalanobis distance using SPSS version 23.0 (IBM SPSS Statistics).

3 Results and Discussion

3.1 Biodiversity of Bryophytes in Plant Communities on the Sandstone Outcrops

In the study of natural sandstone outcrops 98 species of bryophytes were revealed. 72 species of mosses and 26 species of liverworts were identified. Diversity of bryophytes on three locations differs: 34 species—"Saba" (Saba River), 65—"Luga" (Luga River and Rastishenskii Stream,), 51—"Yashchera" (Yashchera River). The comparison of species lists shows low commonality (Table 1). Only 12 species (8 mosses and 4 liverworts) are common for all locations, and 58 species occur uniquely in the one location. For each pair of locations, the Jaccard similarity index was calculated. The results show from medium to low floristic similarity: the Jaccard index range from 0.55 ("Yashchera" and "Luga", 28 common species) to 0.44 ("Luga" and "Saba", 21 common species) and 0.27 ("Saba" and "Yashchera", 15 common species). The greatest floristic similarity is shown for the closest locations ("Yashchera" and "Luga", 15.7 km). It can be explained by the geographical proximity of these locations to each other and the presence of many non-specific multisubstrate species. The degree of similarity of the next pair of locations ("Luga" and "Saba", 31.6 km) decreases disproportionately to the increase in distance due to the appearance of substrate-specific calciphilous species common only for these sandstones.

34 species are found on more than 10% of sample plots so they play a significant role in plant communities. For each species, the occurrence within the outcrop was calculated from its presence in vegetation sample plots. Species with the highest abundance were determined by ranking.

To analyze the relationship between the ecological groups of bryophytes and the chemical characteristics of the substrate, 16 vegetation patches were selected for substrate analyzes. The vegetation features are in Table 2. The main interest was paid to patches dominated by species playing a leading role in the vegetation cover of the outcrop (abundant and frequent). It was also important to analyze the substrate under patches with calcicoles and calcifuges. The ecological characteristics of species have

Table 1 Bryophyte species found on the outcrops

Table 1. Bryophyte species found on the outcrops

Bryophyte species	Separate outcrop on the riverbank			
	Saba River	Rastishenskii Stream	Luga River	Yashchera River
Liverworts				
Bazzania trilobata (L.) Gray				***Sp***
Blasia pusilla L.		R		U
Blepharostoma trichophyllum (L.) Dumort.		**Fr**	Sp	**Sp**
Calypogeia integristipula Steph.		U		R
Calypogeia muelleriana (Schiffn.) Müll. Frib.				U
Cephalozia bicuspidata (L.) Dumort.	U	R	Sp	R
Chiloscyphus polyanthos (L.) Corda		Sp		
Conocephalum salebrosum Szweyk., Buczk. & Odrzyk.	**R**	R	***Sp***	R
Fuscocephaloziopsis lunulifolia (Dumort.) Váňa & L. Söderstr.			U	U
Geocalyx graveolens (Schrad.) Nees	R	U		R
Lepidozia reptans (L.) Dumort.		R	U	***Sp***
Liochlaena subulata (A. Evans) Schljakov		***C***	R	
Lophocolea minor Nees	R		R	
Lophozia guttulata (Lindb. & Arnell) A. Evans				R
Lophozia ventricosa (Dicks.) Dumort.		R		R
Mesoptychia collaris (Nees) L.Söderstr. & Váňa		R		
Mesoptychia heterocolpos (Thed. ex Hartm.) L. Söderstr. & Váňa			***R***	
Mylia taylorii (Hook.) Gray				**Fr**
Pellia neesiana (Gottsche) Limpr.		R		
Plagiochila porelloides (Torr. ex Nees) Lindenb.		**C**	Sp	***C***
Ptilidium ciliare (L.) Hampe				U
Scapania mucronata H. Buch	R	R	R	R
Schistochilopsis incisa (Schrad.) Konstant.		U		Sp
Solenostoma confertissimum (Nees) Schljakov				U
Sphenolobus minutus (Schreb. ex D. Crantz) Berggr.		**Sp**		**Fr**
Trilophozia quinquedentata (Huds.) Bakalin				R
Mosses				
Alleniella complanata (Hedw.) S.Olsson, Enroth & D. Quandt		U		
Amblystegium serpens (Hedw.) Schimp.	R		R	
Athrichum flavisetum Mitt.		R	R	
Atrichum undulatum (Hedw.) P. Beauv.	U	R	R	U
Aulacomnium palustre (Hedw.) Schwägr.				U
Bartramia pomiformis Hedw.		R		
Brachytheciastrum velutinum (Hedw.) Ignatov & Huttunen	***Sp***	R	R	
Brachythecium capillaceum (F. Weber & D. Mohr) Giacom.	U			
Brachythecium mildeanum (Schimp.) Schimp.	U			
Brachythecium rutabulum (Hedw.) Schimp.			R	
Brachythecium salebrosum (Hoffm. ex F. Weber & D. Mohr) Schimp.		R		R
Bryoerythrophyllum recurvirostrum (Hedw.) P. C. Chen	R	R	R	***Sp***
Buxbaumia aphylla Hedw.				U
Campylophyllopsis calcarea (Crundw. & Nyholm) Ochyra	U			
Ceratodon purpureus (Hedw.) Brid.	U			U
Chionoloma tenuirostre (Hook. & Taylor) M. Alonso, M. J. Cano & J. A. Jiménez		R	**Fr**	
Cirriphyllum piliferum (Hedw.) Grout		U		

(continued)

Table 1 (continued)

Bryophyte species	Separate outcrop on the riverbank			
	Saba River	Rastishenskii Stream	Luga River	Yashchera River
Conardia compacta (Drumm. ex Müll. Hal.) H. Rob.	U			
Dichodontium pellucidum (Hedw.) Schimp.		Sp	R	
Dicranella crispa (Hedw.) Schimp.		**Sp**		
Dicranella subulata (Hedw.) Schim				Sp
Dicranum bonjeanii De Not.				R
Dicranum flagellare Hedw.				R
Dicranum montanum Hedw.				U
Dicranum polysetum Sw. ex anon.				U
Dicranum scoparium Hedw.				R
Didymodon fallax (Hedw.) R. H. Zander	U			
Distichium capillaceum (Hedw.) Bruch & Schimp.		R	**Fr**	
Encalypta vulgaris Hedw.	U	U		
Eurhynchiastrum pulchellum (Hedw.) Ignatov & Huttunen			R	
Fissidens adianthoides Hedw.	U			
Fissidens bryoides Hedw.				U
Fissidens bryoides var. *viridulus* (Sw.) Broth.		R		
Fissidens osmundoides Hedw.		R		
Fissidens taxifolius Hedw.		R		
Gymnostomum aeruginosum Sm.	**Sp**			
Gyroweisia tenuis (Hedw.) Schimp.	**Fr**	U	R	
Homalia trichomonoides (Hedw.) Brid.		U	***R***	
Hylocomium splendens (Hedw.) Schimp.			U	U
Hypnum cupressiforme Hedw.			R	
Leptobryum pyriforme (Hedw.) Wilson	**C**	R	Sp	C
Mnium hornum Hedw.		U		
Mnium lycopodioides Schwägr.		U		
Mnium stellare Hedw.	R	R	**Fr**	
Myurella julacea (Schwägr.) Schimp.		***R***	R	
Myurella sibirica (Müll. Hal.) Reimers	U			
Oxyrrhynchium hians (Hedw.) Loeske			U	
Plagiomnium cuspidatum (Hedw.) T. J. Kop.	U	R	**C**	R
Plagiomnium medium (Bruch & Schimp.) T. J. Kop.		R	U	R
Plagiothecium cavifolium (Brid.) Z. Iwats.			Sp	
Plagiothecium curvifolium Schlieph. ex Limpr.		U		
Plagiothecium rossicum Ignatov & Ignatova	R	**C**	R	R
Pleurozium schreberi (Willd. ex Brid.) Mitt.		U		R
Pogonatum urnigerum (Hedw.) P. Beauv.		R	U	**C**
Pohlia cruda (Hedw.) Lindb.	U	***Sp***	**Fr**	***Sp***
Pohlia elongata Hedw.				U
Pohlia nutans (Hedw.) Lindb.	U			R
Pohlia proligera (Kindb.) Lindb. ex Broth.	**Sp**	***C***	R	**Fr**
Polytrichum juniperinum Hedw.	***Sp***			
Pseudanomodon attenuatus (Hedw.) Ignatov & Fedosov			**Sp**	
Pseudoleskeella nervosa (Brid.) Nyholm	R			
Ptychostomum moravicum (Podp.) Ros & Mazimpaka	U	U	U	
Rhizomnium punctatum (Hedw.) T. J. Kop.	U	***C***	***Sp***	R
Saelania glaucescens (Hedw.) Broth.		R	R	U

(continued)

Table 1 (continued)

Bryophyte species	Separate outcrop on the riverbank			
	Saba River	Rastishenskii Stream	Luga River	Yashchera River
Sanionia uncinata (Hedw.) Loeske				R
Sciurohypnum curtum (Lindb.) Ignatov		U	R	
Sciurohypnum oedipodium (Mitt.) Ignatov & Huttunen			U	R
Sciurohypnum populeum (Hedw.) Ignatov & Huttunen	R		R	
Sciurohypnum reflexum (Starke) Ignatov & Huttunen	R			U
Sphagnum capillifolium (Ehrh.) Hedw.				U
Tetraphis pellucida Hedw.				**C**
Tortula lingulata Lindb.	***Sp***	R	***Sp***	

Note. Frequency categories: Fr (frequent) >30% of sample plots on the outcrop, C (common) 20–30%, Sp (sporadic) 10–20%, R (rare) <10%, U (unique) – a single record. The most abundant species of the outcrop are highlighted in **bold** type (top five of liverworts and top five of mosses) and in ***bold italic*** type (next five). Data on outcrops in "Luga" location are marked in gray.

been determined from the literature (Ignatov and Ignatova 2003, 2004; Potemkin and Sofronova 2009; Ellenberg and Leuschner 2010; Ignatov et al. 2017, 2018, 2020).

3.2 *Intact Rock Composition*

SiO_2 is the main of the major components of the sandstone. The SiO_2 concentration exceeds 77% in all substrate samples. The Yashchera deposits show the highest content of SiO_2 and the lowest contents of other oxides. The Saba sandstone is partially enriched in CaO. The substrate samples from "Luga" location show the medium content of CaO, Al_2O_3, Fe_2O_3, K_2O. The average, minimum, maximum and standard deviation of major oxides of sandstone deposits in three locations are represented in Table 3.

Comparison of average contents was carried out using the Mann-Whitney U-test (for each oxide significant differences marked by different letters in Table 3). According to it, there are no significant differences between the composition of substrate from "Saba" and "Luga" locations except SO_3 content. The individual samples from "Luga" location seem to be more homogeneous and comparable to each other in the oxide composition than the samples from "Saba". The "Yashchera" substrate samples are also quite homogeneous in oxide contents but significantly differ from the samples of "Saba" and "Luga". This corresponds to the differences in formation characteristics between sandstone in "Yashchera" location and others (the color of the sandstone deposits, its age and formation conditions).

In several vegetation patches the calcicole bryophytes were discovered and had a great abundance. Calcicole mosses require more calcium ions from the environment for their normal physiology processes, are sensitive to variability in substrate pH and have greater calcium concentrations in their tissues than calcifuges (Bates

Table 2 The description of 16 vegetation patches selected for chemical analyzes of substrate

Location	Vegetation patch	Number of plots	Number of species	Minimal, average and maximum species number in the sample plot	Minimal, average and maximum total projective cover in the sample plot	Dominants of the vegetation patch	Other species in the vegetation patch
"Saba"	S1	3	2	(1)—1.3—(2)	(20)—36.7—(50)	*Pohlia nutans*	***Leptobryum pyriforme***
	S2	4	2	(1)—1.5—(2)	(10)—25.2—(60)	*Polytrichum juniperinum*	***Pohlia proligera***
	S3	3	2	(1)—1.3—(2)	100	***Pohlia proligera***	*Polytrichum juniperinum*
	S4	3	3	(1)—1.7—(2)	(60)—77.3—(91)	***Gyroweisia tenuis***	*Dicranella* sp., *Tortula lingulata*
	S5	3	1	1	(15)—20—(25)	***Gyroweisia tenuis***	
	S6	4	1	1	(25)—53.8—(90)	***Gymnostomum aeruginosum***	
	S7	5	6	(1)—2.6—(4)	(83)—94.6—(100)	***Gymnostomum aeruginosum***, ***Leptobryum pyriforme***	*Conocephalum salebrosum*, *Brachytheciastrum velutinum*, *Tortula lingulata*, *Myurella sibirica*
"Luga"	L1	43	31	(1)—6.1—(12)	(37)—80.3—(100)	***Pseudanomodon attenuatus***, ***Mnium stellare***, *Conocephalum salebrosum*	***Chionoloma tenuirostre***, *Mesoptychia heterocolpos*, ***Plagiomnium cuspidatum***, *Rhizomnium punctatum*, etc.
	L2	26	5	(1)—2.5—(4)	(0.5)—17.3—(70)	*Chiloscyphus polyanthos*	*Dichodontium pellucidum*, *Pelia neesiana*, ***Rhizomnium punctatim***, ***Plagiothecium rossicum***

(continued)

Table 2 (continued)

Location	Vegetation patch	Number of plots	Number of species	Minimal, average and maximum species number in the sample plot	Minimal, average and maximum total projective cover in the sample plot	Dominants of the vegetation patch	Other species in the vegetation patch
	L3	4	10	(3)—4—(7)	(10)—55.6—(95)	*Tortula lingulata*	*Mesoptychia collaris*, ***Leptobryum pyriforme***, *Distichium capillaceum*, *Gyroweisia tenuis*, etc
	L4	3	2	(1)—1.3—(2)	(70)—80—(90)	*Myurella julacea*	*Distichium capillaceum*
"Yashchera"	Y1	3	2	(1)—1.3—(2)	(65)—76.7—(85)	*Pogonatum urnigerum*	***Pohlia proligera***
	Y2	18	19	(1)—3.9—(7)	(55)—89.4—(100)	*Bryoerythrophyllum recurvirostrum*, *Mylia taylorii*	***Pohlia proligera***, *Pogonatum urnigerum*, *Plagiomnium cuspidatum*, *Geocalyx graveolens*, etc
	Y3	9	9	(1)—3—(6)	(40)—82.7—(100)	*Mylia taylorii*	*Tetraphis pellucida*, ***Sphenolobus minutus***, *Dicranella subulata*, etc
	Y4	5	7	(2)—3.6—(6)	(70)—89.4—(100)	***Sphenolobus minutus***, *Bazzania trilobata*	*Lepidozia reptans*, *Plagiochila porelloides*, *Leptobryum pyriforme*, *Pohlia cruda*, *Pohlia nutans*
	Y5	3	4	(2)—2.7—(4)	(60)—68.3—(80)	*Leptobryum pyriforme*, ***Pohlia proligera***	*Blepharostoma trichophyllum*, *Tetraphis pellucida*

Note Dominant species are highlighted in **bold** type (the most abundant frequent, common, and sporadic species of the separate outcrop; abundant frequent and common of the separate outcrop), calciphiles and basiphiles are underlined, acidophiles are double underlined

Table 3 The content of major oxides (wt%) in intact sandstone of three locations: the average content and standard error (first line), minimum and maximum contents (second line)

Oxide	Location		
	"Saba"	"Luga"	"Yashchera"
SiO_2	87.92 ± 1.76 **a**	88.12 ± 0.89 **a**	97.45 ± 0.20 **b**
	77.69 – 92.77	86.01 – 91.17	96.31 – 97.82
TiO_2	0.27 ± 0.07 **a**	0.24 ± 0.02 **a**	0.09 ± 0.02 **b**
	0.12 – 0.59	0.18 – 0.31	0.05 – 0.21
Al_2O_3	6.48 ± 1.13 **a**	6.45 ± 0.54 **a**	1.14 ± 0.21 **b**
	3.63 – 12.91	4.63 – 7.88	0.72 – 2.28
Fe_2O_3	1.67 ± 0.31 **a**	2.08 ± 0.19 **a**	0.79 ± 0.12 **b**
	0.79 – 3.53	1.51 – 2.59	0.41 – 1.34
MnO	0.029 ± 0.007 **a**	0.028 ± 0.001 **a**	0.006 ± 0.001 **b**
	0.008 – 0.067	0.021 – 0.031	0.004 – 0.01
MgO	–	–	–
	–	–	–
CaO	0.60 ± 0.37 **a**	0.16 ± 0.04 **a**	0.01 ± 0.01 **b**
	0.10 – 3.10	0.00 – 0.27	0.00 – 0.04
Na_2O	–	–	–
	–	–	–
K_2O	2.89 ± 0.37 **a**	2.69 ± 0.20 **a**	0.44 ± 0.05 **b**
	1.89 – 4.88	2.09 – 3.29	0.33 – 0.64
P_2O_5	0.05 ± 0.03 **a**	0.08 ± 0.03 **b**	0.00 **a**
	0.00 – 0.19	0.04 – 0.19	–
Loss on ignition	0.04 ± 0.01 **a**	0.04 ± 0.002 **a**	0.03 ± 0.004 **a**
	0.02 – 0.09	0.04 – 0.05	0.02 – 0.05
SO_3	0.05 ± 0.01 **a**	0.12 ± 0.01 **b**	0.05 ± 0.01 **a**
	0.00 – 0.10	0.10 – 0.14	0.00 – 0.07

Note Significant differences calculated with the Mann-Whitney U test are marked by the letters a, b in each line ($p < 0.05$)

1982; Cleavitt 2001). At the same time, bryophytes are more sensitive to a little concentration of elements than tracheophytes (Puczko et al. 2018). The comparison of the content of calcium in the intact sandstone under different bryophyte-dominated patches shows that 0.2% calcium oxide in the substrate composition is already sufficient for the appearance of calciphilous species in plant communities.

3.3 *Features of the Element Composition of the Bryophyte-Dominated Crusts*

The chemical composition of 16 pairs of bedrock and crust samples was compared using X-ray fluorescence spectroscopy with a spectrometer capable of analyzing low-mass samples. A total of 24 elements were detected, but only 11 elements exceeded the detection threshold in all samples. The received data show differences between the chemical composition of the upper layer and intact sandstone. The concentrations of 6 major elements differ noticeably under the bryophyte vegetation (sulfur, calcium, iron, potassium, titanium, manganese) (Fig. 2a, b).

The contents of calcium and sulfur mainly increase in the weathered layer (Fig. 2c). These changes positively correlate (Kendall's tau correlation statistic 0.38,

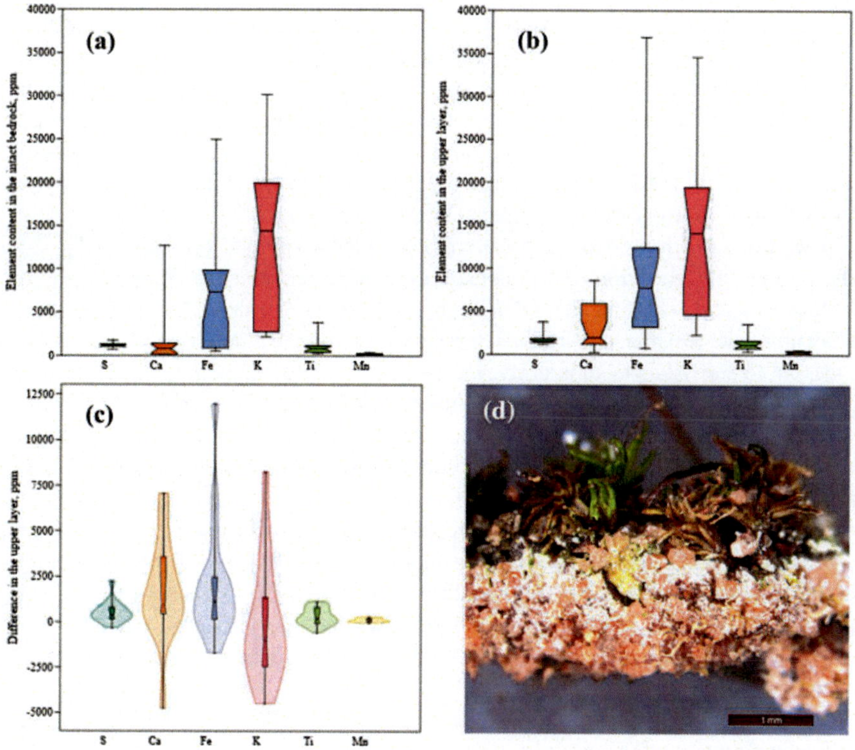

Fig. 2 Element composition of sandstone samples: **a** Box plot showing the element content in the intact bedrock (box—interquartile range, line in box—median, whiskers—extreme values); **b** box plot showing the element content in the upper layer (without one outlier in calcium content (**d**)); **c** Violin (data distribution, wider sections represent more frequently occurring values) and box plot showing absolute difference between the element content in the upper layer and in the intact bedrock (without one outlier in calcium content (**d**)); **d** Formation of calcium mycelium in the upper layer under the vegetation patch L3 (calcium content—48000 ppm)

$p < 0.05$): the sulfur content under bryophyte cushions is predominantly slightly higher than its content in the intact substrate, the calcium content—much higher. The highest calcium content is shown in the upper layer under the vegetation patch L3 with basiphilous *Tortula lingulata* and strongly calciphilous *Mesoptychia collaris* (Fig. 2d)—48000 ppm in the crust sample which is 39 times more than in the intact sandstone sample under this patch. An exception to the general upward trend is the significant decrease of calcium content (4749 ppm lower) in the crust under the vegetation patch S4 dominated by moderately calciphilous *Gyroweisia tenuis*—the calcium content in the intact substrate here was initially high (12690 ppm). The significant positive correlation between the contents of calcium and sulfur can be associated with the formation of salts or complexes containing calcium and sulfur in the crusts. The formation of gypsum ($CaSO_4$) on the surface of non-carbonate sandstones was previously noted in the studies of rock weathering, while gypsum was absent in the bedrock layers (McGreevy 1985; Warke and Smith 2000; Peña-Monné et al. 2022). Gypsum formation seems to be related to rock weathering in conditions of high humidity (the conclusion of the recent study on cave wall weathering (Peña-Monné et al. 2022); two other studies were carried out on coastal sandstones under the similar humidification conditions), and bryophyte cushions improve the water retention. Another possible explanation is a combined increase in the calcium and sulfur concentrations due to its combined accumulation in the same or several organic compounds forming under the mosses (Yamakita et al. 2021).

The content of iron also mainly increases under bryophyte vegetation (Fig. 2c). The greatest concentrations and the greatest increase were noted under the vegetation patches S1 and S3 dominated by *Pohlia* spp. It is interesting that many authors previously noted the resistance of species of the genus *Pohlia* to elevated metal concentrations, but this resistance, apparently, was not associated with metallophilicity (Shaw 1989; Xu et al. 2022). The changes in the concentrations of iron positively correlate with the changes in the contents of other metal elements: potassium (Kendall's tau correlation statistic 0.45, $p < 0.05$), titanium (Kendall's tau correlation statistic 0.41, $p < 0.05$), etc.

For potassium, a very soluble nutrient, leaching processes predominate over accumulation processes in the crusts, but the changes largely depend on element content in the intact fresh rock (Fig. 2c). In conditions of initially low content of potassium, bryophyte vegetation contributes to the accumulation of potassium in the crusts. In conditions of initially high concentrations, plants do not prevent the potassium from leaching from the sandstone surface.

The concentrations of titanium and manganese vary moderately (Fig. 2c). Apparently, the increase in the content can be partially associated with the accumulation of fine earth and the sorption of elements on the surface of fine earth (Cnudde and Boone 2013; Alexandrowicz et al. 2014), while the decrease in the content of elements can be associated with a high degree of moisture and leaching of elements from wet rock. For example, the greatest leaching of titanium, manganese, potassium, iron was noted under the vegetation patch L2 which was located the closest to the water's edge in conditions of constantly high humidity.

Contents of other elements (V, Cr, Ni, Rb, Sr, Zn, Pb, Cl, Co, Cu, As, Zr, Mo, Cd, Sn, Sb, W, U) vary slightly and often do not exceed the detection threshold of the element.

3.4 Comparison of Substrate Transformation Under Different Bryophyte Vegetation

A classification was required to assess the presence of different trends in substrate transformation under activity of different bryophyte vegetation. The substrate samples and vegetation data corresponding to the samples were classified based on the chemical composition of the intact substrate. The classification was checked by discriminant analysis with the Mahalanobis distance in SPSS 23. Four categories were obtained: the substrate under the communities dominated by acidophilous species (Y1, Y2, Y3, Y4, Y5), neutrophilous species (L1, L2), metal-resistant species (S1, S2, S3), and calcicolous species (S4, S5, S6, S7, L3, L4). The names of the categories were given conditionally, based on data on the composition of the substrate and the literature data on the ecology of the species.

The volcano plot (Fig. 3) shows the fold change of the element concentrations between the intact rock and the weathered layer for all categories of the classification. 13 out of 96 values belonging to four categories differ significantly from the initial values. The maximum number of significant fold changes is shown in the category 1—changes in concentrations of 7 elements pass the p value threshold of 0.1 (calcium, titanium, iron, manganese, sulfur, rubidium, strontium). All these elements accumulated in the upper layer of predominantly siliceous sandstone under the communities dominated by acidophilous species. The content of 3 elements (manganese, zinc, zirconium) significantly increased under communities dominated by calcicolous bryophyte species (category 4). In the weathered substrate of all categories, the calcium concentration noticeably increased, but these changes were not always statistically significant. The content of strontium and cadmium decreased in the upper layers under different vegetation, which may be due to high bioavailability of these elements and tissue migration.

3.5 Soil Formation

The composition of the organic matter of the crusts was determined using the methods of analytical chemistry. Organic substances are present in all samples, but in small amounts. The content of total organic carbon (TOC) in crusts (crust extracts with sodium pyrophosphate) does not reach a 1% in all samples (0.10–0.74%). However,

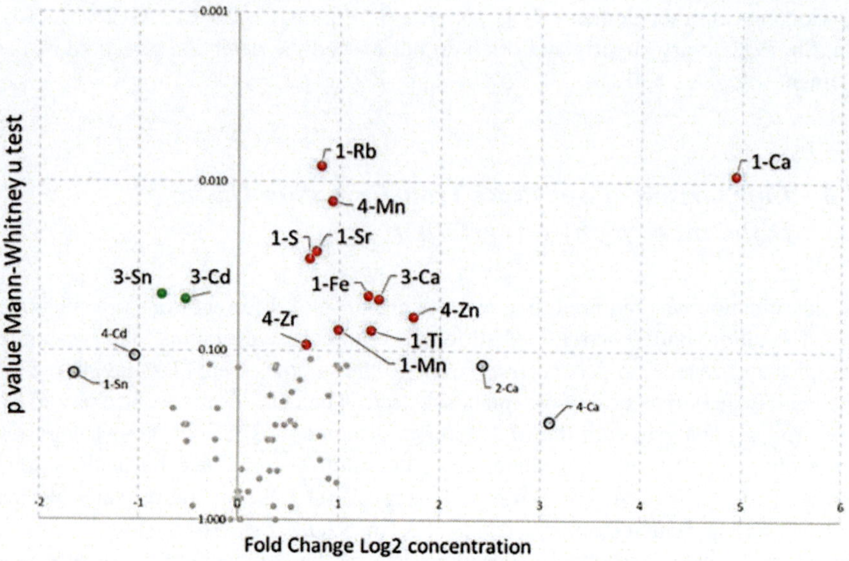

Fig. 3 Fold changes in element concentrations under plant communities dominated by acidophilous (1), neutrophilous (2), metal-resistant (3), and calcicolous (4) bryophyte species; statistically non-significant changes ($p > 0.1$) are in gray, whereas significant increase is in red, significant decrease is in green

the content of TOC in the controls (extracts of the intact sandstone samples) is much lower (0.03–0.05%).

The main part of bryophyte protosoil consists of fulvic acids. Their high content may contribute to the further development of the soil-forming process, the destruction of the minerals, and the increase in availability of nutrients in the substrate (Tan 1978; Kastelan-Macan and Petrovic 1996; Luo et al. 2015). At the same time, it indicates the excess of litter accumulation processes over the rate of decomposition and condensation (Lang et al. 2009; Hagemann and Moroni 2015). The humic acids with low optical density are also present. Low values of optical density (from 3–4 to 8–9) are typical for newly formed humic compounds.

The average content of TOC in four categories of classified samples was compared using the Mann-Whitney U test (Fig. 4). The content of TOC varies a lot between samples in each category, and category averages do not differ. The highest TOC concentrations are noted under the vegetation patches S2 (dominated by *Polytrichum juniperinum*), Y1 (*Pogonatum urnigerum*). The lowest—under the vegetation patches S1, S3, Y5 (dominated by *Pohlia* spp.), L2 (*Chiloscyphus polyanthos*), and L4 (*Myurella julacea*).

When comparing the carbon content of humic acids, differences in the average content for categories 1 and 4 is statistically significant. Thus, humic acids were formed the least intensively under the communities of calcicole bryophytes.

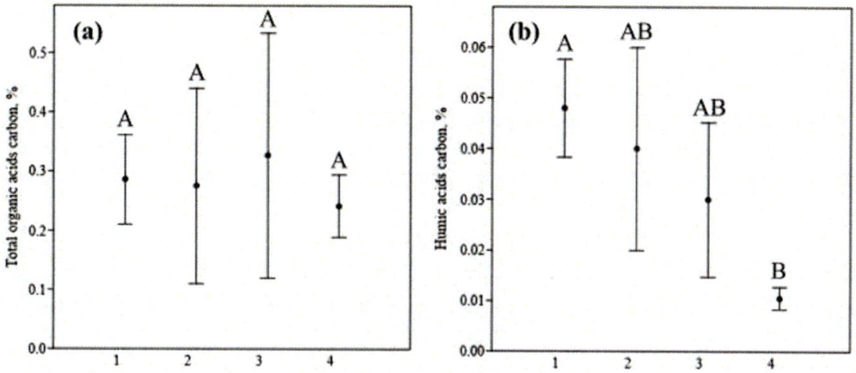

Fig. 4 Average content of organic compounds under plant communities dominated by acidophilous (1), neutrophilous (2), metal-resistant (3), and calciphilous (4) bryophyte species: **a** total organic acid carbon (mean with standard error), **b** humic acid carbon (mean with standard error); significant differences calculated with the Mann-Whitney U test are marked by the letters A, B ($p < 0.05$)

The highest concentrations of humic acid carbon are noted under the vegetation patches Y1 and S2 (dominated by species of Polytrichaceae), L1 (*Pseudoanomodon attenuatus, Mnium stellare, Conocephalum salebrosum*), Y2 (*Bryoerythrophyllum recurvirostrum, Mylia taylorii*). These bryophytes are not the typical epiliths, but occasionally grow on the rocks.

The most efficient formation of humic acids and TOC occurs under Polytrichaceae mosses. These mosses are common on soil outcrops and are regularly mentioned as plants of the first successional stages. The increase in content of TOC and soil acids under *Polytrichum* spp. was recorded previously and was compared to productivity of algae biofoulings and vascular plants (Rahmonov et al. 2010, 2021; Dümig et al. 2014). However, the formation of organic matter under different bryophyte species (bryophyte-dominated communities) is rarely compared (Bahuguna et al. 2012).

4 Conclusion

This research reveals that content of chemical elements in rocky substrate is important for development of epilithic bryophyte vegetation. The transformation of external layers of sandstone takes place with different effects. The initial content of elements in the intact rock and domination of various bryophytes on the sandstone surface appear to be the most important factors affecting the weathering. The bryophyte vegetation also influences the formation and accumulation of total organic carbon and humic acids. In our opinion, the bryophyte vegetation can be divided into four categories: the communities dominated by acidophilous species, neutrophilous species, metal-resistant species, calcicolous species. Calcicolous species have the most specific influence on the underlaying sandstone. Thus, the beginning of the biogeochemical cycle of transformations begins with the mutual contribution of abiotic and biotic components.

Acknowledgements The authors are grateful to L. E. Kurbatova, A. D. Potemkin and I. V. Czernyadjeva (Komarov Botanical Institute of RAS, St. Petersburg) for confirmation of species identification. Analytical work was carried out in Resource Centres of Saint Petersburg State University "Centre for X-ray Diffraction Studies", "Centre for Microscopy and Microanalysis", "Chemical Analysis and Materials Research Centre".

References

Alexandrowicz Z, Marszałek M, Rzepa G (2014) Distribution of secondary minerals in crusts developed on sandstone exposures. Earth Surface Processes and Landforms 39(3):320–335 https://doi.org/10.1002/esp.3449

Altieri A, Ricci S (1997) Calcium uptake in mosses and its role in stone biodeterioration. International biodeterioration & biodegradation 40(2-4):201–204 https://doi.org/10.1016/s0964-8305(97)00047-4

Bahuguna YM, Gairola S, Semwal DP, Uniyal PL, Bhatt AB (2012) Soil physico-chemical characteristics of bryophytic vegetation residing Kedarnath Wildlife Sanctuary (KWLS), Garhwal Himalaya, Uttarakhand, India. Indian journal of science and technology 5(4):2547–2553 https://doi.org/10.17485/ijst/2012/v5i4.25

Bai FY, Chen XP, Huang JZ, Lu YS, Dong HY, Wu YH., Song SL, Yu J, Bai S, Chen Z, Yang TY, Peng XY (2021) Microbial biofilms on a giant monolithic statue of Buddha: The symbiosis of microorganisms and mosses and implications for bioweathering. International Biodeterioration & Biodegradation 156:105106 https://doi.org/10.1016/j.ibiod.2020.105106

Bakalin VA (2005) The monograph of the Lophozia (Dumort) Dumort. s. str. Nauka, Moscow (in Russian)

Bates JW (1978) The influence of metal availability on the bryophyte and macrolichen vegetation of four rock types on Skye and Rhum. The Journal of Ecology 66:457–482 https://doi.org/10.2307/2259146

Bates JW (1982) The role of exchangeable calcium in saxicolous calcicole and calcifuge mosses. New Phytologist 90:239–252 https://doi.org/10.1111/j.1469-8137.1982.tb03256.x

Cao W, Xiong Y, Zhao D, Tan H, Qu J (2020) Bryophytes and the symbiotic microorganisms, the pioneers of vegetation restoration in karst rocky desertification areas in southwestern China. Applied microbiology and biotechnology 104(2):873–891 https://doi.org/10.1007/s00253-019-10235-0

Chen X, Bai F, Huang J, Lu Y, Wu Y, Yu J, Bai S (2021) The organisms on rock cultural heritages: growth and weathering. Geoheritage 13(3):1–17 https://doi.org/10.1007/s12371-021-00588-2

Cleavitt N (2001) Disentangling moss species limitations: the role of physiologically based substrate specificity for six species occurring on substrates with varying pH and percent organic matter. The Bryologist 104(1): 59–68 https://doi.org/10.1639/0007-2745(2001)104[0059:dmsltr]2.0.co;2

Cnudde V, Boone MN (2013) High-resolution X-ray computed tomography in geosciences: A review of the current technology and applications. Earth-Science Reviews 123:1–17 https://doi.org/10.1016/j.earscirev.2013.04.003

Damsholt K (2002) Illustrated flora of Nordic Liverworts and Hornworts. Nordic Bryological Society, Lund

Danin A, Ganor E (1991) Trapping of airborne dust by mosses in the Negev Desert, Israel. Earth Surface Processes and Landforms 16: 153–162 https://doi.org/10.1002/esp.3290160206

Dümig A, Veste M, Hagedorn F, Fischer T, Lange P, Spröte R, Kögel-Knabner I (2014) Organic matter from biological soil crusts induces the initial formation of sandy temperate soils. Catena 122:196–208 https://doi.org/10.1016/j.catena.2014.06.011

Ellenberg H, Leuschner C (2010) Vegetation Mitteleuropas mit den Alpen: in ökologischer, dynamischer und historischer Sicht. Eugen Ulmer KG, Stuttgart https://doi.org/10.1007/bf02853332

Galloway AF, Pedersen MJ, Merry B, Marcus SE, Blacker J, Benning LG, Field KJ, Knox JP (2018) Xyloglucan is released by plants and promotes soil particle aggregation. New Phytologist 217(3):1128–1136 https://doi.org/10.1111/nph.14897

Giordano S, Vingiani S, Adamo P (2012) Bioweathering of a basalt from Etna (Sicily) by the moss *Grimmia pulvinata* (Hedw.) Sm. EGU General Assembly Conference Abstracts: 3324

Hagemann U, Moroni MT (2015) Moss and lichen decomposition in old-growth and harvested high-boreal forests estimated using the litterbag and minicontainer methods. Soil Biology and Biochemistry 87:10–24 https://doi.org/10.1016/j.soilbio.2015.04.002

Hodgetts NG, Söderström L, Blockeel TL, Caspari S, Ignatov MS, Konstantinova NA et al. (2020). An annotated checklist of bryophytes of Europe, Macaronesia and Cyprus. Journal of Bryology 42(1):1–116. https://doi.org/10.1080/03736687.2019.1694329

Ignatov MS, Ignatova EA (2003) Moss flora of the Middle European Russia. Vol. 1. Sphagnaceae — Hedwigiaceae. KMK, Moscow (in Russian)

Ignatov MS, Ignatova EA (2004) Moss flora of the Middle European Russia. Vol. 2. Fontinalaceae — Amblystegiaceae. KMK, Moscow (in Russian)

Ignatov MS, Ignatova EA, Fedosov VE, Ivanova EI, Blom HH, Muñoz J, Bednarek-Ochyra H, Afonina OM, Kurbatova LE, Czernyadjeva IV, Cherdantseva VYa (2017) Moss flora of Russia. Volume 2. Oedipodiales — Grimmiales. KMK, Moscow (in Russian and English)

Ignatov MS, Ignatova EA, Fedosov VE, Zolotov VI, Koponen T, Czernyadjeva IV, Doroshina GYa, Tubanova DYa, Bell NE (2018) Moss flora of Russia. Volume 4. Bartramiales – Aulacomniales. KMK, Moscow (in Russian and English)

Ignatov MS, Ignatova EA, Fedosov VE, Czernyadjeva IV, Afonina OM, Maksimov AI, Kuchera Ya, Akatova TV, Doroshina GYa (2020) Moss flora of Russia. Volume 5. Hypopterygiales – Hypnales (Plagiotheciaceae – Brachytheciaceae). KMK, Moscow (in Russian and English)

Jang KM (2022) Moss on Rocks: Evaluating biodeterioration and bioprotection of bryophytic growth on stone masonry. PhD thesis, University of Oxford

Kastelan-Macan M, Petrovic M (1996) The role of fulvic acids in phosphorus sorption and release from mineral particles. Water Science and Technology 34(7–8):259–265 https://doi.org/10.2166/wst.1996.0630

Lang SI, Cornelissen JHC, Klahn T, van Logtestijn RSP, Broekman R, Schweikert W, Aerts R (2009) An experimental comparison of chemical traits and litter decomposition rates in a diverse range of subarctic bryophyte, lichen and vascular plant species. Journal of Ecology 97:886–900 https://doi.org/10.1111/j.1365-2745.2009.01538.x

Langston G, Neuman CM (2005) An experimental study on the susceptibility of crusted surfaces to wind erosion: a comparison of the strength properties of biotic and salt crusts. Geomorphology 72(1–4):40–53 https://doi.org/10.1016/j.geomorph.2005.05.003

Lisci M, Monte M, Pacini E (2003) Lichens and higher plants on stone: a review. International Biodeterioration & Biodegradation 51(1):1–17 https://doi.org/10.1016/s0964-8305(02)00071-9

Luo Xp, Feng B, Wang Pc., Zhou Hp, Chen Xm (2015) The effect of fulvic acid on the leaching of a weathered rare-earth ore. Metallurgical and Materials Transactions B 46:2405–2407 https://doi.org/10.1007/s11663-015-0425-8

McGreevy JP (1985) A preliminary scanning electron microscope study of honeycomb weathering of sandstone in a coastal environment. Earth Surface Processes and Landforms 10(5):509–518 https://doi.org/10.1002/esp.3290100509

Mitchell RL, Strullu-Derrien C, Sykes D, Pressel S, Duckett JG, Kenrick P (2021) Cryptogamic ground covers as analogues for early terrestrial biospheres: Initiation and evolution of biologically mediated proto-soils. Geobiology 19(3):292–306 https://doi.org/10.1111/gbi.12431

Panova, EG, Vlasov AD, Popova TA, Zelenskaya MS, Vlasov DYu (2015) Biological weathering of granite in an urban environment. Biosphere 7(1):61–79 (in Russian) https://doi.org/10.1007/978-3-319-24987-2_27

Paton J (1999) The Liverwort Flora of the British Isles. Harley Books, Colchester https://doi.org/10.1006/anbo.2000.1047

Peña-Monné JL, Sampietro-Vattuone MM, Báez WA, García-Giménez R, Stábile FM, Martínez Stagnaro SY, Tissera LE (2022) Sandstone weathering processes in the painted rock shelters of Cerro Colorado (Córdoba, Argentina). Geoarchaeology 37(2):332–349 https://doi.org/10.1002/gea.21890

Pérez FL (2010) Biogeomorphic relationships between slope processes and globular Grimmia mosses in Haleakala's Crater (Maui, Hawai'i). Geomorphology 116(3–4):218–235 https://doi.org/10.1016/j.geomorph.2009.11.017

Pérez FL (2020) Growth of Grimmia mosses on volcanic tephra: Geoecological processes of biocrust development in Haleakalā crater (Maui, Hawai'i). Catena 195:104911 https://doi.org/10.1016/j.catena.2020.104911

Popova TA, Vlasov DYu (2015) Specificity of primary soil on the granite embankments in Saint-Petersburg. Izvestia: Herzen University Journal of Humanities & Sciences (173):121–125 (in Russian)

Potemkin AD, Sofronova EV (2009) Liverworts and hornworts of Russia. Vol. 1. Boston-Spektr, St. Petersburg, Yakutsk (in Russian and English)

Puczko K, Zieliński P, Jusik S, Kołakowska A, Jekatierynczuk-Rudczyk E (2018) Vascular plant and bryophyte species richness in response to water quality in lowland spring niches with different anthropogenic impacts. Environmental monitoring and assessment 190(6):1–16 https://doi.org/10.1007/s10661-018-6703-6

Rahmonov O, Kowalski WJ, Bednarek R (2010) Characterization of the soil organic matter and plant tissues in an initial stage of the plant succession and soil development by means of curie-point pyrolysis coupled with GC-MS. Eurasian Soil Science 43(13):1557–1568 https://doi.org/10.1134/s1064229310130144

Rahmonov O, Skreczko S, Rahmonov M (2021) Changes in Soil Features and Phytomass during Vegetation Succession in Sandy Areas. Land 10:265 https://doi.org/10.3390/land10030265

Schljakov RN (1976) Hepaticae of the North of the USSR. 1. Antocerotae; Hepaticae: Haplomitriaceae- Metzgeriaceae. Leningrad, Nauka (in Russian)

Schljakov RN (1979) Hepaticae of the North of the USSR. 2. Herbertaceae-Geocalycaceae. Leningrad, Nauka (in Russian)

Schljakov RN (1980) Hepaticae of the North of the USSR. 3. Lophoziaceae – Mesoptychiaceae. Leningrad, Nauka (in Russian)

Schljakov RN (1981) Hepaticae of the North of the USSR. 4. Jungermanniaceae- Scapaniaceae. Leningrad, Nauka (in Russian)

Schljakov RN (1982) Liverworts of the north of USSR. 5. Lophocoleaceae-Ricciaceae. Leningrad, Nauka (in Russian)

Shaw J (1989) Heavy metal tolerance in plants: evolutionary aspects. CRC-Press, Boca Raton, Florida https://doi.org/10.1017/s0024282991000178

Szweykowski J, Buczkowska K, Odrzykoski J (2005) Conocephalum salebrosum (Marchnatiopsida, Conocephalaceae) – a new Holarctic liverwort species. Plant Systematics and Evolution 253:133–158 https://doi.org/10.1007/s00606-005-0301-0

Tan KH (1978) Effects of humic and fulvic acids on release of fixed potassium. Geoderma 21(1):67–74 https://doi.org/10.1016/0016-7061(78)90005-8

Tiano P (2002) Biodegradation of cultural heritage: decay, mechanisms and control methods Seminar article, New University of Lisbon, Department of Conservation and Restoration; 7–12 January 2002

Wang Z, Shi W, Liu D (2011) Continual erosion of bare rocks after the Wenchuan earthquake and control strategies. Journal of Asian Earth Sciences 40(4):915–925 https://doi.org/10.1016/j.jseaes.2010.07.004

Warke PA, Smith BJ (2000) Salt distribution in clay-rich weathered sandstone. Earth Surface Processes and Landforms: The Journal of the British Geomorphological Research Group 25(12):1333–1342 https://doi.org/10.1002/1096-9837(200011)25:12<1333::aid-esp141>3.0.co;2-6

Warscheid T, Braams J (2000) Biodeterioration of stone: a review. International Biodeterioration & Biodegradation 46(4):343–368 https://doi.org/10.1016/s0964-8305(00)00109-8

Xu Y, Yang R, Zhang J, Gao L, Ni X (2022) Distribution and dispersion of heavy metals in the rock–soil–moss system of the black shale areas in the southeast of Guizhou Province, China. Environmental Science and Pollution Research 29(1):854–867 https://doi.org/10.1007/s11356-021-15335-x

Yamakita E, Moriya S, Nakashima S (2021) Organic boundaries between a moss species and a limestone as analyzed by multiple micro-spectroscopic methods. Catena 204:105426 https://doi.org/10.1016/j.catena.2021.105426

Changes of Trace Element Composition of Dandelion (*Taraxacum Officinale*) in Urban Landscapes: A Case Study of Serpukhov

Natalia E. Kosheleva, Natalia Y. Kuzminskaya, and Olga V. Novikova

Abstract The content of 16 priority pollutant elements was determined in the leaves of dandelion (*Taraxacum officinale*) in Serpukhov (Moscow Oblast) in summer 2016. The greatest changes in the urban plant composition were associated with the accumulation of Bi, Pb, W, V, Cr and As, which are released from mechanical engineering, metalworking, chemical and textile enterprises and motor transport. The maximum excess of Bi in the dandelion leaves was found in the industrial zone (19.6 times that of the background level), the excess of Pb, V, Cr and As was identified in the traffic zone. Such elements as Fe, Ni, Co, Mo, Sr and Sc accumulated in the dandelion leaves in all land use zones, while the essential elements Mn, Cu were depleted. The leading factor that determines the levels of Bi, Cr and Cu in the plant leaves is the type of land use. Plant uptake of Fe, Mn, Cr, Ni, W, V, Bi largely depends on the physicochemical properties of the urban soils. Two associations, Sc–Fe–V–Cr–Ni–Co and Cu–Zn, were formed in the dandelion leaves. The Fe/Mn, Pb/Mn, and Cu/Zn ratios in the dandelion leaves revealed early disturbances in the processes of photosynthesis, respiration, and enzyme synthesis in most of the plants studied. The greatest imbalance of the elements exceeding the physiological optimum by more than 7 times was observed in the industrial and traffic zones, the smallest imbalance was detected in the recreational zone. Almost half (47%) of the examined plants were characterized by high and very high levels of biogeochemical transformation ($Zv = 30–60$). An abnormally high content of Bi, Pb, V, As, Cr, W ($Zv > 80$) was found in the industrial and traffic zones.

Keywords Trace elements · Urban landscapes · Pollution · Bioindicator

N. E. Kosheleva (✉) · O. V. Novikova
Department of Landscape Geochemistry and Soil Geography, Faculty of Geography, Lomonosov Moscow State University, Moscow, Russia
e-mail: natalk@mail.ru

N. Y. Kuzminskaya
LLC Gazprom Geotechnologies, Pervaya Magistral'naya ul. 11/2, 123290 Moscow, Russia

O. V. Frank-Kamenetskaya et al. (eds.), *Biogenic—Abiogenic Interactions in Natural and Anthropogenic Systems 2022*, Springer Proceedings in Earth and Environmental Sciences, https://doi.org/10.1007/978-3-031-40470-2_17

1 Introduction

A study of trace elements in plants is an essential part of environmental and geochemical research in urban landscapes (Kasimov 2013). The development of tolerance mechanisms in some plants allowed them not only to survive in an anthropogenic environment with a high content of bioavailable trace elements, but also to perform an important function for the urban environment associated with biotransformation and dispersion of pollutants. Changes in the chemical composition of plants, as a rule, precede visible morphological disturbances, which explains the use of individual species as indicators of the current conditions of urban landscapes (Bargagli 1998; Djingova et al. 2004; Kabata-Pendias 2011).

One of these species is *Taraxacum officinale* (dandelion), a common herbaceous plant of cities, for which the information on more than 35 elements in the background and urban landscapes of Russia, Poland, Italy, Macedonia, Hungary, Bulgaria, Romania, Czech Republic, Germany, Pakistan, Iran, the USA and Canada has been obtained. In eastern Moscow geochemical features of plants were determined and compared in the following land use areas: industrial, traffic, recreational, agricultural and residential (Vlasov and Kasimov 2015; Kasimov et al. 2016). The seasonal dynamics of the Al, Cd, Cr, Cu, Mg, Mn, Pb, Zn, and Pd accumulation in the aboveground and underground organs of dandelion in various land use zones of Sofia (Bulgaria), in the mid-western United States and in Rome (Italy) was studied (Djingova and Kuleff 1999; Keane et al. 2001; Malizia et al. 2012). It was revealed that for the purpose of bioindication it is better to use the dandelion leaves after the flowering phase (Djingova and Kuleff 1993).

The routes of supply of trace elements into plants have been investigated. The relationship between the total content of elements in the soil and their concentration in dandelion leaves, as a rule, is not revealed (Kabata-Pendias and Dudka 1991; Marr et al. 1999; Keane et al. 2001). However, the positive correlations between the content of As, Ba, Cd, Cu, Zn, Ga, K, Mn, Mo, Na, Ni, Rb, Ni, and Tl ionic forms in the soil and washed leaves indicate the predominant absorption of these elements by the roots. Al, Ce, Fe, Ti, V, Pb and Hg enter the aboveground parts of the dandelion mainly with airborne particles (Czarnowska and Milewska 2000; Winter et al. 2000; Shishlova and Khristoforova 2009; Klinskaya and Khristoforova 2011; Kleckerová and Dočekalová 2014). This is supported by a positive relationship between Pb concentrations in leaves and atmospheric particulate matter $\leq$10 μm (PM_{10}) (Keane et al. 2001).

The role of natural and anthropogenic factors influencing the trace element composition of the dandelion leaves was studied. It was found that the leading factors are physical and chemical properties of soil. The bioaccumulation of Fe, Cd, Mo also depends on their precipitation from the atmosphere. Vehicle emissions are responsible for Pb, Fe uptake. The absorption of Cu, Cd, and Mo by herbaceous plants depends on geochemical load determined by the land use zone of the city (Vlasov 2017). The distribution of trace elements in the aboveground and underground organs of dandelion were evaluated. The studies showed that Cd from polluted soil readily accumulates in the roots and only part of the elements could be transferred to the leaves. For Zn, Pb, Mn, and Fe, an intensive transfer from the roots to the ground organs was observed. Significant accumulation of Cu in roots and weak translocation to leaves was reported for Cu (Hammami et al. 2016; Angelova and Ivanov 2018; Kano et al. 2021). It was found that the soil properties (pH, Corg, Ntot, fungicides content) control the translocation of Zn, Cu, Fe, Mn, and Pb to aboveground plant organs (Królak et al. 2018; Adamczhuk-Szabela et al. 2021).

Recent interest to the trace element composition of *Taraxacum officinale*, in addition to its bioindicative abilities, is also due to the use of the plant parts (roots, leaves and flowers) in traditional medicine (A guide… 2003; Hussain and Khan 2010; Petrova et al. 2013; Jedrejek et al. 2019) and in the daily diet in Western European countries (Giacomino et al. 2016).

The aim of this work—to evaluate the changes in the amount of essential and toxic elements in the leaves of dandelion (*Taraxacum officinale*) with a case study of Serpukhov, a city that is characterized not only by a high traffic, but also by a significant number of industrial enterprises.

The particular tasks of the study include:

- the identification of the background biogeochemical features and the changes in the trace element composition of dandelion leaves in various land use areas of the city;
- the assessment of transformational change of the trace element composition of the plants in urban environment;
- the evaluation of landscape and technogenic factors in the formation of trace element composition of the dandelion leaves;
- the determination of the ecological state of *Taraxacum officinale* in relation to different geochemical loads on landscapes.

2 Study Area

Serpukhov, a city with population of 126.6 thousand people and the square of 32.1 km^2, is located within the Moskvoretsko-Okskaya moraine-erosion plain in the area of wide flat terraces of the Oka river, dissected by the valleys its tributaries. The interfluve area is represented by gently hilly landscapes with absolute heights of 170–175 m. The elevations from 117–122 to 160–165 m are occupied by alluvial terraces. River floodplain elevations reach 117–122 m. The territory belongs to the Central Prioksky soil region with light grey forest soils in the eastern part and sod-podzolic soils in the western part where loamy to clay loamy parent material dominates. Natural plant communities in the city are represented by the patches of coniferous and small-leaved forests, meadows, marshy associations and groups of aquatic plants.

The main contributions to pollution of the urban landscapes are made by industrial enterprises and transport infrastructure. A total of 147 industrial enterprises (including 33 large ones) were registered. The main components of the industrial complex are: mechanical engineering and metalworking, food and chemical industries. The total length of the road network is about 145.2 km, the average density in the city is 4 km/km^2.

3 Methods and Materials

Leaves of dandelion (*Taraxacum officinale*), the most common herbaceous plant in the urban area, were sampled in the beginning of July 2016. Plant material (68 mixed samples of leaves, each collected from 5 to 7 individual plants) was taken on a regular grid at 500 m intervals in different land use zones: industrial, traffic, recreational and residential with high- and low-rise buildings. Due to the absence of dandelion in the Prioksko-Terrasny Nature Biosphere Reserve, the background samples were collected in the vicinity of Serpukhov. Leaves were taken in the phase of vegetation after the complete maturation of tissues, and after a five-day rainless period. Prior to the laboratory analysis they were washed with distilled water to remove the elements deposited on the leaf surface but not participating in plant metabolism. The samples were then dried for 7 h at 70 °C in a drying oven. Bulk concentrations of 33 elements in the samples of the dry plant material were measured by ICP-MS and ICP-AES on Elan-6100 and Optima-4300 analyzers (Perkin Elmer, USA) at the certified laboratory of All-Russian Research Institute of Mineral Raw Materials. Seventeen priority pollutants were selected for the analysis, including elements of hazard classes I (As, Cd, Pb, Zn), II (Co, Ni, Cu, Cr), and III (V, Mn); in addition, the data on the concentrations of Fe, Mo, Sn, Sr, W, Sc, and Bi were used.

The background concentrations of the elements (*Cb*) were compared with the mean global concentrations (*Cglob*) in the annual production of terrestrial plants (Markert 1992; Dobrovol'skii 1998) via calculating the clarkes of concentration *CC* = *Cb/Cglob* and dispersion clarkes *DC* = *Cglob/Cb*. The concentrations of the trace elements in the urban samples *Ci* were averaged for each land use zone and compared with the local biogeochemical background via calculating the contamination *CF* = *Ci/Cb* and the dispersion factors *DF* = *Cb/Ci*. Using the *CF* and *DF* values, geochemical spectra were obtained for the leaves of dandelion grown in different land use zones in order to trace geochemical changes in the plants. The specialization of the background (urban) plants was determined from the formula with the values of *CC (CF)* for concentrating elements in the nominator and the values of *DC (DF)* for deconcentrating elements in the denominator.

Trace element associations in the dandelion leaves were identified using cluster analysis. Dendrogrames were constructed based on the distance functions *D* = *l* – *Pearson r*, where *r* denotes correlation coefficient between element concentrations, and Complete linkage as amalgamation rule. The obtained associations characterize a similar mode in accumulation intensity of different trace elements.

To characterize changes in the trace element composition of the urban plants in comparison with background ones, we proposed an integral parameter: the index of biogeochemical transformation calculated as follows: $Zv = \sum_{1}^{n1} CF + \sum_{1}^{n2} DF \; -$

$(n_1 + n_2 - 1)$, where n_1 and n_2 are the numbers of trace elements for which $CF > 1$ and $DF > 1$, respectively (Kosheleva et al. 2012). It is known that plants can respond to the deterioration of the environment by either accumulation or depletion of trace elements (Bargagli 1998; Dobrovol'skii 1998; Kabata-Pendias 2011). *Zv* index reflects the disturbance of the normal proportions of trace elements in plant organs typical of their phylo- and ontogenetic specialization. This parameter quantitatively expresses the imbalance of chemical elements in plants resulting from the geochemical load. The degree of anthropogenic transformation of the trace element composition of plants is characterized by five gradations: low ($Zv = 10$–20), medium (20–30), high (30–40), very high (40–60), and extremely high (60–80 and higher).

Multivariate regression analysis was used to assess the role of a number of natural and anthropogenic factors in the accumulation of trace elements in dandelion leaves. The method allows one to predict the levels of element content in plants under various combinations of factors and to assess their significance (Kosheleva et al. 2015).

The Fe/Mn, Pb/Mn, and Zn/Cu ratios made it possible to identify the ecological state of the urban plants. Fe/Mn is one of the most informative indicators of the photosynthesis process. Pb/Mn sets the ratio of technogenic to biophilic elements. Zn/Cu characterizes the degree of proportionality in the provision of enzyme synthesis processes with these biometals. The values of the indicators were compared with the average values of the physiological optimum for plants (Arzhanova and Elpat'evskii 1990; Kabata-Pendias 2011; Timofeev and Kosheleva 2016).

4 Results and Discussion

***Characterization of the local biogeochemical background*.** Molybdenum accumulated most intensively in the dandelion leaves of the urbanized background, its concentration was 9 times that in the land vegetation, which is caused by the high need of plants for this vital trace element. The content of Cd, efficiently supplied through the root and foliar pathways, was also markedly higher than the world average (Ertel et al. 1991). Sc, Cu, Zn, Ni accumulated less actively, other elements disperse. Bi, V, As, Pb in the dandelion leaves were in much lesser amounts compared to the reference (background) samples (Table 1). These elements, except for V, are involved in lipid metabolism, but they are not considered as essential, so their absorption from the soil by plants occurs passively (Bargagli 1998). The high level of dispersion of Mn, playing an important role in plant physiology, was associated with its low concentrations in the soils of the city, but might be also caused by antagonism with other elements (Kosheleva et al. 2005; Kabata-Pendias 2011; Kopylova and Yakimova 2011). For Co, W, Cr, Sr the dispersion in the leaves of plants growing in an urbanized background was also observed.

Changes in the trace element composition of dandelion leaves in urban landscapes. Formulas of geochemical specialization of the dandelion leaves in the land use zones of the city characterize the transformation of their trace element composition under

Table 1 Geochemical features of dandelion in the background areas and in the land use zones of Serpukhov

Zone	CF/DF for the different land use zones against the local background	CC/DC for the local background against the clarkes (Dobrovol'skii 1998; Markert 1992)
Industrial	$\frac{Bi_{20}Pb_{6.0}W_{3.0}V_{2.9}Cr_{2.4}Sr_{1.6}As_{1.4}Mo_{1.4}Fe_{1.3}}{Cd_{2.7}Mn_{2.5}Cu_{1.6}}$	$\frac{Cd_{9.1}Mo_{5.2}Sc_{2.2}Cu_{2.2}Zn_{1.2}Ni_{1.1}}{Bi_{33.3}V_{31.3}As_{12.0}Pb_{10.1}Co_{5.0}Mn_{3.3}W_{3.3}Cr_{1.5}}$
Traffic	$\frac{Pb_{12.8}Bi_{9.9}V_{9.2}Cr_{6.0}W_{5.5}As_{5.5}Fe_{3.2}Ni_{2.6}Sc_{2.4}Mo_{1.8}Co_{1.7}Sr_{1.4}}{Mn_{2.6}Cd_{1.9}Cu_{1.3}}$	
Residential (a) high-rise (b) low-rise	$\frac{Bi_{10}W_{8.5}Pb_{6.0}Cr_{5.2}V_{5.1}As_{3.9}Sr_{2.2}Ni_{2.1}Mo_{1.8}Co_{1.6}Sc_{1.6}Fe_{1.5}}{Mn_{2.7}Cd_{2.1}Cu_{1.2}}$ $\frac{Pb_{6.8}Bi_{6.7}V_{5.7}W_{4.2}Cr_{3.6}As_{2.6}Fe_{1.9}Sc_{1.8}Mo_{1.7}Sr_{1.6}Ni_{1.6}Co_{1.5}}{Mn_{2.2}Cd_{1.7}Cu_{1.2}}$	
Recreational	$\frac{W_{11.3}Bi_{6.7}Pb_{5.0}V_{3.4}Cr_{3.1}Mo_{2.0}Fe_{1.7}Sc_{1.7}Ni_{1.4}Co_{1.2}}{Cd_{1.9}Mn_{1.6}Cu_{1.3}}$	

the anthropogenic pressure (Tables 1 and 2). The greatest changes in plants were observed for cationic Bi (*CF* = 20) and Pb (*CF* = 12.4) accumulated in the industrial and traffic zones, respectively. It should be noted that the largest amounts of the mobile forms of Bi (15–26.6%) and Pb (17.1–19.7%) in the soils available for plant uptake were restricted to the same zones (Kuzminskaya 2016). The content of Bi and Pb in other areas of the city also significantly exceeds the background levels: by 6.7–10 times and 5.0–6.8 times, respectively.

Throughout the studied area of the city, the dandelion leaves accumulated anionic W, V, Cr, and As. The plants of the traffic zone were distinguished by the highest uptake of V (*CF* = 9.1), Cr (*CF* = 6.2), W and As (*CF* = 5.5). The residential zone ranks second in terms of the concentration of these elements in the dandelion leaves, exceeding the background level by 3.9–8.5 times. The lowest level of bioaccumulation for As, V and Cr was observed in the recreational zone (*CF* = 1.0–3.4). The content of W in the dandelion leaves showed a significant increase in the recreational zone (*CF* = 11.5), while in other zones its content was 3–8.5 times that of the background level. The accumulation of a group of elements Fe, Ni, Co, Mo, Sr, Sc was 1.2–3.2 times higher than in the urbanized background in all land use zones.

Changes in the trace element composition of the dandelion leaves was manifested not only by a significant accumulation of pollutants, but also by a noticeable decrease in the accumulation of physiologically significant, essential elements. So, in all land use areas of the city, the biological uptake of Mn (*DF* = 1.6–2.7) and Cu (*DF* = 1.2–1.6) was lower compared to the background. This pattern of accumulation is explained by the low availability of the elements in soils where the mobile fractions of Mn and Cu accounted for 3.3–5.7% and 2.1–9.3%, respectively (Kuzminskaya 2016). It should be noted that the bioaccumulation of plant-toxic Cd in all land use zones also decreased by 1.7–2.7 times compared to the urbanized background. This is caused by the intense accumulation of Cd in the roots and its weak transfer to aboveground organs from polluted soils (Angelova and Ivanov 2018).

The features of the geochemical spectra revealed characteristic pollutants and their main sources in the land use areas of the city (Fig. 1). In the *industrial zone*, the dandelion leaves most intensively accumulated Bi, whose large anomalies are confined to radio engineering (RATEP) and Neftegazspets-stroy enterprises. The highest concentrations of Pb and Cr were observed in the vicinity of the RATEP plant, oil depot and cotton wool factory. The main contribution to the supply of W and V is made by enterprises of mechanical engineering and metalworking, chemical and textile industries. The additional source of Pb and Cr in the industrial areas is the transport network. Construction and building materials industries are responsible for the release of Bi, V and W. In the *traffic zone*, the active accumulation was observed for Pb, which occurs in various types of emissions from vehicles. The source of Cr and Ni is tire abrasion, Fe and Cr accumulates in plants with blown road dust and soil particles. The accumulation of W, V, and As in the dandelion in this zone is associated with their supply from neighboring industrial enterprises of mechanical engineering and metalworking, chemical, textile, woodworking, pulp and paper industries, and building materials. In the *residential areas*, the set of priority pollutants (Bi, Pb, W, V, Cr, As) is the same as in the two previous areas. The decrease in the transport

Table 2 Concentrations of trace elements in the leaves of *Taraxacum officinale* in land use zones of Serpukhov

Parameter, mg/kg dry matter (n is the number of samples)	Hazard class								
	I				II				
	As	Cd	Pb	Zn	Co	Ni	Mo	Cu	Cr
Mean for the city (n = 68)	0.031	0.16	0.37	41	0.15	4.0	4.5	14.0	5.0
Min–Max	0.010–0.25	0.04–1.08	0.09–1.58	15–91	0.07–0.38	0.9–15.8	0.8–9.4	6.2–24.9	0.7–23.7
Mean for the industrial zone (n = 10)	0.014	0.12	0.30	32	0,10	2.5	3.5	10.9	2.9
Min–Max	0.010–0.051	0.04–0.49	0,09–0.95	17–60	0.08–0.19	0.9–5.7	0.8–8.9	6.2–15.4	0.7–7.2
Mean for the traffic zone (n = 12)	0.055	0.17	0.62	38	0.17	5.8	4.8	14.2	7.4
Min–Max	0.010–0.25	0.08–0.38	0.19–1.58	25–60	0.11–0.36	2.3–13.2	3.4–7.7	9.7–23.7	2.3–17.0
Mean for the high-rise residential zone (n = 16)	0.039	0.15	0.30	45	0.16	4.6	4.8	15.4	6.2
Min–Max	0.010–0.24	0.07–0.22	0.13–0.80	30–75	0.10–0.35	1.7–15.8	3.0–6.0	9.6–22.8	1.7–23.7
Mean for the low-rise residential zone (n = 25)	0.026	0.19	0.34	43	0.15	3.5	4.5	14.3	4.3
Min–Max	0.010–0.098	0.08–1.08	0.12–0.90	15–91	0.07–0.38	1.3–9.0	2.7–6.5	8.0–24.9	1.4–11.2
Mean for the recreational zone (n = 5)	0.010	0.17	0.25	40	0.12	3.1	5.3	13.5	3.7
Min–Max	0.010–0.010	0.12–0.22	0.16–0.32	22–56	0.10–0.19	2.1–5.0	2.4–9.4	9.9–16.4	2.0–7.1
Background (n = 5)	0.010	0.32	0.05	36	0.10	2.2	2.6	17.8	1.2

(continued)

Table 2 (continued)

Parameter, mg/kg dry matter (n is the number of samples)	Hazard class								
	I				II				
	As	Cd	Pb	Zn	Co	Ni	Mo	Cu	Cr
Clarkes (Dobrovol'skii 1998)	0.12	0.035	1.25	30	0.5	2	0.5	8.00	1.8
Clarkes (Markert 1992)	–	–	–	–	–	–	–	–	–

III				No hazard class is assigned		
Mn	V	Sr	W	Sc	Fe	Bi
27	0.20	42	0.35	0.078	155	0.003
13–79	0.035–1.11	19–137	0.07–3.76	0.015–0.22	52–499	0.0003–0.026
25	0.10	41	0.18	0.043	96	0.006
15–40	0.04–0.30	28–71	0.07–0.41	0.015–0.075	52–186	0.0003–0.026
25	0.32	36	0.33	0.108	244	0.003
19–38	0.035–1.11	23–59	0.11–0.68	0.061–0.22	91–499	0.0003–0.011
23	0.18	55	0.51	0.073	112	0.003
14–33	0.035–0.58	27–137	0.08–3.76	0.015–0.13	88–360	0.0003–0.008
28	0.20	41	0.25	0.080	145	0.002
13–74	0.035–0.75	19–83	0.07–0.80	0.015–0.160	79–332	0.0003 0.007
40	0.12	29	0.68	0.077	132	0.002
16–79	0.035–0.37	22–46	0.07–1.63	0.061–0.130	91–222	0.0003–0.004
62	0.035	25	0.06	0.045	76	0.0003
205	1.5	35	–	–	–	–
–	–	–	0.2	0.02	–	0.01

load caused the decrease in the accumulation of Pb (by 1.8–2 times) and Cr (by 1.2–1.7 times) compared to the traffic zone. The source of other elements is the nearby industrial enterprises. The content of Bi compared to the industrial zone was 2 times lower in the high-rise and 3 times lower in the low-rise zone. The opposite situation was observed for V, W, and As accumulated in the dandelion leaves 1.4–2.8 times more intensively than in the industrial zone. This can be explained by the active transfer of pollutants to residential areas. In the *recreational area*, W, Bi, Pb, V and Cr still remain priority pollutants. Accumulation of Bi, Pb, V and Cr was nearly equal or 1.2–3.0 times lower than in other zones. In the case of W, an intense accumulation in the plants was observed due to the proximity of engineering and metalworking enterprise to one of the urban parks.

Trace elements in the dandelion leaves form two associations which were identified through cluster analysis (Fig. 2). The Sc–Fe–V–Cr–Ni–Co multi-element association is characterized by strong correlations ($r = 0.52$–0.92) and is close in composition to the Fe–Co–V–As–Ni soil association, which indicates the predominantly root absorption of these elements (Kuzminskaya 2016). The accumulation of these trace elements in various land use zones decreased in the following order: traffic ($CF = 1.7$–9.12) > residential (1.5–5.7) > recreation (1.2–3.4) > industrial (1.0–2.9). The Cu–Zn pair ($r = 0.70$) was formed by biogenic elements involved in many enzymatic reactions in plants. There has been no definitive explanation of the interaction between these elements yet. Some authors consider them to be antagonists due to mutual competition for carriers when absorbed by the root system. As a result, one element inhibits the entry of another (Kabata-Pendias 2011). Others point out

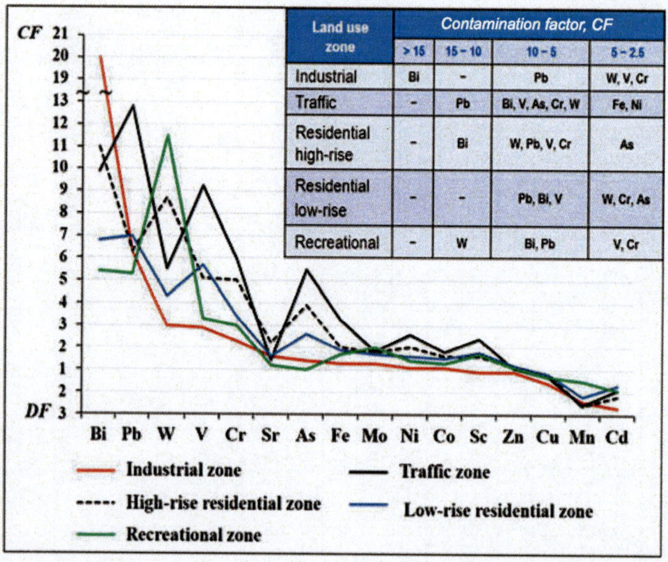

Fig. 1 Geochemical spectra of trace elements in the dandelion leaves in different land use areas of Serpukhov

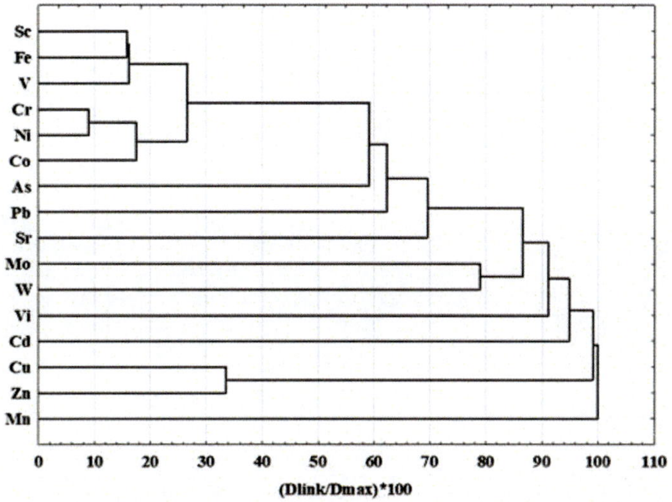

Fig. 2 Dendrogram of cluster analysis and associations of trace elements in the leaves of *Taraxacum officinale* in Serpukhov

a strong correlation between Cu and Zn, due to the predominance of foliar uptake over root uptake under high geochemical load (Timofeev and Kosheleva 2016). In Serpukhov the content of Zn in the urban plants was close to the background level. For Cu, the depletion was observed ($DF = 1.2$–1.6) with a maximal manifestation in the industrial zone.

The remaining elements were not incorporated in associations, although they were intensively accumulated in the dandelion leaves. The elements from this group enter the environment from different spatially isolated technogenic sources and accumulate in the plant leaves in different places. Therefore, the correlations between these trace element concentrations in dandelion leaves were not observed.

Anthropogenic transformation of the trace element composition of leaves. The consequences of anthropogenic impact on plants are conveyed in the index of biogeochemical transformation *Zv* (Fig. 3). According to the average values of the *Zv* index the land use zones form a series: traffic ($Zv = 46$) > industrial (39) > residential high-rise (32) > recreation (31) > residential low-rise (22).

The analysis of the biogeochemical transformation of the dandelions under various anthropogenic load (Fig. 4) showed that the plants with abnormally high ($Zv > 60$) and very high ($Zv = 40$–60) values of the index were found in all land use zones, except for recreational one. They accounted for 9% and 21% of all studied samples, respectively. At the same time, the trace element composition of plants in the industrial ($Zv_{max} = 99$) and traffic ($Zv_{max} = 88$) zones was exposed to the greatest impact. A high level ($Zv = 30$–40) of biogeochemical transformation of dandelions was found in all land use areas, it was observed in 26% of the studied plants. The trace element composition of plants in residential and recreational areas ($Zv < 30$) was least changed. The share of

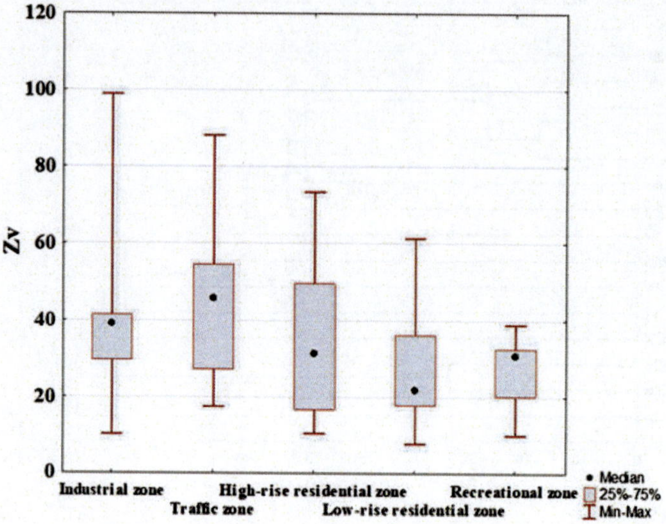

Fig. 3 Biogeochemical transformation of the trace element composition of dandelion leaves in the land use areas of Serpukhov

such plants in these zones accounted for 40–60%, while in the traffic and industrial zones it was only 25–30%.

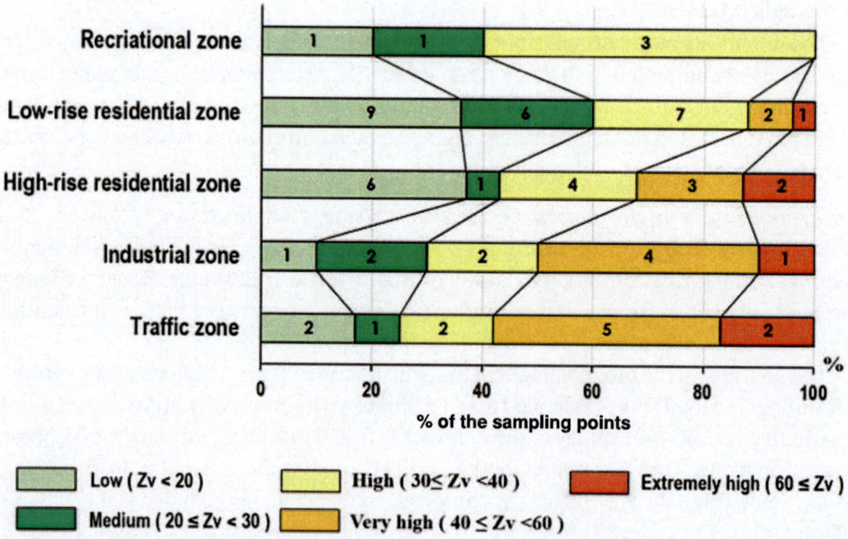

Fig. 4 Levels of biogeochemical transformation of the trace element composition of dandelion leaves in the land use zones of Serpukhov

Within Serpukhov, the trace elements formed several anomalies in the leaves of dandelion (Fig. 5). The center of the extensive *northern anomaly (Zv* = 51–99) is confined to the "Northern Industrial Zone", where machine-building, food, light and construction industries are located. The anomaly affected nearby high-rise quarters: the playground plants with *Zv* = 73 revealed the appearance of spotting on the leaves and morphological changes. An additional source of pollutants in yards is vehicle emissions from neighboring parking lots of large shopping malls. *The western anomaly (Zv* = 83) was formed on Konshinykh Street, which connects Serpukhov to Pushchino, due to high traffic congestion on this road. Emissions from the Serpukhov Cardboard & Paper Manufactory and the Khimvolokno plant might serve as additional sources of the trace elements. *The central anomaly (Zv* = 47–81) was formed on both sides of the Nara River in the vicinity of the industrial enterprises, it also occupies the residential areas adjacent to the industrial zone. The main contribution to the anomaly formation is made by tool, electromechanical, instrument-making, machine-tool, tanneries, as well as clothing and furniture factories. High values of the *Zv* index (49–56) were observed near Konshinykh and Proletarskaya streets, which separate the industrial area from residential areas, and near Chernyshevsky street crossing these two streets and characterized by high-traffic intensity. *The southern anomaly (Zv* = 53–73) was formed due to emissions from the PRANCE metal-working plant, the Serpukhov condenser plant, and the Nara machine-building plant. The anomaly extends to the nearby residential areas on both sides of the Nara River. *The eastern anomaly (Zv* = 42–57) was formed under the influence of the plant "RATEP" producing radio equipment and radio components. It covers the residential areas adjacent to it from the south. The comparison of the biogeochemical data with the soil pollution data (Kuzminskaya 2016) showed a weak relationship between the content of trace elements in the soils and plants, which is due to additional input of the elements from the atmosphere.

Factors of accumulation and dispersion of trace elements in the leaves of dandelion. Tree-based regression models were built depending on the following factors:

- natural: geochemical position (Geochemical landscapes); elevation over sea level (Height); physical and chemical properties of the soil: acid-base conditions (pH), electrical conductivity of water extract (EC), content of humus (Humus), content of physical clay (Clay), total index of soil pollution (*Zc* of soils), bulk content of trace element in the topsoils (0–15 cm) (TE in soil), concentration of mobile forms of TE (in extract of ammonium acetate buffer with EDTA) (TE_{mob} in soil); concentrations of synergistic and antagonist elements in dandelion leaves (TEs in leaves);
- anthropogenic–land use zone (Zone): industrial (1), traffic (2), residential high-rise (3), residential low-rise (4), recreation (5), background area (6).

The results of multivariate regression analysis showed that among the natural factors, the soil physicochemical properties have the most significant effect on the accumulation of Pb, Fe, Mn, Cd, Ni, and W in the dandelion leaves (Table 3). The increase in Pb concentration in urban plants is primarily associated with an increase

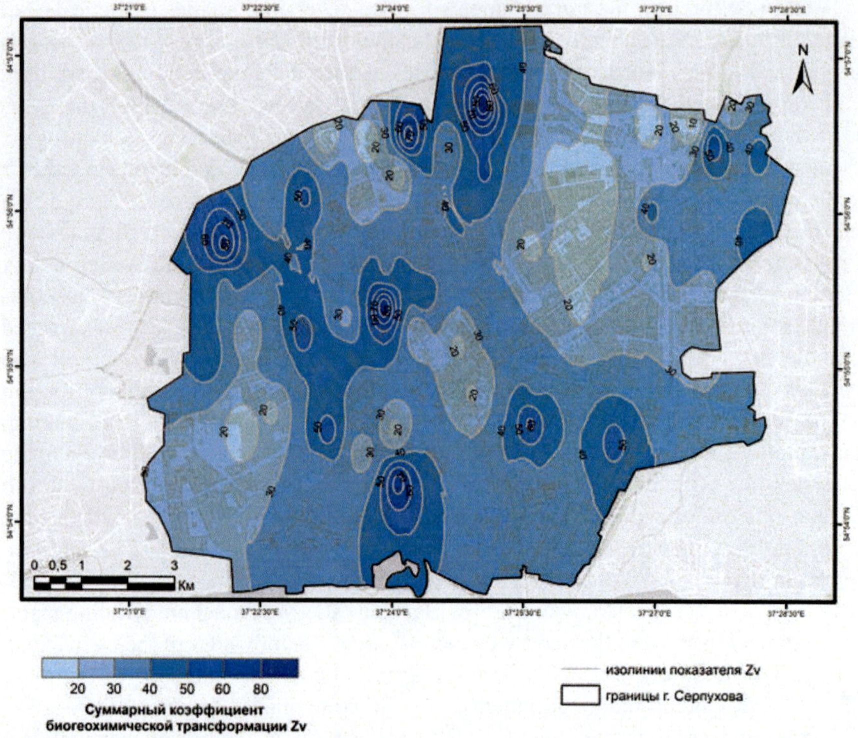

Fig. 5 Distribution of index of biogeochemical transformation *Zv* in the leaves of dandelion across the city

in the proportion of mobile forms of the element in the soil and to a lesser extent it is controlled by EC and the clay content (Fig. 6a). The entry of Fe into aboveground organs of dandelion depends more on its total content in soil and only then on pH. However, acid-base conditions govern Ni and Mn bioaccumulation (Fig. 6b). With an increase in pH, the content of Mn in the dandelion leaves decreases, while the concentration of Ni increases. This is explained by the Ni accumulation in soils at the alkaline geochemical barrier, from where it is subsequently mobilized by plant roots. In turn, cationic Mn in alkaline and neutral media becomes less mobile and thus less available for plant uptake. An increase in the humus content slows down the bioabsorption of Mn, which indicates a strong binding of this essential element by soil organic matter.

The content of physical clay and EC play a leading role in the absorption of Cd and W by dandelion, respectively. Heavier texture of soils associated with increased emission of anthropogenic dust slows down the entry of Cd into the aboveground organs of dandelion, and the increase in EC due to the use of deicing mixtures reduces the mobility of W under the alkaline soil reaction and, as a result, its bioavailability. It should be noted that the interaction between the elements in the plant turned

Table 3 Factors of accumulation and dispersion of trace elements in the leaves of *Taraxacum officinale*

Factors		Concentration in leaves										
		Cu	Pb	Zn	Fe	Mn	Bi	Cr	Cd	Ni	V	W
Land use zone		2					1, 3	1	3		3	
Geochemical landscapes								4				
Altitude			2–	4+	4–			2– 4–			1– 4–	
Soil properties	pH				2+ 3+	1–	5+	2+		1+ 2–	5+	
	EC		3+	4–		4–	4–	3+	4+			1– 4+ 5–
	Clay	3+	4+ 5+	3–	3–				1+		2+	3–
	Humus					2–	2+ 4–				2+	
	Zc						3+	5–	2+	2+		2–
	C_{mob}		1+ 3+	2+				3–				
	C_{total}				1+ 2+	5+	4–					
Concentration in soil	Fe					3+						
	Mn				4+							
	Zn	1+ 2+										
	Cu			1+ 2+								

Note. Ranks 1-5 show a decrease in the significance of the factor; "+" – denotes the growth of the factor corresponds to anincrease in the concentration of the element in plant (highlighted in green), "-"– means its decrease (highlighted in gray); for qualitative indicators, the kind of the relationship is not determined. C_{total} – is the bulk content of the element in soil; C_{mob} – is content of its mobile forms; *Zc* is the total index of soil pollution.

out to be the most significant factor only for the Cu–Zn pair. An increase in the concentration of one of these elements leads to an increase in the concentration of the other. The second most important factor in the accumulation of Zn in the plants is the content of mobile fraction of Zn in soils. Despite the fact that in much of the urban area (62%) the total soil pollution index *Zc* is very high (Kuzminskaya 2016), its effect on bioabsorption is not a key factor for any of the studied elements. A positive significant relationship between *Zc* and the concentration of pollutants in the dandelion leaves can be traced only for Cd, Ni, Bi, and V.

Unlike the aforementioned elements, whose accumulation in the dandelion leaves is primarily controlled by natural factors, the concentrations of Bi and Cr are largely determined by the level of geochemical load, which differs in different land use zones of the city (Fig. 6c). Humus content in soil is the second factor controlling bioaccumulation of Bi. In contrast to biogenic Mn, for technogenic Bi, the relationship with this factor is direct. The elements Cu, Cd, and V discovered a weaker dependence of their accumulation intensity in the plant leaves with the type of land use in a particular area.

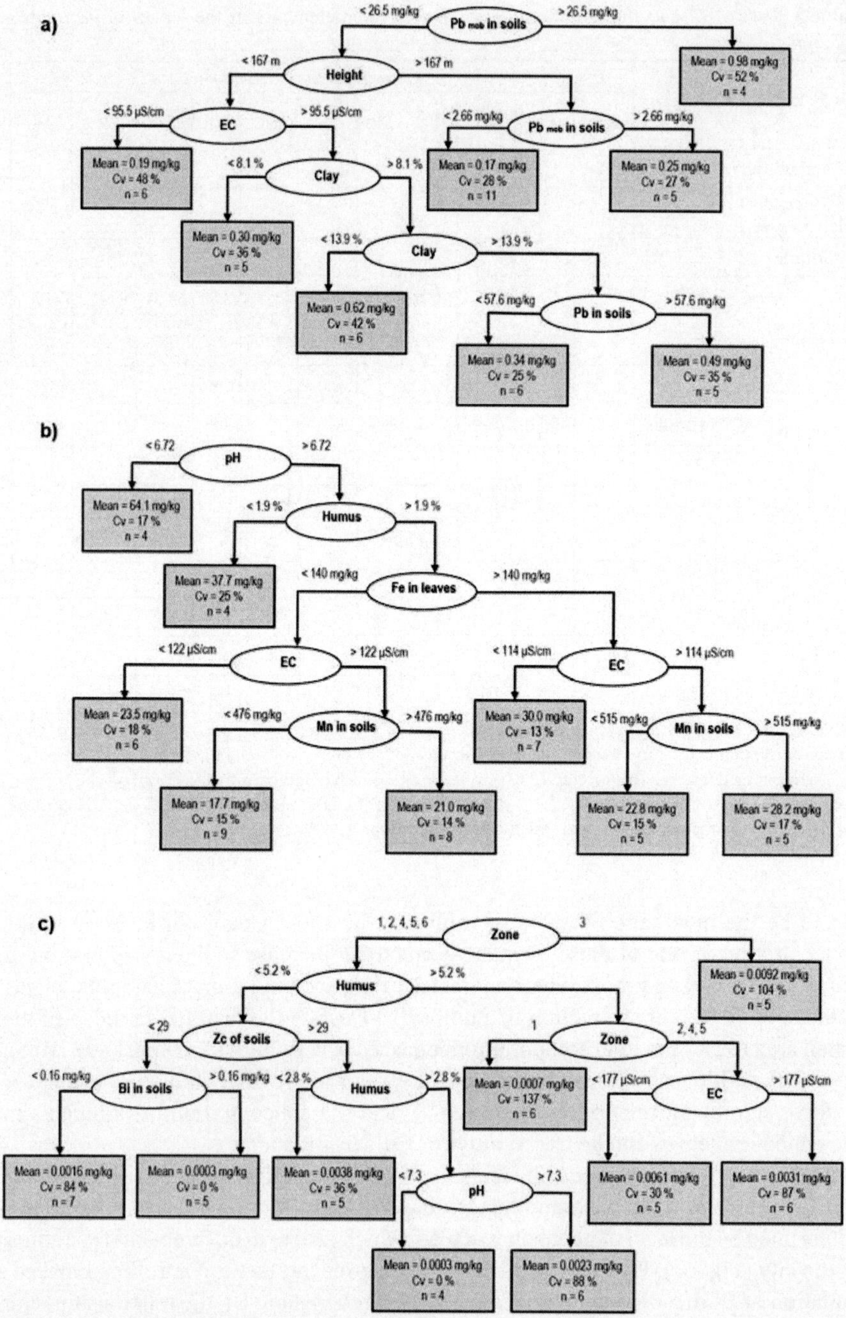

Fig. 6 Distribution of Pb (**a**), Mn (**b**) и Bi (**c**) in the leaves of dandelion under various combinations of anthropogenic and landscape factors in Serpukhov

The heterogeneity of *Zv* index values is determined by functional use of the territory (Fig. 7). In the industrial, recreational and background zones of Serpukhov the *Zv* distribution depends on electrical conductivity on water extract which characterize mineralization of soil solution caused by de-icing agents. In the residential and traffic zones the value of the biogeochemical transformation index *Zv* is related to the altitude. Watershed areas with absolute heights of 170–175 m above s.l. are very gently sloping surfaces slightly inclined towards the Oka and Nara valleys. Terraces above the floodplain are located at elevations up to 160–165 m; these are the areas of removal and partial accumulation of products of liquid and solid runoff and bioaccumulation of trace elements in dandelions. In this geochemical landscape, the total accumulation of trace elements is affected by mobile Cu in soils, with an increase in its content in the soil, the absorption of trace elements by *Taraxacum officinale* increases.

Assessment of the ecological status of plants under technogenic impact. Changes in the trace element composition of plants affect their ecological state. The values of the Fe/Mn, Pb/Mn, and Cu/Zn ratios make it possible to diagnose early disturbances in such important physiological processes as photosynthesis, respiration, and enzyme synthesis (Novikova and Kosheleva 2007; Kasimov et al. 2016; Kosheleva et al. 2016). For these ratios, the following optimal values for plant life were taken: 1.5–2.5 for Fe/Mn, 0.006 for Pb/Mn, and 0.25–0.30 for Cu/Zn (Arzhanova and Elpat'evskii 1990; Kabata-Pendias 2011; Timofeev and Kosheleva 2016). For the dandelions growing within urban background areas of Serpukhov, these ratios were 1.2, 0.001, and 0.5, respectively, which indicates a fairly balanced uptake of the elements.

The values of the Fe/Mn ratio in the plants varied greatly both throughout the city and within individual land use areas (Table 4). The high spatial variability of the ratio is explained by the fact that in areas with anthropogenic pressure it could be affected not only by root absorption, but also by intensive addition of the elements from the atmosphere. Complex synergistic and antagonistic relationships between numerous pollutants in the growth medium might produce various multi-element associations in plants, causing an imbalance between the elements (Kosheleva and

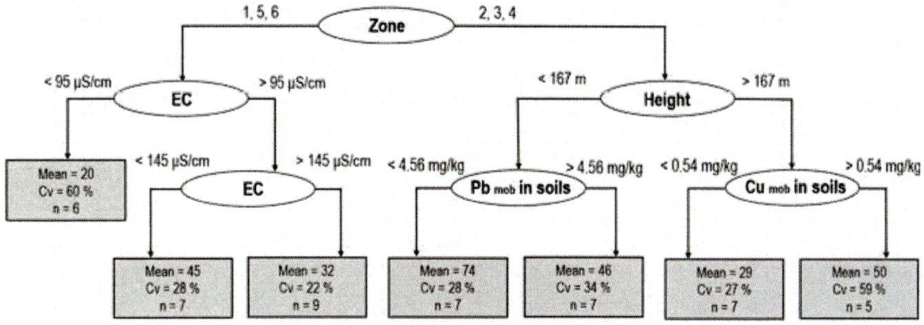

Fig. 7 Distribution of the index of biogeochemical transformation *Zv* for the leaves of *Taraxacum officinale* under various combinations of anthropogenic and landscape factors in Serpukhov

Novikova 2007). The highest values of the index were found in the dandelions of the traffic (16.6) and residential low-rise zones (17.8), where they were 7 times that of the upper limit of the physiological optimum. Such imbalance of the elements is caused by the excessive accumulation of Fe and a sharp deficiency of Mn in plants, which often leads to leaf chlorosis (Lyanguzova 2010). Only in a few plants of the recreational, residential low-rise type and industrial zones, the process of chlorophyll formation proceeds without disturbances and the activity of photosynthesis was not reduced, as evidenced by the values of the indicator not higher than 1.4.

For most plants in the city, the imbalance in the content of Pb and Mn was found. The intensive accumulation of technogenic Pb in the dandelion leaves due to an increase (up to 38%) of its available for plant uptake forms in the soils (Kuzminskaya 2016) inhibits the absorption of biophilic Mn. The average Pb/Mn values for all land use zones of Serpukhov were 1.5–4.3 times higher than the optimum (Table 4). The maximum values of the index (0.042–0.046) in the traffic, industrial, and residential low-rise zones exceeded the balanced ratio of elements by 7–7.7 times, which indicates a deterioration in intracellular respiration and the process of photosynthesis in plants. The Pb/Mn values in the recreation area (0.003–0.018) were closest to the physiological optimum. Within the land use zones both Pb/Mn and Fe/Mn ratios were characterized by high spatial variability.

For most plants in the city, there is a balance in the accumulation of Cu and Zn, which is in agreement with insignificant changes in the accumulation of these elements compared to the background (Table 4). The average values of the index (0.35–0.38) in the dandelion leaves in all land use zones are close to the upper limit of the optimum (Table 4). However, the maximum and minimum values of Cu/Zn for individual plants indicate a violation of the proportional supply of these elements to enzymatic synthesis. Thus, the Cu/Zn maximum in the dandelion leaves in the traffic (0.58) and low-rise residential (0.60) zones exceeds the physiological balance of the elements by a factor of 2 and in other urban areas 1.6 times. The Cu/Zn minimum (0.17–0.19) is 1.3–1.5 times less than the lower limit of the optimum in the recreational and low-rise residential areas.

Table 4 Indicators of the ecological state of *Taraxacum officinale* in land use zones of Serpukhov

Zone	Fe/Mn		Pb/Mn		Cu/Zn	
	Mean	Min–Max	Mean	Min–Max	Mean	Min–Max
Industrial	4.0	1.8–7.1	0.014	0.004–0.043	0.35	0.23–0.48
Traffic	10.0	3.2–16.6	0.026	0.007–0.042	0.38	0.29–0.58
High-rise residential	6.4	4.0–13.2	0.013	0.005–0.028	0.35	0.27–0.48
Low-rise residential	5.9	1.4–17.8	0.015	0.002–0.046	0.36	0.19–0.60
Recreational	4.7	1.2–8.9	0.009	0.003–0.018	0.38	0.17–0.50

5 Conclusions

The greatest changes in the trace element composition of the leaves of the urban plants are observed in the concentrations of two cationic (Bi, Pb) и four anionic (W, V, Cr, As) elements, which are present in the emissions of most enterprises and vehicles. The maximum excess over the background levels was found for Bi (by 19.6 times) in the industrial zone, for Pb, V, Cr and As (by 5.5–12.8 times) in the traffic zone. In all land use zones, the ability of dandelion to indicate the contamination of urban landscapes was also manifested in relation to Fe, Ni, Co, Mo, Sr and Sc. Along with the active bioaccumulation of toxic elements, a decrease of 1.2–2.6 times compared with the background concentrations of essential Mn and Cu was found.

The Sc–Fe–V–Cr–Ni–Co association formed in the dandelion leaves is similar in composition to the Fe–Co–V–As–Ni association in soil, which indicates the predominant absorption of these elements by the roots. The Cu–Zn association is formed by biophilic elements with similar transfer mechanisms in the plant.

Almost half (47%) of the examined plants are characterized by a high and very high level of the biogeochemical transformation index ($Zv = 30$–60). The share of the urban plants with extremely high Zv values ($Zv > 60$) accounts for 9%, mainly in industrial and traffic zones. They form some contrasting multi-element anomalies in different parts of the city. The most balanced relationships between the elements ($Zv < 20$) are observed in plants from the residential areas.

The leading factor determining the accumulation of Bi, Cr and Cu in the plant leaves is the attachment to one or another land use zone of the city. The plant uptake of Fe, Mn, Cr, Ni, W, V, Bi to a great extent depends on the physicochemical properties of the urban soils.

The Fe/Mn, Pb/Mn, and Cu/Zn ratios in the dandelion leaves revealed early disturbances in the processes of photosynthesis, respiration, and enzyme synthesis in most of the plants studied. The greatest imbalance of elements exceeding the physiological optimum by more than 7 times was observed in the industrial and traffic zones, the smallest one was registered in the recreational zone.

Acknowledgements The authors are grateful to N.V. Kuzmenkova and I.S. Kapustina for participation in the field work and to N.B. Zhaksylykov for the help in compiling geochemical map. The study was made within the framework of the Development Program of the Interdisciplinary Scientific and Educational School in the Lomonosov Moscow State University "The Future of the Planet and Global Environmental Changes."

References

A guide to traditional herbal medicines: a sourcebook of accepted traditional uses of medicinal plants within Europe (2003). BHMA, Bristol

Adamczhuk-Szabela D, Lisowska K, Wolf WM (2021) Hysteresis of heavy metals uptake induced in Taraxacum officinale by thiuram. Sci Reports 11:20151

Angelova V, Ivanov K (2018) Heavy metal content in dandelion (Taraxacum officinale Web.). Agric Sci 10(23):55–61

Arzhanova VS, Elpat'evskii PV (1990) Geochemistry of landscapes and technogenesis. Nauka, Moscow (in Russian)

Bargagli R (1998) Trace elements in terrestrial plants. Springer, Berlin

Czarnowska K, Milewska A (2000). The content of heavy metals in an indicator plant (Taraxacum officinale) in Warsaw. Pol J of Environ Stud 9(2):125–128

Djingova R, Kuleff I (1993) Monitoring of heavy metal pollution by Taraxacum officinale

Djingova R, Kuleff I (1999) Seasonal variation in the metal concentration of Taraxacum officinale, Plantago major and Plantago lanceolata. Chem and Ecol 16:239–253

Djingova R, Kuleff I, Markert B (2004) Chemical fingerprinting of plants. Ecol Res 19:3–11

Dobrovol'skii VV (1998) Fundamentals of biogeochemistry. Vysshaya Shkola, Moscow (in Russian)

Ertel J, Prohl G, Paratzke HG (1991) The contamination of plants by cadmium due to root uptake and aerial deposition. Water Air Soil Pollut 57–58:861–874

Giacomino A, Malandrino M, Colombo ML, Miaglia S, Maimone P, Blancato S, Conca E, Abollino O (2016) Metal content in dandelion (Taraxacum officinale) leaves: influence of vehicular traffic and safety upon consumption as food. Hindawi Pub Corp J Chem ID 9842987:9

Hammami H, Parsa M, Mohassel MHR, Rahimi S Mijani S (2016) Weeds ability to phytoremediate cadmium – contaminated soil. Int J Phytorem 18(1):48–53

Hussain I, Khan L (2010) Comparative study on heavy metal contents in Taraxacum officinale. Inter J Pharm Phyto Res 2(1):15–18

Jedrejek D, Lis B, Rolnik A, Stochmal A, Olas B (2019) Comparative phytochemical, cytotoxicity, antioxidant and haemostatic studies of Taraxacum officinale root preparations. Food Chem Toxicol 126:233–247

Kabata-Pendias A (2011) Trace Elements in Soils and Plants, 4th edn. CRC Press, Boca Raton

Kabata-Pendias A, Dudka S (1991) Trace metal contents of Taraxacum officinale (dandelion) as a convenient environmental indicator. Environ Geochem Hlth 13:108–113

Kano N, Hori T, Zhang H, Miyamoto N, Anak DEV, Mishima K (2021) Study on the behavior and removal of cadmium and zinc using Taraxacum officinale and Gazania under the application of biodegradable chelating agents. Appl Sci 11:1557

Kasimov NS (2013) Landscape Ecogeochemistry. Publisher Filimonov M.V., Moscow

Kasimov NS, Vlasov DV, Kosheleva NE, Nikiforova NE (2016) Geochemistry of landscapes of Eastern Moscow. APR, Moscow (in Russian)

Keane B, Collier MH, Shann JR, Rogstad SH (2001) Metal content of dandelion (Taraxacum officinale) leaves in relation to soil contamination and airborne particulate matter. Sci Total Environ 281:63–78

Kleckerová A, Dočekalová H (2014) Dandelion plants as a biomonitor of urban area contamination by heavy metals. Int J Environ Res

Klinskaya EO, Khristoforova NK (2011) Biogeochemical estimation of the condition of the road reserve in the Far East of the Russian Federation by the heavy metal content in dandelion. Uch Zap Komsomolsk-na-Amure Gos Tech Univ 1(5):89–92 (in Russian)

Kopylova LV, Yakimova EP (2011) Specificity of the accumulation of metals by arboreal plants in the urban environment. Uch Zap Zabaikal'skogo Gos Univ Ser Estestven nauki 1(36):183–185 (in Russian)

Kosheleva NE, Kasimov NS, Sorokina OI, Vlasov DV (2012) Assessment of pollution of urban landscapes using integral indicators. In Geochemistry of landscapes and geography of soils (to the 100th anniversary of M.A. Glazovskaya). Reports of the All-Russian scientific conference. Faculty of Geography, Moscow State University, Moscow 174–176 (in Russian)

Kosheleva NE, Makarova MG, Novikova OV (2005) Heavy metals in the leaves of tree species in urban landscapes. Vestn Mosk Univ Ser 5: Geogr 3:74–81 (in Russian)

Kosheleva NE, Kasimov NS, Vlasov DV (2015) Factors of the accumulation of heavy metals and metalloids at geochemical barriers in urban soils. Eurasian Soil Sci 48(5):476–492

Kosheleva NE, Timofeev IV, Kasimov NS, Kisselyova TM, Alekseenko AV, Sorokina OI (2016) Trace Element Composition of Poplar in Mongolian Cities. In: Frank-Kamenetskaya OV, Panova EG, Vlasov DYu (eds) Biogenic–Abiogenic Interactions in Natural and Anthropogenic Systems. Springer International Publishing AG, Switzerland

Królak E, Marciniuk J, Popijantus K, Wasilczuk P, Kasprzykowski Z (2018) Environmental factors determining the accumulation of metals: Cu, Zn, Mn and Fe in tissues of Taraxacum sp. sect. Taraxacum. Bulletin Environ Contamin Toxicol 101:68–74

Kuzminskaya NYu (2016) Heavy metals and metalloids in urban landscapes of the city of Serpukhov. Master's Dissertation, Lomonosov Moscow State University (in Russian)

Lyanguzova IV (2010) Tolerance of the Components of Forest Ecosystems in the Russian North towards Aerotechnogenic Pollution. Dissertation, Komarov Botanical Institute of the Russian Academy of Sciences (in Russian)

Malizia D, Giuliano A, Ortaggi G, Masotti A (2012). Common plants as alternative analytical tools to monitor heavy metals in soil. Chem Centr J 6 (2):S6 10

Markert B (1992) Presence and significance of naturally occurring chemical elements of the periodic system in the plant organism and consequences for future investigations on inorganic environmental chemistry in ecosystems. Vegetation, 103:1–30

Marr K, Fyles H, Hendershot W (1999) Trace metals in Montreal urban soils and the leaves of Taraxacum officinale. Can J Soil Sci 79:385–387

Novikova OV, Kosheleva NE (2007) Ecological and geochemical assessment of the state of woody vegetation in Quito (Ecuador). Vestn Mosk Univ Ser Geogr 6:43–48 (in Russian)

Petrova S, Yurukova L, Velcheva I (2013) Taraxacum officinale as a biomonitor of metals and toxic elements (Plovdiv, Bulgaria). Bul J Agric Sci 19(2):241–247

Shishlova NA, Khristoforova NK (2009) Assessment of the air pollution in Ussurijsk by heavy metal contents in Taraxacum officinale. Vestn SVNC DVO RAN 4:77–80 (in Russian)

Timofeev IV, Kosheleva NE (2016) Environmental and geochemical assessment of woody plants in the nonferrous mining landscape (Zakamensk, Buryat Republic). Problems of Botany of South Siberia and Mongolia. Proc of the 15h Int Sci and Pract Conf (Barnaul, May 23–26 2016). Barnaul: 463–472 (in Russian)

Vlasov DV (2017) Factors of accumulation and depletion of heavy metals and metalloids in dandelion leaves in Moscow. RUDN J Ecol Life Saf 25(2):230–241

Vlasov DV, Kasimov NS (2015) Heavy metals in maple and dandelion leaves from different land use areas in Moscow Eastern District. Geogr Environ Sustain 8(1):53–70

Winter S, Wappelhorst O, Markert B (2000) Löwenzahn *Taraxacum officinale* Web. als (städtischer) Bioindikator. Z Umweltchem Ökotox 12(6):311–321 (in German)

Seasonal and Diurnal Changes of Organic Molecules in Plants

Vladimir P. Chelibanov, Alexander V. Golovin, and Irina V. Shtangeeva

Abstract Circadian rhythms are able to affect numerous physiological processes in plants. Now there is an understanding of circadian biology of model plant species *Arabidopsis thaliana*. The studies of the circadian clock function in other plants, especially perennial weeds, are still limited. In the research, Raman spectroscopy was used to estimate the differences between organic composition of different plant species and influence of climatic conditions on the plant development during vegetation season as well as over a day time. For the field experiment, two widespread plant species, couch grass and dandelion, were selected. It was found that concentrations of carotenoids, phenylpropanoids, aliphatic compounds, cellulose, and carbohydrates in the plant leaves varied both during day and during growing season. Although the plants grew under the same conditions in a small site and were collected at the same time, the variations were often different on different sampling dates, and there were also differences between dandelion and couch grass. No correlation was found between temperature and humidity at the site and temporal changes in the concentration of organic compounds in couch grass leaves collected at different times and in leaves of dandelion, with the exception of the dandelion collected in early summer and mid-autumn. It can be suggested that the main difference in the distribution of the biomolecules in couch grass and dandelion leaves might be because these plant species belong to two different groups that differ in many biochemical and physiological functions.

Keywords Biological clocks · Couch grass · Dandelion · Plant leaves · Carotenoids · Aliphatic compounds · Cellulose · Phenylpropanoids · Carbohydrates · Raman spectroscopy

---

V. P. Chelibanov (✉)
ITMO University, St. Petersburg, Russia
e-mail: chelibanov@gmail.com

A. V. Golovin · I. V. Shtangeeva
St. Petersburg University, St. Petersburg, Russia

O. V. Frank-Kamenetskaya et al. (eds.), *Biogenic—Abiogenic Interactions in Natural and Anthropogenic Systems 2022*, Springer Proceedings in Earth and Environmental Sciences, https://doi.org/10.1007/978-3-031-40470-2_18

1 Introduction

Daily rotation of the Earth is one of the most important drivers of various processes on our planet. Circadian rhythms are widespread in nature and are able to affect numerous physiological processes in all living organisms. The rhythmical variations are among main factors affecting plant development and nutrition (Petersen et al. 2022). The circadian clock is a regulatory mechanism that synchronized with the external diurnal light/dark cycles (Pan et al. 2015). Plants are capable of anticipating the daily cycles and temperature variations to synchronize internal biological events with the external fluctuations (Sing and Sing 2022). It generates periodic physiological, biochemical and developmental processes in plants. Plants have developed an internal clock mechanism that allows them to run biological processes with the daily rhythms (Creux and Harmer 2019).

The first plant circadian oscillator genes were cloned 24 years ago (Harmer et al. 2022). Since that time there has been a serious effort made to identify the basis of circadian rhythms and understand how the circadian system impacts the biology of plants. The well-studied core oscillator of circadian clock is *Arabidopsis thaliana* (Nimmo et al. 2020; Okada and Mas 2022). Now there is a good understanding of its circadian biology. However, the studies of the circadian clock function in plants other than *A. thaliana* are still limited. Only recently new experimental results on circadian rhythms of some other plant species, first of all major crops, have been also published (Wang et al. 2020; Kugan et al. 2021; Singh and Singh 2022). Till now the mechanisms of circadian variations of many other plant species are as yet little understood. Especially little is known about photosynthetic response of perennial weeds.

The relationship between internal fluctuations and plant responses to environmental cues is also relatively poorly studied due to the complexity of the problem and because the circadian oscillator changes over time (Harmer et al. 2022). Although all plants are exposed to the same environmental rhythms throughout a day, it can be assumed that different plant species are able to respond differently to these changes depending on individual needs. It is also suggested that in plants circadian clocks can be different in different tissues, thus resulting in complexity in circadian systems and problems with interpretation of experimental data.

It is generally accepted that light and temperature are two most important environmental factors influencing plant development (Franklin 2009). The intensity, quality and photoperiod affect the development and especially the photosynthesis of plants (Yan et al. 2019). Temperature is a primary factor influencing the rate of plant development (Hatfield and Prueger 2015). These parameters can change throughout the day and growing season and thus will have a significant impact on the plant growth. It can be expected that the fluctuations in light and temperature can also influence the biological clocks.

An understanding the relationships between various biomolecules can help to gain an insight into the knowledge of their contribution to the biochemical processes in plants. Biochemical methods in vitro are more relevant, but unfortunately, they are destructive and time-consuming. There is a need to develop more reliable techniques for non-destructive, accurate and rapid evaluation of various organic compounds in plants. Raman spectroscopy has a great potential for non-destructive analysis of highly complex plant tissues (Saletnik et al. 2021). So far, it has been demonstrated that Raman spectroscopy can be successfully used to study the changes in the most important classes of biomolecules. However, in some cases, a direct comparison of different Raman spectra, even if they were measured under the same conditions, can be difficult due to certain analytical problems (Payne and Kurouski 2021). In particular, the Raman spectra often have a background noise. The spectra are usually dominated by intrinsic luminescence of the sample (Cadusch et al. 2013).

In the research, Raman spectroscopy was used to estimate the differences between organic composition of different plant species and the influence of external factors on the development of plants both during vegetation season and during the day time. The article describes rhythmical changes in the plants growing under the same environmental conditions. For the experimental study, leaves of two widespread plant species, couch grass and dandelion, were selected.

Couch grass *Elytrigia repens* is a rhizomatous perennial grass from the Poaceae family with both vegetative and sexual reproduction. It is a widely distributed weed, mainly in temperate climates (Mitra and Bhowmik 1999). Dandelion *Taraxacum officinale* is a herbaceous perennial plant from the Asteraceae family. The species is native in Europe and Asia, but can be found throughout the world in temperate climates (Berg 2007). The plants belong to different clades. Couch grass is monocot, and dandelion is dicot. On the basis of present knowledge, we can suggest that circadian variations of these two plants can differ and demonstrate different changes of organic compounds in the plant leaves. This research is devoted to comparison of the ontogeny of leaves of monocot and eudicot plants on the example of the two plant species.

2 Materials and Methods

2.1 Study Area

A small (1 × 1 m) site was chosen for the field experiment in the Leningrad region, Russia (60°72′N, 30°06′E). The soil at the site had a loamy texture. The mean soil pH was 6.0. The main plant species found in the site were couch grass *Elytrigia repens* (L.) Desv. ex Nevski, dandelion *Taraxacum officinale* (L.) Weber ex F.H. Wigg, sleepy *Aegopodium podagraria* L., clover *Trifolium* L., and celandine *Chelidonium majus* L.

Table 1 Climatic conditions at the site on different dates of sampling

Time of sampling	Air temperature °C	Air humidity %
4 June, sunrise at 3:45, sunset at 22:03		
06:00	11.0	82
14:00	16.0	65
20:00	14.0	67
20 August, sunrise at 5:23, sunset at 20:40		
06:00	18.2	74
14:00	31.4	33
20:00	24.1	54
26 September, sunrise at 6:45, sunset at 18:55		
06:00	7.9	81
14:00	15.5	48
20:00	9.2	68
1 October, sunrise at 7:01, sunset at 18:34		
06:00	8.6	86
14:00	13.2	71
20:00	8.5	75

2.2 Experimental Design

The field trial was carried out from June to October 2022. The experiment was conducted in a completely randomized design. Couch grass and dandelion were collected on 4 June, 20 August, 24 September and 1 October. On each sampling date, leaf samples were taken three times: at 6:00, 14:00, and 20:00. Table 1 shows fluctuations of air temperature and air humidity at the time of sampling. At each sampling time, 2–3 samples of leaves of each plant species were taken. The differences between measurements of replicates did not exceed 5%. After sampling, the plant leaves were placed in separate plastic containers with a sealed lid (container dimensions: diameter 30 mm, length 50 mm) and stored in a fridge at a temperature of 5 °C.

2.3 Raman Spectroscopy

For the Raman spectroscopy of plant samples, an ORTEC-785Video-M confocal Raman spectrometer was used, including spectrometer CT-RAMAN-785-S703/Horiba, laser 785 nm/CCTX-RAMAN-785-FI, laser power 10 … 50 mW. The spectra were obtained at a given exposure time of 30 s in the range of wave numbers 200…3300 cm^{-1}. The average number of measurements for each spectrum was 100 … 3000. A smoothing algorithm was applied to process the Raman spectra. The spectral characteristics were filtered using the method described by Savitzky and Golay (1964). In Table 2 is given description of the main peaks of the Raman spectra of major vibrational bands in leaves of couch grass and dandelion.

3 Results and Discussion

Main contribution to the spectra of the plant leaves was given by carotenoids, phenylpropanoids, aliphatic compounds, cellulose, and carbohydrates.

Carotenoids are widely distributed pigments that are present in various living organisms (Maoka 2020). In plants, the most important function of carotenoids is to collect light and protect the photosynthetic apparatus from photooxidation. In addition, they serve as precursors of signaling molecules that influence development and biotic/abiotic stress responses (Farré et al. 2010).

Phenylpropanoids are a large group of secondary metabolites. They are widespread in plants and play an important role in the plant development as major components of cell walls that are capable of protecting from strong light and ultraviolet radiation (Deng and Lu 2017). The metabolism of phenylpropanoids also provides plants with lignin for mechanical support, transporting water and nutrients over long distances, and creating physical barriers against water leakage and infection by pathogens (Zhao et al. 2016).

Aliphatic compounds are organic compounds containing C and H linked together in straight chains, branched chains, or non-aromatic rings. These are ethylene, isooctane, acetylene, propene, propane, squalene, and polyethylene. The most famous aliphatic compound is methane (Bhimabhai 2018).

Cellulose is the most abundant biopolymer available in nature. It is one of the main components of the cell walls of many plants (Gupta et al. 2019).

In plants, carbohydrates are sources of energy and carbon skeletons for organic compounds and storage components (Trouvelot et al. 2014). Because carbohydrates interact with diurnal changes, abiotic and biotic stresses, and hormonal signaling, they participate in a complex communication system necessary to coordinate metabolism with growth, development, and response to environmental changes and stresses (Rolland et al. 2002).

Being the main components of the plant cell structure, all organic compounds can be involved in the dynamic processes of plant development. Results of our experiment showed that the concentrations of carotenoids, phenylpropanoids, aliphatic compounds, cellulose, and carbohydrates in the plant leaves varied during day. The variations were often different on different sampling dates. Besides, there were differences between dandelion and couch grass, although the plants grew under the same conditions in a small site and were collected at the same time.

Figures 1 and 2 illustrate the Raman spectra of leaves of the dandelion and couch grass collected on different dates. On all sampling dates, the temperature on the site was the highest and the air humidity was the lowest at 14:00 (Table 1). Since nature gives a complex set of signals during a day, including changes in temperature,

Table 2 Identification of peaks of Raman spectra of major vibrational bands in plant leaves

Raman shift, cm^{-1}	Vibrational mode	Assignment
915	ν(C–O–H) In plane, symmetric	Cellulose, lignin
917	ν(C–O–C) In plane, symmetric	Cellulose, phenylpropanoids
964–969	δ(CH_2)	Aliphatic compounds
1000–1005	In-plane CH_3 rocking of polyene aromatic ring of phenylalanine	Carotenoids, protein
1048	ν(C–O) + ν(C–C) + δ(C–O–H)	Cellulose, phenylpropanoids
1070	CC and CO stretching	Cellulose
1080	ν(C–O) + ν(C–C) + δ(C–O–H)	Carbohydrates
1112	δ(C–OH)	Carbohydrates
1115–1119	Sym ν(C–O–C), C–O–H bending	Cellulose
1155	C–C Stretching; ν(C–O–C), ν(C–C) in glycosidic linkages, asymmetric ring breathing	Carotenoids, carbohydrates
1180	ν (C–C) γ(CH)	Chlorophills
1185	ν(C–O–H) Next to aromatic ring + σ(CH)	Carotenoids
1218	δ(C–C–H)	Carotenoids, xylan
1263	=(CH)	Carotenoids
1265	Guaiacyl ring breathing, C–O stretching (aromatic); –C=C–	Phenylpropanoids, unsaturated fatty acids
1284	δ(phenyl-OH)	Phenolics
1286	δ(C–C–H)	Aliphatic compounds
1301	δ(C–C–H) + δ(O–C–H) + δ(C–O–H)	Carbohydrates
1310	δCH_2 bending	Cellulose
1327	δCH_2 bending	Aliphatic compounds, cellulose, and phenylpropanoids
1328	δ(CH) ν(CN)	Chlorophills, Pyrrole ring br
1339	ν(C–O); δ(C–O–H)	Carbohydrates
1384–1387	δCH_2 Bending	Aliphatic compounds
1435	δ(CH_2), δ(CH_3)	Aliphatic compounds
1443–1446	δ(CH_2) + δ(CH_3)	Aliphatic compounds
1462	δ(CH_2), δ(CH_3)	Cuticle Triterpenoids
1482	δ(CH_2), δ(CH_3)	Cuticle Triterpenoids
1515–1535	–C=C– (in plane)	Carotenoids*
1545	C=O stretching	Flavonoids
1600	C=C (aromatic ring)	Lignin
1606–1632	ν(C–C) Aromatic ring + σ(CH)	Phenylpropanoids
1620	β sheet	Proteins

(continued)

Table 2 (continued)

Raman shift, cm^{-1}	Vibrational mode	Assignment
1654–1660	–C=C–, C=O Stretching, amide I	Unsaturated fatty acids, proteins
1665	β sheet	Amide I
1680	Cycl. [ν(C=O)]	Flavonoids
1682	COOH	Carboxylic acids
1690	β turn	Proteins
1748	C=O stretching	Esters, aldehydes, carboxylic acids and ketones

* Including zeaxanthin (1520 cm^{-1}) and violaxanthin (1523 cm^{-1})

quality and quantity of light, one can expect a certain influence of the fluctuations in the environmental conditions on the concentration of organic molecules in plants. These variations can be related, at least to some extent, to climatic changes.

The concentration of carotenoids, phenylpropanoids, aliphatic compounds, cellulose, and carbohydrates in dandelion leaves collected on 4 June and 1 October were indeed the lowest in the middle of the days and higher in the morning and in the evening. Thus, it can be assumed that there was some relationship between climatic conditions and accumulation of different biomolecules in leaves of dandelion on the dates. However, no direct relationship between external factors (temperature and humidity at the site) and changes in the concentration of organic compounds in couch grass leaves, as well as in dandelion leaves taken on other dates. For example, the concentrations of carotenoids, phenylpropanoids, aliphatic compounds, cellulose, and carbohydrates in leaves of couch grass sampled on 26 September increased from morning to evening (Fig. 2). The circadian clocks allow plants to adapt to the rotation of the Earth and to the changes in the rhythms of light and temperature on the Earth's surface. However, it has been shown that the rhythmical fluctuations are not a direct response to any external factor (Millar 2004).

Both in dandelion and couch grass, the contribution of carotenoids to the Raman spectra was always the largest on all sampling dates. In most cases (except for dandelion leaves collected on 4 June and 1 October), the second most significant contribution to the spectra was made by phenylpropanoids. Other compounds were present in smaller amounts. The comparison of the relative intensity of the carotenoid peaks in the spectra of dandelion and couch grass leaves showed that on 4 June at 6:00 leaves of couch grass exhibited much lower intensities of carotenoid vibrations compared to those of dandelion leaves. Similar situation was also observed in the morning on 1 October. On all other dates, the relative intensities of peaks of carotenoids in leaves of both plant species were similar.

In leaves of the dandelion collected on 4 June and especially on 1 October the concentration of all the organic molecules were much higher at 6:00 than at other times of sampling. It is unlikely that this could be caused solely by changes in

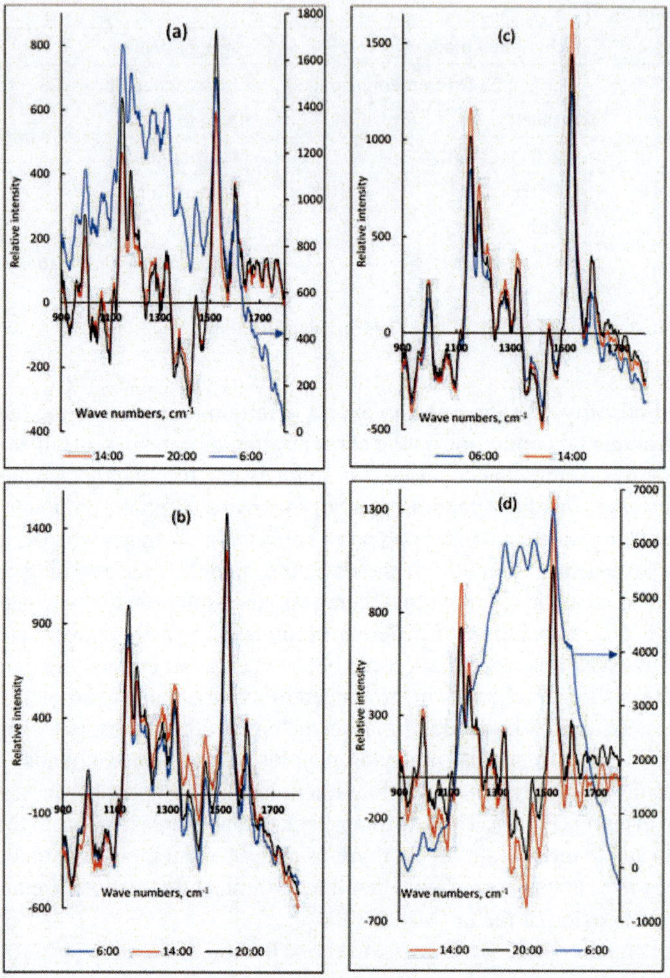

Fig. 1 Raman spectra of dandelion collected on 4 June (**a**), 20 August (**b**), 26 September (**c**), and 1 October (**d**)

climatic conditions at the site on these dates, since similar changes in temperature and humidity were also found on other sampling dates. Besides, in this case, a similar effect on couch grass could be expected, but this was not observed. The spectra of the dandelion leaves obtained on two dates, 4 June and 1 October, demonstrated serious difference compared to the spectra of leaves of the dandelion that were observed on other dates. Especially this concerns morning sampling. On the other hand, the differences between the Raman spectra of the couch grass collected on different dates were not so marked.

It is known that fast-growing plant species have a higher photosynthetic capacity than slow-growing species, which also corresponds to higher maximum rates of

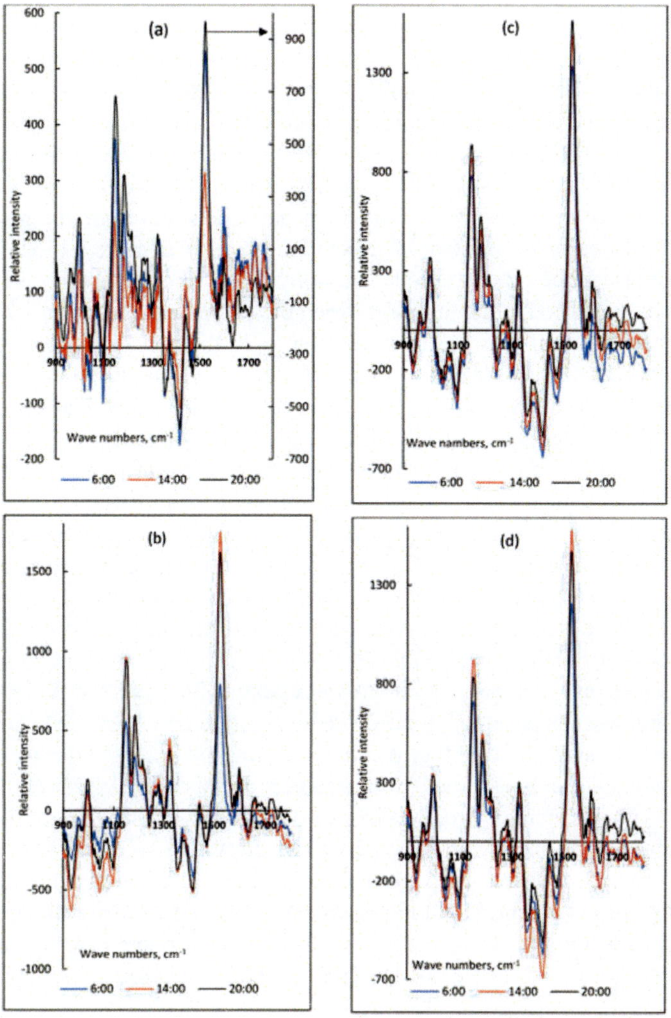

Fig. 2 Raman spectra of couch grass collected on 4 June (**a**), 20 August (**b**), 26 September (**c**), and 1 October (**d**)

photosynthetic electron transport in fast-growing plants (Demmig-Adams et al. 2020). On leaves of dandelion (a fast-growing plant) a significant amount of zeaxanthin (a form of carotenoid, wave number 1520 cm^{-1}) can be actively formed on the front side during the day. This can protect the photosynthetic apparatus from the formed reactive oxygen species (Yamamoto 2008). As was reported (Demmig-Adams et al. 2020), zeaxanthin can oxidize to violaxanthin (this is also a form of carotenoid, wave number 1523 cm^{-1}) in the second part of the day. Violaxanthin plays a role of "photosensitizer" that transfers the triplet excitation energy to chlorophyll molecules. One of the main functions of carotenoids is to collect light and protect the plant photosystem from photooxidation. This function is realized through the xanthophyll cycle involving zeaxanthin and violaxanthin (Kamrani et al. 2022).

In our experiment, the dominance of violaxanthin concentration was observed in dandelion leaves in the morning on 4 June. At the same time, a relatively high level of concentration of aliphatic compounds was also found. A similar behavior of the organic compounds was observed in dandelion leaves in autumn, after 1 October. In this case, the carotenoid zeaxanthin was also converted to violaxanthin with an increase in the concentration of carbohydrates. The highest concentration of carbohydrates occurred in the morning, when the temperature was minimal.

It can be suggested that the main difference in the concentrations of organic molecules in couch grass and dandelion could arise, among others, because the plant species fall into different taxons. Although the mechanisms of leaf development of monocots and dicots are similar, differences in the time, degree, and duration of ontogeny can lead to certain differences in the plant growth and biochemistry. The results of the plant cell cycle studies have confirmed the role of phylogenetically conserved and plant-specific regulatory mechanisms in the control of cell division (Lendvai et al. 2007). It is generally accepted that the coordination of growth processes and the temporal patterns of cell division in leaves of the plants belonging to monocots and dicots differ (Nelissen and DirkInzé 2016).

In addition, it should be noted that a peak of 2875 cm^{-1} was found in dandelion leaves collected on 1 October at 14:00. A weak peak of 2875 cm^{-1} was also detected on 4 June, but was not observed in the dandelion collected in August and September. In the spectra of leaves of couch grass, this peak was not observed during the entire growing season. It can be assumed that the appearance of the peak 2875 cm^{-1} may be associated with producing by roots of dandelion higher concentrations of the protective compound β-D-glucopyranosyl ester of taraxic acid (TA-G). TA-G is known to keep cockchafer larvae from eating dandelion roots (Huber et al. 2021). It can be assumed that plants exposed to biotic stress can activate their own cellular defenses by synthesizing substances that contribute to their survival and development under adverse conditions. In this case, TA-G allows dandelion to adapt to the environment.

4 Conclusions

The results of the experiments showed that concentrations of main organic molecules presented in leaves of couch grass (monocot) and dandelion (dicot) can vary both during vegetation season and during day. The fluctuations are often different for the two plants species. This can result from very different ways of adaptation of monocots and dicots to the Earth's rotation. The effects of external (climatic) conditions on the variations of carotenoids, phenylpropanoids, aliphatic compounds, cellulose, and carbohydrates in the plant leaves was not so significant as it might be expected and was found only for dandelion leaves collected in the very beginning of summer and in the mid-autumn. It was also shown that dandelion can serve as a good model for studying the mechanisms of plant defense against biotic factors.

References

Berg LR (2007) Introductory botany: Plants, people, and the environment. Thomson Brooks/C, Belmont, CA

Bhimabhai K (2018) Comparative study of aliphatic and aromatic compounds. J Pharm Innov 7:175–177

Cadusch PJ, Hlaing MM, Wade SA, McArthur SL, Stoddart PR (2013) Improved methods for fluorescence background subtraction from Raman spectra. J Raman Spectrosc 44(11):1587–1595

Creux N, Harmer S (2019) Circadian rhythms in plants. CSH Perspect Biol 11(9):a034611

Demmig-Adams B, Stewart JJ, López-Pozo M, Polutchko SK, Adams WW (2020) Zeaxanthin, a molecule for photoprotection in many different environments. Molecules 25:5825

Deng Y, Lu S (2017) Biosynthesis and regulation of phenylpropanoids in plants. Crit Rev Plant Sci 36(4):257–290

Farré G, Sanahuja G, Naqvi S, Bai C, Capell T, Zhu C, Christou P (2010) Travel advice on the road to carotenoids in plants. Plant Sci 179(1–2):28–48

Franklin KA (2009) Light and temperature signal crosstalk in plant development. Curr Opin Plant Biol 12(1):63–68

Gupta PK, Raghunath S, Prasanna DV, Venkat P, Shree V, Chithananthan C, Choudhay S, Surender K, Geetha K (2019) An update on overview of cellulose, its structure and applications. Cellulose 201(9)

Hatfield JL, Prueger JH (2015) Temperature extremes: Effect on plant growth and development. Weather and climate extremes 10:4–10

Harmer SL, Fankhauser C, Webb AA (2022) Focus on circadian rhythms. Plant Physiol 190(2):921–923

Huber M, Thomas Roder, Sandra Irmisch, Alexander Riedel, Saskia Gablenz, Julia Fricke, Rahfeld P, Reichelt M, Paetz C, Liechti N, Hu L, Bont Z, Meng Y, Huang W, Robert CAM, Gershenzon J, Erb M (2021) A beta-glucosidase of an insect herbivore determines both toxicity and deterrence of a dandelion defense metabolite. Elife 10:e68642

Kamrani YY, Shomali A, Aliniaeifard S, Lastochkina O, Moosavi-Nezhad M, Hajinajaf N, Talar U (2022) Regulatory role of circadian clocks on ABA production and signaling, stomatal responses, and water-use efficiency under water-deficit conditions. Cells 11(7):1154

Kugan HM, Rejab NA, Sahruzaini NA, Harikrishna JA, Baisakh N, Cheng A (2021) Circadian rhythms in legumes: what do we know and what else should we explore?. Int J Mol Sci 22(9):4588

Lendvai Á, Pettkó-Szandtner A, Csordás-Tóth É, Miskolczi P, Horváth GV, Györgyey J, Dudits D (2007) Dicot and monocot plants differ in retinoblastoma-related protein subfamilies. J Exp Bot 58(7):1663–1675

Millar AJ (2004) Input signals to the plant circadian clock. J Exp Bot 55(395):277–283

Maoka T (2020) Carotenoids as natural functional pigments. J Nat Med 74:1–16

Mitra S, Bhowmik PC (1999) Effect of growth stages on quackgrass (*Elytrigia repens*) control in corn (*Zea mays*) with rimsulfuron. Weed Technol 13(1):37–42

Nelissen H, DirkInzé NG (2016) Leaf growth in dicots and monocots: so different yet so alike. Curr Opin Plant Biol 33:72–76

Nimmo HG, Laird J, Bindbeutel R, Nusinow DA (2020) The evening complex is central to the difference between the circadian clocks of *Arabidopsis thaliana* shoots and roots. Physiol Plantarum 169(3):442–451

Okada M, Mas P (2022) In vivo bioluminescence analyses of circadian rhythms in *Arabidopsis thaliana* using a microplate luminometer. In: Solanas G, Welz PS (eds) Circadian regulation. Methods in molecular biology, vol 2482. Humana, New York, pp 395–406

Pan W-J, Wang X, Deng YR, Li JH, Chen W, Chiang JY, Yang J-B, Zheng L (2015) Nondestructive and intuitive determination of circadian chlorophyll rhythms in soybean leaves using multispectral imaging. Sci Rep 5(1):1–13

Payne WZ, Kurouski D (2021) Raman spectroscopy enables phenotyping and assessment of nutrition values of plants: a review. Plant Methods 17(1):1-20

Petersen J, Rredhi A, Szyttenholm J, Mittag M (2022) Evolution of circadian clocks along the green lineage. Plant Physiol 190(2):924–937

Rolland F, Moore B, Sheen J (2002) Sugar sensing and signaling in plants. Plant Cell 14(Suppl):S185–S205

Saletnik A, Saletnik B, Puchalski C (2021) Overview of popular techniques of Raman spectroscopy and their potential in the study of plant tissues. Molecules 26(6):1537

Savitzky A, Golay MJE (1964) Smoothing and differentiation of data by simplified least squares procedures. Anal Chem 36(8):1627–1639

Singh G, Singh V (2022) Systems scale characterization of circadian rhythm pathway in Camellia sinensis. Computational and Structural Biotechnology Journal 20:598–607

Trouvelot S, Héloir M-C, Poinssot B, Gauthier A, Paris F, Guillier C, Combier M, Trdá L, Daire X, Adrian M (2014) Carbohydrates in plant immunity and plant protection: roles and potential application as foliar sprays. Front Plant Sci 5:592

Wang Y, Yuan L, Su T, Wang Q, Gao Y, Zhang S, Qian J, Guolong Y, Fu Y, Cheng Q, Liu B, Kong F, Zhang X, Song C-P, Xu X, Xie Q (2020) Light-and temperature-entrainable circadian clock in soybean development. Plant Cell Environ 43(3):637–648

Yamamoto HY (2008) A random walk to and through the xanthophyll cycle. In: Demmig-Adams B, Adams WW, Mattoo AK (eds) Photoprotection, photoinhibition, gene regulation, and environment. Advances in photosynthesis and respiration, vol 21. Springer, Dordrecht

Yan Z, He D, Niu G, Zhai H (2019) Evaluation of growth and quality of hydroponic lettuce at harvest as affected by the light intensity, photoperiod and light quality at seedling stage. Sci Hortic-Amsterdam 248:138–144

Zhao S, Zhao L, Liu F, Wu Y, Zhu Z, Sun C, Tan L (2016) NARROW AND ROLLED LEAF 2 regulates leaf shape, male fertility, and seed size in rice. J Integr Plant Biol 58:983–996

Short-Term Variability of Element Concentrations in the Rhizosphere of Couch Grass and Nettle in St. Petersburg, Russia

Irina V. Shtangeeva, Matti Niemelä, Alexander G. Ryumin, and Paavo Perämäki

Abstract The rhizosphere of plants is a very dynamic system that can vary over time depending on the plant physiological state, its developmental stage and environmental conditions. The concentrations of many elements in the rhizosphere soil are also not constant. Until now, more attention has been paid to long-term (seasonal and annual) changes in the root zone. However, short-term (within hours) fluctuations in the concentrations of elements in the rhizosphere soil of different plant species have not been studied enough. In this work, diurnal variations in the concentrations of macro- and trace elements in the rhizosphere of two widespread plant species, couch grass and nettle, growing in the same place under the same conditions, were studied. The short-term variations in the rhizosphere soil of couch grass and nettle were significant and regular. Depending on the time of sampling, the temporal changes of one or another element in the soil taken from roots of the plants were often different. This may be caused by differences between couch grass (monocot) and nettle (dicot), which release various organic compounds into the rhizosphere, and may also be due to natural factors, including climatic conditions.

Keywords Rhizosphere · Macro-nutrients · Trace elements · Daily variations

I. V. Shtangeeva (✉) · A. G. Ryumin
St. Petersburg University, St. Petersburg, Russia
e-mail: shtangeeva@gmail.com; i.shtangeeva@spbu.ru

M. Niemelä · P. Perämäki
University of Oulu, Oulu, Finland

O. V. Frank-Kamenetskaya et al. (eds.), *Biogenic—Abiogenic Interactions in Natural and Anthropogenic Systems 2022*, Springer Proceedings in Earth and Environmental Sciences, https://doi.org/10.1007/978-3-031-40470-2_19

1 Introduction

The rhizosphere is a highly dynamic micro-ecosystem located between plant roots and bulk soil where root exudates and soil microorganisms actively influence various biogeochemical processes (Nannipieri et al. 2008). An additional point to pay attention to is that the soil around the roots of different plant species can vary, even if the plants grow in the same site and under absolutely the same conditions. Similarly, the concentrations of many elements in different plant species grown under the same environmental conditions can vary significantly (Fischerová et al. 2006; Fässler et al. 2010; Tang et al. 2018). In plants, these differences are under genetic control (Clárk 1983). It can be assumed that plants belonging to different classes are capable of releasing different organic compounds into the rhizosphere. These compounds can affect the soil at plant roots in different ways. It can be expected that, despite the same physicochemical parameters of the rhizosphere soil of different plant species, the exchange of root exudates of a particular plant with the soil can lead to formation of dissimilar organic compounds. As a result, the concentrations of macro- and trace elements in the plant rhizosphere can also be different.

The concentrations of many elements in the rhizosphere soil are often variable and usually change over time. The temporal variations can cover a long time, such as a growing season (Hamon et al. 1995), or they can last a relatively short time, for several days or even hours (Shtangeeva et al. 2020). Annual and seasonal variations of different elements in the rhizosphere were observed by many researchers. However, short-term fluctuations are less studied. There are many reasons for the temporal variability of the transport and uptake of elements at the soil—roots of plants border. Seasonal variations can be mainly due to changes in the temperature and, as a consequence, in biological activity of plants and rhizosphere microorganisms (Turpault et al. 2007). Cheng et al. (2003) reported that plant phenology (sampling times) is one of the main factors controlling the amount of soil carbon loss in the rhizosphere.

The aim of this research was to study short-term (in the course of one day) changes in the concentrations of macro- and trace elements in the rhizosphere of two widely distributed plant species, couch grass and nettle, growing in the same place under the same conditions. The plants belong to different clades, and it can be expected that impact of their root exudates on the surrounding soil can also be different.

2 Materials and Methods

2.1 Study Area

A small (1.5 × 1.5 m) site in the south of St. Petersburg, Russia (59°53′N, 30°38′E) was selected for the field testing. The prevailing vegetation (90%) was represented by couch grass *Elytrigia repens* L. and nettle *Urtica dioica* L. The rest of the vegetation was represented by clover *Trifolium repens* and plantain *Plantago major*. Podzols and Umbrisols were the dominant soils in the studied region. The texture of the soil in the site was loamy sand (sand 75%, silt/clay 24%, silt 1%), with clarified quartz grains, roots of herbaceous vegetation, and small anthropogenic inclusions in the form of fragments of brick and glass.

2.2 Experimental Design

The field trial was carried out in May 2021. The experiment was conducted in a completely randomized design. Samples of the couch grass and nettle rhizosphere soil were collected on 3 May, 10 May, and 25 May. On each date, the soil samples were taken three times: at 9:00, 14:00, and 19:00. The soil was shaken off the roots, and then soil residues adhering to the plant roots were collected. At each sampling time, the soil was taken from at least six plants of each plant species. Figure 1 shows variations in air and soil temperature at the time of sampling. The soil samples were air-dried to constant weight and passed through a 1 mm sieve. The pH (1:2.5 H_2O) of the soil samples was measured.

2.3 Elemental Analysis

An elemental analyzer CHN628 (LECO) was used to determine the total amount of C, N, and H. The soil samples (270–300 mg) were placed in a tin foil and burned in a stream of pure oxygen at a temperature of 950 °C. The instrument was calibrated using EDTA with certified values for the concentrations of C, N, and H. The ICP-OES/MS were applied to determine the concentrations of 12 macro- and trace elements. The soil samples (80–100 mg) were weighed into microwave digestion vessels. Nine mL of concentrated HNO_3 were added. The vessels were closed and heated in the Milestone Ethos up (Milestonem Srl, Sorisole, Italy) microwave oven. The programme was the following: heating for 10 min to 180 °C and holding at 180 °C for 9.5 min. Then the vessels were cooled down to room temperature. The samples were diluted to 50 mL with ultrapure water. The ICP-OES with an Agilent 5110 VDV ICP-OES equipped with a SPS 4 autosampler, an U-series concentric glass nebulizer, a cyclonic glass double pass spray chamber, and an 1.8 mmsemi-demountable torch was applied to determine the concentrations of Na, Mg, P, K, Ca, Mn, Fe, Zn, and Ba. Scandium (wavelength 335.372 nm) and Y (wavelength 377.433 nm) were used as internal standards to correct sensitivity drift and matrix effects. The ICP-MS using an Agilent 8900 Triple Quad ICP-MS equipped with a SPS 4 autosampler, a MicroMist glass concentric nebulizer, a quartz Scott double-pass spray chamber, and a 2.5 mm quartz torch was applied to determine the concentrations of Co, Ni, Cu, and Zn. The parameters for ion lenses and torch position were tuned automatically before each ICP-MS measurement. Helium cell gas was used to reduce interferences. In addition, a certified reference material Tomato Leaves NIST 1573a (National Institute of

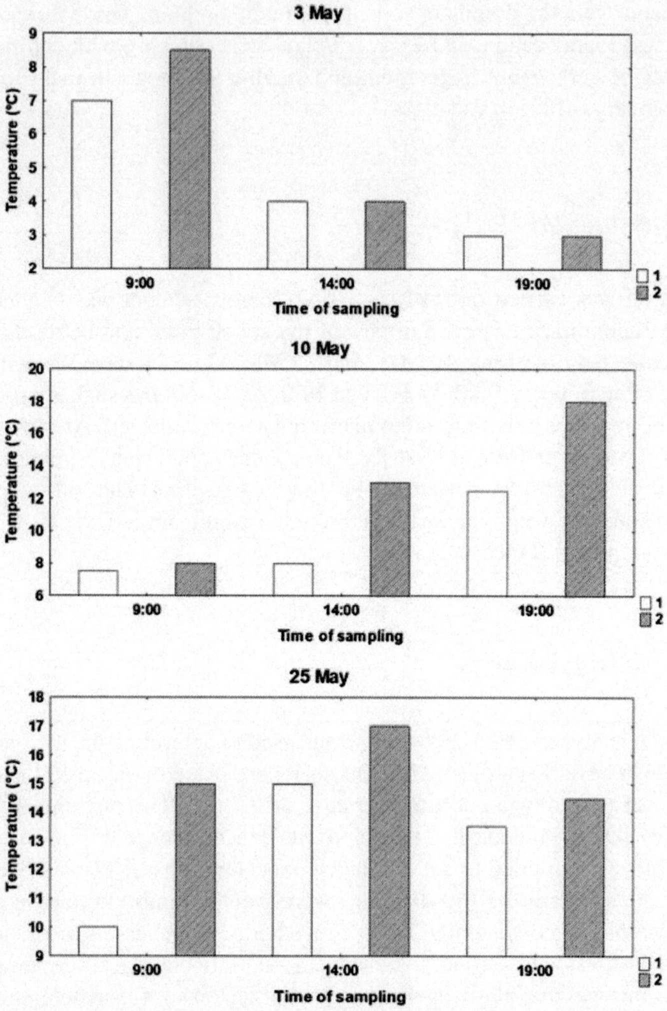

Fig. 1 Soil (1) and air (2) temperature at the time of sampling

Standards and Technology, USA) was measured as a standard. The concentrations of elements determined in the NIST 1573a were in a good agreement with the certified values.

2.4 Data Processing and Statistics

Statistica for Windows 8.0 Software packages (StatSoft, Tulsa, OK, USA) were applied for multivariate statistical analysis of the experimental data. The average concentrations of elements in the rhizosphere soil of couch grass and nettle were calculated, and analysis of variance was carried out to assess the differences between the experimental samples. The level of statistical significance was set at $P < 0.05$. The Shapiro–Wilk test was used to assess the normality of the data. Cluster analysis (CA) and principal component analysis (PCA) were carried out to examine the relationships between elements and between soil samples collected on different dates and times of day. For the CA, Ward's method was applied using the squared Euclidean distances as a measure of similarity. The data for the CA and PCA were normalized to unit concentration to avoid misclassifications caused by different order of magnitudes of variables.

3 Results and Discussion

Although one might expect certain differences between the concentrations of elements in the rhizosphere soil of couch grass and nettle, the mean concentrations of most of the elements were similar (Table 1). The only exception was Mn; its concentration in the soil taken from roots of couch grass was statistically significantly ($P < 0.05$) lower than in the rhizosphere soil of nettle.

Table 1 Mean concentrations (mg g^{-1}) ± SD of elements in the rhizosphere soil of cough grass and nettle calculated from all three sampling dates

Element	Couch grass	Nettle
Na	0.596 ± 0.043	0.594 ± 0.059
Mg	3.31 ± 0.14	3.44 ± 0.20
P	1.92 ± 0.49	2.09 ± 0.39
K	2.20 ± 0.21	2.00 ± 0.21
Ca	25.6 ± 0.7	27.2 ± 3.2
Mn	0.357 ± 0.015*	0.390 ± 0.017
Fe	19.0 ± 0.6	19.8 ± 1.8
Co	0.0049 ± 0.0009	0.0044 ± 0.0005
Ni	0.0125 ± 0.0024	0.0108 ± 0.0026
Cu	0.0228 ± 0.0019	0.0211 ± 0.0021
Zn	0.113 ± 0.020	0.107 ± 0.027
Ba	0.208 ± 0.009	0.216 ± 0.021

* Difference between concentration of the element in the rhizosphere soil of couch grass and nettle was statistically significant ($P < 0.05$)

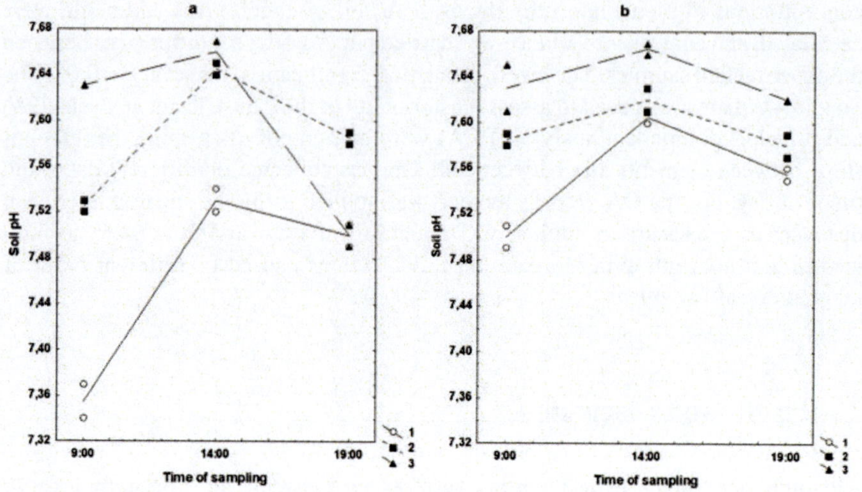

Fig. 2 Daily variations of the pH of the rhizosphere soil of couch grass (**a**) and nettle (**b**) collected on 3 May (1), 10 May (2), and 25 May (3)

3.1 Short-Term Variations of the Soil pH

The pH of the rhizosphere soil of couch grass and nettle varied with time (Fig. 2). As a rule, it increased from morning to mid-afternoon, and then decreased again. This tendency was observed on all dates and was attributed to the soil taken from roots of both plant species. The change in soil pH over time was more pronounced for the rhizosphere soil of couch grass compared to that of nettle. The pH of the soil taken from roots of the plants collected on 3 May was lower than the pH of the soil collected on two other dates. This may be due to lower soil and air temperature on this date. It is known that temperature is an important factor capable of affecting the soil pH, especially in the A horizon (Zhang et al. 2019). Such a relationship between soil temperature and soil pH was reported by other researchers (Ghazali et al. 2020; Tyler et al. 2010).

3.2 Daily Variations of Element Concentrations in the Rhizosphere Soil

The rhizosphere of a plant is a dynamic system that changes over time depending on the developmental stage and physiological state of the plant, as well as on environmental conditions (Selvakumar et al. 2012; Zamioudis and Pieterse 2012). There are various processes that can affect the situation in this zone. The main factors influencing short-term variability might be time-varying activity of root exudates and microorganisms (Chen et al. 2017; Nikolausz et al. 2008). One of the most important reasons of the diurnal fluctuations of element concentrations can be circadian variations, which are characteristic of all living organisms (Harmer 2009; Haydon et al. 2013).

In our experiment, the concentrations of elements in the rhizosphere soil were not stable during the daytime. For example, daily variations of macro-elements such as N, C, and H were regular and quite significant (Fig. 3). Moreover, depending on the date of sampling, certain differences between daily changes of the N, C, and H concentrations in the rhizosphere of couch grass and nettle were observed. On 3 May, the diurnal fluctuations of the element concentrations in the soil taken from roots of couch grass and nettle were similar, while in the rhizosphere soil of the two plants collected on 10 May and 25 May the short-term variations of N, C, and H were different. Root exudates are capable of changing the chemical and physical properties of the rhizosphere soil (Hubbard et al. 2018). It can be suggested that different plant species are able to do this differently via producing different organic compounds that are typical for a particular plant, and thereby affecting the microbial community in the rhizosphere. The differences between daily variations of N, C, and H in the soil probably may also be due to different weather conditions on these dates.

Although couch grass and nettle grew in the same small site and were collected simultaneously, the concentrations of many elements in the rhizosphere soil of the plants varied during the day differently. Figure 4 illustrates short-term changes in the concentrations of essential plant nutrients, K and Mn, in the soil taken from roots of couch grass and nettle. In the soil samples collected on 3 May, the concentration of K increased by the end of the day, while the concentrations of Mn decreased. Differences in the daily dynamics of K and Mn concentrations were also observed in the rhizosphere soil of the plants collected on two other dates. Although the short-term variations of the elements in the soil taken from roots of couch grass and nettle on 3 and 10 May were more or less similar, on 25 May the diurnal change of K and Mn in the rhizosphere soil of the plants exhibited certain differences. It can be assumed that this was because the composition and concentration of root exudates are different for different plant species (Na et al. 2018). Besides, under stable environmental conditions the quantity and chemical composition of root exudates of a given species strongly depend on the plant development stage (Badri and Vivanco 2009).

3.3 Cluster and Principal Component Analysis of Soil Samples

The CA and PCA were conducted to confirm whether the sampling time and vegetation are important factors influencing the concentrations of elements in the soil. The CA was used to identify similarities between the experimental samples. Figure 5 shows results of the CA of the soil collected on different dates at different times. The soil samples taken at 09:00 and 14:00 were grouped similarly (Fig. 5a, b), while the

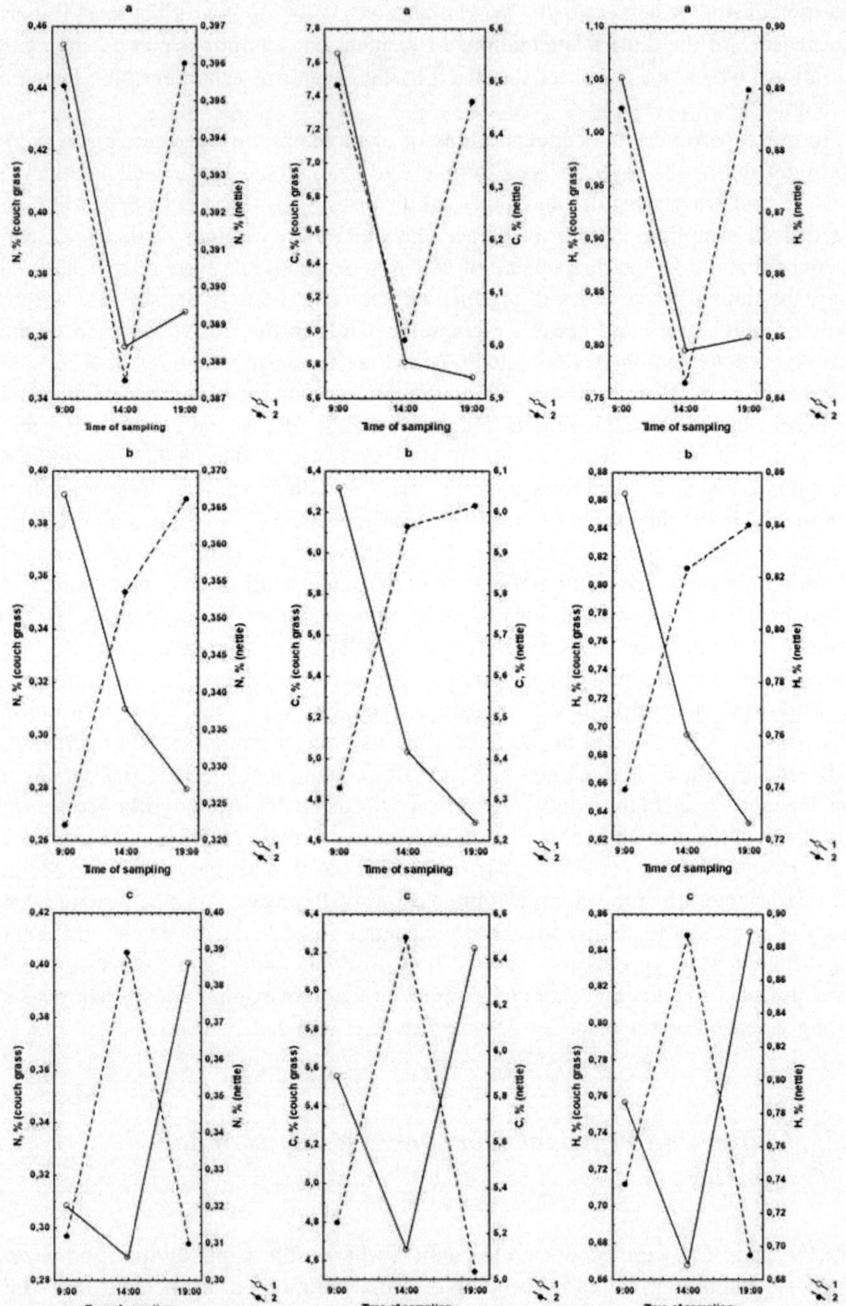

Fig. 3 Daily variations of the concentration of N, C, and H in the rhizosphere soil of couch grass (1) and nettle (2) collected on 3 May (**a**), 10 May (**b**), and 25 May (**c**)

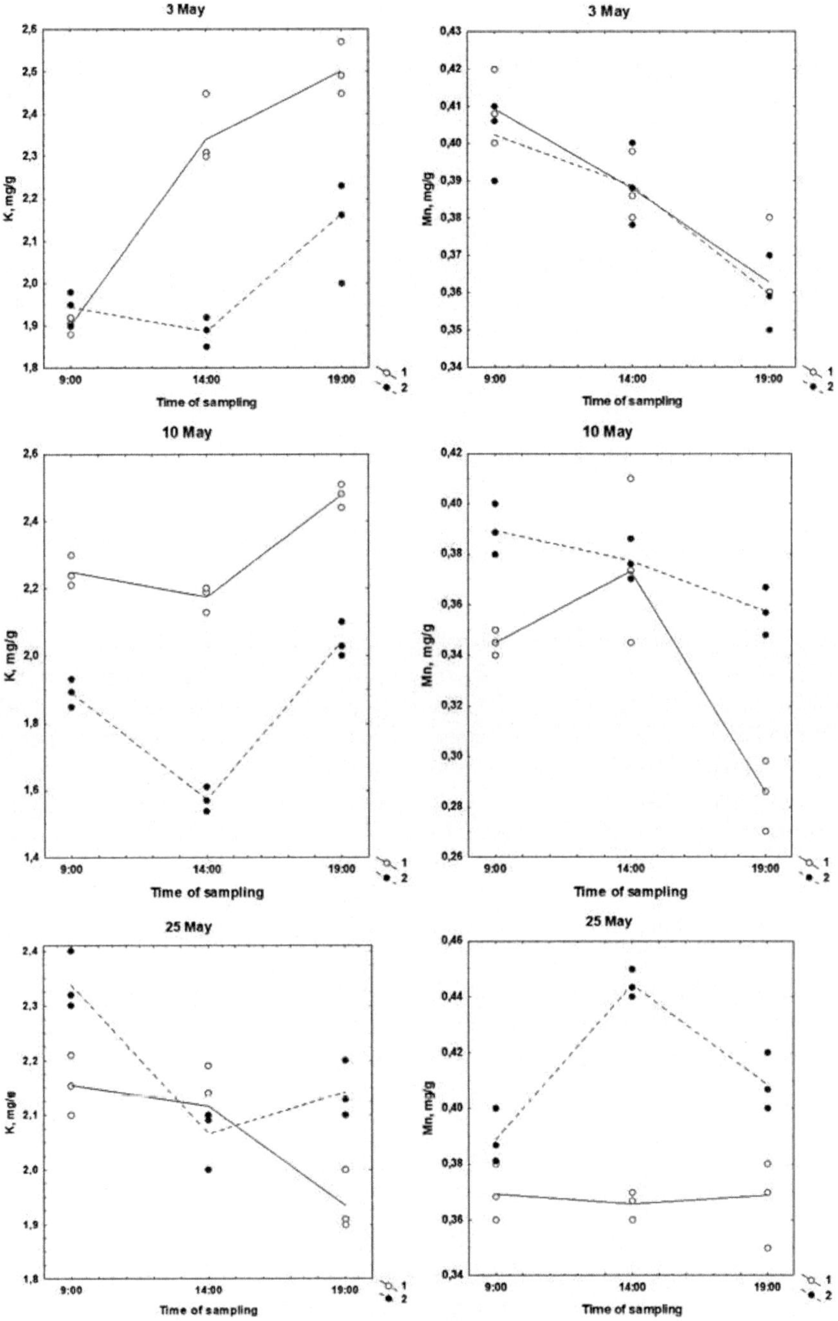

Fig. 4 Daily variations of the concentrations of K and Mn in the rhizosphere soil of couch grass (1) and nettle (2) collected on 3 May, 10 May, and 25 May

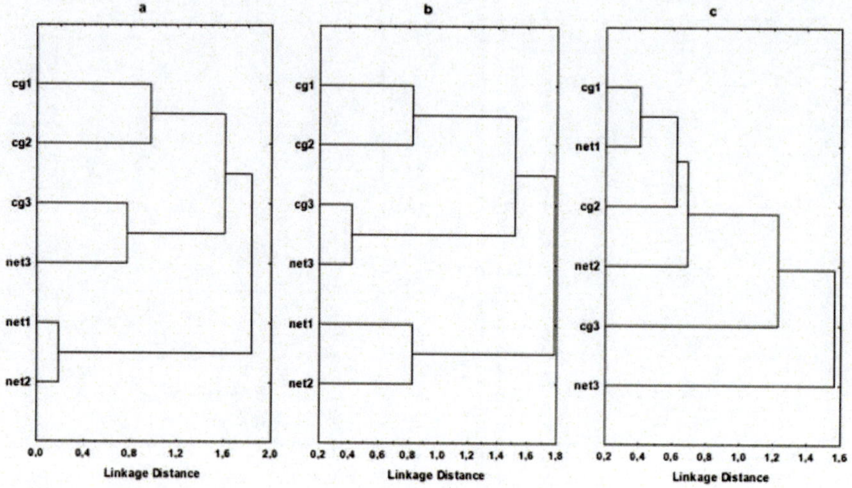

Fig. 5 Cluster analysis (Ward's method) of the soil samples taken from roots of couch grass (cg) and nettle (net) on 3 May (1), 10 May (2), and 25 May (3) at 9:00 (**a**), 14:00 (**b**), and 19:00 (**c**)

samples of couch grass and nettle rhizosphere soil taken at 19:00 were grouped quite differently than in the previous case (Fig. 5c). It is hardly possible to explain this phenomenon by differences in such parameters as soil (and air) temperature and pH of the soil samples taken during these periods. Probably, this was due to the different relationships between elements in the soils collected at 9:00 and 14:00 and at 19:00.

To identify patterns of the relationships between elements and their dependence on the sampling time, the PCA was performed (Table 2). Calcium, Mn and Fe were highly correlated with the PC1 on all sampling dates. This may indicate the general nature of the relationships of these three elements in the soil, regardless of the time of sampling. On both 3 May and 10 May, P, Ca, Mn, and Fe were highly correlated with the PC1, while Na, Ni, and Cu made the dominant contribution to the PC2. As for other elements, there were certain differences between the samples collected on the first two and on the last day. It can be assumed that the biochemical activity of the plant root exudates may change over time as the plants grow.

4 Conclusions

The concentrations of elements in the rhizosphere soil changed over a long (month) and a short (several hours) time. The temporal variability depended on the plant species, age of the plants and differed from plant one to another, even if they grew in the same place and under identical environmental conditions. Considering such short-term variability of element concentrations in the rhizosphere soil, it is necessary to carefully choose the timing of sampling to ensure the correct interpretation of experimental data.

Table 2 Principal component (PC) loadings for 12 elements in the rhizosphere soil collected on different dates. Values in bold indicate variables that most influence each PC

Date	3 May		10 May		25 May	
Element	PC1	PC2	PC1	PC2	PC1	PC2
Na	0.565	**−0.729**	−0.281	**−0.934**	**−0.766**	−0.522
Mg	0.578	−0.503	−0.616	−0.740	**−0.913**	−0.041
P	**0.915**	−0.170	**−0.766**	0.311	0.230	**0.926**
K	−0.693	−0.569	0.514	**−0.817**	−0.327	**−0.918**
Ca	**0.915**	0.374	**−0.964**	−0.203	**−0.753**	−0.181
Mn	**0.943**	0.209	**−0.903**	0.337	**−0.953**	0.265
Fe	**0.898**	−0.001	**−0.917**	−0.268	**−0.908**	0.138
Co	0.580	−0.605	**0.798**	−0.497	**−0.940**	0.173
Ni	0.336	**−0.919**	0.588	**−0.779**	**−0.887**	0.378
Cu	−0.234	**−0.934**	−0.038	**−0.820**	**−0.902**	0.291
Zn	**0.803**	−0.003	−0.433	−0.697	0.330	0.117
Ba	0.669	0.551	−0.555	−0.584	**−0.981**	−0.014
Explained var	6.10	3.71	5.37	4.76	7.43	2.37
% var	50.8	30.9	44.8	39.7	61.9	19.7

Acknowledgements Irina Shtangeeva acknowledges that her visit to the Oulu University was funded by Academy of Finland, grant number 333659.

References

Badri DV, Vivanco JM (2009) Regulation and function of root exudates. Plant Cell Environ 32(6):666–681

Cheng W, Johnson DW, Fu S (2003) Rhizosphere effects on decomposition: controls of plant species, phenology, and fertilization. Soil Sci Soc Am J 67(5):1418–1427

Chen YT, Wang Y, Yeh KC (2017) Role of root exudates in metal acquisition and tolerance. Curr Opin Plant Biol 39:66–72

Clark RB (1983) Plant genotype differences in the uptake, translocation, accumulation, and use of mineral elements required for plant growth. In: Sarić MR, Loughman BC (eds) Genetic aspects of plant nutrition. Springer, Dordrecht, pp 49–70

Fischerová Z, Tlustoš P, Száková J, Šichorová K (2006) A comparison of phytoremediation capability of selected plant species for given trace elements. Environ Pollut 144(1):93–100

Fässler E, Robinson BH, Gupta SK, Schulin R (2010) Uptake and allocation of plant nutrients and Cd in maize, sunflower and tobacco growing on contaminated soil and the effect of soil conditioners under field conditions. Nutri Cycl Agroecosyst 87(3):339–352

Ghazali MF, Wikantika K, Harto AB, Kondoh A (2020) Generating soil salinity, soil moisture, soil pH from satellite imagery and its analysis. Information Processing in Agriculture 7(2):294–306

Hamon RE, Lorenz SE, Holm PE, Christensen TH, McGrath SP (1995) Changes in trace metal species and other components of the rhizosphere during growth of radish. Plant Cell Environ 18(7):749–756

Harmer SL (2009) The circadian system in higher plants. Annu Rev Plant Biol 60(1):357–77

Haydon MJ, Mielczarek O, Robertson FC, Hubbard KE, Webb AAR (2013) Photosynthetic entrainment of the Arabidopsis circadian clock. Nature 502:689–692

Hubbard C, Brock M, van Diepen L, Maignien L, Ewers BE, Weinig C (2018) The plant circadian clock influences rhizosphere community structure and function. ISME J 12:400–410

Na X, Li X, Zhang Z, Li M, Kardol P, Xu TT, Wang M, Cao X, Ma F (2018) Bacterial community dynamics in the rhizosphere of a long-lived, leguminous shrub across a 40-year age sequence. J Soils Sediments 18:76–84

Nannipieri P, Ascher J, Ceccherini MT, Landi L, Pietramellara G, Renella G, Valori F (2008) Effects of root exudates in microbial diversity and activity in rhizosphere soils. In: Nautiyal CS, Dion P (eds) Molecular mechanisms of plant and microbe coexistence Springer, Berlin, Heidelberg, pp 339–365

Nikolausz M, Kappelmeyer U, Székely A, Rusznyák A, Márialigeti K, Kästner M (2008) Diurnal redox fluctuation and microbial activity in the rhizosphere of wetland plants. Europ J Soil Biol 44(3):324–333

Selvakumar G, Panneerselvam P, Ganeshamurthy AN, Maheshwari DK (2012) Bacterial mediated alleviation of abiotic stress in crops. In: Maheshwari DK (ed) Bacteria in agrobiology: stress management. Springer, New York, pp 205–224

Shtangeeva I, Vīksna A, Bērtiņš M, Ryumin A, Grebnevs V (2020) Variations in the concentrations of macro-and trace elements in two grasses and in the rhizosphere soil during a day. Environ Pollut 262, 114265

Tang T, Xu W, Zhou G, Bai Y, Li J, Tang X, Chen D, Liu Q, Ma W, Xiong G, He H, He N, Guo Y, Guo Q, Zhu J, Han W, Hu H, Fang J, Xie Z (2018) Patterns of plant carbon, nitrogen, and phosphorus concentration in relation to productivity in China's terrestrial ecosystems. Proceedings of the National Academy of Sciences 115(16):4033–4038

Turpault MP, Gobran GR, Bonnaud P (2007) Temporal variations of rhizosphere and bulk soil chemistry in a Douglas fir stand. Geoderma 137(3-4):490–496

Tyler JJ, Nederbragt AJ, Jones VJ, Thurow JW (2010) Assessing past temperature and soil pH estimates from bacterial tetraether membrane lipids: evidence from the recent lake sediments of Lochnagar, Scotland. J Geophys Res Biogeosci 115(G1)

Zamioudis C, Pieterse CM (2012) Modulation of host immunity by beneficial microbes. Mol Plant-Microbe Interact 25(2):139–150

Zhang YY, Wu W, Liu H (2019) Factors affecting variations of soil pH in different horizons in hilly regions. PLoS ONE 14(6):e0218563

Biomineral Interactions in Soil

Metagenomic Studies of Chernozem Under Different Type of Land Use

Boris F. Aparin, Yaroslavna V. Valchenko, Elena Yu. Sukhacheva, Maria K. Zakharova, and Evgeny E. Andronov

Abstract This study is a comparative analysis of microbiome of chernozems of different types of land use in the Kamennaya Steppe nature reserve (the Voronezh Region, Russia). The soil transect was crossed over: the mown fallow plot, forest plantation, and wheat crops. The soil samples were collected at the same time from the soils horizons up to a depth of 100 cm (in fallow up to 200 cm). The key plots of soil pit location are characterized by similar parent substrate, climate parameters, and humus-accumulative type of pedogenesis. The history of land use of this territory is known. The age of the fallow land and field-protecting plantations is about 100 years; the arable land use is about 200 years old. That allows us to use these profiles as the possible objects to investigate the specificity of the microbiome structure affected by the type of land use. The microbiome was studied of different taxonomic levels (phylum, class, order, family, genus). The microbiome structure analysis was performed in different aspects: (a) changes in the microbial community along the depth of soil profiles, (b) evaluation of the similarity and differences in the microbial community of the surface and middle soil horizons of different lands, and (c) comparison of the microbocenoses in parent substrate from soils affected by the different land use systems.

Keywords Soil profile · Pyrosequinization · Phylogenetic diversity · Mowed fallow · Forest belt · Cropland · Kamennaya Steppe · Nature reserve

B. F. Aparin (✉) · E. Yu. Sukhacheva
Saint-Petersburg State University, Saint-Petersburg, Russia
e-mail: soilmuseum@bk.ru

B. F. Aparin · Y. V. Valchenko · E. Yu. Sukhacheva · M. K. Zakharova
Central Soil Museum By V.V. Dokuchaev - Branch of the Federal Research Centre V.V. Dokuchaev Soil Science Institute, Saint-Petersburg, Russia

E. E. Andronov
All-Russia Research Institute for Agricultural Microbiology, Saint-Petersburg, Russia

O. V. Frank-Kamenetskaya et al. (eds.), *Biogenic—Abiogenic Interactions in Natural and Anthropogenic Systems 2022*, Springer Proceedings in Earth and Environmental Sciences, https://doi.org/10.1007/978-3-031-40470-2_20

1 Introduction

The use of high-throughput 16S rRNA gene sequencing became a qualitatively new stage in the study of the soil microbial community. Vladimir Vernadsky's statement that "the soil is saturated with life" was confirmed by the quantitative methods. The molecular genetic method opened up unique opportunities for studying metagenome as the main element of soil, which is a system of both biotic and abiotic components. There is a wide spectrum of directions for studying the soil microbial community (MC) (Andronov 2017; Fierer and Jackson 2006; Ananyeva et al. 2021; Blume et al. 2002; Hansel et al. 2008; Ivanova et al. 2015; Schloter et al. 2018; Saccá et al. 2017). Of great interest to researchers is the study of the impact of land use types, the intake of fertilizers and plant protection chemicals, as well as other types of anthropogenic influence on the soil microbiome. Generally, this kind of research is limited to the top soil layer (0–20 cm). The number of works devoted to the description of the full-profile microbiome of soils of different genetic types (Blume et al. 2002; Chernov 2015a, b; Fierer et al. 2003; Aparin et al. 2020; Tikhonovich et al. 2018) is clearly insufficient for understanding the mechanism of the influence of agricultural use on the taxonomic structure of the soil prokaryote community. The aim of this study is to compare the structure of the microbiome of the genetic profile of chernozem in different sites with a known history of land use and located in homogeneous natural conditions.

2 Objects and Methods

The microbiome structure of chernozem was studied in the Kamennaya Steppe Nature Reserve (the Voronezh Region) (N: 51.03137°, E: 40.72780°) (Fig. 1a). The pits were laid on a flat watershed along transects (mowed fallow— forest belt—wheat crops) with a distance of about 140 m between them (Fig. 1b).

Pit 1 is laid in a flat area under steppe grasses.

AU 0–50 cm: black, well-structured, granular, in the lower part of the horizon the structure changes to angular blocky; the slightly moist upper layer (10–15 cm) is densely penetrated by roots; silty clay; the transition to the next horizon is gradual.

AB 50–90 cm: non-uniform in color, brownish gray, light brownish from 85 cm, angular blocky, slightly moist, roots are sparse; soil reacts with 10% HCl with a score of 78 cm; silty clay, the transition to the next horizon is clear.

C 90–200 cm: uniformly colored, light brown, slightly moist, abundant mycelium, from 140 cm there are sparse white nodules, silty clay.

The soil is huperhymic chernozem (WRB 2014), light clay on carbonate brown clays. Morphological signs of agricultural use are absent.

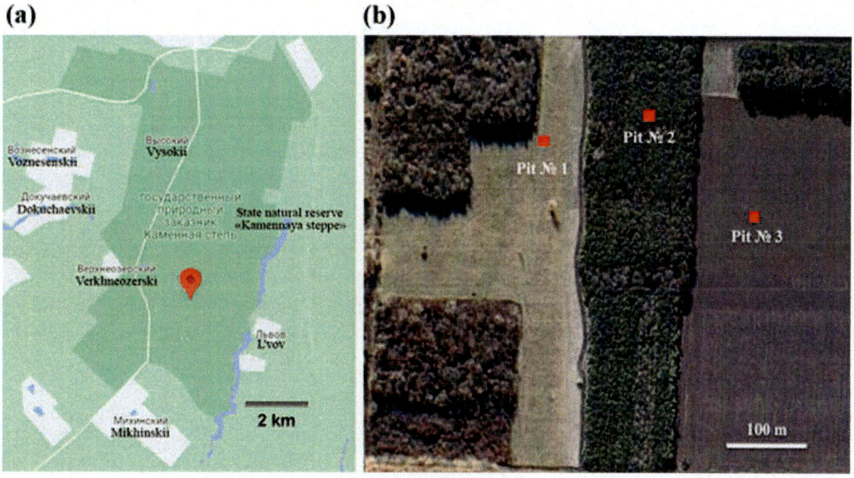

Fig. 1 Soil study areas of different land use: **a** Nature Reserve "Kamennaya Steppe"; **b** Location of the pits

The morphological structure of soil under forest belt (Pit 2) differs from the fallow soil by the presence of sparse undergrowth (up to 1 cm), lighter black colour of the humus horizon and looser topsoil (0–30 cm). The main features of agricultural chernozem (Pit 3) are the presence of a ploughed layer (0–20 cm), a fine-aggregate structure, and a higher depth of reaction to 10% HCl (30 cm).

Samples from the main genetic horizons (A, AB, C) at the depths of 5–10, 50–60 and 90–100 cm were taken for the chemical characterization of soils in triplicate. Simultaneously, in order to study chernozem microbiome, samples were taken from the same depths in three pits laid under different sites according to a special technique (Andronov et al. 2011). In addition, samples without signs of soil formation were taken from a depth of 190–200 cm in the fallow.

The level of pH_{H20}, C_{org}, NH_4 and mobile P and K compounds (P_2O_5, K_2O) was determined using the generally accepted methods (Vorob'eva 1998).

The microbiome was studied according to the methodological guidelines (Andronov et al. 2011; Caporaso et al. 2010) and by DNA and RNA isolation from soil with further 16S rRNA gene sequencing on a Roche GS junior sequencer at the Laboratory of Soil Microbiological Monitoring and Bioremediation of the Institute for Agricultural Microbiology. The comparative analysis of the sequencing libraries was performed using the Q1IME program.

The Shannon index (formula 1) (Chernov 2015b, c), which characterizes the diversity of prokaryotic communities, was calculated on the basis of the data obtained.

$$H = \sum_{i=1}^{S} p_i \log p_i \tag{1}$$

where,

H Shannon index
S total number of species
p_i proportion of a *i* species in a community.

Additionally, samples of the clay fraction were studied. The clay fraction was isolated from two samples of the fallow soil, taken from depths of 5–10 and 190–200 cm, by soaking in distilled water. In the clay fraction samples, microbial communities were isolated and analyzed at the phylum taxonomic level.

3 Results

3.1 Physical and Chemical Characteristics of Soils

All soils are characterized by weak variation in the content of the physical fraction of clay and silt both by horizons and under different land plots at the same depth (Table 1). Arable soils contain 2% less C_{org} in the AU horizon compared to the fallow and forest belt soils. At the depth of 50–60 cm (horizon AB) the difference in C_{org} content in different sites decreases to 0.6%. In the soils under the forest belt and fallow soils, the water extraction reaction is close to neutral (6.5–6.6). In the cropland soil, the values of water extract pH are higher and reach 7.4 in the layer of 5–10 cm and 8.0 in the layer of 50–60 cm deep.

Table 1 Characteristics of soil samples in different land-use areas

Horizon	Sampling depth, cm	HM*, %	pH_{H2O}	C_{org}, %	Fraction content, %	
					<0,01 mm	<0,001 mm
Fallow						
AU	5–10	5,3	6,5	6,70	66	33
AB	50–60	5,5	6,5	3,90	69	46
C	90–100	4,9	8,2	2,70	67	43
Forest belt						
AU	5–10	8,8	6,6	7.40	67	39
AB	50–60	6,1	6,4	2,30	69	43
C	90–100	4,4	8,4	0,90	71	45
Cropland						
PU	5–10	5,4	7,4	4,50	64	40
AB	50–60	4,5	8,0	2,90	66	41
C	90–100	3,6	8,1	0,60	68	41

* Hygroscopic moisture

The phosphorus content in the fallow and forest soils is two–three times lower (4.8 mg/100 g of soil) than in the cropland soil. Below the depth of 60 cm, the phosphate content changes insignificantly. The most exchangeable potassium is contained in the upper layer of the forest belt soil (30 mg/100 g of soil), and its least content is identified in the fallow soil. In the lower horizons, potassium content is the same almost everywhere (3–5 mg/100 g of soil).

3.2 *Taxonomic Composition of the Soil Microbial Community*

The microbiome was studied at different taxonomic levels (phylum, class, order, family, genus). The microbiome structure analysis was carried out in different aspects: (a) MC changes at soil profile depths, (b) evaluation of similarity and difference of MCs of surface and middle horizons of soils in different sites, and (c) comparison of soil-forming microbocenoses of soils in different land plots in question.

3.3 *Microbial Community Composition at the Phylum Level*

28 phyla of microorganisms have been identified and seven of them were dominant: the p. Proteobacteria, p. Actinobacteria, p. Firmicutes, p. Verrucomicrobia, p. Acidobacteria p. Gemmatimonadates and p. Chloroflexi. The proportion of the first three phyla ranged from 70 to 90% (Table 2).

3.4 *Profile Distribution of Phyla*

Three types of changes can be distinguished in the phylum content, as well as other taxa changes, along the soil profiles (horizons AU, PU, AB, and C). The first type is characterized by high taxon content in the upper horizon and its gradual decrease in the middle horizon and the source rock. The second type includes taxa with their minimum content in the upper horizon, with a gradual increase in the middle horizon and reaching a maximum in the parent rock. The third type includes taxa with a higher percentage in the middle horizon compared to the upper and lower horizons.

In forest soils the variation of phyla content by depth is complex by nature. The content of the p. Verrucomicrobia, p. Actinobacteria, p. Chloroflexi and Gemmatimonadetes phyla reaches its maximum in the AB horizon and minimal in the C horizon. The Proteobacteria phyla in horizon B are at a minimum and in horizon C at a maximum. The Acidobacteria and Planctomycetes phyla are distinguished by a high proportion in the humus horizon (A1), and the Firmicutes phyla in the C horizon.

Table 2 Soil microbiome composition at the phylum level (% of phyla content)

The type of land use		Forest belt			Cropland			Fallow			
Horizon		AU	AB	C	PU	AB	C	AU	AB	C	C
Sampling depth, cm		5–10	50–60	90–100	5–10	50–60	90–100	5–10	50–60	90–100	190–200
Phylum	Crenarchaeota	1	1	1	2	1	0	1	0	1	0
	Acidobacteria	8	4	4	8	5	2	7	2	1	2
	Actinobacteria	35	49	15	41	27	14	47	43	12	13
	Bacteroidetes	2	1	2	3	3	1	2	1	2	2
	Chloroflexi	3	7	2	3	2	2	2	5	1	1
	Firmicutes	1	1	26	1	15	30	1	8	26	25
	Gemmatimonadetes	4	5	2	13	4	2	3	4	2	2
	Nitrospirae	2	2	1	1	1	0	1	1	0	1
	Planctomycetes	4	1	1	2	2	1	3	2	1	1
	Proteobacteria	29	9	42	22	34	45	20	26	51	50
	Verrucomicrobia	10	18	4	2	6	3	13	6	3	4
	WS3	2	0	1	0	0	0	0	0	0	0

Overall, these three phyla account for 76–83% of the whole content in the cropland soil. The following features are observed in the microbiome structure change along the profile depth:

(a) Gradual more than twofold decrease of p. Actinobacteria, p. Gemmatimonadetes and p. Acidobacteria with depth;
(b) Two-fold increase in the Proteobacteria phyla and three-fold increase in p. Firmicutes;
(c) Higher phyla content in horizon AB.

In the soil-forming rock, the percentage content of phyla differs from the average values for all samples. Primarily, this applies to Proteobacteria, which predominates in horizon C (about 50%), the second place is occupied by Firmicutes (25–29%) and the third position is taken by Actinobacteria (12–13%). They are followed by p. Verrucomicrobia constituting a much smaller share (2.4–4.2%).

3.5 *Soil Microbiome Composition at the Class Level*

At the taxonomic class level, 16 representatives were identified among the phyla with a proportion exceeding 1%. Four classes are represented by the Proteobacteria and Actinobacteria phyla. Two classes were identified in the phyla Acidobacteria and Gemmatimonadetes, and one class was found in other phyla. These 16 classes represent from 89 to 98% of the total class content. The range of class content is from 1 to 31%. The number of classes with more than 9% content is half of the total number of classes, and in each horizon the content of classes exceeding 9% is 3–4%. In the phyla Actinobacteria in all samples, the class Thermoleophilia prevails significantly. In the phyla Proteobacteria, the maximum share is occupied by c. Gammaproteobacteria and the minimum by c. Deltaproteobacteria (2%), c. Alphaproteobacteria are in the second position (9%).

3.6 *Profile Distribution of Classes*

The first type of profile content variability includes five classes (c. Thermoleophilia, c. Acidobacteria-6, c. Alphaproteobacteria, c. Deltaproteobacteria and c. Acidimicrobia) (Fig. 2). The second type includes the following two classes: c. Bacilli and c. Gammaproteobacteria. The third type includes three classes: [Spartobacteria], Actinobacteria, and MB-A2-108. The remaining three classes of Chloracidobacteria, Sphingobacteria and Gemm-1 are observed in small amounts of 1–2% and have no definite confinement to the horizons.

The considered distribution types of profile classes are found in soils of all land types According to the soil class distribution by profile in soils of different sites, the following features are observed: the first type of distribution prevails in the cropland

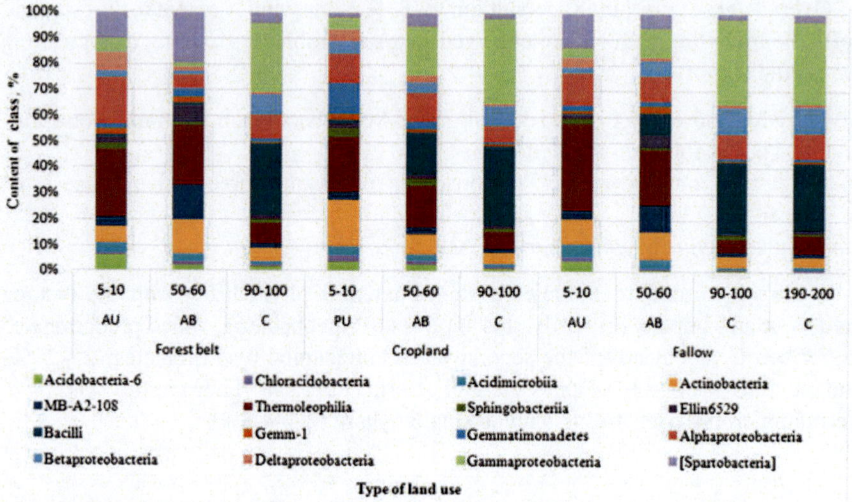

Fig. 2 Class distribution by soil genetic horizons in different land uses

and virgin soils; the first and third types are represented to a different extent in the forest soils; the second type of class share distribution in soils of all site is represented by three classes.

3.7 *Soil Microbiome Composition at the Order Level*

At this level, ten representatives of four phyla (Actinobacteria—5 orders, Proteobacteria—3 orders, Verrucomicrobia—1 order and Firmicutes—1 order) were attributed; their content exceeds 1%. The total proportion of bacteria ranged from 42 to 64%, with minimum values characterizing the soil-forming rock at 42–47%. The following orders are given in order of descending (from 22 to 9%): o. Pseudomonadales—22%, o. [Chthoniobacterales]—18%, o. Solirubrobacterales—17%, o. Gaiellales—14%; o. Actinomycetales, o. 0319-7L14 and o. Rhizobiales—13% each. The order Xanthomonadales rounds off the series with 10%.

There is no uniformity in the share of orders in the soil horizons. One to three orders are found among the predominant orders (with more than 10% content) in horizons.

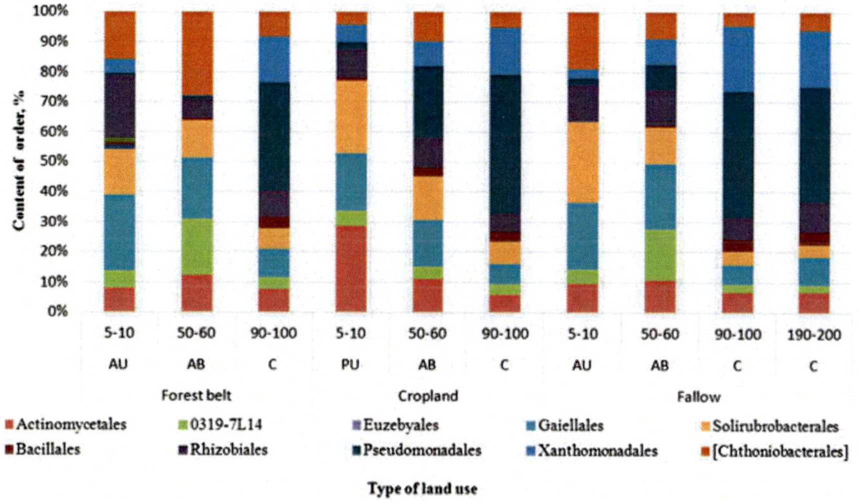

Fig. 3 Distribution of orders in the genetic horizons of different soils

3.8 *Profile Distribution of Microorganisms at the Order Level*

The same three types discussed above stand out in the pattern of changes in the microbial community fractions at the order level (Fig. 3).

The first type includes the orders Gaiellales and Solirubrobacterales. The second type (which content increases with depth) includes the orders Pseudomonadales and Xanthomonadales, but the difference in content of the order Xanthomonadales between horizons is much smaller than that of the order Pseudomonadales. The order Bacillales can be assigned to the same type, with a slight difference in content (1–2%). The order 0319-7L14 belongs to the third type. Two orders ([Chthoniobacterales] and Actinomycetales) have a mixed type distribution (third and second types) depending on the sites.

3.9 *Microbial Community Structure at the Family Level*

A total of 158 families were identified in the soil microbial community, with contents ranging from 0.1 to 14%. The number of families with more than 1% content is an order of magnitude smaller. Six families representing three phyla were attributed among them: two families of Actinobacteria, three families of Proteobacteria and one family of Verrucomicrobia (Table 3). In general, their total share varies from 12 to 25%. The family Gaiellaceae has the maximum content (14%), and the minimum contents are identified for f. Patulibacteraceae (1–4%) and f. Bradyrhizobiaceae (1–2%).

Table 3 Soil microbiome composition at the family and genus level (% of content)

The type of land use		Forest belt			Cropland			Fallow			
Horizon		AU	AB	C	PU	AB	C	AU	AB	C	C
Sampling depth, cm		5–10	50–60	90–100	5–10	50–60	90–100	5–10	50–60	90–100	190–200
Family	Gaiellaceae	14	13	4	8	8	3	13	13	3	4
	Patulibacteraceae	1	2	0	1	1	0	4	1	1	0
	Bradyrhizobiaceae	2	1	1	1	1	1	1	2	1	1
	Hyphomicrobiaceae	8	3	1	2	3	1	6	4	2	3
	Xanthomonadaceae	0	0	7	1	3	7	1	5	10	9
	[Chthoniobacteraceae]	9	18	4	2	5	2	12	5	2	3
Genus	Rhodoplanes	7	2	1	2	3	1	5	3	1	2
	Pseudomonas	0	1	12	1	9	17	1	3	15	14
	Stenotrophomonas	0	0	7	0	3	7	0	5	10	9

According to the character of profile distribution, the families belong to two types: type one of Xanthomonadaceae, type two of Hyphomicrobiaceae, and all others. The exception is the family of Patulibacteraceae, which belongs to the third type with a slight difference in content.

3.9.1 The Composition of the Microbial Community at the Genus Taxonomic Level

The microbial community at this level is represented by three genera of the Proteobacteria phylum, the content of which exceeds 1%. Their total proportion is from 21 to 26%, and it is the highest in the soil-forming rock, and the minimum in horizon A1 (Table 3).

The share of the genus Pseudomonas prevails among the others. According to the depth distribution type, the genus Stenotrophomonas belongs to type 1 and the genus Pseudomonas to type 2. The genus Rhodoplanes in the forest and fallow soils belongs to the second type and in the cropland soils to the third one.

4 Discussions

The history of agriculture in the study area exceeds 200 years. Three developmental trends were revealed in the development trajectory of chernozem after the land-use change 120 years ago. They determined the transition of the soil to the evolutionary stage of development while maintaining the profile-forming process.

Restoration of the natural soil biocenosis began in the arable chernozem at the site of mowed fallow. In the forest belt site, the arable soil began to evolve in accordance with the humus-accumulative eluvial trend. And in modern arable soil, the trend of anthropogenic degradation process remained (dehumification, acidification, carbonate accumulation, degradation of soil structure and the imbalance of natural elemental chemical composition as a result of mineral fertilizers). Therefore, changes in the structure of the soil microbial community of different land-use sites can be considered in conjunction with soil development trends.

4.1 The Impact of Land-Use Type on the Microbiome Structure at the Phylum Level

The influence of the type of land-use on the microbial community is considered in the fallow–cropland–forest sites. The content (share) of phyla serves as the main indicator of changes in the microbial community structure of the humus horizon under the influence of land use. Among the predominant seven phyla, we can distinguish four phyla with significantly different contents in the studied soils. In the sites of different land-use types (fallow–cropland–forest), the share of the p. Actinobacteria decreases and the share of the p. Proteobacteria increases. In the cropland soils, the proportion of p. Gemmatimonadetes increases and the share of p. Verrucomicrobia decreases relative to the soils from other sites.

The situation is different in the middle horizon (AB), here the ratio of phyla and their content changes. The content of three phyla (p. Proteobacteria, p. Firmicutes and p. Acidobacteria) in the cropland soil is significantly higher relative to other soils, and the content of the p. Actinobacteria and p. Chloroflexi phyla is lower. The content of p. Verrucomicrobia increases in the forest soil compared to fallow and cropland soils. In the soil-forming rock, with the same ratio of dominant phyla, the content of the Proteobacteria and Actinobacteria phyla increases in the fallow–cropland–forest sequence and the content of the p. Acidobacteria increases insignificantly.

The taxonomic structure of microbiomes of the seven dominant phyla is similar in the upper horizons of the forest belt and fallow soils. In the cropland soils, differences appear in the third, fifth, and sixth positions: the share of the p. Gemmatimonadetes increases relatively and the share of p. Verrucomicrobia decreases significantly. At the same time, p. Actinobacteria (1), p. Proteobacteria (2) and p. Firmicutes (7) occupy similar positions in the soils of all land use types. Differences in the positions of phyla are observed in the microbiome structure in the middle horizons of different soils.

The comparison of microbiome structures at the depth of 90–100 cm and 190–200 cm in fallow soils showed their complete identity (Fig. 4). This gives grounds to believe that the influence of modern soil-forming processes and land-use types on microbiome formation is practically absent already at a depth of 90–100 cm.

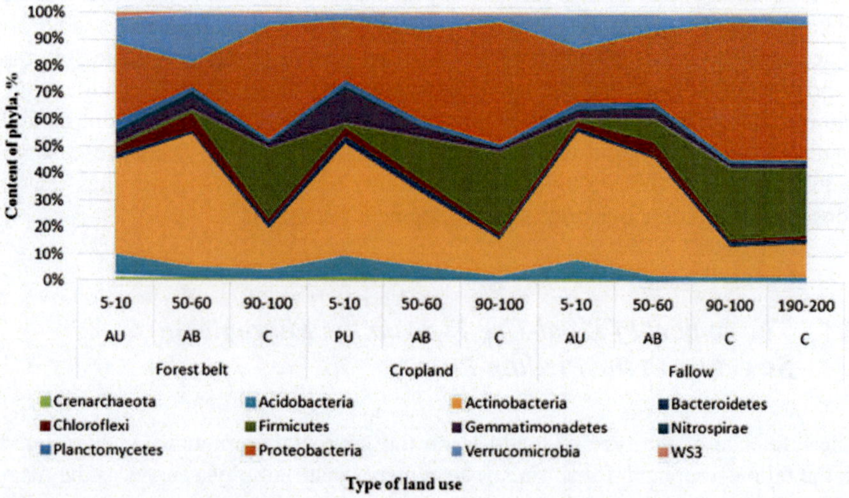

Fig. 4 Distribution of phyla in genetic horizons of soils in different sites

The predominance of the Proteobacteria phyla in the soil-forming rock, judging by the published data (Andronov 2017; Tikhonovich et al. 2018), is typical for soils of different types of loam and clay soils (Umbric Albic Luvisols, Albic Luvic Phaeozems, Calcic Cambisols, Chernozems, Solonetz) (soil names are given according to the World Reference Base for Soil Resources (WRB 2014)). It should be noted that the environmental reaction and carbon content, which are considered to be the most important factors influencing the microbiome structure (Hartman et al. 2009; Lauber et al. 2008, 2009) are in a wide range of pH levels from 5.2 to 8.7, with organic carbon from 0.01 to 2%, and obviously with a wide range of temperature and moisture content. Our data (Aparin et al. 2020) on the microbiome structure of podzolic soil in clay loam under spruce forest and soil in an urban park of Saint-Petersburg also showed dominance of the Proteobacteria phyla. This allows us to hypothesize that this might have been inherited from past epochs of soil formation.

4.2 *Influence of Land-Use Type on the Microbiome Structure at the Class Level*

A pronounced feature of the distribution of classes according to their percentage content by horizon is the absolute predominance of the Bacilli and Gammaproteobacteria classes in the soil-forming rock of the soils in all sites, their share ranging from 25 to 31%. Two classes, Thermoleophilia (19–30%) and Alphaproteobacteria (10–16%), prevail in the upper soil horizons of all sites. The middle soil horizons of all land-use sites are distinguished by a higher content of four classes. At the same time, differences in the combination of classes from the two phyla are observed. In the forest belt soil, these are c. Thermoleophilia, c. Actinobacteria, c. MB-A2-108 and c. [Spartobacteria]. Out of the four classes in the cropland soil, c. Gemmatimonadetes dominate; they are followed by c. Thermoleophilia, c. Bacilli and c. Alphaproteobacteria representing three phyla.

In the middle horizon of virgin soil there are six classes with relatively high content from 8 to 20%. The first position is occupied by c. Thermoleophilia, the second position is occupied by three classes: Actinobacteria, MB-A2-108 and Gemmatimonadetes; the third position is occupied by the Alphaproteobacteria class, and the Bacilli class (8%) closes the series.

In general, the soil-forming rock (horizon C), regardless of the type of site has a very similar composition and shares of classes. This suggests that the change in the content of classes in horizons A and B is related to the influence of soil processes and the land-use type.

The markers of the microbial community of different horizons at the taxonomic level of classes are the content and proportion of [Spartobacteria], Actinobacteria, MB-A2-108, Thermoleophilia and Bacilli representing the three phyla of Verrucomicrobia, Actinobacteria and Firmicutes.

4.3 Influence of Land Type on the Microbiome Structure at the Order Level

The microbial communities of the soil horizons in all sites have individual features in terms of the predominant orders of bacteria. It is not possible to single out any specific soil indicators. Pseudomonadales and Xanthomonadales orders are representing the phyla Proteobacteria, dominate in the soil-forming rock (first and second positions). Horizon C differs strongly from the horizons located above it.

4.4 Influence of Land-Use Types on the Soil Microbiome Structure at the Family Level

In the upper horizon (A) and the middle horizon (AB), regardless of the soil type, the p. Gaiellaceae ranks first in terms of content. This family significantly prevails over the others in both horizons of the cropland soil and in horizon AB in the forest soil. In both horizons of the cropland soil, the content of the f. Gaiellaceae is significantly lower (by almost 1/3) in comparison with the forest and fallow soils. The f. Hyphomicrobiaceae occupies the second place in horizon A of the forest soil and fallow soil. The f. Xanthomonadaceae ranks second in horizon AB of the fallow soils. The Xanthomonadaceae family predominates in the soil-forming rock regardless of the land type, and the content is similar in the soils under forest plantations and croplands. In the fallow soils, the share of this family is one-third higher. The family Gaiellaceae occupies the second position with a twofold decrease only in the forest soils.

4.5 Influence of Land-Use Type on the Soil Microbiome Structure at the Genus Level

In all sites, the genus Rhodoplanes dominates in the humus and median horizons. At the same time, its share in the cropland soils in horizon A1 is significantly lower than in the same horizon of the forest and fallow soils. In horizon AB, there is no difference in the content of this genus in soils of different sites. The genus Pseudomonas prevails in the soil-forming rock, with Stenotrophomonas occupying the second place. The difference in the content of families in different soils is insignificant.

4.6 Diversity of Microbiomes in Soils of Different Land-Use Types

The Shannon index was used to assess the phylogenetic diversity of prokaryotes of soils in different sites, which is not dependent on the sample size and is sufficiently informative (Chernov 2015b, c; Fierer and Jackson 2006; Nannipieri et al. 2020; Pulleman et al. 2014). The analysis of diversity indices revealed a decrease in the abundance of prokaryotic communities with depth in soils of all sites. The Shannon index values varied within a quite wide range from 5.9 to 9.1 (Fig. 5).

The inverse relationship between the diversity index and the structure of the microbial community of the horizons is clearly visible. The diversity decreases as the proportion of p, Protebacteria increases. The maximum diversity of microbiomes is observed in the humus horizon of the cropland soil. This soil is also characterized by the widest range of index values. In soils of all land-use types, the index consistently decreases with depth. The diversity variation between layers from 0–10 to 50–60 cm and from 50–60 to 90–100 cm differs in soils of different sites in question. In the cropland soils the differences between diversity indices of the middle horizon and the soil-forming rock are wider than in the fallow and forest soils. This peculiarity may be related to agricultural use of soils.

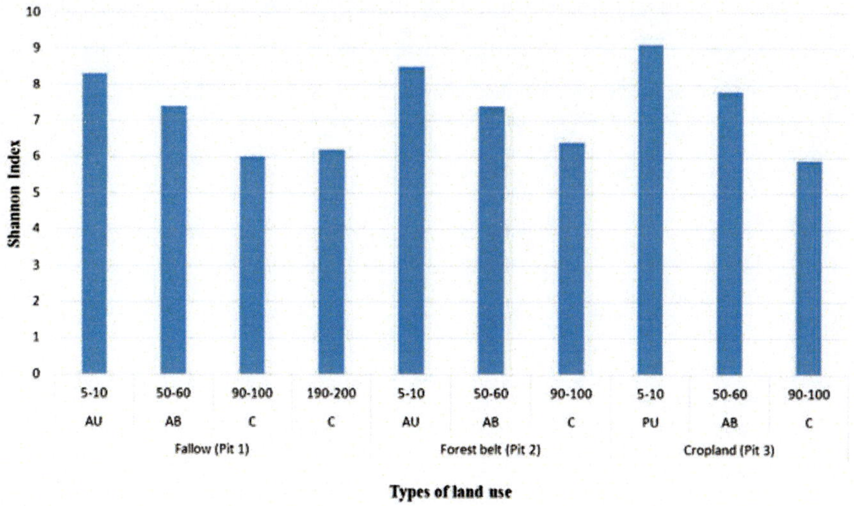

Fig. 5 Biodiversity index of chernozem prokaryotic community (by Shannon)

4.7 Comparative Analysis of the Prokaryotes Content in Bulk Samples and the Clay Fraction of the Fallow Soils

It is known that microorganisms are predominantly located on the surface of the solid soil and are held by sorption forces (adhesion or immobilization of microorganisms) (Saeki and Kunito 2010). The finer the soil particles, the more firmly microorganisms are held on their surface, forming an organic-mineral soil matrix. It can be assumed that the microbial composition of the fine-grained (<1 mm) soil will differ from that of the clay fraction (<0.001 mm).

The clay fraction contained practically the same phyla representatives as in the bulk sample (Table 4). However, the comparison of the samples revealed significant differences in the bacterial community structure. They were manifested both in the ratio of bacteria and in their content. The share of Verrucomicrobia and Actinobacteria in the prokaryotes MC of the clay fraction from the humus horizon (AU) decreased significantly in absolute value, and the share of the Proteobacteria phyla increased. Overall, there is an increase in the proportion of most phyla in the clay fraction. Significant changes were observed in the p. Chloroflexi, p. Firmicutes, p. GAL15 and p. Gemmatimonadetes. The positions occupied by the p. Proteobacteria, p. Actinobacteria and p. Verrucomicrobia were reversed in the clay fraction.

The phyla Proteobacteria prevails in the clay fraction of the soil-forming rock (89%), there is a slight increase in the share of the p. Acidobacteria, p. Bacteroidetes

Table 4 Composition of the soil microbial community of the mowed fallow at the taxonomic phylum level of bulk soil samples and isolated clay fraction samples (% of content)

Horizon		AU		C	
Sampling depth, cm		5–10		190–200	
The type of sample		Bulk sample	Clay fraction	Bulk sample	Clay fraction
Phylum	Crenarchaeota	1	4	0	2
	[Parvarchaeota]	0	1	0	0
	Acidobacteria	7	5	1	2
	Actinobacteria	50	13	12	2
	Bacteroidetes	2	1	1	3
	Chloroflexi	2	6	1	0
	Firmicutes	1	4	26	1
	GAL15	0	3	0	0
	Gemmatimonadetes	3	7	2	0
	Nitrospirae	1	1	1	0
	Planctomycetes	1	1	1	0
	Proteobacteria	19	45	51	89
	Verrucomicrobia	14	1	4	2

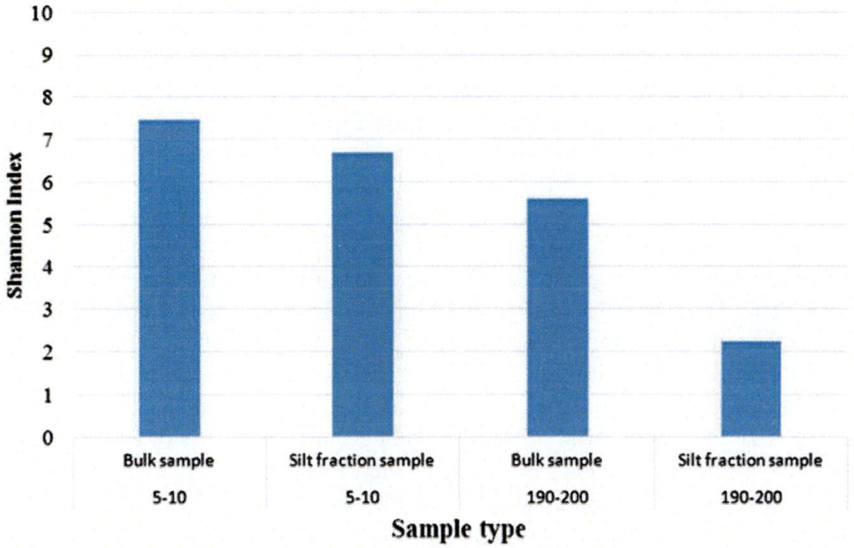

Fig. 6 Shannon index of soil samples from the Kamennaya Steppe Nature Reserve and the isolated silt fraction samples of the fallow (pit. 1)

and p. Crenarchaeota, and a decrease in the content of the p. Firmicutes and p. Actinobacteria compared to the general sample.

The Shannon index calculation showed that the diversity of microorganisms in the bulk samples is greater than in the clay fraction, and the diversity of the microbial community is lower primarily in the soil-forming rock (Fig. 6).

Thus, the study of the clay fraction MC showed that its structure is significantly different from that of the bulk soil sample. An independent study of the MC of the fine-dispersed fraction of soils is important for understanding the formation of soil microbiome.

Significant differences between the microbial communities of the clay fraction and the bulk samples of the humus horizon and the parent rock allow us to hypothesize about the more ancient origin of the organo-mineral matrix of soils.

5 Conclusion

The microbiome structure of chernozems was formed mainly under the influence of humus-accumulative profile-forming process in the organo-mineral matrix of soil-forming rocks. The matrix is characterized by the absolute predominance of three phyla of Proteobacteria, Firmicutes, Actinobacteria, whose total share exceeds 80%.

Changes in the microbiome of soil-forming rocks under the influence of humus-accumulative process of soil formation was expressed in the increase in the phylogenetic diversity of prokaryotes, the substitution of the dominant phylum Proteobacteria by the phylum Actinobacteria, and a decrease in the content of phylum Firmicutes, which moved from the second to seventh position in horizon A. The depth of MC transformation of soil-forming rocks, as follows from the analysis of profile distribution of prokaryotes, does not depend on land-use type and does not exceed 90 cm.

The comparative analysis of the microbiomes of the genetic horizons of soils from different land-use sites showed that the type of land use does not fundamentally change the structure of the chernozem microbiome. In soils of humus horizons AU and PU of all sites, the first two positions among the microbial community are occupied by the phyla Actinobacteria and Proteobacteria with the share of 61–67%. This ratio of the predominant phyla in the humus horizon of soils of all land types characterizes the high stability of the MC of humus horizons.

The identified changes are carried out with agricultural use. The MC structure (the sequence of seven phyla) is completely the same in the forest and fallow soils. Individual features of the MC are observed in the middle horizons (AB) in soils of different sites. In the fallow and forest soils, the MC has more similarity with the humus horizon than with the soil-forming rock.

The obtained data show that the soil-forming rock differs sharply in the absolute predominance of the Proteobacteria phylum. Analysis of the published data (Ananyeva et al. 2021; Chernov 2015b; Chirak et al. 2013; Ivanova et al. 2015; Tikhonovich et al. 2018; Zhelezova et al. 2015) shows that the prevalence of the Proteobacteria phylum is a characteristic feature of soil-forming rocks of loam and clay composition, regardless of natural conditions and the type of soil formation. This is probably a feature inherited from the previous epochs of soil formation.

Studies of the clay fraction of the humus horizon and the soil-forming rock revealed significant differences in the structure of the microbial communities between them. Independent research of MC of silt soil fraction is important for understanding the regularity in the formation of soil microbiome.

The comparison of the microbiome structure of the profiles of the studied soils with the published materials revealed an invariant structure of the microbiome regardless of the soil type. It is formed by eight-nine phyla. Obviously, this invariant structure constitutes the specific structure of the soil microbiome as an interaction between biotic and abiotic components. This is an important result of soil genotypic studies. The microbial community reflects the genetic features of the soil. However, the small number of microbiome profile studies does not allow us to unambiguously determine the influence of typomorphic (genetic) characteristics on soil microbiomes.

Acknowledgements We thank the staff of FGBSI "Voronezh Federal Agricultural Scientific Centre named after V. V. Dokuchaev" for access to the soil archives, their assistance and detailed consultation during the process of soil sampling.

References

Ananyeva ND, Ivashchenko KV, Sushko SV (2021) Microbial indicators of urban soils and their role in the assessment of ecosystem services: a review. Eurasian Soil Science 54(10): 1517–1531

Andronov EE (2017) The preservation of microbial DNA in archived soils of various genetic types. PLoS ONE 12(3):e0173901

Andronov EE, Pinaev AG, Pershina EV, Chizhevskaya EP (2011) DNA extraction from soil samples (methodological guidelines). Federal State Budgetary Scientific Institution — All-Russian Research Institute for Agricultural Microbiology. Saint Petersburg (in Russian)

Aparin BF, Sukhacheva EYu, Kichko AA, Andronov EE, Valchenko VYa (2020) Structure of microbial community in forest and anthropogenic changed soils of megalopolis (St. Petersburg, Russia) Processes and Phenomena on the Boundary Between Biogenic and Abiogenic Nature – Springer, 395–416

Blume E, Bischoff M, Reichert JM, Moorman T et al (2002) Surface and sub-surface microbial biomass, community structure and metabolic activity as a function of soil depth and season. Applied Soil Ecology 20:171–181

Caporaso JG, Kuczynski J, Stombaugh J et al (2010) QIIME allows analysis of high throughput community sequencing data. Nature Methods 7(5):335–336

Chernov TI, Tkhakakhova AK, Kutovaya OV (2015a) Assessment of diversity indices for the characterization of the soil prokaryotic community by metagenomic analysis. Eurasian Soil Science 48(4):410–416

Chernov TI, Tkhakakhova AK, Lebedeva MP, Kutovaya OV (2015b) A comprehensive analysis of the microbiome in the complete profile of virgin light-colored solonetz soil at the territory of Dzanybek experimental station Bulletin of V.V. Dokuchaev Soil Science Institute 77:66–77(in Russian)

Chernov TI, Tkhakakhova AK, Kutovaya OV, Ivanova EA, Turusov VI (2015c) Seasonal dynamics of the microbiome of chernozems of the long-term agrochemical experiment in Kamennaya steppe. Eurasian Soil Science. 48(12):1349–1353

Chirak EL, Pershina EV, Dol'nik AS, Kutovaya OV et al (2013) Taxonomic structure of microbial association in different soils investigated by high through-put sequencing of 16S rRNA gene libraries. Sel'skohozjajstvennaja biologija 3:100–109 (in Russian)

Fierer N, Jackson RB (2006) The diversity and biogeography of soil bacterial communities. Proceedings of the National Academy of Sciences USA 103:626–631

Fierer N, Schimela JP, Holden PA (2003) Variations in microbial community composition through two soil depth profiles. Soil Biology and Biochemistry 167–176

Hansel C, Fendorf S, Jardine P, Francis C (2008) Changes in bacterial and archaeal community structure and functional diversity along a geochemically variable soil profile. Applied and Environmental Microbiology 74:1620–1633

Hartmann M, Lee S, Hallam SJ, Mohn WW (2009) Bacterial, archaeal and eukaryal community structures throughout soil horizons of harvested and naturally disturbed forest stands. Environmental Microbiology 11:3045–3062

Ivanova EA, Pershina EV, Andronov EE, Kutovaya OV, Tkhakakhova AK, Chernov T, Markina LG, Kogut BM (2015) The structure of microbial community in aggregates of a typical chernozem aggregates under contrasting variants of its agricultural use. Eurasian Soil Science 48(11):1242–1256

Lauber CL, Hamady M, Knight R, Fierer N (2009) Pyrosequencing-based assessment of soil pH as a predictor of soil bacterial community structure at the continental scale. Applied and Environmental Microbiology 75(15):5111–5120

Lauber CL, Strickland MS, Bradford MA, Fierer N (2008) The influence of soil properties on the structure of bacterial and fungal communities across land-use types. Soil Biology and Biochemistry 40:2407–2415

Nannipieri P, Ascher-Jenull J, Ceccerini MT, Pietraramellara G, Renella G, Schloter M (2020) Beyond microbial diversity for predicting soil functions: A mini review. Pedosphere 30(1): 5–17

Pulleman M, Creamer R, Hamer U, Helder J, Pelose C, Peres G, Rutgers M (2014) Soil biodiversity, biological indicators and soil ecosystem services – an overview of European approaches. Current Opinion in Environmental Sustainability 4:529–538

Saccá A, Caracciolo AB, Di Lenola M, Grenni P (2017) Ecosystem services provided by soil microorganisms. Soil Biological Communities and Ecosystem Resilience, Sustainability in Plant and Crop Protection. Springer 9–24

Saeki K, Kunito T (2010) Adsorptions of DNA molecules by soils and variable-charged soil constituents. Current Research, Technology and Education Topics in Applied Microbiology and Microbial Biotechnology 1:188–195

Schloter M, Nannipieri P, Sørensen SJ, van Elsas JD (2018) Microbial indicators for soil quality. Biology and Fertility of Soils 54:1–10

Tikhonovich IA, Chernov TI, Zhelezova AD, Tkhakakhova AK, Andronov EE, Kutovaya OV (2018) Taxonomic structure of prokaryotic communities in soils of different bioclimatic zones. Bulletin of V.V. Dokuchaev Soil Science Institute 95:125–153 (in Russian)

Vorob'eva LA (1998) Soil chemical analysis Moscow State University, Moscow (in Russian)

World Reference Base for Soil Resources (2014) International soil classification system for naming soils and creating legends for soil maps. FAO, Rome

Zhelezova AD, Kutovaya OV, Dmitrenko VN, Thakahova AK, Hohlov SF (2015) Estimation of DNA quantity in different groups of microorganisms within genetic horizons of the dark-gray soil Bulletin of V.V. Dokuchaev Soil Science Institute 78:87–98 (in Russian)

Clay Mineralogy as a Marker of Volcanic Biogeosystem Evolution in Laetoli, Tanzania

Andrei B. Rozanov, Sofia N. Lessovaia, Anatoly N. Zaitsev, Gennady P. Kopitsa, Yulia E. Gorshkova, Natalia V. Platonova, Dmitry Yu. Vlasov, Irina Y. Tikhomirova, and Joshua Mwankunda

Abstract The Laetoli hominid footprints dating back some 3.6 million years discovered by Leakey in 1978 is an archaeological site of great importance in understanding the human evolution. The footprints of hominids, animals, and birds cast in the volcanic ash consolidated into tuff deposits are also an insight into the evolution of the whole biogeosystem of this area dominated by volcanic activity. The volcanic ash

A. B. Rozanov (✉)
Department of Soil Science, Stellenbosch University, P.Bag X1, Matieland, South Africa
e-mail: dar@sun.ac.za

S. N. Lessovaia
Institute of Earth Sciences, St. Petersburg State University, Universitetskaya Nab. 7/9, St. Petersburg, Russia

A. N. Zaitsev
Department of Mineralogy, St. Petersburg State University, Universitetskaya Nab. 7/9, St. Petersburg, Russia

Image and Analysis Centre, Natural History Museum, Cromwell Road, London SW7 5BD, UK

G. P. Kopitsa
B.P. Konstantinov Petersburg Nuclear Physics Institute NRC KI, Gatchina, Russia

I.V. Grebenshchikov Institute of Silicate Chemistry RAS, Saint Petersburg, Russia

Y. E. Gorshkova
Joint Institute for Nuclear Research, Joliot-Curie Str. 6, Dubna, Russia

N. V. Platonova
The Research Park, St. Petersburg State University, Universitetskaya Nab. 7/9, St. Petersburg, Russia

D. Yu. Vlasov
Biological Faculty, St. Petersburg State University, Universitetskaya Nab., d. 7/9, St. Petersburg, Russia

I. Y. Tikhomirova
Herzen University, 48 Moika Embankment, St. Petersburg, Russia

J. Mwankunda
Ngorongoro Crater, Ngorongoro Conservation Area Authority, P.O. Box 1, Arusha, Tanzania

O. V. Frank-Kamenetskaya et al. (eds.), *Biogenic—Abiogenic Interactions in Natural and Anthropogenic Systems 2022*, Springer Proceedings in Earth and Environmental Sciences, https://doi.org/10.1007/978-3-031-40470-2_21

deposits consolidated into tuffs are a marker which sets the base line for tracking the further sediment transport in the catchment of this historic site. The surface of the tuff exposure along the Garusi river carrying the footprints shows no signs of weathering and the soils of the area form in subsequent sediments that filled the valley after the deposition of the tuff material. Following the deposition and consolidation of the airfall tuffs, the biogeosystem of the Laetoli gorge and its surroundings experienced a complex evolution which led to formation of the present-day soil cover. The study of clay mineralogy of these soils has provided interesting insights into the evolution of this system.

Keywords Mbuga clay · Alluvial soils · Smectite · Internal pores

1 Introduction

The Laetoli hominid footprints dating back some 3.6 million years discovered by Leakey in 1978 is an archaeological site of great importance in understanding the human evolution (Agnew et al. 1996). The footprints of hominids, animals, and birds cast in the volcanic ash consolidated into tuff deposits are also an insight into the evolution of the whole biogeosystem of this area dominated by volcanic activity. The footprints in airfall tuff were preserved by subsequent sediments overlying the tuff in the Laetoli gorge. The site is located close to the course of the Garusi stream—a tributary of the Kakesio river cutting through the Eyasi plateau and draining into the lake Eyasi. After a few years after excavation, it was observed that the surface containing footprints started to deteriorate due to colonization be vegetation and beginning of soil formation (Agnew and Demas 2016). The preservation attempts continue to this date with an ongoing initiative to create a protective display cover hosting a museum.

The soils of the Laetoli have not been studied in detail yet but were observed and noted (mainly in terms of their colour and texture) during the botanical survey (Andrews and Bamford 2008). A detailed study was undertaken further North-East of the study area (Stonehouse and Duff 1974) and, though outdated in terminology is of substantial value for understanding the soil genesis and structure of soil cover in the Laetoli vicinity. One of the main features of the area is accumulation of thick layers of mbuga clays covering the plain at the foothills of the Lemagrut volcano. It is this surface that was cut by the Garuzi river to expose the Laetoli beds and later sediments underlain by the Precambrian basement. Unfortunately, these clay deposits were not displayed on the geological map of Pickering (1964) unlike on the later maps, e.g. of the Arusha 1:250 000 sheet (Wilkinson 1983). It is the complex interaction of river incision, rock exposure, and sediment transport that seem to define the soil cover patterns of the Laetoli.

The nature and source of the sediments yielding the soils of the valley is not yet well studied and the analysis of clay mineralogy may significantly improve our understanding of landscape evolution in the area so clearly dated by the hominid footprints.

The main aim was to determine a relationship between the soil clay mineralogy of the Laetoli gorge and its recent evolution, establishing the main drivers of soil formation.

2 Study Area

The study area is located close to the Gregory Rift margin on the Eyasi plateau peneplain largely defined by the Precambrian rocks and complicated by the Pliocene volcanism. The geology of the area has been mapped at 1:125 000 scale by Pickering (1964) in the 1958–1961. The bedrock of the area South of Kakesio river is part of an Archaean granite–greenstone terrain of the Tanzanian Craton. According to Westerhof and Koistinen (2005), pegmatites, aplite, dolerite, lamprophyre dykes and quartz veins cut the greenstones of the Nyanzian System. The greenstone belt commonly comprises metavolcanics and sediments (banded iron formations and greywackes, sandstones and shales). Some areas in the Southern part of the area contain kimberlites. These rocks form the base of the old African surface (Burke and Gunnel 2008) which often displays the presence of laterites, calcretes and silcretes on the surface marking the surface in the sense attributed to this term by Partridge and Maud (1987).

The recent sediments of the Pliocene sedimentary sequences of the Laetoli vicinity have been studied in some detail by various authors from the 1940's and summarized by Ditchfield and Harrison (2011). Figure 1 shows the location of our sampling points from the foot of Sadiman volcano across the Eyasi plateau and the Kakesio river to the old African surface. The highest sampling density is in the Laetoli gorge and its vicinity. The main focus of this study are the soils formed on tuffs (Laetoli vicinity), basalts (Lemagrut foothills) and Precambrian metamorphic complex (right bank of the Kakesio river and the shore of the lake Eyasi).

The topography of the study area is dominated by the Sadiman and Lemagrut volcanoes in the North-East. The hydrological system of the Southern Sadiman slopes is separated from the Laetoli basin by the foothills of Lemagrut. The drainage system of the Sadiman is isolated from Laetoli by the Lemagrut foothills and at present cannot contribute to sediment accumulation in the valley. On the other hand the slopes of Lemagrut and the steams draining it may have a material contribution to sediment accumulation in the Laetoli gorge. The river basins of the Kakesio (including the Garusi tributary) and Enaishu rivers form some 20-km wide alluvial plain stretching towards the escarpment between the Lemagrut foothills and the Kakesio batholith in the South West. The plain was probably cut by a much bigger river in the Precambrian gneisses and schists prior to the volcanic activity and may be even prior to rifting, since the current size of the catchment does not justify such a wide valley.

The rainfall measurements are scarce, but the mean annual rainfall map clipped from the WCMC conservation database via ILRI gives some idea of the rainfall distribution within the study area peaking at 1000–1200 m in the highlands of Karatu area and rapidly falling in all directions. The rainfall at Laetoli may be estimated as 800–1000 mm per annum decreasing to 600–700 mm towards Kakesio. Vegetation

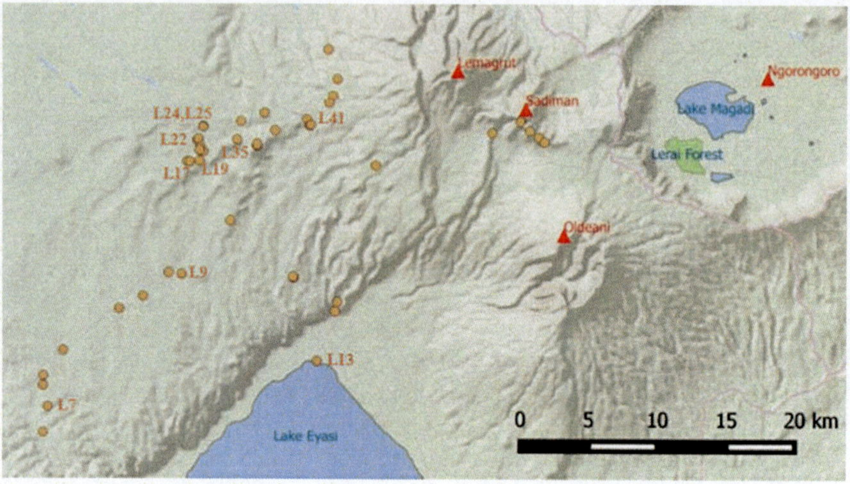

Fig. 1 Sampling points in the Laetoli vicinity (OCM Landscape and OSM Natural Features overlay), L7—Lithic Haplustept, L9—calcrete on the surface, L13—salt pans, L17—Mollic Ustifluvent, L19—Vertic Ustifluvent, L25—Calcic Haplustert, L24—Plinthustox, L22 and L35—Lithic Haplustolls, L41—Typic Rhodustalf

is well studied and was closely linked to soil conditions of the area during a botanical survey largely limited to the Endulen-Laetoli surroundings mapped at 1:70,000 scale (Andrews and Bamford 2008).

3 Methods

3.1 *Field Research and Principle of Soil Sampling*

The cartographic software used in this study is Q-GIS v. 2.10.1-Pisa. DEM: SRTMVF 1 arc-second, approximate resolution: 30 m. Topographic map with 10 m isohypse contour interval (available from OSM as OCM Landscape via QGIS OpenLayers plugin) was used as a background. Geological map (scanned and georeferenced raster of the Quarter degree sheet 52 "Endulen" (Pickering 1964) was used for geological backdrop to maximize the diversity of lithologies covered by soil survey. The 55 soil profiles were studied (Fig. 1). The sequences of soil horizons were described and sampled. The soils were classified using the most recent Keys to Soil Taxonomy (Soil Survey Staff 2014).

3.2 Mineralogy Composition (XRD Data)

The soil samples were ground and sieved to pass through a sieve with 1 mm aperture. The samples were treated according to Mehra and Jackson (1958) method. Then the <1 μm soil fractions were separated by sedimentation and decantation (Gorbunov 1971). For coagulation of the clay fraction, 1 N $MgCl_2$ solution was used. Pretreatment of the fine fraction before XRD examination has included ethylene glycol solvation and heating at 550 °C for 3 h. The soil samples (the <1 μm fractions) were also studied in the 060 area. The mineralogical composition of several rock fragments sampled from the soil profiles was studied in powders. The mineralogical composition of the samples was studied using a Rigaku MiniFlex II (Rigaku Corporation, Tokyo, Japan) X-ray diffractometer (XRD) with Co*Kα* radiation. The minerals identified in the clay fractions of the studied soil profiles according to (Harris and White 2008) and database (PDF-2 2020).

3.3 Specific Surface Area of the Rock Fragments Specific Surface Area of the Rock Fragments

The specific surface area (the total surface area of a material per unit of mass) and the pore size distribution of the rock fragment samples were determined with the low-temperature nitrogen adsorption method, using a QuantaChrome Nova 1200e analyzer (Quantachrome Instruments, Florida, USA). The samples were degassed in a drying chamber at 100 °C for 19 h; then the samples were analyzed by the multipoint Brunauer–Emmett–Teller (BET) method, using 7 points. The pore size distribution was calculated by the Barret–Joyner–Halenda (BJH) method, using the nitrogen desorption isotherm (NOVAWin™—Windows® Based Software for Operation from PC). The method can be used to investigate pores with diameters < 100 nm (Anovitz and Cole 2015).

3.4 Mesostructure and Fractal Properties of the Rock Fragments

The mesostructure (mesostructure data—structural parameters in diapason 1–200 nm) and fractal properties (fractality—self-similarity, which means that the object is exactly or almost the same as a part of itself) of the rock fragment samples were investigated by the small angle X-ray scattering (SAXS) method. SAXS measurements were carried out using the Xeuss 3.0 SAXS/WAXS System (Joint Institute for Nuclear Research, Dubna, Russia), operating in point geometry and using a $GeniX^{3D}$ microfocus generator of X-ray radiation with Cu*K*α(λ = 0.154 nm), operating in the mode of 30W/30 μm. The spectrometer was equipped with a moving detector Eiger2 R 1 M with a sensitive area of 77.1 × 79.7 mm^2 (pixel size 75 μm). Using two distances from the sample to detector (400 and 4500 mm) enabled measurement of the intensity $I(q)$ of X–ray scattering in the range of momentum transfer of $0.025 < q < 6$ nm^{-1} ($q = (4\pi/\lambda) \sin(\theta/2)$, where λ is the incident x-ray wavelength and θ is the scattering angle). Measurements of the samples were carried out in a vacuum at room temperature.

4 Results

4.1 Soils

The footprint tuff 7 is part of the upper Laetoli beds and is exposed in the channel of the Garusi river cutting through the surrounding Eyasi plateau. The left bank of the Garusi river appears as a severely eroded bad land—a dense network of glullies with residual islands of soil standing out like inselbergs held in place by the roots of the bushy acacias. The soils are Mollic Ustifluvents (profile L17) displaying a simple stratification into A (0–20 cm, 10YR 6/2) and C (20–80 cm, 2.5Y 6/2), underlain by the solid tuff. The soils here are the products of fluvial reworking of tuffacesous mainly light in colour and of mixed texture forming layers of sandy and clay alterations with occasional gravel layers. The topsoil is dark (where still present) and the subsoil horizons are strongly calcareous with some cementation, but not cemented into a calcrete pan.

The Laetoli gorge in its upper part is surrounded by an extensive mbuga plain—a thick deposit of black clay where soils are classified as Vertisols. The right bank of the Garusi river is higher and the thickness of alluvial deposits is much greater here, though still severely eroded lies a few meters higher than the left and presents a possibly the youngest alluvial terrace. The soils here have strong vertic features but considering the alluvial nature of the material classified as Vertic Ustifluvents. Here severe erosion also produces residual inselbergs of soil. An example of such soil is profile L19 stratified into Av (0–20 cm, 5YR 3/1)—V (20–80 cm, 5YR 3/1)—C (80–100 cm, 10YR 4/2, an alluvial deposit)—CV (100–120 cm, 5YR 2.5/1) underlain by solid tuff. Mbuga deposit contain high concentration of chalky carbonate neoformations, which may indicate a margin of a water body (lake, swamp or a seasonal wetland) where concentration of carbonates occurred with particularly high intensity due to regular desiccation of the surface.

The right edge of the valley is the watershed between Garusi and Bitin rivers. The latter belongs already to the Olduvai catchment. Here we find two rather different soils in close proximity. One (L24) is a fossilized Plinthustox—a soil with distinct soft plinthic layer containing Fe–Mn concretions. This soil seems to have been further transformed by a different environment, which resulted in formation calcite deposits on the surface of the peds in the prismatic B horizon. The main material is yellow–brown to red and is more common on the surrounding slopes of volcanoes. The other

(L25)—Calcic Haplustert located just a few meters downhill, is a Vertisol formed on complex of various deposits dominated by mbuga clay material containing layers re-deposited carbonate neoformations. It rests on the gleyic horizon, which indicates permanent wetness. The combination of these two profiles may indicate a former wetland margin which existed prior to the formation of the gorge and deep incision of the Garusi river. A shallow Mollisol (Lithic Haplustoll) without vertic properties (profile L22) with a single 20 cm thick ACca horizon, was observed in the lower part of the valley on tuffs. It is distinctly different in appearance and structure from mbuga clays and contains large amounts of calcrete fragments.

The left edge of the valley is defined by the flow of Ogol lavas basaltic in their mineralogy. The top of the Oldonyo Mati hills is covered by shallow Lithic Humustepts with extremely high clay content. The total soil depth barely exceeds 30 cm where the soil merges into fractured basalt. Basalt fragments are also occasionally imbedded into the ground mass and are scattered on the surface. Further downhill the soils remain the same but become deeper—Lithic Haplustoll (L35) until the final transition into the mbuga plain entirely covered by Vertisols. The profile L35 has a well-structured mollic topsoil A (0–18 cm, 5YR 3/2), transitional AB (18–40 cm, 5YR 3/3), underlain by basaltic rock of the Ogol lavas.

The Northern edge of the mbuga plain lies at the foothills of the Lemagrut volcano. The basalts shaping the foothills are topped by the well-developed red soils (Typic Rhodustalf) with pronounced clay accumulation in the subsoil (Argillic B horizon). Typic Rhodustalf on basalt (profile L41) displays a 2 cm layer of sod at the surface, an A (2–20 cm, 7,5YR 3/2) well-structured horizon, a transitional AB (20–33 cm, 7,5YR 3/2)—B1 (33–60 cm 7,5YR 3/4)—subangular blocky B2fe (60–90 cm, 7,5YR 5/4) with Fe–Mn mottles.

A fragment of the Precambrian peneplain was inspected to the South-West of Laetoli along the Kakesio road on the right bank of the Kakesio river. Here the foothills of the granite inselbergs are largely eroded exposing the basement schists and mylonite-cataclasite complex (L7, L9). Profile L7—Lithic Haplustept on mylonite was described in the wall of a gully exposing the Precambrian basement. The AC (0–20 cm) horizon is thin and bleached (10YR 7/2) structure is poorly developed throughout the profile, stones are common. Coarse fragments are not rounded. The soil overlies a silcrete deposit. Location L9 represents a sample of calcrete cropping out in proximity. The shores of the lake Eyasi are covered by salt pans (L13).

4.2 *Soil Clay Mineralogy (XRD Data)*

Pure smectite was identified by the response to ethylene glycol solvation whereby a peak at 15.2–15.4 Å in the air-dry state shifts to a fundamental 001 spacing of ≥17 Å and the appearance of the 002 peak at 8.5 Å. The intensive and symmetrical peaks of smectite were observed on the XRD patterns of the <1 μm fraction only in the Mollic Ustifluvents of the valley bottom (example of L17) (Fig. 2). Besides a predominant smectite, the small proportions of illite and kaolinite were identified. Although the basal 001 peak is symmetrical, the samples show the 060 peaks at 1.50 and 1.54 Å (Fig. 3). The dominant peak at 1.50 Å has demonstrated the dioctahedral nature of smectite. The second smaller peak at 1.54 Å implies the presence of trioctahedral phase (More and Reynolds 1997). Based on the XRD data the same material within predominance of pure smectite and two peaks (1.50 and 1.54 Å) in 060 area was also spotted in the alluvial strata of the mbuga—the light coloured layers without vertic features.

In mbuga soils (L19 and L25) predominance of smectite and small proportions of kaolinite and illite were observed (Fig. 2). In contrast to the alluvial soil, the mbuga has 060 peak at 1.50 Å, which attributes to presence of only the dioctahedral phase (Fig. 3). The smectite 001 peak is broad and asymmetrical. That is particularly well pronounced in the upper horizons (L19) and in the gleyic horizon (L25) where the peak shifts towards low d-values and may be described as "smeared" (Fig. 2). Smectite identified by the "smeared" peak was also observed on the XRD pattern of the Mollisols (L35 and L22). A very different clay mineralogy was found in the Typic Rhodustalf (L41) and Lithic Haplustept (L7)—soils formed on the Northern and Southern bounds of the valley. The profile of L41 at the foot of the Lemagrut volcano shows illite—kaolinite association. In the profile L7 formed from basement rock sepiolite and mica were observed in the clay fraction.

4.3 *Mineralogy of Rock Fragments (XRD Data)*

Calcrete fragments that were sampled from the points L7, L22, and L9 are dominated by calcite. The traces of feldspars (microcline in L7, microcline and albite in L9, and albite in L22) were identified. The L7 also contains the small proportions of dolomite, quartz, chlorite, amphibole, and epidote. The sample from the salt pan on the shore of the Lake Eyasi (L13) the sodium minerals—analcyme, halite, and trona were predominant within small amounts of illite, plagioclases, and calcite. The soft plinthic layer (L24) shows traces of calcite and quartz, the high proportions of feldspars (anorthoclase) and X-ray amorphous phase. In the clay size fraction separated from the sample (L24), a "smeared" 001 peak of smectite was observed on the XRD patterns. The basaltic lava fragments sampled from L35 are dominated by plagioclase and pyroxene within lower proportion of microcline and traces of quartz and dolomite.

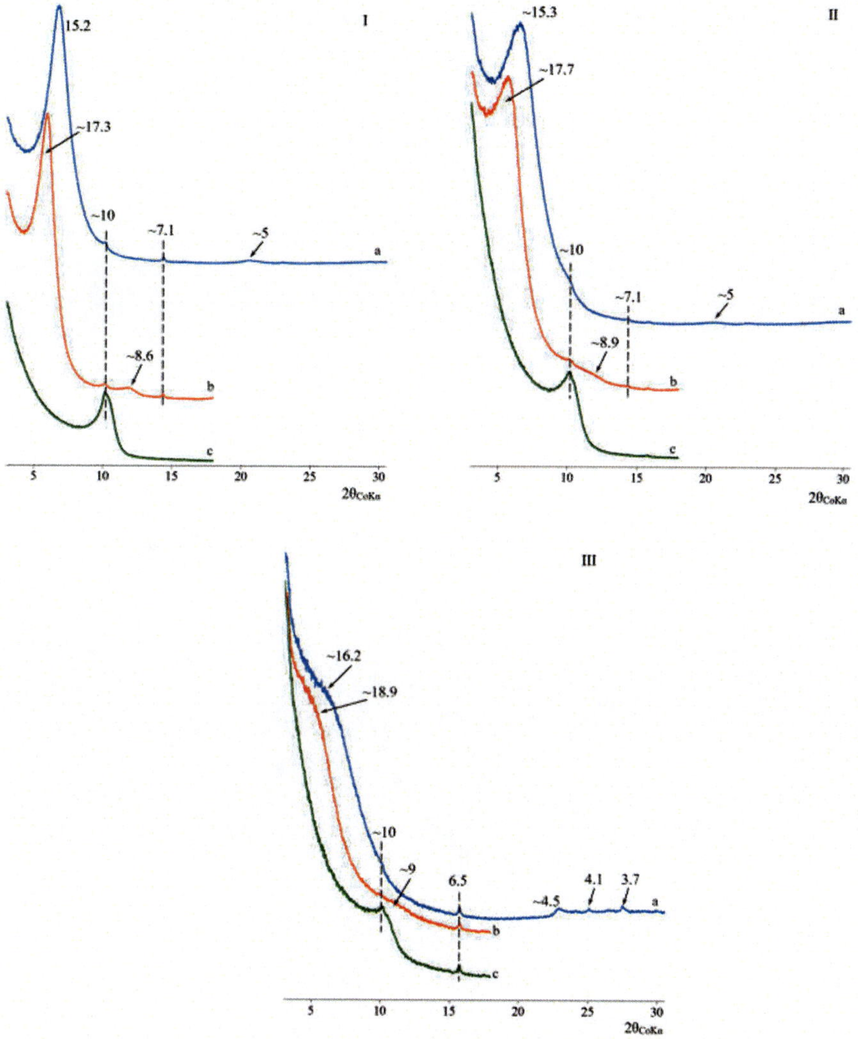

Fig. 2 XRD patterns of the fine size fractions of: I—C horizon (L17); II—V horizon (L25); and III—G horizon (L25); **a** air-dried, **b** ethylene glycol solvated, and **c** 550 °C heated samples; d-values are given in angstroms

4.4 Specific Surface Area of the Rock Fragments and Mesostructure and Fractal Properties of the Rock Fragments

The full nitrogen adsorption—desorption isotherms of all rock fragment samples (Fig. 4) are characterized by expressed capillary condensation hysteresis and refer to

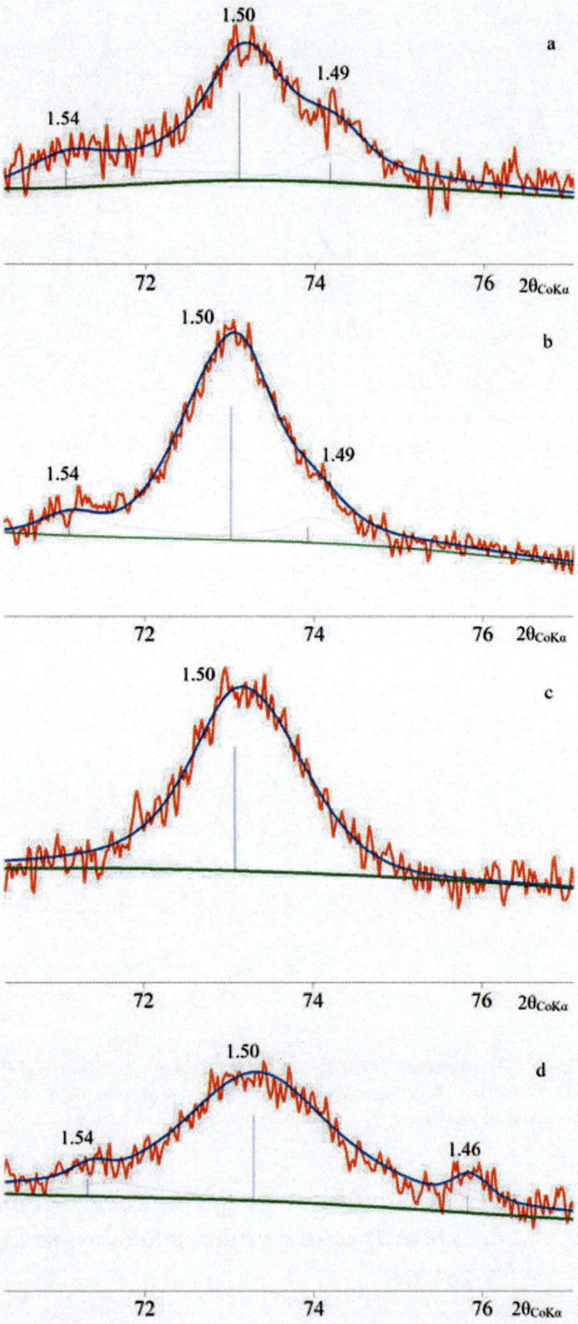

Fig. 3 XRD patterns of the fine size fractions: the 060 peaks and their decomposition into individual components: **a** L17 (C horizon), **b** L19 (the alluvial strata), **c** L25 (V horizon), **d** L35 (AB horizon)

type IV (IUPAC). It corresponds to adsorption on mesoporous (diameter 2–100 nm) materials. The shape of the hysteresis loops for all samples corresponds to the classical type H3 (associated with slit-like pores). Mathematical processing of complete nitrogen adsorption—desorption isotherms within the BJH model allowed us to obtain dV(d) pore size distributions for all samples. For example, the dV(d) distributions are shown for samples L22, L13, and L35 (Fig. 4). For all the studied samples, a narrow maximum is observed on the dV(d) pore size distribution graphs at $d_P =$ 3.9 nm. While the samples L9 and L35, have the smallest values of the specific surface area $S_{BET} \approx 5.9$ and 3.6 m^2/g, are characterized by a normal dV(d) distribution. A sufficiently large number of micropores (< 2 nm) is observed for the sample L7, whereas both micropores and mesopores are present in L22. Samples L24 and L13 are characterized by the presence of a large number of mesopores in the complete absence of micropores (<2 nm).

The texture parameters for rock fragments obtained from the analysis of nitrogen adsorption–desorption isotherms using BET and BJH models are shown in Table 1. Figure 5 shows the experimental dependences of the intensity of $I_S(q)$ SAXS for rock fragment samples. The scattering pattern is typical for scattering on disordered porous systems (solid phase—pore). Two observed types of scattering show inhomogeneities that differ significantly in their characteristic scale R_c. The analysis of the scattering for rock fragment samples was carried out according to the standard procedure that

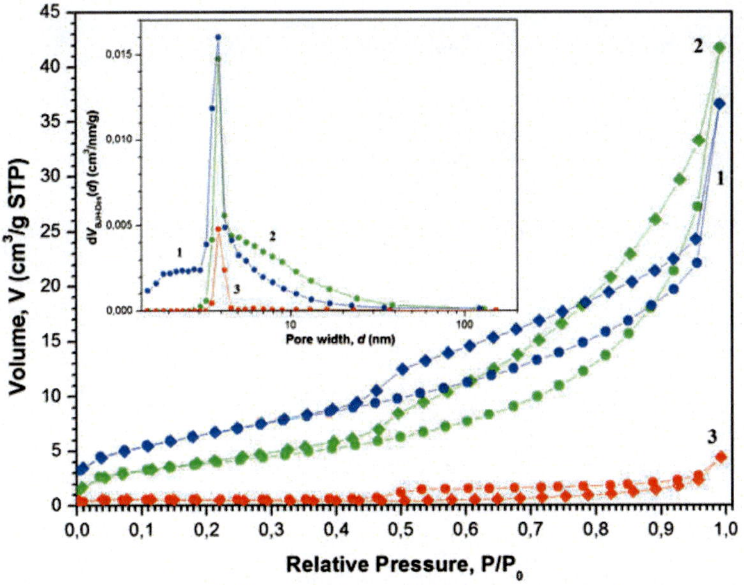

Fig. 4 Full nitrogen adsorption (circles)—desorption (diamonds) isotherms and their corresponding pore size distributions *dV*(*D*) (insert) obtained from the treatment of full nitrogen adsorption–desorption isotherms within the BJH model for the samples: 1—L22, 2—L13, and 3—L35

we previously used in the study of porous materials of various origins (Lessovaia et al. 2020; Ganyushkin et al. 2021; Smyslov et al. 2017; Yorov et al. 2019; Matović et al. 2022), using a generalized unified exponential-power expression that takes into account the presence of several structural levels in the scattering system (Beaucage 1995). The results of the analysis of SAXS data are presented in Fig. 5 and Table 1.

Analysis of the SAXS curves showed that the 1-st type of inhomogeneities, smaller in characteristic scale, for samples L7, L9, L22, and L24 consists of particles with a developed fractal phase interface (solid phase—pore) with dimension $2.66 \leq D_{S1} = 6–n_1 \leq 2.93$ and an upper limit of self–similarity $39 \leq R_{c1} \leq 52$ nm (Table 1). For the L35 sample, the 1-st type of inhomogeneities are volume fractal clusters with dimension $D_{V1} = n_1 = 2.84$ and the upper limit of self-similarity $R_{c1} \approx 48$ nm (Table 1). A special case is the sample L13, which is characterized by the organization of the structure according to the type of logarithmic fractal $D = n_1 = 3$. The volume and surface fractals are self-similar at different scales, whereas the logarithmic fractal changes hierarchically when scaled (Iashina et al. 2017, Yashina and Grigoriev 2019).

The second type of inhomogeneities for all samples consists of large–scale particles having either a "diffuse" phase interface (solid phase—pore) (Schmidt et al. 1991) for samples L7, L9, L24, and L35, or a sharp smooth phase interface for samples L13 and L22. For all samples, the characteristic size of these inhomogeneities is R_{c2} > 120 nm.

Table 1 Texture properties and parameters of mesostructure for rock fragment samples obtained from analysis data of low-temperature nitrogen adsorption and SAXS

Parameters	Samples					
	L7	L9	L22	L24	L35	L13
Low-temperature nitrogen adsorption method						
S_{BET} (m^2/g)	17.3 ± 0.5	5.9 ± 0.1	23.9 ± 0.7	14.7 ± 0.2	3.6 ± 0.1	14.1 ± 0.4
V_T (cm^3/g)	0.034	0.015	0.056	0.076	0.007	0.065
d_P (nm) BJH (des)	3.9	3.9	3.9	3.9	3.9	3.9
SAXS						
R_{c2} (nm)	>120					
n_2	4.25 ± 0.02	n_2	4.25 ± 0.02	n_2	4.25 ± 0.02	n_2
R_{g1} (nm)	40 ± 2	35 ± 1	30 ± 1.5	38 ± 1.2	37 ± 1	33 ± 1
$D_{S1} = 6 - n_1$	2.66 ± 0.01	2.93 ± 0.01	2.88 ± 0.01	2.66 ± 0.01	–	–
$D_{V1} = n_1$	–	–	–	–	2.84 ± 0.01	–
n_1	–	–	–	–	–	3.01 ± 0.01

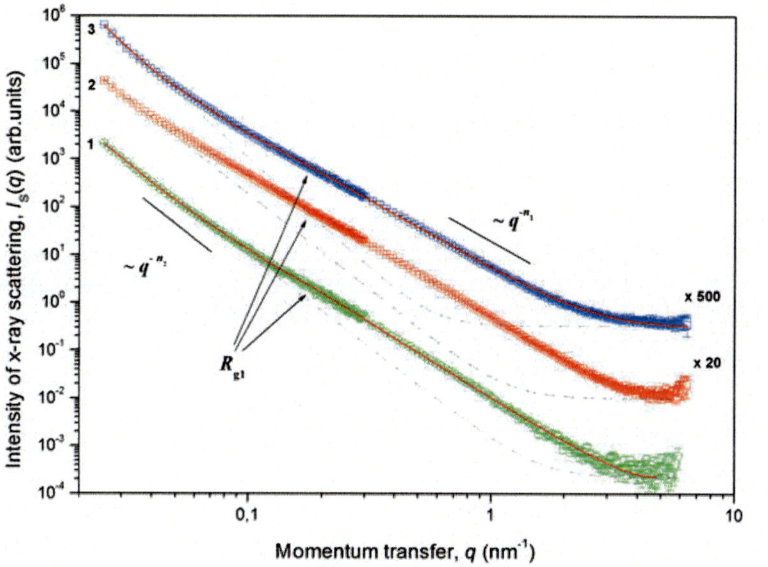

Fig. 5 The experimental dependences of the intensity of $I_S(q)$ SAXS versus momentum transfer q for rock fragment samples: 1—L22, 2—L13, and 3—L35. For the sake of clarity, $I_S(q)$ values for samples 2—L13 and 3—L35 were multiplied by 20 and 500 (corresponding factors are given next to the curves)

5 Discussions

The Laeotli soils may be divided into several groups according to clay mineral association that reflects different sources and/ or transformation pathways of soil parent material: (i) illite-kaolinite (Typic Rhodustalf, profile L41), (ii) mica-sepiolite (Lithic Haplustept, L7), and (iii) smectitic—all the other observation points (alluvial soils—Mollic Ustifluvent, L17), Mbuga soils—L19 and L25, and Mollisols—L35 and L22). The smectite groups in soils are also not uniform clearly seen in 060 peak.

In the alluvial soils (example of L17) two peaks in the 060 area are associated with dioctahedral smectite (1.50 Å) and trioctahedral phase (1.54 Å). According to Deocampo et al. (2009) such a complex asymmetric 060 peak (Fig. 3) shows the presence of Mg-rich phase along with Al-rich phase. The above authors attribute this phase to the change in the octahedral composition as a result of solid-state alteration. This phase shows coexisting dioctahedral and trioctahedral domains within individual sheets, or a homogeneous octahedral sheet of intermediate composition. These conclusions were reached while studying the lacustrine clays of the Olduvai Gorge located in the neighboring catchment West of Laetoli. Previously Hover and Ashley (2003) studying the same deposits proposed that the dissolution precipitation reaction in the lacustrine deposits of Olduvai led to replacement of end-member dioctahedral smectite (montmorillonite) with end-member trioctahedral smectite (stevensite or saponite). Deocampo (2015) suggested that the highly saline and alkaline environment of the Olduvai paleo lake, the dissolved Mg in the presence of Al- or Fe-rich detrital clays may lead to formation of Mg-rich smectite.

Based on the clay mineralogy we suspect that the material of these alluvial soils (L17) as well as the alluvial strata in mbuga soils is unlike the tuff 7, in which only dioctahedral smectite—montmorillonite was found (Zaitsev et al. 2016, 2019; Perova et al. 2022). The Mollic Ustifluvents (L17) of Laetoli gorge are formed predominantly on alluvial tuffaceous material, rather than from the direct weathering of the airfall tuff. The alluvial material of the described soils possibly formed from the Ndolayna beds (Ditchfield and Harrison 2011) represented by tuffaceous materials and calcretes, which overly the Laetoli airfall tuffs.

We can assume that soil formation did not affect the surface of the Footprint Tuff until recently, and the material of this tuff did not contribute to soil formation in the vicinity of the exposure. That was also confirmed by the data of molecular and cultural investigations of fungal and bacterial diversity (Zaitsev et al. 2020) studied in tuffs and soils we identified as alluvial. There are noticeable differences in the microbiome structure of soils and tuffs. The proportion of phylum Actinobacteria in tuff samples was significantly higher than in soils. The proportion of bacteria from the phylum Bacteroidetes was significantly higher in soils. Among Bacteroidetes halophilic bacteria are common. Species associated with aquatic biotopes are also present, which may indicate the influence of these factors in past periods. In relation to fungi, there were also noticeable differences between soil and tuff samples. Thus, tuffs were characterized by the predominance of dark-colored fungi (Dothideomycetes), known for stress tolerance. These fungi often exhibit a conservative development strategy and can exist for a long time on solid substrates, rocks, and minerals. At the same time, representatives of the Eurotiomycetes dominated the soil samples. They are quite typical for developing soils. Genus Aspergillus dominated in soil samples. Some species of Aspergillus are resistant to salinity. Fungi of Sordariomycetes, often associated with plants or plant residues were in a significant proportion of the soil samples. The above study provides additional evidence showing quite different ecological conditions that existed in the tuff and the alluvial sediments.

Generally, the mbuga soils (Vertisols) are dominated by smectite and calcite with traces of mica which has not been altered to smectite (Ngailo 1998). The mbuga clays display 060 peak at 1.50 Å indicating the presence of only dioctahedral phase in the studied samples (L19 and L25). Whereas the 001 peak that is getting smeared in the horizon mostly effected by pedogenesis—top horizons and gleyic horizon. The smectite may be characterized as curved smectite described in Kenyan and Tanzanian soils formed from peralkaline volcanic ash (van der Gaast et al. 1986). The authors suggest that these clays are poorly ordered along the c-axis due to a poor stacking of layers and despite the large proportion in the sample curvature of layers led to poor orientation and weak 001 reflections on the XRD patterns**.** Exclusively dioctahedral smectite is attributed in neighbouring Olduvai gorge to upland weathering which produce dioctahedral smectite through pedogenic processes in the vadose zone, and limited Mg-enrichment by episodic saturation with saline and alkaline waters along the paleo lake margin (Hover and Ashley 2003). Accordingly, mbuga clays did not pass the salt alkaline weathering stage.

The mbuga clays are the most prominent and odd feature of the valley. Until recently the mbuga clays (classified here as Vertisols) were probably covering this part of the Garusi catchement to more-or less the line connecting the two small volcanoes: Emusenge and Endolaunya (Ndolanya). The lavas of these volcanoes could have partially blocked the valley resulting the valley fill by a mixture of tuff and lava material until the river has cut the gorge. This assumption is based on the fact that mbugas do not commonly occur in uplands, but rather occupy circular depressions and in the Arusha area often fill the floors of minor parasitic craters (Wilkinson 1983).

The unexpected results were observed for the clay mineralogy of the Mollisol (L35) formed from the Ogol lavas on the watershed. The smectite here may be characterized as curved smectite, judging by the smearing of the 001 peak and containing Mg-rich phase based on 060 peak. The basalt fragments extracted from the soil has few signs of weathering that confirmed by its texture and mesostructure parameters (Table 1).

The whole landscape from the outcrops of the Precambrian basement in the South to much younger tuffs of the Garusi cachment in the North is strongly calcified. This is a feature of the whole area including the Serengeti plain to the East and the escarpment overlooking the lake Eyasi in the West. The only exceptions are the slopes of the volcanoes like the Lemagrut at the top of the catchment. Calcretes are a common feature of the area. They are present on the African surface along with silcretes and ferricretes on top of the basement complex South of the Kakesio river. Calcretes are also present in the upper layers of Ndolanya beds (Ditchfield and Harrison 2011). Their fragments are present in the alluvial deposits throughout the area. The texture and mesostructure parameters of calcrete fragments (L9, L7, and L22) are not homogeneous. Some contain a large proportion of micropores (L7 and L22) as opposed to L9. The micropores tenaciously retain solutes interacting with the mineral surfaces (Simonyan et al. 2012). The hydromorphic materials—the plinthite (L24) and the sodic sediment from the lake Eyasi shore (L13) have no micropores as opposed to the calcrete fragments (L7 and L22) of the African surface (Table 1).

6 Conclusions

The spatial soil pattern of the Laetoli is largely influenced by the topography and parent material. The Footprint tuff exposures of Laetoli do not show any significant signs of weathering apart from the historically documented deterioration following the excavations of the 1970s. The recent soil formation occurs in the sediments deposited on top of the tuff 7. Clay mineralogy analysis has shown marked differences in the alluvial material from the tuff 7, though all the materials in the valley are dominated by smectites. Only the foothills of Lemagrut at the top of the valley show presence of kaolinite-illite association without smectite. The mica-sepiolite complex was observed in the clay mineralogy of soils formed on the Precambrian basement schists and mylonite-cataclasites. The clay mineralogy of the alluvial valley fill supposes presence of Mg-rich phase in the clays, not observed in the overlying mbuga clay deposits. The latter shows poor orientation of smectite shown by the smearing of the 001 basal peak.

The above observations give us an insight into a complex evolution of the Laetoli gorge biogeosystem strongly influenced by volcanic activity and related change in topography. Valley fill, waterlogging and subsequent erosion resulting in significant linear drainage of the area are the main features of the landscape evolution of this landscape. This evolution can be traced in part by changes in clay mineralogy of exposed deposits. The clay mineralogy indicates that the Footprint tuff exposure is very recent. It's not clear what triggered the erosion, but it seems to have not stabilized yet. Most of the contribution of tuffaceous material to soil formation is limited to alluvial soils at the bottom of the Garusi valley where it is mixed with the mbuga material in various proportions.

Acknowledgements We would like to thank the Tanzania Commission for Science and Technology for granting permission to conduct research (2016-245-NA-2016-181, 2017-234-NA-2016-115 for A.N. Zaitsev). Field work at Laetoli was supported by Peter Rich Architects—GMP Consulting Engineers—Laetoli JV (Arusha), St. Petersburg State University (grants 0.42.955.2016 and 3.42.740.2017) and Ngorongoro Conservation Area Authorities (Tanzania). XRD studies were carried out in the X-ray Diffraction Centre of St. Petersburg State University.

References

Agnew N, Demas M (2016) Fossil tracks and trackways: The dilemmas of preservation. Journal of Paleontological Techniques 15:3–21

Agnew N, Demas M, Leakey MD (1996) The Laetoli Footprints. Science (80) 271:1651–1652

Andrews P, Bamford M (2008) Past and present vegetation ecology of Laetoli, Tanzania. Journal of human evolution 54(1):78–98

Anovitz LM, Cole DR (2015) Characterization and analysis of porosity and pore structures. Rev. Mineral. Geochem. 80:61–164

Beaucage G (1995) Approximations leading to a unified exponential/ power-law approach to small-angle scattering. J. Appl. Cryst. 28:717–728

Burke K, Gunnell Y (2008) The African erosion surface: a continental-scale synthesis of geomorphology, tectonics, and environmental change over the past 180 million years. In: Memoir of the geological society of America 201, Geological Society of America

Deocampo DM, Cuadros J, Wing-Dudek T, Olives J, Amouric M (2009) Saline lake diagenesis as revealed by coupled mineralogy and geochemistry of multiple ultrafine clay phases: pliocene Olduvai Gorge, Tanzania. American Journal of Science 309:834–868 DOI https://doi.org/10.2475/09.2009.03

Deocampo DM (2015) Authigenic clay minerals in lacustrine mudstones. Special Paper of the Geological Society of America 515:49–64·DOI: https://doi.org/10.1130/2015.2515(03)

Ditchfield P, Harrison T (2011) Sedimentology, Lithostratigraphy and Depositional History of the Laetoli Area. In: Harrison T (ed) Paleontology and Geology of Laetoli: Human Evolution in Context: Volume 1: Geology, Geochronology, Paleoecology and Paleoenvironment. Springer Netherlands, Dordrecht

Ganyushkin DA, Lessovaia SN, Vlasov DY, Kopitsa GP, Almásy L, Chistyakov KV, Panova EG, Derkach E, Alekseeva A (2021) Application of Rock Weathering and Colonization by Biota for the Relative Dating of Moraines from the Arid Part of the Russian Altai Mountains. Geosciences (Switzerland) 11(8):342–363

Gorbunov N (ed) (1971) Metody Mineralogicheskogo I Mikromorphologicheskogo Izucheniya Pochv (Methods of Mineralogical and Micromorphological Study of Soils). Nauka, Moscow (in Russian)

Harris W, White NG (2008) X-ray diffraction techniques for soil mineral identification. In: Ulery AL, Drees LR (eds) Methods of soil analysis. Part 5. Mineralogical Methods, SSSA Book Series, No. 5, Madison, Wisconsin

Hover VC, Ashley GM (2003) Geochemical signatures of paleodepositional and diagenetic environments: A STEM/AEM study of authigenic clay minerals from an arid rift basin, Olduvai Gorge, Tanzania. Clays and Clay Minerals 51:231–251 doi:https://doi.org/10.1346/CCMN.2003.0510301

Iashina EG, Grigoriev SV (2019) Large-Scale Structure of Chromatin: A Fractal Globule or a Logarithmic Fractal? Journal of experimental and theoretical physics 129(3):455–458

Iashina EG, Velichko EV, Filatov MV, Bouwman WG, Duif CP, Brulet A, Grigoriev SV (2017) Additive scaling law for structural organization of chromatin in chicken erythrocyte nuclei. Physical Review E. 96(1):012411

IUPAC. The International Union of Pure and Applied Chemistry classification (https://iupac.org/)

Lessovaia SN, Gerrits R, Gorbushina AA, Polekhovsky YuS, Dultz S, Kopitsa G (2020) Modeling biogenic weathering of rocks from soils of cold environments. In: Frank-Kamenetskaya OV, Vlasov DYu, Panova EG, Lessovaia SN (eds) Processes and Phenomena on the Boundary Between Biogenic and Abiogenic Nature. Part of the Lecture Notes in Earth System Science book series (LNESS). Springer Nature Switzerland AG. Part of Springer Nature

Matović B, Gorshkova YuE, Kottsov SYu, Kopitsa GP, Butulija S, Minović Arsić T, Cvijović-Alagić I(2022) Carbon cryogel preparation and characterization. Diamond and Related Materials 121:108727

Mehra OP, Jackson ML (1958) Iron oxide removal from soils and clays by a dithionite citrate system buffered with sodium bicarbonate. Clays and Clay Minerals 7:317–327

Moore DM, Reynolds RC (1997) X-ray Diffraction and the identification and analysis of clay minerals. Oxford University Press

Ngailo JA (1998) Studies on mineralogy, micromorphology and relationships of soils along the Sukuma catena in Maswa district, Tanzania. Tanzania J. Agric. Sc. 1(2):109–120

Partridge TC, Maud RR (1987) Geomorphic evolution of southern Africa since the Mesozoic. J. Geol. 90(2):179–208

PDF-2 2020 (Powder Diffraction File), ICDD (International Centre for Diffraction Data)

Perova EN, Zaitsev AN, Spratt J, Vlasenko NS, Platonova NV, Bubnova OG (2022) Thermodynamic analysis of primary and secondary mineral stability in melilite-nephelinite tuff with Australopithecus Afarensis footprints (Laetoli, Tanzania). Vestnik of Saint Petersburg University. Earth Sciences 67(2):227–242 (in Russian)

Pickering R (1964) Geological Survey of Tanzania Quarter degree sheet 52 Endulen

Schmidt PW, Avnir D, Levy D, Höhr A, Steiner M, Röil A (1991) Small-angle X-ray scattering from the surfaces of reversed-phase silicas: Power-law scattering exponents of magnitudes greater than four. J. Chem. Phys. 94:1474–1479

Simonyan A, Dultz S, Behrens H (2012) Diffusive transport of water in porous fresh to altered mid-ocean ridge basalts. Chemical Geology306–307:63–77

Smyslov RYu, Ezdakova KV, Kopitsa GP, Khripunov AK, Bugrov AN, Tkachenko AA, Angelov B, Pipich V, Szekely NK, Baranchikov AE, Latysheva E, Chetverikov YuO, Haramus V (2017) Morphological structure of Gluconacetobacter xylinus cellulose and cellulose-based organic-inorganic composite materials. Journal of Physics: Conference Series 848(1):012017

Soil Survey Staff (2014) Keys to Soil Taxonomy, 12th ed. USDA-Natural Resources Conservation Service, Washington, DC

Stonehouse HJ, Duff JP (1974) Report on the soils of the Karatu—Oldeani area, Tanzania. Canada Department of Agriculture/CIDA, Ottawa

van der Gaast SJ, Mizota C, Jansen JHF (1986) Curved smectite in soils from volcanic ash in Kenya and Tanzania: a low-angle X-ray powder diffraction study. Clays and Clay Minerals 34(6):665–671

Westerhof AB, Koistinen TJ (2005) Geological Outline of Western Tanzania with Special Reference to Map Legends. Tanzania/MSDTAP/Component B: Geo-Surveys (NDF Credit 277) /Final Report/Westerhof/Koistinen/. Geological Survey of Tanzania

Wilkinson P (1983) Arusha. Quarter degree sheet 55. Geological Survey of Tanzania

Yorov KhE, Kottsov SYu, Baranchikov AE, Boytsova O, Kiskin MA, Varaksina EA, Kopitsa GP, Lermontov SA, Sidorov AA, Pipich V, Len A, Agafonov AV, Ivanov VK (2019) Journal of Sol-Gel Science and Technology 92(2):304–318

Zaitsev AN, Savchenok AN, Vlasov DYu, Zelenskaya MS (2016) Consultancy services for design, construction & supervision of Laetoli State of the Art Museum. Part I. Unpublished Report to Peter Rich Architects—GMP Consulting Engineers Laetoli JV, St. Petersburg

Zaitsev AN, McHenry L, Savchenok AI, Strekopytov S, Spratt J, Humphreys-Williams E, Sharygin VV, Bogomolov ES, Chakhmouradian AR, Zaitseva OA, Arzamastsev AA, Reguir EP, Leach L, Leach M, Mwankunda J (2019) Stratigraphy, mineralogy and geochemistry of the Upper Laetolil tuffs including a new Tuff 7 site with footprints of Australopithecus afarensis, Laetoli, Tanzania. Journal of African Earth Sciences 158:103561

Zaitsev AN, Vlasov DYu, Zelenskaya MS, Zaitseva OA, Pavlova O, Chakhmouradian AR, Savchenok AI, Leach L, Leach M, Mwankunda J (2020) Microbiology of the Laetolil Tuff 7 with 3.66 Ma *Australopithecus Afarensis* Footprints, Ngorongoro Conservation Area, Tanzania. In: Frank-Kamenetskaya OV, Vlasov DYu, Panova EG, Lessovaia SN (eds) Processes and Phenomena on the Boundary Between Biogenic and Abiogenic Nature. Part of the Lecture Notes in Earth System Science book series (LNESS). Springer Nature Switzerland AG. Part of Springer Nature

Acid Sulfate Pedogenesis of the Geological Past

Tatiana V. Alekseeva and Andrey O. Alekseev

Abstract The given paper presents the multidisciplinary study of two Devonian paleosols recently discovered on the territory of Voronezh Anteclise (South Russia) in the actively excavated quarries at the border of Proterozoic rocks with Devonian sedimentary cover. First of it was developed from weathered ferruginous quartzite, second—from rhyolitic tuff. Both rocks contain inclusions of pyrite: of metamorphic origin in quartzite and pyrite substituted allochthonous coalified remnants of *Nematophytes* in rhyolite. Originated from mineralogically and nutrients poor, potentially toxic hard rocks, under rootless vegetation the developed paleosols, however, demonstrate the surprisingly deep development. Along with typical for these rocks a quartz dominance, paleosols contain in situ developed kaolinite, Fe-oxides, gypsum, jarosite. Considering all results obtained the discovered paleosols were attributed to acid sulfate soils. The development of paleosols was initiated by pyrite oxidation which played the role of trigger. In such aggressive system the chemical weathering prevailed under physical weathering and biological effect. Being gypseous, paleosols bear characteristics which are typical for paleosol of humid type.

Keywords Paleosols · Devonian · Chemical weathering · Rhyolite · Sulfides · Gypsum · Fe-oxides · Jarosite · Nematophytes

1 Introduction

Acid sulfate soil is formed from sulfidic (typically pyrite containing) sediments. Being exposed to air pyrite oxidizes and produce aggressive sulfuric acid under which pH of the system may decrease to 4 units and less. The final oxidation products of Fe-sulfides are Fe-oxides and sulfates (e.g. gypsum) with different transitional Fe and S—containing metastable phases. Due to diverse possible sources of sulfides in

T. V. Alekseeva (✉) · A. O. Alekseev
Institute of Physical Chemical and Biological Problems of Soil Science, Russian Academy of Sciences, Pushchino Moscow Region, Russia
e-mail: alekseeva@issp.serpukhov.su

O. V. Frank-Kamenetskaya et al. (eds.), *Biogenic—Abiogenic Interactions in Natural and Anthropogenic Systems 2022*, Springer Proceedings in Earth and Environmental Sciences, https://doi.org/10.1007/978-3-031-40470-2_22

nowadays acid sulfate soils, they are widely distributed and represent the azonal group of soils (Krasilnikov et al. 1995; Wilson 2005; Zazovskaya et al. 2015; Mendonca et al. 2021). Over 17 million hectares of the globe is covered with acid sulfate soils which is one percent of the worldwide extent of the cultivated land (Andriesse and Van Mensvoort 2002).

As it is documented, paleosols developed under aggressive acid sulfate weathering are among the first appeared on the Earth (3.7–3.2 Ga) (Retallack et al. 2016; Retallack 2022). The remote studies of Martian deposits (Christensen et al. 2004; Chevrier and Mathe 2007) show that sulphur is a key element of Mars surface. The presence of Mg, Ca, Na and Fe sulfates, different iron oxides (hematite, goethite, ferrihydrite) is documented (Retallack et al. 2023). Recently reported akageneite on Mars (Carter et al. 2015)—one of the product of Fe-sulfide oxidation, which is present on Earth in acid sulfate soils, additionally supports the possible acid sulfate soils/paleosols extraterrestrial occurrence.

This paper gives two examples of acid sulfate paleosols recently discovered on territory of Voronezh Anteclise (South Russia). The age of pedogenesis was roughly estimated as Eifelian (Alekseeva et al. 2021; 2023). Both paleosols developed at the top of Proterozoic deeply weathered rocks which are overlain by Middle and Upper Devonian sediments of different genesis.

2 Materials and Methods

Laboratory investigations of paleosols covered the obtaining of the basic soils characteristics (granulometry, organic carbon content), mineralogy of the whole rock, clay fraction (XRD), Fe-oxides (magnetic susceptibility, Mössbauer spectroscopy), geochemistry (XRF) and sub-microscopy of organic remnants (SEM–EDX).

Devonian paleosol discovered in Stoilensky quarry (town Stary Oskol, Belgorod Region—51°15′N; 37°44′E) was developed from the weathered Pre-Cambrian ferruginous quartzite. The formation of weathering crust took place in the humid climate prevailing in the region in the Ordovician and Lower Silurian (Nikitina 1968). Later, at the transgressive stage the mineral composition of the weathering crust was transformed by carbonatization and pyritization (Nikulin and Savko 2015; Ovchinnikova 2020). Given paleosol presents fragmentary and most probably did not develop a continuous soil cover. The top of quartzite has numerous little caverns and rare voids which go 12–15 cm below the surface. The surface of quartzite is motley: the caverns are bleached, and the mounds are reddish. In the voids fine-grained bleached unconsolidated material accumulated. The former includes rare sand-size unsorted quartz grains. The contours of voids are bleached. The top of paleosol is densely covered with coalified plant remains. They create the mat and partly penetrate into voids (Fig. 1.1, 1.2, 1.3, 1.4).

The second paleosol of Eifelian age (middle Devonian) is discovered in Pavlovsk granite quarry (Voronezh region) (50.38°N, 40.22°E). Its parent material is a rhyolitic tuff. The underlying rock is the weathering crust of Proterozoic granite basement

Fig. 1 Stary Oskol. Photographs of a fragment of the paleosol profile; (1) general view; (2) top surface, partially covered with plant remains, spotty-bleached; (3) cavity with plant penetrations; (4) unconsolidated bleached material and plant remains filling the cavities

with a pronounced relief. Paleosol is overlain by marine clay deposits. Paleosol is exposed in several places over the wall of the quarry and most probably developed the continuous soil cover.

The depth of paleosol varies from 20 till 150 cm and depends on the relief of the basement. Rhyolite tuff is hard, glassy and brecciated. In a glassy mass the fragments of crystalline pyroclastic material of cm-size are included. Their surfaces are covered by clay cutans. The cementing breccia material includes Fe-—nodules, rounded, with a diameter up to 10 mm. Their concentration is larger in the second half part of the paleosol profile. Tuff contains coal particles. They are mostly of mm size, rare—reach 1 cm and larger. Coal particles don't create the accumulation zones as well as are not regularly distributed by size. Based on these finding their allochthonous origin is suggested. Big part of coal is pyritized. The bottom of paleosol profile is tongue-like (Fig. 2.2).

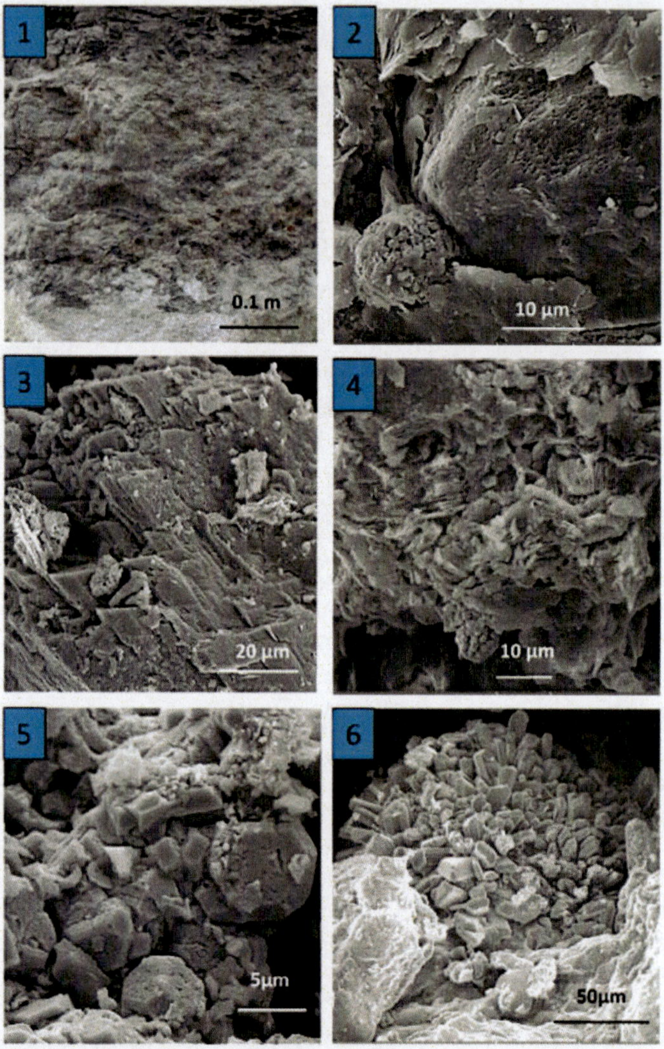

Fig. 2 Paleosol of Pavlovsk quarry: (1)—photo of the profile; (2)–(5)—SEM microphotographs of rhyolite transformation features. (2)—the dissolution of quartz grains surfaces with development of etching pits, clay coatings and a concretions of framboidal pyrite; (3)—dissolution of feldspar; (4)—morphological varieties of newly formed kaolinite—flakes and "booklets"; (5)—dissolution of pyrite crystals; (6)—gypsum "micro rose"

3 Results and Discussion

3.1 Paleosol on Ferruginous Quartzite: mineralogy, Fe-Compounds, Biota

Figure 3 and Table 1 give the obtained data on the mineralogy of ferruginous quartzite and the products of its transformation by pedogenesis.

The quartzite is composed from quartz and siderite, also contains traces of pyrite. Quantitively, 82% of iron is concentrated in siderite and 14%—in pyrite (Mossbauer spectroscopy, Table 1). The topsoil besides quartz and siderite contains neoformed kaolinite and goethite. Mossbauer spectroscopy data show the increase of bulk Fe and input of Fe-oxides: ultrafine goethite (~20 nm crystallite size) and hematite.

The bleached unconsolidated fine material from cavities of the quartzite top is mainly presented by quartz. Additionally, traces of kaolinite, mica and siderite are present (Fig. 3). Kaolinite creates flacks, which cover mineral grains (Fig. 4.3).

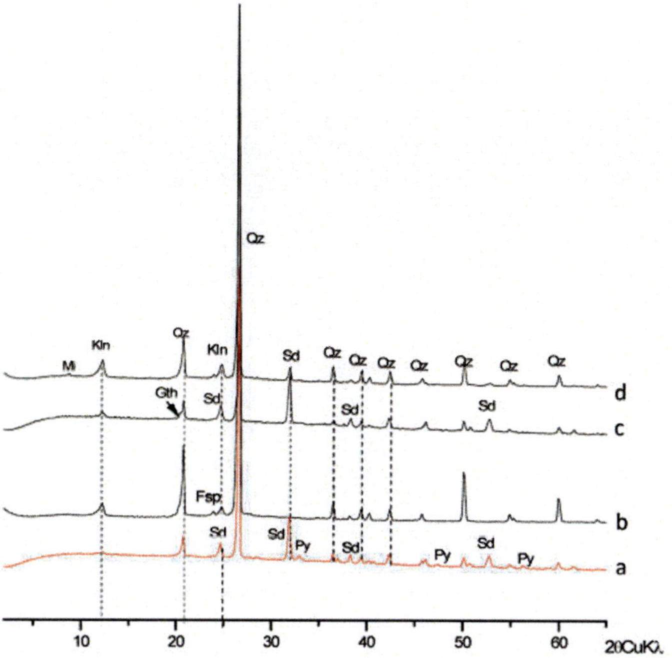

Fig. 3 Stary Oskol. Mineral composition of the paleosol (X-ray diffractometry data): (a)—parent rock—weathered quartzite; (b)—"powdery material" from the cavity—the "rhizosphere" zone; (c)—micromounds on the top of the paleosol; (d)—bleached material of the cavity in the top of the paleosol. Abbreviations: Qz—quartz, Sd—siderite, Py—pyrite, Kln—kaolinite, Mi—mica, Gth—goethite, Fsp—feldspars

Table 1 Magnetic susceptibility (MS × 10^{-8} SI) and mineralogy of iron compounds in quartzite and paleosol according to Mossbauer spectroscopy

Sample	MS	Mineral phases	*IS*, mm/s	*QS*, mm/s	H_{eff}, kOe	Proportion, %	Fe bulk Sq
Ferruginous topsol	48–55	Hematite goethite (20 nm) siderite	0.49 0.47 1.33(85K)	−0.15 −0.17 1.98	520(85K*) 477(85K)	4.43 9.46 86.11	5.47
Rhizosphere material	204	Siderite pyrite	1.23 0.32	1.77 0.64		44.4 55.6	0.23
Quartzite	44	goethite (20 nm) siderite pyrite	0.34 1.33 0.33	−0.33 1.99 0.57	476(85K)	3.7 81.9 14.4	2.83

IS—isomeric shift; QS—quadrupole splitting; Heff—value of the magnetic field strength. The isomeric shift was determined relative to α-Fe, * temperature of spectrum measurement, Sq—square of spectra (relative units).

Siderite forms spherules with a diameter of 1–5 μm (Fig. 4.5). Recently micro-nodules of siderite of similar shape were laboratory obtained as a product of microbiological transformations of chemogenic-sedimentary siderite (Zavarzina et al. 2020). Most quartz and hematite grains show signs of dissolution (Fig. 4.1–4.2). The specific characteristic of the voids filling material is the presence of gypsum, which, most probably, is one of the products of pyrite oxidation (Fig. 4.4). Rare pyrite is present as secondary—framboidal variety (Fig. 4.6).

Bleached unconsolidated material from voids (coarse "rhizosphere material") is mostly composed from quartz grains. Those are angular and unsorted by size: large grains (2–5 mm) are distributed within finer, around 500 μm in size material. Besides quartz, it contains kaolinite, as well as feldspars and gypsum as impurities (Fig. 3). The content of Fe in this material is ten times less of that in the host quartzite (Table 1). It is represented by siderite and pyrite. The observed enhancement of the magnetic signal of the finer part of rhizosphere material (Table 1) allows to propose the formation of biogenic super dispersed magnetite.

Surfaces of quartz grains have signs of dissolution which are present as oriented etching pits and caverns (Fig. 5.1–5.2). Often quartz grains have mineral (kaolinitic) coatings and biofilms. Associated gypsum crystals are very abundant (Fig. 5.3–5.6).

Results of SEM study of plant remains which develop a mat on the topsoil allow to concluded that the mat is composed from monotype bryophyte-like vegetation. Plant remains have good tissue preservation (Fig. 6.1–6.2). Partly they show the features of lysis caused by acidification produced by pyrite oxidation (Fig. 6.4). Rare surviving fragments of tracheids allow to say about the vascular nature of this plant (Fig. 6.3). Recently Xu et al. (in press) have reported the finding of plant with close characteristics of tissues. Based on the complex study these authors went to a

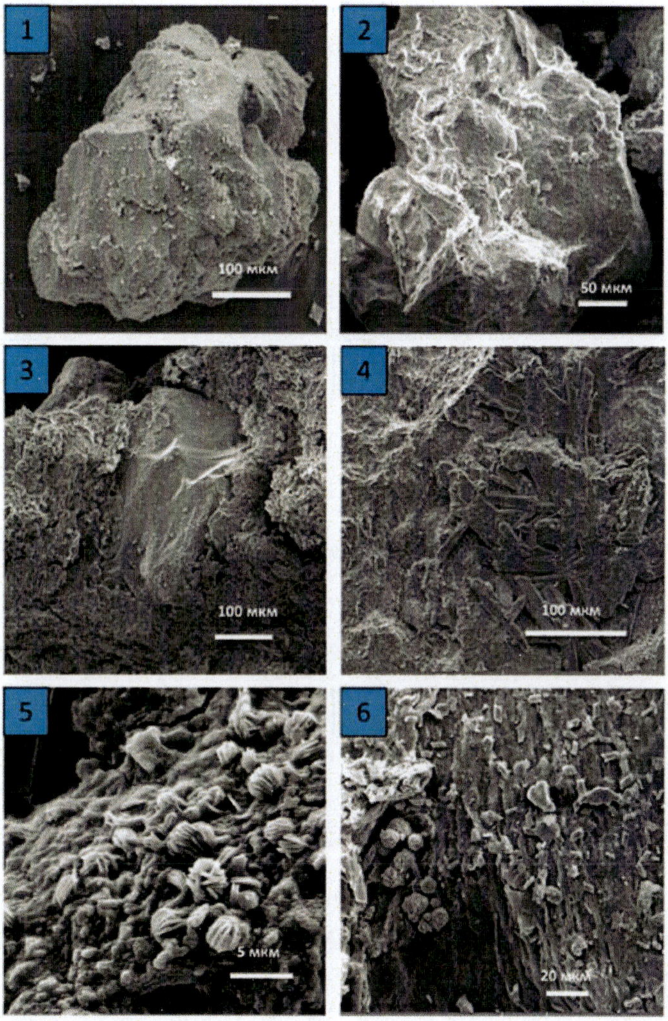

Fig. 4 Stary Oskol. SEM data of unconsolidated material from the top of paleosol: (1) and (2)—quartz grains with cavernous surface; (3)- clay coating on the surface of a quartz grain; (4)—crystallization of gypsum on the surface of quartz grain; (5)—organo-mineral film with inclusions of biogenic spherosiderite; (6)—crystallization of gypsum and formation of framboidal pyrite on the surface of the plant remain

conclusion that this plant is Upper Ordovician and is the earliest vascular land plant described upto now.

The obtained data allow to make the conclusions about the systematic position of the discovered paleosol. Firstly, it is a lithosol. Hard soil-forming rock, its nutrient-poor and potentially toxic due to the presence of pyrite mineral composition, the absence of fine earths, lithophytic vegetation pointed to the extreme conditions of its

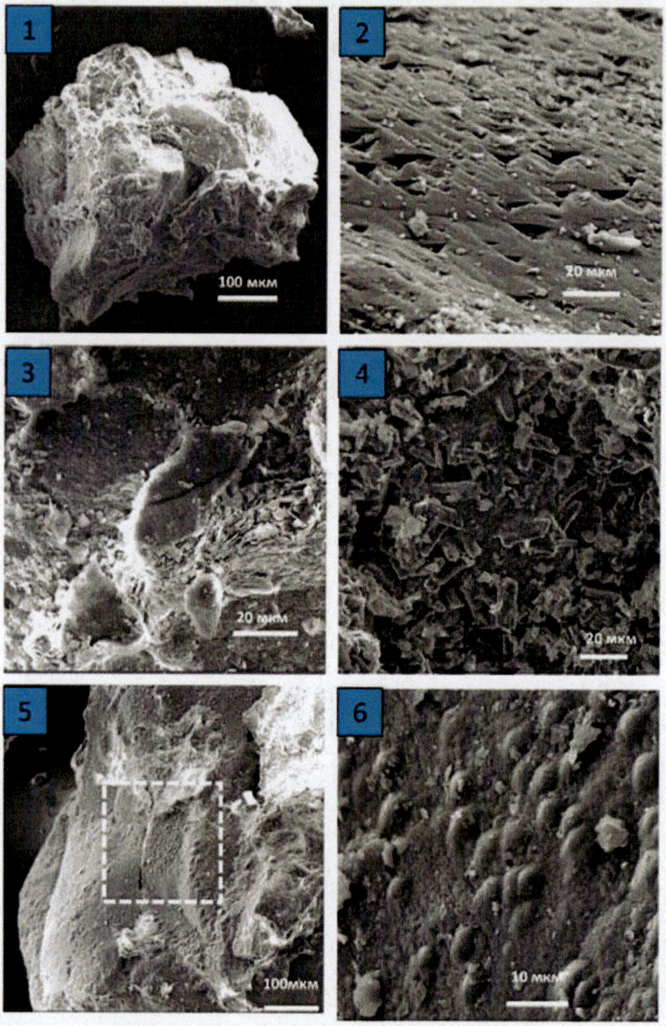

Fig. 5 Stary Oskol. SEM data of unconsolidated material from cavities ("rhizosphere zone"): (1)—quartz grains with cavernous surface; (2) oriented etching pits on the surface of a quartz grain; (3)—leafy aggregates of kaolinite on quartz grain; (4)—crystallization of gypsum on the surface of quartz grain; (5)—colony of cyanobacteria under a film of amorphous silica; (6) fragment of (5)

development (Goryachkin 2022). A unique feature of this paleosol is a macro-scale local subsurface (in voids) transformations against a weak transformations of the top of quartzite. The important role played the vegetation cover which performed some physical destruction of the rock, a protection for the developing fine earth, the improvement of drainage conditions, provided a habitat for microorganisms, etc. These features put the studied paleosol on a line with modern typically microscale

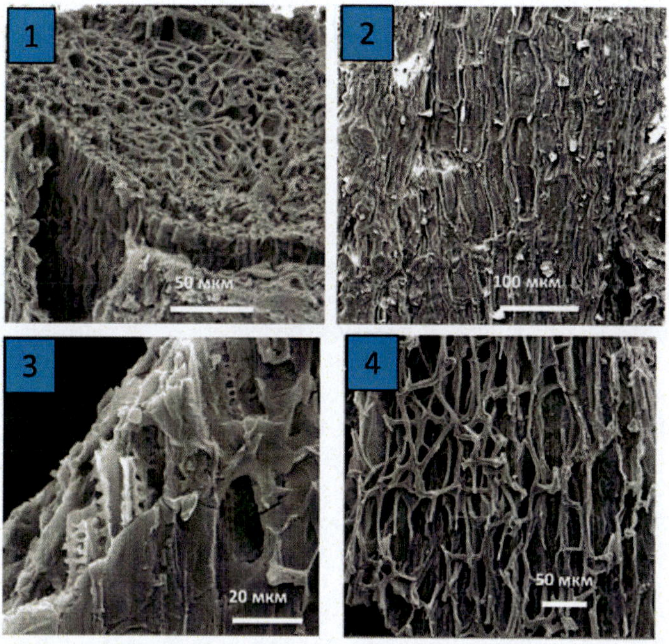

Fig. 6 Stary Oskol. The morphology of plant remains: (1)–(3)—structural elements of the plant fragments; (4)—lysis of plant tissues

epilithic and endolithic systems but being of macroscale instead (Mergelov et al. 2018; Coleine et al. 2021; Mitchell et al. 2021). The observed properties of the paleosol are the results of first of all abiotic processes where large-scale chemical weathering prevailed. The described transformations of quartzite are caused primarily by oxidation of pyrite which created the aggressive acidic conditions. Based on this, the studied paleosol was classified as an acid sulphate paleosoil. Cavities and other features of dissolution on the surfaces of mineral grains, their bleaching accompanied by neosynthesis of kaolinite, Fe-oxides, gypsum, secondary siderite and pyrite—all manifest the deep transformations of parent rock. The aggressiveness of the environment causes the transformation of organic –tissues, their lysis.The participation of biota in the surface and subsurface pedogenesis is confirmed by the presence of colonies of cyanobacteria, neosynthesis of biologically induced framboidal pyrite, spherosiderite, biomagnetite.

3.2 Paleosol on Rhyolitic Tuff: Basic Properties, Mineralogy, Fe-Compounds, Biota

Paleosol is sandy with the mean contents of sand 64%, silt—21% and clay 15%. Upper part shows from 3 to 7% increase in clay. The volcanic origin of parent material gives the specific characteristics to its chemical composition. Thus, large concentrations of Zr provide the small values (2–7) of Ti/Zr ratio and sharply mark out it. The principal chemical element in rhyolite is Si. The mean SiO_2 concentration here is 73%. One more specific characteristic of the rock is the large concentration of sulfur, in some samples it reaches 4%. More clayish upper parts of the paleosol profiles are slightly enriched in Al_2O_3. The behavior of Fe_2O_3 is more complicated. Its mobility and redistribution are visualized via concentration in nodules and accumulation in the bottom of the profile.

Tuff (the whole rock) is composed from predominantly quartz (Fig. 7c–d), which sharply contrast with kaolinitic composition of underlying saprolite (Fig. 7a). The paleosol additionally contains the minor amounts of kaolinite, K-feldspar and gypsum. Clay fraction (<2 μm) separated from paleosols is monomineral—kaolinitic.

Fe-nodules additionally to quartz contain visible amounts of gypsum and jarosite—$KFe_3(SO_4)_2(OH)_6$ (XRD data, Fig. 7b). Based on Möessbauer spectrum the biggest part of Fe (~45%) in nodules is attributed to pyrite. Two Fe—sulfates: jarosite and voltaite—$K_2Fe^{2+}{}_5Fe^{3+}{}_3Al(SO_4)_{12}\cdot 18H_2O$ present 15% and 30% of iron respectively. Both of them are known as products of pyrite oxidation. Additionally, super dispersed goethite with particle size ~ 20 nm presents 8% of iron.

The morphological features of mineral constituents of paleosol were studied by SEM-EDX. The surfaces of quartz and K-feldspar grains show alterations. For quartz—first of all the dissolution with the development of etching pits. Feldspar crystals demonstrate the microcleavage (Fig. 2.2–2.3). In situ formed kaolinite develops mostly flakes but sometimes—aggregates of parallel oriented individual crystals—so-called "booklets" (Fig. 2.4). Pyrite is very abundant. It substitutes coal particles, present as monocrystals of different size and cubic morphologies, framboidal concretions. The listed ones are often coexist (Fig. 2.2 and 2.5). The neoformed gypsum is very abundant both in cement material and Fe-nodules and creates different morphological varieties (Fig. 2.6).

As it was mentioned above, the coal particle inclusions are completely or partly substituted with pyrite (Fig. 8). At the same time some particles have the excellent preservation of tissues. Among different morphological types of tissues tubular structures are prevail. Two principal types of tubular tissues should be noted. Type 1 is composed of thick-walled tubes tightly associated and randomly oriented (Fig. 8.1). Type 2 tissue is composed of two types of tubes. The main part is composed of thin walled tubes (binding tubes), associated and randomly oriented. Most of them have diameter of about 5 μm. Besides, rare thick—skeletal tubes with diameter upto 40 μm are observed. Some of them are filled with banded tubes with a diameter upto 20 μm (Fig. 8.2–8.3). In parallel paleosol material include abundant stick-like

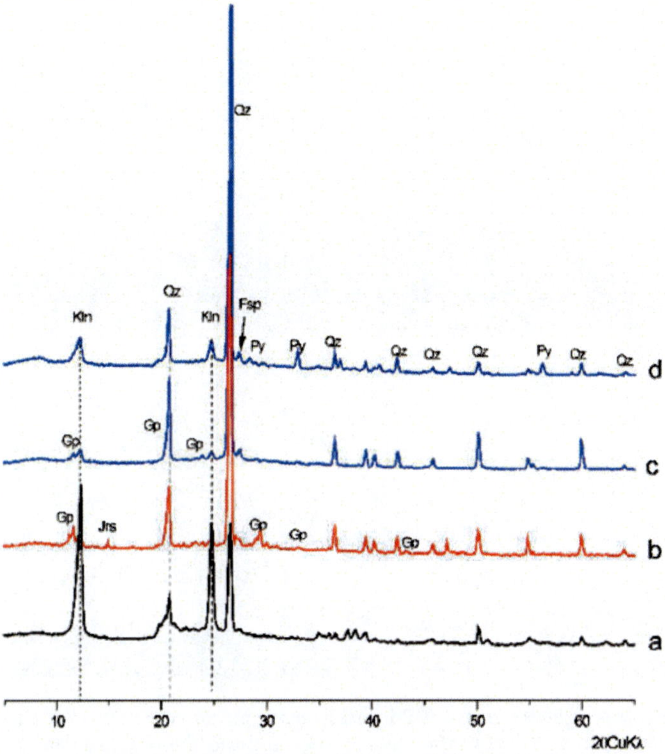

Fig. 7 XRD spectra of paleosol of Pavlovsk quarry (bulk samples): **a** granite weathering crust; **b**—Fe-nodules; **c**—middle part of the profile; **d**—top part of the profile. Abbreviations: Qz—quartz, Kln—kaolinite, Fsp—K-feldspar, Gp- gypsum, Py-pyrite, Jrs-jarosite

coal debris, pyrite substituted. Their size and morphology allow to suppose that they could be mineralized banded tubes described above (Fig. 8.4).

The discovered coal particles with tubular tissues belong to Nematophyte group—enigmatic terrestrial heterotrophic fossils which are interpreted as fungi (Edwards and Axe 2012; Nelsen and Boyce, 2022) or lichens (Honegger 2018). Those are one of the oldest non-marine organisms, described from Late Silurian to Late Devonian (Taylor et al. 2009). The morphology of type 1 fossil looks very similar to the remnants of *Nematophytales* (Edwards and Axe 2012; Filipiak and Szaniawski 2016). In both studies, the specimens were dated as Lower Devonian. Type 2 fossil belongs to *Prototaxites Loganii* (Taylor et al. 2009; Nelsen and Boyce, 2022).

Based on the complex of data the studied paleosol is attributed to acid sulfate soil. An oxidation of pyrite associated with coal particles and microdebris imbedded in rhyolite matrix plays the key role in this paleosol formation being its trigger. Pyrite oxidation is accompanied by dissolution of quartz and feldspars, neoformation of kaolinite, goethite, Fe-sulphates (jarosite, voltaite), amorphous compounds of Si, Al and Fe, gypsum. The formation of this mineral complex is typical for modern

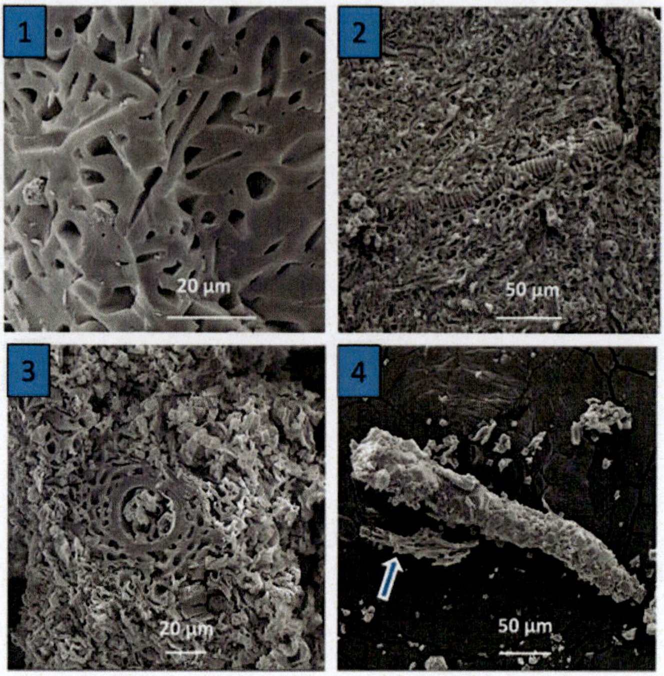

Fig. 8 Paleosol of Pavlovsk quarry. SEM microphotographs of coal: (1)—type 1 tissue, thick-walled tubular of assumed *Nematophytales*; (2)—type 2 tissue, thin—walled tubular of assumed *Prototaxites* with a inclusion of banded tube; (3)—type 2 tissue, showing two types of tubes with different diameters, typical for *Prototaxites Loganii*; (4)—pyrite substituted tube ("banded"?) and fine tubular tissue (arrowed)

acid sulfate soils (Krasilnikov et al. 1995; Bockheim 2015; Zazovskaya et al. 2015; Mendonca et al. 2021). The newly formed gypsum does not develop the distinct salt bearing horizon, but distributed in the soil matrix from the top to the bottom, being associated with mineral matrix, organic fragments, or Fe-nodules.

Mineralogical and chemical poverty of rhyolite and its potential toxicity are matched to extreme pedogenesis (Goryachkin 2022). The sandy texture and brecciated porous matrix of rhyolite providing good drainage, acid pH and exothermic character of reaction of pyrite oxidation (abiotic and/or microbially induced)—all favor chemical weathering and pronounced rhyolite transformation. The complex of obtained data contradicts the aridity of climate which could be expected basing only on salts presence and better corresponds to subtropical environments.

4 Conclusions

Considering all results obtained and discussed the following main conclusions can be made. Both discovered Devonian paleosols are attributed to acid sulfate soils. Being developed from different rocks with different source of sulfides they at the same time show plural similarities of transformational trends and products: pyrite oxidation, deep chemical weathering of parent rock including dissolution of quartz and other rock forming minerals, neoformation of kaolinite, iron oxides and sulfates, gypsum, mineral and organo-mineral films, secondary pyrite. These characteristics could be the reflections of both the aggressiveness of the environment and favorable climate conditions. Azonal character of modern acid sulfate soils which are distributed everywhere from cold and arid Antarctica, Canada, north Russia to warm and humid Brazil and Australia, supports the idea that the presence of sulfides is the key factor of their development. Sulfides oxidation provided the surprisingly deep development of paleosols under the extreme environment: hard rocks, their mineral poverty and potential toxicity, rootless vegetation. In both cases the chemical weathering prevailed under physical weathering and biological impact. Deep transformation of parent rocks had the restricted local area or "explosive" character and was not accompanied by formation of distinct salt-bearing soil horizons.

Acknowledgements The investigation was supported by the Russian Science Foundation (Grant N∘ 22-27-00370).

References

Alekseeva TV, Alekseev AO, Mitenko GV (2021) A paleosol on a Pre-Cambrian ferruginous quartzite weathering crust (Stary Oskol, Belgorod Region, Russia). Paleontolog. J. 55: 1476–1490

Alekseeva T, Kalinin P, Malishev V, Alekseev AO (2023) Sulfide oxidation as a trigger for rhyolite weathering and paleosol formation in Devonian (Voronezh High, South Russia). Catena 220, Part A: 106712

Andriesse W, van Mensvoort MEF (2002) Acid sulfate soils: distribution and extent. In: Lal R (ed). Encylopedia of Soil Science CRC Press: Boca Raton, FL, USA, 14–19

Bockheim JG (ed.) (2015) The Soils of Antarctica. Springer International Publishing, Switzerland

Carter J, Viviano-Beck Ch, Loizeau D, Bishop J, Le Deit L (2015) Orbital detection and implications of akaganeite on Mars. Icarus 253: 296-310

Chevrier V, Mathe PE (2007) Mineralogy and evolution of the surface of Mars: A review. Planetary and Space Science 55: 289–314

Christensen PR, Wyatt M B, Glotch T D, Rogers A D, et al. (2004) Mineralogy at Meridiani Planum from the Mini-TES experiment on the Opportunity Rover. Science 306 (5702): 1733–1739

Coleine C, Stajich J E, Rios A, Selbmann L (2021) Beyond the extremes: rocks as ultimate refuge for fungi in drylands. Mycologia 113(1): 108–133

Edwards D, Axe L (2012) Evidence for a fungal affinity for *Nematasketum*, a close ally of *Prototaxites*. Botanical Journal of the Linnean Society 168: 1–18

Filipiak P, Szaniawski H (2016) Nematophytes from the Lower Devonian of Podolia, Ukraine. Review of Palaeobotany and Palynology 22:109–120

Goryachkin SV (2022) Geography of extreme soils and soil-like systems. Vestnik of Russian Academy of Sciences 92 (6): 564-571 (in Russian)

Honegger R (2018) Fossil lichens from the Lower Devonian and their bacterial and fungal epi- and endobionts. In: Blanz P. (ed). Biodiversity and Ecology of fungi, lichens and mosses. Kerner von Marilaun Workshop 2015 in memory of Josef Poelt. Biosystematics and Ecology Series vol. 34. Verlag der Österreichischen Akademie der Wissenschaften, Wien, 547–563

Krasilnikov PV, Safonova VM, Sedov SN (1995) Acid sulfate weathering in the soils of North Karelia. Pochvovedenie 6:740–746 (in Russian)

Mendonca SKG, Moraes E M V, Otero XL, Ferreira TO, Correa MM, Sousa JES, Nascimento CWA, Neves LVMW, Souza Junior VS (2020) Occurrence and pedogenesis of acid sulfate soils in northeastern Brazil. Catena 196 (2): 1-11

Mergelov N, Mueller CW, Prater I, Shorkunov I, Dolgikh A, Zazovskaya E, Shishkov V, Krupskaya V, Abrosimov K, Cherkinsky A, Goryachkin S (2018) Alteration of rocks by endolithic organisms is one of the pathways for the beginning of soils on Earth, Sci. Rep. 8 (1): 3367

Mitchell RL, Strullu-Derrien C, Sykes D, Pressel S, Duckett JG, Kenrick P (2021) Cryptogamic ground covers as analogues for early terrestrial biospheres: Initiation and evolution of biologically mediated proto-soils. Geobiology 19 (3): 292-306

Nelsen MP, Boyce CK (2022) What to do with Prototaxites? International Journal of Plant Sciences 183 (6): 556-565

Nikitina AP (1968) Drevnyaya kora vyvetrivaniya kristallicheskogo fundamenta Voronezhskoy Anteklizy i yeye boksitonosnost' (Ancient weathering crust of the crystalline basement of the Voronezh Anteclise and its Bauxite Content). Moscow, Nauka (in Russian)

Nikulin II, Savko AD (2015) Iron ore weathering crusts of the Belgorod region of the Kursk magnetic anomaly. Trudy Nauch.-Issled. Inst. Geol. Voronezh. Gos. Univ. 85: 1–102 (in Russian)

Ovchinnikova MYu (2020) Evolution of weathering crust of rocks of the Lower Proterozoic ferruginous-siliceous-shale formation of the Kursk Magnetic Anomaly (KMA) in the Paleozoic. Dissertation, Voronezh University (in Russian)

Retallack GJ, Krinsley DH, Fischer R, Razink JJ, Langworthy KA (2016) Archean coastal-plain paleosols and life on land. Gondwana Research 40: 1-20

Retallack GJ, Jepson S, Broz A (2023). Petrogypsic paleosols on Mars. Icarus 394; 115436

Retallack GJ (2022) Soil salt and microbiome diversification over the past 3700 million years. Palaeogeography, Palaeoclimatology, Palaeoecology 598: 111016

Taylor TN, Taylor EL, Krings M (2009) Paleobotany and the evolution of plants. Academic Press, Burlington Massachusetts

Wilson BP (2005) Elevations of sulfurous layers in acid sulfate soils: What do they indicate about sea levels during the Holocene in eastern Australia? Catena 62: 45-56

Xu H, Wang K, Huang Z, Tang P, Wang Y, Liu B, Yan W (in press) The earliest vascular land plants from the Upper Ordovician of China. Reseach Square https://doi.org/10.21203/rs.3.rs-1672132/v1

Zavarzina DG, Chistyakova NI, Shapkin AV, Savenko AV, Zhilina TN, Kevbrin VV, Alekseeva TV, Mardanov AV, Gavrilov SN, Bychkov AYu (2020) Siderite-based anaerobic iron cycle driven by autotrophic thermophilic microbial consortium. Scientific Reports 10(1)

Zazovskaya EP, Fedorov-Davydov DG. Alekseeva TV, Dergacheva MI (2015) Soils of Queen Maud Land. In: Bockheim J. (ed) The Soils of Antarctica. World Soils Book Series Springer, Berlin, pp. 21–44

Can Earthworm Invasions from Rudny Altai (Kazakhstan) in the South of Western Siberia Change the Amount of Humus in Meadow Chernozem (Calcic Chernozem) Soils?

Elena V. Golovanova, Daria P. Unru, Kirill A. Babiy, Stanislav Yu. Kniazev, and Olga A. Golovanova

Abstract Since the second half of the twentieth century, significant losses of humus have been noted in the soils of the south of Western Siberia. This period also became favorable for the spread of invasive earthworm species in both agrocenoses and natural ecosystems as a result of soil plowing and climate change. In this study, we aim to determine whether earthworm invasions in the south of Western Siberia could alter the amount of humus and further aggravate soil fertility. To this end, we analyzed soil samples from field experiments in mesocosms. Earthworm invasions in Western Siberia were simulated in experiments on meadow-chernozem soils. We studied the influence of the native species *Eisenia nordenskioldi* and invasive species *E. ventripappilata*, E. nana and their cohabitation for 92 days on soil parameters. A significant effect of invasive species and all combinations of earthworms on the amount of soil humus was identified. *E. ventripappilata* was found to have the greatest impact on soil humus. This species contributed to a decrease in the amount of humus. Thus, earthworm invasions in the south of Western Siberia may lead to even greater soil degradation and soil fertility loss.

Keywords Soil organic matter · Carbon · Humus · Mesocosms · Invasions · Species Interactions · Lumbricidae · Endogeics

E. V. Golovanova (✉) · O. A. Golovanova
Omsk F. M, Department of Inorganic Chemistry, Dostoevsky State University, Prospekt Mira 55a, Omsk 644077, Russia
e-mail: nilseb@omgpu.ru

E. V. Golovanova · D. P. Unru · K. A. Babiy · S. Yu. Kniazev
Research Laboratory of Systematics and Ecology of Invertebrates, Omsk State Pedagogical University, Tukhachevsky Embankment 14, Omsk 644099, Russia

O. V. Frank-Kamenetskaya et al. (eds.), *Biogenic—Abiogenic Interactions in Natural and Anthropogenic Systems 2022*, Springer Proceedings in Earth and Environmental Sciences, https://doi.org/10.1007/978-3-031-40470-2_23

1 Introduction

The quantitative and qualitative composition of organic matter is the most important parameter of soil fertility. Changes in the composition of soil organic matter can eventually lead to cascade transformations in natural ecosystems (Wall et al. 2012) and agrocenoses, disturbing the performance of ecosystem functions (Briones 2018). Humus is part of soil organic matter, represented by a combination of specific and non-specific organic substances of the soil, with the exception of compounds that make up living organisms and their residues (GOST 27593-88). The arable land of the Western Siberia was found to have a stable negative balance of humus state (Demin and Eremina, 2022).

Monitoring of humus content in the soils of Western Siberia from 1955 to 2018 showed a significant decrease in the amount of humus in forest-steppe soils (Reinhard 2009; Bobrenko et al. 2021), particularly in southern regions. The humus content decreased from 1955 to 1985 by an average of 14.9% (Reinhard 2009), and by another 12% over the next 30 years (Bobrenko et al. 2021). This trend prevailed in ordinary chernozems and meadow chernozems (Reinhard 2009; Bobrenko et al. 2021). During this period, invasive earthworm species spread to their maximum extent (Golovanova 2019; Golovanova et al. 2020; Kniazev et al. 2022). Thus, *Aporrectodea caliginosa* (Savigny, 1826) began to inhabit agrocenoses. The species *Eisenia ventripappilata* Perel, 1985 and *E. nana* Perel, 1985, which had been previously unable to live under the conditions of deeply freezing soils, spread widely beyond their natural habitats (Golovanova 2019; Golovanova et al. 2020, 2021; Kniazev et al. 2022).

Earthworms are regarded as soil engineers (Lavelle et al 2006) capable of affecting the physical, chemical, and biological properties of soils. Their influence was shown to be both climate- and region-dependent (Wall et al. 2008). Simulation experiments on Siberian Lumbricidae species found that the water-soluble part of earthworm coprolites depends on soil characteristics and earthworm species (Babiy et al. 2021b). The interaction of invasive and native species of Lumbricidae may promote the influence of earthworms on soil characteristics (Babiy et al. 2021a, Babiy et al. 2022). Does this effect extend to soil humus, and can earthworm invasions in Western Siberia change the amount of humus? To answer these questions, we considered the following hypotheses:

1. earthworms affect the amount of humus in meadow-chernozem soils in comparison with the control;
2. invasive *Eisenia ventripappilata, E. nana*, and native earthworm species affect the amount of humus to a different extent;
3. interactions between earthworm species affect the content of soil organic matter.

2 Materials and Methods

2.1 Experimental

In order to test the formulated hypotheses, we analyzed soil samples from two field simulation experiments in mesocosms. The experiments simulated two actual situations of earthworm invasions in the south of Western Siberia. We investigated soils and the species inhabiting them, the composition and amount of litter common in model habitats, taking the natural density of earthworms into account.

The study was carried out on meadow-chernozem soils (Calcic chernozem) with forb-grass meadows (Fig. 1a, d) at the field research station of the Omsk State Pedagogical University (Omsk Oblast, N55°02′38.9″, E73°22 ′52.4″). The depth of groundwater was 3.5 m. We determined the soil parameters that have the greatest impact on the humus content, including the particle size distribution, pH, and the ionic composition of aqueous soil extracts.

Fig. 1 Experimental site and mesocosms. **a**—A soil section of meadow-chernozem soils at the experimental site; **b**—The upper part of an open mesocosm with the remains of the litter; **c**—An open mesocosm with soil; **d**—Experimental site with several excavated mesocosms

The granulometric composition was measured by a pipette method according to the state standard 12,536-2014 (GOST 12,536-2014). The pH values of the samples were determined in a mixture of soil with deionized water (ratio 1:5) using an ESK-10301 combined electrode on Anion-4100 (NPP Infraspak-Analit, Russia). The ionic composition of soil aqueous extracts was analyzed using a Capel-104 capillary electrophoresis installation. Preparation of soil samples and soil aqueous extracts, as well as their analysis, was carried out according to the certified methodology of the Environmental Regulation F 16.1:2:2.2:2.3.74–2012 (PND F 16.1:2:2.2:2.3.74–2012). The soils were medium loamy, slightly alkaline, with the predominance of sodium and calcium, as well as nitrates and sulfates, in the ionic composition of soil aqueous extracts (Table 1).

The meadows at the experimental site were dominated by *Poa pratensis* L., 1753, *Filipendula ulmaria* (L.) Maxim., 1879, *Trifolium pretense* L., 1753. The earthworm fauna was represented by *A. caliginosa* (25–113 ind./m^2). *E. nordenskioldi* (5–17 ind./m^2), *Dendrobaena octaedra* (3–5 ind./m^2) (Savigny, 1826), and *A. caliginosa* (9–35 ind./m^2) inhabited the adjacent forest area.

Samples were collected in tube containers, 25 cm in diameter and 50 cm in length (Fig. 1d). The soil of the humus accumulative horizon A and the next horizon AB was sifted through soil sieves with a mesh size of 4 mm to remove the soil macrofauna before placing into mesocosms, for an approximation to natural conditions (Fig. 1a, c). A piece of cloth was attached at the top and bottom to prevent earthworms from escaping and macrofauna from entering into the containers (Fig. 1d).

The experiments simulated interactions between native and invasive earthworm species in Kazakhstan. The species were identified according to the T.S. Vsevolodova-Perel classification based on (Perel 1985); the names were specified according to R. Blackmoore (2008). The species were also clarified by a genetic analysis. The genetic sequences for *E. nordenskioldi* are available in the GenBank under the numbers of MK618513, JX531526, JX531527, JX531528, JX531529, JX531530, JX531531, JX531532, JX531533, JX531534, JX531535, JX.531. The genetic analysis data are presented in previous publications (Shekhovtsov et al. 2013, 2017, 2020a, 2020b). Photographs of the species are shown in Fig. 2.

In total, experiments were performed in eight variants and five repetitions: three monovariants (native species *E. nordenskioldi*; invasive species *E. nana* and *E. ventripappilata*), three divariants of pairwise combinations of species, and one trivariant. E. nana and E. n. nordenskioldi are epigeic (Perel 1979, 1985; Golovanova et al. 2021), E. ventripappilata is endogeic (Perel 1985; Golovanova et al. 2021). Each mesocosm contained 12 individuals. The density of individuals corresponded to the maximum density of species in the natural conditions of the Omsk Oblast. Increasing the density to the next multiples of two and three values would have led to the oppression of populations as a result of intraspecific competition. Experiments were carried out during the growing season from June to September 2019. To exclude the transition of earthworms into diapause, each mesocosm was watered with distilled water in a volume of 450 ml once a week. Irrigation was started provided that the average soil moisture decreased below 40% at a layer depth of 15 cm. The survival rate of mature individuals in mesocosms comprised 83 ± 9%, 74 ± 18%, and 72 ± 5%

Table 1 Ionic composition in aqueous soil extracts from experimental plots, mg/kg, n = 5

NH^{4+}	K^+	Na^+	Mg^{2+}	Ca^{2+}	Cl^-	$SO_4{}^{2-}$	$NO_3{}^-$	F^-	$HCOO^-$	pH
0.33 ± 0.08	4.5 ± 0.6	89 ± 6	8.8 ± 0.4	39 ± 1	16.5 ± 0.9	47 ± 3	118 ± 12	0.95 ± 0.11	1.26 ± 0.27	7.52 ± 0.03

Fig. 2 Earthworms used in simulation experiments. **a**—*Eisenia nordenskioldi*, **b**—*Eisenia nana*, **c**—*Eisenia ventripapillata*. Side view. Symbol: tb—tubercle; cl—clitellum. Scale bars = 5.0 mm

for *E. nordenskioldi*, *E. nana*, and *E. ventripappilata*, respectively. No significant differences between the species were observed in terms of survival rate and biomass change. A significant increase in the number of populations was noted only for the native species (Golovanova et al. 2022). Leaf litter of *Populus tremula* (L., 1753) dry leaves weighing 12.0 ± 0.1 g was placed in each mesocosm. The final amount of litter at the end of the growing season (Fig. 1B) in each experimental variant is presented in Table 2.

Table 2 Humus contents and the amount of litter in various experimental variants

Treatment	C_{org}, %			Humus, %	Litter, g		
	Mean	N	SE	Mean	Mean	N	SE
Eisenia nordenskioldi	4.08	32	0.12	7.10	4.4	5	0.19
E nana	4.21	35	0.12	7.33	2.8	5	0.19
E ventripappilata	3.24	35	0.10	5.64	6.5	5	1.27
E nana + *E nordenskioldi*	3.83	34	0.11	6.66	2.6	5	0.21
E nordenskioldi + *E ventripappilata*	3.14	35	0.14	5.46	4.5	5	0.50
E nana + *E ventripappilata*	3.16	35	0.15	5.50	3.6	5	0.32
E nana + *E nordenskioldi* + *E ventripappilata*	4.61	35	0.23	8.02	3.6	5	0.21
control	4.40	34	0.08	7.65	8.9	5	1.63
Total	3.83	275	0.06	6.66	4.6	40	1.07

2.2 *Quantitative Analysis of Humus*

In soil samples, we determined humus according to the Tyurin method in the modification of the Central Research Institute of Agrochemical Services (GOST 26,213–2021). In the meadow-chernozem soils in the south of the Omsk Oblast, humus is characterized by a humate composition. C of humic acids/C of fulvic acids in the upper horizon varies from 2.3 to 3.2 (Mishchenko 1991). The humus composition is dominated by the second fraction of humic acids associated with calcium. These soils have a high content of humic acids associated with stable sesquioxides and clay minerals (Gradoboev and Bogdanov 1968). Since humus is of the humate type in meadow chernozem soils throughout the entire depth of the humus profile, the standard coefficient of 1.724 was used to recalculate the percentage of carbon. A quantitative analysis of humus was performed only for the 0–5 cm layer of the upper soil horizon, where all three species of earthworms are active. We determined the content of humus in each mesocosm in seven replicates. Each variant of the experiment was analyzed 35 times.

2.3 *Statistical Data Processing*

Data normality was verified using the Shapiro–Wilk test at the $P = 0.05$ level (Statistica 13). Outliers in the sample were removed. A three-factor ANOVA was employed to process the results followed by the post hoc Tukey HSD test (Statistica 13). A correlation analysis was also carried out.

3 Results

The data obtained followed a normal distribution. All types of earthworms and their combinations, except for the native species, had a significant impact on the amount of humus in the soil layer of 0–5 cm of the humus horizon (Tables 2, 3).

According to Tukey's test, the humus content in mesocosms with *E. nordenskioldi* (Table 2) did not differ significantly from the control (Table 4). The amount of soil humus in the *E. nordenskioldi* and *E. nana* variants was also similar; however, its value was 1.3 times lower ($p < 0.001$) in the *E. ventripappilata* variant compared to that with the native species. Pairwise interaction of species in all variants of the experiment reduced the humus content (Tables 2, 4), with this trend being most pronounced when interacting with *E. ventripappilata.* During the interaction of all species in the three-species version of the experiment, the amount of humus remained at the level of the control.

The maximum content of humus was observed in the three-species variant, while all other variants with the participation of *E. ventripappilta* showed its minimum

Table 3 Results of a multivariate analysis of variance of the influence of earthworm species and their combination on the humus content in mesocosm soils

Effect	One-dimensional significance test for % (Humus) Sigma-restricted parametrization Hypothesis decomposition			
	SS	MS	F	p
E nordenskioldi	1.798	1.798	2.690	0.102176
E nana	3.894	3.894	5.826	0.016468
E ventripappilata	24.252	24.252	36.286	0.000000
*E nordenskioldi*E nana*	9.364	9.364	14.010	0.000223
*E nordenskioldi*E ventripappilata*	18.053	18.053	27.011	0.000000
*E nana*E ventripappilata*	14.377	14.377	21.511	0.000006
*E nordenskioldi*E nana*E ventripappilata*	11.200	11.200	16.757	0.000056

value. No significant correlations between the total amount of litter and humus content were found, as well as between the total number of earthworms in the mesocosm and humus content.

4 Discussion

The conducted quantitative analysis of humus content confirmed our assumption about the influence of invasive earthworm species on this indicator. Earthworms affect the humus content in soils, although not equally. Thus, the invasive species *E. ventripappilata* was found to have the greatest impact. The presence of the species in the monovariant and two-species experiments significantly reduced the humus content in soils.

Literatures presents conflicting data on the role of earthworms in the accumulation of carbon, or humus. Researchers frequently associate the activity of earthworms with humification of plant residues (Edvards and Lofty 1972, Aleksandrova 1980, Haimi and Huhta 1990, Burtellow et al. 1998, Holdsworth et al. 2012, Griffiths et al. 2021) and carbon accumulation in coprolites (Sokolov 1956; Zrazhevsky 1957; Pulleman et al. 2005) and soil (Frouz et al. 2014). However, earthworms are both consumers and producers of humus, including humic acids (Chekanovskaya 1960). Earthworms are typically classified into two morpho-ecological types: those feeding on the soil surface and those feeding on humus (Perel 1979)—or epigeic, anecic, and endogeic, and combinations thereof (Bouche 1977). Worms belonging to the former morpho-ecological type are capable of increasing the content of humus (Perel et al. 1966) and carbon in the soil (Huang et al. 2020; Geraskina 2021). Worms that consume humus may either reduce soil carbon or have no effect on soil carbon (Huang et al. 2020; Geraskina 2021). At the same time, even worms belonging to the same

Table 4 Significance level of the study (p) in pairwise comparison of variants of experiment 2 using Tukey's test (only significant values are presented)

Treatment	Enn	Ena	Ev	Enn*Ena	Enn*Ev	Ena*Ev	Enn*Ena*Ev	Control
E nordenskioldi (Enn)	–	–	0.0006	–	0.00009	0.0001	–	–
E nana (Ena)	–	–	0.00005	–	0.00003	0.00003	–	–
E ventripappilata (Ev)	0.0006	0.00005	–	–	–	–	0.00003	0.00003
E nordenskioldi **E nana* (Enn*Ena)	–	–	–	–	0.01	0.02	0.002	–
E nordenskioldi * *E ventripappilata* (Enn*Ev)	0.00009	0.00003	–	0.01	–	–	0.00003	0.00003
*E nana*E ventripappilata* (Ena*Ev)	0.0001	0.00003		0.02	–	–	0.00003	0.00003
E nordenskioldi **E nana*E ventripappilata* (Enn*Ena*Ev)	–	–	0.000032	0.002	0.00003	0.00003	–	–
Control	–	–	0.000032		0.00003	0.00003	–	–

ecological group can have different effects on soil carbon (Andriuzzi et al. 2016). The participation of earthworms in the process of humus formation and humification is a species-specific process (Sokolov 1956; Zrazhevsky 1957), which also depends on climatic conditions (Perel 1979; Wall et al. 2012) and litter quality (Perel 1979; Zrazhevsky 1957; Holdsworth et al. 2012). In our study, endogeic earthworm species reduced the amount of humus, which agrees well with the data of other authors.

In addition to ecological categories, it is of great importance that earthworm species can be classified as invasive. According to Fragoso et al (1993), the distribution of invasive species in the Mexican tropics was associated with land disturbance, with the humus content having decreased from 7.26% to 4.05%. In a study by Alban and Berry (1994), the 14-year joint effect of two invasive endogeic and epi-endogeic species annually reduced the total soil carbon in Northern Minnesota. In the northeastern United States, the invasive Amynthas hawayanus increased the proportion of active and readily mineralized soil organic matter (Burtelow et al. 1998) on light alluvial soils. A. hawayanus is a group of earthworm species belonging to different ecological categories (Zhao et al. 2009). Unfortunately, the authors did not specify to which ecological category the studied species belonged. A study by James (1991) in the tall grass prairies of Kansas in areas with native species of the genus *Diplocardia* and invasive endogeics *A. caliginosa* and *Octolasion cyaneum* found that, despite their high abundance, invasive species left less coprolites and processed nutrients than native species. Thus, invasive species can either increase or decrease soil organic matter, both in comparison with control and native species. This process can demonstrate different dynamics in different regions (Wall et al. 2012), depending on climate, soil type, particle size distribution (Barthod et al. 2020), plant community, past agricultural activity (Ma et al. 2013), type of applied organic matter (Zhang et al. 2011), ecological category (Huang et al. 2020), and earthworm species. In our study that simulated invasions in the south of Western Siberia, the invasive epigeic E. nana did not significantly affect the amount of humus, while the endogeic *E. ventripappilta* significantly reduced its proportion in the soil. The interaction of species also had a significant effect on the amount of humus in our experiments.

Interactions within earthworm species are known to affect their survival, including in winter (Uvarov, 2021), the number of cocoons, and the number of communities (Uvarov 2009, Uvarov et al. 2019). Species, such as *Aporrectodea trapezoides* and Lumbricus rubellus, have an advantage over other species in areas with high resource availability; however, in less productive communities, these species can either deplete all available resources and go extinct (Golovanova et al., 2018) or not reach the reproductive age (Winsome et al. 2006). All these factors affect soil parameters. Thus, in experiments on Siberian earthworm species, the influence of species interaction on the ionic composition of soils was established (Babiy et al. 2021a, 2021b). Obviously, this influence also extends to the amount of soil humus. The highest humus content was noted in the three-species version of the experiment. This variant included worms of three ecological groups, i.e., epigeic, epi-endogeic, and endogeic. Apparently, this combination is most favorable for the accumulation of humus in soils, which is supported by a number of publications (Huang et al 2020; Geraskina 2021). Could the spread of invasive earthworm species in the forest-steppe of Western Siberia lead to a decrease in humus, which is noted in the works of Reinhard (2009), Demin and Eremina (2022)? The authors attribute the intense degradation of soils in the south of the Omsk Oblast to their plowing and a decrease in the application of organic fertilizers. It is known that plowing of lands contributes to the introduction of invasive species therein (Hendrix et al. 2008; Ma et al. 2013), at the same time as depriving the soil of epigeic earthworms. The combination of these two factors exacerbates the loss of humus in soils.

5 Conclusion

In this work, we identified a significant effect of invasive species and all combinations of earthworms on the amount of humus. E. ventripappilata had the greatest impact on soil humus, which contributed to a decrease in its content. Thus, earthworm invasions in the south of Western Siberia may lead to even greater soil degradation and soil fertility loss.

Acknowledgements We thank E.V. Lisunova for the analysis of 80 soil samples for pH, E.A. Lozhnikova and E.A. Alekhina for their assistance in the analysis of the content of organic matter in the samples, and S.V. Shekhovtsov for conducting genetic studies. We also express appreciation to the administration of the Omsk State Pedagogical University represented by I.I. Krott, the Rector, and I.P. Gerashchenko, a Vice-Rector, for providing access to the Agrobiological Station of the Omsk State Pedagogical University for simulation experiments.

The study was supported by the Russian Science Foundation (grant No. 22–14-20,034) and the Ministry of Industry and Scientific and Technical Development of the Omsk Oblast (grant no. 32-C dated 06/22/2022).

Author Contributions Elena V. Golovanova—project supervision; research concept and experiment design; soil type determination; description of the research site; collection of litter; collection and identification of earthworms; conducting experiments; soil sampling; statistical data processing; photographs of worms; manuscript draft.

Daria P. Unru—sample preparation; photometric analysis of samples for the content of soil organic matter in terms of carbon.

Kirill A. Babiy—production of mesocosms; collection of earthworms; experiment design development; conducting experiments; maintaining soil moisture in mesocosms; soil sampling; analysis of water extract for ionic composition by electrophoresis; preparation of graphic materials.

Stanislav Yu. Kniazev—development of a mesocosm model; production of mesocosms; collection of earthworms and litter; conducting experiments; maintaining soil moisture in the mesocosm; soil sampling.

Olga A. Golovanova—administration of soil chemical analysis; approbation of the study; revision and editing of the manuscript.

References

Aleksandrova LN (1980) Soil Organic Matter and the Processes of Its Transformation. Nauka, Leningrad (in Russian)

Andriuzzi WS, Ngo P-T, Geisen S, Keith AM, Dumack K, Bolger T, Bonkowski M, Brussaard L, Faber JH, Chabbi A, Rumpel C, Schmidt O (2016) Organic matter composition and the protist and nematode communities around anecic earthworm burrows. Biol Fertil Soils 52:91–100. https://doi.org/10.1007/s00374-015-1056-6

Babiy KA, Yu. Kniazev S, Golovanova EV, Solomatin DV, Bel'skaya LV (2021b) What Determines Ion Content of Lumbricid Casts: Soil Type, Species, or Ecological Group? Polish Journal of Ecology 69:. https://doi.org/10.3161/15052249PJE2021.69.2.003

Babiy KA, Kniazev SYu, Golovanova EV, Abramenko AS (2021a) Influence of exotic *Eisenia nana* (Opisthopora, Lumbricidae) on the cationic composition of three types of soils in South Western Siberia (Experiment in microcosms). RJEE. 6(3) https://doi.org/10.21685/2500-0578-2021-3-4

Babiy KA, Yu. Kniazev S, Golovanova EV (2022) Influence of invasive Eisenia nana and Eisenia ventripapillata (Opisthopora, Lumbricidae) on soil cation composition In: Tiunov AV, Gongalsky KB, Uvarov AV (Eds) Biota, genesis and productivity of soils: materials of the XIX Russian meeting on soil zoology. Publisher: Buryat Scientific Center of the Siberian Branch of the Russian Academy of Sciences, Ulan-Ude (in Russian) https://www.elibrary.ru/item.asp?id=49438799

Barthod J, Dignac M-F, Le Mer G, Bottinelli N, Watteau F, Kögel-Knabner I, Rumpel C (2020) How do earthworms affect organic matter decomposition in the presence of clay-sized minerals? Soil Biology and Biochemistry 143:107730. https://doi.org/10.1016/j.soilbio.2020.107730

Blakemore RJ (2008) Checklist of USSR/Russian Federation taxa updated from Perel (1979, 1997) in A series of searchable texts on earthworm biodiversity, ecology and systematics from various regions of the World. 3rd Edition http://www.annelida.net/earthworm/ (accessed on 17 January 2023).

Bobrenko IA, Matveychik OA, Bobrenko EG, Popova V (2021) Changes in humus content in forest-steppe soils of Western Siberia IOP Conf. Ser.: Earth Environ. Sci. 624 012219. https://doi.org/10.1088/1755-1315/624/1/012219

Bouché MB (1977) Stratégies lombriciennes. In: Lohm U, Persson T (Eds) Soil organisms as components of ecosystems. 25. Ecology Bulletin, Stockholm

Briones MJI (2018) The Serendipitous Value of Soil Fauna in Ecosystem Functioning: The Unexplained Explained. Front Environ Sci 6:149. https://doi.org/10.3389/fenvs.2018.00149

Burtelow AE, Bohlen PJ, Groffman PM (1998) Influence of exotic earthworm invasion on soil organic matter, microbial biomass and denitrification potential in forest soils of the northeastern United States. Applied Soil Ecology 9:197–202. https://doi.org/10.1016/S0929-1393(98)00075-4

Chekanovskaya OV (1960) Earthworms and soil formation. Nauka, Moscow (in Russian)

Demin EA and Eremina DV (2022) Balance model of humus state of arable chernozems of the Western Siberia IOP Conf. Ser.: Earth Environ. Sci. 949 012084

Edwards CA, Lofty JR (1972) The role of earthworms in organic matter cycles. In: Biology of Earthworms. Springer US, Boston, MA, pp 141–154. https://doi.org/10.1007/978-1-4899-6912-5_6

Fragoso C, Barois I, Gonzalez C, Arteaga C and Patron JC. (1993) Relationship between earthworms and soil organic matter levels in natural and managed ecosystems in the Mexican tropics. In: K. Mulongoy and R. Merckx (eds.): Soil Organic Matter Dynamics and Sustainability of Tropical Agriculture. 231–239.

Frouz J, Špaldoňová A, Fričová K, Bartuška M (2014) The effect of earthworms (Lumbricus rubellus) and simulated tillage on soil organic carbon in a long-term microcosm experiment. Soil Biology and Biochemistry 78:58–64. https://doi.org/10.1016/j.soilbio.2014.07.011

Geraskina AP (2021) Impact of earthworms of different morpho-ecological groups on carbon accumulation in forest soils. Forest Science 3(3):1–20. https://doi.org/10.31509/2658-607x-202141g15

Golovanova EV, Kniazev SYu, Karaban K (2018) Are there any advantages for a native species of earthworms compared to invasive species in Western Siberia? In: Problems of soil zoology.

Proceedings of the XVIII All-Russian Conference on Soil Zoology. 60–61. (In Russian) https://www.elibrary.ru/item.asp?id=36442522

Golovanova EV, Kniazev SYu, Babiy KA, Tsvirko EI (2020) Distribution of the alien earthworm species Aporrectodea caliginosa in natural habitats of the Omsk region. Omsk State Pedagogical University, 299–302 (In Russian) https://www.elibrary.ru/item.asp?id=44322710

Golovanova EV, Kniazev SYu, Babiy KA, Tsvirko EI, Karaban K, Solomatin DV (2021) Dispersal of earthworms from the Rudny Altai (Kazakhstan) into Western Siberia. Ecol Monten 45:48–61. https://doi.org/10.37828/em.2021.45.9

Golovanova EV, Kniazev SYu, Babiy KA (2022) Interaction between a species of earthworm native to Western Siberia and invasive species from Kazakhstan In: Tiunov AV, Gongalsky KB, Uvarov AV (Eds) Biota, genesis and productivity of soils: materials of the XIX Russian meeting on soil zoology. Publisher: Buryat Scientific Center of the Siberian Branch of the Russian Academy of Sciences, Ulan-Ude (in Russian) https://www.elibrary.ru/item.asp?id=49438814

Golovanova EV (2019) Alien species of earthworms in Western Siberia In: Ushakova KI, Veselkin DV, Vasiliev AG (Eds) Ecology and evolution: new horizons: materials of Intern. symposium dedicated. 100th anniversary of Academician S.S. Schwartz, Humanitarian University, Yekaterinburg (in Russian) https://www.elibrary.ru/item.asp?id=42357515

GOST 26213-2021. (202) Soils. Methods for determination of organic matter. Russian Institute for Standardization, Moscow (in Russian)Alban DH, Berry EC (1994) Effects of earthworm invasion on morphology, carbon, and nitrogen of a forest soil. Applied Soil Ecology 1:243–249. https://doi.org/10.1016/0929-1393(94)90015-9

GOST 27593–88. (2008) Soils.Terms and definitions. Decree of the USSR State Committee for Standards of February 23, 1988 N 326. Standartinform, Moscow (in Russian)

Gradoboev ND and Bogdanov NI (1968) Agrochemical characteristics of soils in the Omsk region in: Agrochemical characteristics of soils in the USSR (regions of Western Siberia). Nauka, Moscow (in Russian)

Griffiths HM, Ashton LA, Parr CL, Eggleton P (2021) The impact of invertebrate decomposers on plants and soil. New Phytologist 231:2142–2149. https://doi.org/10.1111/nph.17553

Haimi J, Huhta V (1990) Effect of earthworms on decomposition processes in raw humus forest soil: A microcosm study. Biol Fertil Soils 10:178–183. https://doi.org/10.1007/BF00336132

Hendrix PF, Callaham MA, Drake JM, Huang C, James SW, Snyder BA, Zhang W (2008) Pandora's Box Contained Bait: The Global Problem of Introduced Earthworms. Annu Rev Ecol Evol Syst 39:593–613. https://doi.org/10.1146/annurev.ecolsys.39.110707.173426

Holdsworth AR, Frelich LE, Reich PB (2012) Leaf Litter Disappearance in Earthworm-Invaded Northern Hardwood Forests: Role of Tree Species and the Chemistry and Diversity of Litter. Ecosystems 15:913–926. https://doi.org/10.1007/s10021-012-9554-y

Huang W, González G, Zou X (2020) Earthworm abundance and functional group diversity regulate plant litter decay and soil organic carbon level: A global meta-analysis. Applied Soil Ecology 150:103473. https://doi.org/10.1016/j.apsoil.2019.103473

James SW (1991) Soil, Nitrogen, Phosphorus, and Organic Matter Processing by Earthworms in Tallgrass Prairie. Ecology 72:2101–2109

Kniazev SYu, Kislyi AA, Bogomolova IN, Golovanova EV (2022) Territorial Heterogeneity of the Earthworm Population (Opisthopora, Lumbricidae) of Omsk Oblast and Environmental Factors: A Quantitative Assessment of the Relationship. Contemp Probl Ecol 15:484–493. https://doi.org/10.1134/S1995425522050079

Kurcheva GF (1971) The role of soil animals in the decomposition and humification of plant residues. Nauka, Moscow (in Russian)

Lavelle P, Decaëns T, Aubert M, Barot, S, Blouin M, Bureau F, Margerie P, Mora P, Rossi J-P (2006) Soil invertebrates and ecosystem services. European Journal of Soil Biology 42:S3–S15. https://doi.org/10.1016/j.ejsobi.2006.10.002

Ma Y, Filley TR, Johnston CT, Crow S E, Szlavecz K, McCormick M K (2013) The combined controls of land use legacy and earthworm activity on soil organic matter chemistry and particle association during afforestation. Organic Geochemistry 58:56–68. https://doi.org/10.1016/j.orggeochem.2013.02.010

Mishchenko LN (1991) Soils of the Omsk region and their agricultural use. Omsk Agricultural Institute, Omsk (in Russian)

Perel TS (1979) Range and regularities in the distribution of earthworms of the USSR fauna. Nauka, Moscow (in Russian)

Perel, TS (1985) Specific Features of the Earthworm Fauna (Oligochaeta, Lumbricidae) in Altai Refugia ofNemoral Vegetation. Doklady Akademii Nauk SSSR, 283(3): 752–756 (In Russian)

PND F 16.1:2:2.2:2.3.74–2012. (2012) Method for measuring the mass fraction of water-soluble forms of cations in soils, soils, clay, peat, sewage sludge, activated sludge, bottom sediments by capillary electrophoresis using the Kapel capillary electrophoresis system. Moscow (in Russian)

Pulleman MM, Six J, Uyl A, Marinissen J C Y, Jongmans A G (2005) Earthworms and management affect organic matter incorporation and microaggregate formation in agricultural soils. Applied Soil Ecology 29:1–15. https://doi.org/10.1016/j.apsoil.2004.10.003

Reinhard JR (2009) Soil degradation of ecosystems in the south of Western Siberia Loz, Poland (in Russian)

Shekhovtsov SV, Golovanova EV, Peltek SE (2013) Cryptic diversity within the Nordenskiold's earthworm, Eisenia nordenskioldi subsp. nordenskioldi (Lumbricidae, Annelida). European Journal of Soil Biology 58:13–18. https://doi.org/10.1016/j.ejsobi.2013.05.004

Shekhovtsov SV, Berman DI, Golovanova EV, Peltek SE (2017) Genetic diversity of the earthworm Eisenia nordenskioldi (Lumbricidae, Annelida). Vestn VOGiS 21:589–595. https://doi.org/10.18699/VJ17.24-o

Shekhovtsov SV, Golovanova EV, Ershov NI, Poluboyarova TV, Berman DI, Szederjesi T, Peltek SE (2020a) Phylogeny of the Eisenia nordenskioldi complex based on mitochondrial genomes. European Journal of Soil Biology 96:103137. https://doi.org/10.1016/j.ejsobi.2019.103137

Shekhovtsov SV, Shipova AA, Poluboyarova TV, Vasiliev GV, Golovanova EV, Geraskina AP, Bulakhova NA, Szederjesi T, Peltek SE (2020b) Species Delimitation of the Eisenia nordenskioldi Complex (Oligochaeta, Lumbricidae) Using Transcriptomic Data. Front Genet 11:598196. https://doi.org/10.3389/fgene.2020.598196

GOST 26423–85. (1985) Soils. Methods for determining the specific electrical conductivity, pH and dense residue of aqueous extract. Decree of the USSR State Committee for Standards of 08.02.1985 No. 283. Standards Publishing House, Moscow (In Russian)

Sokolov AA (1956) Significance of earthworms in soil formation. Publishing House of the Academy of Sciences of the Kazakh SSR, Alma-Ata (in Russian)

Uvarov AV, Ilieva-Makulec K, Karaban K, Yakovenko N S, & Uchmański, J (2019) Effects of Intra- and Interspecific Interactions in Earthworm Assemblages: A Comparative Study. Biol Bull Russ Acad Sci 46:475–482. https://doi.org/10.1134/S106235901905011X

Uvarov AV (2009) Inter- and intraspecific interactions in lumbricid earthworms: Their role for earthworm performance and ecosystem functioning. Pedobiologia 53:1–27. https://doi.org/10.1016/j.pedobi.2009.05.001

Uvarov AV (2021) The Overwinter Survival of three Earthworm Species in Mono- and Multispecific Assemblages. Biol Bull Russ Acad Sci 48:821–828. https://doi.org/10.1134/S1062359021130069

Vsevolodova-Perel TS (1997) The earthworm of the fauna of Russia: Cadaster and key. Nauka, Moscow (in Russian)

Wall DH, Bradford MA, St. John MG, Trofymow J A, Behan-Pelletier V, Bignell D E, Dangerfield J M, Parton W J, Rusek, J, Voigt W, Wolters V, Gardel H. Z, Ayuke, F O, Bashford R, Beljakova O I, Bohlen P J, Brauman A, Flemming S, Henschel J R … Zou X (2008) Global decomposition experiment shows soil animal impacts on decomposition are climate-dependent: SOIL FAUNA IMPACTS ON GLOBAL DECOMPOSITION. Global Change Biology 14:2661–2677. https://doi.org/10.1111/j.1365-2486.2008.01672.x

Wall DH, Bardgett RD, Behan-Pelletier V, Herrick JE, Ritz K, Six J, Strong D, van der Putten WH (eds) (2012) Soil Ecology and Ecosystem Services. Oxford University Press. https://doi.org/10.1093/acprof:oso/9780199575923.001.0001

Winsome T, Epstein L, Hendrix PF, Horwath WR (2006) Competitive interactions between native and exotic earthworm species as influenced by habitat quality in a California grassland. Applied Soil Ecology 32:38–53. https://doi.org/10.1016/j.apsoil.2005.01.008

Zhang J, Hu F, Li H, Gao Q, Song X, Ke X, Wang L (2011) Effects of earthworm activity on humus composition and humic acid characteristics of soil in a maize residue amended rice–wheat rotation agroecosystem. Applied Soil Ecology 51:1–8. https://doi.org/10.1016/j.apsoil.2011.08.004

Zhao Q, Sun J, Qiu J (2009) Three new species of the *Amynthas hawayanus* -group (Oligochaeta: Megascolecidae) from Hainan Island, China. Journal of Natural History 43:1027–1041. https://doi.org/10.1080/00222930902767433

Zrazhevsky AI (1957) Earthworms as a factor in the fertility of forest soils.: Publishing House of the Academy of Sciences of the Ukrainian SSR, Kyiv (in Russian)

Geochemical Features of the Waste Processing Plant Landfill Soil

Elena G. Panova, Tatiana V. Lemanova, and Irina Yu. Tikhomirova

Abstract Recycling, neutralization and storage of solid household waste is an urgent problem due to the constant growth of their accumulation. When they are stored outdoors, as the result of mechanical, chemical and biological weathering, chemical elements become mobile and migrate to surrounding areas, creating soil, hydrochemical and biological anomalies. Some of the elements form secondary minerals (salts, oxides-hydroxides, chlorides, phosphates) or are part of the organic component of landfill soil. In the young soils (2 years old), accumulate high values of total carbon, humus, N, and H. The number of microbial cells in them is less than in the old ones (20 years old), but their diversity is greater. In young landfill soils, high concentrations of most chemical elements (Zc = 74) recorded in comparison with the settled soils (Zc = 32). The degree of weathering of 20-year-old soils is 10–100 times greater compared to young landfill soils. Most of elements accumulate in the soil filtrate of young sediments (from 2 to 15 times) compared to the settled soils.

Keywords Landfill soil · 2 and 20 years old soils · Macro- and microelements · Water-soluble forms · Secondary minerals · Biofilms

1 Introduction

Waste storage areas from the processing of solid household waste (municipal solid waste, MSW) are one of the most interesting objects of bioinert interactions. The problem of processing, neutralization and storage of MSW is essential due to the constant growth of its accumulation (Vilar and Carvalhod 2004; Zhang et al. 2010; Tunesi 2011; Assamoi and Lawryshyn 2012; Basha et al. 2016; Hadinata et al. 2018; Speier et al. 2018). Mechanical, chemical and biological decomposition of

E. G. Panova (✉) · T. V. Lemanova
Saint Petersburg State University, St. Petersburg, Russia
e-mail: e.panova@spbu.ru

I. Yu. Tikhomirova
The Herzen State Pedagogical University of Russia, St. Petersburg, Russia

O. V. Frank-Kamenetskaya et al. (eds.), *Biogenic—Abiogenic Interactions in Natural and Anthropogenic Systems 2022*, Springer Proceedings in Earth and Environmental Sciences, https://doi.org/10.1007/978-3-031-40470-2_24

MSW stored in the open air takes place in the result of weathering processes. So chemical elements and compounds can transform into mobile forms and migrate to the surrounding areas, forming soil, hydro-chemical and biological anomalies.

The main morphological composition of MSW is presented by different authors (Dixon and Jones 2005; Chen et al. 2008, 2020; Jo and Jang 2021). They are: 1. Food waste—20–50%; 2. Cardboard and paper—9–26%; 3. Polymers—10–25%; 4. Glass—5–20%; 5. Scrap metal—6–9%.

The composition of municipal solid waste depends on natural zones and climatic conditions, as well as the scale of megacities, diet and culture of the population, income level and infrastructure. There is the trend: the higher level of development of the country shows the less volume of food waste and at the same time the higher level of recyclable dry waste, such as paper and cardboard, plastic, rubber, glass and metals. In developing countries, more than 50% of the total MSW is food waste.

Landfill soil (LS) formed during MSW storage has a porous structure and three-phase state (Chen et al., 2020). Unlike natural soil has chemically and biologically active composition, which is more complex and variable. MSW also has a wider variety of particle shapes and sizes. MSW is an anisotropic system, due to the diverse composition and properties of particles that unevenly distributed within the mass (Zhang et al. 2017).

In the process of storage, solid waste particles are deformed and reduced in volume. During compression or decomposition MSW release liquid and gas, which affect the compression deformation of these particles. Physical, chemical and biodegradation reactions take place within in the body of LS, causing formation of leachate, gases, volumetric deformation and heat release. Porous water squeezes out, gas pressure changes, and substances migrate over the landfill. These processes affect the rate of biochemical reactions.

Physicochemical processes of waste decomposition in landfill soil divides into four stages (Song et al. 2015; Liu et al. 2019; Santos et al. 2022). *The first stage* takes place under aerobic conditions (at a depth of 50–80 cm). It characterizes by the consumption of oxygen and nitrate, which contributes to the hydrolysis and oxidation of food waste containing fats, proteins, and carbohydrates. Phosphorus from organic compounds oxidizes to phosphates, and then precipitates in the course of interaction with metal ions (Ca2+, Mg2+, etc.). Fats hydrolyze under the impact of microorganisms and mold fungi, forming glycerol and fatty acids. Nitrogen-containing organic compounds affected by microorganisms and exoenzymes are hydrolyzed forming polypeptides that break down to amino acids, which, in the presence of ammonifier microorganisms decompose to form ammonia, CO_2 and H_2O. Proteins containing sulfur atoms during decomposition give an oxidizing agent—sulfate ions. The processes of oxidation and decomposition under aerobic conditions accompanies by heat release. This causes rising of temperature within the landfill up to 80 °C. Appearance of oxidizing agents and leachates is the reason of metal corrosion; the resulting acids dissolve the metals, which transfer into the leachate.

The second stage is the anaerobic acidic phase of MSW decomposition. There is a compression of the body of the landfill, hydrolysis increases, anaerobic biological activity increases and acetogenesis occurs. As a result, of this process, salts and esters of acetic acid, carbon dioxide and water form. This stage characterizes by high humidity, acidic pH (3.5–6.5) and a temperature of 25 °C. The appearance of fungi contributes to the appearance of fungi that destroy wood. Humification occurs and landfill soil forms. Zinc, iron, nickel, chromium, cadmium oxidized in an acidic environment and become soluble with the formation of soluble complex salts.

The third stage is the stage of active methanogenesis. This stage characterizes by a decrease in the content of organic acids and methane formation. Microorganisms consume organic matter at a high rate. The drop in sulfate content is associated with its reduction by sulfate-reducing bacteria.

The fourth stage is stable methanogenesis. It marked by maximum decomposition of organic matter. Organometallic compounds of metals gradually leach and pass into the leachate. At this stage, 50–70% of MSW is decomposed. Enzymatic hydrolysis of lignin takes place at this stage in an alkaline environment with the formation of aromatic and fatty acids and chemical destruction of hard-to-destruct polymer materials.

According to different authors, the following forms of heavy metals in LS have been identified: (1) exchangeable—highly mobile and bioavailable; (2) metal carbonates; (3) recoverable—Fe and Mn oxides; (4) oxidizable or organically bound; (5) residual, where metals are firmly connected with the crystal lattice of mineral components.

The aim of the current study is to identify geochemical, physico-chemical and biological features of soils from storing zones of processing products of different age of storage of one of the waste processing plants (MSW) of St. Petersburg. The task of the research is to determine in LS: (1) component composition; (2) chemical composition; (3) physical and chemical properties; (4) mobile forms of chemical elements; (5) biological composition.

2 Materials and methods

The soils divided into two zones: (a) old storages (2000–2008); (b) new storages (2019–2020). More then, 60 samples collected. The study of LS include: analysis of published data; macroscopic description and photo documentation, scanning electron microscopy, granulometric analysis, determination of organic C (550 °C), common C, thermal analysis and CNH analysis. Geochemical features were revealed using X-ray spectral analysis (complete silicate and trace element analyses), ICP MS and mercury metering. Chemical weathering indices and total toxicity index were calculated. The water soil extract obtained at a ratio of sample/water as 1:5. Acidity, electrical conductivity and microelement composition of the water extract carried out. Microorganisms determined by the growth method. Water added to a sample of LS and mixed with Czapek-Dox culture medium. After two weeks, the number of colony units counted and the species composition of microorganisms that had grown on the surface of the nutrient medium analyzed. Biotesting on watercress carried out, aerobic and anaerobic communities of microorganisms were determined. Biotesting carried out in Petri dishes. A disk of filter paper placed on the bottom, on the surface of which 30 cress seeds placed and they covered with the same disc of filter paper. The filter paper moistened with an aqueous extract of their LS. The control sample moistened with distilled water. Petri dishes placed in a thermostatic cabinet for 10 days (temperature of 20–25 °C) periodically moistening. At the end of the experiment, the indicators of the test measured: the length of the main root, the length of the shoot, the average dry weight, germination, germination energy, germination friendliness. Based on the results of the biotesting, the Factor Toxicity Index calculated as the ratio of the value of the test function in the sample with respect to the control.

3 Results and Discussion

3.1 *Physical Chemical Properties of Landfill Soils*

Grain size determined by sieving and weighing. Figure 1 demonstrates that clay fraction prevails in the 20-years settled landfill soils. The proportion of large fractions (more than 1 mm) is typical of 2-years young landfill soils.

Hygroscopic moisture is the amount of water contained in an air-dried LS sample. This characteristic depends on the grain size, chemical and mineralogical composition. Hygroscopic moisture determined by the weight method expressed as a mass percentage of the dry soil. The young landfill soils have shown four times higher this characteristic (3.82%) in comparison to settled LS (0.95%).

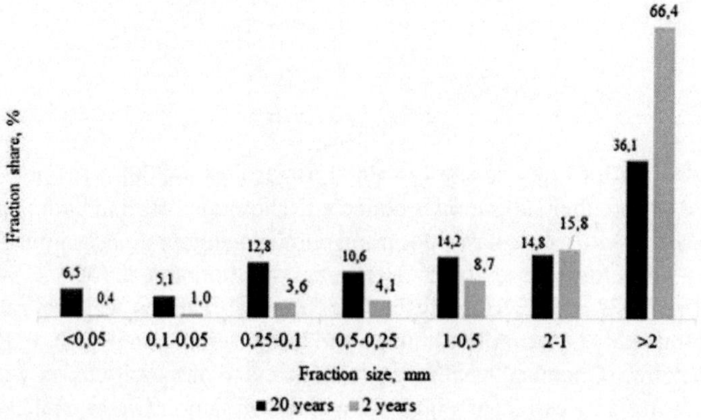

Fig. 1 Fractional composition of 20 and 2-years LS samples, %

3.1.1 Actual Acidity (pH) of Water Soil Leachate

Due to atmospheric precipitation and moisture, a leachate contaminated with organic and inorganic components formed within the body of a landfill. To study this property, pH value was determined in aqueous soil leachates prepared in the proportion soil to water = 1:5. For pH measuring was used the potentiometric method. The experiments have showed a more alkaline environment in the young LS (8.06) and relatively neutral in the settled LS (6.49).

Electrical conductivity of an aqueous soil leachate is determined by the presence of strong electrolytes and evaluates the ability of soil to conduct electric current. Value of electrical conductivity depends on density, grain size, and chemical composition and it is affected by the amount of salts of singly charged, doubly charged cations and anions (Na+, K+, Ca2+, Cl−, SO42-, HCO3-). The results show that the total amount of cations and anions is significantly higher in the young LS (9554 μS/cm) compared to the settled LS (745 μS/cm). The obtained results confirmed by the value of the total mineralization of leachates. Total mineralization of the young LS is (1300–3100 mg/l), and of the settled LS—150–370 mg/l. The results indicate a high degree of physical chemical changes in the settled soils.

3.2 Organic Component of Landfill Soils

The content of organic component determined by ignition losses, the humus component determined by the method of Tyurin. Total content of three elements (C, H, N) was determined simultaneously using an elemental analyzer Euro EA3028-NT. The results of the analysis presented in Figs. 2, 3.

The results of analysis show the enrichment of the young soils with organic components. The content of organic carbon, humus, nitrogen, and hydrogen is approximately two times higher in the young LS in comparison to the settled LS. The humus content according to the Tyurin method in the young soils is 2.82%, and in the settled ones—0.26%.

Fig. 2 Ignition losses at 550 °C and 900 °C

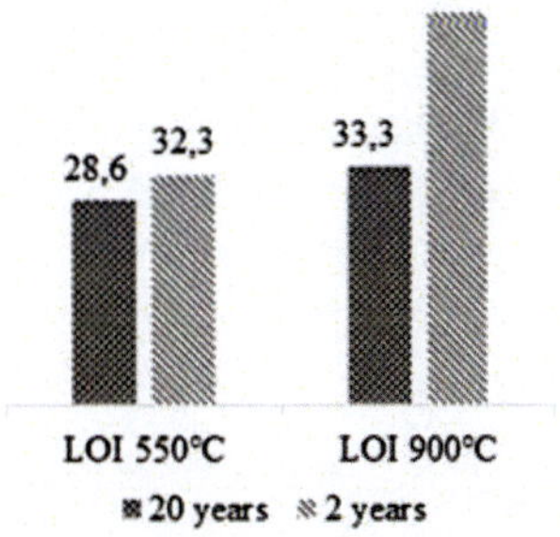

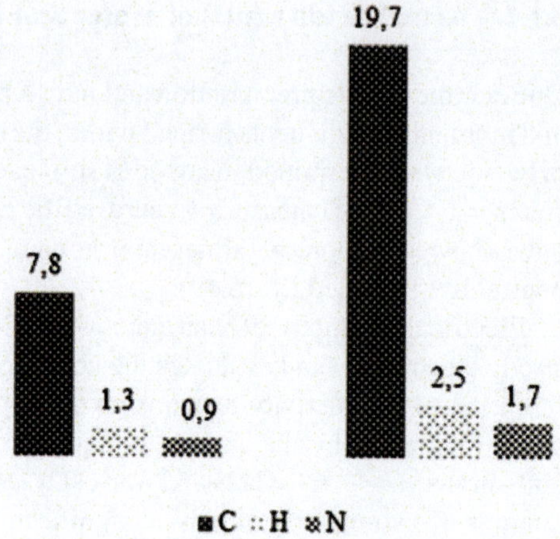

Fig. 3 Results of C–H–N analysis of 2-years and 20-years LS

Thermal analysis reveals the differences in the composition of organic matter in the soils of different ages (Fig. 4). The earliest endothermic effect (about 100 °C) on the thermograms corresponds to the weight loss due to hygroscopic water. This effect is marked in the both types of LS. Dehydration of gypsum corresponds to a maximum about 140 °C and found in the settled LS. Maximum mass loss is typical of the young LS (35–45%) and coincides with the data on high content of C_{org} in them. Oxidation of organic matter corresponds to temperatures of 300–500 °C (Boyanov et al., 2014). The young soils show two peaks at 315 °C and 498 °C. The first exothermic effect associates with the presence of immature organic matter represented by humic acids. More stable functional organic groups decompose at higher temperatures. The settled landfill soils characterized by the presence of one distinct peak at 274 °C and a shoulder at about 400 °C. The thermograms of the studied samples show endothermic effects corresponding to the decomposition of carbonate minerals in the high-temperature interval (600–800 °C) (Olszak-Humienik and Mozejko, 1999). This peak clearly pronounced in the samples of young soils, while in the settled old soils, the peak in this area is wide, which is typical of the decomposition of a mixture of different phases. In some cases, phases represented by newly formed minerals.

Scanning electron microscopy revealed the presence of biofilms and fungal hyphae in LS (Fig. 5). The results of microbiological analysis presented in the Table 1.

All the presented species are typical for the North-West region. The young LS are poorer in the number of microorganisms in comparison with the settled LS, but the microorganisms are more diverse.

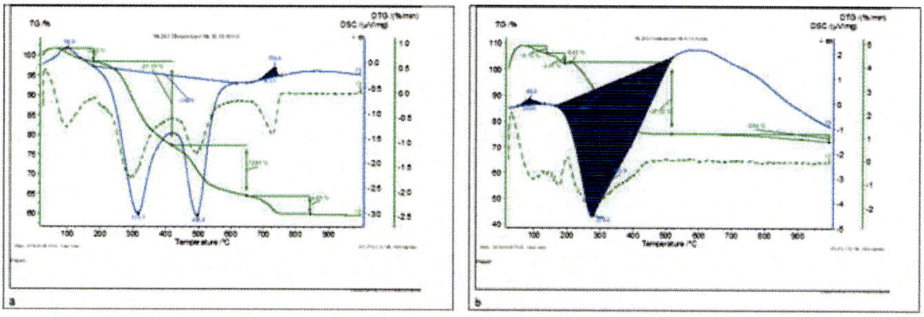

Fig. 4 Thermograms of 2-years soils (**a**) and 20-years soils (**b**)

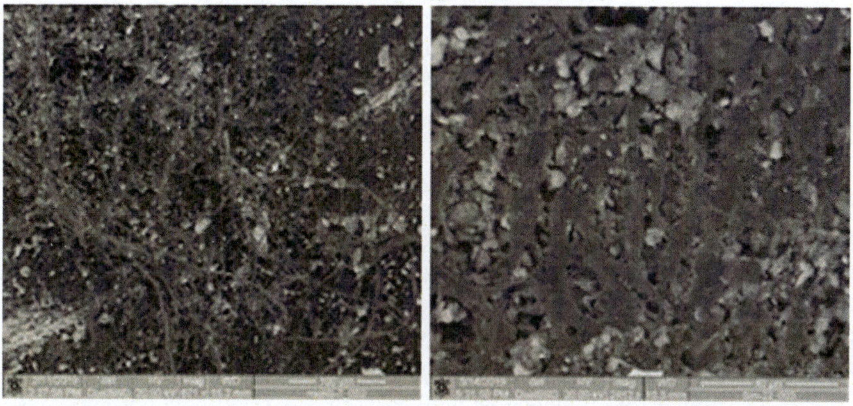

Fig. 5 Biologic component of LS. SEM analysis. **a**—2-years soils; **b**—20-years soils

Table 1 Microbiological biotopes of the different ages landfill soil

Landfill soil	Young Landfill Soil (2 years)	Settled Landfill Soil (20 years)
Number of microbial cells	4000	7000
Microorganisms	Mucor plumbeus Mucor hiemalis Penicillium brevicompactum Penicillium chrysogenum Penicillium spinulosum Fusarium oxysporum Fusarium solani	Mucor hiemalis Mucor ramosissimus Penicillium chrysogenum Penicillium funiculosum Fusarium oxysporum

3.3 Chemical Features of Landfill Soils

Data on macro-element composition of LS used to study the degree of chemical weathering. Indices of chemical weathering calculated basing on the content of

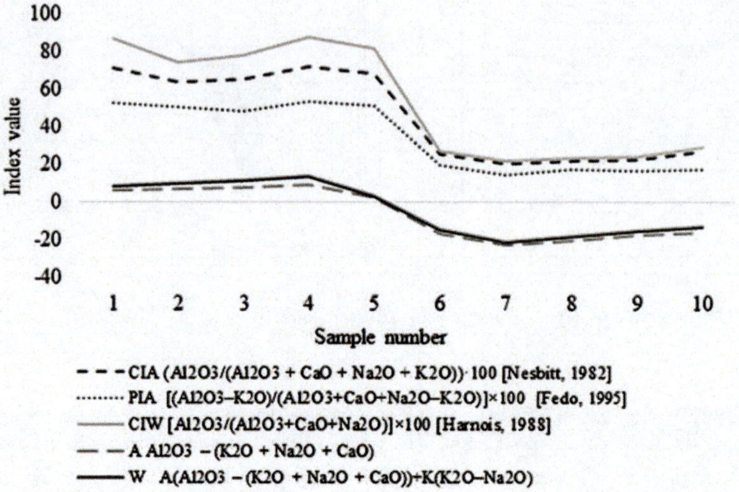

Fig. 6 Indexes of chemical weathering. 1–5—20-years LS; 6–10—2-years LS

petrogenic oxides in the samples. The results show the degree of hypergenesis in the landfill soils. The index values in the settled LS are higher than in the young LS (Fig. 6).

The trace element composition was determined using XRF analysis. Mercury analyzed by Zeeman spectrometry on a mercury analyzer (Fig. 7). Elements of the first (Hg, Pb, Zn, Cd) and second (Cu, Mo, Sb, Ni) toxicity classes accumulate in the soils in amounts up to 2000 Clark Concentrations. Total pollution index (Zc) calculated as the resulting value. All data indicate the accumulation of chemical elements in the young LS compared to the settled soils. Thus, the Z_c index for the young LS is 74, and for the settled ones—32 conventional units. The obtained results are in good agreement with the data of other authors (Hull et al. 2005; Burlakovs et al. 2016; Zhang et al. 2017). The results are in good agreement with the data on the intensity of chemical weathering, the content of C, N, H, and the physico-chemical properties of different ages LS. The results of the study indicate that over time, chemical elements transfer into the soil leachate and washed out of the landfill soils.

To confirm this fact, water soil leachates from the different ages LS analyzed by the ISP MS method (Table 2). The accumulation coefficient calculated as the ratio of the contents of chemical elements in the young LS to the settled LS.

The content of all analyzed chemical elements in the water leachates from the young LS is higher than in the settled ones. Fe and Mn show the highest values of the accumulation coefficient. As this value decreases, the following series of accumulation of chemical elements is formed: Fe (14.9)—Mn (14.1)—Ti (6)—K (4.6)—Pb, Hg, Mo (4.4)—Sr (3.5)—Y (3.3)—Na (3)—Ni (2.8)—Ca, As (2.5)—Rb (2.3)—Mg (2.2)—Sn (2.1)—U (2). For Si, Zn, Th, La, Ba, and Cu the value of the coefficient is less than 2. For Se and V the value is less than 1. It is noteworthy that Ca

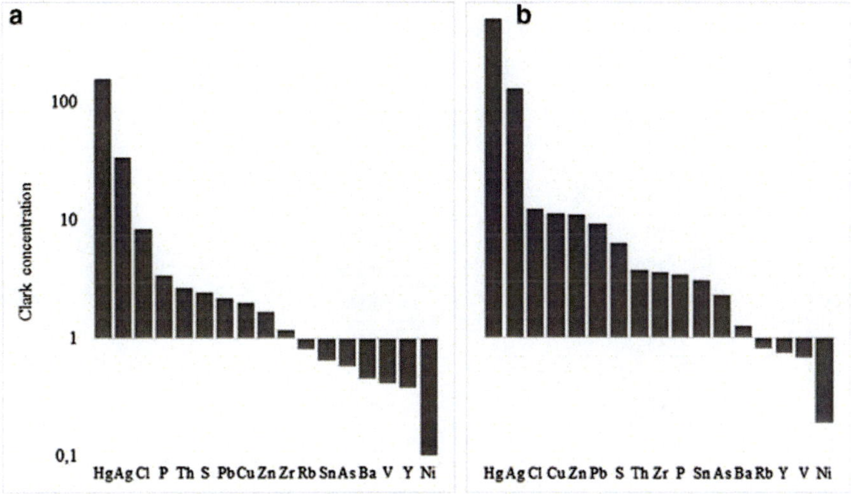

Fig. 7 Geochemical spectra of chemical elements (normalized to clark in soils, according to Dobrovolsky 1993). A—2-years LS; b—20-years LS

and Mg demonstrate the low values. These and other elements can form secondary minerals in the LS (Fig. 8).

Thus, in the process of outdoor storage under the impact of rain and snow, chemical elements can transfer into mobile forms. Some of the chemical elements are washed out of the soils, and some elements form new mineral phases. Scanning electron microscopy made it possible to establish, that the most common are oxides and hydroxides, sulfates, carbonates, phosphates, chlorides of Cu, Zn, Pb, Fe, Mn. Selenium and vanadium can be part of organic biofilms.

3.3.1 Biotesting of Water Soil Leachate.

The method of biotesting has recently received wide recognition and prevalence. Biotesting understood as a procedure for determining the toxicity of the environment with the help of test objects that signal danger, regardless of which substances and in which combination cause changes in the vital functions of objects. Various organisms that react to the presence of harmful substances are widely used for biotesting. Watercress, an annual plant from the cruciferous family (Latin Lepidium sativum), can be used to assess the degree of water pollution.

The germination of seedlings and the length of the roots of watercress are the most sensitive parameters for detecting the general stress effect of toxic substances (tabl. 3).

According to the results of testing, it revealed that the lowest observed in watercress sprouted on water extracts of samples of young LS (70–80%) compared with the settled LS (83–90%). The length of the root of watercress, sprouted on water

Table 2 Chemical elements in the soil water leachates from the different ages landfill soils, mg/l

Element	Young Landfill Soil (2 years)	Settled Landfill Soil (20 years)
Si	14,520	7940
Al	395	311
Ti	2,95	0,49
Fe	2050	145
Mn	232	15,6
Mg	18,740	8500
Ca	143,860	58,290
K	71,320	15,590
Na	87,460	28,820
Rb	62,8	27,7
Ba	161	149
Sr	1220	349
V	0,67	1,55
Se	7,35	9,27
Cu	15,8	15,5
Ni	14,3	5,13
Zn	54,3	32
As	20,8	8,45
Pb	0,84	0,19
Mo	2,22	0,51
Hg	345	78
Sn	0,15	0,07
Y	0,3	0,09
La	0,062	0,054
Th	0,006	0,004
U	1,38	0,69

Table 3 Biotesting results of water LS extracts by the watercress

Sample	Young Landfill Soil	Settled Landfill Soil	Control
Germination, %	70–80	83–90	100
Root length, mm	4,0–4,2	2,5–3,5	2,1
Factor Toxicity Index	1,9–2,0	1,2–1,5	1,0

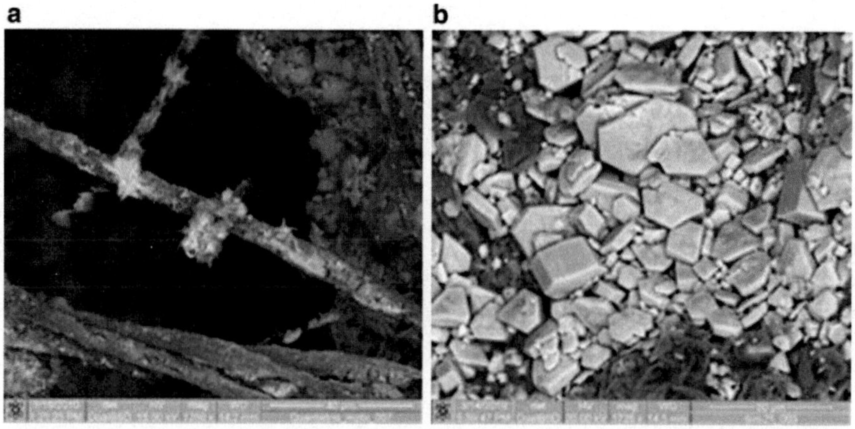

Fig. 8 SEM images of secondary minerals in the settled landfill soils. Cuprite crystals on copper wire (a); gypsum crystals on organic biofilm

extracts of samples of young SL, is 4,0–4,2 mm, in settled LS—2,5–3,5 mm. The Factor Toxicity Index was calculated to assess the aquatic soil extracts for toxicity along the length of the root of the cross salat. Its value for the cross salat sprouted on the aqueous extracts of samples of young LS is 1,9–2,0 compared with the observed LS—1,2–1,5. The value of the Factor Toxicity Index of the young LS corresponds to hazard class VI.

4 Conclusions

(1) High values of total carbon, humus, N, H recorded in the young soils (2-years old). They contain various forms of organic matter (immature and with stable functional groups). The number of microbial cells in the young soils is lesser in comparison with the settled ones (20-years old), but their diversity is greater. The settled soils contain well-formed biofilms.
(2) High contents of most chemical elements recorded in the young soils in comparison with the settled soils. Z_c index for the young LS is 74, and for the settled ones—32 conventional units. 20-year old soils characterized by a greater degree of weathering (10–100 times) compared to the young soils.
(3) Under the atmospheric precipitation, chemical elements transfer into mobile forms and accumulate in the soil leachate. Some of the elements washed out of the soils, while the others form secondary minerals (salts, oxide-hydroxides, chlorides, phosphates) or are included in the organic component of the landfill soils.

Acknowledgements The authors thank M.S. Zelenskaya for the results of microbiological experiments. The studies carried out at the Resource Centers of St. Petersburg State University (Microscopy and microanalysis, X-ray diffraction research methods, Chemical Analysis and Materials Research Centre).

References

Assamoi B, Lawryshyn Y (2012) The environmental comparison of landfilling vs. incineration of MSW accounting for waste diversion. Waste Manag. 32 (5): 1019–1030 https://doi.org/10.1016/j.wasman.2011.10.023

Basha BM, Parakalla N, Reddy KR (2016) Experimental and statistical evaluation of compressibility of fresh and landfilled municipal solid waste under elevated moisture contents. Int. J. Geotech. Eng. 10 (1): 86–98 https://doi.org/10.1179/1939787915Y.0000000018

Boyanov B, Peltekov A, Petkova V (2014) Thermal behavior of zinc sulfide concentrates with different iron content at oxidative roasting. Thermochim. Acta. Vol. 586. 9–16

Burlakovs J, Kaczala F, Vincevica-Gaile Z, Rudovica V, Orup˜old K, Stapkevica M, Bhatnagar A, Kriipsalu M, Hogland M, Klavins M, Hogland W (2016) Mobility of metals and valorization of sorted fine fraction of waste after landfill excavation. Waste Biomass Valorization 7:593–602 https://doi.org/10.1007/s12649-016-9478-4

Chen Y, Zhan TL, Ling W-A (2008) Mechanical properties of municipal solid waste from Suzhou landfill in China. In: GeoCongress 2008: Geotechnics of Waste Management and Remediation, New Orleans, United States, pp. 160–167 https:// doi.org/https://doi.org/10.1061/40970(309)20

Chen Y, Xu WJ, Ling DS, Zhan LT, Gao W (2020) A degradation–consolidation model for the stabilization behavior of landfilled municipal solid waste. Comput. Geotech. 118:103341 https://doi.org/10.1016/j.compgeo.2019.103341

Dixon N, Jones DRV (2005) Engineering properties of municipal solid waste. Geotext. Geomembr. 23 (3):205–233 https://doi.org/10.1016/j.geotexmem.2004.11.002

Dobrovolsky, V (1983) Geography of trace elements: Global scattering. 174 p

Hadinata F, Damanhuri E, Rahardyan B, Widyarsana IMW (2018) Identification of initial settlement of municipal solid waste layers in Indonesian landfill. Waste Manage. Res. 36 (8): 737–743 https://doi.org/10.1177/0734242X18788149

Hull RM, Krogmann U, Strom PF (2005) Composition and characteristics of excavated materials from a New Jersey landfill. J. Environ. Eng. 131:478–490. https://doi.org/10.1061/(ASCE)0733-9372(2005)131:3(478)

Jo Y-S, Jang Y-S (2021) Comparison of waste settlement characteristics for two landfills disposed in long sequential periods. Waste Manage. 131:433–442. https://doi.org/10.1016/j.wasman.2021.07.003

Liu SJ, Xi BD, Qiu ZP, He XS, Zhang H, Dang QL, Zhao XY, Li D (2019) Succession and diversity of microbial communities in landfills with depths and ages and its association with dissolved organic matter and heavy metals Sci. Total. Environ, 651:909-916 https://doi.org/10.1016/j.scitotenv.2018.09.267

Olszak-Humienik M, Mozejko J (1999) Kinetics of thermal decomposition of dolomite. J. Therm. Anal. Calorim. Vol. 56: 829–833

Santos VHJM, Engelmann AdM, Marconatto L, Borge LGdA, Palhano Adl, Augustin AH, Rodrigues LF, Ketzer JMM, Giongo A (2022) Exploratory analysis of the microbial community profile of the municipal solid waste leachate treatment system: A case study. Waste Manag, 141:125–135 https://doi.org/10.1016/j.wasman.2022.01.014

Song L, Wang Y, Tang W, Lei Y (2015) Bacterial community diversity in municipal waste landfill sites. Appli. Microbio. Biotechnol 99:7745–7756 https://doi.org/10.1007/s00253-015-6633-y

Speier CJ, Mondal MM, Weichgrebe D (2018) Evaluation of compositional characteristics of organic waste shares in municipal solid waste in fast-growing metropolitan cities of India. J. Mater. Cycles Waste Manage. 20 (4): 2150–2162 https://doi.org/10.1007/s10163-018-0757-y

Tunesi S (2011) LCA of local strategies for energy recovery from waste in England, applied to a large municipal flow. Waste Manage. 31 (3): 561–571 https://doi.org/10.1016/j.wasman.2010.08.023

Vilar OM, Carvalhod M (2004) Mechanical properties of municipal solid waste. J. Test. Eval. 32, 438–449 https://doi.org/10.1520/JTE11945

Zhang DQ, Tan SK, Gersberg RM (2010) Municipal solid waste management in China: status, problems and challenges. J. Environ. Manage. 91 (8): 1623–1633 https://doi.org/10.1016/j.jenvman.2010.03.012

Zhang C, Feng J, Zhao T, Rong L (2017) Physical, chemical, and engineering properties of landfill stabilized waste. Land Degrad. Dev. 28:1113–1121 https://doi.org/10.1002/ldr.2594

Field Trials of Soil Improvement Technology with a Bacterial Mixture

Darya A. Golovkina, Elena V. Zhurishkina, Jing Xu, and Anna A. Kulminskaya

Abstract Microbially induced calcite precipitation (MICP) is a process of biological cementation that takes place during the metabolic processes of microorganisms under conditions of elevated concentrations of calcium and carbonate ions in the microenvironment. As a result, the geotechnical properties of the soil are improved due to the binding of soil particles by the resulting calcium carbonate minerals. Despite a large number of laboratory studies, data on field trials are scarce. In this work, we present our first attempts of a field technology for soil compaction using microorganisms. Screening of the bacterial strains resulted in a bacterial mixture M2 that showed the highest values for the amount of calcium carbonate formed and for the uniaxial compressive strength when treating soil samples in laboratory studies. After selection of the bacterial preparation storage and transportation methods, field experiments were carried out to improve the geotechnical properties of the soil by bioaugmentation and biostimulation. The experimental soil site treated with the M2 bacterial mix and a cementing mortar in 2 stages yielded 4 cm of compaction depth; filtration coefficient of 53.61 m/day; the concentration of formed calcium carbonate was 2.43%.

Keywords Biomineralization · Soil stabilization · MICP · Field tests · Microbes · Calcium carbonate

D. A. Golovkina (✉) · E. V. Zhurishkina · A. A. Kulminskaya
Petersburg Nuclear Physics Institute Named By B.P. Konstantinov of National Research Center "Kurchatov Institute", Gatchina, Russia
e-mail: golovkina_da@pnpi.nrcki.ru

Kurchatov Genome Center—PNPI, Gatchina, Russia

J. Xu
School of Materials Science and Engineering, Tongji University, Shanghai 201804, PR China

Key Laboratory of Advanced Civil Engineering Materials (Tongji University), Ministry of Education, Shanghai 201804, PR China

O. V. Frank-Kamenetskaya et al. (eds.), *Biogenic—Abiogenic Interactions in Natural and Anthropogenic Systems 2022*, Springer Proceedings in Earth and Environmental Sciences, https://doi.org/10.1007/978-3-031-40470-2_25

1 Introduction

Insufficiently stable mechanical properties of the soil in different regions can cause serious problems in the construction of buildings and roads leading to soil erosion and destruction of slopes (DeJong et al. 2010; Jiang et al. 2017). To prevent this problem and stabilize the soil, chemical or biological methods are usually used. The use of cement or other chemical reagents does not seem an ideal solution due to their high cost and definite difficulties in placing stabilizing materials deep in the soil (Rahman et al. 2020; Chu et al. 2015; Achal and Mukherjee 2015). Thus, there is an urgent need to find alternatives to the above methods. In recent years, there has been growing interest in such a "green" technology, like microbial induction of calcium carbonate precipitation (MICP) resulted in the creation of a new type of building material, biocement (Whiffin 2004). Biocementation is an innovative method for improving the mechanical characteristics of the soil that is based on minerals formed as a result of microbiological action (Chu et al. 2015).

In nature, soil cementation can occur due to natural processes associated with climate (Ivanov and Stabnikov 2017). The most effective process seems to be microbiological cementation (Castanier et al. 1999). In this case, soil compaction occurs due to the induction of calcium carbonate precipitation by bacteria, most often ureolytic ones. Most bacteria are known to produce the urease, the enzyme that catalyzes the degradation of urea. As a result of their activity, the pore space is filled with formed calcium carbonate and due to this, soil particles are bound together (Seifan and Berenjian 2019; Seifan et al. 2020).

The process of biocementation of soil samples the improvement of soil or sand by microorganisms has been studied in sufficient detail at the level of laboratory experiments (Bahmani et al. 2019; Marın et al. 2021; Safdar et al. 2021; Jain et al. 2021). Despite this, there are still few examples of field applications. Several publications by L. van Passen (Van Passen et al. 2010; Zeng et al. 2021) and M. Gomez et al. (2015) reported soil improvements in several patches of sand in the desert. Obviously, there are still many aspects of the soil improvement process in the field conditions that need to be addressed in order to optimize the technology. In our previous work, ureolytic bacteria were screened by pH parameters, catalytic activity of the enzyme, and the amount of $CaCO_3$ formed in a liquid medium, and a comparative analysis of the pathways of ureolysis and metabolism of organic calcium salts was carried out (Golovkina et al. 2020). The aim of the current work was to find the optimal composition of the bacterial mixture to improve the strength characteristics of the soil and attempt to scale the experiment for field conditions.

2 Materials and Methods

2.1 Bacterial Strains and Cultivation Conditions

The bacterial strains used in the work were isolated, identified and maintained in the collection of the Laboratory of Biotechnology of the MRBD NRC Kurchatov Institute—PNPI. For the study, 5 strains were selected that were previously characterized (Golovkina et al. 2020): *Bacillus subtilis* 168, *B. subtilis* K51, *B. cereus* 4b, *B. cereus* BL, *Micrococcus luteus* 6.

To assess the ability of bacteria to stabilize soils in laboratory conditions, several variants of combinations of microorganisms were compiled. Each microorganism was cultivated in 5 mL of LB medium for 1 day at 37 °C with stirring at 120 rpm. The grown bacterial suspension was separated by centrifugation for 15 min at 3000 rpm and resuspended in 2 ml of 0.9% NaCl solution.

To select the optimal method for storage and transporting the bacterial preparation to the work site, several spore storage carriers were studied: storage of spores in PBS; in 0.9% NaCl, in sterilized water, in 1% glucose solution and in the soil. For this, bacterial mixtures were placed in different types of carriers for spore formation and storage. In 100 mL of B4 calcification medium (Golovkina et al. 2020), 1 mL of bacterial spores at a concentration of 4×10^8 cells/mL from buffers after a month of storage was placed (1 g was added from the soil sample). For the formation of minerals, the samples were cultivated at 37 °C and stirring at 120 rpm for 7 days. Then, all samples were filtered from the medium, the precipitate was washed with water and dried in air.

To prepare a bacterial sample for incorporation into the soil in the field conditions, each of the selected bacterial strains for the mix M2 was cultivated in 700 mL of LB medium for 48 h at 37 °C with stirring at 120 rpm. Then the cells were separated by centrifugation at 3000 rpm for 45 min and resuspended in 0.9% NaCl solution. All cultures were combined and diluted in 15 mL of 1% glucose solution. One sample contained about 6–8 g of dry cell suspension for the treatment of 300 g of pre-sterilized soil. The resulting immobilized bacterial preparation was placed at 37 °C for 2 weeks to fix the bacteria in the soil.

3 Soil Form Preparation and Assessment of the Ability of Selected Strains to Stabilize the Soil

Soil samples were picked up in different parts of the Leningrad region, Russia. The soils were collected from a depth of 0.25 m using sterile drill (washed with 70% ethanol) and stored at 4 °C till further analysis. Details on characteristics of the collected samples of soils were made in accordance with the Unified State Register of Soil Resources of Russia and the soil-geographical database of Russia and are presented in the Table 1.

Table 1 Characteristics of soil samples used in the study

Sample No	Soil type	Collecting location	Parameters						
			pH (of soil suspended in distilled water)	Humus content, %	N_2 content, %	$CaCO_3$ content, %	Granulometric composition, partical size, mm		
							1–0,25	0,5–0,01	< 0,001
Sample # 1	Peat boggy gley	A1h soil horizon, SNT Brigantina-2, Leningrad region, Russia	4,53	4,5	0,23	0,7	2	5	24
Sample # 2	Peat podzolic gley	A2Bg soil horizon, Moskovsky city district of St. Petersburg, Russia	5,56	1,15	0,07	0,695	6,7	5,1	11
Sample # 3	Soddy podzolic surface gley	A1A2 soil horizon, Gatchina, Leningrad district, Russia	5,82	1,29	0,06	0,605	1	60	37
Sample # 4	Soddy podzolic surface gley	B1 soil horizon, Gatchina, Leningrad district, Russia	5,4	0,24	0,53	0,55	1	44,5	28,7

Fig. 1 Soil samples treated with CS and bacteria;

Each plastic mold (4.5 × 4.5 × 4 cm in size) was filled with 80 g of soil. The prepared bacterial suspension (4 mL) was injected into the mold along with 20 mL of a cementing solution (CS) consisting of 1 M urea and 0.5 M $CaCl_2$. A sample treated with water was used as a negative control, and a sample treated with CS was used as a positive control. To prevent the samples from drying out, the molds were covered with foam sponges and placed at a temperature of 28 °C. During the first 5 days, 5 mL of CS was added to the molds with experimental samples daily, after which the samples were kept for 2 weeks at a temperature of 37 °C. Then soil samples were placed in a drying oven at 60 °C for 3 days, removed from the molds and, if necessary, dried in air until the water completely evaporated. The final samples are shown in Fig. 1.

4 Unconfined Compression Strength and Determination of the Calcium Carbonate Quantity in Soil Sample

Unconfined strength test was carried out according to the method reported in (GOST 12,248–2010 Soils. Methods of laboratory determination strength characteristics and deformability) in the PSN-0.16.10 M concentrated compression device with lowering of the upper part of the device at a constant rate of 0.1 MPa/s until the destruction of the soil sample.

To estimate the amount of calcium carbonate formed, 1 g of each soil sample was crushed in a mortar to a homogeneous structure and placed in a carbonometer (KM-NT with an MTI manometer 2.5/0.5), followed by 10 mL of 10% HCl (ASTM D 4373 Standard Test Method for Calcium carbonate content in soil) addition. The percentage of carbonate formed in 1 g of the sample was calculated according to the formula (1):

$$CaCO_3 = \frac{P * 100}{m * a} \quad (1)$$

where P is pressure, bar; m is a sample mass, g; a is the correction coefficient equal to 1.4116. All experiments were carried out in 3 repetitions.

5 Field Experiments

For the field experiment, the soil (sample # 4, Table 1) was sampled at a depth of 0.40 m with a sterile drill as described above. Experimental soil plots (60 × 60 × 10 cm) were prepared for field tests by preliminary digging and getting rid of excess vegetation (Fig. 2):

Experiment # 1: Control soil plot treated with 20 L of tap water;

Experiment # 2: Experimental soil plot treated with 20 L of a cementing solution (CS) containing 1 M urea and 0.5 M calcium chloride;

Experiment # 3: Experimental soil plot treated with a bacterial mix M2 (300 g) and CS (20 L);

Experiment # 4: Experimental soil plot treated with a bacterial mix M2 (300 g) and CS (10 L in the beginning of the experiment and 10 L 4 days later).

The climate conditions at the start of the experiments were as follows: temperature was 19 °C and humidity was 72%. Then, the experiments were left for a month before the soil compaction was assessed.

Fig. 2 Experimental soil plots (60 × 60 × 10 cm) prepared for field tests;

6 Soil Improvement Evaluation After Treatment

The following methods were used to assess the soil:

- visual assessment of the $CaCO_3$ layer formation degree;
- the permeability test;
- determination of calcium carbonate quantity formation in 1 g of a soil sample;
- determination of soil sample resistance.

For the visual analysis of a carbonate layer thickness, a soil sample was removed from each experimental plot using glass tubes (2 × 20 cm).

Modified method of the procedure (described in GOST 25,584–2016 Soils Methods of laboratory determination of the permeability coefficient) was used to evaluate the water permeability of the soil. An experimental soil sample (70 g) was placed in a column (14.5 × 4 cm) and the flow rate of passing water was measured. The standardized water filtration coefficient K_{10} (m/day) was calculated using the formula (2):

$$K_{10} = \frac{665 * Q}{S * t} \tag{2}$$

where Q is the volume of water filtered during the testing stage, cm^3; t is the filtration time, sec; S is the cross-sectional area of the cylinder with soil, cm^2; 665 is conversion coefficient.

The method for determination of the calcium carbonate quantity in the soil sample was carried out in accordance with the procedure described in the Sect. 3.

To determine the resistance force of soil samples, a cargo (0.465 kg) was fell down from 0.15, 0.3 and 0.45 m height and the soil resistance force values were calculated using the formula (3):

$$F_r = \frac{m * g * H}{l} \tag{3}$$

where m is the weight of the cargo, kg; g is a coefficient of free fall acceleration, m^2/sec; h is the height of the cargo falling down, m; l is a depth of cargo immersion in the ground, m.

7 Results

7.1 *Strength Characteristics of Soil Improvement by Microorganisms Under Laboratory Conditions*

To assess the ability of previously selected individual strains and their mixtures to improve the mechanical characteristics of the soil, three different ground samples were withdrawn at different parts of the St. Petersburg region: Gatchina and Vsevolozhsk districts, as well as Moskovsky city district of St. Petersburg. After the soil samples were treated with each of the microorganisms used in this study and their mixture, the uniaxial compressive strength parameters were calculated as described in the Materials and Methods section (Fig. 3).

The highest values were found for the soil sample # 1 (Table 1) when strains *M. luteus* 6 and *B. cereus* BL were explored showing a greater ability to strengthen. For the soil sample # 2, treatment with the strains *B. subtilis* 168 and *M. luteus* 6 had led to the highest values while for the soil sample # 3 the highest results were reached with the strains *B. subtilis* K51 and *B. cereus* BL. Interestingly, a bacterial mixture composed of all strains presented also had a significant effect on the change in compressive strength in all soil samples. It has become the decisive argument in favor of using a microbial mixture rather than individual strains for soil compaction in a large-scale field experiment.

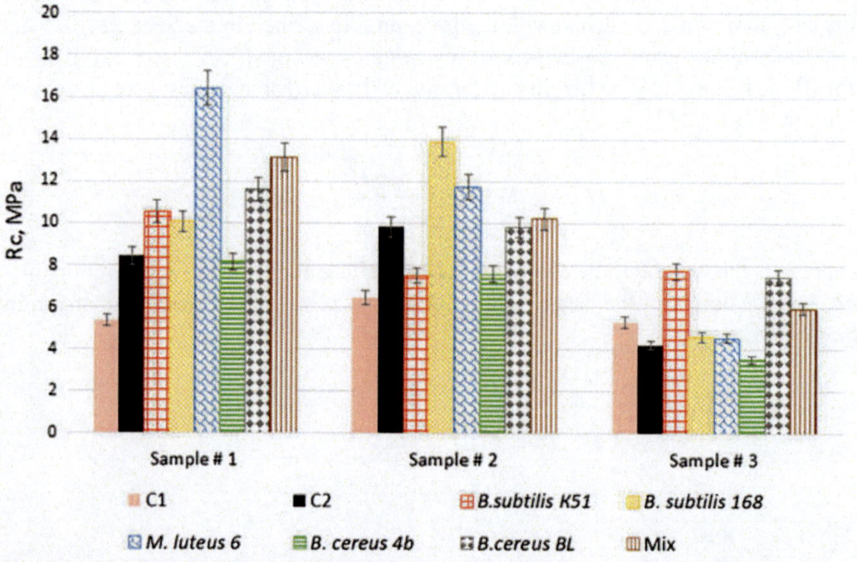

Fig. 3 Values for the compressive strength (Rc) of three types of soil improvement with different bacterial strains and their mixtures. A soil sample treated with water (C1) and a soil sample treated with cementing solution (C2) were taken as controls

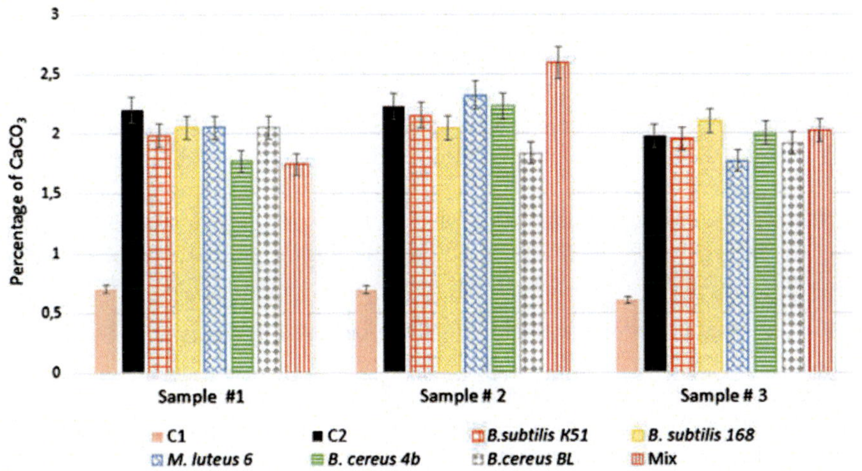

Fig. 4 The percentage of calcium carbonate formed in each sample of soil after the treatment with individual bacterial strains and their mixtures. A soil sample treated with water (C1) and a soil sample treated with cementing solution (C2) were taken as controls

In each of the studied samples (samples # 1–3, Table 1), the percentage of calcium carbonate formed during microbial induction was calculated (Fig. 4).

Figure 4 shows a significant increase in calcium carbonate in each sample compared to the control one treated with water (C1). Surprisingly, in the experimental soil sample # 1, the highest percentage of $CaCO_3$ was detected in the control sample treated with CS (C2). The soil sample # 2 was characterized by the highest quantities of calcium carbonate formed by the bacterial mix and *M. luteus* 6 strains while on the third soil sample all strains acted in a similar way.

Analysis of the determined values for the soil compressive strength revealed the similarity in action for all the strains except for *B. cereus* 4b that was the worst in the soil compaction (Fig. 3). The bacterial strain *B. cereus* 4b was found to form a large amount of calcium carbonate crystals in all cases (Fig. 4), however, they were probably fragile in the structure. Therefore, samples treated with this strain didn't provide sufficient strength when compacting the soil. On the other hand, the bacterial mixture composed of the mixture of strains showed a greater increase in strength characteristics compared to the control sample C1. Therefore, the next step was the selection and preparation of a bacterial mixture, which can later be used in field trials.

8 Selection of the Bacterial Mixture Composition

In order to select the most optimal bacterial composition for the field work, the following combinations of microorganisms were tested:

M1: the bacterial consortium composed of 5 strains: *B. subtilis* 168, *B. subtilis* K51, *B. cereus* 4b, *B. cereus* BL, *M. luteus* 6; the bacterial mixture concentration was equal to 4.29 × 10^8 cells/mL (Wang et al. 2022);

M2: the bacterial consortium composed of strains *B. subtilis* 168, *B. subtilis* K51, *B. cereus* BL, *M. luteus* 6; bacterial mixture concentration was equal to 4.29 × 10^8 cells/mL;

M3: the bacterial consortium composed of *B. subtilis* 168, *B. subtilis* K51, *B. cereus* 4b, *B. cereus* BL, *M. luteus* 6. The concentration of the most potential *B. subtilis* K51 and *B. cereus* BL was twice increased in the mixture.

Key parameters of soil treatment by several bacterial mixtures are presented in Fig. 5.

The analysis of the results revealed that the M2 mix composed of bacterial microorganisms without the *B.cereus* 4b strain showed the highest compressive strength while in the soil sample treated with CS (C2 control sample) the percentage of calcium carbonate was the highest. The other mixes demonstrated similar results. Interestingly, an increase in the concentration of the two most active strains (M3 mix) had led to a worse result that can be explained by the inhibition of other bacteria and the loss of the effect of culturing individual strains. Based on these data, the M2 mix consisting of *B. subtilis* 168, *B. subtilis* K51, *B. cereus* BL, and *M. luteus* 6 at the same concentration was taken as a model of a bacterial mixture for further field work. Experiments have shown that stimulation of soil microorganisms by CS led

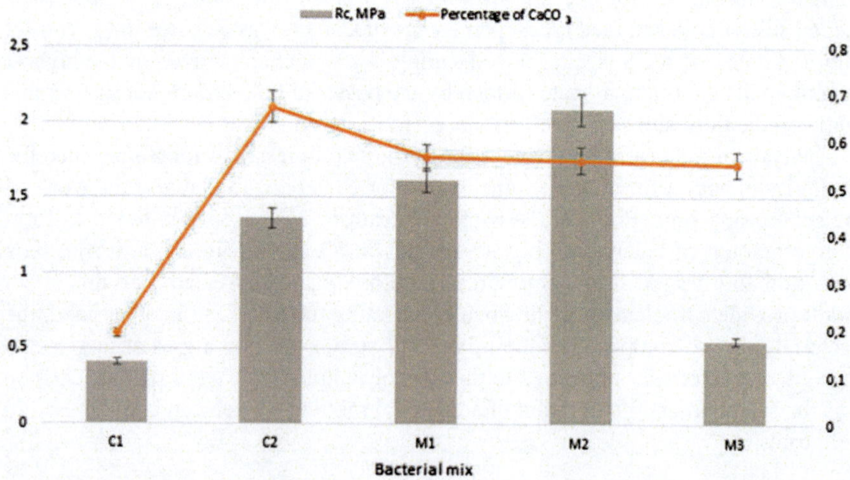

Fig. 5 Parameters of soil treatment by several bacterial mixtures. The diagram shows the uniaxial compressive strength values (Rc); the percentage of calcium carbonate formed in a soil sample using different variants of a bacterial mixtures shown by a curve. Details on soil samples # 1–3 used here are given in the Table 1. A soil sample treated with water (C1) and a soil sample treated with cementing solution (C2) were taken as controls

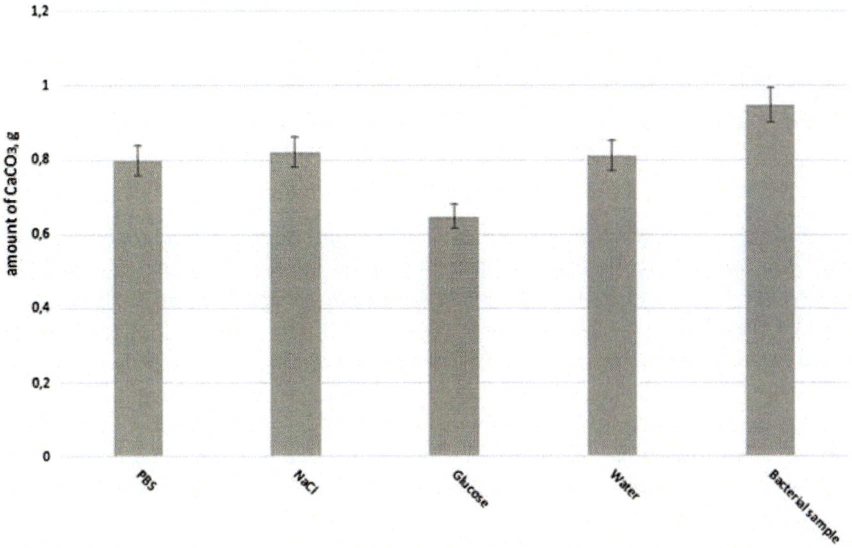

Fig. 6 Quantity of $CaCO_3$ precipitates produced by selected strains in 100 mL of the B4 medium using different spore carriers

to a significant increase in the strength characteristics of the ground sample, which also makes this variant a potential for compaction.

The next step of the study was the development of a system for storage and transport of the bacterial mix M2 to the field test plot. Several media have been investigated for the preservation of bacterial spores. The mineral precipitate obtained after cultivation of the bacteria in the B4 calcification medium was measured. The results are presented in Fig. 6.

The greatest mineral-forming activity was shown by bacterial mix M2 after being stored in soil, water and NaCl solution. Unlike soil as a solid carrier, liquid buffers appeared to be not convenient for transporting the bacterial preparation to the site. Therefore, sterilized and prepared soil was chosen as a carrier and immobilization agent. Such a carrier makes it possible to conveniently store and deliver the bacterial preparation to the place of its use in the field without losing their ability to form $CaCO_3$ crystals.

9 Field Trials of Soil Compaction by Microorganisms

Field studies were carried out in summer (average temperature 20 °C) using 4 experimental soil plots (sample # 4, Table 1):

1. A plot of soil treated with 20 L of tap water as a Control experiment;

2. A plot of soil treated with 20 L of CS to stimulate the existing soil microflora as an Experiment # 1;
3. A plot of soil simultaneously treated with 20 L of CS and 300 gm of the bacterial mix M2 to increase the potential of soil microflora as an Experiment # 2;
4. A plot of soil treated with 300 gm of the bacterial mix M2 and twice treated with 10 L of CS in order to fix the introduced bacteria in the soil as an Experiment # 3.

To study the process of soil stabilization under the above conditions, soil samples were removed twice, after 7 days and after 6 weeks. No changes were found in 7-days-old samples. For this reason, only 6-weeks-old samples were analyzed.

The results of visual inspection of the soil layer thickness (Dhami et al. 2013) reached during the microbially induced compaction processes are presented in Fig. 7.

The results of visual inspection indicate that fractional treatment with the CS leads to the fixation of the introduced bacteria in the soil and, subsequently, to a greater depth of filling the pores with mineralized calcium carbonate. It should be noted that a mineral layer also appeared in the Control experiment, which is probably formed as a result of independent autogenous processes. In addition, despite the fact that in

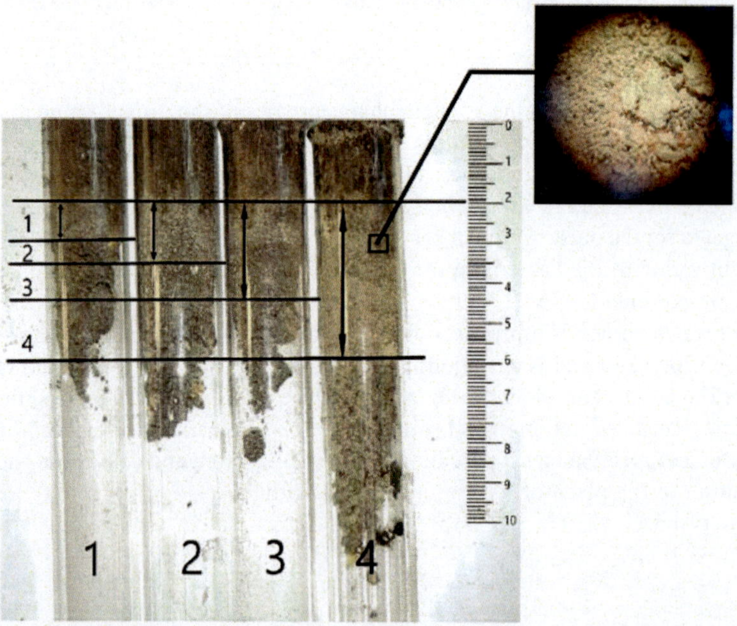

Fig. 7 Thickness of the compacted soil layer measured by height of the ground whitening due to $CaCO_3$ mineralization. Soil sample # 4 (Table 1) was used for the experiment. Tube 1 contains a sample of the control experiment (without treatment); the tube 2: a sample removed from the Experiment # 2 (treated with CS only); the tube 3: a sample from the Experiment # 3 (treatment with CS and the bacterial mix M2); the tube 4: a sample after the treatment with the bacterial mix M2 and CS in 2 stages. Insert shows a view (40x) of formed minerals under an optical microscope Bresser Advance ICD 10x–160x

experiments in the laboratory a high potential was found to stimulate the processes of compaction by the natural microflora of the soil, in the field the same result could not be achieved.

Analysis of the quantity of formed minerals revealed that the most calcium carbonate was formed in Experiments # 2 and # 4 (Table 2) that coincides with the visual assessment (Fig. 7). The permeability coefficient, which characterizes the water permeability of soil samples and the degree of their compaction, turned out to be the lowest in the Experiment # 4 (Table 2).

We found that the soil tends to have a greater resistance to a cargo fall down in the Experiment # 4 compared to the rest of the experimental conditions (Fig. 8).

All tests indicate that the treatment of the soil with a bacterial mixture M2 followed by a two-stage treatment with CS (Experiment # 4) the strength characteristics were improved compared to the control. Stimulation of compaction processes in native soil microflora by the treatment of CS (Experiment # 2) also increased strength compared to control but to a lesser extent.

Table 2 Percentage of formed calcium carbonate and the permeability coefficient K_{10} in field Experimental samples

Experiment No	$CaCO_3$, %	Permeability coefficient K_{10}, m/day
1	1,46	204,40
2	2,10	160,70
3	1,44	156,78
4	**2,43**	**53,61**

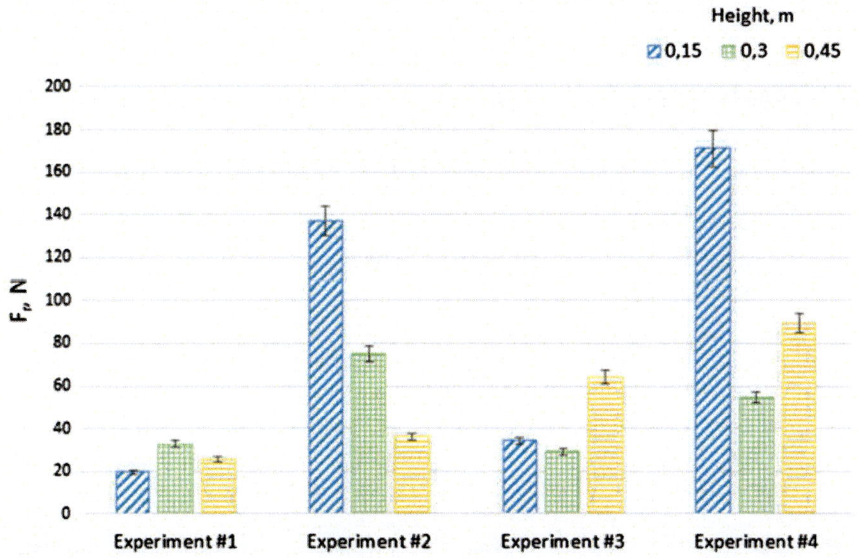

Fig. 8 Dependence of the resistance force (F_r) of soil on the height of the cargo fall down. Soil sample # 4 (Table 1) was used for the experiment

10 Discussion

To date, there are a large number of studies on using MICP to compact soils and sands in order to improve their strength characteristics (Marin et al. 2021; Safdar et al. 2021; Jain et al. 2021). The vast majority of these reports are mainly devoted to the results of laboratory studies while published information on field trials are very scarce. For example, the L. van Passen team (Van Passen et al. 2010), reported a large-scale experiment to compact 100 m^3 of soil. Zeng with co-auth. used a bioaugmentation approach comparing the results of the direct monitoring with a mathematical model of the processes taking place in the field trials (Zeng et al. 2021). Gomez et al. (2015) demonstrated the improvement of several soil areas with a cementing composition during 20 days of continuous treatment. The largest increase in strength value was found in the sample treated with the composition with the lowest concentration of urea and calcium salt (15 and 13 g/L, respectively). Unfortunately, each research group uses different methods to analyze changes in strength characteristics of soils treated by MICP methods depending on the conditions and goals of the experiment that limits a comparative analysis of the results in published reports. In order to develop an efficient and economically viable soil compaction technology, it is necessary to optimize the process parameters to increase the depth of the compacted soil layer, increase the strength values and reduce water permeability. This requires further optimization of cementing solution and the concentration and composition of precipitating microorganisms, modification of the technology itself aiming to reduce costs on the treatment of a large soil area. The success of these studies should optimize the economic aspect of the technology and bring it to the level of large-scale production.

11 Conclusions

In this work, an experiment on soil stabilization by microorganisms was carried out in the real field conditions. For this, bacterial strains were screened for their ability to improve the strength characteristics of several types of soils (boggy peat-gley, peat-podzolic-gley, and soddy-podzolic soils) under laboratory conditions. The treatment of soil samples with the optimal mixture M2 led to the highest values of the uniaxial compressive strength due to the greater amount of calcium carbonate formed. An effective method of storing and transporting the bacterial mixture in the soil to the work site was tested. Subsequent treatment of soil of the soddy-podzolic type under field conditions with cementing solution and M2 mixture in 2 stages showed better results compared to others, resulting in 4 cm of compaction depth; filtration coefficient of 53.61 m/day and 2.43% of formed calcium carbonate.

Acknowledgements This research was funded by the Genome Research Center development program "Kurchatov Genome Center—PNPI" (agreement No. 075-15-2019-1663).

References

Achal V, Mukherjee A (2015) A review of microbial precipitation for sustainable construction. Construct and Build Mater 93:1224–1235

Bahmani M, Fatehi H, Noorzad A, Hamedi J (2019) A Study on Biological Soil Improvement Using New Environmental Bacteria. Environ Geotech 1–16

Castanier S, Le Métayer-Levrel G, Perthuisot JP (1999) Ca-carbonates precipitation and limestone genesis — the microbiogeologist point of view. Sediment Geol 126(1–4):9–23

Chu J, Ivanov V, He J, Maeimi M, Wu S (2015) Ground Improvement Case Histories (ed) Use of Biogeotechnologies for Soil Improvement. Oxford, Butterworth-Heinemann 571–589

DeJong JT, Mortensen BM, Martinez BC, Nelson DC (2010) Bio-mediated soil improvement. Ecol Eng 36(2):197–210

Dhami NK, Reddy MS, Mukherjee A (2013) Bacillus megaterium mediated mineralization of calcium carbonate as biogenic surface treatment of green building materials. World J Microbiol Biotech 29(12):2397–2406

Golovkina DA, Zhurishkina EV, Ivanova LA, Baranchikov AE, Sokolov AY, Bobrov KS, Kulminskaya AA (2020) Calcifying Bacteria Flexibility in Induction of $CaCO_3$ Mineralization. Life 10(12):317

Gomez MG, Martinez BC, DeJong JT, Hunt CE, deVlaming LA, Major DW, Dworatzek SM (2015) Field-scale bio-cementation tests to improve sands. Procs Instit Civ Eng Ground Impr 168(3):206–216

Ivanov V, Stabnikov V (2017) Construction Biotechnology. Green Energy and Technology, Springer Nature

Jain S, Fang C, Achal V (2021) A critical review on microbial carbonate precipitation via denitrification process in building materials. Bioeng 12(1):7529-7551

Jiang NJ, Soga K, Kuo M (2017) Microbially Induced Carbonate Precipitation for Seepage-Induced Internal Erosion Control in Sand–Clay Mixtures. J Geotech Geoenviron Eng 143(3):04016100

Marín S, Cabestrero O, Demergasso C, Olivares S, Zetola V, Vera M (2021) An indigenous bacterium with enhanced performance of microbially-induced Ca-carbonate biomineralization under extreme alkaline conditions for concrete and soil-improvement industries. Acta Biomaterialia 120:304–317

Rahman MM, Hora RN, Ahenkorah I, Beecham S, Karim MR, Iqbal A (2020) State-of-the-Art Review of Microbial-Induced Calcite Precipitation and Its Sustainability in Engineering Applications. Sustainability 12(15):6281

Safdar MU, Mavroulidou M, Gunn MJ, Purchase D, Payne I, Garelick J (2021) Electrokinetic biocementation of an organic soil. Sustainable Chem Pharm 21:100405

Seifan M, Berenjian A (2019) Microbially induced calcium carbonate precipitation: a widespread phenomenon in the biological world. Appl Microbiol Biotech 103:4693–4708

Seifan M, Sarabadani Z, Berenjian A (2020) Microbially induced calcium carbonate precipitation to design a new type of bio self-healing dental composite. Appl Microbiol Biotech 104:2029-2037

Van Paassen LA, Ghose R, van der Linden TJM, van der Star WRL, van Loosdrecht MCM (2010) Quantifying Biomediated Ground Improvement by Ureolysis: Large-Scale Biogrout Experiment. J Geotech Geoenviron Eng 136(12):1721–1728

Wang X, Xu J, Wang Z, Yao W (2022) Use of recycled concrete aggregates as carriers for self-healing of concrete cracks by bacteria with high urease activity. Constr Buil Mater 337:127581

Whiffin V (2004) Microbial $CaCO_3$ Precipitation for the Production of Biocement. PhD Thesis, Murdoch University, Perth

Zeng C, Veenis Y, Hall CA, Young ES, van der Star WRL, Zheng JJ, van Paassen LA (2021) Experimental and Numerical Analysis of a Field Trial Application of Microbially Induced Calcite Precipitation for Ground Stabilization. J Geotech Geoenviron Eng 147(7): 05021003

Biogenic-Abiogenic Interactions in Soils Affected by Wildfires in Russian North-West (Novgorod Region)

Ekaterina Y. Chebykina, Timur I. Nizamutdinov, and Evgeny V. Abakumov

Abstract Studies of postpyrogenic soils in the southern taiga zone in 2022 using the example of the Borovichi forestry territory (quarter 44 of the Pereluchsky district forestry) in the Novgorod region were conducted. The superficial layer of the moss litter was affected by fires in 2021. Three soil pits along the catena (a top of the kama terrace, a slope, and an accumulative position) were taken in the postpyrogenic area and one pit—for control, in the non-burnt area. A similar area of the forest ecosystem with the same type of soil, but not subjected to forest fire was used as a control. The soil at pyrogenic area is represented by illuvial-ferruginous postpyrogenic podzol on kama sandy loams (Podzol—WRB) with the profile structure Qpyr-E-BF-BC-C. It should be noted that postpyrogenic soils poorly fit into the framework of existing soil classifications. Upper part of soils of postpyrogenic succession are represented by surface organogenic horizons that retain traces of pyrogenic impact—burnt moss tirr. The presence of charcoal particles is visually diagnosed at the boundary of organic and podzolic horizons. The lower part of the soil profile has the structure which is typical of Podzols. The paper discussed the dynamics of biogenic elements content in postfire soils and results of the vertical electrical resistivity sounding as well.

Keywords Forest fire · Succession · Podzol · Burned soils · Biogenic elements · Landscape-geochemical position

1 Introduction

Forest fires make fundamental changes in forest ecosystems. If we take into account that, despite a destroying ecosystems, fires lead to the transfer of forests from carbon stocks to sources of greenhouse gases, then in the future there may be a problem with the implementation of international agreements and additional economic losses.

E. Y. Chebykina (✉) · T. I. Nizamutdinov · E. V. Abakumov
Saint-Petersburg State University, Saint-Petersburg, Russia
e-mail: e.chebykina@spbu.ru

O. V. Frank-Kamenetskaya et al. (eds.), *Biogenic—Abiogenic Interactions in Natural and Anthropogenic Systems 2022*, Springer Proceedings in Earth and Environmental Sciences, https://doi.org/10.1007/978-3-031-40470-2_26

Therefore, the importance of studies on fires impact on forest ecosystems has increased significantly in recent years.

A large number of works have been devoted to the nature of pyrogenic changes in soil parameters (Bento-Goncalves et al. 2012; Certini 2005; DeBano 2000; Tarasov et al. 2011a; Fultz et al. 2016; Dymov et al. 2018). A soil organic matter composition is transformed in soils after a fire (Alcañiz et al. 2018; Gonzalez-Perez et al. 2004), a content and forms of compounds of biogenic elements (nitrogen, carbon and phosphorus), which are important indicators of the ecological soil state, change (Covington and Sackett 1992; Knicker 2007; Prieto-Fernández et al. 1998). It should be noted that there are certain differences in various researchers' opinions according a number of issues. Thus, according to some works, organic nitrogen partially volatilizes during a fire, and partially mineralizes and turns into ammonium—into a form available to plants (Certini 2005). At the same time, a number of authors point at the absence of ammonium nitrogen forms on burnt sites (Bezkorovainaya et al. 2005a). As for alkaline hydrolyzable nitrogen, its content decreases as a result of fires due to the fact that most of organic nitrogen compounds are destroyed at temperatures of about 500 °C (Gorbunova et al. 2014).

Contradictory data are given on the content of mobile phosphorus in soils after wildfires. There are studies' results on the loss of mobile forms of this element. As a result of an increase in the pH of soil solution, the mobility of some compounds increases—most of cations, such as phosphorus, potassium, and nitrogen (Gongalsky 2015; Tarasov et al. 2011b; Kulagina 1975; Sannikov 1981). At the same time, some authors found an increase in total and mobile forms of phosphorus in surface horizons (Bezkorovainaya et al. 2005b). The content of P_2O_5 and K_2O increases after fires due to ash formation (Gorbunova and Devyatova 2013; Gorbunova et al. 2014; Borisochkina and Markina 2016; Shcherbov et al. 2015), as well as the content of silicon, iron, aluminum, calcium, and the content of magnesium decreases (Krasnoshchekov et al. 2013; Krasnoshchekov 2009). However, some authors consider the effect of forest fire on phosphorus content to be less than its effect on carbon and nitrogen due to the lower volatility of phosphorus during combustion (Hume et al. 2016). Thus, fires lead not only to the mobilization of many elements contained in soil, but also to the simultaneous entry into the soil of huge amounts of elements that accumulate in plants over a long period.

Therefore, forest fires have an impact on all components of the landscape, but most significantly on soil and vegetation cover. A scientific basis, including data on soils and vegetation state after wildfire, is needed in order to counteract this dangerous phenomenon, to develop programs to combat forest fires and their consequences.

The purpose of the paper is to analyze and evaluate the impact of forest fire on soils properties in the North-West of Russia within a representative area in the Novgorod region.

2 Materials and Methods

Two postpyrogenic sites were selected at the territory of Borovichi forestry in the Novgorod region in order to study pyrogenic soils changes, which burned, according to local foresters, in 2021: the first site is near a lake Shabodro (quarter 44 of the Pereluchsky district forestry), 58°10′18.4"N, 34°41′32.1"E (Fig. 1); the second site is near the village of Zhadiny (64 quarter of the Opechensky district forestry), 58°14′51.7"N, 34°13′13.1"E (Fig. 1).

The climate of study area is temperate continental, close to maritime. The average annual rainfall ranges within 540–750 mm. The maximum precipitation occurs in summer (38%), slightly less—in autumn (27%). The region is located in a zone of excessive moisture. The precipitation amount is unevenly distributed due to features of the relief. The warmest month is July and the coldest month is January. Summer is moderately warm, with an average temperature of + 17– + 18 °C. Winters are characterized by periodic thaws and cloudy weather. The average winter temperature

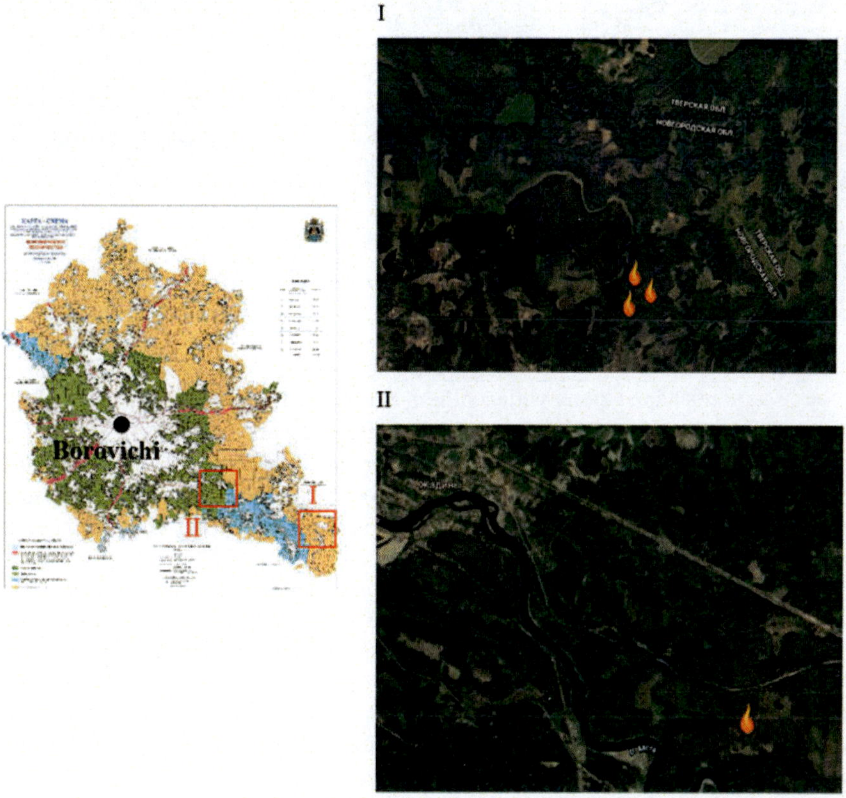

Fig. 1 The study site location. The source of this map is Google Earth Pro (https://www.google.com/intl/ru/earth/)

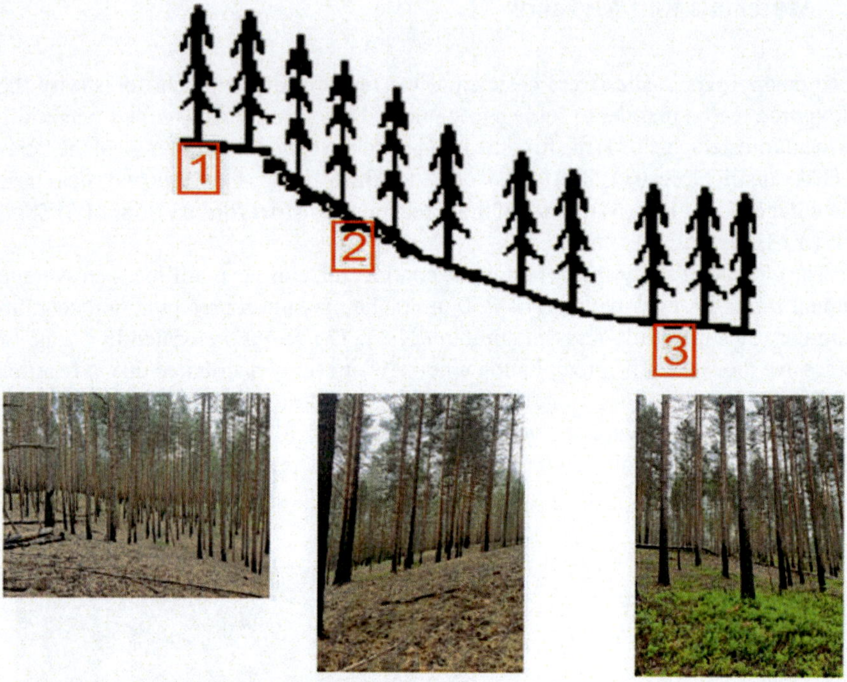

Fig. 2 Catena along the kama slope at the first study area

is −12 °C (Main Directorate of the Ministry of Emergency Situations of Russia for the Novgorod Region https://53.mchs.gov.ru/).

A catena was laid at the first study site: plots were located along lines of water flow on the upper, middle, and lower parts of the slope (Fig. 2). The study site occupies a slope of a ridge-shaped kama with southern exposure. The total length of the slope is about 600 m, the height difference is about 50 m, the slope is within 10°.

Unburned forest areas with the same soil type at a considerable distance from the pyrogenic impact were used as a control. The vegetation before the fire at this area was represented by green moss pine forest with an abundance of *Vaccínium myrtíllus* L. and *Vaccínium vítis-idaéa* L.

The fire broke out at the study area in 2021 and engulfed the entire lower forest canopy and weakened some of tree species. The band of surface fire stretched along the slope has a length of about 2 km and a width of 500 m within altitudes of 140–190 m above sea level (Fig. 3). The total fire area is 13.8 ha. The intensity of burning at study area was medium. The litter-histic horizons are damaged throughout the burnt area, but to varying degree: some where they are charred, in other cases they are completely burnt out.

The area of the surface fire at the second study area was insignificant—about 4.3 hectares, as well as the intensity of the fire was weak. The fire affected only the top of the esker hill, so only two soil profiles were put in this study area—at the burnt

Fig. 3 Strip of burning of surface fire

and control areas. The vegetation of the control plot is represented as at the first plot by green moss-white moss pine forest with dominance of *Vaccínium myrtíllus* L. and *Vaccínium vítis-idaéa* L.

The field soils description and sampling was carried out in August 2022. Soil samples were taken from depths of soil horizons using a volume-weight method to determine the density of undisturbed soils at each sampling point. Soil identification was carried out using the "Classification and Diagnostics of Soils of Russia" (2004) (Shishov et al. 2004) and the World Reference Base for Soil Resources (FAO 2015) (IUSS WRB, 2015).

All samples were dried at the +25 C at the Department of Applied ecology of Saint-Petersburg state university and then sieved through 2-mm sieve.

Key agrochemical characteristics such as mobile phosphorus and potassium were determined using the standard procedures according to GOST (Soils. Determination of Mobile Phosphorus and potassium Compounds 2013), which is based on extracting mobile phosphorus (P_2O_5) and potassium (K_2O) compounds from the soil with a hydrochloric acid solution and quantitatively determining the phosphorus and potassium content. Available forms of nitrogen (NO_3) were determined according to ISO/TS 14256–1:2003, which first extracts exchangeable ammonium ($N\text{-}NH_4$) from the soil with a potassium chloride solution and then photometrically measures the colored solution.

Data normal distribution was verified and a variance analysis (ANOVA) and post hoc test (Fisher's least significant difference) were performed. Differences were considered significant at $p < 0.05$. Statistical data processing and analysis were carried out using standard methods in software packages MS Excel 2016, Past (version 3.20), Statistica 64 (version 10).

3 Results and Discussion

The soils of study area are represented by illuvial-ferruginous-humus Podzol on kame sandy loams (Folic Podzol–WRB) with a typical soil profile structure O–Eh–BHF–BC–C (Fig. 4). The upper layer of moss litter was affected by running surface fire impact. The soil at pyrogenic area is represented by illuvial-ferruginous postpyrogenic podzol on kame sandy loams (Podzol – WRB) with the soil profile structure Qpyr-E-BF-BC-C. It should be noted that postpyrogenic soils poorly fit into the framework of existing soil classifications. A visual survey revealed significant differences in main characteristics of the upper organogenic horizon–forest litter–at the fire-affected and control plots. First of all, its concerns a thickness, which noticeably decreased as a result of partial combustion of the litter, as well as the color of its surface, which became noticeably darker after the fire. Soils of postpyrogenic succession are represented by surface organogenic horizons that retain traces of pyrogenic impact–burnt moss tirr. The moss tirr, forest litter, and upper humus horizon have the maximum impact of fire, as a result of which, according to researchers, there is a loss of organic matter and the formation of "pyrogenic humus" resistant to biodegradation and oxidation. The surface is represented by histic forest litter in the postpyrogenic soil in the accumulative position at the first study site and at the second postpyrogenic site. In the case of the first study site, postpyrogenic succession began, and in the second study plot, a fire of low intensity did not allow the powerful moss tirr to burn completely. The presence of charcoal particles is visually diagnosed at the boundary of organic and podzolic horizons. The lower part of the soil profile has the structure which is typical of podzols. Illuvial-ferruginous-humus horizons lie under the podzolic horizons, and if they contain signs of a pyrogenic effect in the form of coal inclusions, then only from previous fires a long time ago. Differences in the pyrogenic series can also be traced in a change of the podzolic horizon color. It is lighter at postpyrogenic areas due to the fact that erosion processes were activated as a result of fires, which led to an increase in both lateral and vertical migration along the soil profile, and, accordingly, to the removal of dark-colored particles to lower horizons. Burns (1952) spoke about this, also pointing out that soils of burnt areas are more compacted, have low water permeability and are more susceptible to erosion than soils not exposed to fire.

Specific pyrogenic horizons formed in the organic part of the profile after the fire: *pyr*, coals and charred twigs lying fragmentarily on 20% of the soil surface; *Opyr*–charred forest litter, black-brown in color with horizontal divisibility.

The density of the charred litter is higher (Table 1), layering is more pronounced in it, and the proportion of the living phase is lower. The density of upper horizons at postpyrogenic areas ranges from 0.10–0.35 g/cm^3, while for the control area this

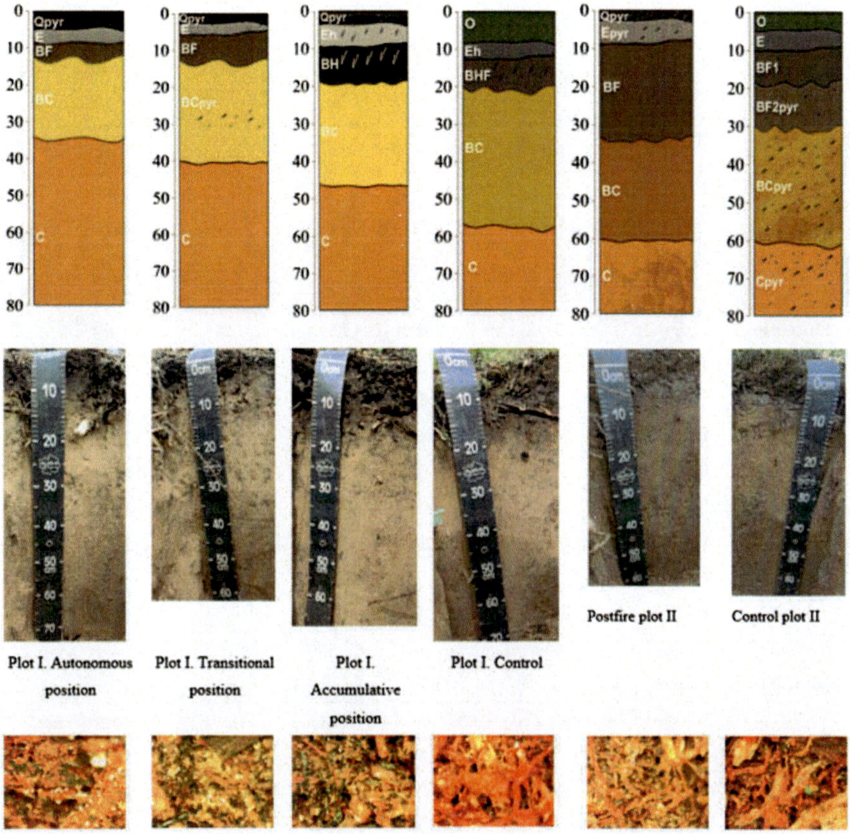

Fig. 4 Soil profiles of studied areas and mesomorphological structure of upper horizons

indicator is 0.04 g/cm^3. Lower horizons do not differ significantly according to this indicator: the podzolic horizon is characterized by a bulk density of 0.6–1 g/cm^3, then in the middle horizon there is a slight compaction of the soil stratum up to 1.3 g/cm^3 and then a further increase of this value with depth.

It would appear reasonable that the noted postpyrogenic changes in main characteristics of the litter had a corresponding effect on its thermophysical properties, on which the features of the inflow, accumulation, and movement of heat in soil depend. For this purpose, the temperature of soil horizons was determined. The most important factor influencing temperature, characterized by the albedo value, is color (Voronin 1986). According to the research data (Sheshukov et al. 1992; Tarasov et al. 2008, 2011a), the albedo of darker burnt-out surfaces is one and a half to two times lower than in the control. It has a corresponding effect on their temperature regime, which is fully confirmed by the results obtained (Table 1). The surface temperature of the control plot was on average 1–2 °C lower than in the burnt areas: postpyrogenic soils were characterized by surface temperature in the range of 18.9–19.8 °C, while

Table 1 Characteristics of studied soils

Plot	Horizon, depth, cm	$N\text{-}NH_4^+$, mg/kg	P_2O_5, mg/kg	K_2O, mg/kg	Bulk density, g/cm^3	Temperature, °C
Postfire plot I. Autonomous position	Qpyr, 0–5	146.7	56.99	180.72	0.24	18.9
	E, 5–8	5.65	27.85	41.88	0.78	16.2
	BF, 8–12	1.35	158.06	11.56	1.22	16.0
	BC, 12–35	1.71	163.35	5.41	1.53	15.0
	C, 35–70	0.35	176.27	3.36	1.46	14.8
Postfire plot I. Transitional position	Qpyr, 0–3	24.0	102.00	133.63	0.35	19.8
	E, 3–6	3.59	122.22	28.15	0.58	18.1
	BF, 6–12	1.12	200.55	10.20	1.19	16.3
	BCpyr, 12–40	0.59	180.36	5.53	1.38	15.7
	C, 40–65	0.53	225.04	4.21	1.48	15.4
Postfire plot I. Accumulative position	Qpyr, 0–4	106.6	61.60	336.55	0.10	19.1
	Eh, 4–9	5.60	122.16	42.79	0.93	16.0
	BH, 9–19	2.89	221.88	21.76	0.98	15.0
	BC, 19–46	0.82	257.42	4.30	1.24	14.6
	C, 46–75	0.47	317.29	2.94	1.35	14.3
Control plot I	O, 0–8	477.0	103.77	433.54	0.04	18.1
	Eh, 8–12	12.2	51.99	41.07	0.90	15.9
	BHF, 12–20	1.71	220.11	13.01	1.12	14.9
	BC, 20–57	0.77	202.56	4.11	1.26	14.2
	C, 57–78	1.65	338.79	3.19	1.43	13.8
Postfire plot II	Qpyr, 0–3	48.3	56.04	239.56	0.23	19.5
	Epyr, 3–8	5.65	124.94	12.10	0.94	16.9
	BF, 8–35	1.12	324.57	5.77	1.28	15.8
	BC, 35–60	0.77	315.84	5.12	1.26	15.3
	C, 60–70	0.47	459.15	6.65	1.33	15.0
Control plot II	O, 0–5	352.8	272.94	554.76	0.08	18.4
	E, 5–10	2.3	21.62	21.91	1.03	15.0
	BF1, 10–18	1.83	188.77	14.70	1.30	14.5
	BF2pyr, 18–30	1.06	150.19	9.00	1.23	14.2
	BCpyr, 20–60	1.53	199.91	6.82	1.09	13.6
	Cpyr, 60–75	0.41	130.60	6.68	1.16	12.7
Post hoc *test* Postfire–Control		$p \ll 0.05$	$p \ll 0.05$	$p \ll 0.05$	$p \ll 0.05$	$p \ll 0.05$
Singnificance of differences		Sign.	Sign.	Sign.	Sign.	Sign.

the control soils were 18.1–18.4 °C. At the same time, the same differences were observed in the lower horizons—the soils of burnt areas warmed up more strongly than control plots.

Two components are visible in the mass of fine earth in upper part of soil profiles after wildfires according to mesomorphological studies: finely dispersed ocherous ash and large carbonaceous particles (Fig. 4). Upper horizons of control soils consist of a large amount of organic residues. Therefore, a pore space in control soils is much larger than in postfire soils. Moisture enters deep into the soil through the layer of forest litter in control soils, and it seeps through the pores and cracks of the pyrogenic horizon in postfire soils, which does not allow lower horizons to become waterlogged.

A radical restructuring of soils occurs in postpyrogenic soil profiles at the morphological (horizontal, aggregate) and molecular levels according to obtained results–the litter burning, the accumulation of ash, carbonaceous material, humons, and other forms of black carbon. There is an activation of erosion processes within landscape-geochemical positions, which lead to a decrease in the horizon thickness and its compaction in the slope position and an increase in thickness with a slight deconsolidation in the accumulative position due to the initialization of postpyrogenic succession. Besides, the activation of erosion processes as a result of fires leads to a depletion of upper soil horizons in ash elements and organic compounds, which will be discussed further.

Many studies show a postfire increase in the nitrogen content in fresh burnt areas and a gradual decrease in the proportion of hydrolysable nitrogen compounds in the future (Sannikov 1981; Tarasov et al. 2011b; Krasnoshchekov 2014; Krasnoshchekov et al. 2014). However, some works, on the contrary, showed a decrease in the nitrogen content in upper soil horizons (Zenkova 2016; Bobrovsky 2010).

Studied samples were analyzed for these indicators (Table 1). The content of nitrate nitrogen did not exceed the limits of detection for studied samples, only traces, therefore, they are not presented in Table 1 and are not discussed. Sushkina (1993) showed that a moderate increase in soil temperature during fire clearing of cutting areas contributes to an increase in the intensity of nitrification.

Difference of ammonium forms of nitrogen between burned plot and control is expressed for all landscape-geochemical positions: the content of ammonium nitrogen decreases after forest fires in the range: top of a kama hill > accumulative position > transit position (146.7 > 106.6 > 24.0 mg/kg compared to 477 mg/kg in control). The same is for podzolic horizon: more or less the same content 5.6 mg/kg is for the autonomous and accumulative position; it is much lower in transitional position–3.6 mg/kg, and for the control soil the content of ammonium nitrogen is the maximum–12.2 mg/kg. The results of the analysis of variance ANOVA showed significant differences in the content of ammonium nitrogen in studied soils. The pyrogenic change in the activity of mineralization processes is due to a change in the dynamics of plant residues entering to the soil surface, a transformation of hydrothermal conditions, and an activity of soil biota (Bezkorovainaya et al. 2007). No significant differences were observed for the middle horizon and underlying ones. The second study area near the village Zhadina shows a decrease in the content of ammonium nitrogen in the surface horizon after the fires, similarly to the first study site, while the podzolic horizon shows a completely different situation: a higher content in the soils of postfire plot is possibly due to the activation of erosion processes and vertical migration to underlying layers increased. Duchaufour (1954) noted an increase in the mobility of nutrients and their removal deep into the soil profile.

Therefore, due to the absence of nitrate nitrogen in studied samples, the main product of pyrolysis of organic nitrogen in soils can be called NH_4^+ forms. Moreover, the influence of the pyrogenic factor on nitrogen soil pool is associated with the loss of volatile nitrogen-containing compounds and structural transformations.

It is known that the ash entering the soil surface during the burning of upper horizons enriches it with nutrients. The source of these compounds is not only burnt litter and aboveground plant organs, but also underground organs of dead plants and mineral and organo-mineral soil particles (for example, feldspars). Combustion is accompanied by an increase in precisely the mobile forms of potassium and phosphorus. A mobile form cannot arise from potassium feldspars, since they are among the most resistant to weathering minerals. Therefore, it arises only from the combustion products of plant components.

Moreover, fires reduce acidity, increase the mobility of a number of elements (in particular, potassium) and contribute to their leaching from upper soil horizons. In particular, a relative increase in the phosphorus content in E horizon is observed in studied samples due to its removal from the upper horizon. The phosphorus content in surface horizons is 1.7 times lower at postfire plots compared to the control (about 60 mg/kg versus 103–272 mg/kg in control, except for the plot at the slope position—there the phosphorus content is at the control level). The surface fire, when the entire ground cover burned down, led to the activation of erosion processes and, thereby, the activation of lateral and vertical migration of substances. There is a relative increase in the content of phosphorus in the podzolic horizon at postpyrogenic plots due to vertical migration down the profile from 20–50 to 120 mg/kg. The content of phosphorus increases towards the parent rock—a distribution nature along the soil profile is eluvial. It should be noted that studied soils are formed on soil-forming rocks rather enriched with phosphorus.

On the other hand, potassium turns out to be more mobile, and an increase in its content in the lower horizons—BH or BF—is observed. In general, the potassium content decreases as a result of fires by 1.3–3 times: from 433.54 mg/kg in the forest litter to 130–330 mg/kg in pyrogenic horizon. An intracatenar differentiation in potassium content is significant at the burnt area, in contrast to phosphorus: the potassium content is 2.4 times lower than the control one at the upper part of the slope, at the lower part—only 1.3 times, and it is the maximum decrease in content at the slope position. This probably indicates its transport down the slope and the greater lability of potassium compared to phosphorus as a result of fires. The potassium content decreases down the soil profile, and its minimum content in the parent rock is from 3 to 6 mg/kg.

Therefore, a loss of available forms of nitrogen, phosphorus and potassium with pronounced lateral and vertical differentiation was revealed at burnt studied areas.

4 Conclusion

Forest fires within the studied territory of the Borovichi forestry in the Novgorod region lead to a significant transformation of surface and subsurface horizons of soils and vegetation.

Litter and histic forest litter with an average total thickness of about 8 cm were destroyed at the most of the studied burnt area, and turned into a pyrogenic horizon Qpyr with a thickness of 3–5 cm, depending on the position in the relief. Pyrogenic changes in morphological soils properties affected the upper part of the soil profile in the form of inclusions of coals and charred fragments of soil organic matter, as well as compaction of the upper horizons. The bulk density of the charred litter is higher, layering is more pronounced there. Depths and thicknesses of genetic horizons have undergone changes in middle and lower parts of soil profiles as well.

The total content of ammonium nitrogen decreased significantly after the fire. The nature of the pyrogenic transformation depends on the catenary position: postfire losses of biogenic elements are more significant at slope positions of the catena, possibly due to more intense lateral removal. The content of labile phosphorus and potassium also decreases after a fire; it is more intensive in upper parts of the catena. There is a vertical migration of these elements down the profile as a result of the activation of erosion processes.

Therefore, interesting data were obtained on the effect of a forest fire on soil properties within a representative area in the Novgorod region, which can serve as a scientific basis for further development of programs to combat forest fires and their consequences.

Acknowledgements The work was supported by the Grant of the President of the Russian Federation for young PhDs No. MK-4596.2022.1.4.

This work is dedicated to the 300th anniversary of Saint Petersburg State University.

References

Alcañiz M, Outeiro L, Francos M, Úbeda X (2018) Effects of prescribed fires on soil properties: A review. Science of The Total Environment 613:944–957. https://doi.org/10.1016/j.scitotenv.2017.09.144

Bento-Goncalves A, Vieira A, Ubeda X, Martin D (2012) Fire and soils: Key concepts and recent advances. Geoderma 191:3–13. https://doi.org/10.1016/j.geoderma.2012.01.004

Bezkorovainaya IN, Ivanova GA, Tarasov PA, Sorokin ND, Bogorodskaya AV, Ivanov VA, Konard SG, Makrae DJ (2005a) Pyrogenic soils transformation in pine forests of the middle taiga of the Krasnoyarsk Territory. Siberian Ecological Journal 1:143-152

Bezkorovainaya IN, Verkhovets SV, Kisilyakhov EK, Kuzmichenko VV (2005b) Influence of controlled burning of silkworms on properties of soddy-podzolic soils in the Lower Angara region. Lesovedenie 2:16-24

Bezkorovainaya IN, Tarasov PA, Ivanova GA, Bogorodskaya AV, Krasnoshchekova EN (2007) Nitrogen fund of sandy podzols after controlled burning of pine forests in Central Siberia. Soil Science 6:775-783

Bobrovsky MV (2010) Forest soils of European Russia: biotic and anthropogenic factors of formation. Association of scientific publications KMK, Moscow: 359 p.

Borisochkina TI, Markina LG (2016) Processes of pyrogenic transformation in soils of forest-park landscapes. Abstracts of the VII Congress of the Society of Soil Scientists named after V.V. Dokuchaev and the All-Russian scientific conference with foreign participation "Soil science - food and environmental security of the country", Belgorod, August 15–22, 2016, part II:69–70

Burns RI (1952) Effect of fire on forest soils in the Pine Barren region of New jersey. Bull. Vale Sch. For. 57.

Certini G (2005) Effects of fire on properties of forest soils: a review. Oecologia 143(1):1-10. https://doi.org/10.1007/s00442-004-1788-8

Covington WW, Sackett SS (1992) Soil mineral nitrogen changes following prescribed burning in ponderosa pine. Forest Ecology and Management 54(1–4):175-191. https://doi.org/10.1016/0378-1127(92)90011-W

DeBano LF (2000) The role of fire and soil heating on water repellency in wildland environments: a review. J. Hydrology 231:195-206. https://doi.org/10.1016/S0022-1694(00)00194-3

Duchaufour P (1954) Note sur e influence de e'incinération sur l'évolution de e'humus. Rev. for. Franc. 6 (5).

Dymov AA, Abakumov EV, Bezkorovaynaya IN, Prokushkin AS, Kuzakov YV, Milanovsky EY (2018) Impact of forest fire on soil properties (review). Theor. Appl. Ecol. 4:13–23 https://doi.org/10.25750/1995-4301-2018-4-013-023

Fultz LM, Moore-Kucera J, Dathe J, Davinic M, Perry G, Wester D, Schwilk DW, Rideout-Hanzak S (2016) Forest wildfire and grassland prescribed fire effects on soil biogeochemical processes and microbial communities: Two case studies in the semi-arid Southwest. Appl. Soil Ecol. 99:118–128

Gongalsky KB (2015) Patterns of communities' restoration of soil animals after forest fires. Dissertation, A.N. Severtsov Institute of Ecology and Evolution of the Russian Academy of Sciences

Gonzalez-Perez JA, Gonza´lez-Vila FJ, Almendros G, Knicker H (2004) The effect of fire on soil organic matter—a review. Environment International 30:855-870. https://doi.org/10.1016/j.envint.2004.02.003

Gorbunova YuS, Devyatova TA (2013) Changes in the enzymatic activity of soddy-forest soil after pyrogenic impact. Problems of regional ecology 2:38-42

Gorbunova YuS, Devyatova TA, Pishchulina AN (2014) Ecological consequences of pyrogenic impact on soddy-forest soils. Proceedings of the International Scientific Conference "Ecology and Biology of Soils", Rostov-on-Don, November 17–19, 2014: 49–51

Hume A, Chen HYH, Taylor AR, Kayahara GJ, Man R (2016) Soil C:N:P dynamics during secondary succession following fire in the boreal forest of central Canada. Forest Ecology and Management 369:1-9. https://doi.org/10.1016/j.foreco.2016.03.033

ISO/TS 14256–1:2003. Soil Quality—Determination of Nitrate, Nitrite and Ammonium in Field Moist Soils by Extraction with Potassium Chloride Solution—Part 1: Manual Method (ISO/TS 14256–1–2003:2003). 2003, p. 14. Available online: https://www.sis.se/api/document/preview/903656/ (accessed on 01 November 2022).

IUSS Working Group WRB. World reference base for soil resources international soil classification system for naming soils and creating legends for soil maps; World Soil Resources Reports No. 106; FAO: Rome, Italy, 2015.

Knicker H (2007) How does fire affect the nature and stability of soil organic nitrogen and carbon? A review. Biogeochemistry 85:91–118. https://doi.org/10.1007/s10533-007-9104-4

Krasnoshchekov YuN (2009) Pyrogenic soils transformation of pine forests in the Central zone of the Baikal natural territory. Productivity and stability of forest soils. III International Conference on Forest Soil Science (Petrozavodsk, September 7–11, 2009). Petrozavodsk: 251–255

Krasnoshchekov YuN (2014) Influence of pyrogenic factor on gray-humus soils of pine forests in the Central Ecological Zone of the Baikal Natural Territory. Siberian Forest Journal 2:43-52

Krasnoshchekov YuN, Evdokimenko MD, Cherednikova YuS (2013) Forest-ecological consequences of fires in cedar forests of the Southern Baikal region. Geography and natural resources. Researches in the Baikal basin 1:33-42

Krasnoshchekov YuN, Evdokimenko MD, Dorzhsuren Ch (2014) Influence of fires on ecosystems of subtaiga larch forests of eastern Khentei in Mongolia. Siberian Forest Journal 3:53–63

Kulagina MA (1975) Influence of a surface fire on the regime of mineral nutrition in a forb-lingonberry pine forest. Protection of Forest Resources of Siberia. Krasnoyarsk: 153–165

Prieto-Fernández A, Acea M, Carballas T (1998) Soil microbial and extractable C and N after wildfire. Biol Fertil Soils 27:132–142. https://doi.org/10.1007/s003740050411

Sannikov SN (1981) Forest fires as a factor in the transformation of the structure, renewal and evolution of biogeocenoses. Ecology 6:23-33

Shcherbov BL, Lazareva EV, Zhurkova IS (2015) Forest fires and their consequences. Academic publishing house “GEO”, Novosibirsk

Sheshukov MA, Savchenko AP, Peshkov VV (1992) Forest fires and fighting them in the north of the Far East. DalNIILKh, Khabarovsk

Shishov LL, Tonkonogov VD, Lebedeva II, Gerasimova MI (2004) Classification and Diagnostics of Soils in Russia. Ojkumena, Smolensk, Russia

Soils. Determination of mobile phosphorus and potassium compounds by kirsanov method modified by CINAO. 2013. Available online: http://docs.cntd.ru/document/1200094361 (accessed on 01 November 2022). (In Russian)

Sushkina NN (1993) Nitrification in forest soils depending on the composition of the forest stand, felling and fire cleaning of cutting areas. Izvestiya Akad. Nauk SSSR 1:111–160

Tarasov PA, Ivanov VA, Ivanova GA (2008) Features of the temperature soils regime in pine forests of the middle taiga, impacted by surface fires. Coniferous boreal zone 3-4:300-304

Tarasov PA, Ivanov VA, Ivanov GA, Krasnoshchekova EN (2011a) Postpyrogenic changes in hydrothermal parameters of soils in middle taiga pine forests. Eurasian Soil Sci 7:795-803

Tarasov PA, Mikhno AS, Sizina AF (2011b) Estimation of the pyrogenic influence on soils of the belt pine forests of Altai. Bulletin of KrasGAU 1:26-30

Voronin AD (1986) Fundamentals of soil physics: textbook. Publishing House of Moscow State University, Moscow

Zenkova IV (2016) Soil and soil fauna of the pyrogenic forest: the consequences of a surface fire in the Khibiny mountain range. Abstracts of the VII Congress of the Society of Soil Scientists named after V.V. Dokuchaev and the All-Russian scientific conference with foreign participation “Soil science—food and environmental security of the country”, Belgorod, August 15–22, 2016, part I:221–222

Soil Transformations in the Littorina Terrace Under the Impact of Urbanization (St. Petersburg, Russia)

Kseniia A. Bakhmatova, Anastasia A. Sheshukova, Elena G. Panova, and Sofia A. Egorova

Abstract The central part of the city of Saint Petersburg is located within the Littorina terrace composed of alluvial-marine deposits. Representative soil profiles of the terrace were studied in the context of increasing anthropogenic impact. The levels of pH_{H2O}, organic carbon, $CaCO_3$, mobile phosphorus compounds, heavy metals (Pb, Cu, Zn) were examined, along with the evaluation of soil texture and characteristics of anthropogenic inclusions (artefacts). It was established that natural soils in the central part of St. Petersburg are buried under anthropogenic (cultural) layers with a total thickness of up to 150–190 cm. The maximum degree of soil transformation was observed in technogenic urbostratozems. These soils were characterized by a large number of artefacts and high alkalinity. The content of heavy metals in soils was found to depend on their land-use history.

Keywords Urban soils · Urbostratozem · Soil artefacts · Heavy metals · Soil chemical properties

1 Introduction

Many existing publications investigate exclusively the surface layer of urban soils with a focus on their contamination via dust deposition or technogenic materials introduction. At the same time, modern earthmoving works undertaken for the purposes of urban landscaping frequently concern depths beyond 1 m; therefore, the characterization of urban soil profiles of up to 1–2 m in depth presents a relevant research task (Lenoir et al. 2019). The redevelopment of urban areas may require extraction of deep soil layers to the surface and unsealing of the soils covered with pavement (Tobias et al. 2018).

K. A. Bakhmatova (✉) · A. A. Sheshukova · E. G. Panova · S. A. Egorova
Institute of Earth Sciences, Saint Petersburg State University, Saint Petersburg, Russia
e-mail: k.bahmatova@spbu.ru

O. V. Frank-Kamenetskaya et al. (eds.), *Biogenic—Abiogenic Interactions in Natural and Anthropogenic Systems 2022*, Springer Proceedings in Earth and Environmental Sciences, https://doi.org/10.1007/978-3-031-40470-2_27

Soil formation in urban areas is largely determined by human activities (Gerasimova et al. 2003; Prokof'eva et al. 2017; Makki et al. 2021). The upper parts of urban soils are cultural layers characterized by the presence of artefacts (bones, charcoal, pottery, construction wastes, etc.) (Aleksandrovskaya and Aleksandrovskiy 2000; Mazurek et al. 2016), which can be classified according to their composition (Semenjuk et al. 2013; Howard and Orlicki 2016). The properties of soil horizons formed with the participation of anthropogenic materials strongly depend on the nature and number of artefacts (Delbeque et al. 2022).

Greinert and Kostecki (2019) studied the chemical parameters of technogenic materials in urban soils. The highest pH (10–12 units) was found in plaster, asbestos-cement roof coating, and the ash after biomass combustion.

Pollution with heavy metals (HM) is typical for urban soils. Thus, the elevated concentrations of HM were found in different cities (Plyaskina and Ladonin 2009; Burt et al. 2014; Abel et al. 2015). In addition, the natural geological background also affects geochemical patterns in urban areas (Burt et al. 2014).

In this paper, we study the relationship between the degree of anthropogenic soil transformation and the characteristics of the soil profile in the Littorina terrace. The natural parent material in the research area is characterized by relatively low HM concentrations and acidity.

2 Materials and Methods

2.1 Research Area and Objects

St. Petersburg, a large Russian city with a population of over 5 million, is situated on the east coast of the Gulf of Finland, at the mouth of the Neva River, covering an area of 1,439 km^2. The city center is located within the terraced Prinevskaya lowland. The lowest terrace (up to 10 m above sea level) referred to as the Littorina terrace (Fig. 1a) is composed of alluvial marine deposits, layered and dominated by sand and silt fractions (Fig. 2).

Six representative soil profiles were selected based on the previous studies (Matinyan et al. 2017; Matinian et al. 2019, 2021; Nadporozhskaya et al. 2000; Rusakov and Novikov 2003; Ufimtseva et al. 2011). A soil map of St. Petersburg compiled by (Aparin and Sukhacheva 2014) was analyzed.

The selected profiles were ranged according to the degree of anthropogenic transformation (the location of the profiles is presented in Fig. 1b).

Profile 1. Stratified dark-humus gley soil (Umbric Gleysol (Arenic)), Elagin Island. Here, a landscape park was created at the end of eighteenth century. Profile 1 (P1) is a slightly altered natural soil similar to the most common soils of the Littorina terrace prior to its anthropogenic transformation.

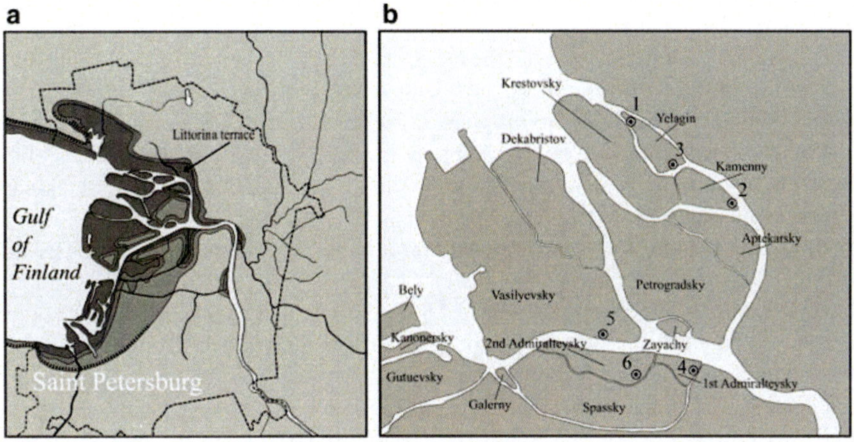

Fig. 1 **a** The Littorina terrace; **b** location of sampling sites on the islands of the Neva River delta

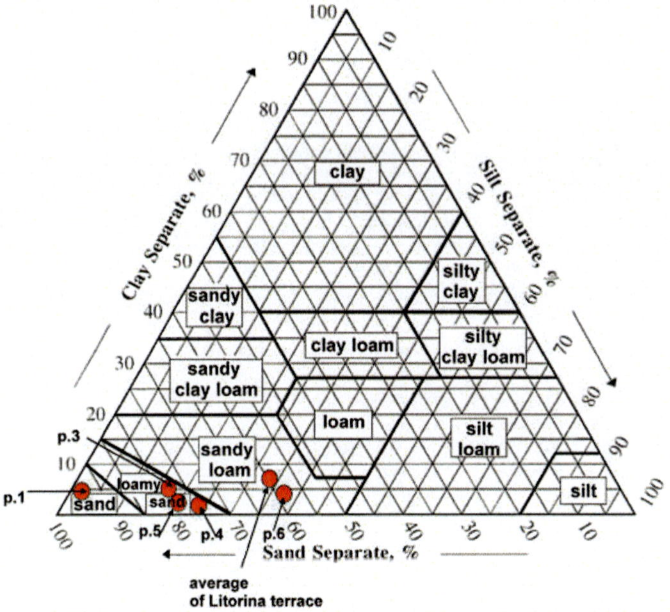

Fig. 2 Soil texture of the Littorina deposits and parent material of the studied soils

Profile 2. Urbistratified gray-humus ferruginous gleyic soil (Umbric Gleysol (Arenic)), Kamenny Island. In this location, a palace and a park were built in 1780.

Profile 3. Urbistratified gray-humus stratozem on a buried gray-humus soil (Hortic Gleyic Phaeozem (Arenic)), Own Garden, Yelagin Island. The Garden located near the Elagin Palace. The soils (P2 and P3) were a result of horticulture in the eighteenth and nineteenth centuries; however, the humus horizon in P3 is thicker than that in P2.

Profiles 4–6. Urbostratozems (urbanozems) with the most complicated morphology due to the high intensity and duration of varied human activity.

Profile 4. Urbostratozem on a buried gray-humus gley post-agrogenic soil (Urbic Technosol (Arenic)) over Umbric Gleysol (Arenic), Summer Garden, the first regular garden in Russia founded in 1703.

Profile 5. Technogenic urbostratozem on a buried gray-humus gley soil (Urbic Technosol (Arenic) over Umbric Gleysol (Arenic)), Rumyantsev Square. Prior to the foundation of the public garden in 1867, the site had experienced changes in the type of land use (market place, construction site, and parade ground) since the beginning of the eighteenth century.

Profile 6. Technogenic urbostratozem, sealed under asphalt, on a buried urbistratified gray-humus gley soil (Ekranic Urbic Technosol (Arenic) over Umbric Gleysol (Arenic), Nevsky Prospekt, 11. This site was part of the Marine Settlement at the beginning of the eighteenth century followed by several changes in ownership. The profile has remained under asphalt pavement in the vicinity of a building, which was constructed in 1802–1804 and then reconstructed in 1898–1900.

The soil morphology is presented in Tables 1–2 and Fig. 3.

Table 1 Morphological characteristics of moderately transformed soils in the Littorina terrace

Horizon, depth, cm	Color	Artefacts, presence and abundance				
		Mortar/lime	Brick	Glass	Ferruginous	Charcoal
P1 Stratified dark-humus gley soil						
AU, 0–18	10YR2/1	–	–	–	–	–
AC, 18–29	10YR3/1	–	–	–	–	–
C, 29–54	10YR4/3	–	–	–	–	–
Cg, 54–90	2.5Y5/2	–	–	–	–	–
P2 Urbistratified gray–humus ferruginous gleyic soil						
RYur,g, 0–25	2.5Y3/1	–	–	+	+	–
AYg, 25–39	2.5Y4/1	–	–	–	–	–
ACg, 39–61	2.5Y5/4	–	–	–	–	–
Cg1, 61–84	2.5Y5/3	–	–	–	–	–
Cg2, 84–110	2.5Y5/4	–	–	–	–	–
P3 Urbistratified gray-humus stratozem on a buried gray-humus soil						
RU1ur,pa,ca 0–14	10YR2/1	+	++	++	++	+
RU2ur,pa,ca 14–41	2.5Y3/1	+	+	+	–	–
[ACur] 41–65	2.5Y3/2	–	–	–	–	–
Cg 65–90	2.5Y5/2	-	–	–	–	–

Table 2 Morphological characteristics of urbostratozems in the Littorina terrace

Horizon, depth, cm	Color	Artefacts, presence and abundance					
		Mortar/ Lime	Bricks	Glass	Slag	Charcoal	Bones
P4 Urbostratozem on a buried gray-humus gley post-agrogenic soil							
UR1ay, 0–20	10YR 3/2	+++	+++	+	+++	–	–
UR2, 20–40	10YR 3/4	+++	+++	–	+++	–	–
UR3g, 40–55	10YR 5/6	+	++	–	++	–	–
UR4, 55–120	10YR 4/3	+	++	–	–	–	–
R1, 121–142	2.5Y 5/4	–	–	–	–	–	–
[AYpa],142–167	5Y 2.5/2	–	–	–	–	–	–
CG, 170–190	2.5Y 6/2	–	–	–	–	–	–
P5 Technogenic urbostratozem on a buried gray-humus gley soil							
RY, 0–25	2.5Y3/1	-	–	+	–	–	–
UR1, 25–41	10YR4/4	+++	+	+	–	–	–
TCH2, 70–80	10YR4/4	++	++++	–	–	+++	–
UR2, 90–100	2.5Y3/3	-	+++	-	-	-	-
TCH3, 100–120	7.5YR4/3	-	++++	-	-	-	-
UR3, 120–140	7.5YR4/3	++	++	–	–	–	+
UR4, 150–160	10YR4/4	++	+++	–	–	–	+
[AYh], 160–170	2.5Y3/1	–	–	–	–	–	–
[AY], 170–185	2.5Y4/2	–	–	–	–	–	–
CG, 185–200	5Y5/1	–	–	–	–	–	–
P6 Technogenic urbostratozem, sealed under asphalt, on a buried urbistratified gray-humus gley soil							
TCH, 15–25	10YR4/2	++++	+	–	–	–	–
UR1, 25–43	2.5Y3/2	++	–	–	–	–	–
UR2, 43–75	2.5Y3/3	+	+++	+	–	–	–
UR3, 75–124	10YR4/3	+++	+++	+	–	+	–
[AYur], 124–131	2.5Y3/3	+	++	–	–	+++	–
[AY], 131–152	2.5Y5/3	+	+	–	–	+	–
CG, 152–160	2.5Y5/3	–	+	–	–	+	–
CG, 160–190	Gley 5/ 10Y	–	–	–	–	–	–

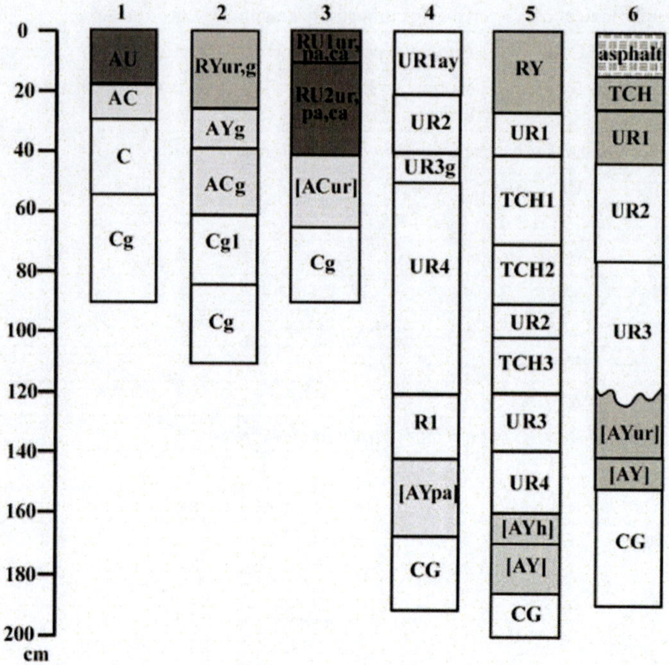

Fig. 3 Profile schemes and the designations of genetic horizons. The soil types are given in the text

Soil Sampling and Description

Fieldwork research was carried out during 2018–2019. Soil profiles 4–6 were described in archaeological sites. Samples were collected from genetic horizons. The diagnostics of buried natural soils was conducted according to the Classification System of Russian Soils (2004). The classification proposals (Prokofieva et al. 2014) and WRB (2014) were used for the diagnostics of urban soils.

Laboratory Analysis

A set of conventional analytical methods was used (Vorobieva 2006). The granulometric analysis was conducted by a pipet method with sodium pyrophosphate used as a peptizer (Rastvorova 1983). The content of organic carbon was determined by dichromate oxidation. The level of pH was electrometrically measured in an aqueous suspension. The content of plant available phosphorus (P) compounds was evaluated by the Olsen method using a 0.5 n solution of $NaHCO_3$ (1:20, pH = 8.5).

The chemical composition of artefacts was studied using a desktop scanning electron microscope TM-3000 microanalyzer (HITASHI, Japan).

The identification of different crystalline phases and determination of their relative concentrations in mixtures was carried out using a Bruker D2 Phaser desktop diffractometer followed by an analysis of the diffraction patterns of powdered samples. Profile 4 was selected for mineralogical analysis as a typical example of urbostratozems in the Littorina terrace without any technogenic layers.

The total content of HM (Pb, Cu, Zn) was determined by a Spectroskan XRF analyzer (Profiles 2 and 4) and by atomic absorption spectrometry (a Kvant-2m1 analyzer). Sample digestion was performed with a mixture of HNO_3, HF, and HCl in a microwave oven. Atomic absorption spectrometry was carried out by an accredited TSK laboratory (St. Petersburg).

The study was performed using the facilities of the research parks of St. Petersburg State University, including the Center for Microscopy and Microanalysis and the Centre for X-ray Diffraction Studies.

3 Results and Discussion

According to the urbanization gradient, the total thickness of anthropogenically-transformed layers and the number of artefacts was found to increase from P1 to P6 (Fig. 3). The mineralogical analysis of P4 showed the domination of quartz and feldspars in the each urbostratozem horizon. The content of clay minerals (chlorite, kaolinite, and mixed-layer formation) slightly varied throughout the profile. The upper two horizons (0–40 cm) are characterized by the higher content of calcite (12–14%) and a lower content of quartz due to admixture of technogenic materials.

Except for the above-mentioned observations, the mineral composition of the studied anthropogenic layers demonstrated no significant differences from that of the buried natural soil and parent material. Therefore, the bulk of the cultural layer was found to consist of local materials (Table 3). This agrees well with a study by Prokofieva et al. (2007), who noted the presence of calcite as a specific feature of urbic horizons.

The soil texture slightly varied in the natural soil (P1) and was characterized by a more prominent variability between different cultural layers (urbic horizons) in urbostratozem (P5) (Fig. 4). In general, all soils have a loamy sand and sandy loam texture. Artefact types were found to be similar in the studied urban soils. The composition of anthropogenic inclusions is quantitatively dominated by brick, mortar, and glass. Charcoal, bone, and metal (ferruginous) fragments were found only in individual horizons. The morphology and composition of artefacts are presented in Fig. 5, as well as Tables 1–2 and 4. It can be seen from Table 4 that charcoal mainly included C; brick contained Si, Al, and Ca; glass comprised Si and Al. In bone and mortar fragments, Ca, P and Ca, Si dominated, respectively. Weathering of artefacts

Table 3 Semi-quantitative phase composition (wt %, Rietveld method), P4 Urbostratozem on a buried gray-humus gley post-agrogenic soil

Horizon, Depth, cm	UR1ay, 0–20	UR2, 20–40	UR3g, 40–55	UR4, 55–65	UR4, 75–85	UR4, 100–120	R1, 121–142	[AYpa], 142–167	CG, 167–190
Minerals									
Quartz	44	49	53	52	60	66	73	63	62
K feldspar	12	10	11	10	9	8	7	7	8
Albite	12	14	18	17	15	15	13	14	10
Chlorite	7	4	5	2	4	4	2	5	7
Kaolinite	2	2	5	<	2	<	<	4	5
Calcite	14	11	<	<	2	<	<	4	5
Pyroxenes	3	4	2	3	3	3	2	<	–
Amphiboles	1	2	<	<	<	<	<	<	<
Mica	4	3	4	3	4	3	<	5	7
Mixed-layer formation or smectite	+ –	+ –	+ –	+	+	+	+	+	+

* < – traces

in soils is known to have an impact on the chemical composition of the solid phase and soil solution (Greinert and Kostecki 2019). Charcoal can be a source of not only C, but also Ca, P, Na, and Mg in soils.

The chemical properties were gradually changing from a slightly transformed soil (P1) to urbostratozems (PP 4–6). The pH in less transformed soils (PP 1–2) was moderately acidic, while it was slightly alkaline in the stratozem (P3) (Table 5). This soil was influenced by horticulture, which might have resulted in the introduction of lime, phosphorous, and organic fertilizers (Burt et al. 2014). The urbostratozems (PP 4–6) have an alkaline reaction (Table 6). Alkaline pH is typical for anthropogenic urban soils (Lehmann and Stahr 2007) due to the presence of technogenic calcareous materials, such as construction debris (Huot et al. 2017). The highest pH values and P-Olsen concentrations were measured in the sealed soil (P6), which can be explained by using the cement and limestone rubble for road foundation (Hulisz et al. 2018). The products of weathering these anthropogenic inclusions accumulated in the soils due to the absence of washout (Prokof'eva et al. 2010). Additionally, alkalinization of the soils under pavement was promoted by the use of sodium chloride for road deicing in winter (Equiza et al. 2017). Increased pH levels in urban soils covered with impervious surfaces in comparison with open sites were also found in Nanjing City, China (Wei et al. 2014).

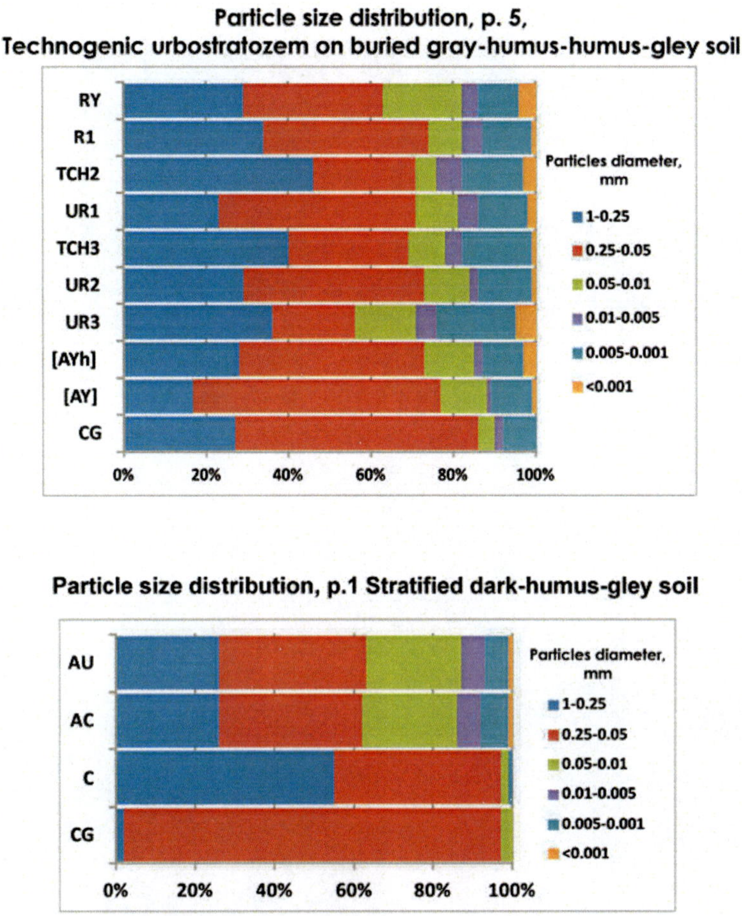

Fig. 4 Soil texture. The soil types are given in the text

In general, higher Olsen-P levels were observed in all the soils studied (Tables 5–6). The P content in urban soils is considered to be an indicator of anthropogenesis (Mazurek et al. 2016). The total P content in urban soils in Nanjing, China, reached 15 g/kg, which is ten times higher than that in agricultural soils, and Olsen-P was 400 mg/kg (Yang and Zhang 2015).

The level of soil organic carbon (SOC) in the studied soils varied significantly depending on soil types and layers. The highest concentration of SOC was observed in the surface horizons of natural and moderately transformed soils. Aparin et al. (2018) found the SOC content of 2–4% as the most common value for surface horizons of St. Petersburg soils, which finding agrees well with our data. Alternative sources of SOC in deeper soil layers comprise ash, slug, and municipal waste (Greinert and Kostecki 2019).

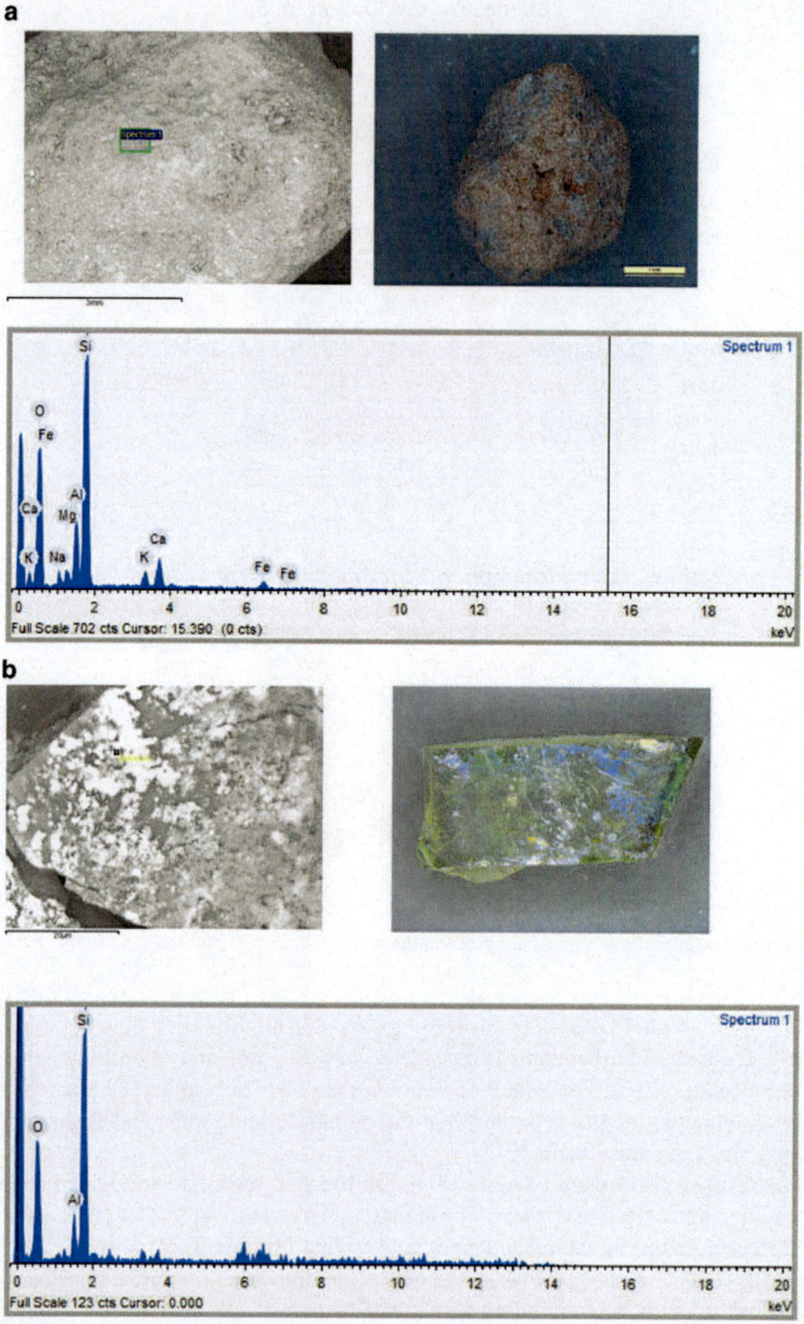

Fig. 5 SEM-image and composition of reference artefacts: **a** brick, **b** glass, **c** charcoal, **d** mortar, **e** bone

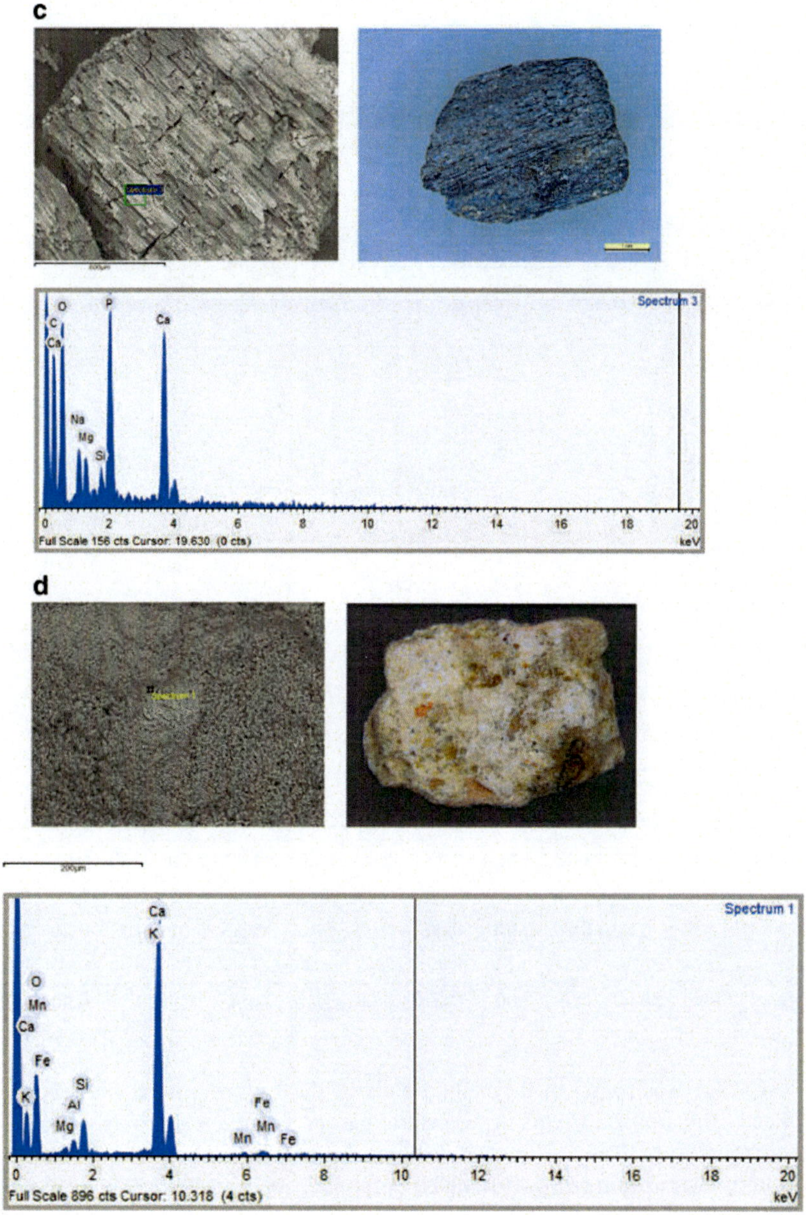

Fig. 5 (continued)

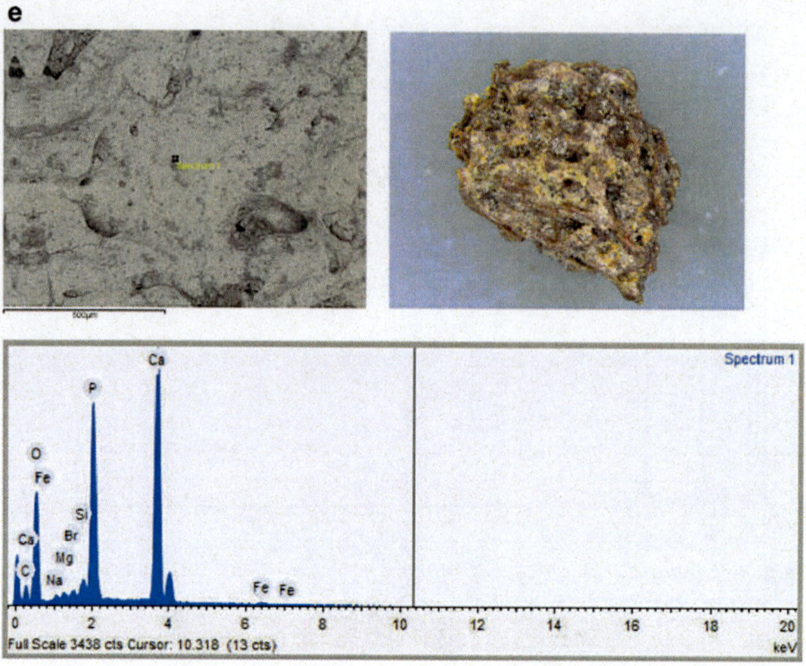

Fig. 5 (continued)

Table 4 Element composition of artefacts, %

Artefacts	Si	Ca	Mg	Na	C	P	O	Fe	Al	K	Br	Mn
Charcoal	0.56	7.45	1.13	1.80	19.57	5.40	64.09	0.00	0.00	0.00	0.00	0.00
Brick	30.72	6.12	1.59	1.39	0.00	0.00	46.53	4.58	6.42	2.65	0.00	0.00
Glass	42.32	0.00	0.00	0.00	0.00	0.00	52.67	0.00	5.01	0.00	0.00	0.00
Bone	0.74	26.26	0.31	0.34	8.22	12.68	50.11	0.66	0.00	0.00	0.66	0.00
Mortar	4.56	56.09	0.79	0.00	0.00	0.00	31.15	3.51	1.59	0.57	0.00	1.75

No $CaCO_3$ was found in the natural soils of the Littorina terrace. Carbonates could enter the urbostratozems in the central part of St. Petersburg only via anthropogenic sources, such as construction debris and urban dust. About 1–2% of $CaCO_3$ was found in stratozem (P3), with higher carbonate concentrations (up to 10–15% and above) observed in urbostratozems (PP 4–6). The presence of carbonates in P3 could be a result of previous liming. Calcium and its compounds from technogenic sources significantly affect biogeochemical processes in urban environments and, consequently, soil properties. The dissolution and recrystallization of technogenic carbonates were found in Moscow soils (Prokof'eva et al. 2021).

Table 5 Chemical properties of moderately transformed soils

Horizon	Depth, cm	pH H_2O	C_{org}, %	$CaCO_3$, %	P_{Ols}, mg/kg
P1 Stratified dark-humus gley soil					
AU	0–18	5.50	4.01	–	19.9
AC	18–29	5.50	1.64	–	2.3
C	29–45(54)	5.80	0.14	–	15.6
Cg	45(54)–90	5.80	0.17	–	9.8
P2 Urbistratified gray-humus ferruginous gleyic soil					
RYur,g	2–25	5.91	5.79	–	32.0
AYg	25–39	5.90	1.34	–	6.7
ACg	39–61	5.85	0.42	–	5.3
Cg1	61–84	5.60	0.11	–	1.3
Cg2	84–98	5.93	0.08	–	2.6
*P3 Urbistratified gray-humus stratozem on a buried gray-*humus *soil*					
RU1ur,pa	0–14	7.50	1.30	2.18	43.0
RU2 ur,pa	14–41	7.60	1.50	1.68	17.0
[ACur]	41–65	7.30	0.69	1.11	2.6
Cg	65–80	7.10	0.15	1.07	14.1

It was previously shown that the present and historical emission of soil pollutants in urban environments may lead to a heterogeneous spatial distribution of HM (Gąsiorek et al. 2017). The concentrations of Pb, Cu, and Zn in the semi-natural soil (P1) were close to the background values (Tables 7–8). The variability of HM concentrations in the studied anthropogenic soils (PP 2–6) was found to independent of the degree of their morphological transformation. A similar observation was made for soils in the city of New York, where HM levels reflected the long-term history of the site rather than followed the urbanization gradient (Burt et al. 2014). The concentration of HMs in the surface horizons of anthropogenic soils (PP 2–6) were also 2–4 times higher than approximate allowable concentrations.

Table 6 Chemical properties of urbostratozems (Urbic Technosols)

Horizon	Depth, Cm	pH H_2O	C_{org}, %	$CaCO_3$, %	P_{Ols}, mg/kg
P4 Urbostratozem on a buried gray-humus gley post-agrogenic soil					
UR1ay	0–20	8.1	1.3	17.2	32.7
UR2	20–40	8.3	0.6	18.7	20.9
UR3g	40–55	8.2	0.1	7.6	36.6
UR4	55–65	7.5	0.7	6.8	28.8
UR4	75–85	7.9	0.9	7.9	42.3
UR4	100–120	7.8	0.9	6.8	28.8
R1	121–142	7.2	0.4	10.1	36.6
[AYpa]	142–167	7.7	1.7	9.0	19.1
AYC	167–170	7.6	1.2	7.5	27.0
CG	170–190	7.9	0.3	6.8	19.1
*P5 Technogenic urbostratozem on a buried gray-*humus *gley soil*					
RY	0–25	7.35	3.20	2.34	26.6
UR1	25–41	8.76	0.72	8.72	32.0
TCH2	70–80	8.50	0.55	3.72	40.0
UR2	90–100	8.17	1.00	10.26	122.7
TCH3	100–120	8.33	0.50	24.49	88.0
UR3	120–140	8.20	1.16	7.16	64.0
UR4	150–160	8.30	0.37	1.56	21.3
[AYh]	160–170	7.05	4.64	-	18.7
[AY]	170–175	6.90	1.63	-	21.3
CG	185–200	7.28	0.09	-	16.0
P6 Technogenic urbostratozem, sealed under asphalt, on a buried urbistratified gray-humus gley soil					
TCH1	15–25	9.10	–	10.45	–
UR1	25–42	9.87	–	10.94	13.3
UR2	38–48	9.80	–	0.81	6.7
UR3	70–100	9.20	–	1.27	36.0
UR3	106–120	8.73	–	1.93	56.0
[AYur]	124–131	8.07	1.33	0.53	82.7
[AY]	144–152	7.87	0.29	0.26	41.3
CG	152–160	7.40	0.89	0.99	48.0
CG	195–200	7.81	0.18	0.19	24.0

Table 7 Concentrations Cu, Pb, and Zn (mg/kg) in moderately transformed soils

Horizon	Depth, cm	Cu	Pb	Zn
P1 Stratified dark-humus gley soil				
AU	0–18	7	14	27
AC	18–29	4	5	11
C	29–45(54)	1	1	7
Cg	45(54)–90	2	2	9
*P2 Urbistratified gray-humus ferruginous gleyic soi*l				
RYur,g	2–25	30	100	120
Cg2	84–98	11	20	24
P3 Urbistratified gray-humus stratozem on a buried gray-humus soil				
RU1ur,pa	0–14	66	51	221
RU2 ur,pa	14–41	27	59	163
[ACur]	41–65	6	18	22
Cg	65–80	2	2	11
Background values, agricultural soils in the North-West Russia (Matinian et. al., 2007)	0–20	9	16	48
Approximate allowable concentrations, sandy and loamy sandy soils (Hygiene standards of Russian Federation 2.1.7.2511–09)		33	32	55

Table 8 Concentrations of Cu, Pb, and Zn (mg/kg) in urbostratozems

Horizon	Depth, cm	Cu	Pb	Zn
P4 Urbostratozem on a buried gray-humus gley post-agrogenic soil				
UR1ay	0–20	5	77	90
UR2	20–40	5	299	131
UR3g	40–55	5	21	67
UR4	75–85	5	58	48
UR4	100–120	5	55	61

(continued)

Table 8 (continued)

Horizon	Depth, cm	Cu	Pb	Zn
R1	121–142	5	14	55
[AYpa]	142–167	5	5	49
CG	170–190	8	15	40
P5 Technogenic urbostratozem on a buried gray-humus gley soil				
R1	0–25	87	175	272
UR1	25–41	46	58	34
TCH2	70–80	18	57	44
UR2	90–100	36	154	209
TCH3	100–120	29	302	118
UR3	120–140	25	183	27
UR4	150–160	16	70	43
[AYh]	160–170	18	79	39
[AY]	170–175	11	25	15
CG	185–200	2	3	10
P6 Technogenic urbostratozem, sealed under asphalt, on a buried urbistratified gray-humus-gley soil				
TCH1	15–25	87	22	254
UR1	25–42	99	26	234
[AYur]	124–131	14	58	17
[AY]	144–152	5	7	20
CG	150–160	6	16	14
Topsoil layer, Berlin, median (Abel et al. 2015)	0–30	29	55	87
34 European cities, median (Luo et al. 2012)	–	46	102	130
21 Chines cities, median (Luo et al. 2012)	–	40	55	109

4 Conclusion

The conducted study of the less transformed soil cover in the marginal parts of the Littorina terrace and the buried soils under cultural layers allowed the authors to determine specific features of natural soils in the terrace (dark- and grey-humus gley soils). The intensive anthropogenic transformation of the Littorina terrace during the evolution of St. Petersburg has led to the appearance of new soil types (stratozems and urbostratozems), different in characteristics from the natural soils. These differences are related to the formation of cultural layers over natural soil horizons, which contain artefacts and are characterized by alkaline pH and the presence of carbonates. The most common types of artefacts in the studied soils were found to be brick, mortar, charcoal, and glass. The properties of urbostratozems strongly depend on the type and composition of artefacts, as well as their variable amounts between different cultural layers.

The majority of modern soils in the Littorina terrace are polluted with Pb, Cu, and Zn. In comparison with semi-natural soils of large urban parks, urbostratozems are characterized by higher HM concentrations. At the same time, HM concentrations are influenced by the history of land use, rather than by the gradient of urbanization.

Full-profile analysis of urban soils should become part of environmental monitoring and city management programs, due to the importance of top and deeper soil layers for plant growth and as a source/sink of technogenic contaminants.

Acknowlegements Great acknowledgements are made to A.B. Galkina for conducting HM determination. The study is dedicated to the memory of N.N. Matinian, a well-known researcher of urban soils of St. Petersburg.

References

Abel S, Nehls T, Mekiffer B, Wessolek G (2015) Heavy metals and benzo[a]pyrene in soils from construction and demolition rubble. J Soils Sediments 15:1771-1780

Aleksandrovskaya EI, Aleksandrovskiy AL (2000) History of the cultural layer in Moscow and accumulation of anthropogenic substances in it. Catena 41:249-259

Aparin BF, Sukhacheva EY, Bulysheva AM, Lazareva MA (2018) Humus Horizons of Soils in Urban Ecosystems. Eur Soil Sci 51(9):1008-1021

Aparin BF, Sukhacheva EY (2014) Principles of soil mapping of a megalopolis with St. Petersburg as an example. Eur Soil Sci 47(7):650–661

Burt R, Hernandez L, Shaw R, Tunstead R, Ferguson R, Peaslee S (2014) Trace element concentration and speciation in selected urban soils in New York City. Environ Monit Assess 186:195-215

Classification System of Russian Soils (2004) Pochv. Inst. im. Dokuchaeva, Moscow (in Russian)

Delbeque N, Dondeyne S, Gelaude F, Mouazen AM, Vermeir P, Verdoodt A (2022) Urban soil properties distinguished by parent material, land use, time since urbanization, and pre-urban geomorphology. Geoderma 413:115719

Equiza MA, Calvo-Polanso M, Sirelli D, Señorans J, Wartenbe M, Saunders C (2017) Long-term impact of road salt (NaCl) on soil and urban trees in Edmonton, Canada. Urban Forestry &Urban Greening 21:16-28

Gąsiorek M, Kowalska J, Mazurek R, Pająk M (2017) Comprehensive assessment of heavy metal pollution in topsoil of historical urban park on an example of the Planty Park in Krakow (Poland). Chemosphere 179:148-158

Gerasimova MI, Stroganova MN, Mozharova NV, Prokofeva TV (2003) Anthropogenic soils (genesis, classification, rehabilitation and use). Smolensk, Oykumena (in Russian)

Greinert A, Kostecki J (2019) Anthropogenic Materials as bedrock of urban Technosols. In: Vasenev V et al (eds) Urbanization: Challenge and Opportunity for Soil Function and Ecosystem Services. Conference proceedings, SUITMA 2017. Book series Springer Geogr. Springer

Howard JL, Orlicki KM (2016) Composition, micromorphology and distribution of microartifacts in anthropogenic soils, Detroit, Michigan, USA. Catena 138:103-116

Hulisz P, Charzyński P, Greinert A (2018) Urban soil resources of medium-sized cities in Poland: a comparative study of Toruń and Zielona Góra. J Soils Sediments 18:358-372

Huot H, Joyner J, Córdoba A, Shaw RK, Wilson MA, Walker R, Muth TR, Cheng Z (2017) Characterizing urban soils in New York City: profile properties and bacterial communities. J Soils Sediments 15:393-407

Hygiene standards of Russian Federation 2.1.7.2511–09 (2009) "Approximate allowable concentrations of chemical substances in soil" (in Russian)

IUSS Working Group WRB (2014) World Reference Base for Soil Resources 2014 International soil classification system for naming soils and creating legends for soil maps. World Soil Resources Reports No. 106. FAO, Rome

Lehmann A, Stahr K (2007) Nature and significance of anthropogenic urban soils. J Soils Sediments 7(4):247-260

Lenoir T, Duc M, Lassabatere L, Bellagh K (2019) Identification of the artifact contribution to two urban Technosols by coupling a sorting test, chemical analyses, and a least absolute residual procedure. J Soils Sediments 19:683-701

Luo X, Yu S, Zhu Y, Li X (2012) Trace metal contamination in urban soils of China. Sci Total Environ 421-422:17-30

Makki M, Thestorf K, Hilbert S, Thelemann M, Makowsky L (2021) Guideline for the description of soils in the Berlin metropolitan area: an extension for surveying and mapping anthropogenic and natural soils in urban environments within the German soil classification system. J Soils Sediments 21:1998-2012

Matinian NN, Reimann K, Bakhmatova KA, Rusakov AV (2007) Background concentrations of heavy metals and As in arable soils of North West Russia. Vestnik of St Petersburg University. Series 3, Biology, 3: 123-134 (in Russian)

Matinian NN, Bakhmatova KA, Sheshukova AA (2019) Anthropogenic and natural soils of urban and suburban parks of Saint Petersburg, Russia. Urbanization: Challenge and Opportunity for Soil Function and Ecosystem Services. Conference proceedings, SUITMA 2017. Eds.: V Vasenev, E Dovletyarova, Z Cheng et al. Book series Springer Geogr. Springer

Matinian NN, Bakhmatova KA, Sheshukova AA (2021) Urban Soils in the Historic Centre of Saint Petersburg (Russia). Advances in Understanding Soil Degradation. Eds.: E Saljnikov, L Mueller, A Lavrishchev, F Eulenstein. Springer Nature

Matinyan NN, Bakhmatova KA, Korenzvit VA (2017) Soils of the Summer Garden (Saint Petersburg). Eur Soil Sci 50(6):637-645

Mazurek R, Kowalska J, Gąsiorek M, Setlak M (2016) Micromorphological and physico-chemical analyses of cultural layers in the urban soil of a medieval city – A case study from Krakow, Poland. Catena 141:73-84

Nadporozhskaya MA, Slepyan EI, Kovsh NV (2000) About soils of historical parts of Saint Petersburg. Vestnik of S.Petersburg University, Biology Series 1(3):115–126 (in Russian)

Pickett STA, Cadenasso ML (2009) Altered resources, disturbance, and heterogeneity: A framework for comparing urban and non-urban soils. Urban Ecosyst 12:23-44

Plyaskina OV, Ladonin DV (2009) Heavy metal pollution of urban soils. Eur Soil Sci 42(7): 816-823

Prokof´eva TV, Gerasimova MI, Bezuglova OS., Bakhmatova KA, Gol'eva AA, Gorbov SN, Zharikova EA, Matinyan NN, Nakvasina EN, Sivtseva N E (2014) Inclusion of soils and soil-like bodies of urban territories into the Russian soil classification system. Eur Soil Sci 47 (10):959-967

Prokof´eva TV, Kiryushin AV, Shishkov VA, Ivannikov FA (2017) The importance of dust material in urban soil formation: the experience on study of two young Technosols on dust depositions. J Soils Sediments 17:515-524

Prokof´eva TV, Shishkov V, Kiriushin A (2021) Calcium carbonate accumulations in Technosols of Moscow city. J Soils Sediments 21:2049-2058

Prokofieva TV, Sedov SN, Kazdym AA (2007) Sources, composition and conditions for the formation of clay material in urban soils. Byulleten Pochvennogo instituta imeni V.V. Dokuchaeva 60:41-55 (in Russian)

Rastvorova OG (1983) Soil physics (Practical guide). Leningrad, Leningrad State University (in Russian)

Rusakov AV, Novikov VV (2003) Biological activity in modern and buried soils of the historical center of Saint Petersburg. Microbiology 72(1):101-109

Semenjuk OV, Gradusova OB, Peleneva MV (2013) Assessment of anthropogenic soils transformation of the estate museum “Arkhangelskoe” on the example of inclusions studying. Vestnik Moscovskogo Universiteta. Ser. 17. 2: 35-42 (in Russian)

Tobias S, Conen F, Duss A, Wenzel LM, Buser C, Alewell C (2018) Soil sealing and unsealing: State of the art and examples. Land Degrad Dev 29:2015-2024

Ufimtseva, M.D., Terekhina, N.V., Abakumov E.V. (2011) Physicochemical description of soils in Central District of Saint Petersburg. Vestnik Sankt – Peterb Universiteta, Seriya Geologiya i Geographiya 4:85-97 (in Russian)

Vorobieva LA (ed.) (2006) Theory and practice chemical analysis of soils. Moscow, MGU (in Russian)

Wei ZQ, Wu SH, Zhou SL, Li JT, Zhao QG (2014) Soil organic carbon transformation and related properties in urban soil under impervious surfaces. Pedosphere 24(1):56-64

Yang JL, Zhang GL (2015) Formation, characteristics and eco-environmental implications of urban soils – A review. Soil Sci and Plant Nutrition 61:30-46

Interactions of Microorganisms with Natural and Artificial Materials

Bacterial Contribution in Biomineralization at the Tomskaya Pisanitsa Rock Art Site

Dmitry Yu. Vlasov, Marina S. Zelenskaya, Alina R. Izatulina, Oksana A. Rodina, Alexey D. Vlasov, Katerina V. Sazanova, Anna A. Vilnet, Irina V. Abolonkova, and Olga V. Frank-Kamenetskaya

Abstract Microorganisms found in various climatic conditions are able to effect the surface of rock art monuments. In this work we consider the influence of bacteria on biomineralization on the unique rock art monument "Tomskaya Pisanitsa" (Western Siberia, Russia). The biomineralization products obtained in vitro and in vivo were investigated by X-ray powder diffraction, binocular and scanning electron microscopy as well as energy-dispersive spectroscopy. Cultures of cyanobacteria and organotrophic bacteria were identified on the basic of the nucleotide sequences of the 16S rRNA gene. The manifestation of bacterial carbonate biomineralization on rock art monument "Tomskaya Pisanitsa" was found. Using experimental modeling it was shown that the formation of biogenic calcite (in some cases together with calcium oxalates) can occur by action cyanobacteria and organotrophic bacteria living on the bedrock. The results of the study can be used for conservation of rock monuments.

Keywords Rock art · Biofouling · Bacterial community · Biomineralization · Carbonates · Oxalates · Experimental modeling

---

D. Yu. Vlasov (✉) · M. S. Zelenskaya
Department of Botany, Faculty of Biology, St. Petersburg State University, St. Petersburg, Russia
e-mail: dmitry.vlasov@mail.ru

A. R. Izatulina · O. A. Rodina · O. V. Frank-Kamenetskaya
Department of Crystallography, Institute of Earth Sciences, St. Petersburg State University, St. Petersburg, Russia

D. Yu. Vlasov · K. V. Sazanova
Komarov Botanical Institute of the Russian Academy of Science, St. Petersburg, Russia

O. A. Rodina · A. A. Vilnet
Avronin Polar Alpine Botanical Garden-Institute of Kola Scientific Center of the RAS, Apatity, Russia

A. D. Vlasov · K. V. Sazanova
St. Petersburg Branch of the Archives of the Russian Academy of Science, St. Petersburg, Russia

I. V. Abolonkova
Kuzbass Tomskaya Pisanitsa Museum-Reserve, Kemerovo, Russia

O. V. Frank-Kamenetskaya et al. (eds.), *Biogenic—Abiogenic Interactions in Natural and Anthropogenic Systems 2022*, Springer Proceedings in Earth and Environmental Sciences, https://doi.org/10.1007/978-3-031-40470-2_28

1 Introduction

Rock art images are an important part of the ancient cultural heritage as they largely characterize the development of prehistoric civilization. Preservation of petroglyphs is one of the important scientific problems that requires an interdisciplinary scientific approach. The study of the petroglyphs state and preservation has been carried out in different countries of the world (Biswas et al. 2013; Neumann et al. 2022; Nir et al. 2022; Rabbachin et al. 2022; Sazanova et al. 2022; Smith et al. 2022, Vlasov et al. 2023). In these works authors focused on the contribution of biological factors in the damage to petroglyphs. It was shown that rock art monuments are colonized by fungi, bacteria, archaea, algae, lichens, seed plants. The use of modern molecular genetic methods has made it possible to establish numerous taxa of microorganisms that settle on rocks with petroglyphs (Esposito et al. 2015; Nir et al. 2022; Rabbachin et al. 2022; Vlasov et al. 2023). Microorganisms have a biochemical and biophysical effect on the stone (Gaylarde and Little 2022), which can be accompanied by the dissolution of calcite, the formation of secondary minerals, and the disintegration of the surface layer due to microbial growth between rock crystals (The effect… 2019; Nir et al. 2022). Cyanobacteria play a special role in the processes of primary colonization of rock art monuments (Crispim and Gaylarde 2005; Wierzchos et al. 2012; Hauer et al. 2015; Del Mondo et al. 2018; González-Gómez et al. 2018). They are able to withstand environmental stressors such as desiccation and radiation (Abed et al. 2010; Perera et al. 2018; Mishra 2020). Cyanobacteria from the genera *Gloeocapsa*, *Nostoc* and *Oscillatoria* are often the cause of the noticeable destructive effect on carbonate rocks (Lian et al. 2008). They can form biofilms with high weathering potential (Mergelov et al. 2018). In many cases, cyanobacteria are the basis of biofilms at archaeological sites. The biofilm provides a protective barrier and improves the survival of microbial cells under nutrient-poor conditions. The composition and structure of biofilms largely determined by external conditions (Del Mondo et al. 2018; Zammit 2019). Bacteria from the phyla Actinobacteria, Proteobacteria, Bacteroidetes, and Chloroflexi also mentioned in the composition of biofilms on rocks with petroglyphs (Esposito et al. 2015; Nir et al. 2019; Vlasov et al. 2023). However, the contribution of these organisms in transformation of the rock substrate is still insufficiently studied. Understanding of biomineralization peculiarities on the surface of the rocks with petroglyphs can be important for rock art monuments preservation. Modeling of lithobiotic systems can help in solving this issue. This experience was widely used earlier by us and other researchers to study the carbonate and oxalate biomineralization on various rocks and minerals (Vlasov et al. 2020). Modeling also used to determine the expected change in the chemical, mineralogical and physical characteristics of the rock surface with petroglyphs under the influence of external factors (Neumann et al. 2022).

This work is a continuation of our study of the interaction of microorganisms with the sandstone of the Tomskaya Pisanitsa site and is aimed at clarifying the contribution of microorganisms in biomineralization on this rock with ancient images. This site is located in the Kemerovo Region on the banks of the Tom River (Western

Fig. 1 The sampling sites on Rock art monument «Tomskaya Pisanitsa»: **a**–rock cracks adjacent to the petroglyphs; **b**–rock cracks filled with a dense substance

Siberia, Russia). The hundreds of ancient images located here on several rocky planes (Kovtun 2021). Previously, we have identified the microorganisms at this site using cultural, morphological and molecular genetic methods (Vlasov et al. 2023).

2 Materials and Methods

The objectives of this study were solved by conducting field studies and modeling.

2.1 Sampling

The previous mineralogical study of the rock on Tomskaya Pisanitsa (Vlasov et al. 2023) showed that it is sandstone with interlayers of siltstones, argellites and their transitional varieties (Fig. 1a). The rock is characterized by significant fracturing and contains numerous inclusions of carbonates. The samples were taken from rock surface (planes) adjacent to the petroglyphs locations (model experiments were carried out using rock fragments ~ 1 × 1 × 0.5 cm). In addition, samples were taken from surface cracks filled with a dense substance (presumably calcite) (Fig. 1b).

2.2 Identification of Bacteria

2.2.1 Cyanobacteria

Cyanobacteria were isolated from rock samples covered by biofilms with carbonate deposits on the surface (Fig. 2 a, b). Initially, a biofilm sample was placed in 100 ml of Z8 culture medium in a flat-bottomed flask to obtain an enrichment culture of cyanobacteria. Cultivation was carried out for a month in a light box with a day/night mode—12/12. After that isolation of cyanobacteria strains was carried out.

The dominant strain (#98) was assigned to the genus *Microcoleus* according to morphological features. Earlier we noted that cyanobacteria of this genus are able to actively develop on rock with petroglyphs in the Tomskaya Pisanitsa (Vlasov et al. 2023). Detailed molecular identification of this strain (#98) was carried out. DNA was extracted from the unialgal culture with the DNeasy Plant Mini Kit (Qiagen, Germany) according to the manufacturer's protocol. Amplification of partial 16S rRNA gene and 16S-23S ITS rRNA region was performed with pair of primers 1 (5’-CTC TGT GTG CCT AGG TAT CC-3’) suggested by Wilmotte et al. (1993) and 2 (5’-GGG GGA TTT TCC GCA ATG GG-3’) by Nübel et al. (1997). PCR was carried out in 20 μl volumes with MasDDTaqMIX (Dialat Ltd, Russia), 10 pmol of each oligonucleotide primer and 1 ng of DNA with the following amplification cycles: 3 min at 94 °C, 40 cycles (30 s 94 °C, 40 s 56 °C, 60 s 72 °C) and 2 min of final extension time at 72 °C. Amplified fragments were visualized on 1% agarose TAE gels by EthBr staining, purified with the QIAquick Gel Extraction Kit (Qiagen, Germany), and then used as a template in sequencing reactions with the ABI PRISM® BigDye™ Terminator v. 3.1 Sequencing Ready Reaction Kit following the standard protocol provided for Applied Biosystems 3730 DNA Analyzer (Applied Biosystems, USA). Additionally for sequence procedure internal primer 3 (5’-CGC TCT ACC AAC TGA GCT A-39’) (Wilmotte, 1994) were used. Newly obtained nucleotide sequence data for strain #98 were assembled in BioEdit 7.0.1. (Hall 1999), and then tested thorough BLAST search (https://blast.ncbi.nlm.nih.gov/) to determine the most similar accessions by 16S rRNA gene and 16S-23S ITS rRNA region.

Also a pure culture of *Phormidesmis* sp. was obtained and identified by characteristic morphological features. This strain was used in experiments for comparison with *Microcoleus vaginatus*, which is dominant in the natural biofilm. The choice of *Phormidesmis* sp. is due to the fact that this cyanobacterium begins to dominate in enrichment culture on Z8 medium after some time of cultivation.

Fig. 2 Cyanobacterial biofilms from Tomskaya Pisanitsa: **a**–calcified biofilm on the rock surface; **b**–enrichment culture of cyanobacteria from calcified biofilm

2.2.2 Organotrophic Bacteria

Two strains of organotrophic bacteria isolated on GMF medium (enzymatic hydrolysate of meat) from the rock with carbonate inclusions (Fig. 3a) were chosen based on the abundance of colonies when inoculating the sample on a nutrient medium (Fig. 3b, c).

Bacteria were identified by the nucleotide sequence of the 16S rRNA gene as described previously (Vlasov et al. 2023). The strains of organotrophic bacteria included in the experiment were identified as *Bacillus* sp. (GB accession number OQ654083) and *Kocuria* sp. (GB accession number OQ654084).

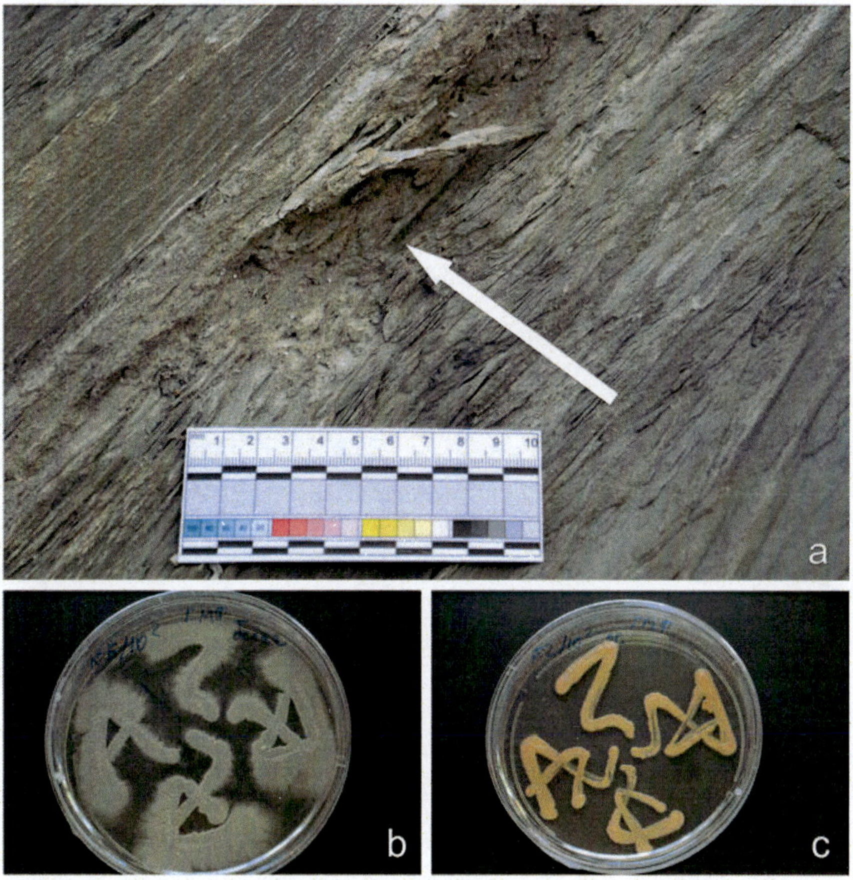

Fig. 3 Organotrophic bacteria isolated from a damaged rock at the Tomskaya Pisanitsa: **a**–sampling place, **b**–*Bacillus* sp., **c**–*Kocuria* sp

2.3 Model Experiments

Enrichment culture of cyanobacterial biofilm as well as pure cultures of cyanobacteria (*Microcoleus vaginatus* and *Phormidesmis* sp) and organotrophic bacteria (*Bacillus* sp. and *Kocuria* sp.) were selected for model experiments on biomineralization on the Tomskaya Pisanitsa sandstone.

The cultures of cyanobacteria *Microcoleus vaginatus* and *Phormidesmis* sp. as well as enrichment culture of natural biofilm with dominance of cyanobacteria were preliminarily ground in a mortar. A homogeneous mass of the initial culture (1 ml) each, a sterile substrate, and 5 ml of the modified Z-8 medium (without the addition of $Ca(NO_3)_2$) were added to a Petri dish d = 6 cm. As a control, cultures were grown in 5 ml of modified Z-8 medium (without addition of $Ca(NO_3)_2$) without substrate. Also, as a control, the rock substrate was placed in 6 ml of modified Z-8 medium (without addition of $Ca(NO_3)_2$) and without cyanobacteria. The growth of cultures was maintained at room temperature under artificial lighting with day/night— 12/12. The experiment lasted for a month without adding the medium in Petri dishes.

The experiments with organotrophic bacteria *Bacillus* sp. and *Kocuria* sp. were carried out in liquid B-4 medium (Boquet et al. 1973) at temperature of + 26°C with pH control of the cultural liquid (initial pH value 7). As a source of calcium for bacterially induced crystallization of calcium carbonate, we used acetate Ca. The rock fragments (~1 × 1 × 0.5 cm) were put on the bottom of plastic Petri dishes and 20 mL of liquid medium was added. The inoculation was carried out by the suspension of bacterial cells. The number of bacterial cells in 1 mL of the initial suspension was 10^8.

The duration of the experiments was 60 days. All syntheses were made in triplicate. The control experiments were carried out without bacteria. The crystallization products and microbial biofilms were investigated directly on the rock substrate.

2.4 Methods

2.4.1 Light Microscopy

Rock fragments with bacterial and cyanobacteria biofilms on its surfaces as well as the products of biomiralization obtained from experiments with were initially examined using a Leica optical stereomicroscope (M205 C) for choosing samples for SEM and XRPD analysis. Small rock fragments taken from rock cracks with signs of natural cementation were also preliminarily examined using a Leica optical stereomicroscope for further study using SEM and XRPD analysis.

2.4.2 X-ray Powder Diffraction (XRPD)

The XRPD method was used for determination of phase composition of biomineralization products in nature and in experiments. The samples were prepared in such a way as to maximally exclude the influence of the underlying rock and study only the mineral component of the formed film. In experiments with organotrophic bacteria, aggregations of crystals in the form of globules, typical of secondary calcite, were selected under a microscope. The measurements were performed by means of Bruker "D2 Phaser" powder diffractometer (CoKα and CuKα radiation). X-ray diffraction patterns were collected at room temperature in the range of $2\theta = 5$–$60°$ with a step of $0.02° 2\theta$ and a counting time of half second per data point. A sample holder of a single crystal silica slice was used to eliminate the background noise. The PDF-2 database was used for phase identifications.

2.4.3 Scanning Electron Microscopy (SEM) and Energy-Dispersive Spectroscopy (EDXS)

SEM and EDXS were used to study the morphology and composition of biomineral crystals and their intergrowths as well for diagnostics of biomineralization products on this base. The investigation was carried out using the TM 3000 (HITACHI, Japan, 2010) microscope. For semiquantitative EDX analysis, the microscope was equipped with the Oxford Inca system (Oxford Instruments, UK), which operated in a low vacuum (60 Pa) mode and at an acceleration voltage of 15 kV. To avoid the charging effect on SEM images, the samples were coated with a thin carbon layer (High Vacuum Carbon Sprayer Q150TE). The EDX spectra were analyzed by means of the EDAX Genesis software package.

3 Results

3.1 Manifestation of Bacterial Biomineralization in Nature

We detected a manifestation of bacterial biomineralization during field work on the Tomskaya Pisanitsa in 2021–2022. We found the rock cracks filled with a dense cementing substance (Fig. 1b, 4a) which suggested carbonate biomineralization. The use of light microscopy, XRPD, SEM and EDXS showed that this substance was formed by calcite and small bedrock particles (Fig. 4b-e). Bacteria of genus *Bacillus* known for the ability to formation of both carbonates and oxalates were isolated from such places. In addition, traces of the cyanobacteria *Microcoleus* sp. (trichomes with characteristic sheaths) are also noted here (Fig. 4f).

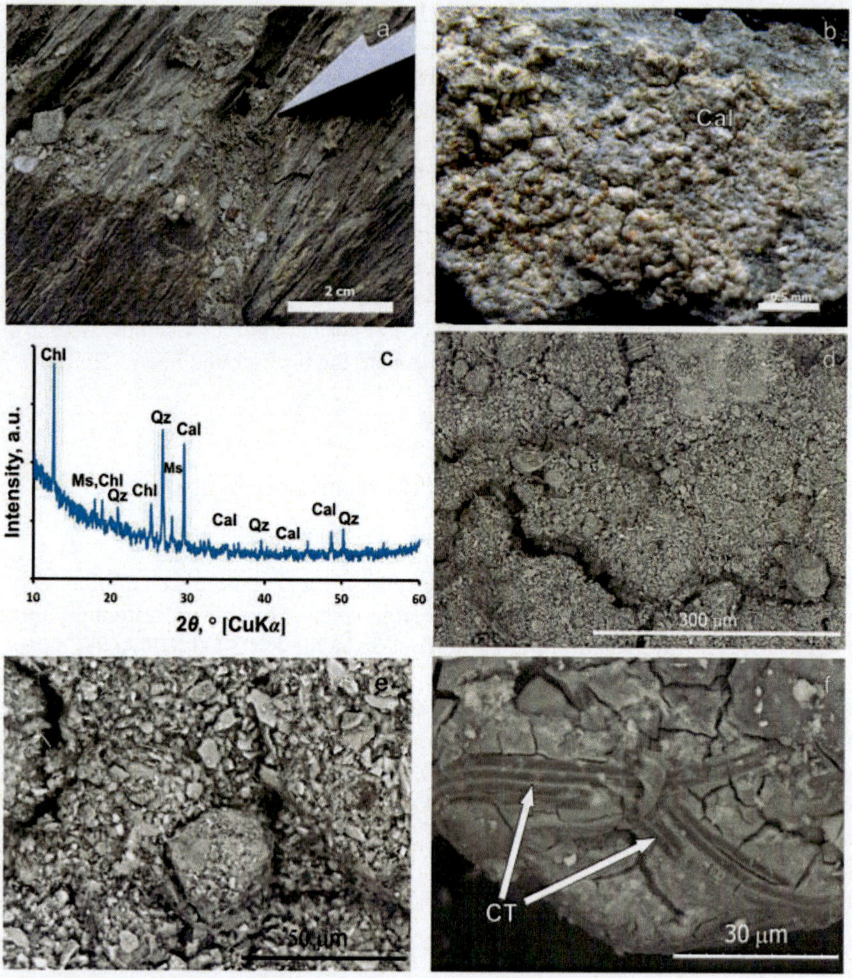

Fig. 4 Newly formed dense substance in the rock cracks in the zone of moisture movement: **a**–general view of the rock cracks filled with a dense substance (photo); **b**–deposits of calcite on a small rock flake from crack (light microscope); **c**–XRPD of dense substance from crack (Cal–calcite; Chl–chlorite-serpentine; Qz–quartz; Ms–muscovite); **d**, **e**, **f**–SEM images: **d**–dense substance in a rock crack (general view); **e**–pieces of rock cemented with calcite; **f**–traces of cyanobacterial trichomes with mucous sheets (CT) *Microcoleus* sp

3.2 *Molecular Analyses of Dominated Cyanobacteria Strain Microcoleus Vaginatus*

Reconstructed by molecular analyses tree topology is shown in Fig. 5 with an indication of bootstrap support values.

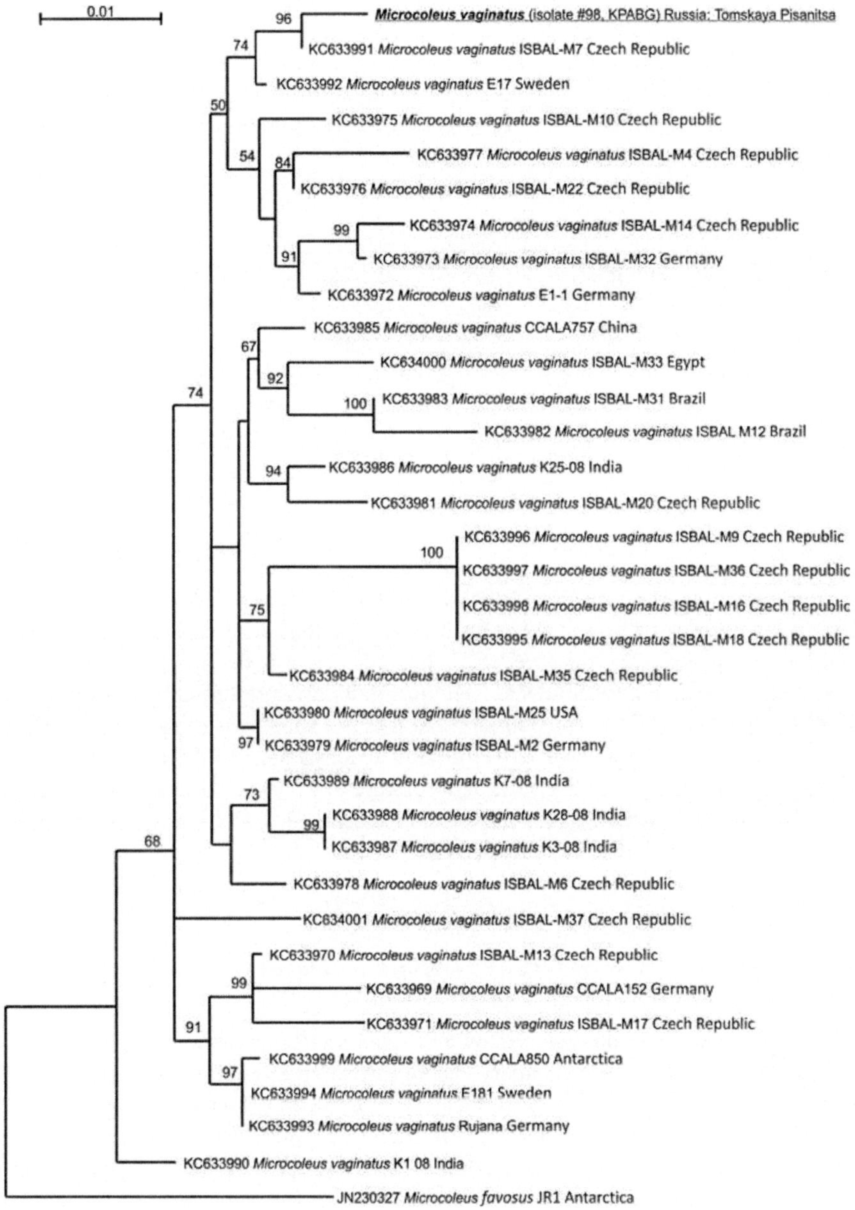

Fig. 5 The phylogram obtained under the maximum likelihood approach for 35 accessions of *Microcoleus vaginatus* Gomont ex Gomont based on 16S rRNA gene and 16S-23S ITS rRNA region

The strain #98 from rock surfaces in Tomskaya Pisanitsa in nested within multiply sampled strains of *Microcoleus vaginatus* gathered from European countries, U.S.A., Brazil, Egypt, China, India and Antarctica. Evidently, strain originated from rock surfaces in Kemerovo Province of Russia belongs to widely distributed cyanoprocariotic species—*Microcoleus vaginatus.*

3.3 Experiments with Cyanobacteria

Thin biofilms formed predominantly by filamentous cyanobacteria and some organotrophic microorganisms were observed on the sandstone surface after a month of cyanobacterial biofilm enrichment culture growth (pH = 8). XRPD and SEM methods clearly shown on the surface of this biofilm the calcite crystals forming clusters (20–100 microns) (Fig. 6). The cavities (channels) remaining in the places of cyanobacteria filaments are surrounded by calcite deposits (Fig. 6a). The diameter of the cavities corresponds to the width of *Microcoleus vaginatus* trichomes. Cyanobacterial EPS is abundantly released in the enrichment culture (Fig. 6b).

In experiments with algologically pure cultures of *Microcoleus vaginatus* and *Phormidesmis* sp. after a month of cultivation at pH = 8.5 local biofilms on the surface of the substrate also were observed. Calcite is formed in both variants of cyanobacteria cultures. In the culture of *M. vaginatus* calcite is present as intergrowths of small globules (20–50 microns) around the mucous sheets of cyanobacteria. Also, dipyramidal single crystals of calcium oxalate dihydrate weddellite between the trichomes of *M. vaginatus* (Fig. 6c). In the culture of *Phormidesmis* sp. calcite globules around cyanobacteria trichomes are not observed. Calcite is represented by collapsing crystals of 20–30 microns.

3.4 Experiments with Organotrophic Bacteria

Results of a study of sediments formed by action *Bacillus* sp. with using XRPD and SEM methods confirm the secondary calcite formation after 60 days of experiment at pH = 8.5 (Fig. 7). The secondary calcite intergrows (globules) were observed on SEM images. Calcite globules (Fig. 7a) ranging in size from 50 to 150 microns are included in bacterial biofilm or located on its surface. On calcite crystals small cavities repeating the shape of a bacterial cells are clearly visible (Fig. 7b, c). Calcite was confirmed by XRPD received for aggregations of crystals in the form of globules (Fig. 7d).

In the experiment with *Kocuria* sp. after 60 days (pH = 6.0) of cultivation calcite was not detected. Secondary products of crystallization are calcium oxalates, primarily dipyramidal-prismatic crystals of weedellite calcium oxalate dihydrate. The dimensions of the prism faces are at least 5 times larger than the dimensions of the dipyramidal faces (Fig. 8a). Weddellite crystals are collected in intergrowths

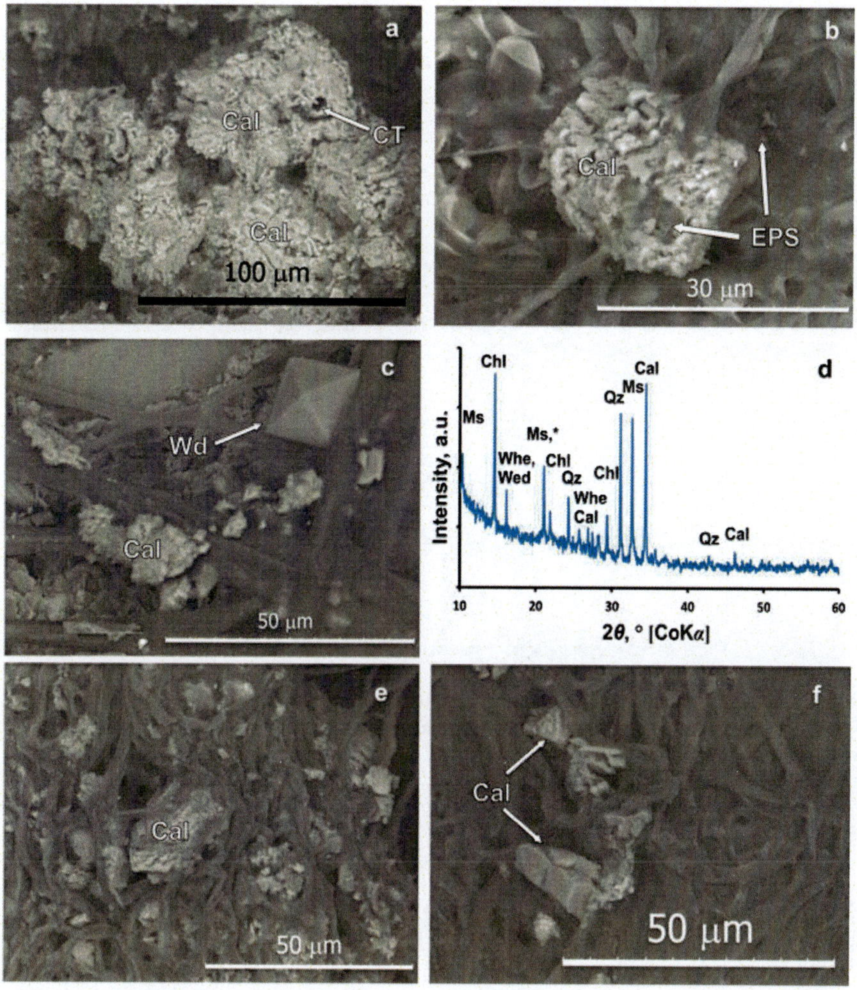

Fig. 6 Crystallization by action of cyanobacteria: **a**–large calcite (Cal) intergrowths with a cavity from the cyanobacterial trichome (CT) formed in enrichment culture; **b**–intergrowth of calcite crystals formed in enrichment culture with abundant EPS; **c**–small calcite intergrowths and large dipiramidal weddellite crystal in biofilm formed by *M. vaginatus*; **d**–XRPD of biogenic oxalates and carbonates (Cal–calcite, Wed–weddellite, Whe–whewellite) and also bedrock minerals (Chl–chlorite-serpentine; Qz–quartz; Ms–muscovite) in *M. vaginatus* culture; **e**, **f**–intergrowths of calcite crystals formed in the culture of *Phormidesmis* sp.

ranging in size from 50 to 80 μm (Fig. 8b). Weddellite crystals can occupy small depressions and cavities on bedrock (Fig. 8c). According to the EDX data, oxalate crystals contain impurities of magnesium and iron ions (Fig. 8d).

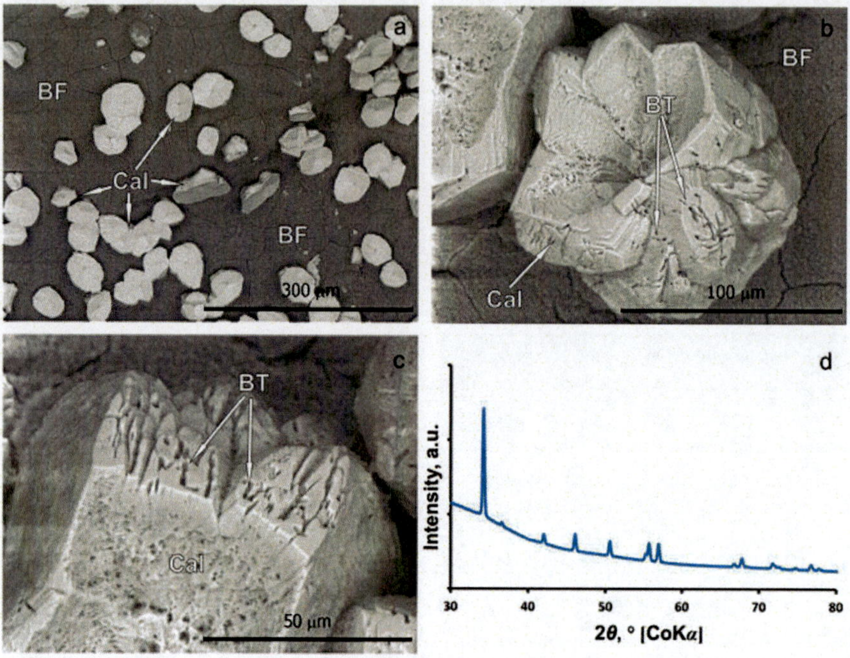

Fig. 7 Calcite crystallization by action of *Bacillus* sp.: **a**–biofilm (BF) with numerous secondly calcite (Cal) crystal intergrows; **b**, **c**–traces of bacterial cells (BT) on calcite crystals; **d** –XRPD diagram of biofilm mineral component. All diffraction peaks are attributed to calcite

4 Discussion

Based on previous studies (Vlasov et al. 2023) we suggested that microbial communities may be involved in carbonate biomineralization on the rock at the Tomskaya Pisanitsa leading to the formation of calcite layers. Cyanobacteria as well as some organotrophic bacteria could take part in it.

The results of this work based on the field researches and modeling clarified the contribution of some dominated in rock with petroglyphs cyanobacteria and organotrophic bacteria in the biomineralization on the rock of Tomskaya Pisanitsa.

For the first time we discovered active formation of calcite in a near-surface cracks. The calcification was also observed in model experiments under action of cyanobacteria: in the enrichment culture and in algologically pure cultures of *M. vaginatus* and *Phormidesmis* sp. dominated on the rock of Tomskaya Pisanitsa. In our opinion, genus *Microcoleus,* which are detected by both cultural and molecular methods, makes the main contribution to biomineralization in the Tomskaya Pisanitsa. This is confirmed by SEM data showing the formation of calcite around the sheets of *M. vaginatus* (Fig. 4f). However, other cyanobacteria also play a significant role. We previously identified cyanobacteria from 17 genera at the Tomskaya Pisanitsa (Vlasov et al. 2023). For example, *Phormidesmis* sp. also dominates in cyanobacterial

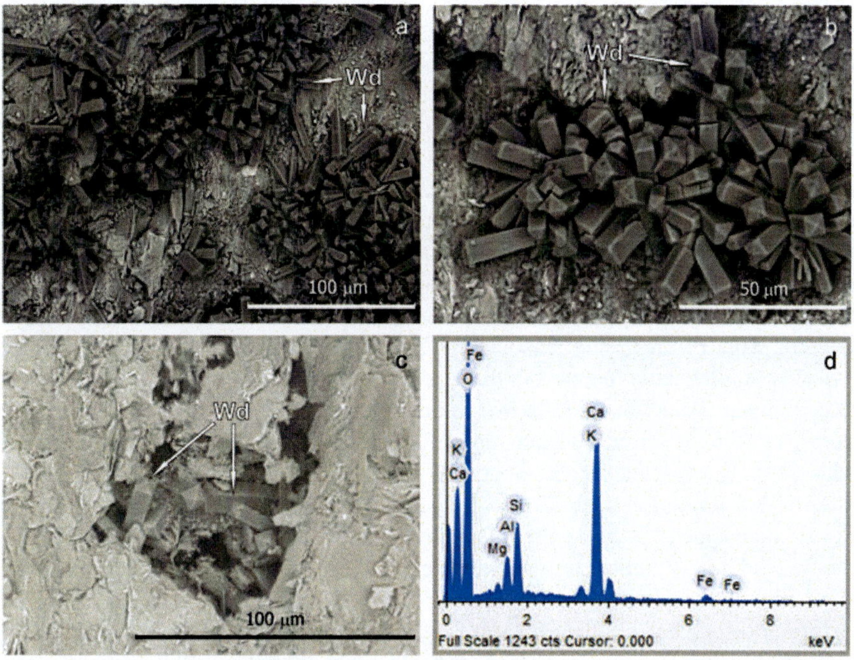

Fig. 8 Oxalate crystallization by action *Kocuria* sp: **a**, **b**–intergrowths of weddellite (Wd) crystals with a strongly developed prism faces; **c**–weddellite crystals in a cavity; **d**–EDX-spectrum of weddellite crystals

biofilms at the Tomskaya Pisanitsa. It is interesting to note that in enrichment culture *Phormidesmis* sp. gradually replaces *Microcoleus* species, which leads to a change in the pattern of calcite crystallization as previously reported (Vlasov et al. 2023). Thus, the use of enrichment culture for controlled syntheses does not always provide answers to questions about the role of specific microorganisms in biomineralization. In this regard the experiments with algologically pure cultures of cyanobacteria which were carried out in this work look most indicative. The deposition of calcite is especially evident around the mucous sheets of *M. vaginatus*, where large aggregates of calcite are formed and cavities from trichomes of cyanobacteria are preserved (Fig. 6a). In the culture of *Phormidesmis* sp. redeposition of calcite is not so active, and calcite crystals of a special morphology are locally formed in the EPS of the biofilm (Fig. 6e, f). Thus, the intensity of carbonate formation depends on the strain of cyanobacteria.

It is interesting to note that the formation of calcium oxalates was also observed in cyanobacterial cultures (Fig. 6c, d). This may be due to the fact that organotrophic bacteria, which are oxalic acid producers, often accompany cyanobacteria. This assumption is based on the data of metagenomic analysis and cultural study of cyanobacterial biofilms from the Tomskaya Pisanitsa (Vlasov et al. 2023).

The formation of oxalate crystals also was shown in experiments with *Kocuria* sp. isolated from the sites of cyanobacterial biofilms development on the Tomskaya Pisanitsa. The metabolism of this bacterium has not been studied. However, the intense oxalate crystallization with participation of this bacteria indicates that under the experimental conditions the main metabolic product of *Kocuria* sp. was oxalic acid, and possibly other organic acids released by organotrophic bacteria (Sazanova et al. 2020). It is possible that intensive production of organic acids by *Kocuria* sp. connected with the concentration of bacterial cells in cavities and depressions on bedrock (Fig. 8c) where clusters of oxalate crystals are noted. In the experiment with this bacterium, the pH of the medium was lower than in experiments with other bacteria. This could be the reason for the inhibition of carbonate crystallization. Probably, a noticeable increase in the size of the prism faces in weddellite crystals formed under the action of *Kocuria* sp. due to the selective adsorption of impurity atoms, in this case, iron and magnesium (Fig. 8d) (Rusakov et al. 2021). Unlike *Kocuria* sp. in experiments with another organotrophic bacterium *Bacillus* sp. at pH 8–8.5 active formation of calcite on the surface of the bacterial biofilm was noted. At the same time, it is known that bacteria genus *Bacillus* are also able to participate in oxalate crystallization (Sazanova et al. 2020). Why do two bacteria behave differently under similar experimental conditions? This question must be answered by further research. At the moment, we see a possible role of organotrophic bacteria of the genus *Bacillus* in carbonate formation at the Tomskaya Pisanitsa, which is consistent with field research data. The results obtained allow us to suggest the participation of bacteria producing secondary calcite in the processes of fracture cementation. Calcium ions for this process come from the rock substrate of the Tomskaya Pisanitsa due to leaching processes. Also, rock particles are washed out from the rock. The minimum size of such rock fragments is 5–10 microns. It is obviously that the production of EPS by cyanobacteria and organotrophic bacteria developing together in rock cracks and fissures promote calcification and bonding of particles. Previously, it was suggested that the presence of autochthonous microorganisms in damaged stone may contribute to the formation of calcite and cementation of the stone (Cardellicchio et al. 2023; Zhang et al. 2023). Obviously, our data on the microbial community composition, the results of model experiments, as well as observations in nature indicate on the complex biomineral interactions in which microorganisms are able not only to damage the rock substrate but also contribute to its strengthening depending on the specific conditions. It is desirable to take this into account when solving the conservation of rock art monument—Tomskaya Pisanitsa.

5 Conclusion

We consider the influence of bacteria on biomineralization on the unique rock art monument "Tomskaya Pisanitsa" (Western Siberia, Russia). The results of the field researches and modeling clarified the contribution of some dominated on rock with petroglyphs cyanobacteria and organotrophic bacteria in the biomineralization. The products of microbial biomineralization (calcite and oxalates) were obtained in vitro and in situ. The formation of biogenic calcite can occur under the action of cyanobacteria and organotrophic bacteria living on the rock of the Tomskaya Pisanitsa. These bacteria are able to participate in the processes of cementation of cracks on the rock with petroglyphs. The results of the study can be used for the conservation of rock monuments.

Acknowledgements This study was supported by the Russian Scientific Foundation (project no. 19-17-00141 "Modern Mineral Formation with the Participation of Microorganisms"). The studies were carried out using the equipment of the X-ray Diffraction Research Methods and Microscopy and Microanalysis resource centers at St. Petersburg State University.

References

Abed RM, Kharusi, SA, Schramm A, Robinson MD (2010) Bacterial diversity, pigments and nitrogen fixation of biological desert crusts from the Sultanate of Oman. FEMS Microbiol Ecol 72(3): 418-428

Biswas J, Sharma K, Harris KK, Rajput Y (2013) Biodeterioration agents: Bacterial and fungal diversity dwelling in or on the pre-historic rock-paints of Kabra-pahad, India. Iranian Journal of Microbiology 5(3): 309-314

Boquet E, Boronat A, Ramos-Cormenzana A (1973) Production of calcite (calcium carbonate) crystals by soil bacteria is a general phenomenon. Nature 246 (5434): 527-529

Bravo D, Braissant O, Cailleau G, Verrecchia E, Junier P (2015) Isolation and characterization of oxalotrophic bacteria from tropical soils. Arch Microbiol 197:65–77 DOI https://doi.org/10.1007/s00203-014-1055-2

Cardellicchio F, Bufo SA, Mang SM, Camele I, Salvi AM, Scrano L (2023)The bio-patina on a hypogeum wall of the matera-sassi rupestrian church "San Pietro Barisano" before and after Treatment with Glycoalkaloids. Molecules 28 (330) https://doi.org/10.3390/molecules28010330

Crispim CA, Gaylarde CC (2005) Cyanobacteria and biodeterioration of cultural heritage: a review. Microbial ecology 49(1): 1-9

Del Mondo A, Pinto G, Carbone DA, Pollio A, De Natale A (2018) Biofilm architecture on different substrates of an Oculatella subterranea (Cyanobacteria) strain isolated from Pompeii archaeological site (Italy). Environ Sci Pollut Res 25: 26079-26089

Esposito A, Ahmed E, Ciccazzo S, Sikorski J, Overmann J, Holmström SJ, Brusetti L (2015) Comparison of rock varnish bacterial communities with surrounding non-varnished rock surfaces: taxon-specific analysis and morphological description. Microbial ecology 70: 741-750

Gaylarde C, Little B (2022) Biodeterioration of stone and metal — Fundamental microbial cycling processes with spatial and temporal scale differences. Sci Total Environ 823: 153193

González-Gómez WS, Quintana P, Gómez-Cornelio S, García-Solis C, Sierra-Fernandez A, Ortega-Morales O, De la Rosa-García SC (2018) Calcium oxalates in biofilms on limestone walls of Maya buildings in Chichén Itzá, Mexico. Environ Earth Sci 77: 1-12

Hall TA (1999) BioEdit: a user-friendly biological sequence alignment editor and analysis program for Windows 95/98/NT. Nucleic Acids Symposium Series 41: 95-98.

Hauer T, Mühlsteinová R, Bohunická M, Kaštovský J, Mareš J (2015) Diversity of cyanobacteria on rock surfaces. Biodivers Conserv 24: 759-779

Kovtun IV (2021) Rock painting of Kuzbass. Research. Researchers. Ideas. Vektor-Print, Kemerovo (in Russian)

Lian B, Chen Y, Zhu L, Yang R (2008) Effect of microbial weathering on carbonate rocks. Earth Sci Front 15(6): 90-99

Mergelov N, Mueller CW, Prater I, Shorkunov I, Dolgikh A, Zazovskaya E, Shishkov V, Krupskaya V, Abrosimov K, Cherkinsky A, Goryachkin S (2018) Alteration of rocks by endolithic organisms is one of the pathways for the beginning of soils on Earth. Sci Rep 8: 3367

Mishra S (2020) Cyanobacterial imprints in diversity and phylogeny. In: Singh PK, Kumar A, Singh VK, Shrivastava CB (eds) Advances in cyanobacterial biology. Academic Press

Neumann JT, Black JL, Hœrlé S, Smith BW, Watkins R, Lagos M, Geisler T (2022) Artificial weathering of rock types bearing petroglyphs from Murujuga, Western Australia. Herit Sci 10(1): 77

Nir I, Barak H, Kramarsky-Winter E, Kushmaro A (2019) Seasonal diversity of the bacterial communities associated with petroglyphs sites from the Negev Desert, Israel. Ann Microbiol 69: 1079-1086

Nir I, Barak H, Kramarsky, Winter E, Kushmaro A, de Los Rios A (2022) Microscopic and biomolecular complementary approaches to characterize bioweathering processes at petroglyph sites from the Negev Desert, Israel. Environ Microbiol 24(2): 967-980

Nübel U, Garcia-Pichel F, Muyzer G (1997) PCR primers to amplify 16S rRNA genes from cyanobacteria. Appl Environ Microbiol 63(8): 3327-3332

Perera I, Subashchandrabose SR, Venkateswarlu K, Naidu R, Megharaj M (2018) Consortia of cyanobacteria/microalgae and bacteria in desert soils: an underexplored microbiota. Appl Microbiol Biotechnol 102: 7351-7363

Rabbachin L, Piñar G, Nir I, Kushmaro A, Pavan MJ, Eitenberger E, Sterflinger K (2022) A Multi-Analytical Approach to Infer Mineral–Microbial Interactions Applied to Petroglyph Sites in the Negev Desert of Israel. Appl Sci 12(14): 6936

Rusakov A, Kuz'mina M, Frank-Kamenetskaya O (2021) Biofilm Medium Chemistry and Calcium Oxalate Morphogenesis. Molecules 26(16): 5030 https://doi.org/https://doi.org/10.3390/molecules26165030

Sazanova KV, Vlasov DY, Zelenskaya MS, Panova EG, Rodina OA, Miklashevich EA (2022) Lithobiotic Communities on the Surface of Rock Art Monuments in the Minusinsk Basin (South Siberia): Conditions for Formation and Biomineral Interactions. Contemp Probl Ecol 15(3): 222-235

Sazanova KV, Frank-Kamenetskaya OV, Vlasov DY, Zelenskaya MS, Vlasov AD, Rusakov AV, Petrova MA (2020) Carbonate and oxalate crystallization by interaction of calcite marble with bacillus subtilis and bacillus subtilis–aspergillus niger association. Crystals 10 (756)

Smith BW, Black JL, Hoerle S, Ferland MA, Diffey SM, Neumann JT, Geisler T (2022) The impact of industrial pollution on the rock art of Murujuga, Western Australia. J Aust Rock Art Res Assoc, 39(1): 3-14

Vlasov DYu, Frank-Kamenetskaya OV, Manurtdinova VV, Zelenskaya MS (2019) Decay of the Monuments. In: Frank-Kamenetskaya OV, Vlasov DYu, Rytikova VV (eds) The Effect of the environment on saint petersburg's cultural heritage. geoheritage, geoparks and geotourism. Springer, Switzerland

Vlasov DYu, Frank-Kamenetskaya OV, Zelenskaya MS, Sazanova KV, Rusakov AV, Izatulina AR (2020) The use of Aspergillus niger in modeling of modern mineral formation in lithobiotic systems. In: Eva Baughan (ed) Aspergillus niger: Pathogenicity, cultivation and uses. Nova Science Publishers, New York

Vlasov DYu, Zelenskaya MS, Sazanova KV, Schigorets SB, Izatulina AR, Rodina OA, Stepanchikova IS, Vlasov AD, Polyanskaya EI, Davydov DA, Miklashevich EA, Pavlova OA, Frank-Kamenetskaya OV (2023) Diversity and Significance of Lithobiotic Communities at the Tomskaya Pisanitsa Rock Art Site. Contemp Probl Ecol 16(2): 173-188

Wierzchos J, de los Ríos, A, Ascaso C (2012) Microorganisms in desert rocks: the edge of life on Earth. Int Microbiol 15(4): 173-183

Wilmotte A, Van der Auwera G, De Wachter R (1993) Structure of the 16S ribosomal RNA of the thermophilic cyanobacterium Chlorogloeopsis HTF ('Mastigocladus laminosus HTF') strain PCC7518, and phylogenetic analysis. FEBS letters, 317(1-2): 96-100

Wilmotte A (1994) Molecular evolution and taxonomy of the cyanobacteria. In: Bryant DA (Ed) The molecular biology of cyanobacteria. Kluwer, Dordrecht, The Netherlands

Zammit G (2019) Phototrophic biofilm communities and adaptation to growth on ancient archaeological surfaces. Ann. Microbiol 69: 1047-1058

Zhang Y, Su M, Wu F, Gu J-D, Li J, He D, Guo Q, Cui H, Zhang Q, Feng H (2023) Diversity and composition of culturable microorganisms and their biodeterioration potentials in the sandstone of Beishiku temple, China. Microorganisms 11(429) https://doi.org/10.3390/microorganisms11020429

Surface Properties of Carbonate Speleothems in Karst Caves Changing Under Biofilms

Oxana A. Sofinskaya, Oleg Y. Andrushkevich, Bulat M. Galiullin, Nataliya E. Gogoleva, Nurislam M. Shaikhutdinov, Eduard A. Korolev, Fedor A. Mouraviev, and Rustem M. Usmanov

Abstract Some surface properties of carbonate speleothems from Kirillov's, Pionerskaya and Yaschik Pandory caves in the Republic of Khakasia, as well as Yuryevskaya Cave in the Republic of Tatarstan were investigated. The types of the speleothem samples such as crusts, drips, corallites, and moon milk were studied. All the samples were collected in cave aphotic zones at a wall temperature not higher than + 10 °C. Differently polished marble onyx, gypsum and glass plates were taken as reference surfaces. The surfaces were processed by polishing, heating, etching chemicals, and adding R2A modified growth media. These modes simulated experimentally common natural processes in the system of "calcium carbonate–chemolithotrophic biofilm". The speleothem samples under the cave microbial community, which can exist successfully in the upper soil layer, were considered against the background of the reference mineral surfaces. The captive bubble method, SEM, XRD, EDX, as well as DTG analyses were carried out to determine wettability, roughness, total organic matter content, and elemental and mineral compositions of the samples. The metagenome of the microbial community was estimated using 16S rRNA sequence analysis. The influence of fresh and long-lived biofilms on the carbonate surface properties is assessed. The assumption that biofilm dynamics affects the carbonate surface properties and its toughness is substantiated. Our work hypothesized that the part of organic matter can enter the gaps in growing carbonate crystals, then, is sealed with a new mineral phase, and later, is assimilated by heterotrophic and organotrophic organisms.

Keywords Wettability · Organic matter · Surface roughness · Biocement toughness · Biofilm dynamics · Extremophiles · Chemolithotrophs · Biomorphs · Pseudomorphoses

---

O. A. Sofinskaya (✉) · O. Y. Andrushkevich · B. M. Galiullin · N. E. Gogoleva · N. M. Shaikhutdinov · E. A. Korolev · F. A. Mouraviev · R. M. Usmanov
Kazan (Volga Region) Federal University, Kazan, Russia
e-mail: ushik2001@mail.ru

O. V. Frank-Kamenetskaya et al. (eds.), *Biogenic—Abiogenic Interactions in Natural and Anthropogenic Systems 2022*, Springer Proceedings in Earth and Environmental Sciences, https://doi.org/10.1007/978-3-031-40470-2_29

1 Introduction

Carbonate formations are promising objects for the design of biocement structures and protective (transport) capsules for microflora, as well as for the refinement of the geologic record (Melim et al. 2016). The surface properties of carbonate formations are closely related to biofilms developing on them (Mouraviev et al. 2006; Banks et al. 2010; Tugarova 2021). Biofilms have a life cycle, at each stage of which they can affect the covered surface differently either destroying or preserving it (Andryukov et al. 2020; Zorina et al. 2019). The interaction between autochthonous or inoculated biofilms and the carbonate leads to the deposition or destruction of minerals and isomorphic replacement in crystal lattices, which should be taken into account when using cements and other carbonate-based materials in geotechnologies (Gray and Engel 2013; Banks et al. 2010; Glazovskaya and Dobrovolskaya 1984; Guvensen et al. 2013; Pronk et al. 2017; Wiseschart et al. 2019; Leonova et al. 2015). Hereby, both lithoautotrophic and organoheterophic modes of microbial metabolism can destroy carbonates. In the first case, this occurs due to biogenic acids production, and in the second case, due to carbon dioxide release (Gray and Engel 2013). Carbonate precipitation requires microbial communities to shift the pH of their environment toward the alkaline side, for example, via alkalis releasing or sulfate reduction (Leonova et al. 2015). Calcium carbonate precipitation passes four stages—amorphous, vaterite, aragonite, and calcite, among which calcite is the only thermodynamically stable phase, while others are metastable. However, the presence of organic matter in the environment is proved to be an inhibitor on the metastable phase transformation and thus, makes them possible to coexist (Myszka et al. 2019).

In our work, we have simulated experimentally some of the mechanical and chemical effects that can be induced by chemolithotrophic biofilms on the carbonate. From the extended version of the DLVO theory, fineness, roughness, hydration, and the type of contact between particles are known to be the factors significant for particle adhesion strength (Andryukov et al. 2020). Carbonate biofilm can affect each of them. Both the abilities to disperse and to stick particles together are peculiar to microbial communities (Gray and Engel 2013; Banks et al. 2010; Glazovskaya and Dobrovolskaya 1984; Pronk et al. 2017). Particles disintegrate via the dissolution of the cement binding them in aggregates. Particle aggregation begins either by a gel forming in the extracellular polymer substances (EPS) or by a cement precipitating in gaps between particles. Surface roughness is similarly formed by biofilms, however, herewith the latter affect only the aggregates' surfaces but not intra-aggregate bonds. The hydration of biofilm-coated particles most strongly depends on the stage of biofilm development: a fresh biofilm can attach to a particle only if the latter has hydrophobic contact. It produces the hydrophobic substances that promote the strongest attachment to the surface (Zorina et al. 2019; Andryukov et al. 2020). Nevertheless, for sufficient nutrition delivered as an aqueous phase, the biofilm is forced to produce hydrophilic compounds. Thus, fluctuations in the hydrophilic-hydrophobic properties of the substrate surface occur. This leads to the alternation of strengthening and weakening contacts between particles. In the presence of a hydrophilic surface, particles form coagulates; and vice versa, in the presence of a hydrophobic one, particles can reversibly aggregate, as well as form direct point contacts, after which the cementation or crystallization contacts are likely.

Based on the dynamic behavior of biofilms concerning the coated mineral phase and considered the previous studies (Melim et al. 2016) it can be hypothesized that a part of the organic matter enters the gaps between growing crystals through selective adsorption and/or capillary condensation. Then, these substances can be sealed with a new mineral phase, and thus preserved. Further, under the acids episodically released by the biofilm and etching a mineral phase, the organic matter is reactivated and can be consumed by heterotrophic and organotrophic organisms. The purpose of this work was to check this hypothesis.

2 Sites and Sampling

Carbonate speleothem samples were purposefully selected from totally different geologic zones (Fig. 1). Karst genesis, the contact of clayey deposits with carbonate rocks, 3–sevenfold excess of the atmospheric partial pressure of CO_2, aphotic zones, and the wall temperatures not higher than + 10 °C were the common features of the sampling sites. The samples were taken from the walls and cornices of caves in such a way as to exclude anthropogenic and zoogenic pollution as much as possible.

Yurjevskaya Cave (Fig. 1a) is a horizontal corridor system in Permian sedimentary gypsum-dolomite rocks with a total length of about 1,5 km. It is located on the right

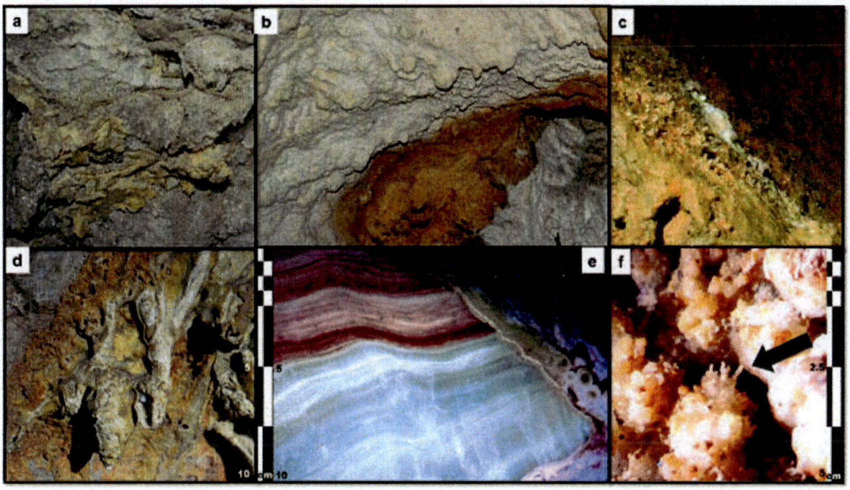

Fig. 1 Some speleothems sampled from caves: **a** Yurjevskaya, crusts; **b** Pionerskaya, dry moon milk; **c** Pionerskaya, fresh moon milk; **d** Yaschik Pandory, drape; **e** Pionerskaya, marble onyx; **f** Yaschik Pandory, corallite with helictites

bank of the Volga River in the Republic of Tatarstan. Crust-type speleothems were selected from the Hall of Organ Pipes located 190 m far from the entrance. This hall is less attended by tourists compared to the neighboring ones. There was a wall temperature of + 4,0 °C, a air temperature of + 13,0 °C, and a relative air humidity of 67%.

Pionerskaya Cave (Fig. 1b, c, e) is a slightly inclined horizontal gallery with a maximum depth of 15 m and a total length of 86 m in Riphean dolomites (Nomokonov 1969) with calcite mines located on the left bank of the Bely Iyus River (the Ob River basin) in the Republic of Khakasia. Sampling points were selected at a distance of 20, 33 and 40 m far from the entrance, which ensured complete coverage of the cave along its length. The speleothem samples were «moon milk» at the different stages of drying, spherulites, and stalactites. There was an average wall temperature of + 0.7 °C. The air temperature was from + 5,2 °C to + 8,0 °C rising from the entrance to the end. A relative air humidity was from 76 to 85%.

Kirillov's Cave is an inclined horizontal gallery with a maximum depth of 36 m and a total length of 320 m in massive Cambrian limestones (Nomokonov 1969) located inside a hill that is 400 m above the right bank of the Bely Iyus River. Somewhere there is an ice floor. Sampling points were selected at a distance of 13, 39 and 68 m far from the entrance of the lower gallery and covered the cave wing, which is the richest in speleothems. The samples collected were small stalactites. A wall temperature increased from −6,0 °C to −0.8 °C from the entrance to the end as the gallery rose; an average air temperature was + 5,9 °C, and a relative air humidity was from 75 to 77%.

Yaschik Pandory Cave (Fig. 1d, f) is a system of vertical and inclined wells with a total depth of 180 m and a total length of about 11 km in the Cambrian rocks, which are represented by argillaceous limestones, sandstones, siltstones, gravelites up to a depth of 40 m, and there appear marls, diabases, porphyrites in the deeper layers (Nomokonov 1969). In some areas there the carbonate rock transition into marble occurs, probably due to the contact metamorphism in the zone of geologic faults. The cave is situated on the left bank of the Bely Iyus River, and is connected hydraulically with it on the lowermost floor. The samples were collected from depths of 0, 100, and 180 m below the entrance. There were spherulites, small stalactites, and corallites. The wall and air temperatures decreased from + 7 °C to + 4 °C and from + 18 °C to + 4 °C, respectively, and a relative air humidity increased from 68 to 90% as descending to the bottom.

Experimental impact on the samples. Freshly sampled speleothems were placed into strong aluminum foils at an initial humidity, a temperature of + 12(±2) °C, and total darkness to preserve the active autochthonous biofilm. These samples participated in our experiment in the following modes: untreated slices, variously polished plates, ones powdered to sizes < 250 microns, heat-treated powders at 105 and 525 °C, as well as chemically treated powders.

The mechanical treatment of a few samples was carried out in order to examine the effect of roughness and surface layer removal on the properties of natural carbonate formations. Marble onyx plates were processed in the following ways:

- *zero polishing*–untreated slices;
- *polishing No. 1*–with a grinding disc grit of 75 microns and the creation of surface roughness at a level of 20–25 microns with oriented hatching;
- *polishing No. 2*–with a grinding disk grit of 20 microns and the creation of surface roughness at a level of 8–12 microns with oriented hatching;
- *polishing No. 3*–with free abrasive SiC M 14 and a uniformly rough surface at a level of 5–8 microns;
- *polishing No. 4*–with free abrasive SiC M 10 and a uniformly rough surface at a level of 2–4 microns.

The samples were heated in order to dehydrate the surface (at 105 °C) and accelerate the removal of chemically bound water and organic matter (at 525 °C).

The chemical treatment of the samples was aimed at the release of the hypothetical organic matter from mineral chambers in two ways: by mineral etching and by the extraction of the organic matter with a solvent. The minerals were etched using oxalic, citric, acetic, hydrochloric, and sulfuric acids, commonly secreted by lithotrophic microorganisms (Glazovskaya and Dobrovolskaya 1984). The organic matter was extracted from the samples into NaOH and ethanol solutions (Golovanova 2022).

The fresh chemolithotrophic biofilm was stimulated by the modified R2-based growth medium intended for psychrophilic oligotrophs (Wiseschart et al. 2019; D'Angeli et al. 2017). The difference from the base R2 medium layed in the addition of an alkaline humus extract and a triple compared to the base medium $CaCO_3$ content. The increased content of $CaCO_3$ was applied to establish the ionic equilibrium between the studied surfaces and the washing solution, as well as to selectively stimulate the organisms adapted to carbonate metabolism. The humus extract was applied as a pH buffer for its natural level of 7.68–8.02, as well as a model for the contact of carbonate formations with the upper soil layer. Thus, the stimulation was aimed at those members of the cave community that can successfully exist in surface soil conditions.

3 Methods

The main surface investigation methods included contact angle measurement, SEM, EDX, TG-DSC, and XRD analyses. The contact angle is an indicator of both hydrophilic-hydrophobic properties and surface roughness. It was determined using the captive bubble method in the author's modification for clay fraction powder specimens (Sofinskaya et al. 2022). The powders were stuck onto flat glass using adhesive tape at a pressure of 30 kPa, after which they were placed into an atmosphere with a relative air humidity of >96% for several minutes to saturate the pores with capillary-condensed moisture. Next, the specimens were immersed into a thermostatic bath of an optical tensiometer with deionized water, and an air bubble was placed and photographed on the powder/water interface. The images obtained were adjusted using the ImageJ program and then contact angles were calculated by the

Contact Angle Option. The representation of the measurements consisted of 30–100 per a specimen depending on its surface stability.

Most of the samples were examined with a FEI XL-30ESEM scanning electron microscope. The SEM was carried out at a low vacuum mode under an accelerating voltage of 20 keV. Elemental analysis was performed using an EDAX energy dispersive spectrometer operating in combination with the SEM, based on which the mineral composition was calculated from the ratio of the weight of elements on the surface.

X-ray diffraction analysis was carried out with a D2 Phaser diffractometer (Bruker, Germany) in order to detail the mineral composition. Analysis mode was an X-ray tube voltage of 30 kV, a current of 30 mA, and the scanning step of 0.02° at a speed of 1 deg/min. The range of scanning angles in the Bragg–Brentano geometry was from 3 to 40°. Standard powder preparations were used. The qualitative and quantitative mineral composition was determined using the DIFFRAC.EVA and TOPAS software.

Thermal analysis was carried out with an STA 449 JupiterF3 device to determine the boiling point of the structural components of the samples. The burning interval was from 30 to 1000 °C, the heating step was 10 deg/min. with continuous air purging. There the reference Al_2O_3, which did not give a thermal effect within the temperature range from 20 to 1200°, was applied.

Taxonomic profiling of the bacterial communities was performed using S-D-Bact-0341-b-S-17 and S-D-Bact-0785-a-A-21 primers (Klindworth et al. 2013) targeting V3 and V4 regions of the 16S rRNA gene. The libraries were sequenced in the "Regulatory genomics" lab of Kazan Federal University (Kazan, Russia) on the MiSeq (Illumina, USA) platform using the 2 × 300 bp paired-end MiSeq Reagent Kit v3. To process the sequencing data, we used the dada2 microbiome data denoising pipeline (version 1.14.0). The dada2 pipeline is based on sequencing error correction algorithms and generates exact amplicon sequence variants (ASVs). The dada2 pipeline implements the Naive Bayes classifier method for taxonomy assignment (Callahan et al. 2016). This classifier compares sequence variants with a training set of classified sequences, the Silva (version 138) rRNA gene database (silva_nr99_v138.1_train_set.fa), which is currently considered the most comprehensive 16S rRNA database of all microorganisms. To visualize and filter the processed data, a package written in the R programming language - phyloseq (McMurdie and Holmes 2013) was used. Data filtering was carried out to remove artifacts generated during the analysis. A set of samples passed through a bioinformatics pipeline with the same parameters.

4 Results

Wetting contact angle (CA). Finely polished marble onyx (*polishing No. 2, 3, 4*) is a hydrophilic matter with a fairly small standard deviation (StDev) of CA fluctuating within the observed range of 28° to 38° (Table 1, see *polishing No. 3*). It is noteworthy that the finest *polishing No. 4* caused a StDev rising (i.e., surface heterogeneity) due to an increase in the range of CA observed at both ends of their distribution, which is probably followed the artifacts introduced after polishing at dimension hatching less than 5 μm.

The *polishing No. 2* and *No. 1* had a larger range of CA variation than finely processed ones, i.e., here macro-roughness was set by irregularities larger than 8 μm. The coarse polishing led to the splitting of one mode of the quasi-Gaussian CA distribution into two, resulting in a bimodal distribution. Herewith, the low-angle mode remained close to surfaces after fine polishing, while the high-angle mode exceeded the maximum CA observed on these surfaces. The *zero polishing* sample also had two CA distribution modes, of which the low-angle one was close to the high-angle mode for a roughly polished surface (*polishing No. 1*), and the high-angle one belonged to the hydrophobic range. That is, hydrophobic properties were set by heterogeneities with a size of more than 25 μm corresponded to the order of biofilm thickness (Wagner and Horn 2017)

The speleothem powders had practically the same CA distribution parameters as the *zero polishing* sample. Nevertheless, after dehydration at 105 °C the hydrophobic heterogeneities and bimodal CA distribution disappeared and the distribution parameters approached surfaces with *polishing No. 1, 2, 3*. Heat treatment at 525 °C led to a strong increase in surface heterogeneity (according to an increase in StDev), but unless the CA distribution mode splitting.

The polished plates treated with the R2 medium demonstrated a decrease in the hydrophilicity and homogeneity of the surfaces as the StDev had risen and the second mode of CA distribution had appeared. At the same time, the first distribution mode

Table 1 Distribution parameters for wetting contact angle on reference and model surfaces (degrees)

	St.dev.est.*	Minimum of CA observed	Mode 1*	Mode 2*	Maximum of CA observed
	Initial/under R2, 1 month / under R2, 2 months				
Glass	6/4	25/23	43/32	–	54/46
Onyx polishing No 3	3/7/7	28/24/19	32/27/27	–/40/46	38/53/52
Onyx polishing No 4	4	23	32	–	42
Onyx polishing No 2	6	23	32	–	51
Onyx polishing No 1	12	22	35	54	67
Onyx zero polishing	15	33	56	102	120
Gypsum smooth slice	7/4	27/30	40/45	–/72	57/81

* estimated by Tikhonov and von Mises

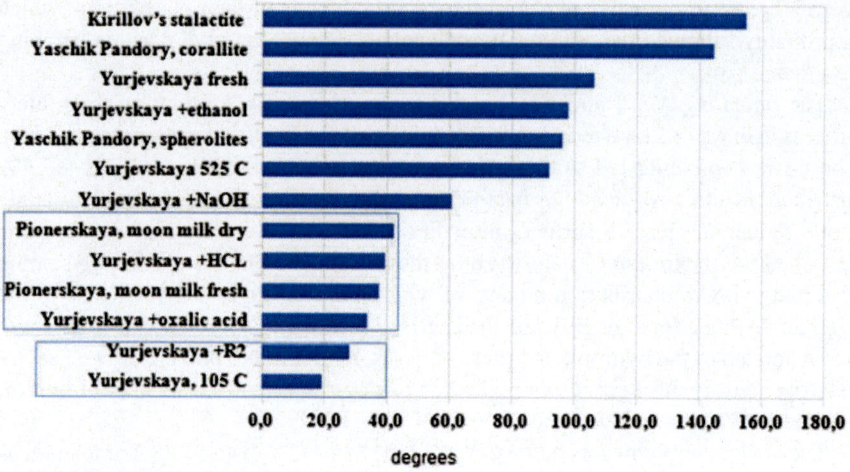

Fig. 2 Euclidean distance between the reference cluster and the samples calculated for wetting contact angle characteristics (min., max., mode 1, mode 2, St.dev.)

and the minimum CA observed shifted toward a more hydrophilic range. This observation is also valid for a gypsum crystal surface. On the contrary, the surface heterogeneity of the calcite speleothem powders under the R2 medium declined, which was indicated by a decrease in the StDev and the disappearance of the hydrophobic CA distribution mode. A similar trend was demonstrated for the silicate glass plate in the homogeneity of the CA distribution and the hydrophilicity of the surface risen under the R2-stimulated biofilm.

CA data was ranged compared to a reference class consisting of flat smooth surface specimens: onyx plates of the *polishing No. 2*, *3*, *4* and a glass plate treated with the R2 medium (Fig. 2). The reference class showed CA fluctuations in the range of 24° to 44° around an average direction of 32° (by Tikhonov and von Mises) with a StDev estimated as a class average of 4°.

Next, cluster analysis was performed using the K-means algorithm for the independent parameters of the empirical CA distribution as a factor space. This resulted in the CA dataset being combined into 3 subgroups with several members and 4 single values. The first combined subgroup included marble onyx plates after treatment with R2 medium, *polishing No. 1*, specimens of Yuryevskaya Cave dehydrated at 105 °C, treated with oxalic and hydrochloric acids, R2 medium, as well as fresh and dry moon milk from Pionerskaya Cave (Fig. 1c and b). The common features of this subgroup included the unimodal CA distribution, the mode of which was higher than in the reference group, hydrophilicity, and roughness. The surface of the specimens from Yuryevskaya Cave treated with NaOH was distinguished by the transition through the threshold of hydrophobicity. The next subgroup included the specimens from Yuryevskaya Cave treated with ethanol and burning at 525 °C, as well as spherulite formations from the entrance grotto of Yaschik Pandory Cave. That subgroup was characterized by an average CA direction of 58° and CA maximum lying in the hydrophobic area. The last combined subgroup included an untreated powder specimen from Yuryevskaya Cave and a *zero polishing* slice of marble onyx. There was the CA distribution split into two modes, a hydrophilic one about 45° and a hydrophobic one about 100° in this group. The differences between the rest powder specimens belonging to Yaschik Pandory Cave (corallites with helictites, Fig. 1f) and Kirillov's Cave (stalactite) did not allow them to be combined into any group, however, all of them were characterized by the low-angle CA mode lied in the hydrophilic area whereas the high-angle CA mode had a superhydrophobic range and individual CA reached 140°–150°.

SEM, EDX, XRD and DTG analyses. In most cases, the images show a film gained with frequent biomorphs and carbon compared to calcite alone (Fig. 3). On crusts filling cracks this film had some differences between the inner smoothed surface adjacent to the host rock (dolomite) and the outer subaerial surface regarding the ratio of elements (Table 2). There was less carbon on the inner side.

When the crusts developed on the bat bones (Fig. 4a) they had the same C/Ca ratio as the outer surface of the crack filler, but a higher Ca/O ratio. A peculiar tubular structure constituted with crystals elongated along one axis allows us to assume confidently aragonite pseudomorphs formed on the initial bone substances. Such an aragonite structure may include the "rose" grown on the gypsum surface (Fig. 4b) structures of micro-rods with notches of moon milk (Fig. 4d), and helictite needles in corallites (Fig. 4c).

Figure 5d shows two synchronously recorded thermal analysis curves, namely, TG–thermogravimetric curve and DSC–differential scanning calorimetry curve. The first indicates the mass loss of the moon milk sample, and the latter shows the thermal effects that appear during its burning. The most noticeable changes in

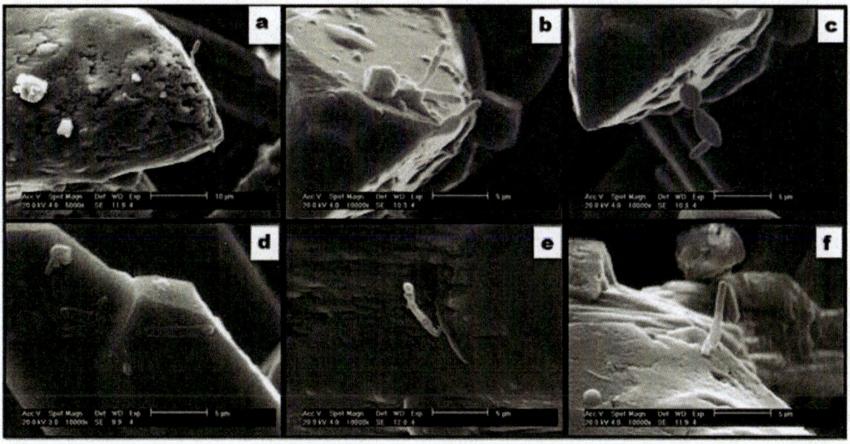

Fig. 3 Biomorphs on calcite speleothems, Yurjevskaya cave: **a**–a continuous biofilm with prominent biomorphs; **b**, **c**, **e**, **f**–cells penetrated the biofilm surface; **d**–monads on the speleothem surface

Table 2 Relative content of elements on the surfaces of speleothems and host rocks

Element, %wt	Speleothem crusts				Other speleothems			Gypsum			Dolomite	
	Inner	Extern	Bat bone	+ R2	Lateral	Apex	Apex	"rose"	Initial	+ R2	Initial	+ R2
C	16	18	22	20	12	12	18	–	–	45	13	18
O	42	50	39	53	38	54	58	60	64	41	48	50
Mg	0,4	0,5	0,3	6,0			0,2			0,5	8,9	0,2
Al	0,3	0,6	0,7	0,8	0,7	0,3		0,3	0,2	0,8	1,1	0,2
Si	0,4	0,5	0,3	1,2	0,5	0,3	0,2	0,4	0,1	0,7	5,3	0,3
P				0,3	0,7	0,5	0,4			4,1		1,5
S	0,7	0,4	0,4	0,3				0,9	16,0	0,9	0,3	0,3
K				0,3						0,8	0,4	
Ca	41	30	37	17	47	32	23	38	19	5	21	30
Mn								0,4			0,2	
Fe		0,3		0,3	0,4	0,3	0,4	0,6	0,5		0,6	
C/Ca	0,4	0,6	0,6	1,2	0,3	0,4	0,8	–	–	9,2	0,6	0,6
Ca/O	1,0	0,6	0,9	0,3	1,2	0,6	0,4	0,6	0,3	0,1	0,4	0,6
Ca/P	–	–	–	54,6	65,4	63,1	58,9	–	–	1,2	–	20,7

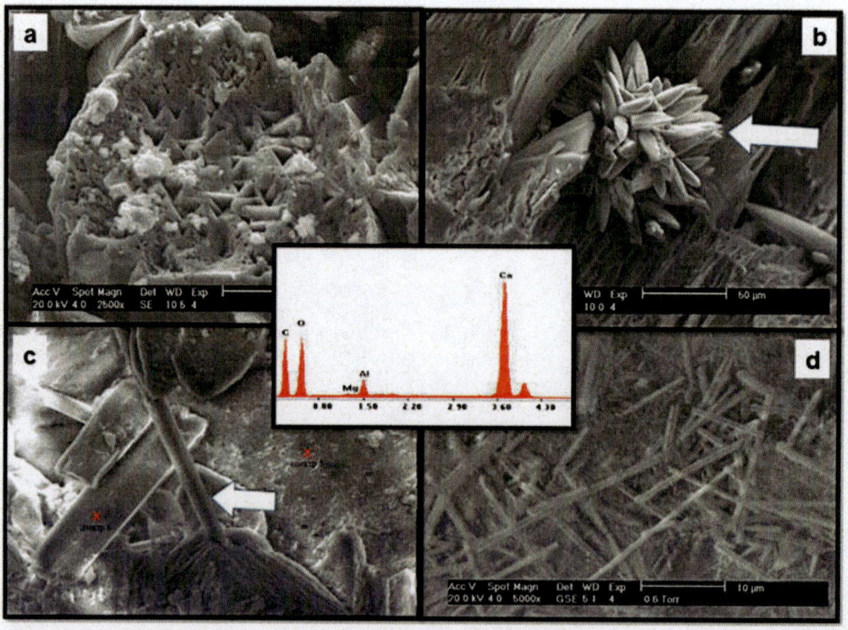

Fig. 4 Pseudomorphoses after aragonite: **a** bat bone from Yurjevskaya cave; **b** "rose" form on gypsum surface from Yurjevskaya cave; **c** micro-rod in corallite from Yaschik Pandory cave; **d** micro-rods in moon milk from Pionerskaya cave

the sample are noted in a burning temperature range of 180–500 °C. In this range, there two exo-effects caused by the thermal dissociation of organic matter are highlighted. These two exo-effects manifested at low temperatures and higher temperatures correspond to the dissociation of low and high molecular weight organic matter, respectively. The total weight loss in this temperature range is 3.5%. The next clear endo-effect is observed at a temperature of 680–780 °C. Its presence is due to the thermal dissociation of calcite according to the reaction.

$$CaCO_3 \rightarrow CaO + CO_2 \uparrow$$

Then, one can suppose the formation of calcite pseudomorphs after aragonite according to XRD and TG-DSC data, since there is present the calcite predominant phase (91–98%) with the admixture of dolomite, celestite, and quartz in these tubes (Fig. 5).

Based on the results of EDX analysis, the lateral part of the tubular crystals reproducibly differed in elemental composition from the apical part and approached organic compounds (Table 2). The presence of phosphorus, which is not common for the speleothems, also indicates organic inclusions in the crystals.

The treatment of calcite speleothems with the R2 medium caused the fixation of Mg, P, and K elements on the surfaces and a threefold increase in the C/Ca ratio (Table 2). The appearance of new threads and films over caverns was visually observed (Fig. 6b, c). The gypsum surface after such treatment was strongly colonized by hyphae-forming microorganisms (Fig. 6f). Herewith high C/Ca ratio, and precipitation of a new phase containing Mg, P, and K clearly increasing the surface roughness were observed (Fig. 6d, g, h). The latter is consistent with the previous investigations that found the phosphate groups released by biofilms to improve their adhesion to the solid (Andryukov et al. 2020). On the of dolomite surface, the R2 environment caused the appearance of a calcite crust (Table 2).

4.1 Taxonomic Diversity

The clone libraries belonged to 6 major phyla that comprised 10 and 17 OTUs in the initial speleothems and in the treated with R2 media one, respectively. Proteobacteria dominated all clone libraries, ranging from 57 to 81% of the clones depending on the option. About one sixth of the OTUs comprised clones from both options. Phyla Actinobacteriota and Nitrospirota being quite abundant in the initial speleothems were suppressed by the R2 medium. Stimulation with the R2 medium was manifested in the phyla Bacteroidota, Firmicutes, Bdellovibrionota (Table 3).

Genera Kribella, Gaiella, and Brevundimonas found in the initial speleothems can be assigned to the typical extremophiles, and genera Pseudoarthrobacter and Nitrospira are obligate chemolithotrophs. R2 media stimulated the most genus Brevundimonas, whose representatives are aerobic consumers of organic matter, inhabit aquatic environments, and have a wide range of resistance to desiccation.

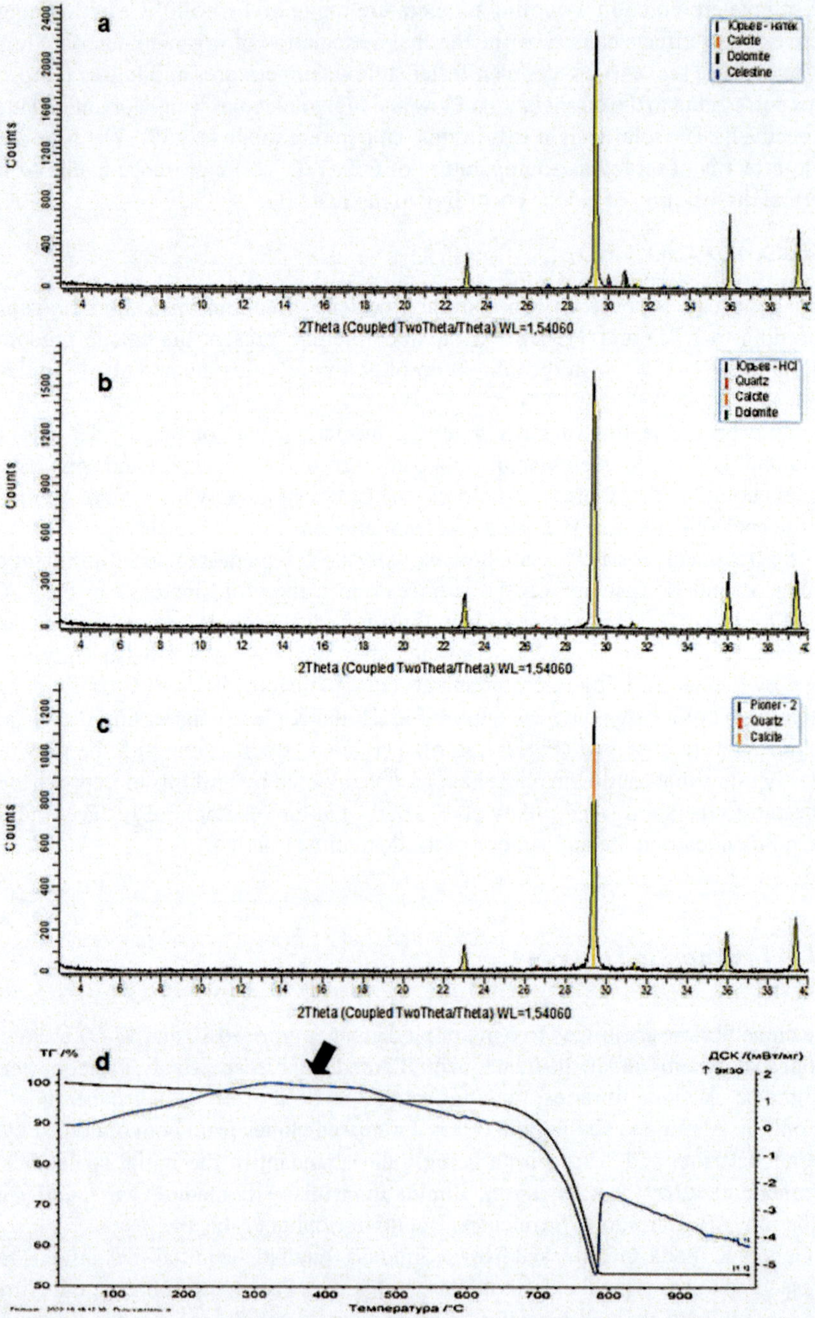

Fig. 5 The analyses of speleothems: **a** XRD of crusts from Yurjevskaya cave; **b** XRD of crusts etched with HCl from Yurjevskaya cave; **c** XRD of moon milk from Pionerskaya cave; **d** DTG and DSC of moon milk from Pionerskaya cave

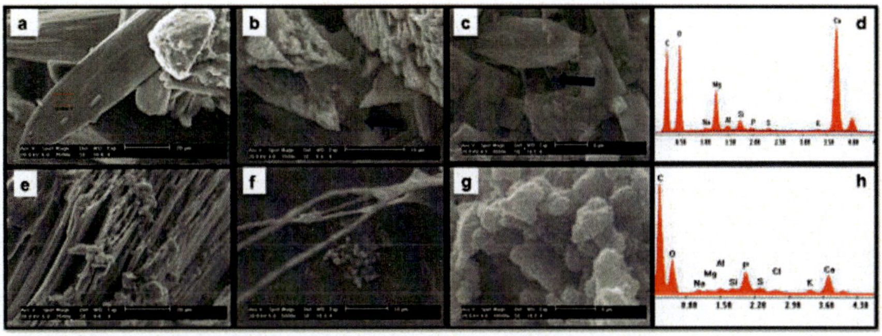

Fig. 6 The surfaces of calcite crusts (**a-d**) and gypsum (**e–h**) from Yurjevskaya Cave before (**a** and **e**) and after the stimulation of biofilm with R2 media: **b** films; **c** threads; **d** EDX analysis of the crusts after the stimulation; **f** hyphae on gypsum slice after the stimulation; **g** precipitants on gypsum slice after the stimulation; **h** EDX analysis of gypsum slice after the stimulation

Table 3 The abundance of bacterial phylla in the crusts of Yurjevskaya cave, %

Option	Initial	+ R2 modified
Proteobacteria	57–77	65–81
Actinobacteriota	16–29	1–1
Nitrospirota	0–4	0
Bacteroidota	0	4–5
Firmicutes	0	2–3
Bdellovibrionota	0	6–7
Total sequence identity	92	96

5 Discussion

The CA distributions demonstrate that the speleothem intact surfaces are covered with a partly hydrophobic film. This film was spread as hydrophobic spots with a size of at least 25 μm. Based on infrequent clear individual biomorphs and a high C/Ca ratio, we can confidently attribute the main part of the film to the EPS.

The ratio of the biofilm elements is approximately the same both on the untreated speleothems and speleothems treated with the R2 medium. The difference is manifested only as more clearly defined threads and films covered cavities in crystal surfaces in the latter case (Fig. 6). Samples of both fresh and hardened moon milk were exceptions not forming hydrophobic spots, but they included more than 3% organic matter in their mass too (Fig. 5). Simultaneously, the experiment showed that the development of a biofilm along with hydrophobization can hydrophilize and smooth the surface, i.e. our observations do not contradict each other.

Based on the comparison of surface microanalysis data, one can see that carbon compounds exceed the classical content for calcium carbonate alone, phosphates sometimes appear, and manganese is almost absent, which in total is characteristic of biogenic deposits. The proportions of elements are closest to calcium oxalate in all samples, but sometimes one can assume the presence of other organic compounds (malates, succinates, etc.).

According to the data presented, the following path of organic matter into the crystalline matrix can be assumed. Biofilms inhabit fresh carbonate formations (along the bone, cave milk, feeding channels of corallites and helictites), which are initially represented by tubular structures of aragonite. Somewhat part of their EPS gets inside the tubes (but not cells because of the tube diameter of fewer than 2 microns), and at the same time, the solution is supersaturated with ions in the tubes (due to selective sorption), which causes precipitation and filling of the inner cavities. Mixing of the organic and crystallizing phases increases the defectiveness of the crystal lattices formed and declines their energy; thus, aragonite is replaced by biochemogenic calcite. This process also occurs without the participation of organic matter (Myszka et al. 2019), but the aragonite structure preservation observed indicates its acceleration under the biofilm.

The roughness could partially reduce the wettability and cause the appearance of a fuzzy bimodality of the CA distribution, but in no case caused the transition of the surface properties through the hydrophobicity threshold by itself. Highly likely, the increase in geometric heterogeneity induced the appearance of preferential inclination angles of the landing sites for air bubbles.

A clear bimodal CA distribution occurs when phases with different surface properties form clusters and spots, due to which there are both hydrophobic and hydrophilic areas on the surface, but few areas with intermediate CA values. We observed that the bimodal distribution of CA appeared only in the following situations: either the samples were not subjected to any processing other than dispersion or the treated surface was repopulated with biofilm. Hereby, when fresh biomass developed, smooth wavy mineral surfaces (calcite and gypsum) able to easily release cations demonstrated an increase in roughness due to the precipitation of a new mineral phase and the formation of hyphae, as well as the appearance of hydrophobized spots. When cations were difficult to release (e.g., glass), the surface could be smoothed (healed) by a biofilm.

More or less, the hydrophobicity of the speleothem specimens was preserved after the treatment, decreasing in the set: ethanol, burning, and alkali. However, at the same time, their bimodality disappeared, i.e., spots of hydrophobic substances spread over the surface. Other modes of treatment formed the following sequence with increasing hydrophilicity: hydrochloric acid, oxalic acid, R2 medium, drying at a temperature of 105 °C, and polishing. Thus, the organic matter extracted from the mineral matrix partially remains as a film sorbed on the upper layer. This residual film can be removed due to the etching of minerals, evaporation, and scraping.

Genus Kribella, Gaiella, and Brevundimonas found in the original speleothems are typical extremophiles, while genera Pseudoarthrobacter and Nitrospira are obligate chemolithotrophs. The latter genera has the potential for the variable consumption of organic matter and/or CO_2, as well as participation in a nitrogen cycle. The modified R2 media contained organic compounds, therefore it stimulated chemoorganotrophic growth (e.g., Brevundimonas) that suppressed other members of the cave community, which most likely have a low chance of adapting to near-surface soil. In this study, just like in the metagenomes studied earlier, many microbial types were partly able to diverge in metabolic pathways, but most of them were heterotrophic (Turrini et al. 2020). The biofilm autotrophic basis could be the genera Nitrospira possessed the potential for variable consumption of organic matter and/or CO_2, as well as participation in a nitrogen cycle. However, the abundance of this genera is not high, therefore, its ability to provide nutrients for the heterotrophic biomass in a cave environment is doubtful. In the stimulated community, the functions of the genera Nitrospira as a nitrifier in an environment depleted in organic matter (Daims et al. 2015) were probably transferred to Pseudomonas as a heterotrophic nitrifier (Trung et al. 2019).

The hyphae-forming members of phylum Actinobacteria are most likely responsible for an increase in surface roughness. The appearance of hydrophobic groups is probably the result of the total biofilm secretions, since both their synthesis during the metabolism of the chemolithoautotrophs and the reverse process of hydrophilization caused by microbial lytics are possible.

6 Conclusion

The biological changes in minerals depended on metabolic precipitants that either smoothed or scratched their surfaces. All the fresh speleothems showed higher carbon in EDX than expected for carbonate alone. Simultaneously, the surface hydrophobicity was detected for these samples in a solid state only, which indicated the hydrophobization of speleothems under long-life biofilms. The spots of a low wettability were removed from the surfaces by the thermal treatment and washing with solvents, whereas acid etching and burning of minerals induced the appearance of such spots. These facts are interpreted as the organic matter is distributed as both spread on the surface and enclosed in mineral channels. The consumption of the enclosed organic carbon compounds by biofilms is possible, since we have reason to believe that the observed cave biofilm community is mostly heterotrophic, with Nitrospira being the only obligate autotrophic among other autochthonous organisms. Nevertheless, there is a lack of organic matter for a heterotrophic nutrition in all the samples, except fresh moon milk. We can conclude that heterotrophs consume the organic matter, which can be released and detected only after the destruction of the crystalline matrix.

Then, the storage of nutrients during the periods of their excess, their uptake and preservation during the crystallization, and the consumption after the mineral destruction by metabolites can be hypothesized. Such a cycle is possible not only in caves, but also in subsoils, as well as technogeneous carbonates. If this dynamic is confirmed, the surface properties of alike formations should be considered as the function of organic matter releasing and preserving cycles. The collection of more complete evidence for this hypothesis requires further research.

Acknowledgements This paper has been supported by the Kazan Federal University Strategic Academic Leadership Program (PRIORITY-2030).

Funding This study was funded by Russian Foundation for Basic Research, project number 20–05-00,151.

References

Andryukov BG, Romashko RV, Efimov TA, Lyapun IN, Bynina MP, Matosova EV (2020) Mechanisms of adhesive-coadhesive interaction of bacteria in the formation of a biofilm. Molekulyarnaya Genetika, Mikrobiologiya i Virusologiya (Molecular Genetics, Microbiology and Virology) 35(4):195–201 (in Russian) https://doi.org/10.17116/molgen202038041155

Banks ED, Taylor NM, Gulley J, Lubbers BR, Giarrizo JG, Bullen HA, Hoehler TM and Barton HA (2010) Bacterial calcium carbonate precipitation in cave environments: a function of calcium homeostasis. Geomicrobiology Journal 27(5):444 — 454. https://doi.org/10.1080/01490450903485136

Callahan BJ, McMurdie PJ, Rosen MJ, Han AW, Johnson AJ, Holmes SP (2016) DADA2: High-resolution sample inference from Illumina amplicon data. Nat Methods 13(7):581-583. https://doi.org/10.1038/nmeth.3869

Daims H, Lebedeva EV, Pjevac P, Han P, Herbold C, Albertsen M, Jehmlich N, Palatinszky M, Vierheilig J, Bulaev A, Kirkegaard RH, von Bergen M, Rattei T, Bendinger B, Nielsen PH, Wagner M (2015) Complete nitrification by Nitrospira bacteria. Nature. 24;528(7583):504–9. https://doi.org/10.1038/nature16461

D'Angeli IM, Serrazanetti DI, Montanari C, Vannini L, Gardini F, De Waele J (2017) Geochemistry and microbial diversity of cave waters in the gypsum karst aquifers of Emilia Romagna region, Italy. Science of the Total Environment 598:538–552. https://doi.org/10.1016/j.scitotenv.2017.03.270

Glazovskaya M A and Dobrovolskaya N G (1984) Geochemical functions of microorganisms. Moscow State University, Moscow (in Russian)

Golovanova OA (2022) Crystallogenesis in the human body. Dostoevsky Omsk State University, Omsk (in Russian)

Gray CJ and Engel AS (2013) Microbial impact on aquifer carbonate geochemistry The ISME Journal. International Society for Microbial Ecology 7:325–337. https://doi.org/10.1038/ismej.2012.105

Guvensen NC, Demir S, Ozdemir G (2013) Effects of magnesium and calcium cations on biofilm formation by Sphingomonas paucimobilis from an industrial environment. Current Opinion in Biotechnology 24(1):S68. https://doi.org/10.1016/j.copbio.2013.05.185

Klindworth A, Pruesse E, Schweer T, Peplies J, Quast C, Horn M, Glöckner FO (2013) Evaluation of general 16S ribosomal RNA gene PCR primers for classical and next-generation sequencing-based diversity studies. Nucleic acids research 41(1): e1. https://doi.org/10.1093/nar/gks808

Leonova LV, Kuz'mina LY, Ryabova AS, Simakova YS, Glavatskikh SP, Cherviatcova OY (2015) Modern nodules: mineralogical investigation and modelling experiments. Vestnik IG Komi SC UB RAS 10:45-51 https://doi.org/10.19110/2221-1381-2015-10-45-51

McMurdie PJ, Holmes S (2013) Phyloseq: An R package for reproducible interactive analysis and graphics of microbiome census data. PLoS ONE 8(4):e61217. https://doi.org/10.1371/journal.pone.0061217

Melim LA, Northup DE, Boston PJ, Spilde MN (2016) Preservation of fossil microbes and biofilm in cave pool carbonates and comparison to other microbial carbonate environments. PALAIOS 31:177–189. https://doi.org/10.2110/palo.2015.033

Mouraviev FA, Vinokurov VM, Galeev AA et al. (2006) Paramagnetism and the nature of dispersed organic matter in the Permian deposits of Tatarstan. Georesources 2(19):40-45 (in Russian)

Myszka B, Schußler M, Hurle K, Demmert B, Detsch R, Boccaccini AR, Wolf SE (2019) Phase-specific bioactivity and altered Ostwald ripening pathways of calcium carbonate polymorphs in simulated body fluid. Royal Soc of Chem Adv. 9:18232–18244

Nomokonov VE (1969) On the stratigraphy of the Upper Cambrian and Cambrian deposits in the region of Yefremkino and Malaya Syja settlements. Izvestiya Tomskogo politekhnicheskogo instituta 196:40-47 (in Russian)

Pronk GJ, Heister K, Vogel C et al. (2017) Interaction of minerals, organic matter, and microorganisms during biogeochemical interface formation as shown by a series of artificial soil experiments. Biology and Fertility of Soils 53:9–22. https://doi.org/10.1007/s00374-016-1161-1

Sofinskaya OA, Kosterin AV, Galeev AA (2022) Heterogeneity of Wetting Contact Angle in Hydrophobized Soils and Parent Rocks. Eurasian Soil Sc. 55:339–347. https://doi.org/10.1134/S1064229322030139

Trung TT, Bott NJ, Dai Lam N, Trung NN, Hoang Thi Dang O, Hoang Le D, Tung Le L, Hoang Chu H (2019) The role of pseudomonas in heterotrophic nitrification: a case study on shrimp ponds (*Litopenaeus vannamei*) in soc trang province. Microorganisms. 29:7(6):155. https://doi.org/10.3390/microorganisms7060155

Tugarova MA (2021) Indicator signs of carbonate microbialites in black shale formations: isotopic composition and biomarkers. Vestnik of geosciences 11(323):55-61. https://doi.org/10.19110/geov.2021.11.5

Turrini P, Tescari M, Visaggio D, Pirolo M, Lugli GA, Ventura M, Frangipani E, Visca P (2020) The microbial community of a biofilm lining the wall of a pristine cave in Western New Guinea Microbiological Research 241:126584

Wagner M and Horn H (2017) Optical coherence tomography in biofilm research: A comprehensive review. Biotechnology and Bioengineering 114(7):1386-1402. https://doi.org/10.1002/bit.26283

Wiseschart A, Mhuantong W, Tangphatsornruang S, Chantasingh D, Pootanakit K (2019) Shotgun metagenomic sequencing from Manao-Pee cave, Thailand, reveals insight into the microbial community structure and its metabolic potential. BMC Microbiol. 19:144

Zorina AS, Maksimova YG, Demakov VA (2019) Biofilm formation by monocultures and mixed cultures of Alcaligenes Faecalis 2 and Rhodococcus Ruber Gt 1. Microbiology 88(2):164-171. https://doi.org/10.1134/S0026261719020140

Examples of Natural and Technogenic Mineral Formation on the Surface of Architectural Objects

Raisa V. Lobzova, Oxana V. Karimova, Anastasia S. Makarova, Alina A. Krotova, Larisa O. Magazina, and Vera N. Smolyaninova

Abstract The influence of structure, mineral and chemical composition of weathering forms on the cultural heritage monuments have been studied by optical microscopy, petrographic analysis and electron microscopy. The objects of the study were architectural, memorial, museum monuments, built and decorated with natural (limestone, dolomite, mica, etc.), or artificial (glass, ceramics and cement) materials. Gypsum, oxides and hydroxides of iron, less often manganese oxides as well as sodium sulfates and halite are the main minerals of the surface formations on the architectural objects. Among the sulfates, we also found minerals of the jarosite group. They form either disseminated phenocrysts in dolomitic limestones and sandstones in association with gypsum and goethite or interlayers in the clays enriched by organic matter. In urban conditions, formations such as streaks, spots and crusts on the architectural objects containing mortars and cement materials differ in the micro-layered structure, zoning of mineral composition and morphology. The resistance of materials to biogenic weathering was evaluated. The surfaces of the glazed tiles are the most stable to biogenic erosion.

Keywords Objects of cultural heritage · Weathering · Efflorescence · Limestones · Ceramics · Glass · Iron oxides and hydroxides · Technogenic jarosite · Muscovite

R. V. Lobzova · A. S. Makarova · A. A. Krotova
State Research Institute for Restoration, Moscow, Russia
e-mail: lobzovarv@mail.ru

O. V. Karimova (✉) · L. O. Magazina · V. N. Smolyaninova
Institute of Geology of Ore Deposits RAS (IGEM RAS), Moscow, Russia
e-mail: oxana.karimova@gmail.com

O. V. Frank-Kamenetskaya et al. (eds.), *Biogenic—Abiogenic Interactions in Natural and Anthropogenic Systems 2022*, Springer Proceedings in Earth and Environmental Sciences, https://doi.org/10.1007/978-3-031-40470-2_30

1 Introduction

The role of microorganisms in the biochemical cycle of matter, especially carbon, nitrogen, sulfur, phosphorus, iron, manganese, etc. is well known and their mineralizing value is great. Of great importance are the studies of biogenic interaction and mineral formation in natural and artificial environments carried out in different directions by a team of scientists, employees, students of St. Petersburg State University. The results of these studies have been published in a number of monographs, articles and materials of International symposia devoted to the bioinert interaction of organic and inorganic matter. In the generalizing monographs: "Examination of stone" (2005), "Monuments. Vector of direction" (2008), "Monuments of the Museum necropolises of St. Petersburg" (2016), and other works by A.G. Bulakh, O.V. Frank-Kamenetskaya, D. Y. Vlasov (Bulakh and Marugin 2013; Bulakh et al. 2005, 2012; Frank-Kamenetzkaya et al. 2010, 2015; Vlasov and Frank-Kamenetzkaya 2006; Vlasov et al. 2016), the authors describe the regularities of the formation of organic minerals on natural objects, cultural heritage monuments and synthesized products. The complex of methods analyzed microbial communities on the modified surface of the stone, selected biocidal compositions, methods and conditions of processing of the monument for each material, developed and applied a qualimetric assessment of the condition of monuments made of stone. The role of microorganisms in the transformation of matter is shown.

Many mineral formations, ore deposits, and various types of iron deposits are directly or indirectly associated with the participation of microorganisms (Chukhrov 1978; Yakhontova and Nesterenko 1983).

Jarosite is one of the minerals in the oxidation zone of iron sulfides. This mineral, previously considered rare in the 19-th century, was subsequently discovered in the oxidation zones of ore deposits. Varieties of the jarosite group were identified in shales, clays, carbonate rocks, coils, etc. Summary information about jarosites is published in reference books (Yakhontova and Zvereva 2007). Jarosites, being a product of the weathering crust, differ both in the content of cations K^+, Na^+, Ag^{2+}, NH^{4+}, Pb^{2+}, Ca^{2+} and others, as well as anionic groups $[SO4]^{2-}$, $[PO4]^{3-}$, $[AsO4]^{3-}$. Gypsum, melanterite, goethite, hydrogoethite (the last two are "limonite"), alunite can be associated with jarosites. They give a various color to pyrite-containing montmorillonite-hydromicas clays and carbonates. It is noted that thion bacteria may participate in the oxidation of pyrite contained in clays. The intermediate product during pyrite dissolution is an easily soluble iron bisulfate—$Fe_2(SO_4)_3$. Decomposition of clays and other terrigenous minerals, formation of iron sulfates is promoted by water, sulfuric and humic acids. Electron microscopy analysis revealed Mn, Sn, Ga, V, Cu, Ba, Sr, Pb and other trace-elements in jarosites.

It is known that minerals of the jarosite group are quite widespread in nature, found in sedimentary and volcanogenic rocks, in different types of clays, coals and coal mining waste. They form pseudomorphoses of pyrite and marcasite nodules, on plant residues, for example, in refractory clays. In this work we consider jarosite formation on different types of cultural heritage monuments.

2 Objects of Research

The objects of research are architectural, memorial, museum monuments, stacked and decorated with natural (limestones, dolomites, sandstones, etc.), or artificial (ceramic and cement) materials from the Ore-Petrographic Museum of IGEM RAS (various jarosites from the oxidation zones of ore deposits), the Lapidarium of the Moscow Kremlin (historical limestone details), restoration limestone from Sant-Danilov Monastery, Bulgarian Historical and Architectural Museum, window panes with mica inserts from the Museum of Pereslavl-Zalessky, window panes with thick glass inserts from the A.V. Shchusev Museum of Architecture, glazed tiles with white and red shards from the collection of the State Historical Museum and the Church of the Novodevichy Monastery. Weathering forms on modern restoration ceramic decor and brickworks were investigated at the Church of the Theotokos of Tikhvin.

3 Research Methods

The samples were studied by optical microscopy using an EZ4 D stereomicroscope (Germany) and an OLYMPUS—BX-51 polarization microscope equipped by Digital Camera (transparently polished thin sections). The most interesting samples were photographed. Chemical composition of the samples were estimated by means of the scanning electron microscopes Hitachi TM 14,000 Plus with EDS (Japan) and Jeol JSM-5610LV (Japan) with EDS Oxford INCA-450. X-ray spectroscopy Quantax 75 (Bruker, Germany) was used to study the features of micromorphology, microstructure and elemental composition of the selected areas.

4 Results and Discussion

4.1 Jarosites from the Collection of the Ore-Petrographic Museum of IGEM RAS

Chukhrov (1950, 1955, 1978) examined jarosite samples from the Musuem collection by various methods. As he pointed out, jarosites are stable at $pH < 3$. Their formation is associated with the initial stages of oxidation iron-containing ores.

We selected jarosites with different cationic composition from the oxidation zones of ore deposits in the Middle Urals, Transbaikalia, Kazakhstan, and Kyrgyzstan. Jarosite samples were marked by V.N. Smolyaninova according to the museum catalog. Macroscopically the jarosite samples differ in hardness, color and chemical composition. Jarosite M-785 is of lemon shade, greasy to the touch, kidney-shaped. M-799 is of mottled color with gray and white inclusions. M-971 is homogeneous fine-grained with an orange tinge, M-970 has a brownish tinge and a granular chip,

M-1017 is brownish with shale and white inclusions. M-882 is a powder of rich yellow color with a brown tinge. Iron oxides and hydroxides, quartz, and sulfide relics associate with jarosites.

SEM studies revealed differences in the morphology and composition of the selected samples (Fig. 1). The SEM images in back scattered electrons reflect various sizes and shapes of jarosite crystals: lamellar, pseudohexagonal, pseudocubic, and film.

EDS analysis estimated that sample M-785 is potassium jarosite; samples M-1017, M-970 and M-799 are sodium containing jarosite. The iron content is the lowest in the samples M-799 and M-882. At the same time, lead, arsenic and strontium impurities were found in them. The highest lead content (19.3 wt.%) was determined in the sample M-882. Phosphorus was detected in the M-970.

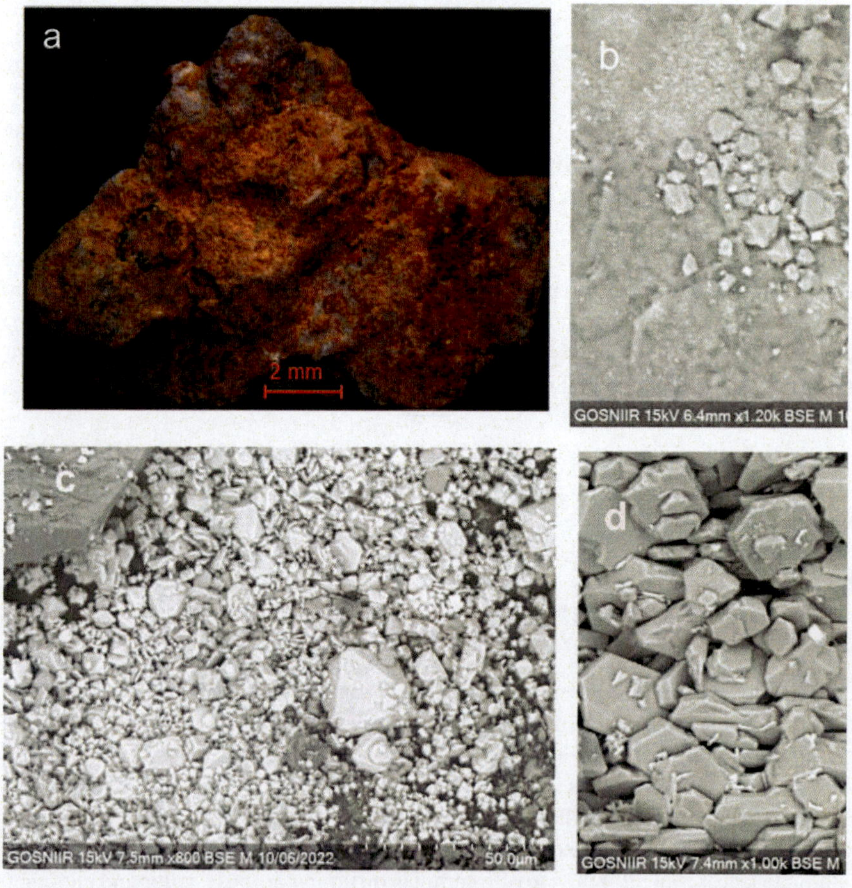

Fig. 1 Jarosites from the Ore-Petrographic Museum of IGEM RAS: (**a**) macrophotographic image of the sample M-799, SEM back scattered images of the samples M-799 (**b**), M-882 (**c**), M-1017 (**d**)

4.2 *Isolation Aggregates of Jarosite on Limestones*

The characteristics of the white stone parts stored in the Lapidary of the Moscow Kremlin are given in the work of Podyapolsky et al. (1997) and the collection of articles (Yarosh et al. 2007). One of the main factors influencing the destruction of white stone is considered to be salt formation. In addition to the predominant calcium and magnesium sulfates, the presence of sodium and potassium chlorides and iron oxides was revealed. During investigations of efflorescences on different architectural objects decorating by white stone we discovered jarosite—potassium sulfate of iron.

4.2.1 The Embedded Plate with a Latin Inscription Originally Located on the Spasskaya Tower of the Moscow Kremlin

Extensive literature is devoted to the history of the plate, the deciphering of the inscription. According to Grashchenkov (2006, 2010), a plate with a Latin inscription (Fig. 2a) was carved in honor of the laying of the Spasskaya Tower in 1491. Its author is considered to be Pietro-Antonio Solari (Grashchenkov et. al., 2006, 2010). According to architect A.I. Khamtsov, when the whitewash was washed away during the restoration of the Spasskaya Tower in 1949, it turned out that the unique monument was badly damaged.

Fig. 2 The plate with a Latin inscription originally located on the Spasskaya Tower of the Moscow Kremlin. Macrophotographic image (**a**) (Grashenkov 2010), microphotographic image of thin film: area with remains of organisms in a fine-grained interlayer (**b**), ferruginous interlayer (**c**), ferruginous halos along sulfide inclusions (**d**)

At the end of the 1990s, when studying the embedded slab of biogenic detritus limestone, yellow needle-like efflorescences were collected on its reverse side and according to X-ray phase analysis conducted by I.M. Marsiy diagnosed as jarosite. During long-term storage of the sample, they turned into a yellow powder. Currently, jarosite is found in the form of small inclusions of different concentrations in the detachments on the back side, which gives a yellow color to organogeno-detritus limestone (Fig. 2). The material of the slab is Myachkovsky limestone with spots of green crust and rare large brachiopod flaps. Peeling crusts on the reverse (back) side of the plate were studied. In addition to yellow and black separations, the presence of greenish inclusions is observed. Calcite, jarosite, melanterite, and glauconite were identified in the composition of the crusts.

4.2.2 The modern restoration block of the Myachkovsky limestone was used for facing the northern stairway and the basement of the facade of the Trinity Cathedral of St. Daniel's Monastery in 2022. In this block yellow spots and primers were revealed, which were concentrated along the cleavage crack in the interlayer containing a large organic residue of the reticulated mossy (Fig. 3). Near the fragment of the mossy, dotted black secretions of biogenic origin are visible. On the surface of the fouling of the flaps with small calcite crystals, a partial filling of the cavities with a yellowish fine-grained mineral is visible.

According to electron microscopy data, the following minerals were determined: gypsum and jarosite, black mineral (melanterite), black microorganic restit matter. Sulfates jarosite and gypsum are considered to be the products of sulfides oxidation.

The results of the EDS analysis of the limestone revealed a high content of potassium, the presence of sodium and magnesium in some spectra, which is associated with the presence of sulfates. The total iron content varies widely from 2.0 to 6.6, and at the point up to 27.8%. The presence of silica (1,7–4,2) and alumina (0,4–1.8) is noted, which corresponds to the admixture of clay substance. Black inclusions of organic matter are confined to the cavities of organic residues and are localized in the oxidation sites of iron-sulfide contamination. Apparently they create a favorable environment for the preservation of jarosite.

4.2.3. Fragments from the Collection of the Bulgarian State Historical and Architectural Museum-Reserve (Republic of Tatarstan). A Fragment of a Reservoir

The monuments are fragments of a single reservoir with a recess for a font. The material is limestone. On the surface there are common dust and soil pollution, traces of biological origin ("bird droppings", cobwebs). The monument is a subject to biological contamination, there are spot traces of yellow lichens, black dot inclusions, brown spots up to 0.5 cm in diameter on two adjacent blocks. The inscription on the limestone is in Arabic, the font is Kufi embedded. Microscopic observations revealed layered impurities of brown color, stains of biological origin. Electron microscopic examination showed an inhomogeneous structure, oolite-like separations are present in the fine-grained mass, in their composition are accretions of lamellar crystals (Fig. 4). Gypsum, potassium–sodium - ferruginous salt formations were revealed.

Fig. 3 Macrophotographic image of the limestone fragment with a piece of the reticulated bryozoan and ferruginous area (**a**), microphotographic image of the weathering forms in a bryozoan cavity (**b**), SEM backscattered image of the salt forms (**c**)

EDS analysis showed the presence of iron, sodium, potassium, sulfur, as well as the presence of phosphorus and zinc.

Of the alkalis, potassium predominates, sodium is present only in some areas. On the maps of the distribution of elements, there is a clear correlation of Ca, K and Cl, as well as Al and P, Si and Mg, sulfur in the form of inclusions. The breed rock corresponds to dolomite (wt.%) CaO−63.1, MgO−26.6 SiO2−6.2 Al2O3−2.0 K2O−0.7, Na2O−0.5, Cl−0.3. When removing the layers with restoration materials, the yellow–brown mottled color of the stone changed into purple tones, in other cases (using compresses with oxalic acid), the dark crusts acquired a light yellow color. The former was removed by washing with water, the latter were preserved, which is due to the technogenic transformation of iron-containing minerals. In nature, iron sulfates in association with gypsum, goethite are found in dolomites and sandstones in the

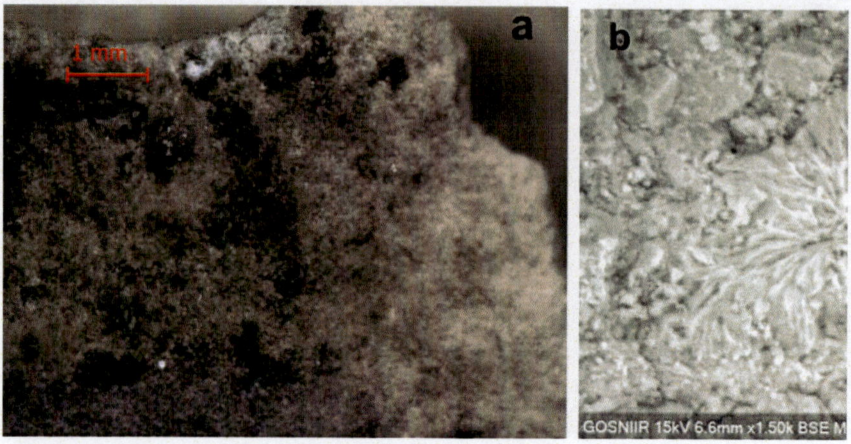

Fig. 4 Biogenic contamination on the limestone fragments from the collection of the Bulgarian State Historical and Architectural Museum-Reserve: macrophotographic image (**a**) and SEM backscattered image (**b**)

form of scattered inclusions or in interlayers enriched with clay matter. Melnikovite, an aqueous sulfate of iron, is developed under the soil cover in the masonry areas.

4.2.4. Sandstone

A memorial cross made of homogeneous light sandstone in St. Daniel's Monastery was installed about 30 years ago at the reburial of the monastery remains. Over time, rounded dark formations appeared in the sandstone, revealing the texture of the sandstone. In the Donskoy monastery cemetery there is a tombstone of Olga Vasilyevna Pushkina, made of light gray sandstone. When removing black spots and crusts by using compresses with a set of different solvents, it led to a change in the color of black crusts through brown to lemon-yellow and greenish tones. Traces of man-made impact during restoration work have repeatedly appeared on the sandstone facing slabs of the basement of the Grand Kremlin Palace, the stairs of the temples of the Danilovsky, Donskoy monasteries, in the slabs of the tombstones of the monastery cemetery of the Novodevichy Monastery. Colored streaks of neoplasms on sandstone have been preserved to the present day, which is clearly seen on the example of sandstones, often used as steps, bases and memorial products. Organic acids—oxalic, citric, and other substances are used to remove the spots of green crust on masonry materials. The use of acids, ammonia and alcohol solutions of various concentrations cause the transformation of the original minerals and contribute to the deposition of newly formed minerals. These acids take part in the migration of iron. The color of the spots varies from black through brown to lemon tones, forming streaks. this is clearly seen in the example of sandstones, which are often used as steps, bases and memorial products. Therefore, it is necessary to approach the clearing of materials with green crust carefully.

4.3 Weathering Forms on Silicate Rocks

4.3.1 Mica (Muscovite) Window Inserts of the Second Half of the XVII Century (Pereslavl-Zalessky)

Muscovite from the pegmatite veins of Karelia began to be exported to Europe at the end of the XV century. During the Tsar Peter I government mica was produced in large quantities for the needs of the fleet construction. And in 1706 even a decree was issued that secured all the extracted mica to the state. An exhibit from the Pereslavl-Zalessky Museum collection was received for restoration at GOSNIIR (Fig. 5). Mica inserts are decorated with colorful paintings of various subjects. Some areas of muscovite are deformed, have cleavage along cleavage planes, hardening on the front surface and the smallest black spots of biota (Fig. 5).

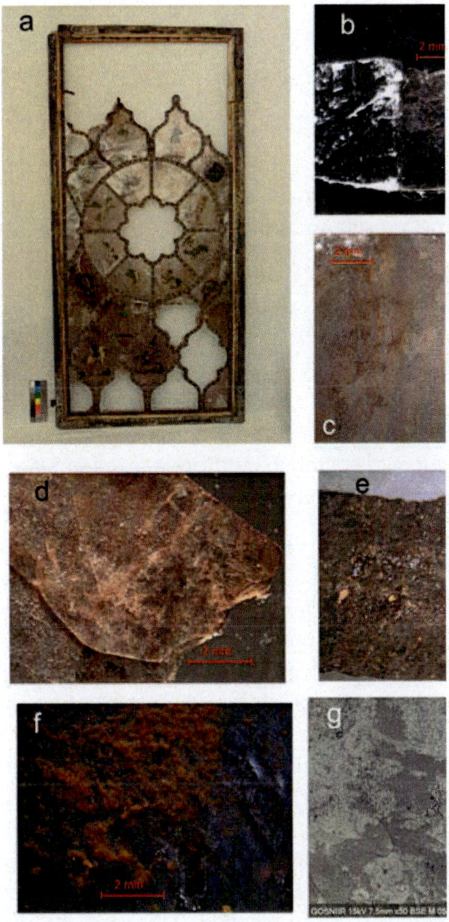

Fig. 5 Macrophotographic image of the mica window (**a**), Microphotographic images. Макрофотография оконницы (**a**). Микрофотографии: muscovite deformation and surface coloring (**b**), crack and spot pigmentation (**c**), ferruginization (**d**), fastening corrosion and jarosite (**e**), jarosite coloring by rust remover (**f**), SEM image of weathering forms (**g**)

Fig. 6 Macrophotographic image of a window fragment with thick glass and neoplasms (**a**), micrograph of glass destruction, clouding and transformation into opal (**b**)

Electron microscopic studies estimated mineral composition of the weathering forms as a mixture of iron oxides and hydroxides, and sulfate—jarosite.

4.3.2 Window made of thick glass from the A.V. Shchusev Museum of Architecture. There are various layers (calcareous, iron-containing, and organic) on the destroyed glass of the exhibit. In some areas, redeposited oxidation products of iron fasteners were observed. In some places they have a rhythmically layered structure of the type of Lysegang rings (Fig. 6).

Opal has been detected in some areas. The ferruginous lattice is corroded, the cement is painted in different tones in places. Rounded inclusions of yellow jarosite were found in oxidized iron fasteners.

4.3.3 Surface changes on the glaze were studied on smooth tiles with a white shard and blue painting on white watering from the collections of the State Historical Museum and relief tiles with a red shard and green glaze from the collection of the Church Museum of the Novodevichy Monastery. On the fragments of the surface of smooth blue-white painted tiles with a white shard of the XVIII century, neoplasms in the form of crusts have different colors, thickness and mineral composition. Under mineral neoplasms and brown impurities (Fig. 7), blue-blue cobalt colorful layers remain on the lead–tin glaze, but the white background often becomes cloudy and acquires creamy and brownish shades. In it, halos of change are observed around metal inclusions, and in the blue glaze, halos of white washing of color are observed.

On a fragment of a relief medallion tile with a red shard and lead green (chromophore copper and iron) glaze under layering, some changes in the surface of the fragment were observed. Traces of the old clearing exposed the surface layer of

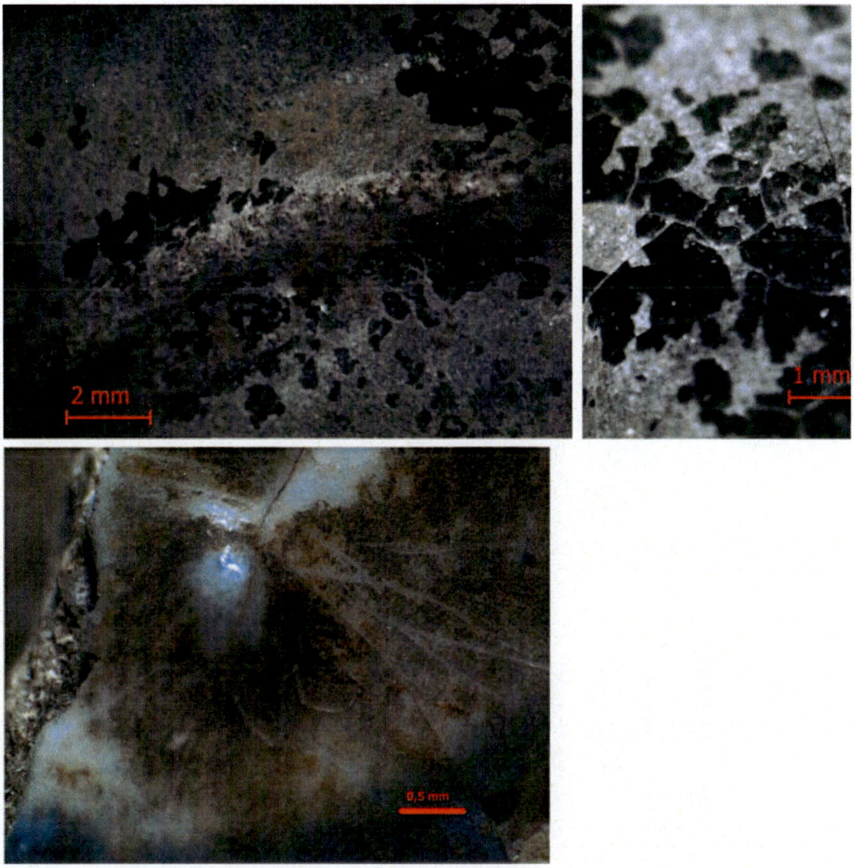

Fig. 7 Macrophotograph of glaze contamination

the glaze with effect optical "raduga" (irisation) caused thin-plate peeling of the glaze. The author's layer was removed in places to an unchanged dark green layer. Therefore, when removing mineral layers on a glazed shard during restoration, it is necessary to examine and pre-strengthen the superficially damaged areas.

5 Modern Restoration Decor

Neoplasms and destruction of ceramic decor, brickwork and efflorescence. The new formations of modern decor are considered by us at the object of cultural heritage of the restored Church of the Theotokos of Tikhvin. The study concerns mainly the destruction observed on the fence of the Church. The temple, built in 1911–1912 (architect N. Martyanov) was closed in 1933, transferred to various organizations, underwent destruction and was not used for its intended purpose for a long time. It was transferred to private ownership in 2004, the restoration was carried out at the private expense of its owner K.V. Akhapkin. The temple is decorated with "porcelain" polychrome domes, platbands, inserts and mosaic icons. The brick fence of the temple is decorated with glazed blue tiles (base), brown tiles and rollers (upper part). Fence posts have polychrome relief inserts.

During the prolonged restoration on the fence, except for cracks, detachments, potholes, etc., there have been significant losses of ceramic decor. In addition to peeling, efflorescences appeared on the brick of the fence posts. In the upper part of the fence posts on the inclined plane there are layers of different textures of white and yellow colors of different shades and intensity. In the lower and middle part of the pillar there are loose white deposits or thin crusts (Fig. 8).

Fig. 8 The fence of the Tikhvin temple. Macrophotograph of the post with neoplasms on the brickwork (**a**), algal fouling of restoration plaster (**b**), efflorescence (**c**), SEM image of neoplasms on the brickwork (**d**)

The micromorphology of efflorescences is different: films, isolated thin-crystalline clusters, tubular fouling, etc. White kidney-shaped crystalline efflorescences in the inner part have a yellow color of different shades and intensity (Fig. 8).

The presence of alternating fine-grained microlayers is noted, which indicates the repeated deposition of efflorescences. On the suture solution, deposits of corrosion products have a different color and structure. composition and layered distribution. Individual yellow spots in the white tile shard with a blue coating are also represented by jarosite, which was part of the cement mortar clinker.

The abnormally hot summer of 2022 affected the state of the ceramic decoration of the temple and the brickwork of the fence. Colored efflorescences were observed in all parts of the fence posts. Significant losses also appeared on the platbands of the temple on the south side. Especially significant changes in the surface of materials occurred in the autumn season after the hot summer of 2022. In addition to peeling and peeling off the surface of the brick, under which white fluffy efflorescences were exposed, colored kidney-shaped neoplasms "grew" on all inclined parts of the fence. The wet warm autumn contributed to the formation of algae films on the white shard of the loss of brown decor. Algae fouling of the roofs of the pillars covered the exposed parts of the lost dark blue paint of the restoration plaster layer. In addition to algae, mosses formed in the corners, and in the horizontal cracks of the brown glaze even rare sprouts of the highest representatives of the biota. The loss of the ceramic decoration of the fence was caused by processes: first mechanical destruction, then salt precipitation, dust pollution and biogenic fouling. After a snowy season with frequent temperature transitions through the zero mark (2022–2023) significant losses of glaze coating on tiles and tiles of ceramic decor and brickwork, especially fences, are expected.

6 Conclusion

Based on the results of Chukhrov's research (1950, 1955, 1978), we can conclude that the formation of jarosite crystals on various objects (architectural objects, monuments and museum samples) indicates an acidic weathering environment. The source of iron and sulfur are layers in limestone enriched with ore minerals—iron sulfides. Organic acids derived from clay and organogenic matter provide an acidic conversion medium for the oxidation of iron sulfides.

Thus, in addition to the well-known facts of the formation of gypsum during weathering and destruction of cultural heritage objects, our study testifies to the formation of another sulfate—jarosite.

Our research also has revealed the presence of neoplasms—various sulfates, including jarosite, on natural and man-made materials of cultural heritage monuments. The features of efflorescence and the formation of crusts on different parts of the fence posts revealed their differences in color, structure, degree of crystallinity, as well as the scale of the appearance of algal fouling. The latter on a silicate-based material are insignificant compared to the intensive development of algae films on lime-plaster "restoration" covering the roofs of pillars. The processes of their formation are comparable to those established for minerals in nature.

Acknowledgements This study was performed in the frame of Fundamental Research Program of State Research Institute for Restoration. The study of natural and tachnogenic jarosites was supported by the Russian Academy of Sciences, Program of Fundamental Research of IGEM RAS.

References

Bulakh AG, Marugin VM (2013) Assessment of the state of monuments of architecture of St.Petersburg andmonumental sculpture before and after restoration. Evaluation of the performance of work according to qualimetry. St. Petersburg University Publishing, Saint Petersburg (in Russian)

Bulakh AG, Vlasov DYu, Zolotarev AA, Marugin VM, Morozov MV, Savchenok AI, Fitzner BB, Frank-Kamenetskaya OV, Kheynrikhs KA, Shigoretz SB (2005) Examination of stone in the monuments of architecture. Nauka Saint Petersburg (in Russian)

Bulakh AG, Vlasov DYu, Frank-Kamenetsky OV (eds) (2012) Open-air Museum. Problems of preservation of monuments made of stone and bronze. Soyuz-Design Saint Petersburg (in Russian)

Chukhrov FV (1950) Formation of jarosites in oxidation zones. DAN USSR 78(4): 781-783 (in Russian)

Chukhrov FV, Lyalikova NN, Gorshkov AI (1978) On the role of microorganisms in the formation of jarosites. DAN USSR 241(4): 929-932 (in Russian)

Chukhrov FV (1955) Colloids in the Earth's crust. Publishing House of the USSR Academy of Sciences (in Russian)

Frank-Kamenetskaya OV, Marugin VM, Vlasov DYu, Egorova VV (2010) Complex qualimetric monitoring of the state of stone monuments. Qualimetric monitoring of construction objects. Polytechnic Saint Petersburg (in Russian)

Frank-Kamenetskaya OV, Polyanskaya EI, Manurtdinova VV, Kurulev KS (2015) Decorative facing stone in the non-fields of the museum of urban sculpture. Open-air museum: modern approaches to the preservation of sculpture. Voronezh (in Russian)

Gracshenkov AV (2006) Plate with a Latin inscription from the Spasskaya Tower and the title of sovereign of all Russia. Questions of epigraphics. 1: 16-25 (in Russian)

Gracshenkov AV (2010) Architectural details and fragments of buildings of the XIV early XX century. Catalogue of the collection. State Historical and Cultural Museum-Reserve «Moscow Kremlin» (in Russian)

Podyapolsky SS, Grashchenkov AV, Lobzova RV, Nikolaenko TZ, Postnikova ON, Skopin VB, Skrynnikova EV, Yarosh VN (1997) Identification of data on the decorative decoration of the facades of the Arkhangelsk and Annunciation Cathedrals of the Moscow Kremlin. Conservation and restoration of historical and cultural monuments. To the 850th anniversary of Moscow. Express information 6: 1-9 (in Russian)

Timofeev VN (ed) (2008) Monuments. Direction vector Collection of articles on sculpture restoration and monitoring of the state of monuments in the urban environment. Saint Petersburg: Soyuz-Design (in Russian)

Vlasov and Frank-Kamenetskaya, 2006Vlasov DYu, Frank-Kamenetskaya OV (2006) Destruction of natural stone in the urban environment. Proceedings of SPbOE 96: 156-169 (in Russian)

Vlasov DYu, Rytnikova VV, Frank-Kamenetskaya OV (eds) (2016) Monuments of the museum necropolises of St. Petersburg. Existence, materials, safety, diagnostics. Saint Petersburg VVM Publishing House (in Russian)

Yakhontova LK, Nesterenko LG (1983) The zone of hypergenesis of ore deposits as a biokos system. Moscow State University (in Russian)

Yakhontova LK, Zvereva V (2007) Minerals of the zone hypergenesis handbook. Vladivostok Dalnauka (in Russian)

Yarosh VN, Lobzova RV, Grashchenkov AV (2007) Investigation of the material of the embedded plate of the XVII century. White stone in architectural monuments of the Moscow Kremlin of the XVI-XVII centuries. Russian Research Institute of Cultural and Natural Heritage named after DS Likhachev. Moscow (in Russian)

Microorganisms and Stages of Stromatolite Formation

Tatiana V. Litvinova

Abstract Since the sixties of the last century, stromatolite buildings have been successfully used for the purpose of stratigraphic dismemberment of the dumb strata of the Precambrian. The age of the stromatolites was determined on the basis of the morphology of buildings associated with a specific period of geological time. Such a formal classification made it possible to distinguish large stratigraphic divisions of erathems and suberathems in Precambrian strata, and much more fractional within stromatolite provinces. It was not possible to identify the remains of microorganisms in stromatolites for a long time. Now, with the help of an electron microscope and a special technique, it has become possible to establish numerous fossilized microfossils in these rocks. However, the reasons for the appearance of various morphological structures (stratiform stromatolites, differently branching and non-branching column-type stromatolites, etc.) and the attachment of each of them to a certain time remain unclear. Currently, it has become possible to study the role of microorganisms in this process. But the scale of stromatolite formation in the Precambrian makes it difficult to compare cyanobacteria in different stromatolite structures. Unlike the powerful Precambrian reefs stretching for kilometers, Phanerozoic stromatolites have a fairly limited distribution and are characterized by small capacities. The remains of cyanobacteria in them are preserved better, and the intensity of secondary processes affects these rocks to a lesser extent. Small incisions in the Phanerozoic make it possible to compare the role of certain cyanobacteria that took part in the construction of stromatolites of different shapes. To determine the age of rocks in the Phanerozoic, numerous remains of fauna were used, rather than rare stromatolite buildings. Therefore, they have not been practically studied before. In this work, for the first time, on the example of stromatolites from the Middle Permian, the restructuring of the organic microcosm is traced, illustrating its evolution. During the formation of the Permian stromatolite structure, three successive phases are distinguished: initial, intermediate and final. The work has established with the participation of which microfossils (coccoid, filamentous, etc.) one or another form of construction is now observed. The features of the vital activity of the cyanobacteria community

T. V. Litvinova (✉)
Geological Institute, Russian Academy of Sciences, Moscow, Russia
e-mail: litvinova-geo@rambler.ru

O. V. Frank-Kamenetskaya et al. (eds.), *Biogenic—Abiogenic Interactions in Natural and Anthropogenic Systems 2022*, Springer Proceedings in Earth and Environmental Sciences, https://doi.org/10.1007/978-3-031-40470-2_31

are considered. The study of microorganisms in Phanerozoic stromatolites and their comparison with the Precambrian cyanobacterial community will greatly simplify the interpretation of data on ancient reef structures.

Keywords Stromatolites · Morphology of buildings · Microfossils · Scanning microscope · Reefs · Permian · Evolution · Cyanobacteria

1 Introduction

Powerful strata of Precambrian stromatolite reefs stretches for kilometers. Their structure has attracted the attention of researchers for a long time (Grotzinger and Knoll 1999; Krylov 1963, 1975; Raaben 1986, 2000; Semikhatov 1985; Serebryakov 1975; Semikhatov and Raaben 1996 and other). In the last century, a classification of stromatolite buildings was compiled, which was subsequently refined and refined many times (Korolyuk 1960; Krylov 1963; Raaben 1986, 2000; Semikhatov and Raaben 1994, 1996). It combined most of the taxa of stromatolites selected from Precambrian rocks of Russia, India, China, Africa, America and Australia, and differentiated them into more than 250 genera and 1,500 species (Raaben 2000).

The classification of stromatolites was formal. It was based on the morphology of buildings, including their volumetric shape (Krylov 1975) and microstructure (Komar 1989). The huge work that was carried out on the study of Precambrian stromatolites of the world was completed, and by the end of the XX century this line of research practically ceased to exist. Although the scanning electron microscope has been known for a long time, it was not possible to identify the remains of fossilized microorganisms in sections of these carbonate rocks either with the help of an optical or with the use of an electron microscope. Thus, the old methods were exhausted, but the new ones never appeared.

The study of microorganisms in stromatolites using a scanning electron microscope (SEM) is possible only when working with volumetric fragments (5 × 5 mm) of stromatolite (Litvinova 2009). They should include an organogenic layer (Litvinova 2009). The first studies showed that the microstructure of stromatolites really significantly depends on the composition of organisms and the peculiarities of the vital activity of cyanobacterial community. So, if filamentous cyanobacteria took the main part in their structure, a ribbon microstructure arose, if, for example, it was formed by separate large particles-colonies including various cyanobacteria, now a clot microstructure is observed in this rock. (Litvinova 2014). It is also possible to interpret a more complexly constructed microstructure, in the formation of which a community of microfossils of various shapes took part (Litvinova 2018). It is more difficult to identify the role of organisms in the construction of columnar stromatolites, since these are quite large objects. However, the need for this is obvious. It was due to the participation of organisms in their formation that it was possible to create a formal classification of these rocks and successfully use it in the stratigraphic division of the Precambrian (Korolyuk 1960, 1963; Semikhatov 1974, 1985; Raaben 1986, 2000; Serebryakov 1975; Semikhatov and Serebryakov 1983 and others).

The generalization of numerous works and the experience of the author, who has studied Precambrian stromatolites and their biogenic ultramicrostructures for a long time (Litvinova 2014, 2016, 2018 and others), show some regularity of the location in the context of these large buildings in the Precambrian. The initial stage of reef formation is often characterized by a wide spreading of stratified stromatolites, which are subsequently replaced by a columnar type. In its upper part there are often layers with oncoliths. However, the reef structure can be much more complicated. Layers or lenses with oncoliths may also occur in the sole of the incision or in its central part.

The main attention of stromatolite researchers was usually focused on Precambrian sediments, where age determination was possible for a long time only with their help. However, the huge scale of stromatolite formation in the Precambrian (Fig. 1a), which formed extended reefs, makes it technically difficult to compare the remains of microorganisms in morphologically different structures and interpret their evolution. Stromatolite structures in the Phanerozoic sediments have a small capacity—2–5 m, unlike many kilometers of Precambrian stromatolite reefs with a capacity of up to 200–500 m. It made no sense to use them for the purpose of approximate stratigraphic gradation. This work is successfully and more accurately carried out with the help of numerous fossils of organisms, so the stromatolites of the Phanerozoic have not been studied practically. The determination of microorganisms in Phanerozoic stromatolites has also not been carried out, with the exception of rare individual works in recent years (Naugolnykh and Litvinova 2014 2021). However, the remains of cyanobacteria in them are preserved much better than in ancient stromatolites, and the intensity of secondary processes affects them much less. In small-area and volume sections of Phanerozoic stromatolites, it is easier to study the composition of microorganisms and compare it with the shape of a particular building. Sometimes, even within a single sample, it is possible to observe (Fig. 1b) such a miniature reef (Naugolnykh and Litvinova 2014). The successive changes in the structure of the geological outcrop in Precambrian and Phanerozoic stromatolites are in general identical. Stratified stromatolites likewise turn into closely contiguous columnar stromatolites up to 16–23 mm in diameter. Each of the layers of the column is rounded, forming an ever higher arch up the section. Subsequently, the surface of large columns serves as a springboard for the formation of mini-columns from 10 to 25 pieces on each, their size is up to 1.5–1.8 mm wide. Gradually they begin to degenerate, decreasing to 1–2 mm in diameter. The distance between these columns increases noticeably, being filled with carbonate material. The small size of Phanerozoic buildings on the area and accessibility almost always make it possible to trace their evolution step by step, as well as to study the role of microorganisms in the construction of the morphology of stromatolites. Such observations are much more difficult to make within the ancient Precambrian reefs.

For this purpose, Permian stromatolites are considered in this paper. Within its limits, three stages of the formation of biogenic buildings (Naugolnykh and Litvinova 2021) were identified, in each of which certain microorganisms were identified

Fig. 1 Stromatolite buildings. **a**—stromatolite reef in the Precambrian ($R_{1\text{-}2}$), Russia, Ural, Bakal. Scale bar 20 m. **b**—stromatolite biogenic edifice of the Phanerozoic (P_1). Russia, Perm Urals, Alabastrovo. Scale 10 mm. Naug. et Litvinova (2014)

and the features of the community's vital activity were analyzed. Similar studies have never been conducted before. The excellent preservation of cyanobacteria in Phanerozoic rocks, compared with the Precambrian, allows us deepen our understanding of microfossils and their vital activity using the method of actualism. Moreover, the conservativeness of these organisms over time is known (Golubic and Hofmann 1976). The purpose of this work was not only to identify the remains of microorganisms in Phanerozoic stromatolites, but also to establish their role in the construction and formation of buildings of different shapes. A consistent comparison of their composition and community function at each stage of growth of the currently observed structure demonstrates the role of microorganisms in creating their morphological features. The factual material discussed in the paper will make a significant contribution to the understanding of complex biogenic-sedimentation processes occurring during the formation of stromatolites. So far, these issues have been discussed only presumably, based on theoretical data, since cyanobacteria have not been identified in them. The obtained material will help to most rationally direct the research of microorganisms in various morphological types of Precambrian stromatolites in order to establish the causes of their formation.

2 Materials and Research Methods

The work uses the material collected in the Solikamsk horizon of the Ufa tier ($P_{1\text{-}2}$) of the Middle Urals (Russia), the Uspenka section (Naugolnykh and Litvinova 2021). The study of microbial residues in stromatolites was carried out on a scanning electron microscope model Tescan MV 2300, with an energy-dispersed prefix (microanalyzer) Cambridge Instrumente INCA 200. The samples were a specially selected rock fragments with a binocular magnifier measuring about 5×5 mm. Gold spraying of the samples was carried out to determine whether the carbon content in the presumably biogenic structures was increased. Organisms and biogenic ultramicrostructures in stromatolites are characterized by a significant increase in the amount of carbon compared to the carbonate rocks containing them, and are accompanied by microdoses of some biophilic elements.

As a result of a detailed study of the Solikamsk horizon of the Ufa tier of the Uspenka section (Chusovaja River Basin, Perm region) of the Permian system, 3 stages of stromatolite formation were identified: initial, intermediate and final (Naugolnykh and Litvinova 2021 Fig. 5). Each of them differs significantly from the others in the composition of microorganisms, their number and distribution features in the host rock. Figure 2 shows the remains of microorganisms established at each stage. The material in the photo table (Fig. 2) is constructed as follows. The vertical columns represent the I (a, d, g, j, m), II (b, e, h, k) and III (c, f, I, l) stages of stromatolite formation, and the horizontal columns show similar or identical remains of microorganisms in each of these stages.

1. The initial stage (Fig. 2I). Areas of uneven accumulation of several identical coccoid microfossils (Fig. 2a). The correct transverse notch along the circumference of the coccoid microfossilium indicates an equal-sized transverse binary division, leading to the formation of two identical cells (Fig. 2d). The sustained recess along the circumference of the coccoid microfossilium (Fig. 2d) indicates an equally equal in magnitude transverse binary division, leading to the formation of two identical cells (Fig. 2a). Sometimes these organisms are accompanied by relatively small fragments of individual thin bacterial films (Fig. 2j). Filamentous Cyanobacteria are extremely rare, in groups of several very short filaments (Fig. 2g). Their length does not exceed 1 micron. The main area of the sample is occupied by the host rock—limestone (Fig. 2m).
2. Intermediate stage (Fig. 2II). Coccoid microfossils are rare, they are single cells, they do not form clusters (Fig. 2b, e). The number of bacterial films increases significantly, they are dense, with a wrinkled surface (Fig. 2e), often found in groups of two or three fragments (Fig. 2h). Bundles of relatively large filamentous cyanobacteria have been identified within their limits (Fig. 2k). They have a length of 20–23 microns, a width of 1.5 to 3 microns. Biogenic material makes up a significant part of the total area of the rock (Fig. 2k1, k2)
3. The final stage (Fig. 2III). Coccoid microfossils are very rare, their size decreases to 5 microns in diameter (Fig. 2c, f). Numerous fragments of bacterial films tightly cover a significant part of the space of the organogenic layer (Fig. 2f, I: l). Large filamentous microfossils abound both surrounded by bacterial films and without them (Fig. 2:l). Their length can be already more than 100 microns, width up to 8 microns, the shape is comparable to a flattened tubular formation. At the end of such tubes, an oval bud–like growth is sometimes observed (calyptra, Fig. 2:11). The samples also contain chains of very small (0.5 microns) coccoid formations connected by a thin thread (Fig. l). Their length is about 6 microns. Similar "beads" with 5–8 "balls" are almost always located on large tubular shapes, overlapping them from above approximately perpendicular to the length. The

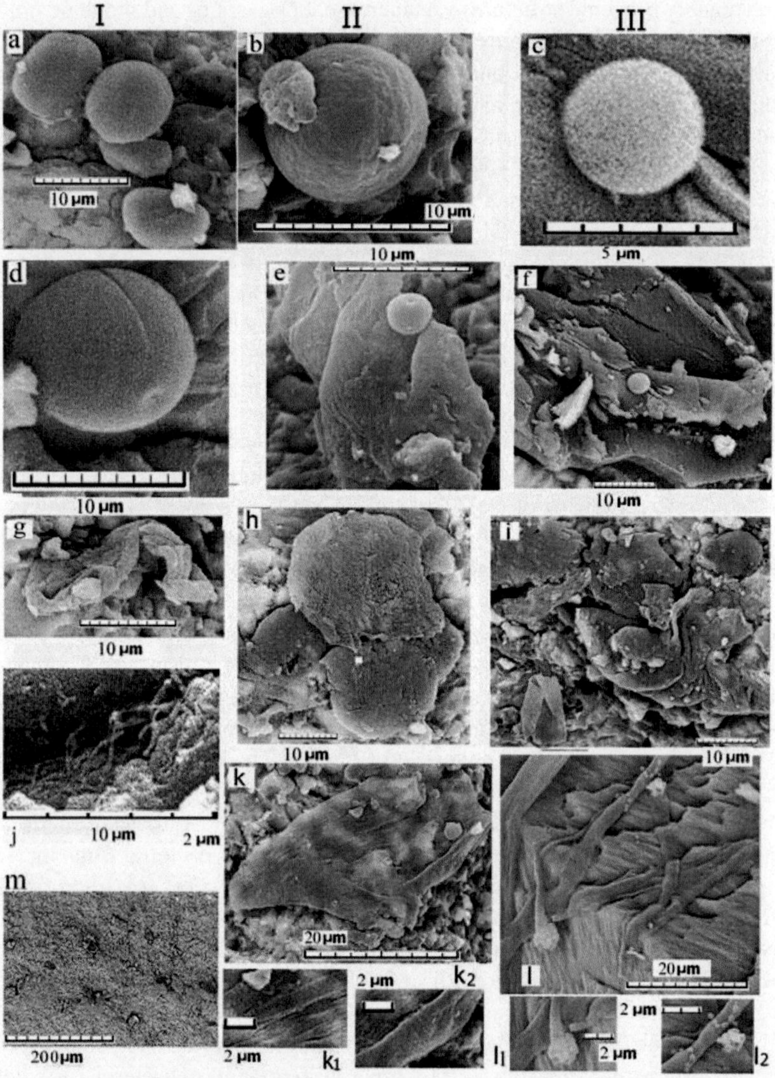

Fig. 2 Microorganisms and their role in the formation of the microstructure of stromatolites on the example of the Solikamsk Horizon (P_1) of the Uspenka section (basin of the Chusovaya River, Perm Territory, Russia). Vertical columns I, II, III—stages of stromatolite formation: I (a, d, g, j, m)—initial stage, II (b, e, h, k)—intermediate, III (c, f, I, l)—final; **a**–**f**—coccoid microfossils at different stages of stromatolite formation, a—accumulation of coccoid microfossils; d—transverse notch on the surface of coccoid microfossils dividing it into two equal halves; **e**, **f**—single coccoid microfossils together with bacterial film; **g**, **h**, **i**—bacterial films: j—accumulation of filamentous microfossils up to 1 microns in size; **k**—bacterial film with bundles of filamentous microfossils, k_1, k_2—enlarged fragments **k**; **l**—accumulation of various filamentous microfossils together with bacterial film, l_1, l_2 (enlarged fragments of l): l_1—bud–like formation (calyptra) at the end of tubular microfossils, l_2—a chain of several coccoid formations connected by a single thread

biogenic components of the rock, modified to one degree or another, occupy a much larger part of the sample than the carbonate material containing it (Fig. 2м).

The three stages of stromatolite evolution differ significantly. The composition, shape, area distribution and number of cyanobacteria in each of them shows the determining role of organisms in the formation of the morphology of the building. At the first stage, a rock resembling gravelite appeared. The peculiar unevenly distributed dark inclusions of organogenic material observed in the light carbonate mass of the rock were created by separate clusters of coccoid microfossils and rare fragments of bacterial films. The widespread distribution of filamentous microfossils together with numerous bacterial films in the second and third stages indicate a change in the composition of microorganisms and the growth of a community that is developing an increasingly large surface on the precipitate (Fig. 2II, III). The size, density, number of bacterial films and relatively large filamentous formations increases significantly in each subsequent stage (Table 1, I: a; II: б; III: c). Accordingly, the size and number of coccoid cyanobacteria decreases in direct proportion. At the second stage, a significant increase in the number and size of filamentous taxa and, accordingly, bacterial films leads to the formation of organogenic columns with a massive structure. The organogenic material in them occupies approximately 35% of the sample (Table 1: II). The maximum increase in the number of cyanobacteria at the final stage of stromatolite formation leads to an even greater increase in the amount of organogenic material. Filamentous cyanobacteria become diverse in shape and size (Fig. 2: l). The remains of microfossils and bacterial films already make up 80% of the sample (Table 1: III). There is a noticeable differentiation of organogenic and mineral materials inside the columns. The organogenic layers are almost completely composed of biogenic material.

Table 1 Quantitative ratio of biogenic ultra microstructures (bacterial films, fossilized remains of organisms and their waste products) and the carbonate rock including them at different stages of stromatolite formation

Stages of building formation	Biogenic and abiogenic components of the rock				The shape of the building, texture
	Host rock	Coccoid microfossils	Bacterial films	Filamentous microfossils	
I initial	89	7,4	1	0, 6	Gravelites
II intermediate	65	4	20	9	Columnar stromatolites
III final	18	2	43	37	Columnar layered stromatolites

3 Discussion

Currently, it is known that the accumulation of cyanobacteria inside a semi-liquid nutrient medium (carbonate sludge) occurs as a result of the reproduction of one or more bacterial cells. Their division is carried out very quickly. They are isolated from outside penetration, creating a spatial isolation of the community in the form of integral "structured colonies" (Oleskin and others 1999). It is due to this that the columns are formed. Nevertheless, the growth of a colony originating from 1–2 individuals is not an obstacle to the emergence of such different shapes and sizes of cyanobacteria. Of course, the microfossilium residues currently observed in rocks may represent different stages of degradation of cyanobacteria residues during their fossilization and lithification of sediment. These processes do not have strictly sequential stages, but depend on many reasons, even on accidents. As a result of various microconditions, many morphotypes can arise (Knoll and Barghoorn 1975), which are different stages of postmortem changes (Hoffman, 1976). Nevertheless, morphological features of cyanobacteria are an important diagnostic feature widely used in Precambrian stratigraphy. In this case, we clearly observe, among other things, different taxa of microfossils (Fig. 2).

The community is able to regulate the number and composition of cyanobacteria. Many researchers believe that it represents a single "superorganism" (Zavarzin 1993; Oleskin 1994; Oleskin and others 1999). Besides, such a multicomponent organized system has enormous advantages that increase the community's ability to survive. For the formation of stromatolite columns, that is, a structure isolated in space, certain conditions are necessary. First of all, it is the presence of a quorum, that is, a significant concentration of microorganisms and a high population density are required (Shapiro 1995; Gray 1997; Kaprelyants and others 1999). Such a multicomponent organized system has enormous advantages that increase the community's ability to survive.

In the initial stage of the formation of Permian stromatolites, rare disparate groups of coccoid cyanobacteria cannot assemble a quorum. In the initial stage of formation of Permian stromatolites, rare disparate groups of coccoid cyanobacteria cannot assemble a quorum. It takes a lot of bacteria to form a biofilm capable completely isolate a community. Only in this case they are able to secrete a large amount of mucosal extracellular matrix consisting of proteins, polysaccharides and nucleic acids. Mucus, to one degree or another, blocks the penetration of other organisms into the colony (Shapiro 1995). In the intermediate stage, as a result of the active reproduction of filamentous microfossils, it is already possible to isolate the community in the area. The mucus tightly envelops the colony, resulting in columns. They are often fenced off from the host rock by a wall, which was originally a tightly compressed mucous material. In the process of sedimentation and lithification of the rock around the columns, mucus and carbonate sediment form a sedimentation-organogenic separation area. At the final stage, the number of microorganisms increases even more, new taxa of filamentous microfossils appear.

The bacterial colony multiplies radially on the sediment surface. Young cells mainly develop along the periphery. This is due to the peculiarity of the intake of nutrients from the medium (Pirt 1975). As a result of the saturation of bacteria in the central part of the building, a small elevation appears. The processes of sedimentation, cyanobacteria vital activity and sediment lithification eventually lead to maximum growth of the central part of stromatolite columns. Due to this, a rounded arch is gradually formed up the section on each stromatolite column (Fig. 1b). Bacteria are equipped with molecular signaling systems used in a process called quorum sensing. The intensity and frequency of the signals received by bacteria depends on the number of cells. If there is a quorum, that is, a sufficient number of cells, the frequency of signaling can reach a threshold that causes changes in the regulation of genes. In response to fluctuations in the density of the cell population, bacteria are able to regulate gene expression (sense of quorum). This leads to the appearance of numerous and diverse filamentous forms of cyanobacteria. There is a reorganization of cellular equipment for the formation of an increasing amount of mucus (glycocalyx). It is observed in the rock using a scanning electron microscope in the form of flattened dried bacterial films. Their number increases markedly already in the intermediate stage, but reaches its maximum only in the final stage of stromatolite formation. And then the participation of different types of cyanobacteria working together contributes to the compaction of biogenic material.

The mucous mass forms barriers both around the colony and on its surface, which leads to a noticeable differentiation of mineral and biogenic materials in the sediment at the third stage of stromatolite formation. When precipitation occurs, the most viable and mobile filamentous microfossils (filamentous and tubular forms) are selected on the surface. Rapid and active reproduction of organisms in a short period of time again allows gain a quorum. An increase in the pH of the environment as a result of the vital activity of organisms accelerates the processes of sedimentation. On the surface of each successively precipitating carbonate sediment, the colony is restored again. Trichomes and mucus formed a thin film covered with sedimentary material. Numerous fragments of bacterial film also indicate that extracellular mucus (glycocalyx) served as a barrier to the penetration of other organisms, isolating each of the colonies with a dense cover. Large filamentous formations involved carbonate material in biogenic activity. Due to the alternation of carbonate sediment and a densely populated bacterial-algal colony, an alternation of organogenic and mineral materials was formed in the columns. In the process of lithification, thin dark and light layers replacing each other, respectively, emerged from them. Thus, organogenic and biogenic material participated in the construction of stromatolite columns with a thin transverse layering.

4 Conclusions

Comparison of the three stages of stromatolite formation shows the essential role of cyanobacteria, reflected in the morphology of buildings.

1. The first stage. Individual clusters of coccoid microfossils scattered over the area and rare fragments of bacterial films formed gravelite texture.
2. The second stage. Filamentous forms of cyanobacteria actively multiply and displace coccoid. Accordingly, the amount of extracellular mucus (glycocalyx), now observed in the form of numerous fragments of bacterial films, increases. The slime isolates the colony in an area. It becomes a barrier to the penetration of other organisms and limits the community by the area. Cyanobacteria actively involve carbonate material in biogenic activity. As a result, a column fenced off by a wall is formed in the future.
3. The final stage. Active reproduction of filamentous microfossils (filamentous and tubular) on the surface of the precipitate is accompanied by even more mucus. Complete isolation of the colony leads to a noticeable differentiation of mineral and biogenic materials. However, this does not prevent the restoration of the colony on the surface of successive precipitation. In the process of its lithification, a thin layer is formed in the stromatolite columns due to their alternation.

The analysis of the dependence of the shape of stromatolite structures on the composition and vital activity of microfossils in Permian stromatolites is only a precursor to the topic under consideration. Questions about the morphology of buildings, their distribution over the area in the reef and up the section, the change of one type of buildings to another, the reasons for branching columns in one way or another, the timing of each of the buildings to a certain period of time remain open. It is obvious that a complex biogenic system is being rebuilt depending on the conditions of existence. If the community is able to cover a significant surface area of the reservoir, stratiform stromatolites arise. In the case of the Permian stromatolites under consideration, this did not happen. It is known that in Phanerozoic, the niche of stromatolite builders is actively occupied by other organisms, including coral reefs. Each community was forced to isolate itself in a small area, creating a dense protective barrier.

Despite the huge interest in stromatolites, at present they have practically ceased to be studied. This is due to the fact that the old methods were exhausted, and the new ones did not appear for a long time. Determination of the age of Precambrian strata began to be carried out by geophysical and paleontological methods. Researchers have learned to recognize microfossils in flint rocks and isolate them from shales, but there are incomparably more remains of organisms in stromatolites. The study of cyanobacteria with the help of SEM allows observe them at a higher magnification than with an optical microscope, and in volume. With their help, it is possible not only to carry out paleontological and stratigraphic studies. The study of the vital activity of cyanobacteria in stromatolite reefs, widespread in the Precambrian, is also of interest for solving more global problems of the evolution of our planet.

Acknowledgements The research was funded within the framework of the state assignment of the Geological Institute of the Russian Academy of Sciences "Periodization of the Proterozoic and the role of global environmental and climate changes in the evolution of the Precambrian biosphere" (FMMG-2023-0004)

References

Golubic S, Hofmann HJ (1976) Comparision of holocene and mid-Precambrian Entopysalidaceee (Cyanophyta) in stromatolitic algal mats: cell division and degradation. Journ. of Paleontol. 50 (6):1074–1082

Gray KM (1997) Intercellular communication and group behavior in bacteria. Trends Microbiol. V 5 (5): 184-188

Grotzinger JP, Knoll AH (1999) Stromatolites in precambrian carbonates: evolutional columns or environmental indexes? In: Science (ed). Annular review Earth and Planetary 27: 313–358

Hoffman HJ (1976) Precambrian Microflora, Belcher Islands, Canada: Significance and Systematics. Journal of Paleontology. 50: 1040-1073

Knoll AH, Barghoorn ES (1977) Archaean microfossils showing cell division from the Swaziland System of South Africa. Science. 198: 396-398

Kaprelyants AS, Mukamolova GV, Kormer SS, Weichart DH, Young M, Kell DB (1999) Intercellular signalling and the multiplication of prokaryotes Microbial Signalling and Communication. Society for General Microbiology Symposium 57. In: England, G Hobbs, Bainton N, Roberts D McL (ed) Cambridge: Cambridge University Press 33–69

Komar VA (1989) Classification of the microstructures of the Upper Precambrian stromatolites. Himalayan Geol 13: 229–238

Korolyuk IK (1960) Stromatolites of the Lower Cambrian and Proterozoic of the Irkutsk Amphitheatre. In: Proceedings of the Institute of Geology and Development of Combustible Minerals. USSR Academy of Sciences, Moscow (In Russian) V 1: 112–167 (In Russian)

Korolyuk IK (1963) Upper Precambrian stromatolites. Stratigraphy of the USSR vol. 2 Upper Precambrian. Gosgeoltehizdat (In Russian)

Krylov IN (1963) Columnar branching stromatolites of the Riphean deposits of the South Urals and their significance in stratigraphy of the Upper Precambrian. Academy of Sciences of the USSR Proceedings of the GIN, Moscow (In Russian)

Krylov IN (1975) Stromatolites of the Riphean and Phanerozoic in the USSR. Science, Moscow (In Russian)

Litvinova TV (2009) New Data on the structure and composition of stromatolite buildups (northern Anabar region), Lithol Miner Resour Soc Notes. 4: 389–398

Litvinova TV (2014) Ultramicrostructures of the stromatolite reef in the Northern Anabar Region and their origin. Lithol Miner Resour Soc Notes 5:416–426

Litvinova TV (2016) Biogenic–Abiogenic Interactions in Stromatolitic Geosystems and Their Mineralization. Abiogenic Interactions in Natural and Anthropogenic Systems. Chapter 6. Springer.

Litvinova TV (2018) Possible role of microorganisms in the formation of structural and textural features of stromatolites Lethaea rossica. 16: 1-16 (in Russian)

Naugolnykh SV, Litvinova TV (2014) Stromatolites from the Permian deposits of the Perm Cis-Urals: a new form-genus Alebastrophyton Naug. et Litvinova, Gen. nov. Palaeontology in museum practice GIN RAS, Moscow (In Russian).

Naugolnykh SV, Litvinova TV (2021) Conditions of forming of the deposits of the Solikamskian Horizon (Ufa tier, Middle Permian) and new date on Permian stromatolites of the Middle Cis-Urals (Chusovaja Rifer Basin, the Perm region). Procedings of the paleontological society, VSEGEI (In Russian)

Oleskin AV (1994) Social behaviour of microbial populations. Basic Microbiol. V. 34. (6): 425–439 (In Russian)

Oleskin AV, Botvinko IV, Tsavkelova EA (1999) Colony organization and intracellular communication in microorganisms Biology Department, Moscow State University Moscow (In Russian)

Pirt SJ (1975) Principles of Microbe and Cell Cultivation. Blackwell Scientific, Oxford, London, Edinburgh, Melbourne

Semikhatov MA (1974) Stratigraphy and geochronology of Proterozoic. Science, Moscow (In Russian)

Semikhatov MA (1985) Stromatolites in the Precambrian stratigraphy: analysis'84. Izv. AN USSR, Ser. Geol., 4: 3–21. (In Russian)

Semikhatov MA, Serebryakov SN (1983) The Riphean Hypostratotype in Siberia. Nauka, Moscow (In Russian)

Semikhatov MA, Raaben ME (1994) Dynamics of the global diversity of Proterozoic stromatolites. Article 1: northern Eurasia, China and India. Stratigr Geol Correl 2: 492–513

Semikhatov MA, Raaben ME (1996) Dynamics of the global diversity of Proterozoic stromatolites. Article II: Africa, Australia, North America, and general synthesis. Strat Geol Correl 4:24–50

Serebryakov SN (1975) Peculiarities of the Formation and Distribution of Riphean stromatolites in Siberia,: Nauka, Moscow. (In Russian)

Shapiro JA (1995) The significances of bacterial colony patterns. BioEssays. V. 17 (7): 597-607. (In Russian)

Raaben ME (1986) Topical problems of stromatolite systematics. In: Topical problems of the recent Paleoalgology. Naukova Dumka, Kiev 137–143 (In Russian)

Raaben ME (2000) Proterozoic stromatolite taxonomy and biostratigraphy. In: Riding RE, Awramik SM (eds) Microbial sediments. Springer-Verlag, Berlin.

Zavarzin GA (1993) Development of microbial communities in through Earth's history. In: Science (ed) Problems of preanthropogenic evolution of the biosphere. Moscow (In Russian)

Influence of Ionizing Radiation on Microfungi at Polar Latitudes

Irina Yu. Kirtsideli, Galina N. Zvereva, Andrej A. Vasilev, Natalya A. Kuzora, Lilit G. Vaganyan, Fedor A. Pak, Vadim A. Iliushin, Eduard M. Matchs, and Aleksandr I. Khalikov

Abstract We investigated the survival of extremophile microfungi (*Exophiala xenobiotica, Aureobasidium pullulans, Cladosporium herbarum, Rhodotorula colostri, Pseudogymnoascus pannorum)* under proton and gamma radiation in a dose range of 0–200 Gy. An increase in survival probability observed for all species of microfungi at doses less than 50 Gy may be explained in terms of hormesis. The increased survival rate was the highest for *Cladosporium herbarum* (140%) and the lowest for *Rhodotorula colostri* (107%). The latter may be due to the presence of carotene in *Rhodotorula colostri*, which blocks free radicals involved in hormesis. An increase in survival probability in *Cladosporium herbarum* observed across a wider range of doses up to 75 Gy may be a result of the presence of melanin in the cell wall. At doses of 100–200 Gy, the mycelial fungi *Pseudogymnoascus pannorum* had the lowest survival probability of 25%, while the corresponding probability of *Cladosporium herbarum* was 48%. At the same time, the probability of survival with gamma-irradiation was about 15% higher than with proton irradiation. Conversely, in the case of yeast fungi, the opposite effect was observed: the probability of survival under gamma irradiation was lower than under proton irradiation.

Keywords Microfungi · Polar region · DNA · Wall cells · Antioxidants · Astrobiology · Gamma- and Proton radiation

I. Yu. Kirtsideli (✉) · V. A. Iliushin · E. M. Matchs
V. L. Komarov Botanical Institute, Russian Academy of Sciences, St. Petersburg, Russia
e-mail: microfungi@mail.ru

G. N. Zvereva
State University of Civil Aviation, St. Petersburg, Russia

S. I. Vavilov State Optical Institute, St. Petersburg, Russia

A. A. Vasilev · N. A. Kuzora · L. G. Vaganyan · F. A. Pak · A. I. Khalikov
Petersburg Nuclear Physics Institute Named By B.P. Konstantinov, NRC "Kurchatov Institute", Gatchina, Russia

O. V. Frank-Kamenetskaya et al. (eds.), *Biogenic—Abiogenic Interactions in Natural and Anthropogenic Systems 2022*, Springer Proceedings in Earth and Environmental Sciences, https://doi.org/10.1007/978-3-031-40470-2_32

1 Introduction

Microfungi at polar latitudes are constantly exposed to low temperatures, physiological desiccation, and increased doses of UV radiation. As such, they can be considered as organisms having a high potential level of adaptability of life to extreme conditions of existence in outer space (Dadachova and Casadevall 2008; Onofri et al. 2012; 2015; Selbmann et al. 2017; Sazanova et al. 2019; Graham et al. 2020).

Resistance of microfungi to ionizing radiation is associated with increased content of sulfhydryl (SH-) groups and higher levels of endogenous thiol compounds (Tugai et al. 2007). resistance is promoted by the synthesis of melanin, a pigment with a high molecular weight, which may have a variety of biological functions (e.g., protection against predators, heavy metal toxicity, and physical influences) that confer an advantage in environmental survival (Saleh et al. 1988; Jacobson 2000; Nosanchuk and Casadevall 2003; Dadachova et al. 2007; 2008; Shuryak et al. 2014).

It is hypothesized that the increased resistance to unfavorable environmental conditions on the part of dormant spores is associated with species conservation (Khramtsov and Stefanovich 2011; Feofilova et al. 2012). In comparison with dormant spores, however, germinating spores are more sensitive to the action of ionizing radiation.

The study of tolerance to radiation, UV-, X-ray, gamma- and high-energy particle bombardment provides an understanding of the mechanisms of cell resistance in fungi (Zhdanova and Vasilevskaya 1982; Zvereva et al. 2018; Kirtsideli et al. 2017, 2018; Zvereva and Kirtsideli 2020, 2021).

The space environment exposes polar biota to both electromagnetic radiation, such as gamma rays, and bombardment by particles, such as atomic nuclei. There are three natural sources of cosmic radiation: (1) trapped radiation, i.e., particles captured by the Earth's magnetic field (trapped radiation); (2) corpuscular radiation from the Sun, i.e., particles ejected into space during solar flares (solar particle events or SPE); (3) galactic cosmic rays (GCR) (Picozza 2013).

The primary lethal damage in the genomes of cells exposed to ionizing radiation consists of DNA strand breaks. Such damage is caused either by direct interaction of DNA (target molecule) with high-energy particles or indirect interaction with radiation-induced free radicals (indirect effects due to reactive particle formation), such as oxygen radicals (Kirtsideli et al. 2017; Zvereva and Kirtsideli 2021).

Within the Earth's atmosphere, collisions of SPE or GCR particles with nuclei produce secondary energetic particles, nucleons, and nuclear fragments, as well as gamma rays and X-rays. When penetrating into substances, ionising radiation can affect biological mechanisms both directly, if the energy excites electrons inside biomolecules, and indirectly, if the radiation primarily interacts with water, generating highly reactive oxygen species (HRA).

While a number of investigations into the effects of radiation on microfungi have been carried out, the nature of the effect of radiation on polar fungi concerning disturbances in the biochemical processes of irradiated cells remains insufficiently studied. In addition, very little is known about the effect of radiation on the stimulation of growth processes at low doses. Since the stimulating effect of radiation depends both on the radiation dose and on the organism and type of radiation, understanding the characteristics of the radiation sources and the radiation conditions under which this effect is observed is of particular importance.

In particular, data on the effects of secondary particles and relativistic GCR protons on microfungi of polar latitudes have been poorly reported. Protons, representing the most common type of ionizing radiation in space, are capable of causing damage to cells regardless of their development cycle and oxygen saturation (McKenna-Lawlor et al. 2011). Therefore, understanding the interaction of environmental space parameters limiting survival over long periods of time with living organisms is essential both in the study of interplanetary transport of microorganisms due to natural collisions (lithopanspermia) and to investigate the consequences of human activities in space (including for planetary protection purposes (NASA 2005; Nicholson 2009; Nicholson et al. 2009) and for the search of biosignatures.

The aim of the present work is to investigate the survival features of microfungi of polar regions under proton and gamma irradiation at doses up to 200 Gy. Fungi with a wide range of biological characteristics were exposed: mycelial and yeast species, pigmented with various pigments (melanin, carotene), at different phases of colony development (stationary, exponential) and mycelial development (conidia and growth tubes).

2 Materials and Methods

2.1 *Fungi Cultivation*

Cultures of microfungi having both yeast and mycelial forms and containing different types of pigments (melanin, carotene) in both exponential and stationary phases of development were used in the experiments (Table 1).

All cultures of microfungi were isolated from Arctic and Antarctic substrates (collection of the Komarov Botanical Institute). Microfungi cultures were grown on standard media (Chapek) at the temperature of 18–20 °C, for 7–10 days (unless other

Table 1 Characteristics of the research subjects

Species	Growth form	Pigments	Object of irradiation
Exophiala xenobiotica	Yeasts	Melanium	Cells (propagules)
Aureobasidium pullulans	Yeasts	Melanium	Cells (propagules)
Cladosporium herbarum	Mycelial growth	Melanium	Conidia, growth tubes
Rhodotorula colostri	Yeasts	Carotene	Cells (propagules)
Pseudogymnoascus pannorum	Mycelial growth	Carotene	Conidia, growth tubes

parameters are specified). Microfungi were obtained from colonies at the exponential phase of their growth.

When studying the effect of age of microfungi on their survival, cultures were grown for a longer period (up to 2 months). A suspension of propagules was obtained by flushing from the surface by distilled water. For Cladosporium herbarum and Pseudogymnoascus pannorum, the suspension was divided into spores and mycelium by filtration. Suspensions of microfungi were placed in ependorphs for further studies (0.5 ml).

Following irradiation, 10–100 μl of fungal suspension samples were transferred to nutrient medium in Petri dishes. The microfungi were cultured in agar media at a temperature of 14 °C. After 10–15 days of cultivation, the number of colonies of fungi was counted and the survival probability (ratio of the number of colonies formed by irradiated spores to the number of colonies formed by unirradiated spores) was determined.

To determine the proportion of spore germination, the spore suspension was placed in liquid Chapek nutrient medium at 20 °C, and 100-field microscopy was performed after 24, 48, and 72 h in a Goryaev chamber. For micro examination, microscopy by Carl Zeiss AxioImager A1 was employed.

2.2 *Irradiation*

Prepared suspensions of fungal propagules were exposed to a monoenergetic proton beam of the medical channel of the Synchrocyclotron SC-1000 at the St. Petersburg Institute of Nuclear Physics (PIAP, Gatchina, Russia). This constitutes a primary proton source having an energy of 1 GeV, a linear energy transfer (LET) of ≈0.3 keV/μm, and an average relative biological effectiveness (RBE) value of 0.86 ± 0.05 (Konnov 1982; Elsayed 2020).

The prepared samples were arranged in a horizontal row in the direction of the radiation flux and irradiated evenly. The irradiation was performed in the irradiation hall of the laboratory of medical physics of PNPP (LMF). SC-1000 protons propagate in a strictly straight line and pass through the beam. The beam width at the half-height of radial dose distribution was chosen according to the size of the eppendorf and was 11 mm. The dose monitoring system (DMS) of the irradiation hall is represented by two plane-parallel ionization chamber monitors, which provide information on the dose rate in real time and its variation over time. Dose delivery was carried out continuously.

In addition to the effect of proton radiation, the biological effect of 6 meV photons induced by electron-induced braking radiation was studied. Photon beams formed in the radiation head of the linear gas pedal COMPACT by ELEKTA were given the form of a narrow pencil beam by braking a low-energy directional flow of electrons formed by an electron gun in a tungsten target. The generated high-energy photon radiation was then passed through a smoothing filter. The field irradiated by the symmetrical beam has dimensions of 30 cm × 30 cm at a source-surface distance

(RIS) of 100 cm. Within the irradiation field, the dose of photons generated by the target is uniformly distributed apart from a pronounced peak along the central beam axis. The solvation model based on density (SMD) is represented by two ELEKTA ionization chambers, which are constantly present in the photon beam and used to continuously monitor the dose rate and dose accumulation in monitor units (MU).

Before irradiating the microfungal propagules, the beam quality and parameters (energy, size, smoothing, symmetry) were checked to determine the absorbed dose. IBA dosimetric equipment was used to determine the absorbed dose of braking radiation photons in water: Blue Phantom 2 water phantom; two cylindrical cameras CC13 for relative dosimetry; a cylindrical camera FC65-G for absolute dosimetry; Dose 1 electrometer; general control unit CCU; OmniPro-Accept 7 data acquisition software.

For microscopic examination of the propagule surface, an atomic force microscope (AFM) NTegra Aura (NT-DT) was used. Surface analysis was carried out in semicontact mode.

DNA of fungi was isolated from control and irradiated spores using a commercial Phire Plant Direct PCR Master Mix kit (Thermo Scientific™); electrophoresis was performed using a Bio RAD Sub-cell Model 96 camera.

3 Results and Discussion

Data were obtained on the radiobiological response of propagules of microscopic fungi to the effects of proton and gamma radiation.

The research was carried out on the example of yeast fungi *Exophiala xenobiotica, Aureobasidium pullulans, Rhodotorula colostri.* Figure 1 shows the obtained survival probability curves of Rhodotorula colostri.

In *Rhodotorula colostri* the surfaces of propagules both of control (unirradiated) and of those exposed to different doses of proton and gamma radiation were studied using the AFM method. This confirmed that the dose of 200 Gy caused changes in the surface of irradiated *Rhodotorula colostri.* Here, the more visually discernible changes in the surface of spores was caused by the effect of protons. Exposure to gamma radiation gave a smaller effect at a dose of 200 Gy. The effect of protons and gamma radiation on the survival rate of propagules (yeast cells) was studied using the example of the microfungis *Rhodotorula colostri.* The effect of the dose up to 200 Gy (both proton and gamma radiation) was shown to lead to the death of fewer than 50% of propagules. Exposure to 200 Gy of gamma radiation had a relatively greater lethal effect on *Rhodotorula colostri* propagules.

Electrophoresis of DNA extracted from irradiated and control propagules showed that partial DNA destruction already occurs at a dose of 100–150 Gy. At a dose of 200 Gy, DNA destruction was more significant both with proton and gamma exposure.

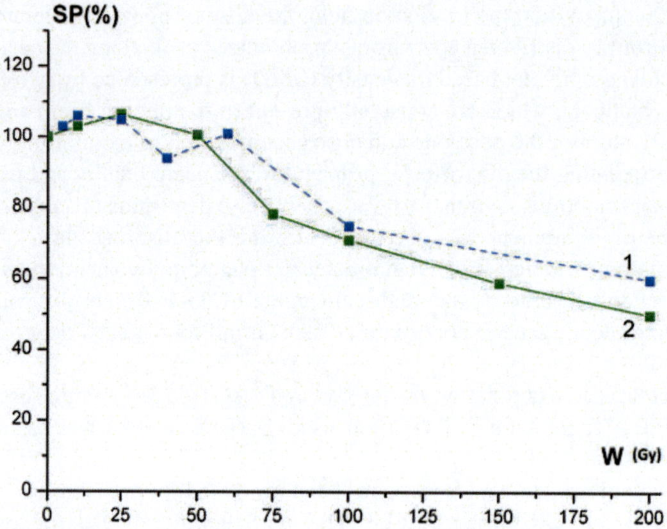

Fig. 1 Survival probability of *Rhodotorula colostri* propaguls by influence of proton (1) and gamma radiation (2). On the ordinate axis—survival rate (%), on the abscissa axis—irradiation dose (Gy)

Figures 4 and 5 depict the obtained survival probability curves of *Aureobasidium pullulans and Exophiala xenobiotica* due to the influence of proton and gamma radiation. When considering the effect of protons and gamma radiation on the survival rate of propagules of microfungi *Aureobasidium pullulans* and *Exophiala xenobiotica.* (Figs. 4 and 5), exposure to doses up to 200 Gy (both proton and gamma radiation) is shown to result in the death of fewer than 30 and 20% of propagules. As compared to proton irradiation, exposure to 200 Gy gamma radiation had a stronger lethal effect on both *Aureobasidium pullulans* and *Exophiala xenobiotica.* The latter is probably a more melanized species than *Aureobasidium pullulans* in the early stages of its development.

Figure 6 shows the influence of the development phase on the survival rate of *Exophiala xenobiotica* propagules under the influence of gamma irradiation. Cultures in a stationary phase of development are shown to have a greater proportion of propagules, which maintain viability even under the influence of a dose of 500 Gy, as well as a smaller peak in the region of 0–50 Gy. This is probably connected with the fact that, despite the lower share of viable propagules in the initial culture, the propagules which maintained viability are transformed into chlamydospores, thus developing a thicker shell in which melanin is accumulated.

Although demonstrating similar behavior patterns, the yeast fungi *Aureobasidium pullulans, Exophiala xenobiotica, Rhodotorula colostri* exhibited different tolerances to gamma and proton radiation.

According to the results shown in Figs. 1, 4, 5 gamma irradiation in the dose range of 0–200 Gy leads to the death of a greater number of cells than proton irradiation in all the studied species. The hierarchy of resistance to irradiation can be

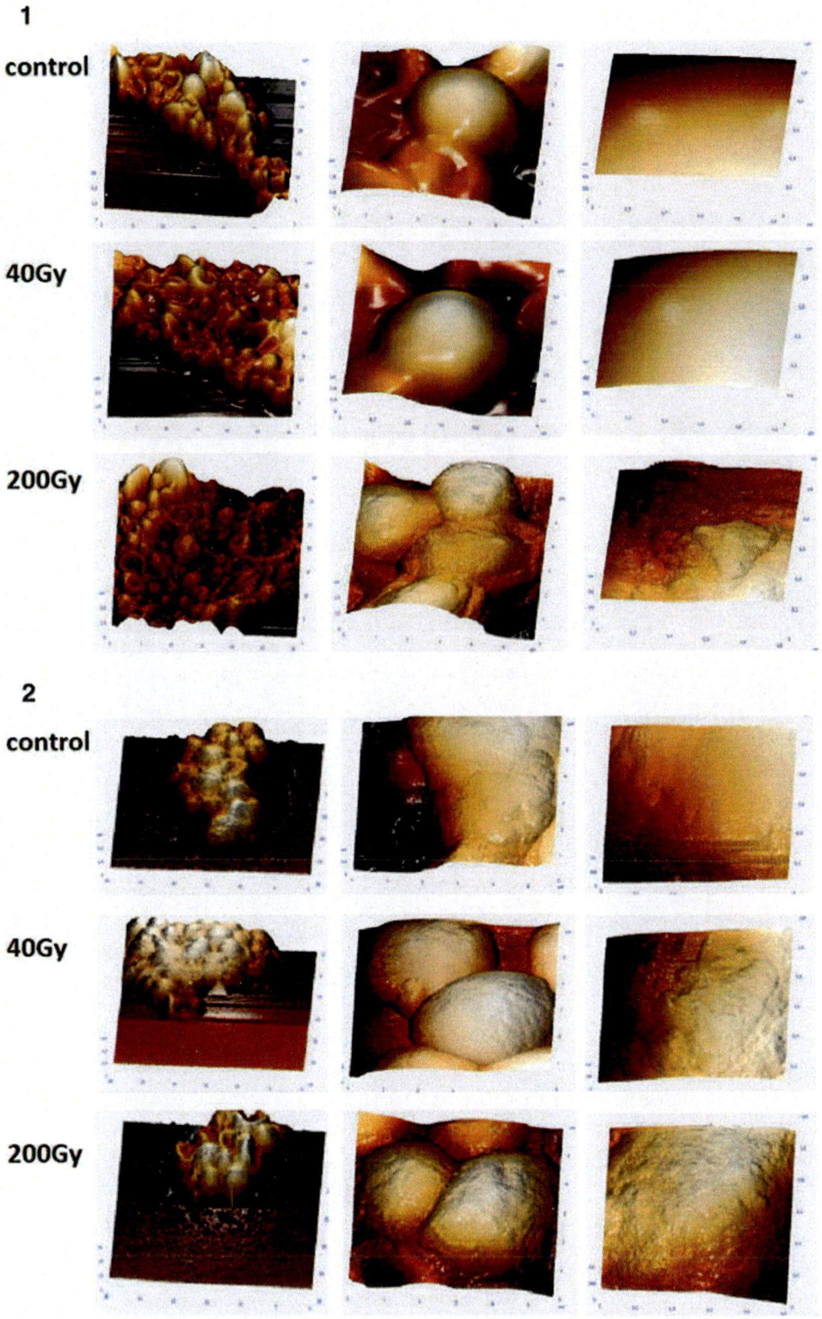

Fig. 2 Influence of proton (Fig. 2-1) and gamma (Fig. 2-2) radiation on *Rhodotorula colostri* surface

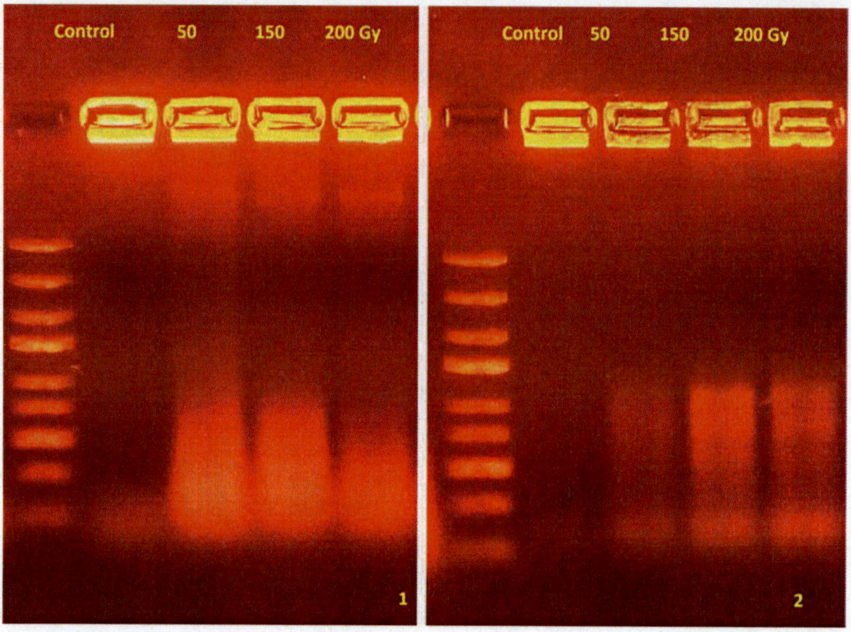

Fig. 3 DNA electrophoresis of *Rhodotorula colostri* by influence of gamma and proton radiation

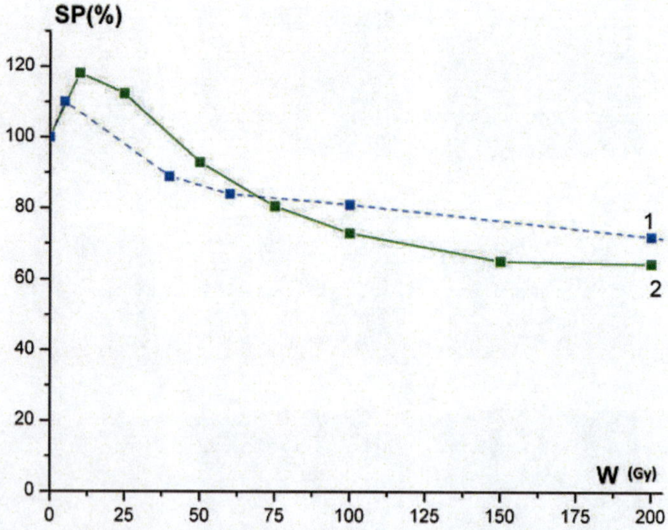

Fig. 4 Survival probability of *Aureobasidium pullulans* propaguls by influence of proton (1) and gamma radiation (2). On the ordinate axis—survival rate (%); on the abscissa axis—irradiation dose (Gy)

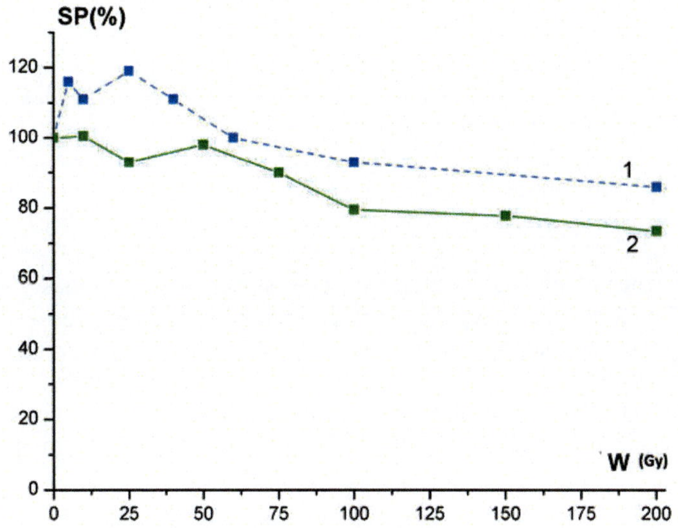

Fig. 5 Survival probability of *Exophiala xenobiotica* propaguls by influence of proton (1) and gamma radiation (2). On the ordinate axis—survival rate (%); on the abscissa axis—irradiation dose (Gy)

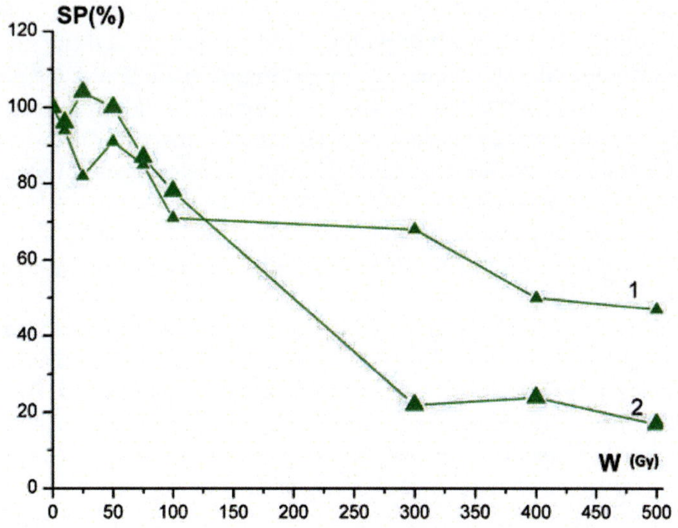

Fig. 6 Survival of *Exophiala xenobiotica* propaguls of different ages by influence of gamma radiation: 1—stationary phase (age 90 days); 2—exponential phase (age 8 days); on the ordinate axis—survival rate (%); on the abscissa axis—irradiation dose (Gy)

established in the following order: *Rhodotorula colostri* < *Aureobasidium pullulans* < *Exophiala xenobiotica.* This hierarchy may be related to different pigmentation: *Rhodotorula colostri* contains carotene, while *Aureobasidium pullulans* and *Exophiala xenobiotica* contain melanin.

Apart from the possibility of the observed resistance to proton and gamma irradiation being enhanced by the melanized cell wall of the studied fungi, this could also be promoted by the varying hypobiosis or anhydrobiosis capability of different strains.

The presence of stimulation of growth processes at doses less than 50 Gy was observed for all investigated species of yeast forms of microfungi. This is explicable in terms of the well-understood phenomenon of hormesis. The greatest increase of survival rate (SP = 120%) was achieved in *Aureobasidium pullulans* and *Exophiala xenobiotica.* The lower stimulation effect in *Rhodotorula colostri* can be explained by the presence of carotene, which blocks free radicals triggering hormesis.

3.1 Survival Probabilities of Mycelial Microfungi

The effect of protons and gamma irradiation on the survival of mycelial fungi was considered on the example of *Cladosporium herbarum* and *Pseudogymnoascus pannorum* (Figs. 7, 8). The effect of proton irradiation led to the death of a greater proportion of propagules as compared to yeast fungi. Irradiation at the germ tube stage increased the death rate of propagules as compared to the effect on spores.

According to the results presented in Figs. 7 and 8, mycelial fungi have lower survival probability in the dose range of 100–200 Gy as compared to yeast fungi (*Pseudogymnoascus pannorum*—25%; *Cladosporium herbarum*—48%). At the same time, the survival probability under gamma-irradiation exceeds the probability of survival under proton irradiation by about 15%, which is usually observed in experiments of this type.

In the dose range up to 50 Gy, an increase in the probability of survival in micelium and yeast fungi is observed. In this case, SP reaches higher values than in yeasts: in *Cladosporium herbarum*—SP = 140% in proton irradiation; in *Pseudogymnoascus pannorum*—SP = 130% in gamma-irradiation. In *Cladosporium herbarum,* increased survival probability was observed across a wider range of doses up to 75 Gy. This may be explained by the presence of melanin in the cell wall.

3.2 Delay in the Process of Spore Germination

The effect of exposure to protons and gamma radiation on the delay of spore germination (time of formation of germ tubes) of microfungi *Pseudogymnoascus* and *Cladosporium* was also considered for the case of mycelial fungi (Figs. 9 and 10).

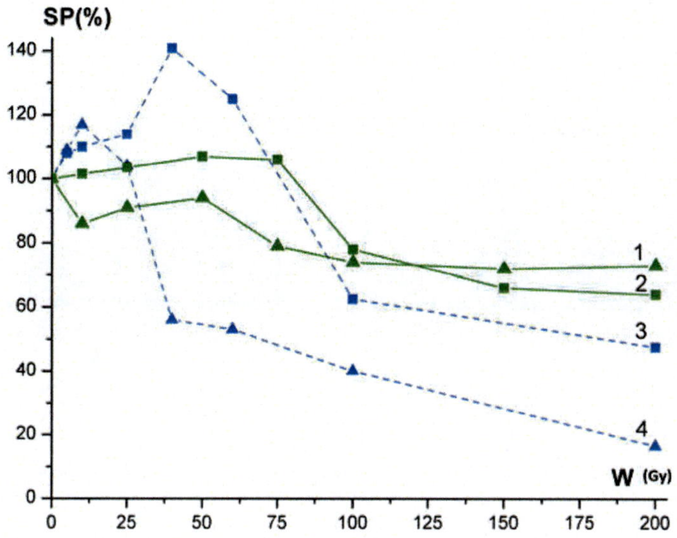

Fig. 7 Survival of *Cladosporium herbarum* by influence of proton and gamma radiation: 1—gamma radiation germ tube; 2—gamma radiation spores; 3—proton-radiation spores; 4—proton-radiation germ tube. On the ordinate axis—survival rate (%); on the abscissa axis—irradiation dose (Gy)

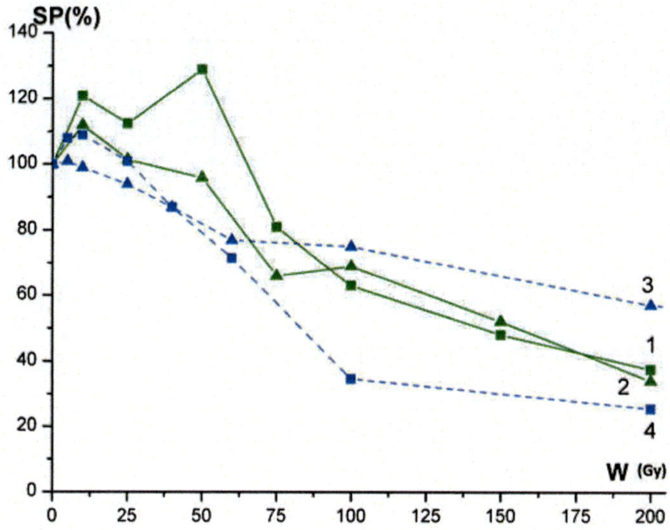

Fig. 8 Survival of *Pseudogymnoascus pannorum* by influence of proton and gamma radiation: 1—gamma radiation germ tube; 2—gamma radiation spores; 3—proton-radiation spores; 4—proton-radiation germ tube. On the ordinate axis—survival rate (%); on the abscissa axis—irradiation dose (Gy)

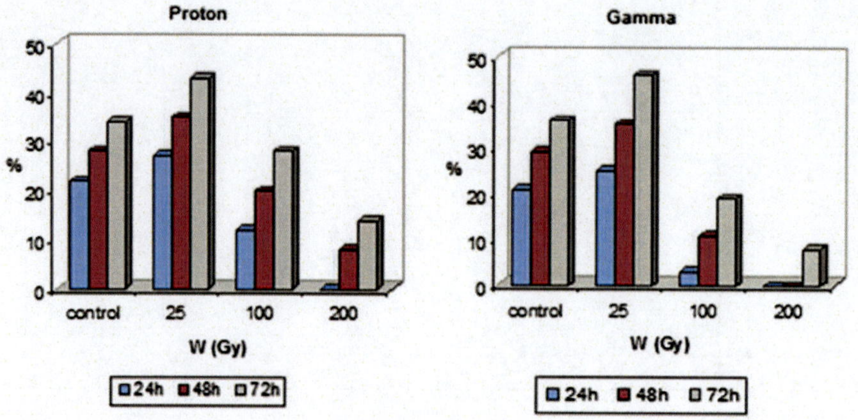

Fig. 9 Spores germinated (%) of *Pseudogymnoascus pannorum* after proton and gamma irradiation

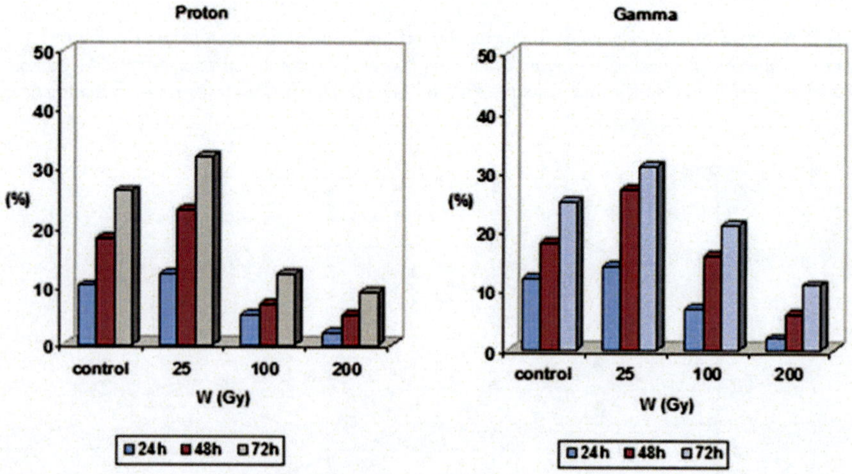

Fig. 10 Spores germinated (%) of *Cladosporium herbarum* after proton and gamma irradiation

Our investigation into the viability of fungi after proton and gamma irradiation showed that proton and gamma irradiation reduce the viability of fungal spores (in vitro). At the same time, a 200 Gy dose did not lead to 100% inhibition of growth of the species in this study. In the case of *Pseudogymnoascus pannorum* proton irradiation, 22% of the spores germinated 24 h after irradiation, germinating 27% of the spores at a dose of 25 Gy. Doses of 100 and 200 Gy resulted respectively in delayed germination (12%) and the complete absence of germination (0%) after 24 h. After 72 h, the % of germinated spores increased to 34% in the control; when irradiated with 200 Gy, the corresponding figure was 14%. Similar trends were observed in the study of growth tube formation under gamma irradiation. The surviving *Pseudogymnoascus pannorum* showed a 24-h delay in spore germination (germ tube formation) under proton irradiation and a 48-h delay in germination under gamma irradiation at the same dose of 200 Gy. The germ tube length of *Pseudogymnoascus pannorum* and *Cladosporium herbarum* was observed to decrease when exposed to proton and gamma radiation.

While a delay in spore germination was observed in *Cladosporium herbarum*, a complete cessation of germ tube formation was not observed even at the 24-h time limit. This indicates that *Pseudogymnoascus pannorum* is more sensitive to irradiation than *Cladosporium herbarum*, possibly due to the effect of melanin, which has also antioxidant properties.

Small radiation doses (up to 50 Gy) not only did not cause a temporary inhibition of growth tube formation, but positively stimulated the rate of conidia germination. Exposure to radiation doses less than 50 Gy resulted in growth activation, along with an increase in the number of germinated conidia and in the stimulation of growth processes in all fungal species. Spores only began to lose the ability to germinate at the dose of 200 Gy.

Individual fungal responses to irradiation differed markedly among the microfungi species of polar region at this study. Fungi, especially melanized fungi, are very radioresistant. Cells may be more radioresistant in the G2 phase of the cell cycle. Resistance may be increased by the presence of two homologous copies of DNA. Our current results (in vitro) and those of (Saleh et al. 1988) clearly indicate that the response of these microfungi is directly related to cellular mechanisms. That is, the radiosensitivity of fungi depend not only on the individual response of the fungis to radiation, but also the phase of development. A significant decrease in the number of germ tubes and a decrease in the length and width of growth tubes all indicate a general inhibition of fungal growth due to irradiation. The increased germination rate and formation of growth tubes suggest that radiation has a stimulating effect with respect to the control germ tubes. In addition, the change in sensitivity to radiation may be related to the life cycle and the ability to recover.

The in vitro results show a decrease in the viability of all investigated fungi at the limit of 200 Gy. Differences in the response of fungi to irradiation, reflected in the delay in germination of irradiated fungal spores and their reduced spore viability, depend on the dose and morphological features (presence of melanin). The radiosensitivity of fungi reflects species specificity, which may have been related to individual adaptive strategies acting under stress conditions.

When studying the effect of these radiation sources on the process of conidia germination in the fungal strains in our study, some regularities were observed. Under the influence of gamma and proton irradiation, both activation of the conidia germination process and suppression of this process were noted. This confirms that extremophile fungi have different adaptive mechanisms in the realization of responses to different doses and types of radiation.

The data obtained can be considered within various current theoretical frameworks. We hypothesize that reactive oxygen species play a major role in the observed effects. Actions on metabolic processes inside the cell to influence on the level of information transfer inside the cell culture and microorganisms may be diverse. Accordingly, in this experimental model, it is possible that processes leading to the formation of a new generation of growth-promoting metabolites, activation of enzymatic systems responsible for this process take place in the process of formation of reactive oxygen species, as well as their interaction with the components of the growth medium of the fungal culture.

4 Conclusions

In this work we obtained survival probabilities of extremophile microfungi of *Aureobasidium pullulans, Cladosporium herbarum, Exophiala xenobiotica, Pseudogymnoascus pannorum, Rhodotorula colostri* species under proton and gamma radiation at at dose interval of 0–200 Gy.

The increase in survival probability observed at doses less than 50 Gy for all species of fungi may be related to the phenomenon of hormesis. The increased survival rate was the highest for *Cladosporium herbarum* (140%) and the lowest for *Rhodotorula colostri* (107%). The latter may be explained by the presence of carotene in Rhodotorula colostri, which blocks free radicals involved in hormesis. In *Cladosporium herbarum*, the increased survival probability observed in a wider range of doses up to 75 Gy can be explained by the presence of melanin in the cell wall.

At doses of 100–200 Gy, the mycelial fungi *Pseudogymnoascus pannorum* had the lowest survival probability (25%), while that of *Cladosporium herbarum* was 48%. At the same time, the probability of survival with gamma-irradiation was about 15% higher than with proton irradiation. In the case of yeast fungi, the opposite effect was observed: the probability of survival under gamma irradiation was lower than under proton irradiation.

The survival probabilities of microfungi spores under proton- and gamma irradiation were obtained. The results of the work indicate the existence of several parallel mechanisms, which lead to spore inactivation by proton and gamma irradiation. These include: destruction of the cell wall and DNA by means of direct absorption of radiation, as well as indirectly by means of reactive products of photolysis.

SP is dependent on the presence of melanin and carotin in the spore cell wall. The inactivation effect can be can be caused by two mechanisms: membrane damage and DNA destruction. Electrophoresis of DNA isolated from irradiated spores shows the formation light fragments. The influence of age, life stage of microfungi and the presence of Al_2O_3 on the survival of microfungi, as well as an effect related to the origin of the isolate, have been demonstrated.

Acknowledgements This study was carried out as part of the state assignment according to the thematic plan of the Botanical Institute of the Russian Academy of Sciences (theme No. AAAA-A19-119020890079-6). The research was carried out using equipment of the "Cell and Molecular

Technologies in Plant Science" Core Facilities Center at the Komarov Botanical Institute RAS (St.-Petersburg, Russia).

References

Dadachova E, Bryan RA, Howell RC, Schweitzer AD, Aisen P, Nosanchuk JD, Casadevall A (2008) The radioprotective properties of fungal melanin are a function of its chemical composition, stable radical presence and spatial arrangement. Pigment Cell & Melanoma Research. 21(2): 192 – 199.

Dadachova E, Bryan RA, Huang X, Moadel T, Schweitzer AD, Aisen P, et al. (2007) Ionizing radiation changes the electronic properties of melanin and enhances the growth of melanized fungi. PloS one. 2:e457. https://doi.org/10.1371/journal.pone.0000457

Dadachova E, Casadevall A (2008) Ionizing radiation: how fungi cope, adapt, and exploit with the help of melanin. Current opinion in microbiology. 11(6):525–531. https://doi.org/10.1016/j.mib.2008.09.013

Elsayed KOM (2020) Effect of the high-energy proton component of cosmic rays on the structure of DNA. Dissertation, St. Petersburg

Feofilova EP, Ivashechkin AA, Alyokhin AI, Sergeeva YE (2012) Spores of fungi: rest, germination, chemical composition and importance for biotechnology (review). Applied biochemistry and microbiology. 48(1): 5–17 (in Russian)

Graham K, Shunk-Xavier R, Gomez NJH, Averesch A (2020) Self-Replicating Radiation-Shield for Human Deep-Space Exploration: Radiotrophic Fungi can Attenuate Ionizing Radiation aboard the International Space Station. ASM Press; Washington.

Jacobson ES (2000) Pathogenic roles for fungal melanins. Clin.Microbiol. Rev. 13: 708–717.

Kirtsideli IYu, Parfenov VA, Zvereva GN, Petrov A, Grigor'eva NO (2017) Effect of ultraviolet radiation from a pulsed ArF laser on the viability of microfungi. Journal of Optical Technology (A Translation of Opticheskii Zhurnal) 84(9): 593–597. https://doi.org/10.1364/JOT.84.000593

Kirtsideli IY, Vlasov DY, Novozhilov YK, Abakumov EV, Barantsevich EP (2018) Assessment of Anthropogenic Influence on Antarctic Mycobiota in Areas of Russian Polar Stations. Contemporary of Ecology. 11(5): 449–457. https://doi.org/10.1134/S1995425518050074

Khramtsov AK, Stefanovich AI (2011) Mycology: guidelines for a special course on the section. Ecology of fungi and fungus-like organisms. Minsk. (in Russian)

Konnov BA (1982) The Use of a Proton Beam with the Energy of 1,000 MeV for Radiation Therapy. Dissertation, Leningrad. (in Russian)

McKenna-Lawlor S, Gonçalves P, Keating A, Reitz G, Matthiä D (2011) Overview of energetic particle hazards during prospective manned missions to Mars. Planet Space Sci. 63–64:123–132

NASA (2005) Planetary protection provisions for robotic extraterrestrial missions. National Aeronautics and Space Administration. Washington.

Nicholson WL (2009) Ancient micronauts: interplanetary transport of microbes by cosmic impacts. Trends Microbiol. 17:243–250.

Nicholson WL, Schuerger AC, Race MS (2009) Migrating microbes and planetary protection. Trends Microbiol. 17:389–392.

Nosanchuk JD, Casadevall A (2003). The contribution of melanin to microbial pathogenesis. Cell Microbiol. 5: 203–223.

Onofri S, de la Torre R, de Vera JP, Ott S, Zucconi L, Selbmann L (2012) Survival of rock-colonizing organisms after 1.5 years in outer space. Astrobiology. 508–516.

Onofri S, de Vera JP, Zucconi L, Selbmann L, Scalzi G, Venkateswaran KJ (2015) Survival of antarctic cryptoendolithic fungi in simulated martian conditions on board the international space station. Astrobiology. 1052–1059.

Picozza PMB (2013) Multi messenger astronomy and CTA: TEV cosmic rays and electrons. Astroparticle physics. 43: 163-170.

Saleh YG, Mayo MS, Ahearn DG (1988) Resistance of some common fungi to gamma irradiation. Appl. Environ. Microbiol. 54:2134–2135.

Sazanova KV, Senik SV, Kirtsideli IY, Shavarda AL (2019) Metabolomic Profiling and Lipid Composition of Arctic and Antarctic Strains of Micromycetes Geomyces pannorum and Thelebolus microsporus Grown at Different Temperatures. Microbiology (Russian Federation). 88(3): 282–291

Selbmann L, Pacellia C, Zucconi L, Dadachova E, Moeller R, Vera JP, Onofri S (2017) Resistance of an Antarctic cryptoendolithic black fungus to radiation gives new insights of astrobiological relevance. 22(6):546-554. https://doi.org/10.1016/j.funbio.2017.10.012.

Shuryak I, Bryan RA, Nosanchuk JD, Dadachova E (2014) Mathematical modeling predicts enhanced growth of X-ray irradiated pigmented fungi. PloS one. 9(1):e85561. https://doi.org/10.1371/journal.pone.

Tugai TI, Zhdanova NN, Zheltonozhsky VA, Sadovnikov LV (2007) Manifestation of radioadaptive properties in microfungi long staying in the territories with elevated background radiation after the Chernobyl accident. Radiation Biology. Radioecology. 27(5): 543 – 549. (in Russian)

Zhdanova NN, Vasilevskaya AI (1982) Extreme ecology of fungi in nature and experiments. Kiev. (in Russian)

Zvereva G, Kirtsideli I, Machs E, Vangonen A (2018) Mechanisms of the effect of VUV radiation on the microfungi. Proc. SPIE 10614, International Conference on Atomic and Molecular Pulsed Lasers XIII, 106141S. https://doi.org/10.1117/12.2303532.

Zvereva GN, Kirtsideli IYu (2020) Features of the Effect of VUV radiation on microfungi from polar regions. Chapter 32. 2020. https://doi.org/10.1007/978-3-030-21614-6_32.

Zvereva GN, Kirtsideli IYu (2021) Irradiation of microorganisms by vacuum ultraviolet radiation. Journal of Optical Technology. 88(8): 454–459. https://doi.org/10.1364/JOT.88.000454

Medical Geology

Medical Geology: Status and Prospects of the Science in Russia and in NIS

Iosif F. Volfson, Marina V. Prozorova, Olga I. Yakushina, Igor G. Pechenkin, Elena V. Kremkova, and Iskhak M. Farkhutdinov

Abstract Medical geology is a relevant interdisciplinary branch of natural science that studies the influence of geological materials, objects, processes and phenomena on the health of the population, animals and biota in general. Medical geology is a dynamically developing scientific field that unites the geological, geochemical, biological and medical communities in solving of wide range of environmental problems. Medical geology, in fact, is a response to one of the most acute challenges of the time—the existence of man as a biological species. Among the priority tasks of medical geology is the integration of various branches of biomedical and geological sciences into a complex system of knowledge about the health of living organisms.

Keywords Medical geology · Environmental geochemistry · Biogeochemical provinces · Biogeochemical endemicity · Selenium · Iodine · Fluorine · Manganese

I. F. Volfson (✉)
Institute of Mineralogy, Geochemistry and Crystal Chemistry of Rare Elements (FSBI IMGRE) st. Veresaeva, h.15, Moscow 121357, Russia
e-mail: imgre@imgre.ru

M. V. Prozorova · O. I. Yakushina · I. G. Pechenkin
All-Russian Institute of Mineral Resources (FSBI VIMS) Staromonetny per., h.31, Moscow 119017, Russia

E. V. Kremkova
Pirogov Russian National Research Medical University, Ostrovitianov st. 1, Moscow 117997, Russia

I. M. Farkhutdinov
Vernadsky State Geological Museum, Mokhovaya st., 11, building 11, Moscow 125009, Russia

O. V. Frank-Kamenetskaya et al. (eds.), *Biogenic—Abiogenic Interactions in Natural and Anthropogenic Systems 2022*, Springer Proceedings in Earth and Environmental Sciences, https://doi.org/10.1007/978-3-031-40470-2_33

1 Introduction

Medical geology is a unique socially oriented project, the scientific and methodological basis of which is the knowledge and skills of number of disciplines of natural sciences, both geological and biomedical. The purpose of this project was initially to create a community of scientists who study the impact of geological objects of natural and technological origin and geological processes and phenomena on the health of population and animals, on the condition of plants. This work began about a quarter of a century ago, when working group on medical geology (IWGMG) was created under the auspices of UNESCO, and then the International Union of Geological Sciences (IUGS). IWGMG consisted of some famous scientists—geologists, geochemists, toxicologists, representing universities, research institutes and geological prospecting organizations from Sweden, the USA, England and first of all Olle Selinus, Robert Finkelman, Jose Centeno, Edward Derbyshire. In the late 1990s IWGMG activities crowned with great success and led to the joining of scientists from more than 50 countries, including Russia. In 2004, at the XXXII Session of the International Geological Congress held in Florence, Italy the International Medical Geological Association (IMGA www.medicalgeology.org) was officially established.

Scientists and specialists from the Russian Federation and Newly Independent States (NIS), play a significant role in IMGA activities. They have been united in the IMGA Russia-NIS Regional Chapter under NGO Russian Geological Society (ROSGEO) umbrella.

2 About the History of Medical Geology

Medical geology has its own history in Russia and the NIS. The foundation of modern research in the field of basic disciplines of medical geology—mineralogy, geochemistry, geodynamics, geoecology, hydrogeology, volcanology, biogeochemistry, soil sciences, biology, microbiology, pathology, epidemiology, balneology, etc.—was laid by the works of outstanding predecessors—Avicenna, Biruni, M.V. Lomonosov, D.I. Mendeleev, V.I. Vernadsky, A.P. Vinogradov, A.I. Perelman, A.N. Sysin, V.V. Kovalsky, A.P. Avtsyn, Yu.E. Saet, P.G. Tsarfis, N.P. Yushkin, A.V. Vikulin and many others. Currently, various aspects of medical geology are being developed by representatives of research organizations and universities of the geological and medical profile in Russia and the NIS such as National Research Tomsk Polytechnic University (Tomsk), Institute of Earth Sciences of St. Petersburg State University, Institute of Mineralogy, Geochemistry, Crystal Chemistry of Rare Elements (Moscow), Institute of Problems of Industrial Ecology of the North Federal Research Center "Kola Scientific Center of the Russian Academy of Sciences", Institute of Geology of the Komi Scientific Center of the Ural Branch of the Russian Academy of Sciences, Center of Noospheric Research of National Academy of Sciences of Republic of

Armenia and others. Scientists and practitioners of Russia and NIS study the sources and haloes and scenarios of accumulation of toxic elements in the components of the environment—the near-surface geosphere, hydrosphere and atmosphere (Volfson et al. 2006; Orlov et al. 2018).

2.1 Essentials of Medicinal Geology

The interpenetration of geology and medicine goes deep into human civilization history. Ancient Chinese, Egyptian, Greek, and Islamic chronicles describe considerable therapeutic significance of various rocks and minerals as well as harm they do to health. Such scholars of the antiquity as Ge Hong, Avicenna, Hippocrates, James Cook, Georgius Agricola, Paracelsus and others were pioneers in studying the interrelation between the geological features and human health. Many centuries ago minerals and rocks were used as medicinal agents by people living in mountain areas, while in the lowland areas various plants were predominant. In accordance to G.A. Pelymsky, Professor of Lomonosov Moscow State University, different areas of medicine were developing through ages, accumulating their own selection of medicinal products and traditions of treatment reflecting geographical and geological aspects of the areas, for example of Tibet, East and West Europe and so on (Pelymsky 2008).

People living in the territory of Russia and the NIS both accumulated their own experience and borrowed that of the people of the Near East, Middle East and Far East and West Europe. Research carried out by the outstanding scholars of the Middle Ages, such as Avicenna, Biruni and others, involved detailed studies into the properties of minerals, including their physical condition, chemical qualities and healing properties. Avicenna in his well-known *The Canon of Medicine* (eleventh century) gave recommendations on the use of more than thirty curative minerals and described the recipes of many medicinal preparations based on them. Later, another remarkable scientist Biruni summarized the early period of studying and using gems in his treatise *The Gems*, which contained data on 450 minerals and naturally occurring compounds. Also, Biruni represented the data of research of semiprecious stones possessing healing qualities.

According to the data of academician N.P. Yushkin, the history of the theory of curative properties of minerals in Russia started with the introduction of Christianity in Ancient Rus in the tenth century, with the epoch of Vladimir the Great. The first reported data on the medicinal properties of a number of minerals can be found in *The anthologies* by Sviatoslav compiled in 1073, in *The Vetrogrady* (fifteenth-seventeenth centuries), in *The Trading Book* (sixteenth-seventeenth centuries) and in *The Azbukovnik* (seventeenth -nineteenth centuries) (Yushkin 2003, 2004).

The pharmacopeia began developing in Russia in the early seventeenth century during the reign of Peter the Great, but as early as in the sixteenth century there were pharmacies where one could buy medicinal preparations based on such minerals and chemical compounds as saltpeter, potash, rosin, sulfur, borax, mercury, verdigris, wax, and also precious stones—sapphire and ruby which were used in making medicines (Volfson et al. 2010a, b; Volfson 2016).

The traditions of medical mineralogy that have been establishing through years are living today and are developing both in Russia and in the NIS. Natural minerals and their compounds are being actively studied and used. Materials based on the known geological formations—minerals and rocks, also combined with medicinal herbs, possess essential properties and are used in cosmetic products, for the prevention and treatment of different diseases, ranging from skin diseases to a broad array of visceral diseases, consequences of injuries and fractures.

Currently medicine is making use of approximately 500 known mineral forms, including zeolites, shilajit, shungites, flints, amber, different clay minerals, minerals of ores of nonferrous and rare earth metals, etc. Such mineral formations as shungites, clay minerals, muds, moor have been successfully used in SPAs (Volfson 2013a, b).

2.2 *Mineral and Thermal Waters*

In Russia and the countries of the NIS and neighboring countries, invaluable experience has been accumulated in water and mud therapy, based on the traditions of the late Middle Ages. Examples are the urban spa resort founded in Moscow by Dr. Justus Christian Loder in the 18-nineteenth century, the Caucasian mineral water resorts—Essentuki, Pyatigorsk, Kislovodsk, Zheleznovodsk, Archman in Turkmenistan, Borjomi in Georgia, Arzni in Armenia and many others. Currently, 273 deposits of mineral waters are known on the territory of the Commonwealth countries. Of these, 57 objects are used for the treatment of cardiovascular diseases (carbonate waters), diseases of the joints and spine, as well as diseases of the peripheral nervous system (sulfide waters)—38 objects (Tsarfis 1985; Tsarfis 1991; Adilov and Razumov 2010; Volfson 2013a, b; Pechenkin et al. 2022; Zhidkova et al. 2023) (Table 1).

3 Geochemistry and Human Health

It seems important to note that systematic studies of the influence of natural factors on human health in Russia and the NIS are currently sporadic. On the one hand, the reason may be the low population density in many regions and, accordingly, the difficulties encountered in conducting statistically reliable epidemiological studies. On the other hand, the intensive industrialization of the economy has led to a high concentration of the population in industrial centers, where people's health is exposed to the products of technological processes, waste of various genesis and composition, derivatives of building materials. All of the above requires significant efforts

Table 1 Health benefits of hot spring bathing (after Finkelman 2008; reproduced with his kind permission; additional credits to M. Takeuchi, National Institute for Advanced Industrial Science and Technology, Japan Volfson 2013a, b)

Type of spring disease susceptible to treatment	Simple hot	Chloride	Hydrogen carbonate	Sulfate	Carbon dioxide	Iron	Sulfur	Acid	Radon
Neuralgia	+	+	+	+	+	+	+	+	+
Muscular pain	+	+	+	+	+	+	+	+	+
Movement paralysis	+	+	+	+	+	+	+	+	+
Frozen shoulder	+	+	+	+	+	+	+	+	+
Joint diseases	+	+	+	+	+	+	+	+	+
Bruise	+	+	+	+	+	+	+	+	+
Chronic digestive diseases	+	+	+	+	+	+	+	+	+
Hemorrhoids	+	+	+	+	+	+	+	+	+
Catarrhal diseases	+	+	+	+	+	+	+	+	+
Skin cracks		+	+	+	+		+		
Skin burns		+	+	+	+				
Chronic skin disease		+	+	+				+	+
Gynecological diseases		+					+		+
Arteriosclerosis				+	+		+		+
Hypertension					+		+		+
Menstrual disorders						+			
Diabetes							+		
Gout									+
Chronic cholecystitis									+
Cholelithiasis									+

in determining the background concentrations of chemical elements in the components of the environment and, therefore, their sources of origin, forms of occurrence, the creation of methodological documents to update the classification of chemical elements by toxicity, and, conversely, essentiality, from the standpoint of new medical and environmental data research. The more important for understanding the relationship between man and geological objects are the results of research conducted in the last century on the territory of the former USSR (Saet et al. 1990). Then invaluable information was obtained on the geological conditions of the presence of fluorine, iodine, selenium, strontium, arsenic and other biologically active chemical elements in the components of the environment and the consequences of their impact on the human body were studied. The outstanding domestic geochemist A.P. Vinogradov established a relationship between the concentrations of certain chemical elements in natural environments and human and animal health problems. He put forward the hypothesis of biogeochemical provinces and biogeochemical endemicity (Vinogradov 1964). The data given in the works of the Biogeochemical Laboratory, headed by A.P. Vinogradov, and later, in the 1970s of the last century, by his follower V.V. Kovalsky, indicate that relatively small excesses of elements in natural environments—2–3 times higher than Clarke concentrations—can cause serious health problems among the local population (Kovalsky 1974). In 2004 A.A. Golovin proposed the concept of geochemical endemicity (Golovin et al. 2004) (Fig. 1). It depends on the parameters and characteristics of a particular region and manifests itself against the background of an excess or, conversely, a decrease in the contents of elements in natural environments relative to their Clarke concentrations. In particular, the cause of Kashin-Bek's disease (or "Urov's disease") is the geochemical features of the territories of its manifestation. This is an endemic disease (deforming osteochondral arthritis), which is inherited, is found in some areas of Transbaikalia, Tajikistan, China and Korea (Rikhvanov et al. 2021). A number of researchers believe that the cause of its occurrence is strontium, phosphorus and manganese in excess present in soils that suppress calcium. This leads to a critical decrease in the values of the Ca/P ratio in the bone tissue. However, the results of more recent studies highlight the role of selenium in the occurrence of this disease. It has been proven that selenium is responsible for the functions of iodine, phosphorus and calcium in the human body. It is precisely when it is deficient in food chains that the balance of the Ca/P ratio in the bone tissue is disturbed, which, ultimately, is the cause of urologic disease, as well as the appearance of foci of Keshan disease (cardiomyopathy), the signs of which also appear in regions with low selenium levels in components of the natural environment (Essentials 2005). According to Professor B.A. Revich (Revich 2001), more than 90 percent of the Russian population do not receive fluorine in the required amount. Especially characteristic is the lack of this microelement for surface sources of drinking water supply in the northern territories: the Kola-Karelian region, the Arkhangelsk and the Leningrad regions, the Republic of Komi. Fluorine deficiency in water is a factor in the increased incidence of about 60% of residents with dental caries. On the other hand, in a number of territories there are examples of fluorine contamination of water supply sources with the development of severe forms of

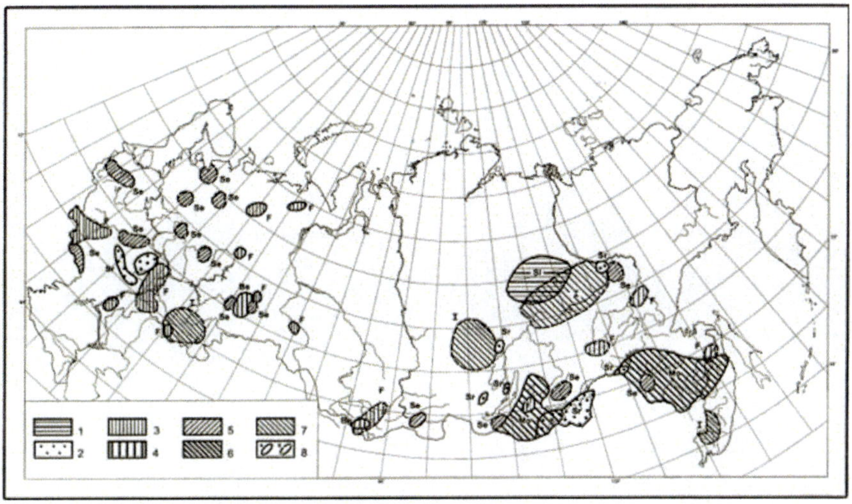

Fig. 1 Areas of some endemic diseases in Russia and NIS. Endemic diseases: *1*—silicotic urolithiasis (Si), *2*—Urov disease (Sr), *3*—dental fluorosis (F), *4*—endemic enteritis (Be), 5—Se toxicosis or white-muscular disease (Se), *6*—Mo gout, molybdenosis (Mo), *7*—endemic wen (I); chemical elements concentration areals: **a** excessive, **b** deficient (Golovin et al. 2004; Volfson 2010a, b)

dental and skeletal fluorosis in the population living near mining and metallurgical enterprises.

More than 60% of the population of the countries of the former USSR lived in areas with iodine deficiency. These are the territories of a number of regions in the center of the European part of Russia, the Upper and Middle Volga regions, the Urals, Siberia, the North Caucasus, Ukraine, and Kazakhstan. The incidence of goiter in such regions, according to epidemiological studies, ranges from 17 to 40% (Revich 2001; Golovin et al. 2004; Volfson 2013a, b) (Figs. 1 and 2).

An assessment of the relationship between the hydrogeochemical characteristics of drinking water in wells of underground water supply and diseases of population of the Smolensk region, carried out at the Institute of Water Problems of the Russian Academy of Sciences with the use of nonparametric statistical method of rank correlation, revealed a statistically significant relationship between an increased content of stable strontium, manganese and iron in drinking water in wells of underground water supply, and fractures of extremity in the child population, diseases of the nervous system and atopic dermatitis. In the Smolensk region, a close relationship has also been established between the level of general mortality and mortality from coronary heart disease, acute myocardial infarction with total water hardness and calcium content (Elpiner 2011).

One of the reasons for the established pathologies is the geochemical specialization of rock complexes for certain elements, the effects of which on the human body can be established over the course of several generations. An indicative example is manganese, which is common in the sedimentary rock complexes of the Paleozoic

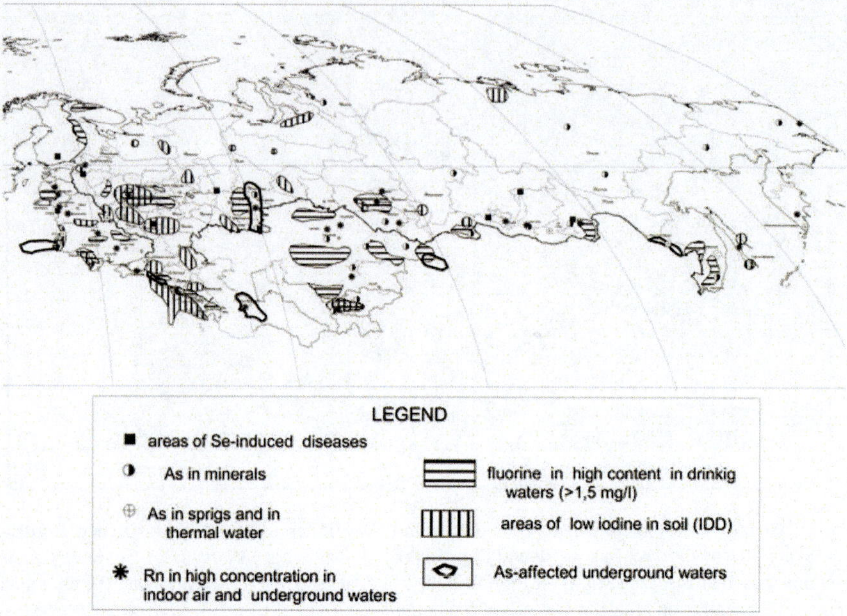

Fig. 2 Occurrence of documented health problems of population related to geological factors in Russia and NIS (F, Se, I, and proposed area of As-induced diseases, Volfson 2010a, b; 2013a, b)

of the Russian Platform. Manganese is actively absorbed by plants and, thus, is able to enter the human body through food chains. Manganese is one of the biologically essential trace elements for both animals and humans. It plays an important role in the formation of connective tissue and bones, growth of the body, in carbohydrate metabolism and is necessary for the implementation of the reproductive function of the body. Unexpected results of studies on the chemical composition of the bone remains of people who lived in the Middle Ages in Moscow and its environs were obtained by archaeologists and anthropochemists E.I. and A.L. Alexandrovsky. According to Aleksandrovskaya and Aleksandrovskiy (2010), in the Middle Ages, aspen and birch firewood, active concentrators of manganese, was used for stove heating of dwellings. It was found that manganese accumulated in the bone tissue of the inhabitants of medieval Moscow, and among commoners the concentration of manganese in the bone remains was 30–70 mg/100 g, among the nobility up to 10 mg/100 g. Such a significant, three-fold difference in the contents of manganese was due, among other things, in connection with the then widespread way of taking hygiene procedures—“black bath”. On the basis of the data obtained, with a high degree of probability, we can talk about common factors in the occurrence of some endemic diseases (conditions) both in our ancestors and in the current inhabitants of the Central regions of Russia and, in particular, about one of them—an imbalance of manganese in the human body and the influence of this trace element as a cofactor on the occurrence of diseases of the endocrine system. The use of modern methods of processing geological, geochemical and geophysical information makes it possible to identify new conditions for the concentration of geochemical elements in natural environments. In particular, the use of geographic information systems—GIS in the retrospective analysis of helium-hydrogeochemical survey materials made it possible to link the boundaries of the distribution of helium halos, gravitating to the intersection nodes of deep fault structures, and halos a number of chemical elements contained in the Devonian-Permian aquifers of the East European Platform—fluorine, strontium, boron, lithium, bromine, etc., which are important in establishing the causes of fluorosis, osteoporosis, digestive system and mental disorders among the population of a number of territories. This circumstance confirms the wide prospects for using GIS technologies in identifying, studying and monitoring hazardous medical and environmental conditions associated with geological objects and processes (Wolfson 2022).

The same tasks are served by geological models of the formation of dangerous medical and environmental conditions on the earth's surface in connection with the manifestation of deep geological processes, built by the doctor of geological and mineralogical sciences Igor G. Pechenkin on the basis of studying the features of epigenetic ore formation within the boundaries of oil and gas and ore-bearing sedimentary basins of Central Asia: catagenetic (stadial), exfiltration and infiltration. Each of them forms its own specific set of geological and geochemical settings in near-surface geospheres and their corresponding epidemiological responses (Pechenkin and Pechenkin 1998; Pechenkin et al. 2005; Volfson 2013a, b) (Fig. 3). Based on the studies carried out, a forecast was made for the occurrence of endemic diseases within the distribution haloes of such elements as U, Se, As, F, Br, B, Rn. At the same time, the use of models of epigenetic ore formation within the sedimentary basins of young platform areas makes it possible to provide forecasts for the identification of deposits of hydromineral and balneological raw materials—therapeutic mud, clays and mineral waters of various compositions and to explore the possibilities of their application in medical practice (Zhidkova et al. 2023).

4 Conclusion

The contemporary anthropogenic technocratic society has a significant influence on the environment and changes the way of life of almost every person. In the early twentieth century Russian geochemist and mineralogist Vladimir I. Vernadsky scientifically proved the inevitability of the transition of the bio-geosphere to the "noosphere"—a new sphere, the sphere of human knowledge, when it becomes an effective geological force (Josephson et al. 2013). Today we are aware of the importance of the threats and consequences of environmental challenges for the future of Earth. Awareness of the role of anthropogenic impact as a huge geological force that changes the conditions of existence of the bio- and geosphere leads to new interdisciplinary studies.

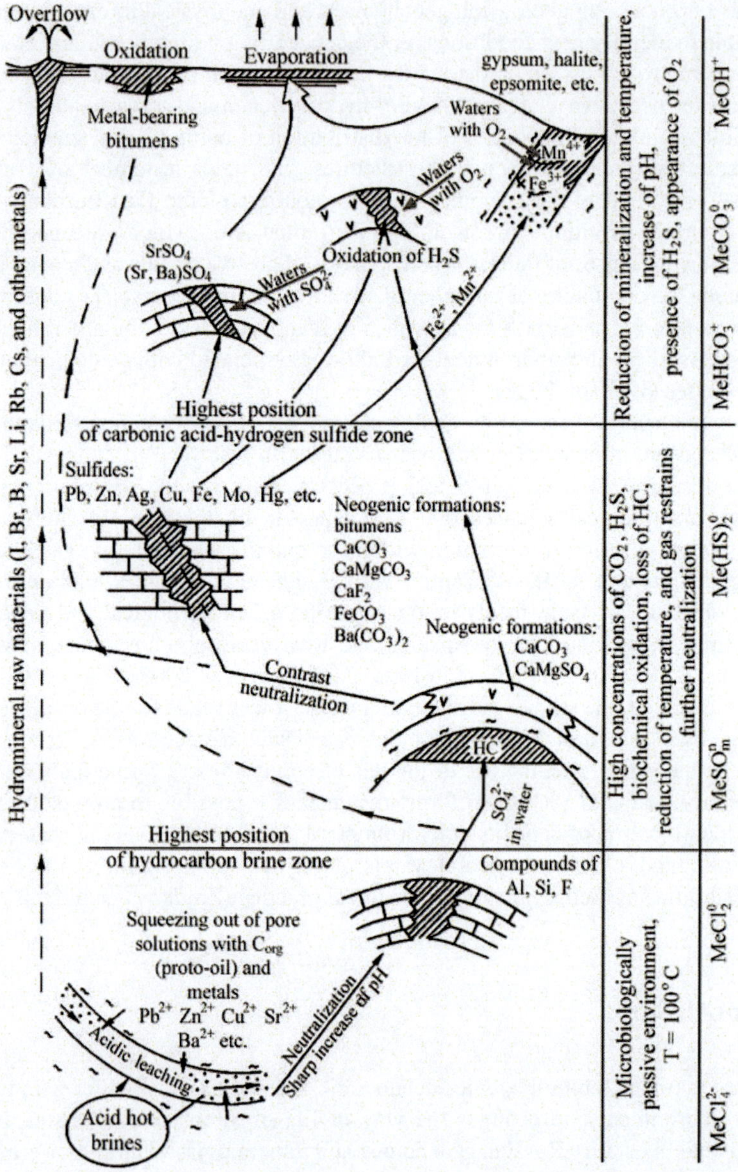

Fig. 3 Shematic diagram illustrating gaseous liquid fluid transformation in the exfiltration ore-forming system *(Metalliferous brines, as well as metalliferous (e.g., Fe, Ag, and Cu) and non-metalliferous occurrences and deposits are observed at the boundaries of currently active zones of geofluid discharge. They are accompanied by high concentrations of toxic elements (e.g., Cd, Tl, As, and Pb) and marked by the presence of rare elements, Hf, W, and Sb. Concentrations of precious metals, such as Au and Pt, are also observed. The deepest areas of the trough served as the principal zones for the accumulation of geofluids enriched in gaseous and liquid phases)* (Pechenkin and Pechenkin 1998; Pechenkin et al. 2005; Volfson 2013a, b)

Human being is a subject of the ecosystem interacting with nature. The consequences of such interaction may not manifest quickly. One needs to consider the fact that the period between exposure to environmental hazards and its effect on public health may have a delay. The "exposure-outcome" link can't always be easily detectable. This period between exposure and visible impact on morbidity may last years. The studies in medical geology are aimed at promoting the development of more effective environmental protection policies and measures in healthcare at the federal and regional levels, which can help to minimize the impact of negative environmental factors on public health.

Medical geology is a strategically important socially oriented discipline. The study of the conditions, factors and mechanisms of the impact of geological objects and processes on human health makes it possible to develop preventive measures necessary for successfully solving current and planning long-term tasks of the economics of managing and implementing various social projects based on the interests of all segments of the population.

Scientific analysis of the established facts and patterns of formation of geochemical halos of various elements allows a reasoned approach to solving a number of economic and social problems, the main ones being the protection of public health, improving the quality of life and the implementation of the constitutional human right to a safe environment (Yakushina et al. 2018).

Prospects for medical geology in Russia and the CIS countries, as well as any developing sphere of human activity, in qualified personnel. One of the priority tasks for the near future is the creation of an interdisciplinary project with the participation of representatives of educational authorities, the geological and medical community, within which the conceptual foundations of the educational subject "Medical Geology: Man and the Geosphere" should be laid.

References

Adilov VB, Razumov AN (2010) Natural healing resources of Russia and problems of their use. Medical geology: state and prospects. Ed. IF Volfson. M.: ROSGEO, 2010: 44–52 (in Russian)

Aleksandrovskaya EI, Aleksandrovskiy AL (2010) Anthropochemical factor in the evolution of European civilization. History and Modernity. 1. March 2010: 148–177 (in Russian)

Elpiner LI (2011) Groundwater quality and public health. Proceedings of the international scientific-practical conference "Drinking groundwater. Study, use and information technologies". VSEGINGEO. Pos. (Settl.) Zeleny, The Moscow region. April 18–22, V 1: 40–55 (in Russian)

Essentials of Medical Geology. Impacts of the Natural Environment on Public Health (2005) Selinus O, Lindh U, Fuge R, Centeno J, Alloway B, Smedley P, Finkelman R (Eds.) Elsevier Academic Press, printed in China

Finkelman RB (2008) Health Benefits of Geologic Materials and Geologic Processes (Lecture). Short course in medical geology, Oslo, Sunday, August 10, 2008 (within the frame of XXXIII Session of the International Geological Congress. Oslo, Norway. August 06–14, 2008)

Golovin AA, Krinochkin LA and Pevzner VS (2004) Geochemical specialization of bedrock and soil as indicator of regional geochemical endemicity. Geologija, 48: 22–28

Josephson P, Dronin N, Mnatsakanyan R, Cherp A, Efremenko D, Larin A (2013) An environmental history of Russia. Cambridge University Press, New York: 54-57

Kovalsky VV (1974) Geochemical ecology. M.: Nauka (in Russian)

Orlov VP, Farrakhov EG, Wolfson IF, Alekseev VM, Prozorova MV (2018) Current state and prospects of medical geology (to the results of the VII conference of the International Medical Geological Association MedGeo-2017). Exploration and protection of mineral resources (Razvedka I Ohrana Nedr - RION). 1: 3–7 (in Russian)

Pechenkin IG, Pechenkin VG (1998) Dynamic models of ore formation in the sedimentary cover of the Turan Plate. Lithology and minerals. 4: 383–396

Pechenkin IG, Sysoev AN, Pechenkin VG, Grushevoy GV, Volfson IF (2005) Metallogeny of the uranium-bearing sedimentary basins. Eds. Jingwen Mao, Frank P. Bierlein. Mineral Deposit Research: Meeting the Global Challenge. Proceedings of the Eighth Biennial SGA Meeting Beijing, China, 18–21 August 2005. Springer. Chapter 3–23: 307–310

Pechenkin IG, Zhidkova ES, Kremkova EV (2022) Essentuki salt-alkaline springs: discovery, study, use. Exploration and Protection of Mineral Resources (Razvedka I Ohrana Nedr - RION). 7. 2022: 53–61. https://doi.org/10.53085/0034-026X_2022_07_53 (in Russian)

Pelymsky G.A. (2008) Minerals presented in the exposition of the Museum of Earth Science on the History of Medicine. Proceedings of the scientific conference “Lomonosov Readings. Museum Studies Section. Museum of Geosciences of Moscow State University Moscow. April 22–24, 2008: 34–37 (in Russian)

Revich BA (2001) Environmental pollution and human health. Introduction to ecological epidemiology. Moscow. MNEPU publishing house (in Russian)

Rikhvanov LP, Soktoev BR, Baranovskaya NV, Ageeva EV, Belyanovskaya AI, Deriglazova MA, Yusupov DV, Ekaterina S. Epova ES et al. (2021) Comprehensive geochemical research of the environmental components in endemic areas of Transbaikalia. Bulletin of the Tomsk Polytechnic University. Geo Assets Engineering. V. 332. 2: 7–25. https://doi.org/10.18799/24131830/2021/02/3039 (in Russian)

Saet YE, Revich BA, Yanin EP (1990) Geochemistry of the environment. Moscow, Nedra (in Russian)

Tsarfis PG (1985) Nature and health: Treatment and rehabilitation by natural factors. Moscow, USSR: Mir Publishers (in Russian)

Tsarfis PG (Ed.) (1991) Resorts, 2 vols. Moscow, USSR: Profizdat (in Russian)

Vinogradov AP (1964) Provinces biogéochimiques et leur rôle dans l’évolution organique. In U. Colombo and G.D Hobson (Eds.), Advances in organic geochemistry: Proceedings of the International meeting, Milan, Italy, 1962: 317–338. New York, NY: Macmillan

Volfson IF, Pechenkin IG, Kremkova EV (2006) Medical geology today: goals, objectives and ways to solve them. *In* (Eds.) Gavrilenko VV, Panova EG: Bioinert interactions: Life and Rocks. Proceedings of SPBOE, series 1, vol.96: 107–124 (in Russian)

Volfson IF, Farrakhov EG, Pronin AP, Beyseev OB, Bogdasarov MA et al. (2010a) The medical geology community in Russia and the NIS In Medical geology. A Regional synthesis (Eds.) Selinus O, Fikelman R, Centeno J Springer Dordrecht Heidelberg London New York: 221–258

Volfson IF, Paul W, Pechenkin I (2010b) Geochemical anomalies: Sickness and health. *In* Man and the Geosphere (Earth Sciences in the 21st Century). Editor: Florinsky IV. Nova Science Publishers, Inc: 69–114

Volfson I (2013a) Non-metallic mineral resources in solving applied geoecological problems. Exploration and Protection of Mineral Resources (Razvedka I Ohrana Nedr - RION). 7. 2013a: 62–67 (in Russian)

Volfson IF (2013b) Hydromineral resources in the medical practice. Mining Journal (Gorniy Zhurnal). 7. 2013b: 94–97 (in Russian)

Volfson IF (2016) On the history and practice of medicinal mineralogy. In Amber in the History of Medicine: Proceedings of the International Conference. Eds. I.A. Polyakova, Ch.J. Duffin, T.I. Suvorova; Kaliningrad Regional Amber Museum.-Kaliningrad (September 14–16, 2015), 2016.-272 p. ISBN 978-5-903920-43-3:9-15: 9–16.

Wolfson IF (2022) Medico-ecological aspects of multi-purpose geochemical mapping. Exploration and protection of mineral resources (Razvedka I Ohrana Nedr - RION). 2: 56–61 (in Russian)

Yakushina OI, Guliaeva YV, Volfson IF, Dasaeva LA, Farrakhov EG (2018) The impact of ecological factors on the wellbeing of the population in the Tyumen and Kaliningrad territories RF Exploration and Protection of Mineral Resources (Razvedka I Ohrana Nedr - RION). 1. 2018: 18–25 (in Russian)

Yushkin NP (2003) Human health effects of "table" minerals. In N. Kamata (Ed.), *Proceedings, 1st International symposium of the Kanazawa University 21st century COE program, Kanazawa, Japan, Mar. 17–18, 2003, Vol. 1: 261–265.* Kanazawa, Japan: Kanazawa University

Yushkin NP (2004) Mineral world and human health. Bulletin of the Earth Sciences Department of the Russian Academy of Sciences. Electronic scientific and information journal section: scientific publications - scientific publication. 1(22)'. http://www.scgis.ru/russian/cp1251/h_dgggms/1-2004/scpub-1.pdf (in Russian)

Zhidkova ES, Kremkova EV, Pechenkin IG (2023) The role of medical geology in solving the environmental problems of the foothills planes and health of their population. Exploration and Protection of Mineral Resources (Razvedka I Ohrana Nedr - RION). 1. 2023. https://doi.org/10.53085/003026X_2023_01_29 (in Russian): 30–38

Drinking Water Influence on the Chemical Composition of Gastric Juice: Monitoring and Modeling

Svetlana I. Mazukhina, Svetlana V. Drogobuzhskaya, Pavel S. Tereshchenko, Andrey I. Novikov, Anna A. Shirokaya, Yuliya A. Kalashnikova, Sergei S. Sandimirov, and Andrey M. Zolnikov

Abstract The paper presents the results of "model" pH values of gastric juice and data on the sampling and study of the chemical composition of the biological fluid of the stomach of patients from different population groups of the Apatity-Kirovsk region (Murmansk region, Russia) depending on HCl content. Deviations from the ideal model are associated with the incidence of patients of different nature and severity. Analysis of the results of the chemical composition of gastric juice of patient 1 before and after treatment indicates its change in the course of treatment: concentrations of sodium, potassium, chlorine, magnesium, and phosphorus decreased several times. Using thermodynamic modeling, it was shown that the result of using mineral

S. I. Mazukhina (✉) · S. S. Sandimirov
Institute of North Industrial Ecology Problems—Subdivision of the Federal Research Center "Kola Scientific Center of the Russian Academy of Sciences", 14 a, Akademgorodok, Apatity 184209, Russia
e-mail: simazukhina@mail.ru

S. S. Sandimirov
e-mail: s.sandimirov@ksc.ru

S. V. Drogobuzhskaya · A. I. Novikov · A. A. Shirokaya
Tananaev Institute of Chemistry—Subdivision of the Federal Research Center "Kola Scientific Center of the Russian Academy of Sciences", 26 a, Akademgorodok, Apatity 184209, Russia
e-mail: s.drogobuzhskaia@ksc.ru

A. A. Shirokaya
e-mail: a.shirokaia@ksc.ru

P. S. Tereshchenko
OPTEC JSC, Maly Prospekt V.O.58, Letter A, Room 20N, Saint Petersburg 199178, Russia
e-mail: tereshchenko_pash@mail.ru

Y. A. Kalashnikova · A. M. Zolnikov
State Regional Budgetary Health Institution "Apatity-Kirovsk Central City Hospital", Kirovsk 184250, Russia
e-mail: lokadoktor@mail.ru

A. M. Zolnikov
e-mail: doctorzol@yandex.ru

O. V. Frank-Kamenetskaya et al. (eds.), *Biogenic—Abiogenic Interactions in Natural and Anthropogenic Systems 2022*, Springer Proceedings in Earth and Environmental Sciences, https://doi.org/10.1007/978-3-031-40470-2_34

waters of the Krasnodar region coincides with the results of drug treatment. The use of mineral waters should be considered as an effective medical resource for the prevention and treatment of stomach diseases. This approach opens up new prospects for research in the field of ecology, health and medicine.

Keywords Modeling · Mineral water · Gastric juice · Forms of element migration · ICP-MS · Mineral exchange · Geochemical barrier

1 Introduction

In the Murmansk region of the Russian Federation, the incidence of the adult population associated with eating disorders and metabolic disorders, according to the Federal Information Fund, exceeds the national average, the incidence of the child population is also higher than the national average by 30–40%, which has been observed for many years (http://mpr.gov-murman.ru). The highest excess of the average Russian level in the incidence of the adult population in the Murmansk region is noted in the Apatity-Kirovsk district and in the city of Apatity, where these figures exceed the national average by more than 2 times. In this work (http://mpr.gov-murman.ru), it is emphasized that the preservation of the health of the nation is the primary task of the state, and the very value of health turns into a factor in the stability and well-being of society (Razumov 2013). The relevance of this task was especially acute in 2020–2021.

The simplest and most promising options for improving the health of the population are non-drug, non-invasive methods of exposure, which include the intake of mineral waters (Lyubchiket al. 2013; Dovgan' 2015; Lyubchik et al. 2016; Ponomarenko et al. 2014; Dupont et al. 2019). A person in the process of life consumes from one to three liters of water per day. Together with water, chemicals enter the body and it is important which and in what quantity. Food intake (water) is a source of energy for the processes occurring in the body, the value of the free energy of chemical compounds (Lyubchik et al. 2016). Previously, we considered the Karelian Marcial waters, drinking waters of the Murmansk region and their impact on the human body (Mazukhina et al. 2020). From the point of view of the effect of medicinal waters on the digestive tract, natural mineral waters of the Crimea, Krasnodar and Stavropol regions are of interest, which have a versatile effect on the gastrointestinal tract, as well as on the urinary and cardiovascular systems of the body, the function of the endocrine glands, on the processes of metabolism and detoxification (Lyubchik et al. 2016).

2 Materials and Methods

The Psekup mineral waters, which come to the surface in the valley of the Psekups River, are the pride of the resort town of Goryachiy Klyuch (Krasnodar region (https://xn-8sbaaidghcy3ac0ckpe8r.xn-p1ai/treatment/mineral-water/). In the very center of the city, right in the sanatorium complex, there is the Drinking Gallery, which receives water from three mineral springs through a special pipeline, which differ from each other both in temperature and in their chemical composition: No. 21, No. 58 and No. 104. Water from them is prescribed for oral use, that is, only for drinking. The temperature of the mineral waters coming to the surface ranges from 38 to 45 °C. Mineral waters from sources No. 58 and No. 104 are classified as low-thermal waters and are used at temperatures up to 40 °C. In case of chronic gastritis occurring with secretory insufficiency, mineral water from spring No. 58 is recommended. In case of stomach disease, occurring with normal and increased secretion, mineral water from spring No. 104 is recommended.

The presented work is a continuation of the study of the chemical composition of natural waters and their transformation in the human body (Mazukhina et al. 2020). The work presents the results of the study of chemical composition of the gastric juice of the population of the Apatity-Kirovsk region and an assessment of changes in the parameters of gastric juice as a result of the use of medicinal waters.

The aim of the work is to obtain new data on the state of the chemical composition of gastric juice of residents of the Apatitsko–Kirovsky district and the influence of medicinal waters on the chemical composition of gastric juice using physico-chemical modeling.

3 Experimental Part

The work applies an integrated approach that combines the study of chemical composition of natural waters and biological substrates, physico-chemical (thermodynamic) modeling of natural waters and the results of their interaction with human biological media. The objects of the study were the waters of the Goryachiy Klyuch spring from wells No. 58 and No. 104 and the natural waters of the Apatity-Kirovsk district of the Murmansk region (Fig. 1), which the population uses as drinking (Mazukhina et al. 2020), as well as the gastric juice of a number of patients.

Elemental analysis of water samples and biosubstrates (gastric juice) in the course of investigation was carried out using inductively coupled plasma mass spectrometry (ELAN 9000 DRC-e, Perkin Elmer, USA), potentiometry with ion-selective electrodes (liquid analyzer Expert-001(F. Econiks-Expert, Russia) and titrimetry (Metodika opredeleniya 2003; Seregina et al. 2019). For thermodynamic modeling, the software package "Selector" software package was used (Chudnenko 2010). A physical and chemical model (PCM) of water–rock interaction includes 42 independent components, 1062 dependent components, including 435 in aqueous solution,

Fig. 1 Map-scheme of water sampling points

76 in the gas phase, 111 liquid hydrocarbons, 440 solid phases, organic and mineral substances. The solid matter phases of the multisystem have been shaped by the mineral composition of the rocks of the Baltic Shield (Sandimirov et al. 2022).

The study of the chemical composition of samples of biological fluids was carried out among the population of the Apatity-Kirovsk region (2020).

4 Results and Discussion

Chemical composition of groundwater is the most important geochemical and environmental factor, since the geochemical properties of groundwater have a direct biochemical effect on the human body and its physiological functions. Groundwater contains a very large number of elements that, in high concentrations, can have a toxic effect on the human body. At the same time, it is important to know that not only the excess of any element, but also its deficiency is of biochemical importance (Krainov et al. 2012). Table 1 presents chemical composition of samples from the Tsentralny water intake (Kirovsk), obtained under the grant of the RFBR and the Ministry of Education and Science of the Murmansk Region 17–45-510,640 ("Geochemical assessment of groundwater in the Khibiny massif, age and identification of the origin of groundwater, chemical composition, forms of migration of elements") and gastric juice of the inhabitants of the Apatity-Kirovsk region.

Analysis of the results (Table 1) shows the presence in the chemical compositions of gastric juice samples of the population of the same set of chemical elements as in natural waters, but in much higher concentrations.

Study of the chemical composition of the gastric juice of the population of the Apatity-Kirovsk region (twenty patients, a random sample) showed low average concentrations (relative to the ideal model) of the elements Na, Cl, K and a high concentration of phosphorus (Table 1). Lanthanides are present in the composition of gastric juice, which, in turn, may lead to the formation of a persistent tendency to the emergence of pathological conditions in the body.

Since the study of this type is being conducted for the first time and there is no standard sample of the chemical composition of a healthy stomach of residents of the Apatity-Kirovsk region, a comparison was made with an ideal model compiled according to the literature data. The works (Mazukhina et al. 2020) present changes in the Eh value, pH of the gastric juice model depending on the HCl content. On Fig. 2 presents the results of modeling and analytical data on the parameters of gastric juice of twenty patients.

Analyzing the anamnestic data of the patients included in the study group, we note the following diseases: oncology (points 1,2,6,7,16 in Fig. 2 (25% of the total number of diseases)), according to the analytical data, the presence of low concentrations of Na ($99.6 < Na < 340$), K ($57.2 < K < 158.7$) - except for point 16, where oncology was detected for the first time), low Zn content (from 0.22 to 1.27 mg/l);cholecystitis (point 3–5%)—Na concentration—302.6; Al—7.72; P—94.4; K—36.8; Ca—46.7 mg/l;anemia (point 8–5%), Na—298.5, K—72.5, respectively;TORCH infection (point 5–5%), Na content—342.9, K—59.7 mg/l;damage to the osteoarticular apparatus (points 4,14,15–15%)—they correspond to a high content of calcium $12 < Ca < 32.6$, phosphorus 92.4 < P—114;damage to the cardiovascular system (point 11–5%)—the content of boron (11.9 mg/l), lithium (0.024 mg/l), low content of sodium (513.7 mg/l), potassium (116.4 mg/l);gastrointestinal disease (points 20,19,18,17,13,12,10,9—40%, including in the acute stage (points 9,12,13), after treatment (point 10) and before treatment (point 20).

It is noteworthy that a number of patients have low levels of stomach acid (hypochlorhydria), which functions to break down food, absorb certain nutrients such as protein and vitamin B12; destruction of bacteria and other pathogens in the stomach. Hypochlorhydria can be caused by age (the stomach produces less acid as

Table 1 Chemical composition of water intake "Central" (5 water intake wells) and gastric juice of residents of the Apatity-Kirovsk region

Element	The chemical composition of the waters of the water intake "Centralny", mg/l			The chemical composition of the gastric juice of the inhabitants of the Apatity-Kirovsk region, mg/l		
	min	max	avg	min	max	avg
Li	0.0004	0.0010	0.0007	0.0031	0.053	0.0138
Be	0.0002	0.0012	0.0006	0.0008	0.0075	0.0035
B	0.0150	0.0169	0.0158	0.28	24.6	7.17
Na	11.5	29.2	19.1	99.6	1199	548.0
Mg	0.002	0.042	0.020	1.50	12.5	5.54
Al	0.35	1.07	0.79	0.057	7.72	1.47
Si	1.65	1.91	1.79	1.02	4.83	2.22
P	0.015	0.032	0.023	3.9	114.0	50.5
K	1.17	4.15	2.27	36.8	316.7	158.7
Ca	0.008	0.24	0.10	4.13	46.7	15.8
V	0.0006	0.0066	0.0037	0.0035	0.035	0.0133
Cr	0.0002	0.0017	0.0009	0.0071	0.074	0.041
Mn	0.0000	0.0004	0.0002	0.0022	0.087	0.0204
Fe	0.005	0.021	0.015	0.134	8.92	1.25
Co	0.00004	0.00016	0.00008	0.00013	0.00046	0.00027
Ni	0.00002	0.00019	0.00009	0.0108	0.36	0.080
Cu	0.00010	0.00029	0.00017	0.130	40.7	5.53
Zn	0.0034	0.0061	0.0048	0.223	14.9	2.81
Sr	0.0002	0.0024	0.0011	0.0202	0.099	0.046
Y	0.00001	0.00005	0.00003	0.00001	0.00038	0.00009
Zr	0.00001	0.00008	0.00005	0.000043	0.0021	0.00054
Se	0.0006	0.216	0.072	0.013	0.398	0.077
Nb	0.00006	0.00030	0.00020	0.000018	0.00084	0.00022
Mo	0.0032	0.0115	0.0061	0.00218	0.0112	0.0054
Ag	0.0006	0.0018	0.0013	0.00027	0.0258	0.0036
Cd	0.000014	0.000038	0.000025	0.0003	0.0258	0.0035
Ba	0.00003	0.00014	0.00009	0.0008	0.0079	0.0045
La	0.00003	0.00018	0.00008	0.00001	0.00029	0.00009
Ce	0.00001	0.00009	0.00006	0.000001	0.00017	0.00007
U	0.00011	0.00103	0.00043	0.0001	0.0023	0.0007
Cl^-	1.25	2.58	2.03	440.0	4939	2180
pH	8.97	9.63	9.37	1.43	7.35	4.44

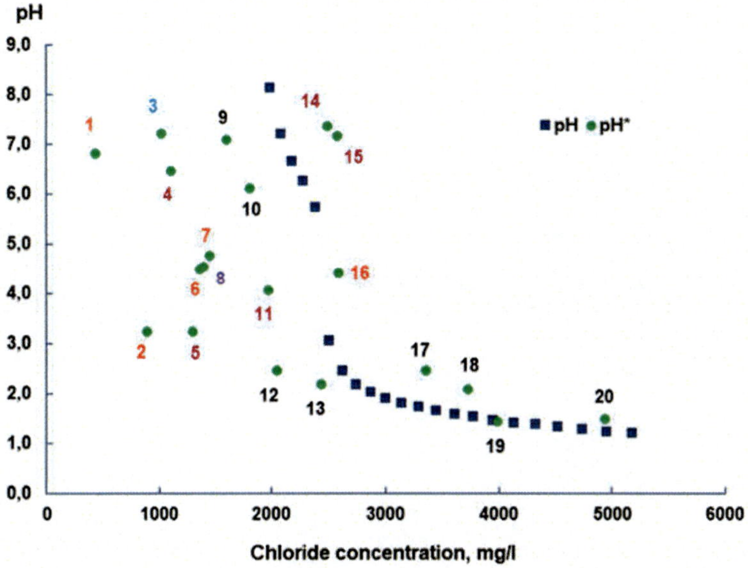

Fig. 2 Comparison of real and model data of gastric parameters (■—simulation results; O—results of gastric juice monitoring

a result of aging), drugs, or bacterial infection (Kulikovsky 2018). Without going into discussions of what is primary and what is secondary, we note these factors in the anamnesis of patients.

To identify and monitor nosological forms of diseases, as well as to correct their course, dispensary observation of patients from different population groups is necessary. We conducted the study of the chemical composition of the biological fluid of the stomach before and after the treatment of patient 1 (Table 2). In addition, modeling of the processes that would occur in the patient's stomach when taking mineral water was carried out. Table 2 presents the results of monitoring (AD) and modeling results (RM) of mineral water samples No. 58, No. 104 and gastric juice of patient 1 before (Fig. 2, point 20) and after treatment (Fig. 2, point 10).

Analysis of the results of the chemical composition of the gastric juice of patient 1 indicates a change in the course of treatment: the concentrations of sodium, potassium, chlorine, magnesium, and phosphorus decreased several times. As a result, the pH values shifted from acidic (1.48) to slightly acidic (6.09).

Comparison of AD and SR of the presented mineral water samples No. 104 and 58 shows a weakly alkaline environment, reducing conditions (negative Eh value) and a difference in the concentrations of the main elements (Na, Ca, Mg, Sr and Cl) at fairly close values of the concentrations of bicarbonate and sulfate. Attention should be paid to the content and chemical forms of iron in water samples No. 104 and No. 58. Note the comparability of the concentrations of Na and Ca of sample No. 58 and gastric juice according to the literature data (Tables 1 and 2). The concentration of iron in water samples No. 104 and No. 58 is quite low, it is present in the ionic form in

Table 2 Results of monitoring and modeling of water samples No. 58, No. 104 and gastric juice of patient 1 before and after treatment

Parameters	Psekup mineral waters «Goryachiy Klyuch»				Patient 1, gastric juice		
	Sample No104		Sample No58		Before treatment		After
	AD	SR	AD	SR	AD	SR	AD
pH	8.10	8.09	8.00	8.00	1.48	1.13	6.09
Eh,V		−0.292		−0.254	0.312	0.099	0.047
T,°C		20		20		38	
Element	Concentration, mg/l						
Ca^{2+}	0.99	0.816	5.10	5.09	12.3	11.9	14.8
Na^{+}	537.0	536.0	1558	1560	821.9	813.0	223.1
K^{+}	4.10	4.10	9.40	9.39	275.7	277.0	60.2
Mg^{2+}	0.19	0.163	2.00	2.00	6.95	6.65	2.05
Fe	0.065	0.065	0.057	5.69E-02	0.257	0.257	0.49
Fe^{2+}		6.31E-02		5.38E-02		2.45E-01	
$FeSO_4$		7.96E-05		1.78E-03			
$FeCl^{+}$		1.88E-04		1.92E-03		1.95E-02	
$FeOH^{+}$		2.32E-03		1.66E-03			
Mn	0.0037	3.70E-03	0.0146	1.46E-02	0.023	1.85E-02	0.013
Mn^{2+}		3.69E-03		1.41E-02		1.73E-02	
$MnSO_4$		4.29E-06		4.32E-04			
$MnCl^{+}$		1.04E-05		4.80E-04		1.90E-03	
Cl^{-}	162.3	162.0	1892	1890	4940	4880	1930
P					46.3	46.6	8.13
$SO_4{}^{2-}$	34.5	1.54E+00	45.8	37.8			
$HCO_3{}^{-}$	1074	1.12E+03	905	842		4.88E-03	
$CO_3{}^{2-}$		6.14		5.05			
CO_2		1.35E+01		1.25E+01		4.11E+02	
H_2S		8.81E-01		6.53E-04		3.25	
HS^{-}		1.04E+01		6.30E-03		7.74E-06	
CH_4		1.19E-02		2.17E-07		1.19E+01	
Sr^{2+}	0.064	5.32E-02	0.377	3.25E-01	0.030	2.87E-02	0.059
$SrCO_3$		2.80E-03		1.12E-02			
$SrHCO_3{}^{+}$		1.45E-02		6.88E-02		5.11E-08	
Zn			0.0065	6.49E-03	11.2	1.13E+01	1.63
Zn^{2+}				3.92E-03		9.07	
$ZnOH^{+}$				3.03E-03		4.61E-06	
B	0.74	7.40E-01	0.0039	3.90E-03	6.01	6.05	13.7

(continued)

Table 2 (continued)

Parameters	Psekup mineral waters «Goryachiy Klyuch»				Patient 1, gastric juice		
	Sample No104		Sample No58		Before treatment		After
	AD	SR	AD	SR	AD	SR	AD
$B(OH)_3$		3.96E+00		2.09E-02		3.46E+01	
BO_2^-		1.89E-01		9.31E-04		2.86E-07	

Note AD—analytical data, SR—simulation results, Eh—redox potential

the +2 oxidation state, since there is no oxygen in the system. The simulation results made it possible to determine the presence of carbon dioxide, hydrogen sulfide and methane in these waters. It can be assumed that the concentrations of sodium and chlorine, as well as the values of Eh and pH should affect the normalization of the function of the stomach.

According to the presented data (Table 2), when different types of water enter the stomach on an empty stomach (sample No. 104, sample No. 58, it will contain: Na (mg/l)—537.0; 1558; Cl^-—162.3; 1890, respectively. Considerpossible forms of element migration obtained in the course of physicochemical modeling (Mazukhina et al. 2020a) as a result of the interaction of natural water with gastric juice (depending on the water content in the stomach).

Figure 3 shows changes in Cl^-, Na^+, K^+, pH and Gibbs free energy in the "water-gastric juice" system, where water is the water of springs No. 58 and No. 104, gastric juice is the chemical composition of the gastric juice of patient 1, taken before treatment. Straight lines represent the concentrations and parameters of the chemical composition of gastric juice after treatment.

The Gibbs potential of the system is calculated on the basis of the Gibbs free energy of the components in equilibrium, and in nature can be compared with the possibility of a spontaneous flow of a chemical process in a given direction. The analysis of changes in the Gibbs potential in the calculation with constant temperature and pressure serves as a measure of the most favorable level of the state from the energy point of view, achieved in closed systems.

Analysis of the results shows a change in acid–base conditions during the interaction of water with gastric juice (Fig. 3). With a volume of more than 320 ml, mineral waters form a slightly acidic environment in the stomach with a decrease in the concentration of K^+ and Cl^- and achieve the results obtained as a result of treatment (water No. 104). Increasing the volume of these waters to 1000 ml in the stomach will contribute to the achievement of the results obtained after treatment under the influence of the waters of the source No. 58 in terms of chlorine with an increase in the concentration of sodium. In real conditions, a large volume of highly mineralized water cannot simultaneously enter the body, but changes in the parameters in the stomach with high acidity occur already when 30 and 100 ml of mineral water No. 104 and No. 58 are consumed, respectively (Fig. 3b, c).

Figure 3 shows the change in the Gibbs free energy (potential) relative to different states of the system, where $\Delta G_{58} = \Delta G$ (gastric juice + water from well No. 58),

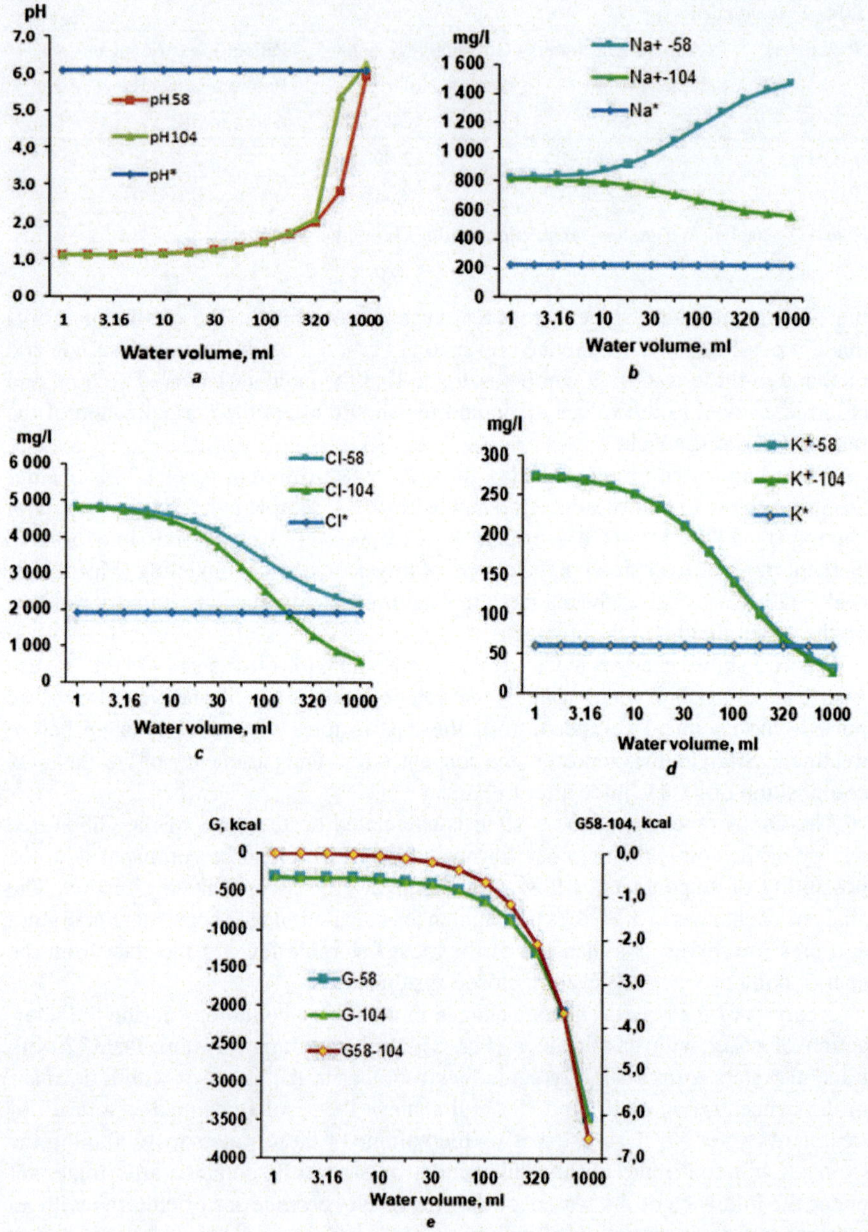

Fig. 3 The results of modeling changes in the parameters of the stomach during the interaction of various waters with 100 ml of gastric juice: **a**—pH, **b**—Na^+, **c**—Cl^-, **d**—K^+, **e**—ΔG, kcal. Straight lines—parameters obtained as a result of treatment (protocols)

$\Delta G_{104} = \Delta G$ (gastric juice + water from well No. 104). The difference between the Gibbs free energies relative to the states of the system "gastric juice—water No. 104", as well as "gastric juice—water No. 58" is presented as $\Delta G_{58\text{-}104}$. The difference in the states of the system begins to appear after the introduction of 10 ml of water. The system with water from well No. 58 is characterized by a large change in parameters, which is expressed in a larger value of ΔG.

Chemical compositions of the waters presented in Tables 1 and 2 indicate the comparability of the northern and Psekup waters (Table 2) in terms of the concentrations of potassium, magnesium and differences by 2–4 orders of magnitude for such elements and compounds as sodium, chlorine, HCO_3^- and sulfate ion (SO_4^{2-}). Thus, northern waters contain very low concentrations of sodium and chlorine, which are a regulator of water metabolism and osmotic pressure in tissues and have a stimulating effect on the secretory function of the digestive glands (Barashkov et al. 2001; Lyubchik et al. 2016).

Analyzing the daily intake of chemical elements in the human body, adults/children, mg: K 2000–5500/530; Na 1100–3300/260; Ca 800–1200/420; Mg 300–400/60; Zn 15/5; Fe 10–15/7,0; Mn 2,0–5,0/1,3; Cu 1,5–3,0/1,0; Cl^- 3200/470; PO_4^{3-} 800–1200/210; SO_4^{2-} 10 (Baranovskaya et al. 2015) and chemical composition of the Tsentralny water intake (Table 1), it can be stated that the concentrations of vital elements in the intake waters are low, and that the residents of the North can get these elements only with food and water.

Knowledge of the exact quantitative elemental composition of a person, taking into account regional and other features, can contribute to a more effective correction of human and animal health and early prevention of various diseases in accordance with the concepts of "Microelementoses", "Geochemical ecology" and "Geochemical ecology of diseases".The study of chemical elements in the human body should become an obligatory geochemical basis for the development of medical-environmental and medical-geochemical studies, which are of great socio-economic importance (Rikhvanov and Baranovskaya 2013).

In this study, we made an attempt to contribute to the study of the elemental composition of human biosubstratum, which is relevant, first of all, in connection with the current difficult environmental situation, when the humanity is destroying itself, while the elemental composition of the human body remains poorly understood (Baranovskaya et al. 2015).

5 Conclusions

Analysis of the information provided allows us make the following conclusions:

1. For the first time, the presence in the chemical compositions of samples of biological fluids of the population (gastric juice) of the same set of chemical elements as in natural waters, but in much higher concentrations, was established.

2. The presence of extremely dangerous: Be and P; highly hazardous: Li, Cd, Sr, Br, Co, B, Nb, Se and Mo; dangerous: Ni, Cu and Zn elements and lanthanides in biological fluids may indicate the presence of pathologies.
3. Weakly mineralized waters (northern waters) do not have the necessary elemental composition to activate the physiological mechanisms of the stomach, which can be an incentive for the development of pathological processes.
4. Mineral water from spring No. 58 (Goryachiy Klyuch) is recommended for use in chronic gastritis occurring with secretory insufficiency, which is confirmed by modeling results, and is especially recommended for residents of regions with low-mineralized water (Northern regions).

   The use of mineral waters can make it possible to reduce the dosage of drugs and reduce the duration of treatment for many chronic pathological conditions and to carry out preventive measures, which will positively affect the quality of life of people.
5. The use of mineral waters should not be considered as a low-budget approach to be used in cases where traditional methods of treatment have not helped or are contraindicated, but as an effective medical resource for the prevention and treatment of stomach diseases.
6. In view of the presence of minimal concentrations of elements Na and Cl in the drinking waters of the North, a salt-free diet is not relevant for this region and may be harmful to the health of the population.
7. Note that GOST(State Standard) and SanPiN (Sanitary Standards) normalize the upper limit of element concentrations, without taking into account the geochemical environment of its location and its environment.

References

Baranovskaya NV, Rikhvanov LP, Ignatova TN, Narkovich DV, Denisov OA (2015) Essays on human geochemistry: monograph. Tomsk Polytechnic University. Publishing House of Tomsk Polytechnic University, Tomsk (in Russian)

Barashkov VA, Koposova TS, Belykh AI et al. (2001) Khimicheskiye elementy v organizme cheloveka: spravochnyye materialy. Pomor. gos. un-t im. M.V. Lomonosova, Arkhangel'sk (in Russian)

Chudnenko KV (2010) Termodinamicheskoye modelirovaniye v geokhimii: teoriya, algoritmy, programmnoye obespecheniye, prilozheniya. Akademicheskoye izd-vo "Geo", Novosibirsk (in Russian)

Dovgan' IA (2015) Sanatorno-kurortnoye lecheniye i meditsinskaya reabilitatsiya v Sakskom voyennom klinicheskom sanatorii im. N.I. Pirogova. Novyye tekhnologii. Perspektivy razvitiya. Vestnik fizioterapii i kurortologii. 2:45-48 (in Russian)

Dupont C, Constant F, Imbert A, Hébert G, Zourabichvili O, Kapel N(2019) Time to treatment response of a magnesium- and sulphate-rich natural mineral water in functional constipation. Nutrition. 65:167-172. https://doi.org/10.1016/j.nut.2019.02.0188

Kraynov SR, Ryzhenko BN, Shvets VM (2012) Geokhimiya podzemnykh vod. Teoreticheskiye, prikladnyye i ekologicheskiye aspekty. TsentrLitNefteGaz, Moskva (in Russian)

Kulikovsky VO Hypochlorhydria: causes, symptoms and treatment https://medicalinsider.ru/news/gipokhlorgidriya-prichiny-simptomy-i-lechenie

Lyubchik VN, Buglak NP, Slobodyan YeI, Mel'tseva YeM, Golubova TF, Gnatenko TV (2013) Lechebnoye primeneniye natural'nykh mineral'nykh pit'yevykh vod Kryma u vzroslykh i detey. Metodicheskiye rekomendatsii. AO «Yevpatoriyskaya gorodskaya tipografiya», Yevpatoriya (in Russian)

Lyubchik VN, Buglak NP, Kaladze NN (2016) Lechebnoye primeneniye natural'nykh mineral'nykh pit'yevykh vod respubliki Krym. FGAOU VO «KFU im. V.I. Vernadskogo», Simferopol' (in Russian)

Mazukhina SI, Maksimova VV, Chudnenko KV, Masloboyev VA, Sandimirov SS, Drogobuzhskaya SV, Tereshchenko PS, Pozhilenko VI, Gudkov AV (2020) Kachestvo vod Arkticheskoy zony Rossiyskoy Federatsii: fiziko-khimicheskoye modelirovaniye formirovaniya vod, formy migratsii elementov, vliyaniye na organizm cheloveka.Izdatel'stvo FRC KSC RAS, Apatity (in Russian)

Metodika opredeleniya mikroelementov v diagnostiruyemykh biosubstratakh mass – spektrometriyey s induktivno svyazannoy argonovoy plazmoy (ISP-MS). Metodicheskiye rekomendatsii (2003) Moscow (in Russian)

Mineral water in the "Foothills of the Caucasus" sanatorium. https://xn-8sbaaidghcy3ac0ckpe8r.xn--p1ai/treatment/mineral-water (Date of access:30.06.2021)

Ministerstvo prirodnykh resursov i ekologii Murmanskoy oblasti, http://mpr.gov-murman.ru (Date of access:14.03.2017)

Ponomarenko GN, Zolotarova TA, Radomkin MV, Leshchev AL, Oleshko AYa, Bakholdina YeI (2014) Vliyaniye mineral'nykh vod na stress indutsirovannuyu endogennuyu intoksikatsiyu organizma (kliniko-eksperimental'noye issledovaniye). Fizioterapiya, bal'neologiya i reabilitatsiya. 2:(4–12) (in Russian)

Razumov AN (2013) Kontseptual'noye obosnovaniye mesta i roli novogo nauchno-prakticheskogo napravleniya vosstanovitel'noy meditsiny v sistemnoy organizatsionnoy strukture zdravookhraneniya Rossiyskoy Federatsii. Voprosy kurortologii. 1:10-19 (in Russian)

Rikhvanov LP, Baranovskaya NV (2013) Khimicheskiye elementy v organizme cheloveka kak osnova dlya razvitiya meditsinskoy geologii. Mezhdunarodnaya molodezhnaya shkola-seminar «Geokhimiya zhivogo veshchestva». Posvyashcheno 150-letiyu so dnya rozhdeniya V.I. Vernadskogo (1863–2013), Tomsk (in Russian)

Sandimirov SS, Pozhilenko VI, Mazukhina SI, Drogobuzhskaya SV, Shirokaya AA, Tereshchenko PS(2022)Chemical composition of natural waters of the Lovozero massif, Russia. Model. Earth Syst. Environ. 8:4307–4315. https://doi.org/10.1007/s40808-022-01362-1

Seregina IF, Osipov K, Bol'shov MA, Filatova DG, Lanskaya SYu (2019)Matrix Interference in the Determination of Elements in Biological Samples by Inductively Coupled Plasma–Mass Spectrometry and Methods for Its Elimination. J Anal Chem 74:182–191. https://doi.org/10.1134/S1061934819020114

Environmental Quality of the Kola Region: Impact on Human Elemental Status

Svetlana V. Drogobuzhskaya, Irina P. Kremenetskaya, Svetlana I. Mazukhina, and Yuliya A. Kalashnikova

Abstract The population of the Kola region lives in an industrialized area. This determines the increased man-made load on the environment and humans. A modern precision method—mass-spectrometry with inductively coupled plasma is used to evaluate the elemental composition of dust emissions, soils, plants, groundwater, sediments and biosubstrates. The state of the central part of the Kola region is determined by enterprises operating on its territory. They produce a large amount of dust emissions that transform the entire environment. Ecotoxicants migrate into soil and water. This affects the elemental composition of organisms in the area. The study shows the sources of industrial pollution, transport of elements in the environment and their influence on the elemental status of a human.

Keywords Man-made pollution · Dust emissions · Ecotoxicants · Biosubstrates · Human · Kola region

---

S. V. Drogobuzhskaya (✉) · I. P. Kremenetskaya
Tananaev Institute of Chemistry—Subdivision of the Federal Research Center, Kola Scientific Center of the Russian Academy of Sciences", 26 a, Akademgorodok, Apatity 184209, Russia
e-mail: s.drogobuzhskaia@ksc.ru

I. P. Kremenetskaya
e-mail: i.kremenetskaia@ksc.ru

S. I. Mazukhina
Institute of North Industrial Ecology Problems—Subdivision of the Federal Research Center "Kola Scientific Center of the Russian Academy of Sciences", 14 a, Akademgorodok, Apatity 184209, Russia
e-mail: simazukhina@mail.ru

Y. A. Kalashnikova
State Regional Budgetary Health Institution "Apatity-Kirovsk Central City Hospital", Kirovsk 184250, Russia
e-mail: lokadoktor@mail.ru

O. V. Frank-Kamenetskaya et al. (eds.), *Biogenic—Abiogenic Interactions in Natural and Anthropogenic Systems 2022*, Springer Proceedings in Earth and Environmental Sciences, https://doi.org/10.1007/978-3-031-40470-2_35

1 Introduction

The quality of the environment is evaluated based on the detection of man-made pollution in the atmosphere, water and soil. The population of the Kola Peninsula lives in an industrialized area. This determines the increased man-made load on the environment and people (Fig. 1).

Contaminated zones have formed (including those formed due to dust emissions) around large mining and processing enterprises over the years of their activity. Elevated concentrations of certain elements are noted (Evdokimova et al. 2012; Tarasova et al. 2020; Saltan et al. 2021); migration of these elements into water is observed (Kremenetskaya et al. 2013, 2015). In addition, large areas are occupied by overburden dumps and ore dressing wastes of various minerals (Pashkevich et al. 2019). Intensive development of non-ferrous metallurgy in the center of the Kola Peninsula has led to the degradation of terrestrial ecosystems in the zone of aerotechnogenic impact and the emergence of wastelands as its extreme manifestation. Most of the cities and towns are also concentrated there (Monchegorsk, Olenegorsk, Apatity, Kirovsk, Lovozero and Revda).

Geoecological state of the territory and the degree of its transformation as a result of the ingress of ecotoxicants affects the elemental composition of organisms

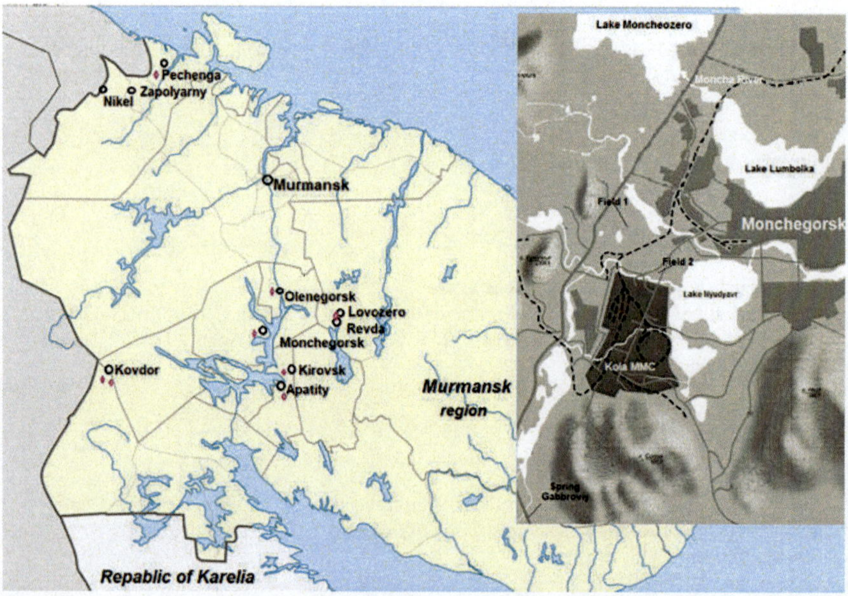

Fig. 1 ◊ Enterprises of the Kola region: JSC «Kola MMC» (Monchegorsk, Zapolyarny and Nickel sites), Olcon (Olenegorsk mining and processing plant), Kovdor mining and processing plant, Kirovsk branch of Apatit, North-Western Phosphorus Company, «Kovdorsluda» mining and processing plant, Lovozero mining and processing plant; insert in fig.—the land plot 700 m from the plant (Monchegorsk site)—experimental fields of the ICT KSC RAS

located on the territory: plants, animals and humans. This impact is exacerbated by the climatic features of the region. The imposition of natural and man-made factors leading to the entry of certain elements into the body forms complex combinations. This can lead to a change in its functional state, therefore, human body can be considered as an indicator that reflects changes in the natural environment. Interdisciplinary research and the study of not only chemical composition of various natural environments but also the elemental composition of the biosubstrates of human body are essential for deep understanding the relationship between the environment and the health of people living in the Arctic.

Mainly food and water supply the human body with macro- and microelements. However, the pathways for the entry of elements are more diverse and are associated both with the environment and with the external environment (Fig. 2). We have been studying the elemental composition of dust emissions from enterprises, the composition of technogenically modified soils and plants growing on these soils, the composition of surface and groundwater for ten years. To evaluate their impact on human body, a detailed analysis of human biosubstrates is required. Elemental analysis of blood, gastric juice and other biological media helps to evaluate the elemental status in real time, and analysis of hair, nails, bone tissue indicates the accumulation of a number of elements in human body.

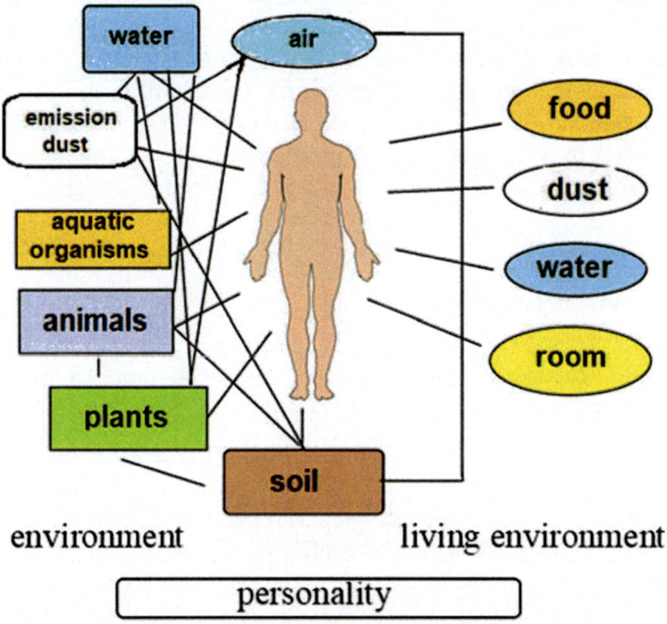

Fig. 2 Pathways by which chemicals enter the human body

The goal of the study is to show a number of sources of man-made pollution, ways of heavy metals (HM) entry into the biosphere and their impact on the elemental status of a human using the example of emissions from a number of enterprises in the center of the Kola Peninsula.

2 Experimental Procedure

2.1 *Methods*

Inductively coupled plasma mass spectrometry (ICP-MS) was used in the study. The potentiometric method with ion-selective electrodes (Expert-001 liquid analyzer, Econiks-Expert, Russia) was used and titrimetry was applied. Elemental ICP-MS analysis (ELAN 9000 DRC-e, Perkin Elmer, USA) of samples of water, soil, dust, plants and biosubstrates (gastric juice) was carried out using available methods (CV 3.18.05 2005) and methods developed in the ICT KSC RAS (Slukovskaya et al. 2020; Saltan et al. 2021). Analysis of soils, plants, and dust was carried out after autoclave microwave decomposition of samples (MW 4, Berghof, Germany); samples of biological media were prepared according to the procedure (Method 2003; Seregina et al. 2019). Quality control of the results of the analysis was carried out using the analysis of international and Russian reference materials certified for the content of the determined elements. The limit of detection of elements was in the range from 0.001 to 0.n ppm, depending on the determined element and the object of study. Study of the chemical composition of samples of biological fluids was carried out among the population of Apatity-Kirovsk district of the Murmansk region. Aspiration of gastric juice was carried out according to existing protocols, pH was determined in vitro. Samples ($n = 20$) were taken from patients suffering from diseases of the gastrointestinal tract.

3 Results and Discussion

1. *Sources of industrial pollution, routes of entry and migration of toxic elements into the biosphere*

Gas and dust emissions from the divisions of Kola MMC are the source of formation of Monchegorsk man-made wasteland and the territory in the Zapolyarny area. Man-made wastelands are open landscapes with polluted and eroded soils, almost completely devoid of vegetation due to the influence of aerotechnogenic load. Significant amounts of metallurgical dust which, according to the current technology, are mainly captured and returned to circulation are formed during pyrometallurgical processing of sulfide copper-nickel raw materials. However, more than a thousand tons of copper and nickel enters the environment per year with uncaptured dust from

Kola MMC. Phase analysis of the air in the working area of nickel production helped to identify the following groups of nickel compounds in dust particles: metallic nickel, soluble nickel sulfates and chlorides, sulfide compounds in the form of Ni_3S_2, NiS, Ni_3S_4, $(Ni,Fe)_9S_8$, nickel arsenides, selenides and tellurides, and oxides and silicates - NiO, $NiFe_2O_4$, Ni_2SiO_4. All air pollutants have a negative impact on human health. These substances mainly enter human body through the respiratory system. The most dangerous are insoluble nickel compounds, since, settling in the lungs, they are not excreted from the body naturally (Nikanov et al. 2018; Chashchin et al. 2014).

In addition, a number of products accumulating microimpurities were identified, first of all, these are dusts formed during ore electric smelting and conversion of raw materials, processing of nickel and copper concentrates. To evaluate the toxicity of dust from copper-nickel production, an extended analysis of fine dust taken from electrostatic precipitators of various technological stages of Kola MMC was carried out. Table 1 shows the results of the analysis.

Thus, the dust can pollute the environment. The dust from nickel production contains 40–70% Ni, 2–7% Cu, 1–2% Co, 2–6% Fe and 0.1–10% S, and the dust from copper production contains 20–50% Cu, 0.5–3% Ni, 0.5–7% Fe, 0.01–0.5% Zn, as well as Cd, As, Hg, Pb, Tl, Sn, Sb, Se and Te.

The territory is “acidified”, in addition to the influx of a large amount of heavy metals, due to the ingress of acid-forming gases into the atmosphere, therefore, areas with a catastrophic and severe degree of damage are stood out on the territory of Monchegorsk man-made wasteland. Anthropogenic pollution of the environment largely associated with heavy metals, causes serious concern for its negative consequences for the health of various population groups (Salnikova et al. 2013). Accumulation of Cu and Ni in the hair of children living in similar conditions in the area of emissions from the Karabashmed is shown by the authors of the work (Notova et al. 2017). Children’s hair contain a very small amount of the main osteotropic chemical elements Ca, Mg, P, B and Sr, which are heavy metal antagonists, and the elements Pb, Cd and As can accumulate in their bone tissue. Their antagonist is P.

Harsh climatic conditions of the Arctic cause both an increase in the level of air pollution and its negative impact on health (Saltykova et al. 2020). The simultaneous action of cold and air pollution, which are synergists, accelerates the development of diseases and aging of human body in high latitudes and affects the circulatory and respiratory systems to the greatest extent.

Thc furthcr nature of the migration of ecotoxicants entering the environment will depend on the phase composition of various types of dust and will determine the form of their presence in natural objects. Chemical analysis of snow samples taken at several sites near the metallurgical plant (Monchegorsk site) is shown on Table 2. The magnetic fraction from the solid phase of snow samples contains 0.3% Cu and 1–3% Ni, nickel is in the form of a spherical metal particle, and copper is in the form of CuFeS.

HM released into the environment with dust emissions then migrate in the system of water and soil objects of the geochemical zone. The results of hydrochemical studies indicate a high level of pollution of water sources, including the entry of HM into them. Concentration of heavy metals in water increases during floods. This

Table 1 Composition of products of roasting of nickel concentrate in fluidized bed furnaces, fine converter dust and dust of reflective smelting of copper concentrate

Element	Products of roasting of nickel concentrate					Heavy metals in fine converter dust			
	Cyclone dust	Dust from electrostatic precipitators				Dust from electrostatic precipitators			
		Prechamber	Field I	Field II	Field III	Gas distribution grid	Field I	Field II	Field III
		Content of elements (weight fraction), %							
Ni	67.9	61.6	63.7	59.1	41.7	–	12.9	25.1	3.35
Cu	2.67	2.36	2.28	2.21	1.70	–	6.74	2.83	1.85
Co	1.53	1.39	1.33	1.28	0.93	–	0.41	0.16	0.10
Fe	2.68	2.24	1.91	1.84	3.08	–	13.1	6.39	5.90
Sn	0.0011	0.89	0.0012	0.0007	0.0016	0.025	0.75	2.02	1.51
Ge	0.0045	0.0045	0.0053	0.0037	0.0029	0.017	0.0261	0.027	0.026
Sc	0.023	0.035	0.0079	0.0043	0.038	0.020	0.0308	0.028	0.017
Cd	0.0002	0.016	0.0014	0.0012	0.0014	0.018	0.24	0.24	0.21
Pb	0.14	0.40	0.70	0.80	0.60	0.47	8.51	12.9	10.5
Te	0.057	0.21	0.38	0.33	0.27	0.020	0.24	0.32	0.28
Se	0.0001	0.0024	0.0039	0.024	0.357	0.22	0.22	0.11	0.22
Ag	0.06	0.13	0.2	0.2	0.22	0.0011	0.0093	0.027	0.0112
As	0.82	5.4	3.95	5.74	4.9	0.61	2.35	3.39	4.6
Zn	0.0039	0.206	0.0052	0.0057	0.204	0.142	1.5	2.44	2.27
Content of elements (weight fraction), ppm									
Hg	0.15	0.45	1.4	0.72	2.7	2.7	1.03	0.40	0.66
Tl	4.8	28.1	19.2	21.0	19.7	19.9	211.0	227.4	219.2
In	0.50	14.6	0.44	0.36	0.45	3.6	26.6	60.8	59.5
Mo	61.3	40.0	47.2	39.8	41.0	20.5	67.6	122.4	95.4

Dust from electrostatic precipitators	Dust of reflective smelting of copper concentrate						
	Cu	Ni	As	Se	Sn	Sb	Te
	Content of elements (weight fraction), %						
Prechamber	25.2	0.90	0.022	0.075	0.026	0.0003	0.0018
1 field	35.0	2.06	0.050	0.030	0.0073	0.0004	0.0019
2 field	33.1	2.29	0.065	0.043	0.0070	0.0007	0.0020
3 field	42.1	2.63	0.055	0.047	0.0027	0.0006	0.0017
4 field	37.1	2.30	0.057	0.044	0.0040	0.0006	0.0016

Table 2 Characteristics of solutions formed during melting of snow samples and the content of components in the solid phase

Element	Snow sampling site			
	Spring Gabbroviy	Moncha river	Experimental site	
			№ 1	№ 2
	Elements concentration, mg/L			
pH	5.8	5.8	5.6	6.5
Cu	0.025	0.023	0.025	0.042
Ni	0.035	0.040	0.260	0.150
Ca	0.23	0.17	0.07	0.53
Mg	0.035	0.027	0.016	0.055
Element	The content of components in the solid phase of snow samples, wt%			
Cu	0.06	0.23	0.18	1.1
Ni	0.25	0.82	1.7	2.2
Ca	3.5	2.8	2.1	2.2
Mg	2.0	1.8	1.2	3.3
Fe	4.0	7.1	4.3	6.2
Al	6.7	5.5	4.7	4.1
Si	23.4	16.9	15.7	13.8
C	11.9	30.1	38.5	29.7
S	0.6	1.8	2.4	2.2

* Insert in Fig. 1 shows experimental fields

is due to the fact that snow cover accumulates HM, and ecotoxicants come from contaminated soil (Kremenetskaya et al. 2015).

Table 3 shows the gross composition of technogenically polluted soil (sampled from Monchegorsk site, insert in Fig. 1) and concentration of mobile forms of elements that affect the plants growing on such soil. The highest values of the fraction of mobile forms in relation to the total content are observed for chemical elements of anthropogenic origin in the group of heavy metals: copper (54%), lead (32%), selenium and cadmium (more than 25%), nickel and antimony (about 20%). The extremely high content of the priority ecotoxicants in the impact zone of the copper-nickel enterprise should be noted, and concentration of toxic metals in peat is even higher.

Figure 3 shows data on the total content of heavy metals in one of the peat samples and shows the forms of these elements: in a mobile form and associated with organic matrix and hydroxides of iron and manganese. Recent experiments have shown that concentrations of nickel and copper in some areas can reach 1–2%.

Experiments carried out on experimental reclamation sites (insert in Fig. 1) helped to analyze the mobility of elements. They were accumulated by herbaceous plants during growth. The TM content was determined in the samples previously washed

Table 3 Gross composition of the averaged sample of man-caused soil, mg/kg

Element	Na	Mg	Al	Si	P	K	Ca	V	Cr	Mn
Gross composition	3,160	1,623	10,665	64,155	1,166	2,202	3,376	44.4	131.9	265.1
Movable form	55	103	85	30	53	536	718	0,4	7	15
Element	Fe	Co	Ni	Cu	Zn	As	Se	Rb	Sr	Mo
Gross composition	26,420	82,5	2,344	5,955	93.1	27.4	7.32	2.43	24.7	7.23
Movable form	2,553	3	440	3,200	4	2	2	–	–	–
Element	Ag	Cd	Sn	Sb	Te	La	Hg	Pb	Bi	Th
Gross composition	0.14	0.38	2.19	1.15	1.01	3.12	0.18	18.6	0.6	0.42
Movable form	–	0.10	0.06	0.20	–	–	–	6	–	–

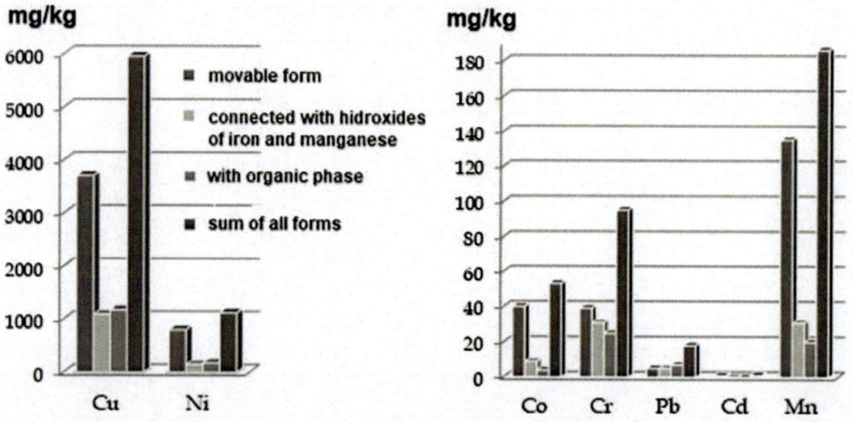

Fig. 3 Results of determining the forms of components in man-made peat

with water and dried in laboratory conditions, and in the samples without pretreatment (Table 4). This procedure helped to estimate and separate the concentration of metal on the surface as a part of dust and metals directly absorbed by the plants. Table 4 shows data on the content of Cu and Ni in the plants sampled near the plant and at a distance of 7 and 31 km from the plant, concentration of copper decreases in more than 40 times, and nickel—in 150 times.

Studies conducted in different parts of the world show that there is a certain relationship between the content of a number of trace elements in water and soil and the incidence of certain diseases. However, a human is affected by a number of factors, and microelements are only part of this series (Salnikova et al. 2019).

2. *Influence of geochemical characteristics of the territory on the composition of drinking water and the intake of elements into human body*

There is a need to study geographical and ecological pathology caused by an excess or deficiency of trace elements for abnormal geochemical regions of the country. Daily intake of chemical elements in the body of an adult is (mg): K 2000–5500; Na

Table 4 Content of elements (mg/kg) in plant material

Element	Planting option				Clark values*	*Avenella flexuosa* L.	
	Direct sowing		Carpet turf			Distance**, km	
	Unwashed	Washed	Unwashed	Washed		*7*	*31*
Cu	278.4	178.5	203.5	188.9	10	170	3.9
Ni	316.1	152.1	242.1	180.2	2	360	2.3
As	7.21	3.10	6.05	2.50			
Se	4.76	2.3	3.01	2.6			
V	7.71	3.70	6.99	3.80			
Cr	209.0	35.9	133.2	38.1			
Mn	265.1	165.8	253.9	266			
Fe	1104	390	728	302			
Co	22.5	11.2	22.2	12.5			
As	7.2	3.1	6.1	2.5			
Se	4.76	2.3	3.01	2.6			
Sb	0.24	0.13	0.18	0.03			
Te	0.49	0.30	0.41	0.17			

* Lebedev et al. (2018), ** distance from a plant

1100–3300; Ca 800–1200; Mg 300–400; Zn 15; Fe 10–15; Mn 2.0–5.0; Cu 1.5–3.0; Mo 0.075–0.25; Cr 0.05–0.2; Cl^- 3200; PO_4^{3-} 800–1200; SO_4^{2-} 10 (Baranovskaya et al. 2015). The lack of vital macro- and microelements in food and drinking water, and in some cases an excess of toxic substances, leads to disruption of vital processes, a decrease in the adaptive capacity of the body and a deterioration in the health of population. Studies of the biological role of chemical elements presented in organisms in micro- and submicroquantities are relevant (Salnikova et al. 2013; Krainov et al. 2012).

When studying the effect of the mineral composition of water on the morbidity of population, it is necessary to take into account the significant contribution of water to the total intake of mineral substances into the body. The real contribution of drinking water to the incidence of population is not always possible to evaluate, since the intake of minerals in the body is determined by many social factors and individual habits. However, most of population consumes water from centralized water supply systems. Water is supplied from ground or underground sources after additional water treatment. Water sources can also be subject to anthropogenic pressure.

The main source of water supply for Monchegorsk is the Moncheozero which is located in the zone of impact of Kola MMC. The waters of the lake contain elevated concentrations of Ni, Cu and Cd (Moiseenko et al. 2010). Concentration of Cu and Ni in the river Moncha flowing out of the lake does not exceed background concentrations, but exceeds the maximum allowable concentration for fishery reservoirs (Kremenetskaya et al. 2015). The water intake of Olenegorsk is affected by the metallurgical plant to a lesser extent, because it is located in a zone of low pollution. Apatity is supplied with water from Lake Imandra. Its northern part is polluted by industrial emissions from the plant, but the concentration of the main ecotoxicants in it is lower due to the flow and large volume of the lake. Concentration of heavy metals in the waters supplied to the population does not practically change after the cycles of technological preparation. All surface waters are low-mineralized.

As an alternative source of drinking water, water packaged in containers can be considered. However, the authors of the work (Khramov et al. 2019) argue that the use of drinking water enriched with Ca and Mg by residents of environmentally polluted cities of the North (for example, Monchegorsk) creates a threat of accumulation of toxic metals in the body. The presence of a direct correlation of the concentrations of Ca and Mg in children's hair and in drinking water and an inverse correlation of the concentration of Cu, Ni, Fe and Sr in hair from the concentration of these metals in drinking water has been established. The study of the accumulation of toxic metals in hair most qualitatively reflects the technogenic geochemical situation in places where people live (Abdelbagi et al. 2017; Skalny et al. 2018; Krivosheev 2003). The method has become widespread, especially in the study of the accumulation of rare earth elements (REE) in recent years (Li et al. 2013).

Table 5 shows the data on chemical composition of the waters of the Tsentralny water intake supplied to the population of the city of Kirovsk and the gastric juice of the inhabitants of the Apatity-Kirovsk district. The water intake is located in the zone of influence of another large plant: Kirovsk branch of the "Apatit" and its tailing damp. The territory is polluted due to the fact that the wind blows solid particles from dusty surfaces (Pashkevich et al. 2019). The work demonstrates the ways in which particles enter the atmosphere and then soils and groundwater in the areas affected by man-made massifs. Analysis of the results (Table 5) shows the presence in the gastric juice samples of the same set of chemical elements as in the underground natural waters, but in higher concentrations. Studies of chemical composition of the gastric juice of the population (20 patients by random sampling) showed low average concentrations (relative to the literature data) of Na, Cl, K and high concentration of P. Rare elements and REE are present in gastric juice. In recent decades, the threat has increased due to the fact that, along with traditional metals-pollutants (Cr, Mn, Be, Cd, Hg, Pb), human exposure to rare and REE (In, Ga, Zr, Nd, etc.), the toxicity of which has not been sufficiently studied, is increasingly observed, therefore they cannot be considered harmless.

The pH value of the underground waters of the Tsentralny is high, the concentration of aluminum exceeds the maximum allowable norms up to five times, at the same time, a very low concentration of Ca and Mg is noted. The waters are poorly mineralized, concentrations of vital elements in the waters of the intake are low, these elements can only be obtained by the residents of the region with food or bottled water (Mazukhina et al. 2020). Mineral composition of whole blood in 25 healthy teenagers aged 18 to 22 years old living in the central part of Kirovsk was studied in the work (Nikanov et al. 2021). It was found that the concentrations of aluminum in the blood of patients are within the physiological norm, high concentrations of HM are not detected, calcium content is below the average value for healthy people, and

Table 5 Chemical composition of the waters of the water intake "Tsentralny" and the gastric juice of the inhabitants of Apatite-Kirovsk region

The chemical composition of the waters of the water intake "Tsentralny", mg/L					The chemical composition of gastric juice of the inhabitants of the Apatite-Kirovsk region, mg/L				
Element	min	max	M, n = 5	SD p = 0,05	Literature data*	min	max	M, n = 20	SD p = 0,05
B	0.015	0.017	0.016	0.00,090	–	0.28	24.6	6.65	5.390
Na	11.5	29.2	18.6	7.150	1,160	99.6	1,199	537.8	299.1
Mg	0.002	0.042	0.016	0.0174	–	1.50	12.5	5.40	2.98
Al	0.35	1.07	0.82	0.279	–	0.057	7.72	0.57	1.69
P	0.015	0.032	0.024	0.0070	3.18	3.88	114.0	49.6	31.7
K	1.17	4.15	2.12	1.239	590	36.8	316.7	156.9	96.9
Ca	0.008	0.24	0.093	0.0921	5.03	4.13	46.7	14.8	10.0
V	0.0006	0.0066	0.0037	0.00059	–	0.0035	0.035	0.013	0.0075
Cr	0.0002	0.0017	0.0008	0.00017	–	0.0071	0.074	0.041	0.019
Mn	0.00001	0.00044	0.00016	0.00017	–	0.0022	0.087	0.018	0.019
Fe	0.0051	0.021	0.016	0.0068	–	0.13	8.92	0.92	1.93
Co	0.00001	0.00016	0.00006	0.000056	–	0.00013	0.0005	0.0003	0.0001
Ni	0.00001	0.00019	0.00007	0.000072	–	0.011	0.36	0.070	0.090
Cu	0.0001	0.00029	0.00016	0.000081	–	0.13	40.7	4.05	8.91
Zn	0.0034	0.0061	0.0048	0.00102	–	0.22	14.9	2.33	3.87
Sr	0.0002	0.0024	0.0010	0.00093	–	0.020	0.099	0.045	0.019
Y	0.00001	0.00005	0.00003	0.000017	–	0.00001	0.00038	0.00007	0.00011
Zr	0.00001	0.00008	0.00005	0.000025	–	0.00004	0.0021	0.00049	0.00048
Nb	0.00006	0.0003	0.0002	0.00012	–	0.00002	0.00084	0.00019	0.00021

(continued)

Table 5 (continued)

The chemical composition of the waters of the water intake "Tsentralny", mg/L					The chemical composition of gastric juice of the inhabitants of the Apatite-Kirovsk region, mg/L				
Element	min	max	M, n = 5	SD p = 0,05	Literature data*	min	max	M, n = 20	SD p = 0,05
Mo	0.0032	0.0115	0.0056	0.00348	–	0.0022	0.011	0.0053	0.0027
Ag	0.0006	0.0018	0.0014	0.000493	–	0.00027	0.026	0.0026	0.0055
Cd	0.00001	0.00004	0.00002	0.000011	–	0.00001	0.00024	0.0001	0.00008
Ba	0.00003	0.00014	0.00009	0.000045	–	0.0008	0.0079	0.0045	0.0022
La	0.00003	0.00018	0.00007	0.000062	–	0.00001	0.00029	0.00008	0.000066
Ce	0.00001	0.00009	0.00007	0.000031	–	0.00001	0.00017	0.00007	0.000057
U	0.0001	0.0010	0.0004	0.00040	–	0.0001	0.0023	0.0006	0.00064
pH	8.97	9.63	9.40	0.260	–	1.43	7.35	4.44	2.081
Cl^-	1.25	2.58	2.03	–		422.5	5,864	2,685	1,618

* Mazukhina et al. (2020)

a high content of boron (2.45 mg/l) is detected (the cause is unknown). The concentration of boron measured by us in the gastric juice of patients living in this region was in the range of 0.28–24.6 mg/l. It is many times higher than its concentration in the water (Table 5).

Chemical pollution of the water bodies in Murmansk region amounted to 33% for the waters of category I, 37%—for category II. 20% of the non-centralized water supply sources and 27% of the centralized water supply sources, 36% of the underground water sources and 29% of the surface water sources do not correspond to hygienic standards for sanitary and chemical indicators. Exceeding the maximum allowable concentration for metals (Ni—21%, Fe—24.2%, Mg—4.6%, Al—16.2%) is observed in the drinking water of the centralized water supply in the whole of the Murmansk region (Dushkin et al. 2015). Murmansk region (springs are popular with the population) also causes concern: 46.2% of samples do not correspond to the standards for sanitary and chemical indicators (2012 data).

4 Conclusions

The geoecological state of the Kola region is determined by the enterprises operating on its territory. They produce a large amount of dust emissions that change the entire environment. More than a thousand tons of Cu and Ni enter the biosphere per year. Dusts also contain Fe, Zn, Cd, As, Hg, Pb, Tl, Sn, Sb, Se, Te, etc. HM that came with dust emissions then migrate in the system of water and soil objects of the geochemical zone, which affects the elemental composition of organisms located in this territory. The results of hydrochemical studies indicate a high level of pollution of the water sources, including the sources of drinking water supply. Anthropogenic pollution of the natural environment with heavy metals is a matter of serious concern. First of all, anthropogenic pollution is dangerous for its negative consequences for the health of various groups of the population. Knowledge of the elemental status of a human, taking into account regional characteristics, can contribute to a more effective correction of human health and the prevention of various diseases.

Funding This work was financially supported by the Ministry of Science and Higher Education Russian Federation scientific topics FMEZ 2022-0018 and FMEZ-2022-0010 (1021051803680-5), and partially supported by the Russian Science Foundation grant 21-77-10111.

References

Abdelbagi AM, Gilani MA, Sharf Eldeen AE (2017) Concentrations of trace elements in human hair as a biomarker expose to environmental contamination. International Journal of Scientific Research and Innovative Technology 4(2):1–11

Baranovskaya NV, Rikhvanov LP, Ignatova TN, Narkovich DV, Denisov OA (2015) Essays on human geochemistry: monograph. Publishing House of TPU, Tomsk (in Russian)

Chashchin VP, Gudkov AB, Popova ON, Odland YuO, Kovshov AA (2014) Characteristics of the main risk factors for health disorders of the population living in areas of active nature management in the Arctic. Ekologiya cheloveka 21(1):3–12. https://doi.org/10.17816/humeco17269

CV 3.18.05-2005. FR.1.31.2005.01714 (2005) Water quality. Determination of elements in water. Method by ICP-MC. Center for research and control of water. St. Petersburg (in Russian)

Dushkin EV, Dudarev AA, Sladkova YuN, Zachinska IYu, Chupakhin VS, Gushchin IV, Talykova LV, Nikanov AN (2015) The content of metals in water sources and drinking water in industrial cities of the Murmansk region. Medicina truda i promyshlennaya ekologiya 2:29–34 (in Russian)

Evdokimova GA, Kalabin G V, Mozgova NP (2012) Assessment of the state of the soil cover in the process of reducing the impact of air emissions from the Severonickel plant on the environment. Cvetnye metally. 11:53–58 (in Russian)

Krainov SR, Ryzhenko BN, Shvets VM (2012) Geochemistry of underground waters. Theoretical, applied and environmental aspects. CentrLitNefteGaz, Moscow https://medicalinsider.ru/news/gipokhlorgidriya-prichiny-simptomy-i-lechenie

Khramov AV, Kontrosh LV, Pankratova MYu, Vezhenkova IV (2019) Influence of the chemical composition of drinking water on the level of accumulation of toxic metals in the human body. Ekologiya cheloveka 6:11–16. https://doi.org/10.33396/1728-0869-2019-6-11-16

Kremenetskaya IP, Kubachina EE, Slukovskaya MV, Drogobuzhskaya SV, Lashchuk VV (2013) Influence of the geomorphology of objects of the hydrogeological system of the Monchegorsk technogenic wasteland on the formation of “minerals” of natural water

Kremenetskaya IP, Drogobuzhskaya SV, Lashchuk VV, Ivanova TK (2015) Influence of climatic, hydrological and geomorphological conditions on the formation of the chemical composition of water bodies located in the impact zone of a copper-nickel plant in subarctic conditions. In: Ohrana okruzhayushchej sredy ot negativnogo vozdejstviya hozyajstvennoj deyatel’nosti (ed. by Eliseev DV). Publisher «Sibak». Novosibirsk (in Russian)

Trudy Fersmanovskoj nauchnoj sessii GI KNC RAN. 10:257–260. ISSN: 2074-2479 (in Russian)

Krivosheev YuK (2003) Comparative assessment of the content of chemical elements in the hair of children in Monchegorsk and other regions of Russia. Byulleten’ Severnogo Gosudarstvennogo Medicinskogo Universiteta 2:133–134

Lebedev AN, Khazov MV, Potseluyev OM, Lebedeva MA, Siromlya TI (2018) Influence of biological products on the elemental chemical composition of lawn plants during the growing season. Agrochemistry. 2:79-87

Li X, Chen Zh, Chen Zh Q, Zhang Y (2013) A human health risk assessment of rare earth elements in soil and vegetables from a mining area in Fujian Province, Southeast China. Chemosphere 93(6):1240–1246. https://doi.org/10.1016/j.chemosphere.2013.06.085

Mazukhina SI, Maksimova VV, Chudnenko KV, Masloboev VA, Sandimirov SS, Drogobuzhskaya SV, Tereshchenko PS, Pozhilenko VI, Gudkov AV (2020) Water quality in the Arctic zone of the Russian Federation: physical and chemical modeling of water formation, forms of element migration, impact on the human body. Izdatel’stvo FIC KNC RAN, Apatity ISBN 978-5-91137-437-2 (in Russian)

Method for determining trace elements in diagnosed biosubstrates by inductively coupled argon plasma mass spectrometry (ICP-MS) (2003). Guidelines, Moscow (in Russian)

Moiseenko TI, Megorsky VV, Gashkina NA, Kudryavtseva LP (2010) The impact of water pollution on the health of the population of the industrial region of the north 37(2):194-203. https://doi.org/10.1134/S0097807810020077

Nikanov AN, Bykov VR, Talykova LV, Dorofeev VM (2018) Occupational risk at enterprises engaged in the extraction and processing of copper-nickel ores. Collection of articles on mater. int. scientific and practical. conf. "Environmental, industrial and energy security". Sevastopol State. Univ, Sevastopol ISBN 978-5-6040112-3-2

Nikanov AN, Gudkov AB, Popova ON, Smolina VS, Chashchin VP (2021) Mineral composition of the blood of residents of the Arctic region with low water salinity in centralized water supply systems. Ekologiya cheloveka 3: 42–47. https://doi.org/10.33396/1728-0869-2021-3-42-47

Notova SV, Kireeva GN, Zhukovskaya EV, Grabeklis AR, Kiyaeva EV, Skalny AV, Deryagina LE (2017) Influence of anthropogenic and geochemical environmental factors on the elemental status of children in the Chelyabinsk region Ekologiya cheloveka 11:23–28

Pashkevich MA, Matveeva VA, Danilov AS (2019) Study of Pollutant Migration from the Territories of Technogenic Massifs of the Kola Peninsula. Gornyj zhurnal 1:17-21. https://doi.org/10.17580/gzh.2019.01.14. (in Russian)

Salnikova EV, Osipova EA, Skalny AV, Burtseva TI, Boldyreva OI (2013) Influence of microelement intake from the biosphere on the elemental status of a person. Vestnik Orenburgskogo gosudarstvennogo universiteta 10(159):21-24 (in Russian)

Salnikova EV, Burtseva TI, Skalny AV (2019) Regional features of the content of trace elements in the biosphere and the human body. Gigiena i sanitariya 98(2):148–152. https://doi.org/10.18821/0016-9900-2019-98-2-148-152

Saltan N, Slukovskaya M, Drogobuzhskaya S, Shirokaya A, Kremenetskaya I, Gorbov S, Mikhaylova I, Khvostova A, Zarov E, Skripnikov P (2021) Assessment of soil heavy metal pollution by land use zones in small towns of the industrialized Arctic region, Russia.

Springer Geography. 100–110. https://doi.org/10.1007/978-3-030-75285-9_10

Saltykova MM, Bobrovnitsky IP, Balakaeva AV (2020) Impact of atmospheric air pollution on the health of the population of the Arctic region: a literature review Ekologiya cheloveka 4: 48–55. https://doi.org/10.33396/1728-0869-2020-4-48-55

Seregina IF, Osipov K, Bolshov MA, Filatova DG, Lanskaya SYu (2019) Matrix Interference in the Determination of Elements in Biological Samples by Inductively Coupled Plasma Mass Spectrometry and Methods for Its Elimination J. Analyt. Chem. 74:182–191. https://doi.org/10.1134/S1061934819020114

Skalny AV, Tinkov AA, Voronina I, Terekhina O, Skalnaya MG, Kovas Y (2018) Hair Trace Element and Electrolyte Content in Women with Natural and In:Vitro Fertilization-Induced Pregnancy. Biological Trace Element Research 181(1):1–9. https://doi.org/10.1007/s12011-017-1032-0

Slukovskaya M. V., Kremenetskaya I. P., Drogobuzhskaya S. V., Novikov A. I. (2020) Sequential Extraction of Potentially Toxic Metals: Alteration of Method for Cu-Ni Polluted Peat Soil of Industrial Barren // Toxics. No 8, 39. https://doi.org/10.3390/toxics8020039.

Tarasova E, Brykov VA, Dovletyarova EA, Paltseva AA, Drogobuzhskaya SV, Tapia-Pizarro F, Neaman A, Morev DV, Slukovskaya M, Navarro-Villarroel C (2020) Vermiculite-lizardite industrial wastes promote plant growth in a peat soil affected by a Cu/Ni smelter: a case study at the Kola peninsula, Russia. Journal of Soil Science and Plant Nutrition. 20(3):1013–1018. https://doi.org/10.1007/s42729-020-00188-z

Mobility and Peculiarities of Trace Elements Content in the Blood Serum of the Population of the Mining Region: Case of Bashkortostan Republic

Irina N. Semenova

Abstract The object of the study is teenagers and adult population of the industrial center of the mining province of the Republic of Bashkortostan—Sibay. Objective: to study the incidence and the microelement composition of blood serum in different age groups of the Sibay city population. Chemical and analytical studies of blood were carried out in a licensed laboratory of the Autonomous Non-Profit Organization "Center for Biotic Medicine" (Moscow) by mass spectrometry with inductively coupled argon plasma. Statistical analysis was performed using the Statistica 6.0 program, and at $p < 0.05$, the differences were evaluated as statistically significant. As a result of the examination of adolescents and adults of the Trans-Ural zone of Bashkortostan, an imbalance of trace elements in the blood serum was revealed. Significant deviations from the physiological norm in adolescents were the following: in less than 10% of individuals, a reduced content of zinc, manganese, copper and cobalt was registered, and an increased content of iron was also detected in 30% of adolescents; in 23%—cadmium; in 16–17%—copper, zinc and manganese. Adults (23%) were found to be deficient in cobalt, as well as an excess of cobalt (32.5%), cadmium (32.5%), arsenic (20%), and zinc (20%). The results obtained indicate that the population of this region is at risk of developing both metal toxicosis and a deficiency of vital trace elements. This fact requires strengthening control over the level of heavy metals in atmospheric air, soil, water, and food, and taking measures to reduce the risk to public health, including correcting the microelement status.

Keywords Heavy metals · Microelement profile · Imbalance of microelements · Adolescents · Adults

I. N. Semenova (✉)
Sibay Institute of Ufa University of Science and Technology, Sibay, Russia
e-mail: alexa-94@mail.ru

O. V. Frank-Kamenetskaya et al. (eds.), *Biogenic—Abiogenic Interactions in Natural and Anthropogenic Systems 2022*, Springer Proceedings in Earth and Environmental Sciences, https://doi.org/10.1007/978-3-031-40470-2_36

1 Introduction

Among the factors that are of particular importance for maintaining human health, the most important role belongs to vital macro-and microelements. The lack of essential macro-and microelements in the food chain, and in some cases an excess of toxic microelements, contribute to disruption of vital processes, a decrease in the adaptive capabilities of the body, and a deterioration in the health of both individuals and the population as a whole (Avtsyn et al. 1991).

The imbalance of chemical elements in the environment determines not only the variability of food chains and metabolism, but also the ability of an organism to adapt to environmental conditions, as well as the probability of developing biogeochemical endemias (Kovalsky 2009; Skalny and Rudakov 2004; Bech et al. 2020).

Studies of a number of scientists in the Southern Uralshave identified biogeochemical provinces that are natural and anthropogenic in origin, formed as a result of the formation of the Earth's crust and pollution of the biosphere by industrial emissions (Belan 2005, 2006; Opekunova et al. 2020). As a result, geochemical anomalies were formed under the influence of natural and man-made factors with an excessive accumulation of some elements—zinc, copper, nickel, lead, cadmium, and more rarely cobalt, and a deficiency of others-selenium and iodine (Gribovsky 2000). The formation of anomalies negatively affects human health, in particular, there is a violation of mineral metabolism due to changes in the concentration and ratio of elements in the environment, drinking water and food products (Kovalsky 2009). Experimental and clinical studies have confirmed that a deficiency of some trace elements and an excess of other trace elements contributes to an increase in the frequency of malignant neoplasms, lymphoproliferative diseases, infectious pathology, autoimmune and degenerative diseases, and congenital anomalies (Avtsyn et al. 1991; Boev et al. 2001).

The development of ore deposits is the main factor of chemical pollution of the environment in the Southern Urals. One of the most significant sources of technogenic impact in the region is the mining complex formed in the Trans-Urals of Bashkortostan, where the extraction and processing of copper-zinc-pyrite deposits is carried out (Belan 2005, 2006; Opekunova et al. 2020; Opekunov et al. 2021). Thus, the population living in a given territory in a geochemical province with an imbalance of chemical elements in environmental objects is a group at risk of developing eco-dependent diseases, which requires the development and adoption of appropriate management decisions.

In the Republic of Bashkortostan, work has been underway to study the relationship of various diseases with trace elements of the living environment since the middle of the last century. In the works of many authors, the trace element composition of soil, water, food and its role in the development of various diseases are analyzed (Askarova 1969; Starova 2003; Farkhutdinova 2006). A number of works highlight the importance of trace elements of rocks for the health of the population on the example of the Trans-Urals (Belan 2005; 2006).

Sibay city with a population of 61 thousand is located in the Trans-Ural zone of the Republic of Bashkortostan and is an industrial, cultural and educational center of the mining region, with the presence of environmental problems caused by the pollution of environmental objects with heavy metals, such as Cu, Zn, Fe, Mn, Cd, Ni, Co, Pb. The main mining production is concentrated in the Sibay branch of the Uchaly Mining and Processing Plant, which includes quarries, underground mines, dumps, tailings dumps and a processing plant. This company has been operating since 1948. The main products of the plant are copper concentrate, zinc concentrate, limestone crushed stone, lime. In 2019, the volume of emissions of pollutants into the atmospheric air amounted to 0.498 thousand tons. The Sibay city also has a thermal power plant, which was put into operation in March 2004. Mining activities contribute to the accumulation of toxic chemical elements in the environment (Semenova et al. 2021a, b).

The study of environmental risks in connection with the emissions of toxic substances, mainly sulfur dioxide, into the atmosphere in 2018–2019 as a result of a fire in the Sibay quarry has become particularly relevant for Sibay. Work in the quarry was stopped in 2003; its sides began to crumble, as a result of which individual inclusions of ore with high sulfur content were exposed, and in November 2019, its ignition occurred. Unfavorable meteorological conditions contributed to the fact that SO_2 began to rise to the surface from the bottom of the quarry. In order to eliminate the fire, the flooding of the quarry was carried out.

Objective: to study the incidence and the microelement composition of blood serum in different age groups of the Sibay city population.

2 Objects and Methods

Assessment of the content of chemical elements in the body of the examined was carried out according to the results of determining their level in the blood serum. The content of 8 chemical elements in the blood serum of 40 men aged 29 to 62 years and 30 boys aged 15 to 17 years was studied.

Blood was drawn from the ulnar vein in the morning on an empty stomach in a volume of at least 5 ml. Serum at room temperature after clot formation was centrifuged at 1500 rpm for 15 min, transferred to a micro tube and stored at -70 °C. The elemental composition of blood serum was studied in the accredited testing laboratory of the Center for Biotic Medicine (Moscow, Russia, ISO 9001: 2008, certificate 54Q10077 dated 21.05.2010), using a combination of atomic emission spectrometry and inductively coupled plasma mass spectrometry (ISP-AES and ICP-MS). As reference values for the concentrations of elements, the norms specified in the form of the laboratory of the Center for Biotic Medicine were used. The obtained data were processed using the Microsoft Excel 2003 and STATISTICA 6.0 application programs.The arithmetic mean (M), standard error of the mean (m) and median (Me) were calculated. The 25th and 75th percentiles were used as measures of dispersion of parameters with a distribution different from normal and the presence of extreme values. Statistically significant differences between groups were calculated using the nonparametric Mann–Whitney method for two unrelated groups (Rebrova 2006). The critical significance level (p) was taken as 0.05 when testing statistical hypotheses.

To analyze the morbidity of the population, a database was used, formed on the basis of the materials (2000–2020) of the official reports of the Medical Information and Analytical Center of the Republic of Bashkortostan.

3 Results

Table 1 shows the content of chemical elements in the serum of the examined individuals.

The distribution of the content of most of the studied chemical elements in the blood serum did not correspond to the normal type. Almost all average concentrations of chemical elements were within physiologically normal values. The exception was the median levels of arsenic, iron and manganese in the blood serum of adolescents (exceeding the upper limit of normal), as well as cadmium (upper limit of normal).

Table 2 shows the frequency of occurrence of deviations from the reference values of the concentration of chemical elements in the blood serum of the subjects.

Significant deviations from the physiological norm of adolescent blood were the following: in less than 10% of individuals, a reduced content of zinc, manganese, copper and cobalt was registered, and an increased content of iron was also detected in 30% of adolescents; in 23%—cadmium; in 16–17%—copper, zinc and manganese.

Cobalt deficiency was found in adults (23%), as well as an excess of cobalt (32.5%), cadmium (32.5%), arsenic (20%) and zinc (20%).

Analysis of the correlation relationships of the levels of the studied chemical elements in serum showed the presence of statistically significant relationships between some indicators ($p < 0.05$). Thus, in the group of adolescents, the relationship between the content of zinc and iron ($r = 0.77$), zinc and nickel ($r = 0.52$), cobalt and manganese ($r = 0.54$) was revealed (Table 3). In the adult group, the relationship between the content of the following chemical elements was revealed: copper and iron ($r = 0.56$), cadmium and zinc ($r = 0.54$), cobalt and arsenic ($r = 0.43$), cobalt and cadmium ($r = 0.46$) (Table 4).

The analysis of statistical reports showed an increase in both the general and primary morbidity of adolescents over the period 2000–2020, both in the Republic of Bashkortostan as a whole and in Sibay. At the same time, the increase in the incidence of adolescents in Sibay was significantly ahead of the average in the Republic (Table 5).

The first three places in the structure of morbidity of adolescents in 2020 were occupied by respiratory diseases, injuries and poisoning, diseases of the digestive system.

Comparison of the average primary morbidity of the adult population of the mining region for 2016–2020 with the average indicators for the Republic showed its excess

Table 1 Content of chemical elements in the blood serum of the population of Sibay, mkg / ml

	Adolescents	(n = 30)	Adults	(n = 40)	Reference values
	M ± m	Me (Q25-Q75)	M ± m	Me (Q25–Q75)	
As *	0.026 ± 0.004	0.019 (0.013–0.045)	0.002 ± 0.0002	0.002 (0.0016–0.0022)	0–0.010
Cd	0.0002 ± 0.0001	0.0002 (0.0001–0.0002)	0.0003 ± 0.0001	0.0001 (0.0000–0.0002)	0–0.00015
Co*	0.0005 ± 0.0001	0.0005 (0.0004–0.0006)	0.0019 ± 0.0004	0.0010 (0.0007–0.0014)	0.00045–0.0010
Fe*	2.78 ± 0.52	2.23 (1.70–3.38)	1.77 ± 0.02	1.80 (1.74–1.83)	0.65–2.00
Mn*	0.007 ± 0.001	0.005 (0.005-0.009)	0.0029 ± 0.0002	0.0027 (0.0021–0.0031)	0.0015–0.0040
Ni*	0.0073 ± 0.0004	0.0075 (0.0061–0.0086)	0.0037 ± 0.0002	0.0036 (0.0031–0.0040)	0–0.010
Zn	1.221 ± 0.078	1.15 (1.02–1.48)	1.07 ± 0.02	1.05 (1.00–1.13)	0.75–1.5
Cu*	0.83 ± 0.04	0.80 (0.72–0.90)	0.98 ± 0.04	1.02 (0.87–1.10)	0.75–1.5

Note *Statistically significant difference ($p < 0.05$) between adolescent and adult indicators

Table 2 Frequency of deviations of the content of chemical elements in the blood serum of the population of Sibay from normal values, %

	Level	Cu	Zn	Fe	Mn	Ni	Co	Cd	As
Adolescents (n = 30)	Reduced	5.0	6.3	6.3	7.5	0	10	0	0
	Increased	16.6	16.6	30.0	15.5	2.5	12.5	23.3	50
Adults (n = 40)	Reduced	5.0	7.5	7.5	7.5	0	23.3	0	0
	Increased	12.5	20.0	12.5	12.5	2.5	32.5	32.5	20

Table 3 Correlation coefficients (r) between concentrations of chemical elements in the blood serum of the adolescents of Sibay (n = 30, critical value r = 0.36, significant at the level of $p < 0.05$ is highlighted in bold)

	Fe	Zn	Cu	Mn	Co	As	Ni
Zn	**0.77**						
Cu	0.08	0.17					
Mn	0.40	0.36	0.03				
Co	0.38	0.25	0.04	**0.54**			
As	0.16	0.07	0.05	−0.02	−0.11		
Ni	0.37	**0.52**	0.46	0.26	0.32	−0.01	
Cd	−0.37	−0.35	−0.03	0.31	0.15	−0.33	0.18

Table 4 Correlation coefficients (r) between concentrations of chemical elements in the blood serum of the adults of Sibay (n = 40, critical value r = 0.31, significant at the level of $p < 0.05$ is highlighted in bold)

	Fe	Zn	Cu	Mn	Co	As	Ni
Zn	−0.11						
Cu	**0.56**	0.06					
Mn	0.03	0.13	0.08				
Co	−0.04	0.31	0.05	−0.20			
As	0.01	−0.03	0.06	−0.02	**0.43**		
Ni	0.06	−0.02	0.13	−0.01	−0.01	−0.15	
Cd	−0.12	**0.54**	0.17	−0.12	**0.46**	−0.07	0.18

Table 5 The increase in the incidence of adolescents aged 15–17 years for the period 2000–2019, per 100 thousand population

	General morbidity			Primary morbidity		
	2000	2019	Growth, %	2000	2019	Growth, %
Sibay	253,293.9	462,984.4	82.8	175,760.1	238,619.2	35.8
Republic of Bashkortostan	190,010.7	249,885.1	31.5	104,925.4	145,143.6	38.3

Table 6 Comparison of the average primary morbidity of the adult population of the mining region for 2016–2019 with the average republican indicators per 100 thousand populations

	Mining region	Republic of Bashkortostan
Total diseases	98,289	71,498
Diseases of the blood, hematopoietic organs and individual disorders involving the immune mechanism	1431	345
Anemia	1409	326
Diseases of the endocrine system, eating disorders and metabolic disorders	1591	1188
Diseases of the thyroid gland	402	274
Diseases of the nervous system	2596	1560
Diseases of the circulatory system	9262	4375
Asthma, asthmatic status	160	60
Diseases of the digestive system	8888	4692

in many nosologies. Particularly high morbidity rates were found for atopic dermatitis (exceeding the average values by 4.9 times), blood diseases (4.1 times), asthma (2.7 times), diseases of the digestive system (2.9 times), diseases of the circulatory system (2.1 times), the level of neoplasms (exceeding 1.4 times), diseases of the endocrine system, eating disorders and metabolic disorders (1.3 times), diseases of the thyroid gland (1.5 times), nervous system (1.6 times), eye and its appendage (1.6 times), ear and mastoid process (1.6 times), bone- muscular system and connective tissue (2.5 times), genitourinary system (1.4 times) (Table 6).

4 Discussion

It is known that blood has a more stable composition of trace elements than hair, so deviations in the content of both essential and toxic elements in the blood from the norm can contribute to the development of various pathological conditions in humans (Rikhvanov et al. 2015).

In our studies, deviations from the standard values were more often detected in adolescents than in adults. This applies primarily to such chemical elements as arsenic, iron, manganese and cadmium.

The median arsenic content in the blood serum of adolescents was 10 times higher than that of adults and 2 times the upper limit of normal.

Arsenicis one of the most powerful and dangerous poisons. In the presence of oxygen, it quickly forms a very toxic arsenic anhydride. In oral poisoning, a high concentration of arsenic is observed in the stomach, intestines, liver, kidneys and pancreas; in chronic poisoning, it gradually accumulates in the skin, hair and nails. Due to the inhibition of various enzymes, it disrupts metabolism. In the process of poisoning, the axons are the first to suffer, which leads to peripheral neuropathy and paralysis of the limbs. Entering the body, arsenic disintegrates/destabilizes more than 200 enzymes involved in various types of cell metabolism and DNA synthesis, activates peroxidation processes, binds thiol or sulfhydryl groups in tissue proteins, which leads to multisystem lesions involving the cardiovascular and nervous systems, liver, lungs, kidneys, spleen, and gastrointestinal tract. Arsenic is considered carcinogenic to humans (Skalny and Rudakov 2004).

Millions of people around the world are exposed to chronic exposure to high concentrations of arsenic due to the geochemical characteristics of the territory of residence. At present, large provinces of copper-arsenic ores have been found in Italy, the United States, Sweden, Norway, Japan, Georgia, and Kazakhstan; arsenic-cobalt ores in Canada; arsenic—tin ores in Bolivia, Chile, and England; and gold-arsenic ores in the United States and France. In Russia, geochemical arsenic provinces have been found in Yakutia, Siberia, Transbaikalia, Primorye, the Urals, and Chukotka (Kovshov et al. 2019). Data from clinical studies conducted in various regions of the Russian Federation indicate that in the territories of arsenic geochemical provinces, the incidence of diseases of the endocrine, genitourinary, nervous, cardiovascular, gastrointestinal and skin systems is 1.3–3.8 times higher than the national average, and the mortality rate due to oncological pathology is twice as high (Dormancheva 2000). Increased arsenic concentrations in drinking water were significantly associated with the development of skin diseases, immune system diseases, and cardiovascular diseases in the population (Bogdanova et al. 2017).

In our studies, the median serum iron content in adolescents was 1.2 times higher than that of adults and 1.1 times the upper limit of normal.

Adolescence is a critical period of ontogenesis, during which chronic diseases worsen and progress. Iron metabolism disorders are not only known iron deficiency conditions, but also poorly understood iron overload. Free iron leads to the formation of free radicals, which stimulates fibrogenesis, intoxication, and dysfunction of many systems (Ramm and Ruddell 2005). Elevated levels of hemoglobin and serum iron often escape the attention of doctors. One of the causes of iron excess in adolescent boys is physiological hyperadsorption of dietary iron under the influence of androgens during the period of musculinization (Testa, 2002).

The median serum manganese content in Sibay adolescents was 1.8 times higher than that of adults and 1.3 times the upper limit of normal.

One of the most possible ways of getting manganese into the human body is drinking water with an increased level of this element. Therefore, manganese is included in the list of priority pollutants in water bodies recommended for systematic control (Decision of the European Parliament and of the Council of the EU No. 2455/2001/EC) (Burlibayev et al. 2010). In terms of its effect on the body, manganese when ingested with drinking water is classified as moderately dangerous substances in accordance with Sanitary Rules and Regulations 2.1.4.1074—01 (3rd hazard class). Along with the general resorptive effect, it is characterized by the development of specific damaging effects, primarily on the part of the central nervous system, the blood system, which are aggravated by an imbalance of oxidative-antioxidant processes, which can lead to an increase of diseases in the population, including children.

Manganese is stably present in groundwater and open water in the form of suspended and colloidal particles. The main sources of manganese entering open water are discharges from ferrous and non-ferrous metallurgy, mechanical engineering, and metalworking enterprises, which contribute about 20–30% to Russia's Gross Domestic Product (Rakitskii and Yudina 2005). Significant amounts of manganese are obtained because of leaching of ferromanganese ores and other minerals containing manganese, during the decomposition of aquatic animals and plant organisms, especially blue-green, diatoms and higher aquatic plants (Teplaya 2013).

Manganese is an essential trace element necessary for performing many functions in the body, such as regulating various types of metabolism, including bone and connective tissues, blood clotting, is a cofactor for such enzymes as transferases, hydrolases, lyases, superoxide dismutase, arginase, glutamine synthetase, and participates in the synthesis and exchange of neurotransmitters (Skalny and Rudakov 2004).

Mazunina's review considers the negative effects of increased manganese content entering the body with drinking water and posing a potential danger to public health (Mazunina 2015).

With increased exogenous intake, manganese has a toxic effect, which is expressed in the development of negative effects (Basova and Khamitova 2008).

It was shown that in children who consume drinking water with a high content of manganese at the level of up to 3.3 MPC, the concentration of manganese in the blood exceeds on average 1.3 times the reference level and the indicator in the comparison group. At the same time, negative effects on the central nervous system were revealed, expressed in an increased frequency of imbalance of neurotransmitters regulating the processes of excitation and inhibition in the central nervous system, initiated by the activation of oxidative damage to cellular and subcellular structures, when the resources of the antioxidant defense system are depleted (Zaitseva et al. 2014).

The content of serum cadmium in the blood of Sibay adolescents corresponded to the upper limit of normal.

Chronic cadmium poisoning destroys the liver and kidneys, leading to severe renal impairment. Excess cadmium disrupts the metabolism of metals, especially iron and calcium, disrupts the action of zinc and other metal enzymes, blocks sulfhydryl groups of enzymes, and disrupts DNA synthesis. Cadmium easily replaces metal-flavoprotein complexes, where iron and molybdenum play a dominant role, disrupting the two-stage oxidation process (Skalny and Rudakov 2004).

Cadmium has carcinogenic, gonadotropic, embryotropic, mutagenic and nephrotoxic effects. Non-ferrous metallurgy enterprises are known to be a source of anthropogenic cadmium in the environment (Teplaya 2013). Previously, it was shown that the content of cadmium in the hair of the population of the Trans-Urals region of the Republic of Bashkortostan in the conditions of environmental pollution with heavy metals in some individuals exceeds the national average (Rafikova and Semenova 2020).

The imbalance of chemical elements in the human body serves as a starting point or contributes to the development of pathology; at the same time, the absence of clinical manifestations of the disease in children does not exclude the presence of pre-nosological and functional disorders.

Changes associated with puberty in adolescence are among the most significant, since during this period there is a morphological differentiation of tissues, organs and systems. It is generally recognized that a growing organism that is constantly under the influence of technogenic factors reacts acutely to these impacts (Boev et al. 2001).

Adolescents are carriers of toxic trace elementsof toxic elements, which is most dangerous for a growing body, since it is during this period that the functional system of the body is laid. In the future, such a carrier can negatively affect fertility, metabolic processes, and the formation of immunity.

It is shown that a complex of factors contributes to the formation of violations of elemental homeostasis in adolescents. It has been proved that the accumulation of Pb in the body of children above the 75th percentile scale index is accompanied by the development of significant changes in health status: deficiency of essential trace elements (Zn, Fe, Cu) and macronutrients (Ca, Mg), increased excretion of d-aminolevulenic acid in urine to 29.8 mmol/g creatinine, Pb—up to 100.8 mcg/l, a decrease in the level of physical development, an increase in the frequency of acute respiratory morbidity, disorders of the functional state of the urinary, nervous, digestive, cardiovascular systems, changes in red blood indicators (Trankovskaya 2004).

The work of Rafikova (2010) is devoted to the study of the state of health and trace element composition of the hair of workers of the Sibay branch of the Uchalinsky Mining and Processing Plant. An in-depth survey of workers revealed that 52.7% of male workers were healthy and 47.3% had health abnormalities. Diseases of the cardiovascular system and circulatory organs (17.4%), digestive system (7.7%), respiratory diseases (5.8%) prevailed in the structure of morbidity. In the structure of morbidity of women, healthy accounted for 18.6%, with deviations in health—81.4%. Among the revealed pathology, the first place was occupied by diseases of the blood (17.9%), the genitourinary system (17.6%), the cardiovascular system and circulatory organs (16.3%). A noticeable excess in the frequency and proportion of diseases of the cardiovascular system, urinary system, blood diseases, respiratory organs suggests a causal relationship with the effects of metal ions, which are present in a huge number in all environmental objects.

Rafikova (2010) showed that workers of different sexes react differently to prolonged stay in production with harmful working conditions. In male workers, in all cases, the content of the studied elements in the hair exceeded the normal physiological level. At the same time, the accumulation of metals occurred quite quickly and practically did not depend on the duration of work at the enterprise. The exception was the content of toxic elements: lead, and, to a lesser extent, cadmium, for which a clearly expressed linear dependence on work experience was shown. In women, the metal content also exceeded the physiological level, with the exception of manganese. But women, unlike men, have a more pronounced dependence of these indicators for copper, zinc and cadmium on the duration of their work in harmful working conditions. It is possible that the body of men is saturated with metals more quickly, compared with the body of women, and adapts better to pollution, forming a relatively stable equilibrium. The body of women accumulates metals more slowly, especially in the case of essential elements of copper and zinc and toxic—cadmium.

An imbalance of chemical elements in the human body can cause certain diseases. The excess of the content of all the studied metals in the hair of healthy male workers was found in comparison with similar indicators of the control group. Consequently, the accumulation of metals in the hair is primarily influenced by the production factor. In healthy female workers, an excess of iron and cadmium content in the hair was observed. In all other cases, the metal content was in the range of 25–75-centile values taken as a conditional norm.

With an increase in work experience, the content of lead in hair increases in men with diseases of the cardiovascular, genitourinary and musculoskeletal systems, and in women with diseases of the circulatory and digestive systems—cadmium, with diseases of the genitourinary system—copper and zinc.

An analysis of the metal content in the workers' hair depending on their health group was also carried out, which revealed that with the deterioration of health, an increase in cadmium and lead occurred in men's hair, copper and lead in women's hair. The largest amount of these metals was found in the hair of persons with a deterioration in their health. Therefore, it cannot be excluded that an increased level of metals such as cadmium, lead and copper in the body plays a certain role in the development of a number of diseases.

Thus, living in conditions of a high degree of chemical pollution of the environment leads to excessive intake of toxic trace elements in the body of children. An excess of heavy metals makes it difficult to assimilate vital trace elements and macronutrients, especially in conditions of their insufficient content in the diet. Conversely, the lack of minerals with food creates additional conditions for the accumulation of toxic metals. Individual risk factors contribute to the development of an elemental imbalance.

5 Conclusion

As a result of the examination of adolescents and adults of the trans-Ural zone of Bashkortostan, an imbalance of trace elements in the blood serum was revealed. Significant deviations from the physiological norm in adolescents were the following in less than 10% of individuals, a reduced content of zinc, manganese, copper and cobalt was registered, and an increased content of iron was also detected in 30% of adolescents; in 23%—cadmium; in 16–17%—copper, zinc and manganese.

Cobalt deficiency was found in adults (23%), as well as an excess of cobalt (32.5%), cadmium (32.5%), arsenic (20%) and zinc (20%).

The results obtained indicate that the population of this region is at risk of developing both metal toxicosis and a deficiency of vital trace elements. This fact requires a comprehensive medical, biological, clinical and functional study of the population, as well as strengthening control over the level of heavy metals in atmospheric air, soil, water, food and measures to reduce the risk to public health, including the correction of microelement status.

Funding Sources The study had no external funding.

Declaration of Interest The author confirms that there is no conflict of interest.

References

Askarova YaN (1969) Regional features of the content and ratio of some trace elements (iodine, copper, lead and manganese) in the external environment and their significance in the development of goiter endemia in Bashkiria: Dissertation, L. (in Russian)
Belan LN (2005) Medical and biological features of mining areas. Bulletin of Orenburg State University 5:112-117 (in Russian)
Belan LN (2006) On the ecological danger of pyrite deposits. Bulletin of Orenburg State University 4:115-120 (in Russian)
Dormancheva EI (2000) Estimation of morbidity and mortality from malignant diseases in the population of an area with abnormal levels of arsenic. Vyatka Medical Bulletin 1:68–70 (in Russian)
Gribovsky YG (2000) Scientific substantiation of a complex of measures to reduce the negative impact of nickel on the body of domestic animals and the sanitary quality of livestock products in the natural and technogenic provinces of the Urals: Dissertation, Troitsk. (in Russian)
Kovalskiy VV (2009) Problems of biogeochemistry of microelements and geochemical ecology: selected works. Moscow: Russian Agricultural Academy (in Russian)
Mazunina DL (2015) Negative effects of manganese in chronic intake with drinking water. Human ecology 3:25-31 (in Russian)
Rafikova YS (2010) Microelement status of the population of Sibay in conditions of technogenic pollution: Dissertation, Togliatti (in Russian)
Rikhvanov LP (2015) Essays on human Geochemistry. Tomsk (in Russian)
Starova NV (2003) The content of chemical elements in human hair. In: Problems of ecology: Principles of their solution on the example of the Southern Urals. M.: Nauka (in Russian)
Teplaya GA (2013) Heavy metals as a factor of environmental pollution. Astrakhan Bulletin of Environmental Education 1(23):182-192 (in Russian)
Testa U (2000) Resent developments in the understanding of iron metabolism. Hematology J.3:63-89
Trankovskaya LV (2004) The role of chemical element imbalance in the formation of children's health disorders Dissertation. Vladivostok (in Russian)
Basova OM, Khamitova RYa (2008) Health risks for children of small towns from ingestion of heavy metals. Kazan Medical Journal 89(2):203-206 (in Russian)
Rakitskii VN, Yudina TV (2005) Antioxidant and trace element status of the body: the modern problems of diagnosis. Bulletin of the Russian Academy of Medical Sciences 3:33-36 (in Russian)
Ramm GA, Ruddell RG (2005) Hepatotoxicity of iron overload: Mechanisms of iron-induced hepatic fibrogenesis. Sem. Liver Diseases 25(4): 433–449\
Skalny VL, Rudakov IA (2004) Bioelements in medicine. Moscow: Onyx 21st century Publishing House: Mir (in Russian)

Avtsyn AP, Zhavoronkov FF, Rish MA, Strochkova LS (1991) Human microelementosis: etiology, classification, organopathology (in Russian)
Bech J, Korobova E, Pérez-Sirvent C (2010) Selenium and iodine anomalies in soils and health Preface. Fuel and Energy Abstracts. 107. https://doi.org/10.1016/j.gexplo.2010.10.004
Boev VM, Utenina VV, Bystroykh VV, Utenin VV, Perepelkin SV, Setko AG, Kuksanov VF (2001) Disbalance of microelements as a factor of ecologically conditioned diseases. Hygiene and Sanitation 5:68 (in Russian)
Bogdanova VD, Kislitsina LV, Kiku PF (2017) Sanitary-hygienic assessment of drinking water quality in Khasanskiy district of the Primorye. Vladivostok:124–126 (in Russian)
Burlibaev MZh, Neronova TI, Saidov II, Mirkhashimov IKh, Kaidarova RK, Kalinin IYu, Sadvokasova SK (2010) Prospects of harmonization of water quality standards and norms in Central Asian countries and the European Union Water Framework Directive. Almaty: NGO OST- 21st century (in Russian)
Farkhutdinova LM, Speransky BB, Gilmanov AG (2006) Trace elements of hair in patients with goiter. Clinical Laboratory Diagnostics 8:19-21 (in Russian)
Kovshov AA, Novikova YuA, Fedorov VN, Tikhonova NA (2019) Diseases risk assessment associated with the quality of drinking water in the urban districts of Russian Arctic. Bulletin of the Ural Medical Academic Science 16 (2):215–222 (in Russian)
Opekunov AY, Opekunova MG, Janson SY, Somov VV, Kukushkin SY, Bychinskii VA, Papyan EE. (2021) Mineral and geochemical characteristics of soils and bottom sediments in the area affected by mining dumps (a case study of the Sibay ore deposit). IOP Conference Series: Earth and Environmental Science. Orenburg: 012078. https://doi.org/10.1088/1755-1315/817/1/012078
Opekunova MG, Opekunov AY, Somov VV, Kukushkin SY, Papyan EE (2020) Transformation of metals migration and biogeochemical cycling under the influence of copper mining production (the Southern Urals). Catena 189:104512. https://doi.org/10.1016/j.catena.2020.104512.
Rafikova YuS, Semenova IN, Khasanova RF, Suyundukov YaT (2020) Levels of cadmium and lead content in the hair of the population of the Trans-Ural zone of the Republic of Bashkortostan. Human ecology 1:17-24. 10.33396 / 1728-0869-2020-1-17-24 (in Russian)
Semenova IN, Ilbulova GR, Rafikova YS, Khasanova RF, Suyundukov YaT, Suyundukova MB. (2021a) Ecological risk assessment of heavy metals in soils of the Trans-Urals zone of the Republic of Bashkortostan (Russian Federation). IOP Conference Series: Earth and Environmental Science. VIII Congress of the Dokuchaev Soil Science Society. Syktyvkar: 012028 https://doi.org/10.1088/1755-1315/862/1/012028
Semenova IN, Rafikova YuS, Khasanova RF (2021b) Metal content in environmental components of the mining region. Trace elements in medicine 22 (1):58-59 https://doi.org/10.19112/2413-6174-2021-S1-29 (in Russian)
Zaitseva NV, Zemlyanova MA, Mazunina DL (2014) Peculiarities of changes in biochemical and hematological parameters in children consuming drinking water with high manganese content. Public Health and Environment. 2(251):41-44 (in Russian)

Physiological Adequacy Assessment of Potable Water in Lovozero District (Murmansk Region, Russia)

Svetlana I. Mazukhina, Sergey S. Sandimirov, and Svetlana V. Drogobuzhskaya

Abstract Findings of a physiological assessment of the potable water of the Lovozero Massif consumed by residents are presented according to its main macro- and microelement content. Our findings indicate that concentrations of calcium and magnesium are significantly lower than those recommended for different water categories, including those based on the criterion of physiological adequacy; the maximum permitted concentration (MPC) standards for Sr, Na^+, HCO_3^-, Cl^-, SO_4^{2-} and NO_3^- for physiologically adequate and bottled water need to be revised for the study area. The water also contains trace elements (Zr, La, Ce) for which no MPCs exist and whose effect on the human body has not yet been studied. Our findings show that the potable water consumed by the residents of Lovozero District, both in its natural state and in tap water, lacks a balanced composition in terms of the content of the main macro- and microelements (magnesium, calcium, hydrocarbonates) and cannot be considered as physiologically adequate in terms of total mineralization. The concept of physiological adequacy of potable water needs to be developed, taking into account regional differences. The study of the role of mineral processes in the human body is important in both environmental science and medicine.

Keywords Potable water · Chemical composition · Thermodynamic modeling · Lovozero Massif · MPC

S. I. Mazukhina (✉) · S. S. Sandimirov
Institute of North Industrial Ecology Problems, Subdivision of the Federal Research Center Kola Scientific Center, Russian Academy of Sciences, 14 a, Akademgorodok, Apatity 184209, Russia
e-mail: simazukhina@mail.ru

S. V. Drogobuzhskaya
Tananaev Institute of Chemistry - Subdivision of the Federal Research Center Kola Scientific Center of the Russian Academy of Sciences, 26 a, Akademgorodok, Apatity 184209, Russia

O. V. Frank-Kamenetskaya et al. (eds.), *Biogenic—Abiogenic Interactions in Natural and Anthropogenic Systems 2022*, Springer Proceedings in Earth and Environmental Sciences, https://doi.org/10.1007/978-3-031-40470-2_37

1 Introduction

Environmental factors affecting the condition and development of the human body, both at the cellular- and macro-levels, include the degree of physiological adequacy of potable water.

The outstanding Russian chemical biologist N. K. Koltsov, who proposed the concept of physiological adequacy of potable water in 1912, conceived it as reflecting a set of cations and anions that are physiologically adequate for the human body at all levels of its functioning. "Water should be a source of macro- and microelements necessary for the human body…"; in particular, these include calcium and magnesium, representing chemical elements that control the hardness of water and whose deposition on household appliances is a pressing issue for consumers.

While stopping short of the claim that water is the main source of essential elements, the World Health Organization (WHO) recognizes the significant contribution of potable water in sustaining human life (Goncharuket al. 2021). A review of Russian and international research and medical literature and regulations (Belyaeva 2013; Drozdova et al. 2017; DSTU 7525–2014; WHO 2004, 2009; SNiP 2012; SanPiN 2.1.3684–21; SanPiN 2.1.4.1074–01 2002; SanPiN 2.1.4.1116–02 2002; SanPiN 1.2.3685–21 2021; National 2009; Mazukhina et al. 2020) demonstrates the relevance of water adequacy studies in terms of its main macro- and microelement content and its impact on health. In Russia, mandatory standards for potable water only exist at the federal level; there are no enforceable regional or local standards.

Conversely, in the USA, in addition to the federal National Primary Drinking Water Regulations (NPDWR), advisory state-and even local water quality standards are applied. According to the NPDWR, each substance contained in the water is represented by two standards: the Maximum Contaminant Level (MCL) and the Public Health Goal (PHG). This enables drinking water supply professionals to focus on achieving optimal water quality indicators when designing and conducting drinking water treatment as constrained by the MCL. However, for a number of indicators—for example, antimony, barium, beryllium—MCL and PHG have the same values.

While approaches to regulating macro- and microelements in potable water from the point of view of its physiological adequacy are discussed in scientific and popular science literature, current EU regulatory documents only state the maximum permissible levels of certain elements such as sodium, manganese, selenium, zinc, fluorine, and iron. Moreover, for most of these elements, the standard is set according to the organoleptic indicator of harmfulness, rather than according to the overall effect on health; there are no stated requirements applying to water mineralization. Thus, the criteria for the physiological adequacy of water are not included in the regulations: only the safety of water for human health is controlled, but not its adequacy. The concept of the physiological adequacy of potable water has, however, been introduced in regulatory documents in Russia describing public health standards for bottled water (SanPiN 2.1.4.1116–02 2002), in Belarus (SniP 2012), and in Ukraine (DSTU 7525–2014).

The purpose of the present study was to assess the chemical composition of potable water from the natural sources available in Lovozero District, Murmansk Region, in terms of the criteria for physiologically adequate potable water.

2 Experimental Procedure

2.1 Materials and Methods

Monitoring of surface and ground water included a determination of the full hydrochemical water composition using inductively coupled plasma mass spectrometry as a precision analysis method. Fluorine concentrations obtained using thermodynamic calculations (Mazukhina 2012) are comparable with earlier analyses carried out in the natural systems of the Khibiny Massif (Mazukhina et al. 2020). Studies of surface water in 2001 in the catchment area of Lake Imandra and monitoring of groundwater in wells 45 and 46 in the Kuniyok River valley (Khibiny Massif) showed the following: concentration of fluorine 0.79, 1.55, 0.76 mg/L, respectively; in 2010, the fluorine content in the wells was 2.5, 1.1, respectively (Mazukhina 2012). Moreover, the results of groundwater monitoring in 2017 from well 1 in the Malaya Belaya River valley (Khibiny Massif) showed a fluorine content of 4.2 mg/L (Mazukhina et al. 2020). According to the data collected by the 2020 expedition, the concentration of fluorine in the water of Lake Lovozero was 0.18 mg/L.

Thermodynamic calculations were performed using the method of physicochemical (thermodynamic) modeling implemented in the software suite Selector (Chudnenko 2010).

A physicochemical model (PCM) of interactions in the water–rock system was developed, adapted to the conditions of Murmansk Region (Mazukhinaet et al. 2020).

Figure 1 shows the Lovozero Tundra Mountains and water sampling stations in 2020 (October) and 2021 (November).

Located at the foot of the Lovozero Tundra among the Quaternary glacial, eluvial-diluvial and diluvial-solifluction sediments, the Sergevan group of loparite placers covers an area of more than 300 km^2. The length of individual placers is up to 10 km, with a width of up to 72 m and a thickness of up to 72 m, while the loparite grade is up to 3 kg/m^3(Pozhilenko et al. 2002). In addition to loparite, the placers contain a large quantity of magnetite and eudialyte. This circumstance must be taken into account when analyzing water samples taken from the Parkovy spring (station 4) or Lake Sikir (station 5).

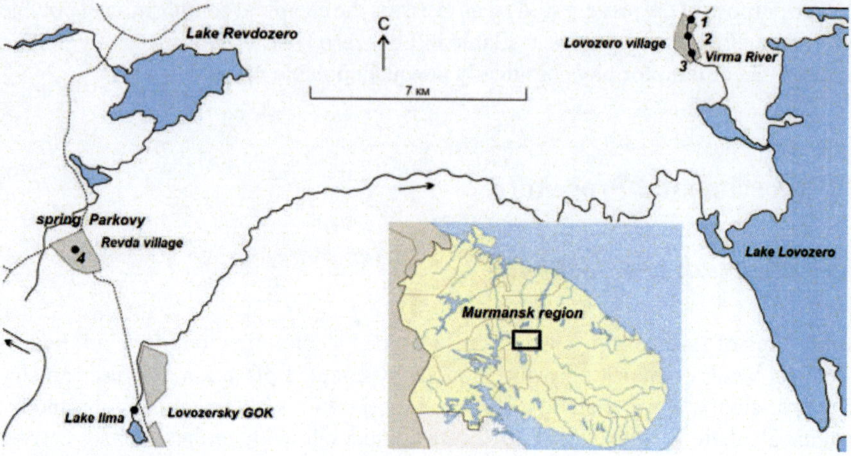

Fig. 1 Water sampling stations: 1–Virma River (water intake of Lovozero); 2–a diner in Lovozero; 3–a well in Lovozero; 4—spring "Park"

3 Results and Discussion

Table 1 presents the chemical analyses of the water of the Lovozero water intake (Virma River) before and after water treatment (diner in Lovozero) and the Parkovy spring popular with the residents of Revda.

According to the chemical composition, the water from the Virma River is characterized by a high natural iron content, color, and turbidity. Based on the water studies carried out by GOUP Olenegorsk vodokanal and FGUZ Center for Public Health and Epidemiology in Murmansk Region in Monchegorsk, Olenegorsk and Lovozersky District, the following conclusion was reached concerning the water sampled in the water intake and distribution system of Lovozero: "Water from the water intake (Virma River) does not meet the standards determined in the Potable Water Quality Guidelines (GOST 56,237–2014). At present, the wear of the water distribution system is 93%".

Considering the concentrations and MPCs of some elements as indicated in the table, *sodium* is within the specified range for all water categories. In the Murmansk Region, high concentrations of Na (starting at 20 and above 200 mg/L) are found only in industrial water; this is also true of HCO_3^-, Cl^-, $SO_4^{2-}$and NO_3^-(Mazukhina 2012). Although *calcium* and *magnesium* are the most important elements for human health, their concentrations are more than an order of magnitude lower than those recommended for different categories of water, including according to the criterion of adequacy. According to the SanPiN standards, *strontium* should not exceed 7 mg/L; following the logic of Kravchenko, calcium should be two orders of magnitude higher, i.e., 700 mg/L, or the MPC of strontium should be within 0.25–0.8 mg/L for complete and bottled water. Microelements (Zr, La, Ce)

Table 1 Chemical composition of the water of the Lovozero Massif and the criteria for the adequacy of water and MPCs for different categories of water

Chemical composition of water				Adequacy criteria of water and the chemical composition of different categories of water				SanPiN 2.1.4.1074-01	SanPiN 2.1.4.1116-02	SanPiN 1.2.3685-21		WHO	EU	USEPA	
Element	Virma	Diner in Lovozero	Spring Parkovy	Adequacy	1st category	Top category	Hazard Class	MPC for potable water	MPC for bottled water, I cat	MPC for different types of water	Hazard class according to	Top category		PHG	MCL
Li	0.0003	0.0005	0.0002		0.03	0.03	2	0.03	0.03	0.03	2	0.03			
Be	0.0007	0.0010	0.0007		0.0002	0.0002	1	0.0002	0.0002	0.0002	1	0.0002		0.004	0.004
B	0.0095	0.0092	0.012		0.5	0.3	2	0.5	0.5	0.5	2	0.3	0.3	1	
Na	1.88	19.87	8.64	2–20	200	20	2	200	200	200	2	20	200	200	
Mg	1.28	1.31	0.92	5–65	65	5–50	-			50	3			50	
Al	0.057	0.29	0.0012		0.2	0.1	2	0.5	0.2	0.2	3	0.1	0.2	0.2	0.05–0.2
Si	5.45	5.31	5.24		1.02	4.83	2	10		2.5–25	2				
P	0.018	0.001	0.001					0.0001		0.0001	1				
K	0.31	0.49	3.30	2–20	20	2–20	–							12	
Ca	1.54	3.07	3.44	25–130	130	25–80	–							100	
V	0.0007	0.0008	0.00089					0.1		0.1	3				
Cr	0.0001	0.0002	0.00005		0.05	0.03	3	0.05	0.05	0.05	2	0.03	0.05	0.1	0.1
Mn	0.047	0.028	0.0011		0.05	0.05	3	0.1	0.05	0.1	3	0.05	0.5	0.05	0.05
Fe	0.76	0.25	0.0083		0.3	0.3	3	0.3	0.3	0.3	3	0.3	0.3	0.2	0.3
Co	0.00010	0.00010	0.00007		0.1	0.1	2	0.1	0.1	0.1	2	0.1			
Ni	< 0.00001	< 0.00001	< 0.00001		0.02	0.02	3	0.1	0.02	0.02	2	0.02	0.02	0.02	
Cu	0.0003	0.0016	0.00017		1	1	3	1	1	1	3	1	2	1.3	1.3

(continued)

Table 1 (continued)

Chemical composition of water				Adequacy criteria of water and the chemical composition of different categories of water				SanPiN 2.1.4.1074-01	SanPiN 2.1.4.1116-02	SanPiN 1.2.3685-21		WHO	EU	USEPA	
Element	Virma	Diner in Lovozero	Spring Parkovy	Adequacy	1st category	Top category	Hazard Class	MPC for potable water	MPC for bottled water, I cat	MPC for different types of water	Hazard class according to	Top category		PHG	MCL
Zn	< 0.00001	< 0.00001	< 0.00001		5	3	3	5	5	5	3	3	3	5	5
Sr	0.033	0.042	0.167		7	7	2	7	7	7	2	7			
Y	0.00017	0.00009	0.00014												
Zr	0.00080	0.00031	0.00020												
Se	0.00023	0.00039	0.00007		0.01	0.01	2			0.01	2			0.05	0.05
Nb	0.00001	0.00002	0.00005					0.01		0.01	2				
Mo	0.0003	0.0003	0.0006					0.25	0.07	0.07	3	0.07	0.07		
Ag	0.0013	0.0013	0.0013		0.0025	0.0025	3	0.05	0.025	0.05	2	0.0025		0.01	0.1
Cd	0.00003	0.00004	0.00005		0.001	0.001	2	0.001	0.001	0.001	2	0.001	0.003	0.005	0.005
Ba	0.0047	0.0056	0.0014		0.7	0.1	2	0.1	0.7	0.7	2	0.1	0.7	2	2
La	0.00015	0.00008	0.000081												
Ce	0.00023	0.00009	0.000055												
U	0.00005	0.00003	0.00009							0.015	1			0	30
Cl^-	2.09	10.9	3.30		250	150	4	350	250	350	4	150	250	250	250
NO_3^-	0.01	0.078	0.928		20	5	3	45	20	45	3	5	50	10	10
SO_4^{2-}	4.76	2.03	1.88		250	150	4	500	250	500	4	150	250	250	250
HCO_3^-	15.7	50.3	28.7	30–400	400	30–400	-								
F^-	0.246	0.15	1.20	0.5–1.5	1.5	0.6–1.2				1.5	2		1.5	1.5	4
pH	6.88	6.67	7.98		6.5–8.5	6.5–8.5	-	6.5–8.5	6.5–8.5			6.5–8.5	6.5–8.5	6.5–8.5	6.5–8.5

Note PHG – Public Health Goal, (mg/L), MCL – Maximum Contaminant Level, (mg/L) were found in the studied potable water, for which no MPCs exist and whose effect on the human body has not yet been studied (Mazukhina et al. 2020).

Table 2 presents the modeling results of the water of the Parkovy spring (Revda) and water from a well in Lovozero. Since all natural waters of the Kola Peninsula have a low total mineralization, they can be considered as fresh xenohalobic (salinity up to 0.1 g/kg according to GOST 17.1.2.04–77); the total mineralization of tap water varies in the range of 58.7–61.6 mg/L. All potable waters are classified in Alekin's carbonate and sodium group classifications (Alekin 1970). In terms of hardness, the water is very soft–no more than 1.5 mg-eq/L. By determining the ratio between ions, all waters are characterized as Type 1–$HCO_3^- > Ca + Mg$. Water of this type forms with the significant participation of the constituents of Quaternary loose rocks and underlying sediments, which contain large amounts of Na^+ or K^+; however, as a rule, it has low mineralization.

The existing water treatment for the central water supply in Lovozero includes chlorination, coagulation and filtration (https://docplayer.com/28705206). Disinfection takes place with liquid chlorine in accordance with the requirements of GOST and coagulant aluminum polyoxychloride (AQUA-AURAT), water is stabilized with a solution of soda ash with a concentration of 50 mg/L. By adding to the Virma model (per 1000 L of original water) NaOCl 0.316; Na_2CO_3 0.158; $AlCl_3$ 0.0050 mol, the following concentrations are obtained (mg/L): Al 0.192; Na 17.2; HCO_3^-35.7; pH 7.67. This is comparable with the concentrations in the well (Table 2) and diner water (Table 1).

The oxygen content in the well water matches that of bottled water, which should be at least 5 mg/L for water of category 1 and 9 mg/L (saturation close to optimal at a temperature of 20–22 °C) for water of the top category. Bottled water with an oxygen content above 15 mg/L must undergo toxicological safety assessment (SanPiN 2.1.4.1116–02 2002). In 2018, we developed a model of the stomach environment based on literature sources (PC "Selector"). With the help of modeling, the qualitative and quantitative composition of the solution–crystalline substance system was studied, taking into account environmental conditions and physiological parameters of the human body, where the solution is natural drinking water, gastric juice (GJ), a mixture of drinking water and gastric juice, while the crystalline substance is newly formed phases in equilibrium with the solution. The interaction of gastric juice with water can be seen in terms of the emergence of a geochemical barrier, i.e., involving such physicochemical parameters as occur when there is a sharp change in conditions and concentrations of elements. By applying physicochemical modeling to the natural water–gastric juice system under conditions of low ($pH = 6.24$, $Eh = 0.218$ V) and high ($pH = 2.02$, $Eh = 0.044$ V) acidity, it was possible to evaluate the pH and Eh of the gastric juice model to demonstrate their changes, i.e., to explore the chemical forms of element migration and determine the composition of mineral phases that can be transported from the stomach to other organs and tissues (Mazukhina et al. 2020).

In our study, within the framework of the solution-crystalline substance system, the forms of migration of lanthanides under the conditions of the human body obtained using physicochemical modeling in PC "Selector" are presented.

Table 2 Analytical data (AD) and modeling results (MR) of ground water and drinking water

Index	Parkovy Spring, mg/l		Well in Lovozero, mg/l		Index	Parkovy Spring, mg/l		Well in Lovozero, mg/l	
	AD	MR	AD	MR		AD	MR	AD	MR
Eh		−0.197		0.770	VO_4^{3-}		3.10E−14		4.00E-09
pH	6.88	6.87	7.68	7.69	HVO_4^{2-}		8.76E-08		1.68E-03
Is (ionic strength)		0.000843		0.001381	H_3VO		1.35E-09		5.99 E-07
Al	0.0012	1.16E-03	0.176	1.76E-01	Ba	0.0014	8.03E-03	2.83E-03	2.83E-03
$Al(OH)_2^+$		6.61E-06		1.17E-04	Ba^{2+}		1.46E-03		2.83E-03
$Al(OH)_2F$		2.47E-03		5.37E-03	$BaCO_3$		1.46E-03		2.38E-06
$Al(OH)_2F_2^-$		2.65E-05		7.18E-06	$BaCl^+$		1.07E-07		3.05E-07
AlO_2^-		2.58E-04		1.94E-01	$BaOH^+$		4.74E-08		2.56E-09
$HAlO_2$		1.26E-04		1.43E-02	Si	5.44	5.44	5.46	5.46
$Al(OH)_2^+$		9.82E-07		2.73E-06	SiO_2		3.86		3.86
$Al(OH)_3$		1.09E-04		1.24E-02	$HSiO_3^-$		8.81E-03		5.77E-02
$Al(OH)_4^-$		3.53E-04		2.65E-01	H_4SiO_4		12.4		12.4
Al^{3+}		1.08E-08		4.73E-09	Sr	0.167	1.67E-01	0.033	3.28E-02
Ca	3.44	3.44	3.96	3.96	Sr^{2+}		1.66E-01		3.24E-02
Ca^{2+}		3.41		3.91	$SrOH^+$		4.09E-08		5.08E-08
$CaOH^+$		3.05E-06		2.23E-05	$SrCO_3$		2.34E-05		5.25E-05
$CaCO_3$		2.10E-03		2.77E-02	$SrHCO_3^+$		1.73E-03		6.01E-04
$CaHCO_3^+$		3.94E-02		8.05E-02	$SrCl^+$		1.11E-05		7.19E-06
$CaHSiO_3^+$		1.09E-05		7.99E-05	SrF^+		1.62E-05		3.83E-07
$CaCl^+$		2.77E-04		1.06E-03	Zn	<0.00001		<0.00001	
$CaCl_2$		1.81E-08		2.33E-07	Cd	0.00005	4.55E-05	0.00003	2.88E-05

(continued)

Table 2 (continued)

Index	Parkovy Spring, mg/l		Well in Lovozero, mg/l		Index	Parkovy Spring, mg/l		Well in Lovozero, mg/l	
	AD	MR	AD	MR		AD	MR	AD	MR
CaF^+		1.40E-03		1.94E-04	Cd^{2+}		4.51E-05		2.79E-05
$CaSO_4$		4.65E-02		3.15E-02	$CdCl^+$		5.12E-07		1.05E-06
B	0.012	1.22E-02	9.75E-03	8.03E-03	CdO		5.92E-12		1.50E-10
H_3BO_3		6.97E-02		4.48E-02	$CdOH^+$		2.19E-08		8.60E-08
BO_2^-		1.82E-04		7.69E-04	Ni	<0.00001		<0.00001	
Fe	0.0083	8.30E-03	0.292	2.92E-01	Pb	0.00008	7.75E-05	0.00007	7.93E-05
Fe^{2+}		8.24E-03		3.92E-12	Pb^{2+}		1.33E-05		2.51E-06
$FeSO_4^+$				2.79E-12	$PbOH^+$		6.94E-05		8.30E-05
$Fe(OH)_3$		7.25E-12		4.16E-02	PbO		5.06E-09		3.91E-08
$Fe(OH)_4^-$		2.96E-13		1.11E-02	$HPbO_2$		3.07E-13		1.55E-11
$FeOH^{2+}$		1.29E-14		1.80E-06	Cu	0.00017	1.73E-04	0.0012	1.19E-03
$FeOH^+$		2.68E-05		8.11E-14	Cu^+		1.69E-04		1.95E-14
FeO^+		2.54E-11		2.25E-02	Cu^{2+}		1.59E-10		8.45E-04
$FeSO_4$		6.04E-05		1.69E-14	$CuCl^+$		5.46E-14		9.70E-07
$HFeO_2$		6.81E-11		3.91E-01	$CuOH^+$		1.28E-11		4.33E-04
FeO_2^-		9.67E-14		3.63E-03	CuF^+		-		2.65E-07
$FeCl^+$		8.12E-07		1.28E-15	$CuCl_2^-$		8.12E-07		1.09E-15
FeF^+		1.76E-05		1.01E-15	$HCuO_2^-$		-		1.46E-08
FeF^{2+}		–		5.38E-11	P		5.87E-03	0.0166	3.26E-04
F^-		1.20		1.51E-01	PO_4^{3-}	0.018	1.97E-08	0.051	1.75E-08

(continued)

Table 2 (continued)

Index	Parkovy Spring, mg/l		Well in Lovozero, mg/l		Index	Parkovy Spring, mg/l		Well in Lovozero, mg/l	
	AD	MR	AD	MR		AD	MR	AD	MR
HF		2.24E-04		1.49E-01	HPO_4^{2-}		2.34E-07		7.65E-04
HF_2^-		6.31E-09		4.24E-06	$H_2PO_4^-$		1.25E-02		2.48E-04
K	2.33	2.33	6.30E-01	6.30E-01	Co	0.00007	7.22E-05	0.00009	8.67E-05
K^+		2.33		6.30E-01	Co^{2+}		7.22E-05		8.64E-05
KCl		1.03E-06		9.34E-07	CoO		3.75E-10		1.85E-08
$KHSO_4$		8.79E-13		2.20E-14	$CoCl^+$		3.84E-08		1.53E-07
KOH		5.88E-08		1.03E-07	$HCoO_2^-$		5.66E-16		1.82E-13
KSO_4^-		2.03E-03		3.34E-04	$CoOH^+$		4.44E-08		3.38E-07
Mg	0.9	9.00E-01	1.92	1.92E+00	Cl	3.30	3.3	11.3	11.3
Mg^{2+}		8.90E-01		1.90E+00	Cl^-		3.30		11.3
$MgOH^+$		1.49E-05		2.02E-04	HCl		9.31E-08		4.87E-08
$MgCO_3$		3.59E-04		8.78E-03	Zr	0.00020	2.04E-04	0.00029	2.95E-04
$Mg(HCO_3)^+$		1.39E-02		5.28E-02	$HZrO_3^-$		2.39E-04		4.32E-04
$MgCl^+$		1.41E-04		1.00E-03	ZrO_2		6.56E-05		1.81E-05
MgF^+		2.12E-03		5.46E-04	U	0.00009	9.42E-05	0.00003	2.87E-05
$MgSO_4$		2.32E-02		2.92E-02	HUO_4^-		8.11E-15		1.77E-06
$MgHSiO_3^+$		7.48E-06		1.01E-04	UO_2^{2+}		5.02E-16		4.06E-10
Mn	0.0011	1.06E-03	0.0161	1.61E-02	UO_2OH^+		1.76E-14		9.04E-08
Mn^{2+}		1.05E-03		1.61E-02	UO_2		1.07E-04		-
$MnOH^+$		1.60E-07		1.56E-05	UO_3		9.86E-13		3.28E-05

(continued)

Table 2 (continued)

Index	Parkovy Spring, mg/l		Well in Lovozero, mg/l		Index	Parkovy Spring, mg/l		Well in Lovozero, mg/l	
	AD	MR	AD	MR		AD	MR	AD	MR
MnO		1.99E-12		1.25E-09	Li	0.00018	1.76E-04	0.00037	
$HMnO_2^-$				1.46E-14	Li^+		1.76E-04		3.68E-04
$MnSO_4$		7.17E-06		6.45E-05	LiOH		6.96E-11		9.38E-10
MnF^+		6.38E-07		1.18E-06	Ce	0.000055	5.47E-05	0.00005	4.67E-05
$MnCl^+$		9.84E-08		4.99E-06	Ce^{3+}		2.65E-05		3.54E-05
CO_3^{2-}		8.56E-03		1.04E-01	CeF^{2+}		2.52E-05		4.03E-06
HCO_3^-	28.7	26.5	49.1		CeF_2^+		2.01E-06		3.89E-08
HCO_3^-		26.5		4.87E+01	CeF_3		2.76E-08		6.58E-11
HCO_2^-		2.87E-06		-	CeF_4^-		1.88E-10		5.57E-14
SO_4^{2-}	4.76	3.42	2.19	2.14	$CeHCO_3^{2+}$		1.21E-06		2.83E-06
HNO_3		4.11E-09		5.49E-11	CeO_2H		1.84E-11		6.33E-09
NO_3^-	0.93	9.27E-01	0.08	8.14E-02	$CeSO_4^+$		5.32E-06		4.07E-06
Na^+	8.64	8.63	18.6	1.86E+01	$CeOH^{2+}$		4.50E-07		3.77E-06
NaOH		4.78E-07		6.65E-06	La	0.000081	8.10E-05	0.00006	
$NaAlO_2$		2.24E-08		3.55E-05	La^{3+}		2.43E-05		2.77E-06
NaCl		3.22E-04		2.35E-03	$LaCO_3^+$		6.27E-05		7.99E-05
NaF		9.30E-05		2.46E-05	LaF^{2+}		9.70E-06		1.32E-07
$NaSO_4^-$		7.37E-03		9.65E-03	LaO^+		3.15E-10		1.44E-09
$NaHSiO_3$		3.30E-04		4.60E-03	LaO_2H		2.78E-13		8.17E-12
O_2		-		7.74	$LaOH^{2+}$		2.48E-07		1.78E-07

(continued)

Table 2 (continued)

Index	Parkovy Spring, mg/l		Well in Lovozero, mg/l		Index	Parkovy Spring, mg/l		Well in Lovozero, mg/l	
	AD	MR	AD	MR		AD	MR	AD	MR
CO_2		5.91		1.66	$LaSO_4^+$		4.89E-06		3.20E-07
V	0.00089	8.93E-04	0.000 7	7.39E-04	Ag	0.0013	4.58E-03	0.001 4	
VO^+		1.14E-03			Ag+		1.22E-10		1.43E-09
VO^{2+}		1.88E-06			$AgNO_3$		1.63E-03		1.66E-03
VOH^{2+}		1.96E-06			$Ag(HS)_2^-$		5.72E-03		-
$VOOH^+$		3.26E-05							

Figure 2 shows the forms of migration of lanthanum and cerium during the interaction of 100 ml of gastric juice and water from the water supply system of Lovozero (diner) with a change in the volume of water in the system from 1 to 1000 ml at normal stomach acidity. An analysis of the results (Fig. 2, Table 2, natural water) shows that, under the local conditions of the human stomach, the following changes take place: *Eh* and *pH*; concentrations of lanthanum; cerium migration. The increased content of $LaCl^{2+}$ by several orders of magnitude coincides with a decrease in the concentration of fluorides; the emergence of $CeCl_3$ is also noted.

Figure 3 shows the forms of migration of lanthanum and cerium during the interaction of 100 ml of gastric juice and water from the water supply system in Lovozero (diner) with a change in the volume of water in the system from 1 to 1000 ml of water at initial GJ values, i.e., at a low stomach acidity: $pH = 6.24$; $Eh = -0.218$ V; C(HCl) = 2284.6 mg/L; V(GJ) = 100 mL; T = + 38°C; P = 1 bar.

The results (Fig. 3a) show that when drinking water having a neutral *pH*, the *pH* value of the stomach shifts to slightly acidic under reducing conditions. Compared to natural water, the content of $LaCl_2^+$ increases by several orders of magnitude (Fig. 3b). The possibility of formation of solid phases with the first sips of water was also demonstrated (Fig. 3c) with fluorapatite predominating.

The effect of zirconium (Zr) on biological systems remains a mystery. It is ubiquitously present in nature in quantities greater than most trace elements. The delay occurs first in the soft tissues, and then slowly in the bones. This element is able to cross the blood–brain barrier and the placental barrier, i.e., it accumulates in the brain and passes into milk (Ghosh et al. 1992). The simulation results show a fundamental difference between the forms of migration of zirconium in natural water ($HZrO_3^-$) and in the human body, depending on the deficiency (ZrO_2) or normal level ($ZrOH^{3+}$, ZrO^{2+}) of hydrochloric acid, which depends on the individual characteristics of the organism. Research results from 2020 to 2021 showed the presence of zirconium in milk (0.0014 mg/L) produced in the study region, whose concentration was an order of magnitude higher than in natural water (Table 1).

Thus, by examining the composition of groundwater obtained from a well (spring) and tap water (diner), we found that, under the conditions of the human body, with the appearance of a geochemical barrier, a change occurs in the forms of migration of such elements as lanthanides, zirconium, etc.

4 Conclusions

The results of our research indicate the following:

- concentrations of *calcium and magnesium* (the most important elements) are more than an order of magnitude lower than those recommended for water of different categories, including according to the criterion of adequacy;
- the need to revise the MPC standards for Sr^{2+}, Na^+, HCO_3^-, Cl^-, SO_4^{2-}, NO_3^- for high-grade and bottled water in the study region;

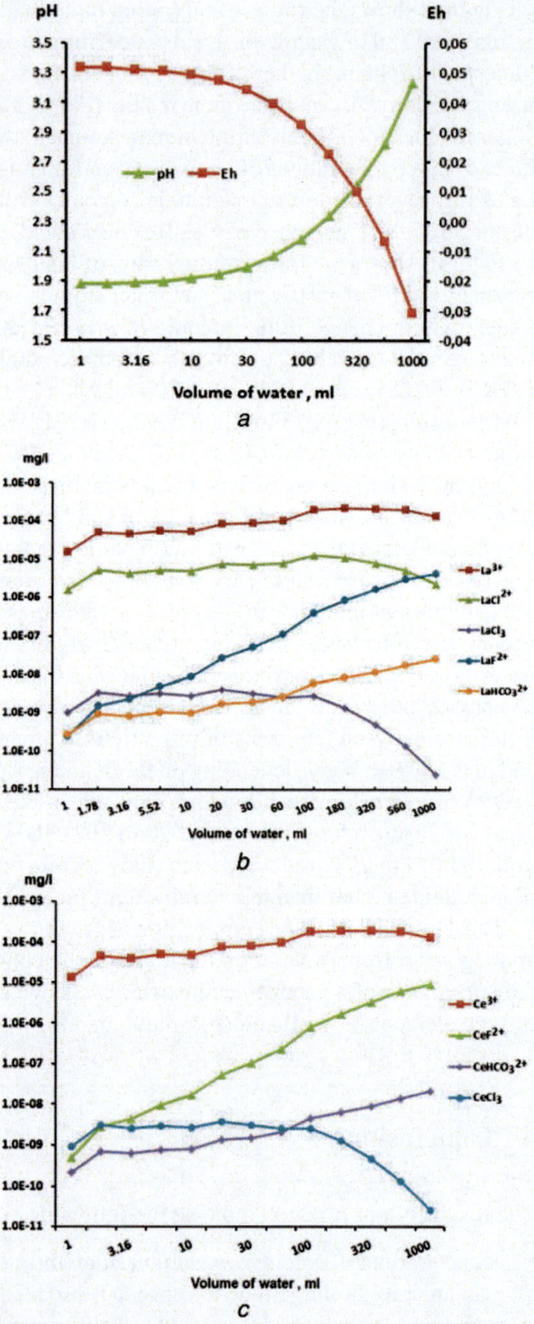

Fig. 2 Change in the parameters of the system–*pH* (**a**), *Eh* (**b**), and concentrations of chemical elements (**c**) when changing the volume of water in the table water–gastric juice system *V*(GJ)–100 ml, T + 38 °C, $P = 1$ bar at initial values of GJ: *pH* 2.02, $Eh =$ + 0.044 V, C (HCl) = 2872.3 mg/L

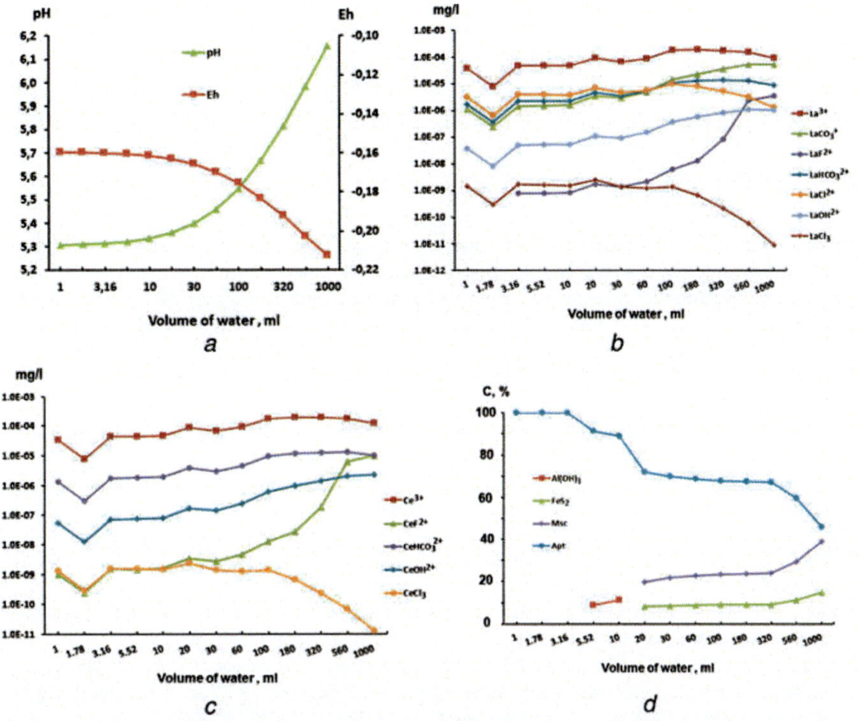

Fig. 3 Change in the parameters of the *pH* system (**a**), *Eh* (**b**), and concentrations of various forms of La, Ce (**c**) and the formation of solid phases (**d**) with a change in the volume of water in the table water–gastric juice system. Volume of gastric juice $V(GJ) = 100$ ml; $T = +38$ °C;$P = 1$ bar. $Al(OH)_3$–gibbsite; Msc—$KAl_3Si_3O_{10}(OH)_2$—muscovite (hydrated phase); Apt–$Ca_5(PO_4)_3F$–apatite

- microelements (Zr, La, Ce) were found for which no MPCs exist and whose effect on the human body has not yet been studied;
- under the conditions of the human body, with the appearance of a geochemical barrier, there will be a change in the forms of migration of such elements as lanthanides, zirconium, etc.

The applied approach within the framework of a single multisystem makes it possible to predict the composition of stones and forms of migration of elements depending on the individual characteristics of the organism, while most other works contain only statements of facts.

Our results indicate that the potable water supplied to the residents of Lovozero District, both in its initial natural state and at the time of supply to the consumer, does not have a balanced and complete composition and cannot be recognized as physiologically adequate. The main conclusion is that it is necessary to develop the concept of physiological adequacy of water to take regional characteristics into account.

The study was carried out as part of the research topics 1021051803677-1, 1021051803680-5, FMEZ-2022-0018.

References

Belyaeva YuN (2013) Diseases of the digestive system as a medical and social problem. Bulletin of Medical Internet Conferences 3(3):566–568 (ISSN 2224–6150) (in Russian)

Calcium and Magnesium in Drinking water (2009) Public health significance, WHO, Geneva.

Chudnenko KV (2010) Thermodynamic modeling in geochemistry: theory, algorithms, software, applications. Academic Publishing House Geo, Novosibirsk (in Russian)

Council Directive 98/83/EC (1998) of 3 November 1998 on the quality of water intended for human consumption P.15

Detsina AN (2017) Water resources and health.Physiologically adequate drinking water.Publishing House Circulation,Yekaterinburg(in Russian)

Drozdova EV, Buraya VV, Surovets TZ, Firago A V, Girina VV (2017) Assessment of drinking water consumed by the residents of the Republic of Belarus, according to macro- and microelement composition. Occupational Medicine and Human Ecology1: 44-49 (in Russian)

DSTU 7525–2014 (2014) Potable water.Ministry of EconomicDevelopmentofUkraine,Kiev (in Ukrainian)

Ghosh S, Sharma A, Talukder G. Zirconium (1992)An abnormal trace element in biology. Biol Trace Elem Res 35(3):247–71. DOI: https://doi.org/10.1007/BF02783770

Goncharuk VV, Zui OV, Melnik LA, Mishchuk NA, Nanieva AV, Pelishenko AV (2021) Assessment of the physiological adequacy of drinking water by biotesting. Chemistry for Sustainable Development1: 35-41. DOI: https://doi.org/10.15372/KhUR2021275 (in Russian)

GOST 56237–2014 (2014) Potable water.Sampling at water treatment plants and pipeline distribution systems.(in Russian)

GOST 17.1.2.04–77(1978) Nature protection.Hydrosphere.Idices of state and regulations for valuation survey of fishery waters. 62 c

Guidelines for ensuring the quality of drinking water (2004), 1: 3rd edn. World Health Organization, Geneva.

Mazukhina SI (2012) Formation of surface and underground water of the KhibinyMountain-Range.Publishing House of the KSC RAS, Apatity (in Russian)

Mazukhina SI, Maksimova VV, Chudnenko KV, Masloboev VA, Sandimirov SS, Drogobuzhskaya SV, Tereshchenko PS, Pozhilenko VI, Gudkov AV (2020) Water quality in the Arctic zone of the Russian Federation: physical and chemical modeling of water formation, forms of element migration, impact on the human body: monograph. Publishing House of the Federal Research Center of the KSC RAS,Apatity.ISBN 978-5-91137-437-2 (in Russian)

National Primary Drinking Water Regulations.(2009) EPA 816-F-09–004.P. 7.

Pozhilenko VI, Gavrilenko BV, Zhirov DV, Zhabin SV (2002) Geology of ore districts of Murmansk Region.Publishing House of the KSC RAS, Apatity (in Russian)

SanPiN 2.1.4.1074–01 (2002) Potable water. Public health requirements for water quality of centralized drinking water supply systems. Quality control.Public health and epidemiological rules and regulations.Ministry of Health of Russia, Moscow (in Russian).

SanPiN 2.1.4.1116–02 (2002) Potable water.Public health requirements for the quality of water packaged in containers.Quality control.Ministry of Health of Russia, Moscow(in Russian).

SanPiN 1.2.3685–21 (2021) Sanitary rules and regulations. Hygienic standards and requirements to ensure the safety and (or) harmlessness of environmental factors for humans". 1025 p. (in Russian)

SanPiN 2.1 3684–21 (2021) Sanitary rules and regulations. Sanitary and epidemiological requirements for the maintenance of the territories of urban and rural settlements, for water bodies, drinking water and drinking water supply to the population, atmospheric air, soils, residential premises, operation of industrial and public premises, organization and conduct of sanitary and anti-epidemic measures.75 p. (in Russian).

SNiP "Requirements for the physiological adequacy of drinking water" (2012) Minsk(in Belarus)

U.S. Environmental Protection Agency https://www.epa.gov/, (Date of access: 10.01.2022)

Water supply and sanitation plan of the community of Lovozero, Lovozero District, Murmansk Region for 2014–2018 and for the period up to 2028(2014) Energo Consult LLC https://docplayer.com/28705206 (accessed 10.03.2022)

Philosophical Aspects

The Ratio of Simple and Complex in Evolution

Evgeniya L. Sumina and Dmitry L. Sumin

Abstract On the usually given examples of increasing complexity, it is shown that the complexity of objects increases only in total. These examples also show something else—each subsequent level of organization is simpler than the previous one. Complication occurs as a result of the fact that elements are more complex than systems and impart their complexity and energy to them. The general mechanism of complication is the mutual-organization of elements, and not the self-organization of systems from simple components.

Keywords Complexity · Self-organization · Mutual-organization · Crystals · Cyanobacteria · Behavior · Morphogenesis

1 Introduction

The complication of matter as a whole and its individual objects is an observable fact. Simplification or destruction as such are subordinated of this process and included in it. From this observation, it seems, an unambiguous conclusion follows—the origin of the complex from the simple is a universal law of development, in fact, the development itself. Further research can only concern the concretization of the mechanisms of this process.

Such a universal circumstance can be demonstrated in any observed cases of change, so we begin with the closest and most familiar to everyone—the development of tools. Their complication goes from raw stone, simply taken in hand, through simple composite tools like a stone ax and tools from artificially obtained materials—an iron ax—to modern autonomous robotic complexes. However, their direct development does not take place—they do not exchange information and matter,

E. L. Sumina (✉)
M.V. Lomonosov MSU, Geological Faculty, Department of paleontology, Moscow, Russia
e-mail: stromatolit@list.ru

D. L. Sumin
Network of Independent Scientists (NISEEB), Moscow, Russia

O. V. Frank-Kamenetskaya et al. (eds.), *Biogenic—Abiogenic Interactions in Natural and Anthropogenic Systems 2022*, Springer Proceedings in Earth and Environmental Sciences, https://doi.org/10.1007/978-3-031-40470-2_38

there is no direct continuity between them. The person is who give them complexity, bringing it closer and closer to his own, including by striving to endow them with the ability for the self-reproduction and self-development. This textbook example demonstrates the opposite of what they usually illustrate—the real process does not go from simple to complex, but from complex to simple. In this case, from a person to the objects of his labor.

Hence, an important observation follows—the theoretical constructions are based on a fictitious process that does not occur and cannot occur. The real process with observable patterns, in which everyone takes part, is ignored. For now, we will only note this fact in order to be considered below.

2 Organisms

The next well-known example of complication is the evolution of organisms both within individual levels of organization and leading to their emergence. Basically, the diversity of organisms consists of extinct groups, that is, those that have experienced simplification to the point of losing the properties of the living body. However, this is only the outer shell of the really ongoing process of complication. Apart from the fossil remains of increasingly complex organisms, this process can be judged by its reflection in individual development and the surviving living transitional forms.

The origin of multicellular from unicellular is usually considered on the example of sponges and volvox. It is believed that sponges do not have organs and tissues, since they have not yet been formed in the process of evolution. However, sponges in individual development go through the stage of gastrulation, and their gastrulae are qualitatively indistinguishable from the gastrulae of true multicellular animals, including chordates. That is, the evolution of sponges obviously went through the stage of at least two-layer organisms, which later experienced simplification. Volvox, which is not an animal at all, but a green algae, also has gastrulation (Desnitskiy 2018). Consequently, its ancestors were multicellular animals no simpler than two-layer ones, having experienced secondary simplification. Thus, these most often mentioned examples, illustrating the stages of the origin of complex multicellular organisms from simple unicellular organisms, actually demonstrate the reverse process.

In ontogeny, the formation of a multicellular organism is indeed observed as a result of the fragmentation of one cell—the zygote. But the history of free-living adult organisms and the history of embryonic stages, obviously not viable in the environment, cannot be directly compared, much less identified.

The data currently available suggest that the formation of multicellular organisms and one of eukaryotic cells are the conjugated processes that began as early as in prokaryotes. Like any other, this process creates predominantly unstable forms, which subsequently experience decay. Most likely, this explains the similarity of both free-living unicellular organisms with multicellular cells, and a number of bacteria with semi-autonomous organelles of eukaryotic cells (Gusev and Gokhlerner 1980).

The current ideas about the evolution of multicellular organisms are based on the provisions set forth by Charles Darwin in his work "The Origin of Species by Means of Natural Selection: Or, the Preservation of Favored Races in the Struggle for Life" (2011). The problem with all theoretical constructions of this kind is that the event of the origin of the species was not observed either by Darwin himself or by the later researchers. Therefore, it is impossible to reasonably conclude anything about its causes and mechanisms. It is assumed that the complication of organisms occurs as a result of the influence of two deliberately simplifying factors. Firstly, a random change in heredity and, secondly, the selection of the resulting variants by the influence of the external environment. In no other cases the complication due to random changes and deliberately lower organized external influences is known. Therefore, this assumption especially needs to be confirmed by a direct observation.

Especially it is necessary, since the observable fact is the predominant survival of the most environmentally independent organisms and the extinction of the fittest (specialized). For animals, this is a change from cold-blooded to warm-blooded, spawners—oviparous and then placental. For plants—a change of spore-bearing gymnosperms, and then angiosperms. The emergence of these groups on land is also an event during which independence from the environment has fundamentally increased.

The change in multicellular organisms in the course of both natural evolution and selection is regarded as a direct change in the organisms themselves. The fundamental disadvantage of this point of view is that the structures of multicellular organisms in the series of their changes do not directly communicate with each other. They are separated by the sexual process, in which only cellular and subcellular structures are successive, and in a destabilized form.

Thus, in any evolutionary series of multicellular organisms, the impact of the environment can only be taken into account in the form of information about it. The corresponding change in the shape of organisms can also be only an expression of newly created information that corresponds to information about the impact of the environment. Since only cellular and deeper structures are successive, this information can be formed and transformed only in them.

Nine years after the publication of the work "The origin of species…". Ch. Darwin published the treatise "The emperor's new mind: concerning, computers, minds, and the laws of physics" (1970). In this work, based on the study of the observed selection process, he has formulated the theory of pangenesis, taking into account the above circumstances. This theory assumed the accumulation in germ cells of information about the current existence of the organism as a result of their communication with somatic cells with the help of special carrier particles—gemmules. For a long time, this theory was considered speculative, not substantiated by facts. His initial point of view, which contradicted the observed facts, was accepted as justified.

At the time of Darwin, the development of microscopic technology only allowed to initiate the study of the structures of the dimensional class of cells. Gemmules were supposed to be much smaller, so discussions on related topics continued throughout the twentieth century, acquiring particular urgency in the middle of it.

By the beginning of the twenty-first century, the research technique had acquired sufficient complexity for the substantive solution of such issues. The work was published in 2014 (Cossetti et al. 2014), in which the authors showed that the long-known exosome particles secreted by cells into the blood perform the functions proposed by Darwin for gemmules. In the experiment, they transferred the genetic information that determines the synthesis of a fluorescent protein from modified human malignant cells implanted in laboratory mice into the germ cells of animals. Thus, the names "gemmules" and "exosomes" can be considered synonymous. The first of them was introduced to denote functions, and the second to denote structures.

Thus, the historical change of multicellular organisms is the result of a complex behavior of cells and structures of deeper levels of organization.

Important point for the question under study is that the behavior of the cells of multicellular organisms is immeasurably more complicated than the behavior of the organisms themselves. And in this case, we observe the formation of a simple from a complex, but never vice versa.

Despite the fact that cells have a microscopic size, which in most cases is not comparable with the size of an organism, their behavior, including morphofunctional specializations, covers not only the entire organism as a whole, but also its external relations. Its positioning is generally provided by a single description of the whole organism in the DNA of each of them. In specific situations, it is the activity of the cytoskeleton, which is a structure not of a cellular, but of an organismic scale (Penrose 1989), and exosomes.

The actions of cells in ontogeny are precisely the behavior that in most cases uses a developed program, and not unambiguous adherence to such a historically emerged program. This is demonstrated in the course of embryonic regulation, during which a whole series of atypical equifinal ontogeny may be constructed (Beloussov 2015). They often originate from such damaged stages that further development in nature would be impossible and, accordingly, the reaction to such damage could not be evolutionarily fixed.

Even in relation to an individual variability, the influence of the environment, leading to a change in the organism, is not the actual influence of the environment in the literal sense. For example, with an increase in the volume of muscles, the influence of the environment must be evaluated, taken into account and converted into a corresponding effect of the organism on its structure. The loads themselves cannot lead to such a result—their direct impact is limited only by the injuries.

Direct change in multicellular organisms as a result of cell behavior is difficult to observe. Exceptions are the embryonic regulation and regeneration. If these stages were present in evolution, they have long been replaced by the following ones. However, evolutionary processes are non-linear and, as a rule, create sister groups that implement similar processes in parallel. Filamentous cyanobacteria turned out to be such a group for eukaryotes (Gusev and Gokhlerner 1980; Sumina and Sumin 2013). The fundamental difference between them is that a eukaryotic cell is a voluminous multicellular prokaryotic organism, while a filament of cyanobacteria is a linear multicellular prokaryotic organism. Them both are the objects of the same level of organization. This implies a general similarity of evolutionary events in these groups and fundamental differences in their course.

The volumetric layout of prokaryotic components in eukaryotic cells allows them to achieve a high degree of specialization in the formation of mitochondria, plastids, nuclei and cytoplasm. In the future, this creates the possibility for deep specialization of the cells themselves in multicellular organisms.

In filamentous cyanobacteria, the multicellular prokaryotic organism is arranged linearly, which prevents the specialization of its cells, although its certain intention is also observable. Respectively, specialization is also impossible in the associations of threads, despite the fact that during the interaction of threads, morphogenesis of an organismic nature occurs, but does not have equifinality and stereotypy. These circumstances hinder their evolutionary development at the initial stages and, in contrast to multicellular eukaryotes, make them observable under normal conditions, and not only during the restoration of damage.

Let us give one of the examples of the creation of structures of an organismic plan by the threads of cyanobacteria, parity in terms of the level of organization of multicellular eukaryotes. In some cases, they form mobile multi-beam aggregates up to five millimeters in size, vaguely reminding of the texture and nature the movement of small spiders. These textures are formed as a result of the directed movement of the threads and, thanks to it, they acquire a certain permanently preserved texture for the duration of their existence. The main function of multi-beam aggregates is the implementation of fast, in relation to the capabilities of a single thread, movement both in the volume of liquid and on the surface of the substrate (Fig. 1).

Another example of this kind is the development of stromatolites, communities of filamentous cyanobacteria attached to the bottom of calcareous structures known

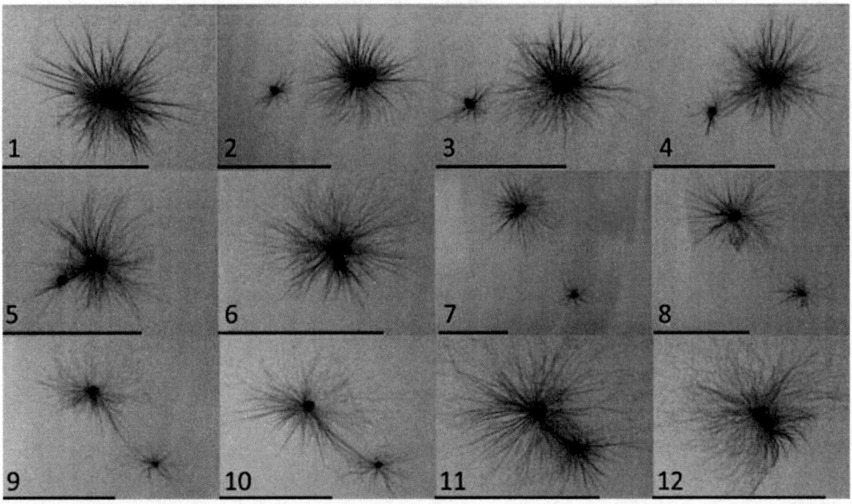

Fig. 1 Movement and merging of multipath aggregates. 1–12—a single series of events (within 4.5 h). Scale bar 5 mm

since the Archean (3.5 Ga). For a long time, these formations were not recognized as organisms, based on the properties attributed to them. However, in geological time they showed all the properties inherent in organisms, what formed the basis for their use in stratigraphy (Komar et al. 1965). Stromatolites are capable of directionally changing their structure, which can be compared with the ontogeny of multicellular eukaryotes, with the difference that the change in stromatolites does not have a certain direction, stereotype, and discontinuity.

This process may last for many thousands of years. But unlike multicellular eukaryotes, every season the community breaks up into separate threads in order to overcome the layer of sediment that has fallen on it. Next, the threads are again combined into a film. Consequently, in the conditions of continuous disintegration and restoration of the community, only individual threads that implement it in their behavior can be interpreted as the carriers of information about the structure of the community and the direction of its change (Fig. 2). These observations are consistent with the ability of some sponges to rebuild their structure after being rubbed through a sieve. In this case, only individual cells can also be pointed as the carriers of information about the structure of the organism. In both cases, this behavior is determined not by the local relationships of the thread or cell, but by the properties of the structure of the organismic plan in the formation of which they participate.

A number of structures that have finally formed in multicellular eukaryotes are in the process of formation in cyanobacteria. Therefore, they are a convenient object

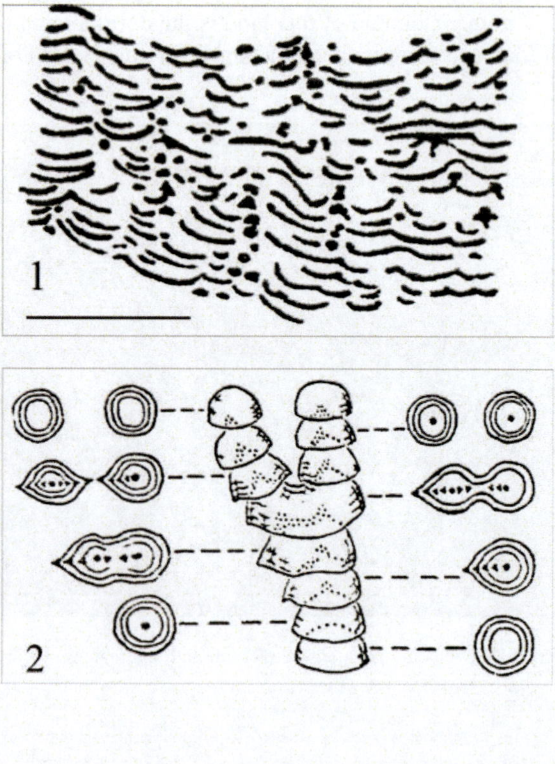

Fig. 2 1—restoration of the structure of the stromatolite layer (film) with a system of reference points after the threads have overcome the sediment layer. Scale bar 10 mm. 2—controlling the division of the top of the building using a system of control points (from Vlasov 1970). GCPs correspond to multibeam aggregates (Sumina et al. 2022)

for considering the general direction of movement of complexity in evolutionary processes. In general, during the formation of organisms at each level of organization, the following stages of complication can be distinguished: the initial stage is the emergence of functional differentiation at the level of behavior of elements, the middle stage is the emergence of a constant spatial arrangement of elements, and the final stage is the emergence of a constant morphofunctional differentiation of elements.

Individual cells of cyanobacteria are formed organisms, the structures of which have both a constant spatial position and a high specialization of elements. The filaments made up of cells have managed to acquire only a permanent arrangement of elements, and the structure of the community of filaments is maintained solely due to the behavior of the elements.

Thus, we observe that the complexity of the structure grows deeper into the organization and each time the elements turn out to be more complicated than the system. Moreover, they deliberately communicate their complexity to it.

The same phenomenon is noted for "colonial" two-layered eukaryotes—hydroid polyps, which are modular organisms. Marfenin (2016) directly notes that "both anatomically and functionally, a modular organism turns out to be simpler compared to a unitary one". Here again, the elements are more complex than the system, and the complexity is given to it by the elements.

Returning to the role of cells in giving complexity to multicellular organisms, it should be noted that multicellular organisms are also modular.

Another common example of increasing the complexity is the origin of life from inanimate. It has not been yet noted a single such case, and the reverse irreversible transition of the living into the inanimate is constantly observed. The origin of life is a purely historical reconstruction. The facts established within the framework of the natural history disciplines testify to the primacy of life as a physical process and the origin of the inanimate as a secondary simplification. This is the main content of bioinert interactions. (Sumina and Sumin 2016).

3 Crystals

Another often considered example of increasing complexity is the formation of crystals.

In this part it should be noted that the type of crystal lattice, shape and other properties of a crystal are determined by the properties of the atoms and molecules that form it, and not vice versa. These properties can, depending on the conditions, manifest themselves in several ways, which is also an intrinsic property of a particular substance.

If, for example, we give an artificial shape to a diamond crystal or provoke a change in the crystal lattice with the formation of graphite, it will not have an effect on the properties of carbon atoms in any way.

The bond energy of the crystal lattice is the energy of the electromagnetic field created by electrons. Thus, electrons have sufficient complexity to assimilate energy and transform it into force fields, while crystals do not have such complexity.

The last point needs to be clarified. It is generally accepted that artificial field generators consume energy corresponding to the characteristics of the field, while natural ones do not consume any energy. In this case, the law of conservation of energy appears and disappears arbitrarily in the reasoning. Also, for the time being, we note this circumstance as a fact to which we will turn later.

This contradiction resolves the discovery of the accelerated expansion of the Universe (Riess et al. 1998). It follows from this that there is a continuous influx of energy into the Universe, which we identify as time (Sumin and Sumina 2020). Bodies exist due to their assimilation of this energy and its dissipation, including in the form of fields.

The bond energy also increases, going deeper into the structure of the system. In order to destroy the crystal lattice, a simple mechanical action is sufficient. In order to destroy the nuclear bonds of its constituent atoms, an accelerator with the appropriate energy and design complexity is necessary.

The same regularity can be traced in the external relations of crystals—they are more complex than aggregates consisting of them, and even more so rocks. For example, pyrite forms cubic, pentagondodecahedral, and other distinctly shaped crystals. Further, it can form spherulites having a certain shape and texture, but simpler than that of the crystals that compose them. This is followed by the level of the breed, which does not have its own form and with very little definable complexity of the texture.

The crystallization process itself is often characterized as "stochastic" or random. How the naturally arising structure of the crystal lattice and the shape of crystals can be characterized as randomly arising cannot be definitely said. Moreover, the concept of «stochastic comes from the ancient Greek "στόχος", which means "goal; conjecture"», and how it can be translated as an accident, is also unclear. This fact, together with the previous ones, will be left for consideration below.

4 Social Processes

In this series of examples, social processes can also be considered as the most observable. Usually, social processes are opposed to the natural ones. The only difference is that the researcher is the direct participant, and not an external observer. In addition, the researcher is the only object given to him in sensation, and he perceives other objects only as a reflection in these sensations. Thus, social processes differ from others not in their nature, but only in the position of the observer.

Society takes place in the series of development of organisms, at the moment closing it, but is not considered applied to the organisms. In a number of organisms, organismal properties unfold fractally, passing through the levels of prokaryotic cells, eukaryotic cells and multicellular organisms. Further, this process captures communities of organisms, including society. However, at this point it is at the first setout (in relation to proper time). Because of this, neither humanity as a whole, nor individual societies can be attributed to any stable biological structures, since they have not acquired sufficient integrity and complexity. At this stage of development, a peculiar archaic variety of types of social structure will arise.

All the complicating influences on society are based on the activities of specific individuals. This is related to the fact that individuals, unlike society, have not only their own integrity necessary for this, but also much greater complexity. If we compare them by the number of elements, then in the human body as a cellular community, about 100 trillion individuals act in concert, and humanity has only 8 billion individuals. The organizational capabilities of an individual organism quantitatively exceed the organizational capabilities of mankind by 4 orders of magnitude. The qualitative superiority in the coherence of cell interactions compared to the interaction of people is so great that at the moment it cannot be definitely characterized.

At the moment, the prevailing opinion about the low organizational capabilities of cells is due to the simplicity of their structure. The very same opinion about the structure of cells changed along with the development of means of observation in the direction of increasing complexity. This indicates that the properties of the research technique were transferred to the object of observation and the idea of its simplicity is not related to it itself. However, the behavior of free-living cells was observed regardless of their structure. When discovered, they were called "animalcules" or "animals", and "bacteria", that is, "stick-animals" in literal translation. Their behavior was immediately recognizable as not inferior in complexity to the behavior of animals.

How complex is the behavior of cells in the human body? If, for example, we consider the organization of the surface of integumentary structures, then the "pattern" of the fingertips is a polar positioning system for tactile information. Most of the skin surface is covered by a data positioning system based on triangulation. That is, the surface of the body to control it is described in the same ways that are now used by society in advanced modeling technologies.

Cyanobacteria also use triangulation to manipulate surfaces (Sumina et al. 2022). The time of application of these technologies in the case of an organism has an evolutionary scale—at least hundreds of millions of years, and in the case of a society—hundreds of years. However, this is precisely the organization of the surface. The organism itself is a three-dimensional system, which is controlled by the cytoskeleton with the help of three-dimensional triangulation. For humanity, these methods have become available only in the most recent time. In addition, the inner surfaces of cells are also controlled by the cytoskeleton through triangulation.

Triangulation is the most convenient object for observation. Operations with information in this case acquire a visible and easily identifiable form of a geometric fractal. This makes it possible to trace the movement of complexity from the molecular level through the cellular to the social. With an increase in size, there is a delay in time and a drop in complexity (Sumina and Sumin 2017).

In speech, this circumstance is expressed quite definitely—here are the corresponding dictionary entries: «Engineer. Comes from the French. ingénieur, further from the Latin ingenium "innate features", further from in- "to" + gignere "to give birth", further from the pra-Indo-Hebrew. *gen- "to give birth"», «The subject (Latin subjectum "underlying; underlying") is the bearer of activity, consciousness and cognition».

Currently, efforts in engineering are focused on the creation of artificial intelligence and neural networks. Here the catch-up position of society in relation to the structure of organisms is clearly expressed. Also, in terms of computing power (only a quantitative characteristic), the total technical power of mankind equaled the capabilities of the individual human brain approximately in 2007–2010.

In this case, the direction of movement of complexity from the side of the elements to their system is also obvious.

This pattern is also traced historically. As soon as society acquires sufficient organization and stability to exert a decisive influence on the formation of individuals, it inevitably collapses. Most often with the extinction of the population involved in it. Fundamentally, these events do not differ from technical disasters. They occur when the products of human activity begin to function according to their own internal laws and extend their properties to the subject of their existence. Thus, in this case, the elements are being simplified by their system.

As a result, human thinking occupies a dual position. On the one hand, it is a product of highly organized matter that is inside the body and creates its structure. On the other hand, it is formed when communicating with society as a system that is poorly organized and does not have integrity.

This circumstance is also reflected in the contradictory structure of the conceptual apparatus. It is based on fundamental concepts that do not have clear definitions—time, space, life, number and, accordingly, simplicity and complexity. For this reason, the discussion of the subject has started without introducing its definition. Despite the fact that these concepts are indefinable, in other words, they cannot be adequately expressed by a combination of words, they are understood unambiguously, since they are identified individually. Definitions, on the other hand, are a product of social activity that does not have the structural complexity and integrity for operating with fundamental concepts.

In this point we also observe an already traced regularity. Elements are more complex than their system—individual words have a clear meaning, but are too complex for public understanding. Their combinations, on the contrary, most often have neither clear connection with the subject, nor among themselves. Earlier we drew attention to several examples of this pattern, now its internal causes have become clear.

Fundamental concepts are integral characteristics that are not defined, since it is impossible to reach their limit and compare with objects that do not have such characteristics. For example, it is impossible to compare objects that have complexity and those that do not. Accordingly, only the current understanding of these characteristics can be determined.

Complexity and simplicity are recognized individually, regardless of their definitions. Therefore, the perception of their individual properties also occurs independently of the definitions. On this basis, it is possible to single out particular, but correlated characteristics of complexity: the number of elements, the number of their types, the energy of connections, the certainty of connections (their correlation, orderliness), the ability to transform, the direction of control, etc. The examples of complexity change analyzed above reflected some of these characteristics depending on how they are expressed in the object of consideration.

5 Conclusion

The emergence of the complexity of objects of each next level of organization is most often considered as a process of "self-organization" of simpler elements, closed in itself (Nicolis and Prigogine 1977). For this reason, their influence on the formation of the complexity of system properties can be neglected. At the same time, the environment cannot be the source of the emerging complexity of the object, since it is easier organized. With such a formulation of the question, the emerging complexity of the object has neither its structural sources nor sources of energy.

Apparently, the idea of a greater complexity of systems compared to their elements arose historically due to the limitation of the means of observation by the capabilities of ordinary vision. Where the structure of objects was not visible, it was considered either simple or absent. As a result, a simple logical connection was constructed—"the bigger, the more complex", which turned out to be only partly true. In total, systems are certainly more complicated than their elements, since they add their own level of connections to the structure. However, being considered separately, the elements turn out to be qualitatively more complex than their systems (Fig. 3).

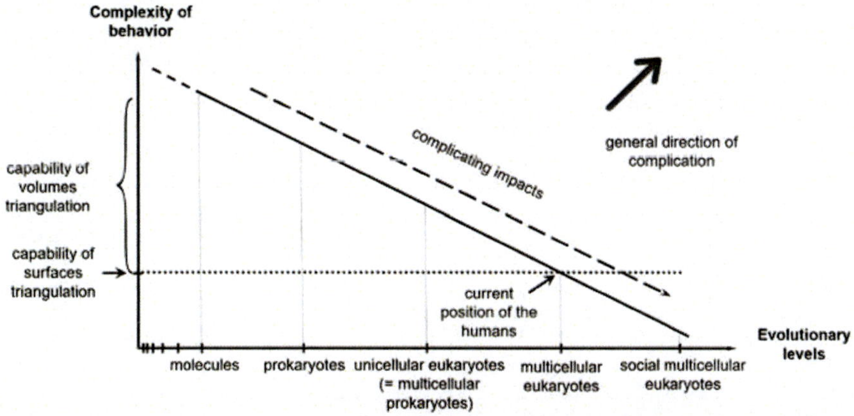

Fig. 3 Principal scheme for changing the complexity of the behavior of biological objects

Thus, the general complication of matter occurs as a result of the spread of complexity and energy from deeper and more ancient levels of its structure to the more superficial and later ones. Therefore, development as such is the origin of the simple from the complex. This corresponds to the principle of energy propagation along a gradient, otherwise called the second law of thermodynamics. The gradient arises due to the fact that the energy of bonds falls in proportion to their length. For the same reason, if the complexity of the elements is removed from the system, it is not clear what will constitute its own complexity.

Raising the question of "the origin of order from chaos" turned out to be groundless. When penetrating deep into the matter, chaos is not observed, but on the contrary, the degree of its orderliness increases.

This leads to the necessity to introduce the concept of "mutual-organization", in contrast to "self-organization", which takes into account these circumstances. The observed processes show that systems do not arise due to local and random interactions of simple elements. This happens because each element carries out behavior, the dimension of manifestations of which determines not only internal, but also external connections of the entire system. In biological systems, this property is most pronounced.

References

Beloussov LV (2015) Morphomechanics of development. Springer, Cham Heidelberg New York Dordrecht, London

Cossetti C, Lugini L, Astrologo L, Saggio I, Fais S, Spadafora C (2014) Soma-to-germline transmission of RNA in mice xenografted with human tumour cells: possible transport by exosomes. PLoS One 9 (7): e101629

Darwin Ch (1998) Variation of Animals and Plants Under Domestication Ser.: The Variation of Animals and Plants under Domestication. Johns Hopkins University Press, Baltimore

Desnitskiy AG (2018) Comparative Analysis of Embryonic Inversion in Algae of the Genus Volvox (Volvocales, Chlorophyta). Russian Journal of Developmental Biology 49: 129–133

Gusev MV, Gokhlerner GB (1980) Free Oxygen and Cell Evolution). Moscow University Press, Moscow (In Russian)

Komar VA, Raaben MA, Semikhatov MA (1965) Conophytons of the Riphean of the USSR and their stratigraphic significance. Proceedings of the Geological Institute of the USSR Academy of Sciences 131: 72 (In Russian)

Marfenin NN (2016) A decentralized organism on the example of colonial hydroids. Biosphere 8 (3): 315-337 (In Russian)

Nicolis G, Prigogine I (1977) Self-Organization in Nonequilibrium Systems. Wiley-Interscience, New York Berlin

Penrose R (1989) The emperor's new mind: concerning, computers, minds, and the laws of physics. Oxford University Press, Oxford

Riess A, Filippenko A, Challis P (1998) Observational evidence from supernovae for an accelerating universe and a cosmological constant. Astron J. 116:1009–1038

Sumin DL, Sumina EL (2020) Time and space of biological morphogenesis. In: Processes and Phenomena on the Boundary Between Biogenic and Abiogenic Nature. Lecture Notes in Earth System Sciences. Springer, Dordrecht Netherlands

Sumina EL, Sumin DL (2013) Morphogenesis in a Community of Filamentous Cyanobacteria. Russian Journal of Developmental Biology 44 (3): 158-176

Sumina EL, Sumin DL (2016) About the properties of relations between living and nonliving. In: Biogenic-Abiogenic Interactions in Natural and Anthropogenic Systems. Lecture Notes in Earth System Sciences. Springer International Publishing, Switzerland Heidelberg New York

Sumina EL, Sumin DL (2017) Morphogenesis as a fractal organization of behavior. Evolutionary and comparative psychology in Russia. Cogito Center, Moscow (In Russian)

Sumina EL, Sumin DL, Kharitonov AN, Grechenko TN (2022) Morphogenetic Activity in Modern and Fossil Communities of Cyanobacteria. Biology Bulletin Reviews 12 (1): 1–17

Vlasov FYa (1970) Anatomy and morphology of stromatolites of the early and Middle Proterozoic of the Southern Urals. In: Materials on paleontology of the Urals, Sverdlovsk (In Russian)

The Concept of Species in Biology and Mineralogy: A Comparative Study

Sergey V. Krivovichev and Evgeny A. Borovichev

Abstract Both descriptive mineralogy and descriptive biology are based upon the concept of species. The definition of species in mineralogy reflects the dialectic unity of chemical composition and crystal structure, two interdependent features of a crystalline substance. Despite some issues that still remain unresolved, the definition of a mineral species is rather strict and depends upon the mutual agreement reached by researchers on the borders that separate one mineral species from another. In biology, there are different approaches to the concept of species. Yet, the biological species is understood as a group of organisms with common morphological, physiological, biochemical and behavioral features, capable of crossbreeding. The modern definition of a biological species is based upon combination of morphological and molecular genetic features. The basic difference between mineral and living organism is the existence of a molecular carrier of genetic information in the latter (DNA). Though mineral kingdom develops in time (mineral evolution), this development has little in common with the development of the domains of life (biological evolution). Thus, the concept of a mineral species has no evolutionary aspect, in contrast to the concept of biological species, where each species is the result of evolution governed by different biological and environmental mechanisms. The morphological and genetic diversity of living organisms is by far larger than the mineralogical diversity, by at least three degrees of magnitude. The biological classification of species, at least at some level of organismal complexity, reflects their evolutionary development, whereas classification of minerals is based purely upon their constitution, i.e., the chemical composition and crystal structure. In both mineralogy and

S. V. Krivovichev (✉)
Nanomaterials Research Centre, Kola Science Centre, Russian Academy of Sciences, Apatity, Russia
e-mail: s.krivovichev@ksc.ru

Department of Crystallography, Institute of Earth Sciences, St. Petersburg State University, St. Petersburg, Russia

E. A. Borovichev
Polar-Alpine Botanical Garden-Institute, Kola Science Centre, Russian Academy of Sciences, Apatity, Russia

O. V. Frank-Kamenetskaya et al. (eds.), *Biogenic—Abiogenic Interactions in Natural and Anthropogenic Systems 2022*, Springer Proceedings in Earth and Environmental Sciences, https://doi.org/10.1007/978-3-031-40470-2_39

biology, the definition of a new species proceeds along specific protocols governed by special codes identified by respective international scientific organizations.

Keywords Mineralogy · Biology · The concept of species · Diversity · Classification · Falsification · Mineral evolution · Biological evolution

1 Introduction

Throughout the whole history of science, the difference between living and non-living was a subject of extensive and considerable debates, due to the relevance of this issue to various concepts in philosophy, theology, sociology, and natural sciences sensu stricto. Aristotle was probably the first who raised the problem in his classification of beings, resulting in the formulation of 'the Great Chain of Being' that was at the core of scientific development till the end of the XIXth century. The history of the views on the similarities and differences between living and non-living is well beyond the scope of this paper, but it is noteworthy that the serious modern analysis of the problem is rarely done on the purely scientific level (with the exception of seminal works by Vladimir Vernadsky in the middle of the XXth century (Vernadsky 1939)), without specific prejudices external to science.

In this work, we would like to abstract ourselves from any external philosophical or theological influences and put the analysis of the problem onto a scientific basis. Our approach is multidisciplinary and is based upon the comparative analysis of basic phenomena that are of high relevance to the fields dealing with living and non-living matter. In this particular contribution, we shall concentrate on the concept of species, which is at the core of descriptive mineralogy (considering mineral as a basic unit of the solid Earth) and biology (where the concept of species is at the center of any scientific exploration). We are well aware that, in biology, the concept of species is tightly linked to the principle of evolution, since any biological specie is the result of an evolutionary development. Following Vernadsky (1939), one may point out that '...there are no changes in non-living natural bodies... that are analogous to the evolutionary processes of living substances'. However, recent advances in the science of mineral evolution (Hazen et al 2008, 2021; Hazen 2013; Hazen and Morrison 2020; Morrison and Hazen 2020, 2021) that was first suggested back in the 1980s (Zhabin 1981; Yushkin 1982) indicate that the kingdom of minerals changed in the course of geological time, though the mechanisms of mineral evolution and its driving forces differ from those observed in the biota. It is our intention to consider the differences in the concept of evolution in mineralogy and biology in the subsequent papers of the series. Other points of interest are also symmetry, complexity and modularity, which also deserve a separate analysis.

Herein we adopt a rather pragmatic view on the concept of species, the one at use by a practical mineralogist or biologist willing to establish a new natural species, be it a new mineral from volcanic fumaroles or a new genus of liverworts. Our analysis is intended to reveal similarities and differences in the concept of species employed in descriptive mineralogy and biology viewed from the eyes of practitioners involved in the studies of mineralogical (S.V.K.) or biological (E.A.B.) diversity. The paper is organized as follows. First, we provide the definitions of species existing in current mineralogy and descriptive biology, with all its limitations and borders that outline a separate mineralogical or biological entity. In the second section, general formal procedures are described for the establishment of new species in modern mineralogy and biology, including listing of the necessary data; the concept of cryptic species is considered in this section as well. Third section is devoted to the mineralogical and biological diversity, their current state and different approaches to their estimation and prediction. The problem of classification of species is touched upon in the fourth part of the paper, though we do not go deep into its philosophical implications; rather we concentrate upon the current methodology used in both fields of scientific enquiry. In concluding section, we try to summarize basic findings of our analysis and to outline similarities and differences in mineralogical and biological concepts of species that are upheld in modern science.

2 Definition of Species

2.1 Mineralogy

As many other concepts of species, the definition of species in mineralogy changed in the course of time. The current definition of a mineral species established by the International Mineralogical Association (IMA) (Nickel and Grice 1998, 1999) is based upon the concept of a mineral substance, which is defined as '…a naturally occurring solid that has been formed by geological processes, either on Earth or in extraterrestrial bodies' (Nickel and Grice 1998). Since solids can be either amorphous or crystalline, a mineral species is defined as a crystalline substance with well-defined chemical composition. The related question 'what is a crystal' is answered in the definition given by the International Union of Crystallography (IUCr): 'a solid is a crystal if it has essentially a sharp diffraction pattern' (Authier and Chapuis 2014; Grimm 2015). The overwhelming majority of solids with sharp diffraction patterns are those based upon three-dimensional lattice with elementary unit cells described (in general) by six independent parameters: three linear parameters, *a*, *b*, *c* (usually provided in ångströms, Å), and three angular parameters, α, β, γ (usually given in degrees). Crystals are characterized by their crystallographic symmetry, which is given as a space group symbol. There are in total 230 space groups derived by E.S. Fedorov (1891) and independently by A. Schoenflies (1891) at the end of the XIXth century. Majority of minerals are crystals with three-dimensional periodicity. Recently, there have been several reports on minerals with structures without three-dimensional periodicity (quasicrystals, which have sharp diffraction patterns with five-, seven- or tenfold symmetries prohibited in three-dimensional lattices) (Bindi et al 2009, 2011, 2015); however, their natural origin had been questioned (Ivanova et al. 2017), despite their approval by the IMA.

The mineral species (or one may say, an archetype of a mineral species (Hawthorne et al. 2022)) is defined by its end-member formula (idealized chemical composition) and crystal structure. In turn, the latter is defined by its unit-cell parameters, space-group symmetry, and the finite set of atomic sites occupied by different chemical elements. In principle, each site is predominantly occupied by a specific chemical element: if '…at least one structural site…' in a potentially new mineral is '… predominantly occupied by a different chemical component than that which occurs in the equivalent site in an existing mineral species', such a mineral may be considered as a new mineral species (Nickel and Grice 1998). In practice, however, this rule can be avoided and several sites may be grouped and their total predominant occupancy is considered as a species-defining parameter (this is an accepted procedure, e.g., in eudialyte-group minerals (Johnsen et al 2003)).

In the former times, physical appearance of a mineral (e.g., color, luster, crystal morphology, cleavage, etc.) have been considered as major mineral-defining characteristics. However, with the advent of X-ray diffraction methods and electron-microprobe analysis, chemical composition and crystal structure became most important, though the provision of data on physical properties, etc. is formally required for the recognition of mineral species as new (see below). All minerals are grouped into classification taxa according to their chemistry and structure. A particular group of minerals may have its own nomenclature rules (within general guidelines) for the definition of a mineral species. Such nomenclatures are elaborated by experts and are given at the site of the Comission of New Minerals, Nomenclature and Classification of the IMA (CNMNC IMA; see below).

Minerals with the same crystal structure and different end-member formula are always considered as different (though they may belong to the same classification taxon). However, minerals with the same end-member formula and different crystal structures may be or may not be considered as different species. The basic criterion in this case is the degree of similarity between their crystal structures. If the structures possess different atomic coordination numbers and different building principles, they are considered as different mineral species. For example, quartz and stishovite are two different polymorphs of SiO_2. In quartz, the coordination numbers of Si and O are 4 and 2, respectively, and the crystal structure is based upon a three-dimensional framework of corner-sharing SiO_4 tetrahedra. In contrast, in stishovite (a high-pressure polymorph of SiO_2) Si and O atoms are 6- and 3-coordinated, respectively, and the crystal structure is a framework of edge- and corner-sharing SiO_6 octahedra. The drastic differences between the two structures determine the definition of quartz and stishovite as two different mineral species. However, there are examples of minerals with crystal structures that have the same coordination numbers for identical chemical elements and are different in the stacking of structural modules only. For instance, polytypes are crystal structures that are based upon the same type of two-dimensional layers but with different stacking sequences. Such structures are considered as belonging to the same mineral species that occurs in different polytypic modifications. In this case, special suffices are applied to different polytypes that nevertheless belong to the same mineral. The example is graphite, a layered carbon polymorph that exists in two polytypes, graphite-2*H* and graphite-3*R*. The nomenclature of polytypes is governed by the rules outlined by the IMA and IUCr (Guinier et al. 1984). Note that polytypes generally have different unit cells and space-group symmetries.

The key issue for the establishing mineral species as new is its origin in the course of natural geochemical or cosmochemical processes without a direct intervention by a human. There are obviously borderline cases, when the absence of human intervention cannot be easily established. The good example are crystalline phases that form as combustion products of burning coal dumps. The IMA policy regarding the status of these phases as mineral species changed over time. Initially, these phases were considered as true mineral species (see, e.g., (Chesnokov et al. 1989, 1990)), whereas later it was adopted that '…substances formed by combustion are not generally regarded as minerals' (Nickel and Grice 1998, 1999). Very recently, this stance was re-considered again and the conclusion was reached by the CNMNC IMA that '…minerals crystallizing on burning coal dumps… are now acceptable as minerals, if it can be proven that the fire responsible for the formation of these phases is the result of natural events' (Parafiniuk et al. 2020).

More difficult issues with regard to the natural vs. technogenic origin of crystalline phases in nature arise due to the problem of contamination. The human activity throughout the history of humankind resulted in the huge amounts of synthetic compounds introduced into surface or subsurface geological environments that can mistakenly be considered as being of natural origin and, therefore, as minerals. The examples are recent discoveries of so-called super-reduced mineral phases (Griffin et al. 2016, 2018a, b, 2019; Lian et al. 2017; Xiong et al. 2017), including diamond, metal carbides, borides, nitrides and oxides, many of which have been approved by the CNMNC IMA as new minerals. These discoveries have been criticized due to the highly possible occurrence of these phases as derived from technical equipment, e.g., from abrasive materials, where they occur as microinclusions in corundum (Litasov et al. 2019a, b; Ballhaus et al. 2021). The topic of technogenic origin of several new mineral species is under current debate in mineralogical community.

In addition to inadvertent misconsideration of technogenic phases as mineral species, there have been several fake mineral discoveries in the history of mineralogy. In fact, mineralogical discoveries can easily be falsified intentionally, taking into account that the person involved is smart enough to emulate natural origin of crystalline phases. The most famous example in mineralogy is that of texasite that was originally described as a pure oxosulfate of praseodymium, $Pr_2O_2(SO_4)$ (Crook 1977). This 'discovery' was soon exposed as a fake, along with several others made by the same author, since rare earths (REs) almost never occur in nature as pure elements, but always with the admixtures of other REs (Peacor et al. 1982).

2.2 Biology

The problem of the definition of biological taxa and the main one of them, the species, is causing heated discussion today, as it was three hundred years ago. Pavlinov (2018), discussing the problem of the complexity of defining a species, notes that one of the paradoxes of science is that the more general and fundamental a certain concept is, the less strictly and unambiguously it can be defined and derived from the traditional logic of definitions, supplemented by the epistemological principle of incompleteness.

In general, the species is understood as a group of organisms (individuals) with common morphological, physiological, biochemical and behavioral characteristics, capable of crossbreeding, which gives fertile offspring in a number of generations, naturally distributed within a certain area and similarly changing under the influence of environmental factors. Grant (1981) recognizes five types of species—taxonomic, biological, micro-species, successional and biosystematic species. The concept of a biological species, which is usually, though sometimes reluctantly, accepted by evolutionary biologists, views species as reproductively isolated populations (Mayr 1963; Grant 1981). A biological species is reproductively isolated from all other species, and therefore cannot interbreed with them. While a lot species concepts and definitions are currently used, only a limited subset might be applied to bacteria due to their asexual reproduction.

Most often, in taxonomy, the concept of a taxonomic species is used, which is based on the difference in the morphology of the systematized individuals. Zavadsky (1968) pointed out that the taxonomic volume of a species depends on its structure. "The smallest imaginable volume of the species (the "ideal monotype") (homozygous and homogeneous population consisting of phenotypically indistinguishable individuals) does not exist in nature. Any species has a polymorphic nature, is a system and includes a certain number of similar or different types of subdivisions." This objective phenomenon is refracted through the subjective view of the researcher and results in a different breadth of understanding of the species. He also suggested that the following should be considered common features of the species: (1) number of individuals, (2) type of organization, (3) reproduction, (4) discreteness, (5) ecological certainty, (6) geographical certainty, (7) diversity of forms, (8) historicity, (9) sustainability, (10) integrity.

In the scientific literature, there are many "concepts of species" at the research and practical level. Mayden (1997) identified 22 different concepts of species, whereas Wilkins (2007) has identified 26. Pavlinov (2018) recognizes the following main categories into which certain concepts fall, depending on what is considered as a key in defining a species (species unit) in its general understanding:

- a species as a community by similarity; the concepts that belong to this group are: typological (goes back to the beginning of the XIX century), phenetic, genetic, "integrative" (a combination of traditional morphological-geographical and genetic);
- a species as an isolated reproductive community, i.e., the biological concept (Mayr 1963, 2000);

- species as a historical community, including the classical generative concept (Komarov 1940), concepts of phylogenetic or genealogical species in cladistics;
- a species as a specific ecological community, including the concept of an eco-species;
- species as an ontological natural unit (Mahner 1993);
- species as a logical classification unit, allocated on a general basis within the framework of the generic scheme (Mahner and Bunge 1997).

Discussing different concepts of the species, we are faced with two different solutions: "monism" and "pluralism" (Ghiselin 2001, 2002; Pavlinov 2018). Vavilov (1931) formulated the concept of different types of species, putting forward the idea of the diversity of species, the relativity of taxonomic criteria and the theory of peripheral accumulation of recessive traits. He suggested that "Linneaons" and "Jordanons" correspond to successive stages of speciation. Great importance in the concepts of species and intraspecific taxa is often attached to the geographical isolation of one from the other. However, if two geographically separated taxa come into interaction, then their imaginary uniqueness is called into question.

Most authors today conceptually agree that species are entities (units) that arise in the process of speciation as separately developing segments of metapopulation lines (Mayden 1997; Hey et al. 2003; Bock 2004; Hey 2006).

It is still popular among taxonomists to distinguish intraspecific taxa. The smallest category is a form, which is a small deviation that has or does not have a genetic basis, but is used to indicate a random deviation that does not form local or regional facies (Rietz 1930). Later, Crundwell (1970) clarified this formulation and limited the form exclusively to inherited deviations.

The larger categories are subspecies and variety. Touw (1979) identified several approaches to the use of these terms: (1) the use of the category of subspecies for geographically isolated aggregation, and varieties for sympatric deviations; (2) the use of both categories to unite single-level groups with a gap between them; (3) some researchers use the term subspecies in the first or second sense, but subordinate the variety to it; (4) distinguish only one of the categories and use the terms 'subspecies' and 'variety' to describe both sympatric and allopatric intraspecific deviations. And, in the Touw's opinion, the use of the subspecies category is possible only in two cases. First, for the separation of satellites in large combined taxa, when it is doubtful to recognize the level of the species for each. The subspecies, in this case, will show a close relationship and balance the classification system more than assigning all deviant taxa the rank of species. Secondly, when a subspecies is considered a taxon that unites a group of subordinate varieties of both sympatric and allopatric distribution.

The recent trend is to refuse to assign the rank of a subspecies. In the rank of the variety, rather persistent deviations are considered, however, connected by transitions to the main variety along this (or part of) the extent of the range.

Since speciation is not always accompanied by morphological transformations, the true number of biological species is likely to be greater than the current number of known species, most of which are determined mainly by morphological features

(Fišer et al. 2018). The massive use of molecular genetic methods in the last few decades, and their significant reduction in price, has made it possible to establish the presence of cryptic species in all groups of animals, plants, fungi and microorganisms (Bickford et al. 2007; Heinrichs et al. 2009, 2010; Adams et al. 2014; Renner 2020). These are species not distinguishable by morphological features, but clearly distinguished by molecular data. As further studies have shown, for most of the species described by molecular data as cryptic, small morphological differences were found, and these species are considered almost cryptic or semicryptic (Vilnet et al. 2004; Fišer et al. 2018; Renner 2020). Cryptic species are also increasingly taken into account in biodiversity studies (especially in microorganisms and fungi), but not as named species, but as clusters of sequences called molecular operational taxonomic units (OTU, also called MOTU) (Fišer et al. 2018; Ryberg 2015). Some authors consider all prokaryotes are species-less, because in their evolutionary speciation the important factors are horizontal gene transfer and homologous recombination. As a result, we deal with highly recombinant lineages. Thus, the main issues in prokaryotic systematics are whether or not lineages necessarily represent a genetic continuum. At the end of this section, we would like to emphasize that the species concept is a theoretical notion of the species, which would be ideally applicable to all organisms, and the species definitions are a set of rules used for practical identification of species.

3 Establishment of a New Species: Requirements and Procedure

In this section, we review basic requirements and procedures for the establishing new mineral species in mineralogy and biology. Though this topic is not directly related to the concept of species sensu stricto, the similarities and differences are of interest and highlight the nature of the concept of species in various subdisciplines of natural sciences.

3.1 Mineralogy

The procedure of consideration and approval or disapproval of new mineral species is governed by the International Mineralogical Association (IMA) founded in 1958 at the founding meeting in Madrid, Spain. The IMA established a special Commission of New Minerals, Nomenclature and Classification (CNMNC) consisting of 35 national representatives and 4 officers (more details about the Commission are available at the site: http://cnmnc.units.it). The procedure to apply for a new mineral discovery is as following. The applicant(s) should fill out a special sheet of data (checklist) and contains essential characteristics of a mineral, including: (i) description of geological occurrence and locality where the mineral was found; information of associated minerals should be provided as well; (ii) appearance and physical properties; this includes description of visual appearance of the mineral, its crystal morphology, colour, streak, luster, hardness (defined according to the Mohs' scale), tenacity, cleavage (the ability of mineral crystals to cleave along particular crystallographic planes), parting, fracture, and density; in some cases (especially when crystals are nanosized), these characteristics are difficult to determine; (iii) optical properties (optical sign, refraction indices, etc. for transparent minerals, reflectance spectra for non-transparent (ore) minerals); sometimes these properties are difficult or even impossible to determine as well; (iv) chemical data (incl. raw chemical analyses, empirical and idealized (end-member) chemical formula; (v) powder X-ray diffraction pattern (usually eight strongest lines are identified); (vi) crystal-structure data (in the past, these data were not always required, however, since the advent of new experimental methodologies, e.g., area detectors, crystal-structure data become more or less mandatory for new mineral descriptions); (vii) additional data, including relation to other species and position in the overall classification scheme of minerals (see below). The mineral receives its name from the authors, which is the topic of a special consideration by the CNMNC members, who vote separately for a mineral and its name. Sometimes, the mineral species is accepted as new, whereas mineral name is not, and the authors are asked to rename the mineral or to modify the chosen name. It is mandatory to indicate the place where the holotype specimen of the mineral is deposited (usually mineralogical museums or open collections) and its catalogue number. It is implied that any researcher who want to re-investigate the mineral may request the part of the holotype specimen for its own study.

After receiving the checklist for a new mineral discovery, the Chairman of the CNMNC checks it for omissions and errors and, after communications with authors, compiles the data and sends them to the CNMNC members for voting. The procedure of voting usually takes two or three months, after which the Chairman collects the votes and sends the results of voting to the authors.

3.2 *Biology*

Until recently, the description of new taxa of biological species and intraspecific rank was based on morphological features only. Within the last few ten years, it occurs most often on the basis of a combination of both classical and molecular genetic methods.

The rules for naming new species are regulated by the codes of nomenclature. The formation and use of scientific names of animals, plants (including fungi) and bacteria are regulated respectively by the International Code of Zoological Nomenclature (https://www.iczn.org/the-code/the-code-online/), the International Code of Nomenclature of Algae, Fungi and Plants (https://www.iapt-taxon.org/nomen/pages/intro/title_page.html) and the International Code of Nomenclature of Prokaryotes (International Code…, 2019). These three codes differ in their approach to the problem and in their scope, but the main content consists of a number of rules or articles, and some of them are supplemented with recommendations (tips). The provisions of the codes have no legal status at the state or international level (law). Their use and implementation are based solely on the voluntary agreement of taxonomists to comply with the rules of the codes. The codes differ in details, but their most important features are common to all three codes. They require all scientific names to be Latin in form, written in letters of the Latin alphabet. The Code also establishes the relevant rules that must be observed in the formation and the use of scientific names in order to ensure their uniformity, if possible.

Let us analyze some features on the example of the code of botanical nomenclature. The Code is changed by the decisions of the Nomenclature Section of the International Botanical Congress, which is open to all participants of the Congress. The leading botanical institutions have a certain number of votes on this section (from 1 to 7), which are transferred by proxy to the members of the section, but none of them can have more than 15 votes, including their own. Proposals to amend the code are made by all interested specialists and published in the journal "Taxon"—the official body of the International Association for Plant Taxonomy, and can also be made directly at the meeting of the section by at least 4 of its participants. The decisions of the Nomenclature Section are approved at the plenary session of the International Botanical Congress. This is the essential difference between the botanical code and the zoological one, which is currently being developed and adopted by a rather narrow composition of the International Nomenclature Bureau and approved by the International Union of Biological Sciences (Geltman and Sokolova 2013).

The general principles include: the independence of botanical nomenclature from zoological and microbiological, the use of names of taxonomic groups by means of nomenclature types, the principle of priority, the presence single correct name for each taxonomic group with certain boundaries, position and rank (except in specially specified cases), consideration of scientific names as Latin regardless of their origin, retroactive force nomenclature rules, unless they are specifically restricted.

The correct name of the species must comply with the rules of nomenclature, be typified, be effectively and truly publicized, possess priority and be legitimate. The name of a species is a binary combination consisting of two words: a generic name and a specific epithet. Grammatically, a specific epithet is most often expressed by an adjective in the nominative singular case, consistent with the generic name in the genus. A specific epithet can speak of some morphological or ecological feature, can be derived from personal names and surnames, from geographical names of countries or localities.

At the first publication, the description of a new species begins with its name in Latin and an indication of novelty—sp. nov. (new species), subsp. nov. (new subspecies), var. nov. (the variety is new). Effective publication is publication in a fairly widespread journal that is accessible to a wide range of specialists. Publication is effected, under the Code, by distribution of printed matter to the general public or at least to scientific institutions with generally accessible libraries. Publication is also effected by distribution on or after 1 January 2012 of electronic material in PDF in an online publication with an International Standard Serial Number (ISSN) or an International Standard Book Number (ISBN) (https://www.iapt-taxon.org/nomen/pages/main/art_29.html). The dissemination of electronic material in any other way, as well as in preliminary versions of publications, does not constitute effective publication (Knapp et al. 2011a). Thus, botanists did not follow the path of the zoological code, according to which the dissemination of indelible information on electronic media, such as disks, in 5 or more libraries is considered publication, and dissemination of information only via the Internet, on the contrary, is not considered publication (Geltman and Sokolova 2013).

For a valid publication, the name of the species must be accompanied by a diagnosis or description to which it relates, or a reference to an earlier diagnosis or description before January 1, 2012 in Latin, after that date in Latin or English. In zoological nomenclature, descriptions in any language are allowed. When describing a new species, a protologue or primary description is given—everything that accompanies the new name when it is actually published: description, diagnosis, illustrations, references, geographical data, a list of studied samples. The nomenclature type must be indicated—the herbarium specimen with which this taxon name is inextricably linked, assigned to it, these are standards. For some species described in the XVIII century, the nomenclature type may be an image. The most important type of type specimen—holotype (holotypus)—is a specimen according to which a diagnosis of a new species is made, the label indicates the place of collection, growing conditions, date of collection, field diary number, surname and initials of the collector, date of determination, surname and initials of the author. Other specimens that have the same label as the holotype and are duplicates of it are called an isotype. All samples cited in the protologue are called authentic. It is quite common to find a topotype—a herbarium specimen collected in the same place as the holotype, in the so-called "classical habitat"—locus classicus. Since January 1, 1990, when publishing the name of a new species or intraspecific taxon, it is necessary to indicate the only herbarium or institution where its type is stored, otherwise the name is not considered to be really published.

4 Diversity of Species

4.1 Mineralogy

As of January 2023, the IMA recognizes 5900 currently valid mineral species; the list is available online and is updated several times per year (Pasero 2023). The problem of the maximal number of mineral species was under intensive debates, especially in the Russian mineralogical literature (Fersman 1938; Povarennykh 1966; Urusov 1983; Khomyakov 1996). The quantitative approach to the problem is due to Hazen et al (2015a, b) and Hystad et al (2015a, b, 2019a, b), who used different statistical models to obtain numerical estimates for the Earth's mineralogical diversity. Hystad et al. (2015a) investigated the relationships between the number of minerals and the number of localities where they have been found. They demonstrated that the frequency distribution of minerals corresponds to that of words in a book: some words are very common, whereas most words are rare. This distribution known as a Large Number of Rare Events (LNRE) frequency distribution was used by Hazen et al. (2015b) to predict that the approximate total number of mineral species is 6394. Taking into account that ~ 100 mineral species are currently discovered every year, that would mean that mineralogical discoveries would end up in *ca.* 6–7 years. Later Hystad et al. (2019a, b) reconsidered these estimates and extended it to the total of 9308 mineral species. Despite the fact that many active mineralogists are skeptical about this number too, this estimate remains to be the only prediction in the field based upon rational mathematical modeling. All other predictions are based upon 'gut feeling' or intuition not supported by any numerical methods.

4.2 Biology

No one knows how many biological species there are on Earth. It is noteworthy that we do not know the true number of species on Earth, even to the nearest order of magnitude (Wilson 1985). According to the most accurate estimate, as of 2011, the number of described species of living organisms was approximately 1.2 million on land and about 200 thousand in the ocean (Moral et al. 2011). The total number of species living on Earth is estimated in different ways: the number of 8.7 million is estimated, of which 2.2 million are marine. According to other estimates, the total number of species may range from 11 to 12 million (Chapman 2009). Estimates vary for different groups of organisms. So, for fungi, it is assumed that there are more than 5 million species (despite the fact that only about 100 thousand are described now) (Blackwell 2011).

According to some estimates, the number of extinct species is about 250 thousand species (Raup 1986), according to others about 500 million (Mayr 1969).

5 Classification of Species

5.1 Mineralogy

It is rather surprising, but, in contrast to biology (see below), there is no single and widely accepted classification of mineral species. Since mineral is a dialectic unity of chemical composition and crystal structure, it is most natural to base the classification of minerals upon combination of chemical and structural features. In fact, such a systematics is employed in the current classification schemes in mineralogy. There are several classifications that combine chemistry and structure as basic criteria. Most popular are Strunz-Nickel (Strunz 1996; Strunz and Nickel 2001), Hey (Clark 1993), and Dana (Gaines et al 1997) classifications. The primary criterion is chemical composition. For instance, in the Strunz system, ten chemical classes are subdivided: (1) elements (metals and intermetallic alloys; metalloids and nonmetals; carbides, silicides, nitrides, phosphides), (2) sulphides and sulphosalts (sulphides, selenides, tellurides; arsenides, antimonides, bismuthides; sulpharsenites, sulphantimonites, sulphbismuthites, etc.), (3) halides, (4) oxides (hydroxides, vanadates, arsenites, antimonites, bismuthites, sulphites, selenites, tellurites, iodates), (5) carbonates (nitrates), (6) borates, (7) sulfates (selenates, tellurates; chromates, molybdates, wolframates), (8) phosphates, arsenates, vanadates, (9) silicates (germanates), (10) organic compounds. The structural principle is used for classification purposes for silicates and borates only, which are subdivided into subclasses according to the degree of polymerization of SiO_4 or BO_n ($n = 3, 4$) coordination polyhedra. For instance, silicates are subdivided into neso-, soro-, cyclo-, ino-, phyllo-, and tektosilicates (with and without H_2O).

Mills et al. (2009) recommended subdivision of the following mineralogical classification taxa: class, subclass, family, supergroup, group, subgroup or series. Ideally, the proposal is reasonable, but there are many points that remain to be unaddressed at present. For instance, borosilicates that contain polymerized units of Si- and B-centered coordination polyhedra belong to both classes of silicates and borates. The present nomenclature for the garnet supergroup (Grew et al. 2013) contains the majority of silicate garnets, but also hydroxides (katoite), halides (cryolithionite), oxides (elbrusite), arsenates (berzeliite), and vanadates (palenzonaite). The reason for the inclusion of non-silicates into the garnet supergroup is obviously due to their crystal chemical features since all these minerals crystallize in the garnet structure types. The same is true for the apatite supergroup that includes phosphates, arsenates, vanadates, sulfates, and silicates (Pasero et al. 2009). Therefore, at least for the garnet and apatite supergroups, the principle of hierarchical classification is violated, since supergroups may not contain species from different chemical classes. It should be recognized that the present nomenclature proposals are based upon structural features rather than on chemical principles employed traditional classifications since Lima-de-Faria (1994, 2001, 2003, 2004, 2012) attempted to provide a comprehensive structural classification of all minerals based upon the principle of close packings, but his scheme was of little or even no use in mineralogical community. Hierarchical structural classifications have been developed for particular mineral classes (Povarennykh 1972; Bokii and Gorogotskaya 1969; Sabelli and Trosti-Ferroni 1985; Christ and Clark 1977; Rastsvetaeva and Pushcharovsky 1989; Moore 1974; Hawthorne 1986, 1990, 1998; Pushcharovsky et al. 1998; Eby and Hawthorne 1993; Hawthorne et al. 1996, 2000; Hawthorne and Huminicki 2002; Burns et al. 1996; Burns 2005), but the overall classification of minerals that would encompass both chemical and structural principles, if ever possible, is still absent.

5.2 Biology

Biologists have been trying for a long time to develop a natural classification that would not be a human invention, but existed objectively, thanks to the laws of living nature. Now almost everyone recognizes that such a natural classification in biology should be an evolutionary classification based on the kinship of living organisms with each other. Modern classifications of living organisms are based on a hierarchical principle. Living organisms are grouped with their closest relatives, then we combine these groups into larger groups with more distant relatives, and so on, building a hierarchy of groups or taxa by kinship. Most often, such a classification will look like an evolutionary tree or graph in which the branches will be groups of living beings, and the separation of branches will indicate the evolutionary separation of these groups that occurred at some point in time.

Currently, the group of the highest rank, the largest taxon is called a domain. Now there are three different domains (Woese et al. 1990). Eukaryotes are organisms whose cells have a nucleus: plants, fungi (including lichens), animals. Bacteria are single–celled organisms without a nucleus. They can be classified on the basis of cell structure, metabolism, as well as differences in the chemical composition of cells (the presence or absence of certain fatty acids, pigments, antigens, quinones). Archaea are unicellular microorganisms that do not have a nucleus, as well as any membrane organelles. Previously, they were combined with bacteria into a common group—prokaryotes, but it was found that archaea have their own independent evolutionary history and are characterized by many biochemical features that distinguish them from other forms of life (Pace 2006). Currently, modern classification within domains is mainly based on molecular phylogenetics.

Within the domains, taxa of lower rank are distinguished: kingdoms, types (in plants—divisions), orders, families, genera. The basic unit of biological systematics is the species. It is important that when determining the rank of a taxon, a correct assessment of the significance of taxonomic features is very important. A important character for the division of the kingdom will not be important when dividing into families or genera. This applies to both traditional and integrative taxonomy.

6 Conclusions

Both descriptive mineralogy and descriptive biology are based upon the concept of species. The definition of species in mineralogy reflects the dialectic unity of chemical composition and crystal structure, two interdependent features of a crystalline substance. Despite some issues that still remain unresolved (Hawthorne 2023), the definition of a mineral species is rather strict and depends upon the mutual agreement reached by researchers on the borders that separate one mineral species from another. In biology, there are different approaches to the concept of species. Yet, the biological species is understood as a group of organisms with common morphological, physiological, biochemical and behavioral characteristics, capable of crossbreeding. The modern definition of a biological species is based upon combination of morphological and molecular genetic features. The basic difference between mineral and organism is the existence of a molecular carrier of genetic information in the latter (DNA). Minerals are not born, do not crossbreed and do not die in the biological sense of the word, though mineral individuals do appear and disappear, and their microstructure and defects reflect their geological history (Grigoriev 1961). Though mineral kingdom develops in time (mineral evolution), this development has little in common with the development of the domains of life (biological evolution). Thus, the concept of a mineral species has no evolutionary aspect, in contrast to the concept of biological species, where each species is the result of evolution governed by different biological and environmental mechanisms. The morphospace explored by organisms is by far larger than the morphospace of crystalline compounds and, in particular, minerals (Krivovichev 2023), which results in the enormous difference between biological and mineralogical diversity, by at least three degrees of magnitude. The biological classification of species, at least at some level of organismal complexity, reflects their evolutionary development, whereas classification of minerals is based purely upon their constitution, i.e., the chemical composition and crystal structure. It is worthy to note that there are some recent attempt to provide an evolutionary classification of minerals, which is, however, based upon the concept of natural kinds that can be viewed as complementary to the traditional concept of mineral species (Hazen 2019; Hazen et al. 2022). Yet, despite the relative simplicity of minerals compared to organisms, there is no single universal natural classification of mineral species. Perhaps, the simpler the objects, the more difficult is their classification (at the level of unicellular organisms classification is more difficult than for multicellular species (Zavarzin 2006)).

In both mineralogy and biology, the definition of a new species proceeds along specific protocols governed by special codes identified by respective international scientific organizations. Whereas for biological species, there is almost no danger of scientific misconduct and falsification during description of new species, in mineralogy the situation is different. Minerals are crystalline compounds of geological origin that may or may not have analogues among synthetic compounds. Thus, mineral discoveries can be intentionally or unintentionally falsified. The intentional falsification can be done by a person looking for recognition in the scientific community, whereas unintentional falsification is due to confusion between naturally occurring minerals and compounds produced by a technical activity of a humankind. The latter kind of misunderstanding is continuously expanding due to the increasing penetration of technology into the geological and geochemical environment. In biology, both kinds of falsification are impossible, since, up to now, nobody could reproduce new living organisms, so there is no world of artificial biological species parallel to the world of natural origin. The question as to whether or not this is theoretically and practically possible is beyond the scope of the present contribution.

Acknowledgements This research was supported by the Russian Foundation for Basic Research, grant 21-011-44141.

References

Adams M, Raadik TA, Burridge CP, Georges A (2014) Global biodiversity assessment and hyper-cryptic species complexes: More than one species of elephant in the room? Syst Biol 63:518–533

Authier A, Chapuis G (2014) A Little Dictionary of Crystallography. Chester International Union of Crystallography

Ballhaus C, Helmy HM, Fonseca ROC, Wirth R, Schreiber A, Jöns N (2021) Ultra-reduced phases in ophiolites cannot come from Earth's mantle. Amer Miner 106:1053–1063

Bickford D, Lohman DJ, Sodhi NS, Ng PKL, Meier R, Winker K, Das L (2007) Cryptic species as a window on diversity and conservation. Trends Ecol Evol 22(3):148–155

Bindi L, Steinhardt PJ, Yao N, Lu PJ (2009) Natural quasicrystals. Science 324:1306–1309

Bindi L, Steinhardt PJ, Yao N, Lu PJ (2011) Icosahedrite, $Al_{63}Cu_{24}Fe_{13}$, the first natural quasicrystal. Amer Miner 96:928–931

Bindi L, Yao N, Lin C, Hollister LS, Andronicos CL, Distler VV et al (2015) Decagonite, $Al_{71}Ni_{24}Fe_5$, a quasicrystal with decagonal symmetry from the Khatyrka CV3 carbonaceous chondrite. Amer Miner 100:2340–2403

Blackwell M (2011) The fungi: 1, 2, 3 …5,1 million species? Amer J Bot 98:426–438

Bock WJ (2004) Species: The concept, category and taxon. J Zool Syst Evol Res 42(3):178–190

Bokii GB, Gorogotskaya LI (1969) On the crystal chemical classification of sulfates. Zh Strukt Khim 10:624-632 (in Russian)

Burns PC (2005) U^{6+} minerals and inorganic compounds: insights into an expanded structural hierarchy of crystal structures. Can Miner 43:1839-1894

Burns PC, Miller ML, Ewing RC (1996) U^{6+} minerals and inorganic phases: a comparison and hierarchy of structures. Can Miner 34:845-880

Chapman AD (2009) Numbers of Living Species in Australia and the World, 2nd ed. Canberra Australian Biological Resources Study, pp. 1–80

Chesnokov BV, Lotova EV, Pavlyuchenko VS, Nigmatulina EN, Usova LV, Bushmakin AR et al (1989) Svyatoslavite, $CaAl_2Si_2O_8$ (orthorhombic), a new mineral. Zap Vses Mineral Obshch 118(2):111-114 (in Russian)

Chesnokov BV, Lotova EV, Nigmatulina EN, Pavlyuchenko VS, Bushmakin AF (1990) Dmisteinbergite $CaAl_2Si_2O_8$ (hexagonal) - a new mineral. Zap Vses Mineral Obshch 119:43-5 (in Russian)

Christ CL, Clark JR (1977) A crystal-chemical classification of borate structures with emphasis on hydrated borates. Phys Chem Miner 2:59-87

Clark A (1993) Hey's Mineral Index. London Chapman & Hall

Crook WW III (1977) Texasite, a new mineral: the first example of a differentiated rare-earth species. Amer Miner 62:1006-1008

Croundwell AC (1970) Infraspecific categories in Bryophyta. Biol J Linn Soc 2:221-224

Eby RK, Hawthorne FC (1993) Structural relations in copper minerals. I. Structural hierarchy. Acta Crystallogr B49:28-56

Fedorov ES (1891) Simmetriia Pravil'nykh Sistem Figur (The symmetry of regular systems of configurations). Zap Min Obshch 28:1-146 (in Russian)

Fersman AE (1938) On the number of mineral species. C R (Dokl) Acad Sci URSS 19(4):269–272

Fišer C, Robinson CT, Malard F (2018) Cryptic species as a window into the paradigm shift of the species concept. Mol Ecol 27:613–635

Gaines RV, Skinner HC, Foord EE, Mason B, Rosenzweig A (1997) Dana's New Mineralogy. Wiley

Geltman DV, Sokolova IV (2013) Botanical nomenclature: specificity and modern tendencies of development. Trudy Zoolog Inst RAN, Suppl 2:230-237 (in Russian)

Ghiselin MT (2001) Species Concepts. Encyclopedia of life sciences. DOI:https://doi.org/10.1038/npg.els.0001744

Ghiselin MT (2002) Species concepts: the basis for controversy and reconciliation. Fish Fisher 3: 151–160

Grant V (1981) Plant Speciation. New York Columbia University Press

Grew ES, Locock AJ, Mills SJ, Galuskina IO, Galuskin EV, Hålenius U (2013) Nomenclature of the garnet supergroup. Amer Miner 98:785–810
Griffin WL, Gain SEM, Adams DT, Huang JX, Saunders M, Toledo V et al (2016) First terrestrial occurrence of tistarite (Ti_2O_3): Ultra-low oxygen fugacity in the upper mantle beneath Mount Carmel, Israel. Geology 44:815–818
Griffin WL, Gain SEM, Bindi L, Toledo V, Cámara F, Saunders M et al (2018a) Carmeltazite, $ZrAl_2Ti_4O_{11}$, a new mineral trapped in corundum from volcanic rocks of Mt Carmel, northern Israel. Minerals 8:601
Griffin WL, Huang JX, Thomassot E, Gain SEM, Toledo V, O'Reilly SY (2018b) Super-reducing conditions in ancient and modern volcanic systems: sources and behaviour of carbon-rich fluids in the lithospheric mantle. Mineral Petrol 112:101–114
Griffin WL, Gain SEM, Huang JX, Saunders M, Shaw J, Toledo V et al (2019) A terrestrial magmatic hibonite-grossite-vanadium assemblage: Desilication and extreme reduction in a volcanic plumbing system, Mount Carmel, Israel. Amer Miner 104:207–219
Grigoriev DP (1961) Ontogeny of Minerals. Lvov, Izd Lvov Univ
Grimm U (2015) Aperiodic crystals and beyond. Acta Crystallogr 71:258-274
Guinier A, Bokij GB, Boll-Dornberger K, Cowley JM, Durovič S, Cox DE et al (1984) Nomenclature of polytype structures. Report of the International Union of Crystallography Ad hoc Committee on the Nomenclature of Disordered, Modulated and Polytype Structures. Acta Crystallogr A40:399–404
Hawthorne FC (1986) Structural hierarchy in $^{IV}M_x{}^{III}T_{yz}$ minerals. Can Miner 24:625-642
Hawthorne FC (1990) Structural hierarchy in $M^{[6]}T^{[4]}f_n$ minerals. Z Kristallogr 192:1-52
Hawthorne FC (1998) Structure and chemistry of phosphate minerals. Miner Mag 62:141-164
Hawthorne FC (2023). On the definition of distinct mineral species: A critique of current IMA-CNMNC procedures. Miner Mag, in press. doi:https://doi.org/10.1180/mgm.2023.8
Hawthorne FC, Huminicki DMC (2002) The crystal chemistry of beryllium. Rev Miner Geochem 50:333-403
Hawthorne FC, Burns PC, Grice JD (1996) The crystal chemistry of boron. Rev Miner 33:41-115
Hawthorne FC, Krivovichev SV, Burns PC (2000) Crystal chemistry of sulfate minerals. Rev Miner Geochem 40:1-112
Hawthorne FC, Mills SJ, Hatert F, Rumsey MS (2021) Ontology, archetypes and the definition of 'mineral species'. Miner Mag 85:125-131
Hazen RM (2013) Paleomineralogy of the Hadean Eon: A preliminary list. Amer J Sci. 313:807-843
Hazen RM (2019) An evolutionary system of mineralogy: Proposal for a classification of planetary materials based on natural kind clustering. Amer Miner 104:810–816
Hazen RM, Morison SM (2020) An evolutionary system of mineralogy. Part I: Stellar mineralogy (>13 to 4.6 Ga). Amer Miner 105:627–651
Hazen RM, Papineau D, Bleeker W, Downs RT, Ferry J, McCoy T et al. (2008) Mineral evolution. Amer Mineral 93:1693-720
Hazen RM, Hystad G, Downs RT, Golden J, Pires A, Grew ES (2015a) Earth's "missing" minerals. Amer Miner 100:2344–2347
Hazen RM, Grew ES, Downs RT, Golden J, Hystad G (2015b) Mineral ecology: Chance and necessity in the mineral diversity of terrestrial planets. Can Miner 53(2):295–324
Hazen RM, Morrison SM, Prabhu A (2021) An evolutionary system of mineralogy. Part III: Primary chondrule mineralogy (4566 to 4561 Ma). Amer Miner 106:325–350
Hazen RM, Morrison SM, Krivovichev SV, Downs RT (2022) Lumping and Splitting: Toward a Classification of Mineral Natural Kinds. Amer Miner 107:1288-1301
Heinrichs J, Klugmann F, Hentschel J, Schneider H (2009) DNA taxonomy, cryptic speciation and diversification of the Neotropical-African liverwort, Marchesinia brachiate (Lejeuneaceae, Porellales). Mol Phylogen Evol 53:113–121
Heinrichs J, Hentschel J, Bombosch A, Fiebig A, Reise J, Edelmann M, Kreier HP, Schäfer-Verwimp A, Caspari S, Schmidt AR, Zhu RL, von Konrat M, Shaw B, Shaw AJ (2010) One species or at

least eight? Delimitation and distribution of Frullania tamarisci (L.) Dumort. s.l. (Jungermanniopsida, Porellales) inferred from nuclear and chloroplast DNA markers. Mol Phylogen Evol 56:1105–1114

Hey J (2006) On the failure of modern species concepts. Trends Ecol Evol 21:447–450

Hey J, Waples RS, Arnold ML, Butlin RK, Harrison RG (2003) Understanding and confronting species uncertainty in biology and conservation. Trends Ecol Evol 18:597–603

Hystad G, Downs RT, Hazen RM (2015a) Mineral frequency distribution data conform to a LNRE model: Prediction of Earth's "missing" minerals. Math Geosci 47:647-461

Hystad G, Downs RT, Grew ES, Hazen RM (2015b) Statistical analysis of mineral diversity and distribution: Earth's mineralogy is unique. Earth Planet Sci Lett 426:154-157

Hystad G, Eleish A, Downs RT, Morrison SM, Hazen RM (2019a) Bayesian estimation of Earth's undiscovered mineralogical diversity using noninformative priors. Math Geosci 51:401-417

Hystad G, Morrison SM, Hazen RM (2019b) Statistical analysis of mineral evolution and mineral ecology: the current state and a vision for the future. Appl Comput Geosci 1:100005

International Code of Nomenclature of Prokaryotes. Int J Syst Evol Microbiol 69:S1–S111

Ivanova MA, Lorenz CA, Borisovskiy SE, Burmistrov AA, Korost DV, Korochantsev AV et al (2017) Composition and origin of holotype Al-Cu-Zn minerals in relation to quasicrystals in the Khatyrka meteorite. Meteor Planet Sci 52:869–883

Johnsen O, Ferraris G, Gault RA, Grice JD, Kampf AR, Pekov IV (2003) The nomenclature of eudialyte-group minerals. Can Miner 41:785–794

Khomyakov AP (1996) Why is there more than two thousand? Priroda 5:62-74 (in Russian)

Komarov VL (1940) The doctrine of the species in plants. Publishing House of the Academy of Sciences of the USSR (in Russian)

Krivovichev SV (2023) Structural and chemical complexity of minerals: the information-based approach. In: Bindi L, Cruciani G, eds. Celebrating the International Year of Mineralogy. Springer, in press

Lian D, Yang J, DIlek Y, Wu W, Zhang Z, Xiong F et al (2017) Deep mantle origin and ultra-reducing conditions in podiform chromitite: Diamond, moissanite, and other unusual minerals in podiform chromitites from the Pozanti-Karsanti ophiolite, southern Turkey. Amer Miner 102:1101–1113

Lima-de-Faria J (1994) Structural Mineralogy. An Introduction. Dordrecht Kluwer

Lima-de-Faria J (2001) Structural Classification of Minerals. Vol. 1. Minerals with A, A_mB_n and $A_pB_qC_r$ General Chemical Formulas. Dordrecht Kluwer

Lima-de-Faria J (2003) Structural Classification of Minerals. Vol. 2. Minerals with $A_pB_qC_rD_s$ to $A_pB_qC_rD_sE_xF$ General Chemical Formulas. Dordrecht Kluwer

Lima-de-Faria J (2004) Structural Classification of Minerals. Vol. 3. Minerals with $A_pB_q...E_xF_y..._nA_q$ General Chemical Formulas. Dordrecht Kluwer

Lima-de-Faria J (2012) The close packing in the classification of minerals. Eur J Miner 24:163-169

Litasov KD, Kagi H, Bekker TB (2019a) Enigmatic super-reduced phases in corundum from natural rocks: Possible contamination from artificial abrasive materials or metallurgical slags. Lithos 340–341:181–190

Litasov KD, Bekker TB, Kagi H (2019b) Reply to the discussion of "Enigmatic super-reduced phases in corundum from natural rocks: Possible contamination from artificial abrasive materials or metallurgical slags" by Litasov et al. (Lithos, v.340–341, p.181–190) by W.L. Griffin, V. Toledo and S.Y. O'Reilly. Lithos 348–349:105170

Knapp S, McNeill J, Turland NJ (2011) Changes to publication requirements made at the XVIII International Botanical Congress in Melbourne – what does e-publication mean for you? PhytoKeys 6:5–11

Mahner M (1993) What is a species? J Gen Philos Sci 24:103–126

Mahner M, Bunge M (1997) Foundations of Biophilosophy. Springer Verlag Frankfurt

Mayden RL (1997) A hierarchy of species concepts: the denouement in the saga of the species problem, in Species. The Units of Biodiversity (eds Claridge MF, Dawah AH and Wilson MR), London Chapman & Hall, pp. 381–424

Mayr E (1963) Animal Species and Evolution. Cambridge Harvard University Press
Mayr E (1969) Principles of systematic zoology. N. Y. McGraw-Hill
Mayr E (2000) The biological species concept., in Species concepts and phylogenetic theory: a debate (Wheeler QD, Meier R, eds), Columbia Univ Press, New York, pp 17–29
Mills SJ, Hatert F, Nickel EH, Ferraris G (2009) The standardisation of mineral group hierarchies: Application to recent nomenclature proposals. Eur J Miner 21:1073–1080
Moore PB (1974) Structural hierarchies among minerals containing octahedrally coordinating oxygen. II. Systematic retrieval and classification of edge-sharing clusters: an epistemological approach. N Jb Miner Abh 120:205-227
Moral C, Tittensor DP, Adl S, Simpson AGB, Worm B (2011) How Many Species Are There on Earth and in the Ocean? PLoS Biol 9:e100112
Morrison SM, Hazen RM (2020) An evolutionary system of mineralogy. Part II: Interstellar and solar nebula primary condensation mineralogy (>4.565 Ga). Amer Miner 105:1508–1535
Morrison SM, Hazen RM (2021) An evolutionary system of mineralogy. Part IV: Planetesimal differentiation and impact mineralization (4566 to 4560 Ma). Amer Miner 106:730–761
Nickel EH, Grice JD (1998) The IMA Commission on New Minerals and Mineral Names: Procedures and guidelines on mineral nomenclature. Can Miner 36:913–926
Nickel EH, Grice JD (1999) The IMA Commission on New Minerals and Mineral Names: Procedures and guidelines on mineral nomenclature. Mineral Petrol 64:237–263
Pace NR (2006) Time for a change. Nature 441:289
Parafiniuk J, Hatert F (2020) New IMA CNMNC guidelines on combustion products from burning coal dumps. Eur J Miner 32:215–217
Pasero M (2023) The New IMA List of Minerals. http://pubsites.uws.edu.au/ima-cnmnc/
Pasero M, Kampf AR, Ferraris C, Pekov IV, Rakovan J, White TJ (2010) Nomenclature of the apatite supergroup minerals. Eur J Miner 22:163–179
Pavlinov IY (2018) Foundations of Biological Systematics. History and Theory. Moscow KMK
Peacor DR, Simmons WB, Jr, Essene EJ, Heinrich EW (1982) New data on discreditation of "texasite", "albrittonite", "cuproartinite", "cuprohydromagnesite", and "yttromicrolite", with corrected data on nickelbischofite, rowlandite, and yttrocrasite. Amer Miner 67:156-169
Povarennykh AS (1966) On the abundance of the chemical elements in the Earth's crust and on the quantity of mineral species. Mineral Sb 20(2):178-185 (in Russian)
Povarennykh AS (1972) Crystal Chemical Classification of Minerals. New York London Plenum Press
Pushcharovsky DYu, Lima-de-Faria J, Rastsvetaeva RK (1998) Main structural subdivisions and structural formulas of sulphate minerals. Z. Kristallogr 213:141-150
Rastsvetaeva RK, Pushcharovsky DY (1989) Crystal chemistry of sulfates. Itogi Nauki Tekhn ser Kristallokhim. Vol. 23 (in Russian)
Raup DM (1986) Biological extinction in earth history. Science 231:1528–1533
Renner MAM (2020) Opportunities and challenges presented by cryptic bryophyte species. Telopea J Plant Syst 23:41-60
Rietz GE (1930) The fundamental units of biological taxonomy. Sven Bot Tidskr 24:333-428
Ryberg M (2015) Molecular operational taxonomic units as approximations of species in the light of evolutionary models and empirical data from Fungi. Mol Ecol 24:5770–5777
Sabelli C, Trosti-Ferroni R (1985) A structural classification of sulfate minerals. Per Miner 54:1-46
Schoenflies A (1891) Kristallsysteme und Kristallstruktur. Leipzig (in German)
Strunz H (1996) Chemical-structural mineral classification. Principles and summary of system. N Jb Mineral Mh 10:435-445
Strunz H, Nickel EH (2001) Strunz Mineralogical Tables. Stuttgart Schweizerbart
Touw A (1979) Notes on the use of infraspecific categories by bryologist. In: Bryophyte Taxonomy: Proc IAB Taxon Workshop Meet Geneva, pp. 13–20
Urusov VS (1983) Why is there only two thousand? Priroda 10:82-88 (in Russian)
Vavilov NI (1931) The Linnean species as a system. Leningrad Izd Selskhozgiz (in Russian)

Vernadsky V (1939) Problemy biogeokhimii. II. O korennom materialno-energeticheskom otlichii zhivykh I kosnykh estestvennykh tel biosphery (Problems of biogeochemistry. II. On the basic material and energetical difference between living and non-living natural bodies of the biosphere). Leningrad Izdatelstvo AN SSSR (in Russian)

Vilnet AA, Borovichev EA, Bakalin VA (2014) Frullania subarctica — a new species of the Frullania tamarisci complex (Frullaniaceae, Marchantiophyta). Phytotaxa 173:61–72

Wilkins JS (2007) The dimensions, modes and definitions of species and speciation. Biol Phil 22:247–266

Wilson EO (1985) The biological diversity crisis: A challenge to science. Iss Sci Technol 2:20–29

Woese CR, Kandler O, Wheelis ML (1990) Towards a Natural System of Organisms: Proposal for the Domains Archaea, Bacteria, and Eucarya. Proc Natl Acad Sci USA 87:4576–4579

Xiong Q, Griffin WL, Huang JX, Gain SEM, Toledo V, Pearson NJ et al (2017) Super-reduced mineral assemblages in "ophiolitic" chromitites and peridotites: The view from Mount Carmel. Eur J Miner 29:557–570

Yushkin NP (1982) Evolutionary ideas in modern mineralogy. Zap Vses Miner Obshch 116(4):432-442 (in Russian)

Zavadsky KM (1968) Species and Speciation. Leningrad Nauka (in Russian)

Zavarzin GA (2006) Does evolution make the essence of biology? Her Russ Acad Sci 76:292–302

Zhabin AG (1981) Is there evolution of mineral species on Earth? Dokl Earth Sci 247:142-144

Index

O. V. Frank-Kamenetskaya et al. (eds.), *Biogenic—Abiogenic Interactions in Natural and Anthropogenic Systems 2022*, Springer Proceedings in Earth and Environmental Sciences, https://doi.org/10.1007/978-3-031-40470-2

D

E

F